AF776513

The Oxford Handbook of Climate Action

The Oxford Handbook of Climate Action

Edited by

PAUL ALMEIDA

OXFORD
UNIVERSITY PRESS

OXFORD
UNIVERSITY PRESS

Oxford University Press is a department of the University of Oxford.
It furthers the University's objective of excellence in research, scholarship,
and education by publishing worldwide. Oxford is a registered trade mark of
Oxford University Press in the UK and in certain other countries.

Published in the United States of America by Oxford University Press
198 Madison Avenue, New York, NY 10016, United States of America.

© Oxford University Press 2026

All rights reserved. No part of this publication may be reproduced, stored in a retrieval system, transmitted, used for text and data mining, or used for training artificial intelligence, in any form or by any means, without the prior permission in writing of Oxford University Press, or as expressly permitted by law, by license or under terms agreed with the appropriate reprographics rights organization. Inquiries concerning reproduction outside the scope of the above should be sent to the Rights Department, Oxford University Press, at the address above.

You must not circulate this work in any other form
and you must impose this same condition on any acquirer

Library of Congress Cataloging-in-Publication Data
Names: Almeida, Paul, 1968– editor
Title: The Oxford handbook of climate action / edited by Paul Almeida.
Description: New York, NY : Oxford University Press, [2026] |
Series: Oxford handbooks series |
Includes bibliographical references.
Identifiers: LCCN 2025053863 (print) | LCCN 2025053864 (ebook) |
ISBN 9780197762097 hardback | ISBN 9780197762110 epub | ISBN 9780197762127
Subjects: LCSH: Climate change mitigation
Classification: LCC TD171.75 .O94 2026 (print) | LCC TD171.75 (ebook)
LC record available at https://lccn.loc.gov/2025053863
LC ebook record available at https://lccn.loc.gov/2025053864

DOI: 10.1093/oxfordhb/9780197762097.001.0001

Printed by Integrated Books International, United States of America

The manufacturer's authorized representative in the EU for product safety is
Oxford University Press España S.A., Parque Empresarial San Fernando de Henares,
Avenida de Castilla, 2 – 28830 Madrid (www.oup.es/en or product.safety@oup.com).
OUP España S.A. also acts as importer into Spain of products made by the manufacturer.

MIX
Paper
FSC FSC® C183721

CONTENTS

CONTRIBUTORS

Paul Almeida is a Professor of Sociology and Environmental Systems at the University of California, Merced. Almeida's research centers on the efficacy of collective action at the local, national, and global levels of social and political life. He has empirically examined the timing, distribution, and outcomes of dozens of large-scale campaigns whereby ordinary people and excluded social groups mobilized to protect themselves from the loss of vital necessities such as environmental quality, health care, pensions, water services/utilities, and other socioeconomic rights. His recent work focuses on locally based challenges and responses to climate change. Since 2018, he has implemented representative surveys with partner groups in two dozen towns and cities throughout the San Joaquin Valley on community needs and/or climate action.

Aisha Al-Sarihi is a political scientist with over a decade of research expertise in the governance, policy, and politics of climate and energy transition. Dr. Al-Sarihi is currently a Non-Resident Fellow at both the Middle East Council on Global Affairs (MECGA) and the Arab Gulf States Institute in Washington (AGSIW). Al-Sarihi has published widely in peer-reviewed journals, policy reports, and edited volumes, including articles in *Climate Policy, Renewable Energy, Oxford Institute for Energy Studies*, and *Environmental Policy and Governance*. Her latest edited volume, *Politics and Policy of Climate Change in the Middle East*, is due to be published in October 2025. Al-Sarihi has held various positions at leading think tanks and academic institutions in the United Kingdom, United States, Saudi Arabia, and Singapore, including the London School of Economics and Political Science's Middle East Centre, Arab Gulf States Institute in Washington, Georgetown University's Center for Contemporary Arab Studies, King Abdullah Petroleum Studies and Research Center, and the National University of Singapore's Middle East Institute. Al-Sarihi holds a PhD from the Centre for Environmental Policy at Imperial College London and an MSc and a BSc, with distinction, in Environmental Science from Sultan Qaboos University.

Camila Alvarez is an Assistant Professor of Sociology at the University of California, San Diego. Her areas of expertise are environmental sociology, environmental justice, and quantitative methods. Her research has been published in *Social Science and Medicine*, *Social Problems*, and *Sociological Perspectives*.

Francisco S. Álvarez is a biologist and Research Leader at Fundación Naturaleza El Salvador, as well as a specialist in forest and biodiversity management and conservation. He studies the ecological processes of secondary forests, wildlife ecology, and strategies for biodiversity conservation and landscape restoration in the tropics.

Moises Arce is the Scott and Marjorie Cowen Chair in Latin American Social Sciences and Professor in the Department of Political Science at Tulane University. Professor Arce specializes in conflict processes, democratic governance, and environmental politics. He is the author of *Market Reform in Society, Resource Extraction and Protest in Peru, Social Protest and Democracy, The Roots of Engagement, Popular Politics and Protest Event Analysis in Latin America*, and numerous book chapters and journal articles.

Joshua A. Basseches is the David and Jane Flowerree Assistant Professor of Environmental Studies and Public Policy at Tulane University. His research focuses on the politics of climate and renewable energy policymaking in the United States. His work has been published in peer-reviewed outlets such as *Climatic Change, Politics and Society*, and *Social Science Quarterly*, among others. He is also the author of *Owning the Green Grid*.

Will Bates is a cofounder and director at 350.org. He has played leading roles in many of the largest international climate justice mobilizations and campaigns over the last two decades. He lives in the Pyrenees in Spain with his partner and two daughters.

May Boeve is a next-generation climate and social movement leader. For 15 years she served as the founding Executive Director of 350.org, an international organization that she cofounded in 2008. May is host of the podcast *How We Build This*, and coauthor of *The Oxford Handbook of Climate Action* and *Fight Global Warming Now*. She was featured as a TIME Magazine next-generation leader, is a recipient of the John F. Kennedy New Frontier Award, and has been featured in numerous publications as a thought leader about movement building to fight the climate crisis. She lives in Oakland, California, with her husband and daughter.

Ion Bogdan ("Bodi") Vasi is Associate Professor in the Department of Sociology and (by courtesy) Department of Management at the University of Iowa. His research explores essential questions at the intersection of energy and environment. He published articles in *American Sociological Review, American Journal of Sociology, Social Forces, Social Problems*, and other journals. He has published the book *Winds of Change: The Environmental Movement and the Global Development of the Wind Energy Industry* (Oxford) and is currently working on the book *Turning the Energy Titanic: Collaboration, Conflict, and the Global Transition to Renewable Energy Systems*.

Ingrid Brostrom is an environmental justice attorney, educator, and advocate with decades of legal and policy experience. She is currently the Climate and Rural Economy Program Director at the UC Merced Community and Labor Center. Ingrid previously served as Assistant Director of the Center on Race, Poverty & the Environment (CRPE), where she led the organization's Toxic-Free Communities campaign. As part of that campaign, Ingrid coordinated the People's Senate, a coalition of residents and advocates from communities impacted by toxic waste working to drive reform in California. She also taught environmental justice at the UC Berkeley School of Law. Ingrid is currently a member of the California Board of Environmental Safety.

Tiago Carvalho is a Researcher at the Centre for Research and Studies in Sociology (CIES-Iscte) and Assistant Professor at Iscte—University Institute of Lisbon. He holds a PhD in Sociology from the University of Cambridge. His work explores the dynamics of social movements, political parties, and their interactions, with publications on austerity politics, housing struggles, and environmental mobilizations. He has written extensively on climate activism and transnational diffusion, in particular, comparative studies of Extinction Rebellion in Europe, Africa, and Australia. He contributes to scholarly debates on the role of grassroots activism in shaping political change and climate governance.

Joyee Shairee Chatterjee is an Associate Professor of Gender and Development Studies at the Asian Institute of Technology and Co-Editor-in-Chief of *Gender, Technology and Development* (Taylor & Francis). A communication and gender expert, her interdisciplinary research explores how media, culture, and storytelling drive social and behavioral changes related to gender norms, human rights, and well-being. Her publications focus on issues of equity, social justice, and sustainability, with studies spanning health communication, climate action, and gender-sensitive technology policy. She has led transdisciplinary research and capacity-building projects across South and Southeast Asia, supported by IDRC, UNDP, UN Women, ITU, and ERASMUS.

Victoria Coon is a union organizer with the New York State United Teachers (NYSUT) based out of Buffalo, New York. She holds an MS in Sociology from the UMass Amherst Labor Center and a BA in Urban Studies from McGill University in Quebec.

Hassan El Tinay is a Doctoral Student in Sociology at Boston College, focusing on environmental sociology and political economy. He uses a mix of computational and quantitative methods to address macro-sociological questions concerning ecologically unequal exchange, economic drivers of climate change, and the climate-militarism nexus. His work appears in the *American Sociological Review*, *Environmental Sociology*, and *Ecological Economics*.

Terence Epule Epule is a Professor of Agroclimatology and Adaptation to Climate Change at the University of Quebec in Abitibi–Témiscamingue. His research is at the interface of the interactions between climate and agricultural systems with a goal of developing robust adaptations. His research works cover topics related to climate stressors, vulnerability, risks, readiness for climate change adaptation and climate change performance inter alia using process-based models, machine learning, and artificial intelligence. He is also involved in co-construction through assessments of population perceptions. He has published more than 90 peer-reviewed papers, three books, and several book chapters.

Deborah Barros Leal Farias is an Associate Professor at the University of New South Wales (UNSW–Sydney) School of Social Sciences, where she teaches in the Politics and International Relations stream. She has a rich interdisciplinary background spanning political science, international relations, law, and economics. Her research has covered topics such as populism, global governance, the politics of categorizations, and Brazil's evolving role in international affairs.

Caroline Farrell is an Associate Professor and Director of the Environmental Law and Justice Clinic at Golden Gate University. She trains law students to represent clients to improve environmental conditions where they live, work, and play. Prior to working at the law clinic, Caroline was the Executive Director at the Center on Race, Poverty & the Environment (CRPE), where she represented clients throughout the San Joaquin

Valley on a range of issues from land use, air permitting, pesticide exposure, and hazardous waste.

James V. Fenelon is Professor and Director of Center for Indigenous Peoples Studies, California State University. He authored *Culturicide, Resistance and Survival of the Lakota (Sioux Nation)*; *Indigenous Peoples and Globalization* (with Thomas Hall); *Redskins?: Mascots, Indian Nations, White Racism*; and most recently *Indian, Black and Irish: Indigenous Nations, African Peoples and European Invasions—1492–1790*; and has published numerous journal articles, chapters, and essays. He is Dakota/Lakota (Standing Rock), of Gaelish Irish, French, Scandinavian descent, and advocates for social justice.

Leonardo Figueroa Helland (PhD) (he/him/le'e/el) is Director and Associate Professor of the Environmental Policy and Sustainability Management graduate (MS) program at The New School university in Lenapehoking/Manhattan/New York City. He is also Faculty Associate Director of the Tishman Environment and Design Center, where he leads the Indigeneity and Decolonizing Sustainability program. He is a decolonizing scholar of mestizo (mixed-blood) heritage, with detribalized Indigenous Mesoamerican Maya Tstosil and Zoque, as well as Euroamerican settler Nordic-Scottish-Hispanic ancestries. His work underlines the centrality of Indigenous resurgence and revitalization, in intersection with other transformative and justice-centered paradigms, in addressing planetary crises, achieving climate justice, and materializing systemic change.

Eliana Fonsah is a PhD Candidate in Sociology at the University of California, Merced. Her research focuses on social movements, transnational movements, the African diaspora, democratic mobilization, coalitions and solidarity, environmental justice, just transitions, and political sociology. She was awarded the Chancellor's Fellowship for Inclusive Excellence designed to attract outstanding doctoral students committed to promoting diversity, equity, and inclusion within their academic fields and communities. She is currently a Fellow at the University of California's Institute of Global Conflict and Cooperation. This fellowship supports her research on diasporic mobilization for democracy and social change in Cameroon.

Peter Gardner is a Senior Lecturer in Sociology at the University of York, United Kingdom. His research examines social movements and protest, global environmental activism, postconflict societies, and the sociology of the climate and ecological emergency. He is the author of *Ethnic Dignity and the Ulster-Scots Movement in Northern Ireland, Supremacy in Peril*, coauthor of *Symbolic Objects in Contentious Politics* (with Benjamin Abrams), and Deputy Chief Editor at *Contention: The Multidisciplinary Journal of Social Protest*.

Jennifer Givens is an Associate Professor of Sociology at Utah State University. Her research areas include environmental sociology, the sociology of development including political economic, international, and global perspectives, and interdisciplinary work that addresses mitigation of and adaptation to climate change and related environmental and human well-being issues.

Luis Rubén González Márquez is a Postdoctoral Fellow at the Center for Inter-American Policy and Research (CIPR) at Tulane University. Luis Rubén specializes in environmental and labor mobilization in marginalized territories. His current research examines the political outcomes of opposition campaigns against renewable energy megaprojects, based in a comparative-historical study of Central America. He has also conducted research on labor and popular mobilization in El Salvador. In addition, Luis Rubén has collaborated in publications about anti-austerity protest and protest music in Central America, and climate action and environmental civic engagement in California's San Joaquin Valley.

Kenneth A. Gould is Professor of Sociology and Urban Sustainability at the City University of New York–Brooklyn College, and Professor of Sociology and Earth and Environmental Sciences at the CUNY Graduate Center. His work centers on the political economy of environment, technology, and development contributing to theoretical frameworks, including the Treadmill of Production and Green Gentrification. His field research has focused on North America, Latin America, and the Caribbean. Gould's recent research examines implications of urban greening and climate resilience for inequality, relationships between ecotourism and extraction, and the impacts of climate change on efforts to achieve sustainable development trajectories.

Tommaso Gravante is a Full-Time Researcher of Social Science at the National Autonomous University of Mexico (UNAM). Gravante's research centers on the role of emotions in grassroots activism. His most recent work focuses on the relations between emotions and prefigurative politics.

marilyn grell-brisk is a Black Study Scholar who holds a PhD in Sociology. She is a Visiting Assistant Professor at Pitzer College and specializes in global structural inequality, hierarchy, power, and the connections between exploitative economic systems and climate change, racism, and othering. Using community-based research practices, she engages in Black Study, and she is particularly interested in global-local social movements that affirm Black(ness) and Black futurity.

Jamie Henn is the Cofounder of 350.org and Founder and Executive Director of Fossil Free Media, a nonprofit media lab that leads campaigns and communications for the broader climate movement. Jamie was a driving force behind some of the largest climate campaigns in history, helping lead global days of action, the fight against the Keystone XL pipeline, the fossil fuel divestment movement, the Global Climate Strikes, and more. Fossil Free Media's campaigns include efforts like Clean Creatives, Make Polluters Pay, #StopLNG, and Sun Day. A trail runner and backcountry skier, Jamie lives in New Hampshire with his wife and daughter.

Kathryn Hochstetler is Professor of International Development and Political Science at the London School of Economics. She has published widely on environmental issues in the Global South, with a special focus on Latin America and Brazil. Her articles and books discuss environmental movements, environmental politics and policy, and the roles of emerging powers in international negotiations and regional trade agreements. Her most recent book is *Political Economies of Energy Transition: Wind and Solar Power in Brazil and South Africa*. Current research explores the Amazonian countries and their economic and environmental alternatives to oil and gas production.

Xiaorui Huang is an Assistant Professor of Sociology at Drexel University. His research focuses on environmental sociology, climate change, global political economy, and advanced quantitative methods. His main published work examines the impacts of social driving forces, such as economic development, international trade, and income inequality, on climate change and global climate justice. His work appears in venues including the *American Sociological Review*, *Social Forces*, *Social Science Research*, and *Energy Research & Social Science*.

Nygmet Ibadildin teaches International Relations and Political Economy of Natural Resources at KIMEP University, Almaty, Kazakhstan. He is currently the chair in the Department of Political Science and International Relations. His latest research is on security challenges for Kazakhstan in the context of the Russian invasion into Ukraine.

Andrew Jorgenson is a Professor of Sociology and Founding Director of the Climate & Society Lab at the University of British Columbia, where he is also affiliated with the Institute for Resources, Environment, and Sustainability, and the Centre for Climate Justice. He also serves as a Senior Research Fellow in the Department of Theoretical Economics at Vilnius University. As an environmental macrosociologist trained in global political economy, Andrew conducts research on the human dimensions of global and regional environmental change, with a focus on the societal causes and consequences of the climate crisis.

Carrie Karsgaard is an Assistant Professor in the Education Department at Cape Breton University, where she teaches in the Master of Education program in Sustainability, Creativity, and Innovation. She is Principal Investigator of Learning Collective Worldmaking, a suite of research projects exploring ways to live in creative and relational ways within the planet's boundaries while simultaneously challenging and repairing harmful systems. Her recent book, *Instagram as Public Pedagogy: Online Activism and the Trans Mountain Pipeline*, taps into social media to explore its educative force in climate and antipipeline activism.

Jasmine Kerrissey is a Professor of Sociology and Director of the Labor Center at the University of Massachusetts, Amherst. Kerrissey's research examines the historical and contemporary role of labor movements in shaping working conditions, inequality, and politics. She is coauthor of the award-winning book *Union Booms and Busts: The Ongoing Fight over the US Labor Movement* (Oxford) and many journal articles. She holds a PhD in Sociology from the University of California, Irvine, and a BS in Industrial and Labor Relations from Cornell University.

Farhana R. Khan is a Graduate of the Massachusetts Institute of Technology (MIT). She is currently employed as a senior AI/machine learning engineer at Maximus, a US federal contractor. She cofounded a nonprofit in Boston, the Center for Environmental Peace Building, which uses quantitative modeling approaches using AI/machine learning to promote prevention-oriented environmental peace-building work.

Mizan R. Khan is the Technical Lead of the LDC Universities' Consortium on Climate Change (LUCCC), which is an official program of the 45 LDC governments for building capacity to address climate change. He served as a professor in multiple national universities in Bangladesh. Dr. Khan also served as a visiting fellow/professor in several Western universities, including Brown and the University of Manitoba. Professor Khan has many publications in peer-reviewed journals and magazines, including *Science* and *Nature*. He has three books published by Routledge and MIT Press on climate change economics, politics, and adaptation. Professor Khan has been serving as the Lead Climate Finance Negotiator with the Bangladesh delegation under the UNFCCC process since 2001.

Benjamin Leffel is an Assistant Professor in the School of Public Policy and Leadership at UNLV. His published work demonstrates how cities and corporations successfully fight climate change, alone and in collaboration, as well as the environmental performance dividends of climate networks, private climate consulting, and climate finance in cities globally. His work appears in *Nature Energy, Nature Cities, Environmental Sociology, Energy Research and Social Science*, and other outlets. He has published op-eds in the *Regulatory Review, Harvard Business Review*, and *The Hill*, and he has advised governments on subnational climate leadership across the United States, Germany, and the United Kingdom.

Tammy L. Lewis is Professor of Sociology and Director of the Urban Sustainability Program at Brooklyn College of the City University of New York; and Professor of Sociology and Earth and Environmental Sciences at the CUNY Graduate Center. Her recent work focuses on institutional efforts to adapt to climate change, with a focus on urban coastal areas. She authored *Ecuador's Environmental Revolutions: Ecoimperialists, Ecodependent, and Ecoresisters* and coauthored *Green Gentrification: Urban Sustainability and the Struggle for Environmental Justice*. She has chaired the Environment Sociology Section of the American Sociological Association and received the Fred Buttel Distinguished Contribution Award.

Aldo Madariaga is Associate Professor of Political Science at Universidad Diego Portales (Chile) and director of the Max Planck-UDP Group Challenges of Green Growth. His research interests focus on comparative and international political economy, growth models, and sustainable development. His research has been published in *Environmental Politics, Environmental Impact Assessment Review, Policy Studies Journal, Socio-Economic Review, Governance, Regulation & Governance*, among others. His book *Neoliberal Resilience. Lessons in Democracy and Development from Latin America and Eastern Europe* received honorable mentions for best book by ISA and SASE.

Marcos Magaña is a PhD Student at the UCLA Institute of the Environment and Sustainability. His research explores the lived experiences of marginalized communities in desert regions, reframing heat not as a purely quantitative metric but as a phenomenon shaped by social, cultural, and historical contexts. His work also challenges dominant narratives of deserts as barren landscapes, instead highlighting them as places rich with life, knowledge, and resilience.

Antoine Maillet is Associate Professor at the Universidad de Chile (Facultad de Gobierno), associate researcher at the Centre for Social Conflict and Cohesion Studies (COES), and adjunct researcher at the Centre for Climate and Resilience Research (CR2). He holds a PhD in Political Science from Sciences Po and Pontificia Universidad Católica de Chile. His research mainly deals with environmental politics, from social mobilization to policies and governance.

Joan Martinez-Alier, Emeritus Professor at ICTA, Universitat Autonoma de Barcelona. He has been awarded the Balzan Prize (2020) and the Holberg Prize (2023) for his work on ecological economics, political ecology, and environmental justice.

Cecilia Menjívar is Distinguished Professor of Sociology, Dorothy L. Meier Chair in Social Equities, and Director of the UCLA Center for the Study of International Migration. She is interested in state actions, inactions such as abandonment and neglect, and manifestations of state power through legal frameworks, bureaucracies, institutions, and agents. Empirically, she studies the US legal, institutional, and community contexts that Central American immigrants encounter, and gender-based violence in the lives of Central American women, within the context of structural, symbolic, state-sponsored, and quotidian violence.

Parimita Mohanty is currently the regional lead for the finance unit of climate mitigation at the United Nations Environment Programme (UNEP) and the Asia Pacific regional lead for sustainable cooling under the Cool Coalition. Dr. Mohanty has more than 22 years of experience in sustainable energy, women's economic empowerment, and climate change. She has been a visiting researcher at many recognized universities in the United States and Europe. In her 22-year career, she has initiated more than 50 collaborative projects with reputed national and international institutions, including governments, civil society, and the private sector, and worked on sustainable technologies for communities. She has led and successfully implemented several renewable energy-driven projects on the ground with innovative financing instruments in more than 10 South and Southeast Asian countries. She has authored and edited seven books and more than 25 peer-reviewed publications on renewable energy, e-mobility, energy policies, regulatory business models, climate change, and socioeconomic development. She holds a MTech degree in Energy Science and Technology and a PhD from the Indian Institute of Technology (IIT), Delhi. Aside from her work on climate change, she is also an avid cricket player and has represented Odisha and East Zone in many national level matches in India.

Alesia Montgomery is an Assistant Professor at UCLA's Institute of Environment and Sustainability (IoES). An ethnographer, Montgomery studies the social and environmental justice concerns of low-income, racialized communities. Her book, *Greening the Black Urban Regime: The Culture and Commerce of Sustainability in Detroit*, focuses on battles over the aims and strategies of green redevelopment. Her publications also include articles in the *International Journal of Urban and Regional Research*, *City and Community*, *Ethnography*, *Antipode*, *Sociological Perspectives*, and *Global Networks*.

Ainhoa Montoya is Ramón y Cajal Senior Research Fellow at the Spanish National Research Council (CSIC). She is also Reader in Latin American Studies and Director of the Centre for Latin American and Caribbean Studies at the School of Advanced

Study, University of London. She is the author of *The Violence of Democracy: Political Life in Postwar El Salvador*. Her current research explores cross-border enactments of environmental care in transboundary watersheds of the Americas and Southern Europe.

Kathryn Norton-Smith is an Assistant Professor of Sociology at the University of Arkansas, Fayetteville. Her research explores how inequalities are embodied, experienced, and resisted, as well as processes of coercion and consent. She specializes in the sociology of health and illness, race and racism, social and environmental inequality, and intersectionality. Her work has been published in *Socius* and *Racial and Ethnic Studies*.

Youngmin Oh is a Fulbright Scholar and Director in the Ministry of Climate, Energy and Environment of the Republic of Korea. Since 2009, Dr. Oh has designed and implemented many public policies and projects both domestically and internationally for the Korean Government and international organizations, including the Inter-American Development Bank in Washington, DC. Her areas of expertise encompass sustainable socioeconomic planning, energy transitions, and climate policies. She has served as an official delegate of the Republic of Korea at UNFCCC Conferences of the Parties. Her research interests focus on science–policy interactions, comparative climate actions, and transboundary environmental challenges.

Kajol Patel is a Doctoral Candidate in Sociology at Stony Brook University. Their research bridges the fields of environmental sociology, global political economy, social movements, and law and society. Their dissertation examines the global rise of climate litigation as a form of legal mobilization, showing how lawsuits brought by diverse actors challenge and redefine climate governance, political legitimacy, and public debate over governments' responsibilities in the climate crisis.

Vincent Poirier is a Professor of Soil Science and Director of the Agriculture and Agrifood Research institute of the University of Québec in Abitibi-Témiscamingue (UQAT). Dr. Poirier holds a doctorate in renewable resources (2011) from McGill University, Canada. His research focuses on soil organic matter, carbon storage, and the roles of roots in soil functioning in topsoil and subsoil horizons, production of fodder plants to support livestock farming, and management and production of conventional and organic field crops. He is involved in numerous research projects mainly related to soils in temperate agroecosystems, and some focus on urban and forest soils as well as soils in tropical environments.

Alice Poma is a Full-Time Researcher of Social Science at the National Autonomous University of Mexico (UNAM). Poma's research focuses on the role of emotions in climate and environmental activism, climate and eco emotions, and social movement theory. She has empirically examined which and how emotions mobilize, how emotions are made by activists, and the emotion work activists do and its impact on mobilization. Her work follows a sociocultural approach, which takes into consideration how the violent and unequal Mexican context influences the emotions people feel. Her most recent work focuses on the role of emotions in locally based responses to climate change in Mexico.

Emily Rainsford is Principal Research Associate at Newcastle University, United Kingdom. She is an expert on youth political activism and was part of the UK team in the Caught in the Act of Contestation (CCC) project and has surveyed over 15 demonstrations and adapted the CCC protest methodology to survey other political gatherings.

Joyashree Roy is Distinguished Professor and Founding Director of the SMARTS Centre at the Asian Institute of Technology, Thailand. Formerly as a Professor of Economics in Jadavpur University, India, she designed and conducted field surveys and developed econometric models for empirical investigations in the adoption of energy and water technologies, technology diffusion, understanding of components of transition management for commercializing community-level technology in resource-poor regions, and policy analysis in a developing country context. Her work has been acknowledged by receiving many international and national awards. She is a national fellow of the Indian Council of Social Sciences Research. She was in the IPCC 2007 Nobel Peace Prize–winning panel and coordinating lead author in the fourth, fifth, and sixth assessment cycles of WGIII of IPCC. She is featured in the documentary *Juice: How Electricity Explains the World*, which describes, among other things, how developing countries are trying to bring people out of the dark and into the light and transforming lives.

Jesse Salah Ovadia is an Associate Professor at the University of Windsor, Canada. His research focuses on the political economy of oil, gas, and mineral extraction and development in Africa as well as sustainable energy transitions. He is the author of *The Petro-Developmental State in Africa*. His academic research has been published in journals such as *Third World Quarterly*, *New Political Economy*, *Resources Policy*, *Land Use Policy*, *Development Policy Review*, *Journal of Modern African Studies*, *Journal of Agrarian Change*, and the *Extractive Industries and Society*.

Keiichi Satoh is an Associate Professor of Sociology at Hitotsubashi University in Tokyo, Japan, specializing in political sociology, environmental sociology, and social network theory and analysis. His research focuses on climate policy processes at the local, national, and global levels. In particular, he has empirically examined the collaborative networks of interest groups in the climate policy arena and how these networks have promoted or hindered policy advancement. He has also investigated antinuclear social movements following the Fukushima Nuclear Power Plant accident. Satoh is a core member of the international comparative climate change policy network project, COMPON.

Clare Saunders is Professor of Environmental Politics at the University of Exeter, Cornwall. She is an expert on political participation, particularly environmental activism. Her work is widely published in a range of journals, including the *British Journal of Sociology*, *Journal of European Public Policy*, and *Political Research Quarterly*. Her latest book is *Organising for Change* (with Silke Roth), illustrative of the way she thinks about social change making as a broader enterprise than simply protest. She is currently writing for Bloomsbury on *Environmental Activism, Power, and the State* (with Sam Nadel and Bob Walley).

Kristen Shorette is Associate Professor of Sociology at the State University of New York, Stony Brook. She does cross-national comparative research with a focus on the environment, economy, health, and human rights. Her work focuses on the relationship between national embeddedness in world society and geopolitical position across the world-system.

Jackie Smith is a Professor of Sociology at the University of Pittsburgh. She has authored or coauthored numerous works on global and local social movements, including *Social Movements and World-System Transformation* (coedited with Michael Goodhart, Patrick Manning, and John Markoff), *Social Movements in the World-System: The Politics of Crisis and Transformation* (with Dawn Wiest), and *Social Movements for Global Democracy*. From 2012 to 2020 she served as editor of the open-access *Journal of World-Systems Research*. Smith is co-coordinator of Pittsburgh's Human Rights City Alliance and serves on the national steering committee of the U.S. Human Rights Cities Alliance.

David A. Sonnenfeld is Emeritus Professor of Sociology and Environmental Policy at the State University of New York College of Environmental Science and Forestry (SUNY-ESF), in Syracuse, New York; and Research Associate, Environmental Policy Group, Wageningen University, the Netherlands. His research interests focus on institutional environmental transformation, reform, and retrenchment in comparative perspective, in Asia, Europe, the Middle East, and elsewhere. His books include *Water Crises and Governance: Reinventing Collaborative Institutions in an Era of Uncertainty* and *The Ecological Modernisation Reader: Environmental Reform in Theory and Practice*.

Zoë Sullivan is a Graduate Student in Political Science at Tulane University. Her research focuses on the politics of climate migration and immobility, disaster recovery, international development, South-South cooperation, and gender and development. She is especially interested in the sociopolitical impacts of climate displacement in both the Global South and the US Gulf South, and she is committed to community-engaged research that centers equity, agency, and local voices.

Daniela Tapia is a Political Scientist from Universidad Diego Portales (Chile). She is currently pursuing a master's degree in social research methods at the same institution, where she is working on her thesis on climate finance. Daniela is also a research assistant in the Max Planck Partner Group on the Challenges of Green Growth.

Sirayuth Thongprasert is a Research Associate at the SMARTS Center, Asian Institute of Technology in Thailand. He is interested in understanding and resolving intersectional social inequities and injustices that crosscut dimensions, sectors, and actors through communication media, and how they interact and affect GESI. His current research and publications focus on climate mitigation, demand-side solutions, circularity, sufficiency, well-being, and GESI from the Global South. He has a bachelor of science in International Relations and Development from Rangsit University and a master of science in Gender and Development Studies from the Asian Institute of Technology in Thailand.

Nick Tinoco is a PhD Candidate in the Department of Sociology at UCLA interested in environmental sociology, climate mobilities, and the politics of climate change adaptation. Currently, his work examines how residents in wildland–urban interface communities mobilize in response to wildfire to remain in and defend high-fire-risk communities. He is interested in understanding how residents act collectively to support, challenge, or transform local wildfire governance regimes and shape the preparedness strategies enacted at the neighborhood and city levels. He has previously published research on postwildfire migration decision-making following the 2018 Woolsey Fire in Los Angeles County in the journal *Population and Environment.*

Fernando Tormos-Aponte is an Assistant Professor of Sociology and the Marianne Brown Dietrich Breakthrough Scholar at the University of Pittsburgh. Tormos-Aponte is also the Policy Lead for the Just Transition Alliance. He specializes in social movements, environmental and racial justice, intersectional solidarity, identity politics, social policy, and transnational politics. His research focuses on how social movements cope with internal divisions and gain political influence. Tormos-Aponte also investigates social movement claims about uneven government response across communities. His work in this area examines the causes and consequences of government neglect of socially vulnerable communities during disaster recoveries.

Jale Tosun is Professor of Political Science at Heidelberg University, Germany, and Adjunct Professor in the Department of Political Science at the University of Oslo, Norway, as part of the "Accelerating Climate Action and the State: Getting to Net Zero" (ACCELZ) project funded by the Research Council of Norway. She serves as Editor-in-Chief of the Nature Portfolio journal *Climate Action* and as Associate Editor of *Policy Sciences.* Her research focuses on comparative public policy, international political economy, comparative public administration, and European integration.

Katrin Uba is an Associate Professor of Political Science at Uppsala University, Sweden. Her research explores social movements, political activism, and policy change, with a strong focus on climate- and environment-related mobilization. She has published extensively on how protests, campaigns, and other forms of collective action influence political decision-making, such as research on fossil-free energy policy in Sweden. Her work appears in leading political science and sociology journals and contributes to understanding the dynamics of climate action in democratic societies.

Jamilya Ukudeeva teaches Political Science at West Valley College and Chabot College in California. She is currently developing an Open Educational Resource textbook in International Relations to expand global access to quality learning materials.

Xiaoliang Yang is the China Country Manager for the Global CCS Institute. Prior to the Institute, Xiaoliang served as Technology Director at OGCI China Climate Investments and held various positions at the World Resources Institute. In addition, Xiaoliang worked at the Department of Social Affairs of the UN Headquarters and KAPSARC in Saudi Arabia. Xiaoliang holds a PhD in Environmental Science from the State University of New York College of Environmental Science and Forestry and an MPA from the Maxwell School of Syracuse University. Xiaoliang also obtained a program certificate in Carbon Capture and Storage: Science, Technology, and Policy from MIT.

CHAPTER 1

The Role of Environmental Threat in Motivating Climate Action

PAUL ALMEIDA, ELIANA FONSAH, AND LUIS RUBÉN GONZÁLEZ

Introduction: Environmental Threats

In the collective action literature, opportunities and threats act as two major structural drivers of mobilization. Opportunities encourage collective action by incentives that involve new advantages and benefits if groups mobilize (McAdam & Tarrow, 2019). Threats spur mobilization via the worsening of existing conditions unless collective action is taken. Sociologist Charles Tilly theorized decades ago that threat has the potential to create greater levels of collective action than opportunity (Tilly, 1978, pp. 134–135). This perspective is consistent with prospect theory, whereby losses encourage more collective action than potential gains (Kahneman & Tversky, 2013; Snow et al., 1998). In recent decades, scholars have developed more precise classifications of types and dimensions of threat. These studies examine the role of repressive, economic, political, and environmental threats in relation to social mobilization and the properties of the threat. For the study of climate action in this volume, the focus centers on environmental threat (Johnson & Frickel, 2011)—the impact of deteriorating environmental conditions associated with climate change on moving civil society and institutions into joint efforts to reduce harms and increase resiliency. Ecological threats also drive climate action when they disrupt the subsistence needs of local communities or entire societies (Simmons, 2016).

The variable properties of environmental threat determine both the pace and the forms of climate action (Almeida, 2019a). These variable dimensions include the severity, endurance, and credibility of the threat (Einwohner & Maher, 2011). When climate change-related threats are severe, sustained, and credible, attempts at climate action are more likely. Severity can be conceptualized in multiple ways. Extreme flooding from hurricanes disrupts entire regions shaping several potential forms of climate action from victims demanding immediate relief to out-migration (Gould & Lewis, 2021, forthcoming; also see chapter by Tinoco & Menjivar, this volume). Such disasters also may result in more investment in climate resiliency and adaptation by creating more flood protection infrastructure and evacuation plans, whereby environmental justice communities may also advocate for equity. Severity (and endurance) can also include the loss of and damages to ecosystems and livelihoods of communities over several decades, inducing climate action by poorer countries in demanding reparations and compensation from Global North countries, which have contributed the majority of historical greenhouse gas (GHG) emissions (see chapter by Khan & Khan, this volume). The more groups experience climate-related events and disasters, the more they become credible threats, leading to greater awareness and willingness to act (Fisher, 2024).

Others have emphasized the breadth, visibility, source, and timing of threats (Zepeda-Millán, 2017). In this perspective, climate change drives climate action when threats are wide in scope (affecting multiple groups or large numbers of people), highly visible, occurring from a single source to focus blame, and impending over time to provide an adequate period to prepare climate actions. The breadth of climate change provides a planetary problem and a threat to unify across continents with the transnational climate movement acting as one of the largest social movements in the twenty-first century, mobilizing upward of 7.5 million people during global days of climate action in the 2010s (Almeida, 2019b). Both highly visible and single-source climate change threats include fossil fuel extraction and transportation in regions with vulnerable and marginalized populations (see chapters by Arce and Sullivan, as well as by Ovadia, this volume). Air, soil, and water pollution motivate communities near extraction sites to mobilize for cleanup, ecological restoration, and compensation for health, economic, and environmental damages (see chapter by Martínez-Alier, this volume).

Ecological Threat Related to Institutionalized Climate Action

Institutional climate actions are initiated by established organizations such as governments, city councils, state agencies, corporations, and nonprofit groups. The more routine actions initiated by institutions include climate action planning, green and renewable energy alternatives, carbon and other GHG capture technologies, nature-based climate solutions, carbon markets, climate education, and climate legal actions (Almeida et al., 2023). These forms of climate action maintain a long-term and broad perception of threat whereby there is time to develop and enact measures to mitigate GHG emissions and adapt to shifting environments. Some types of institutional climate actions may address not only threats but also opportunities for new benefits (e.g., cleaner and healthier communities with less pollution from fossil fuels).

Global Climate Action Cooperation and Planning

Climate action planning takes place at international, national, and local levels. Planning was one of the first forms of climate action to respond to the threat of global warming in the early 1990s. Climate action planning has a multiyear or even multidecade view of ecological threat whereby the long-term consequences of climate change must be challenged through rational planning that will contribute to mitigating GHG emissions and preparing adaptation and resiliency strategies. The threat is viewed as global and understood based on the best available atmospheric science and climate forecasting models (Sofaer et al., 2017).

At the international level, climate action planning is guided by the United Nations (UN) and calls for international cooperation. The UN responded to the threatening projections of global warming that became more apparent in the 1980s (Hansen et al., 1981). In 1988, the United Nations Environmental Program (UNEP) and the World Meteorological Organization (WMO) established the Intergovernmental Panel on Climate Change (IPCC). In the early 1990s, the UN convened climate scientists who articulated the ecological threat to establish international bodies to enact global climate action planning. Working against the threat of reaching ecological tipping points, intergovernmental bodies such as the UN and its affiliated agencies coordinated with nation-states to adopt various common climate mitigation and adaptation actions. The long-term international climate action planning of the UN corresponds to the wide scope of the threat of global warming, encompassing the entire planet.

In 1992, global leaders came together at the United Nations Conference on Environment and Development (UNCED, also known as the Rio Earth Summit), to establish the United Nations Framework Convention on Climate Change (UNFCCC). The UNFCCC responds to the climate threat by requiring member states to examine their contributions to global warming and climate change (Bohringer, 2003; Christoff, 2016; Fanning & Hickel, 2023). The UNFCCC led to the formation of the Conference of Parties (COP) with the goals to review and promote climate change discussions and initiatives among UN member nations on an annual basis. Since 1995, 29 annual COP conferences have been held with member state governments engaged in climate action projects addressing long-term threats from GHG emissions.

In terms of international climate planning, the major achievements of COP conferences include the establishment of the 1997 Kyoto Protocol setting mechanisms and targets for climate change mitigation planning, including the need to reduce GHG emissions from fossil fuels (especially by industrialized nations) 5.2% below aggregate 1990 levels in a collective act to stabilize global temperatures (Babiker et al., 2000; Bohringer, 2003). In 2009, at COP 15 in Copenhagen, member states agreed to reduce GHG emissions and pool funds for developing country protections (though the commitments were not binding and many nongovernmental organizations [NGOs] representing civil society were blocked from participating). The 2015 COP 21 Paris Climate Agreement guaranteed the first unanimously accepted and binding climate agreement after many years of disagreements over emission reduction negotiations (Gomez-Echeverri, 2018). During the Paris Climate conference, 174 governments signed the accord to declare their nationally determined contributions (NDCs) to lowering GHG emissions and action plans to decrease global warming (Jenkins et al., 2020).

At the civil society level, the sector is making significant contributions to climate planning globally. For example, the Climate Action Network (CAN) serves as a major force in international climate planning (Hadden, 2015). Formed in 1989, CAN counts 1,900 civic organizations in its membership with representation from 130 countries. This transnational grouping of NGOs advocates during annual COP meetings for binding international agreements pressuring governments to develop and implement more comprehensive national climate plans. CAN structures itself via regional networks in Africa, Asia, Europe, and the Americas. A more equity-centered faction of CAN split off in 2007, to form a new coalition of NGOs called Climate Action Now!, which was active through the early 2010s.

Regional- and National-Level Climate Action Planning

A large variety of climate action planning programs occur at regional and national levels. Under optimal conditions, the global economic transition to low-carbon economies would be guided by the Just Transition framework informed by global bodies such as the UN and its International Labor Organization (ILO) (Ciplet & Harrison, 2020; Jenkins et al., 2020). Just Transitions involve a focus on green energy conversions that incorporate the needs of the displaced labor sector and marginalized communities (see chapter by Kerrissey & Coon, this volume; Cha, 2024). In terms of regional climate planning, Just Transitions have progressed the furthest in the European Union whereby member states approved a Just Transitions Fund in 2021. The European Green Deal is the largest coordinated program for climate action planning across a continent with ambitious emissions and sustainable industry benchmarks with the goal of making the first climate-neutral continent (see chapter by Tosun, this volume).

Climate change threats have pushed several developing nations to take climate action planning seriously as demonstrated in the high level of commitment to the Paris Climate Agreement through the ratification of the accord and pledges to reduce GHG emissions through a variety of strategies. For example, climate-induced threats propelled the African Union (AU) to establish a 10-year strategic climate action plan incorporating regional and global frameworks to facilitate collaboration in addressing global warming, while building climate-resilient communities (African Union, 2022). The AU climate plan established the African Adaptation Initiative (AAI) to enhance climate information sharing, strengthen policies and institutions, promote concrete actions, and facilitate access to resources and promote low-emission pathways in accordance with the Paris Climate Agreement. In terms of food security, the plan's Comprehensive Africa Agriculture Development Program (CAAP) and Accelerated Agricultural Growth and Transformation Program provide guidelines and knowledge on climate change and agricultural practices through initiatives such as the Africa Climate Smart Agriculture Alliance to tackle food insecurity in the region.

Global climate change is exacerbated in sub-Saharan Africa by severe deforestation and desertification, ecosystem disruption, loss of biodiversity, and dependence on firewood and charcoal for cooking and heating (Frankyln, 2009). Forests play important roles in absorbing CO_2 in Africa and the world at large (see chapter by Alvarez, this volume). Forests equally play critical roles in preserving soil quality, biodiversity, and ecosystems. The climate change threats (warming and environmental degradation) caused by deforestation have motivated African countries to engage in national tree planting projects and to incorporate the United Nations REDD+ program in their climate action plans to mitigate the effects of climate change and rehabilitate degraded lands as a nature-based climate solution.

Ethiopia has experienced a reduction in forest coverage from 30% of total land area to 4% (British Broadcasting Corporation, 2019). The acute deforestation of this drought-prone country prompted the government to embark on a massive reforestation program that witnessed the historic planting of 350 million trees in 2019 in a bid to counter climate change effects exacerbating droughts in the country (British Broadcasting Corporation, 2019). In 2021, Rwanda's government embarked on an afforestation project planting 40 million trees (Mlaba, 2021). In other African countries with large forested areas, including Nigeria and Cameroon, governments are working to protect reserves as well as engage communities in reforestation projects to increase canopy coverage and rehabilitate degraded lands (Gakou-Kakeu et al., 2022; Isyaku, 2021).

Other regional climate action planning projects simultaneously deliver environmental solutions and maximize socioeconomic benefits to communities, critical in disadvantaged regions and societies. The African Climate Action Partnership (AfCAP) comprised of seven African countries—Ghana, Zambia, D.R. Congo, Cameroon, Kenya, Mozambique, and Côte d'Ivoire—provides one example. AfCAP partners with REDD+, the African Union, and UN climate action programs to forecast and develop integrated climate modules that drive investments in climate planning (EU-UNEP Africa LEDS Project, 2017a, 2017b). These countries are currently implementing green agriculture (climate-smart agriculture), agricultural waste processing, renewable energy, and clean domestic fuel projects. In terms of smart agriculture, Côte d'Ivoire has programs composting rice straws to produce fertilizers, while rice husks are processed into fuel briquettes for cooking, replacing charcoal. In a similar fashion, Nigeria, Uganda, and Cameroon approach climate change in agriculture under the Ecosystem-Based Adaptation for Food Security Assembly (EBAFOSA) program and provide economic opportunities for communities (EBAFOSA, 2022). To promote the transition from heavy reliance on forests for food processing to solar drying powered systems, large greenhouse community farms and small solar dryers are being established to preserve staple foods. Farm waste is also being used to produce organic fertilizers for agriculture and affordable fuel briquettes that burn efficiently without contributing to deforestation and pollution (Fonjong et al., 2024).

Scholars are increasingly advocating for the environmental threat of plastics to be incorporated into climate action planning at regional levels as the production and waste of synthetic materials contribute significantly to GHG emissions (Noman et al., 2025). In Africa, plastic waste produced by fossil fuels exacerbates existing concentrations of environmental pollution. Discarded plastics constitute up to 10% of mass human waste, and 60 to 70% is deposited in the oceans and other water bodies (Bezerra et al., 2021). On the continent alone, the neglect of plastic waste management has resulted in the disruption of ocean ecosystems (affecting more than 2,240 marine species and interrupting the ability of the oceans to absorb carbon from the atmosphere), killing wildlife on land, impacting soil quality for agriculture, and intensifying health problems. The impacts of plastic waste on the environment have forced African governments and regional blocks such as the Southern African Development Community (SADC comprising of all 16 southern African countries) to massively adopt plastic-use control measures as a climate change mitigation strategy. More than 40 out of 54 African states have either passed laws banning the use of plastics or adopted legislations restricting the use of nonbiodegradable plastics (Bezerra et al., 2021; Greenpeace Africa, 2020; Wanja, 2022). The laws and legislation ban the importation, manufacture, commercialization, and use of plastic packaging bags and some plastic bottles based on their environmental impact assessment measured mostly in terms of biodegradability. Though challenged by issues of effective implementation, the strategy has drastically reduced the quantity of plastic waste causing environmental issues in Africa. It should also be noted that similar forms of international climate planning taking place in Africa also occur in other regions of the Global South (see chapters on Latin America and Asia, this volume).

Local-Level Climate Action Planning

City-level climate planning served as one of the first actions to respond to the long-term environmental threat of climate change with a focus on mitigation to reduce GHG emissions. Cities and local authorities have been at the forefront of climate action planning (see chapter by Leffel, this volume). They engage with diverse actors from business, civil society organizations, and communities to implement strategies to reduce emissions based on GHG inventories. Municipal-level climate planning also involves adopting renewables energy, low carbon public transportation systems, decarbonizing buildings, and implementing nature-based climate solutions. These strategies are often formalized in a climate action plan (CAP). Some of the leading initiatives for CAPs at the local level include the Transition Towns movement, ICLEI, C40 Cities, Energy Cities, and the Global Covenant of Mayors (Tosun & Schoenefeld, 2017). Each program networks across localities sharing information, knowhow, and experiences with an understanding of the wide and global scope of the threat of planetary warming. The ties between cities provide major conduits for the diffusion of climate action planning across the planet (Leffel, 2022).

The International Council for Local Environmental Initiatives (ICLEI; now called Local Governments for Sustainability) formed in the early 1990s and now partners with over 2,500 local and regional governments across the world. ICLEI specializes in providing expertise in developing local climate action plans starting with GHG inventories. The Global Covenant of Mayors collaborates with 10,000 cities in 140 countries with similar goals as ICLEI to develop local measures to reach GHG reduction targets along with other green energy and decarbonization measures. Founded in 2005, C40 focuses on local-level climate planning in the world's largest metropolises. It currently has nearly 100 city members that account for over 20% of the world's economy. C40 member cities include London, Beijing, Santiago, Sao Paulo, Mexico City, Copenhagen, Bangkok, Jakarta, Melbourne, Lagos, Johannesburg, Los Angeles, and New York (see chapter by Leffel, this volume).

Green Energy and Climate Mitigation Technologies

Even though GHG emissions reduction is critical to addressing the ongoing environmental threat of global warming, it cannot be achieved without the decarbonization of the global economy or a transition to net-zero systems. These strategies include the rapid expansion of renewable energy as well as a variety of carbon capture and removal technologies. The most promising renewable energy sources include photovoltaics, turbine, geothermal, and hydrogen power (see chapter by Hochstetler, this volume). Nuclear energy continues to suffer from the dangers of cooling and hazard waste disposal, while a new generation of small modular reactors (SMRs) is purported to pose relatively fewer risks. Advocates of renewable energy as a replacement to fossil fuels maintain a long-term environmental threat time horizon similar to climate action planning, whereby there is a perceived need to act within a sufficient period to develop new technologies and overcome political and economic obstacles (see chapter by Vasi, this volume). A majority of the world's countries have now reached at least 30% renewable energy usage (see Hochstetler, this volume). Even in western Africa, countries such as Senegal have established clean energy hubs that promote green industrial production by supporting firms and businesses in strengthening their capacities to pursue energy efficiency and solar energy development (The African Climate Foundation, 2025). In Namibia, national climate resilience strategy involves investing petroleum profits into green hydrogen production for global export, relying heavily on local wind and photovoltaics to power the electrolyzers (DeBoom, 2025).

Another related set of technologies involves carbon management to reduce GHG emissions and remove historical releases. There exists a wide variety of these technologies and they include direct air capture of carbon (DAC), carbon capture and storage (CCS), biomass carbon removal and storage (BiCRS), bioenergy with carbon capture and storage (BECCS), and dairy digesters, to name some of the most prominent (Pett-Ridge et al., 2023; see chapter by Yang, this volume). Some experts estimate the need to remove a thousand gigatons of carbon to avoid surpassing the 1.5°C increase in average global temperature threshold (Buck, 2019). Many carbon capture technologies have been developed in collaboration with the oil and gas industry and are highly subsidized by national governments (especially CCS). A range of nature-based climate solutions also provide a means for CCS, including biochar, cover crops in agriculture, algae/seaweed farms, reforestation, and afforestation. The future of both green/renewable energy and carbon management industries depends on the level of commitment by governments, with climate skepticism and misinformation by powerful states emerging as a serious impediment in the 2020s (see chapter by Barros Leal Farias, this volume).

Other Forms of Institutional Climate Action

The list of additional threat-driven institutional climate actions would be difficult to exhaust. Some of the most prominent include climate education, cap and trade, carbon taxes, climate emergency declarations, and lawsuits and referendums. Climate education programs are driven by a long-term view of the climate threat, where there is enough time to raise awareness and skill levels to respond to the changes. The most promising approach includes climate curriculum in public school systems as well as community, labor, and state organizations reaching marginalized and excluded populations. The market-based approaches of cap-and-trade programs and carbon taxes are consistent with an ecological modernization view of climate action (see chapter by Satoh, this volume). Another type of institutional action closely related to pressures from civil society is the declaration of a climate emergency. Commonly used in cities, such declarations aim to bring awareness on the seriousness of the climate crisis and help governments to mobilize resources for addressing it. To date, 2,364 jurisdictions have declared a climate emergency (Climate Emergency Declaration, 2024). The vast majority of those declarations came from local governments, with only 18 from national governments. These declarations highlight environmental threats as urgent and immediate concerns by labeling them as "emergencies" (Rilling & Tosun, 2021).

Civil society groups have used the democratic process to introduce *lawsuits* linked to climate-related losses and damages. Employing the growing jurisprudence on environmental harms, protections, and reparations in the last decades, sociopolitical coalitions are aiming to receive the recognition and ruling of courts regarding intergenerational justice and the fair distribution of hazards and benefits. Such lawsuits also incorporate a long-term threat time horizon whereby there is sufficient time to act. Columbia University's Sabin Center for Climate Change Law has registered 1,961 court cases related to climate change in the United States between 1986 and 2025, and 1,039 in the rest of the world (in 55 countries) between 1994 and 2024 (Sabin Center for Climate Change Law, 2025a, 2025b). Besides lawsuits, civil society groups (Indigenous, environmental, youth, labor, and women's movements) and NGOs have employed democratic participation mechanisms to attempt to bring about institutional action in relation to climate change (Weller & Tran, 2022). This has been the case of the failed climate referendum in the city of Berlin, Germany (to achieve carbon neutrality by 2030) and the successful one in Switzerland (for reaching net-zero emissions by 2050) (British Broadcasting Corporation, 2023; Thomeczek, 2024). The successful Ecuadoran referendum in 2023 to forbid oil exploration and drilling in the Amazonian Yasuní National Park stands apart, as it was the result of a long-standing mobilization from Indigenous communities, environmental NGOs, and their transnational allies (The Associated Press, 2023; Lewis, 2016).

Environmental Threat and Noninstitutional Climate Action

Noninstitutional climate actions include the use of strategies and tactics outside of routine and institutionalized political behavior (Almeida et al., 2024). Such strategies provide one of the defining features of social movements (Almeida, 2019c). Examples include strikes, street marches, rallies, occupations, boycotts, and other unconventional forms of collective action. In relation to the environmental threat of climate change, two major forms of noninstitutional climate action include the international climate social movement and local struggles against fossil fuel extraction, transportation, processing, and nearby GHG emissions.

Transnational Climate Social Movements

The international climate movement is driven largely by the threat of the acceleration of the climate crisis and the lack of advancement by institutional forms of climate action (Fisher, 2024).

Transnational climate protests took off in the mid-2000s after over a decade of institutionalized climate action guided by the UN and city-level climate action planning. In the mid- to late 2000s, climate movements coordinated days of international action, usually co-occurring with the annual UN-sponsored COP summits. These first-wave mobilizations received endorsement by the Assembly of Social Movements within the World Social Forum. The Global Climate Coalition coordinated the protests as they focused on the failure of the Kyoto Protocol and making global climate treaties binding in preparation for COP 15 in Copenhagen in 2009. In the buildup to COP 15, several new international climate movement organizations and campaigns entered the scene, such as Tck, Tck, Tck, 350.org, and Avaaz (see chapter by Bates, Boeve, & Henn, this volume). The first-wave global climate mobilizations of the late 2000s also benefited from the establishment of an international infrastructure of progressive movements built in the previous years by the global economic justice movement, the World Social Forum, and the anti–Iraq war movement (Almeida, 2019b; Chase-Dunn & Almeida, 2020; Hadden, 2014).

After Copenhagen, a second wave of global climate activism occurred in the early 2010s in the leadup to COP 21 in Paris. In September 2014, 350.org and AVAAZ helped coordinate a massive climate march of an estimated 400,000 people in New York City coinciding with the United Nations Climate Summit. At the same time, an additional 2,600 protest events occurred across the planet in 160 countries. Climate activists repeated this same level of mobilization in late November 2015 with global days of climate action reaching 1,900 cities around the world. These protest events and rallies likely assisted in the signing of the Paris Accords in December 2015 (during COP 21) with binding national commitments to prevent global average temperatures from increasing past 1.5°C.

Following COP 21, there was a relative pause in massive transnational climate protest until late 2018. A third wave of global climate activism gained momentum throughout 2019, bringing climate street mobilizations to new heights. The third wave of climate movement energy was brought about by emerging groups such as Fridays for Future (FFF) (see chapter by Uba, this volume) and Extinction Rebellion (XR) (see chapter by Cavalho & Gardner, this volume). FFF launched a series of school strikes across Europe in March and throughout the world in May 2019. By the fall of 2019, FFF was peaking. The high point of FFF occurred during the United Nations Climate Action Conference in New York City in 2019. Between September 19 and 26 of that year, some 6,000 protest events took place with an estimated 7.6 million participants reaching 185 countries, including a march of 250,000 led by Greta Thunberg on September 19. XR and the Sunrise Movement also participated in these actions (Chase-Dunn & Almeida, 2020). The third-wave climate protests influenced key components of the European Green Deal and the Biden administration's climate investments in the United States.

Since the COVID-19 pandemic of 2020 and 2021, the global climate movement has yet to recover in terms of the level of street mobilization. More climate activism moved online, and new organizations emerged in the 2020s, such as Just Stop Oil, Third Act, and the expansion of XR into the Global South (see chapter by Carvalho & Gardner, this volume). One of the largest challenges for the international climate movement involves incorporating an environmental justice focus (see chapter by Tormos-Aponte & Smith, this volume). Extraction struggles provide one site of climate conflict where environmental justice goals often predominate.

Environmental Threats of Resource Extraction and Noninstitutional Climate Action

Antiextraction mobilization involves communities and environmental movements opposed to the localized impacts of the global capitalist economy's increasing demand for raw materials (Gudynas, 2020). Antiextraction protests conform to "glocal" actions—global production and consumption of resources harming local communities (Auyero, 2001; Svampa, 2019). They become a form of noninstitutional climate action when the extraction involves fossil fuels or the transition to green energy development (see chapters by Martínez-Alier, Ovadia, and Arce & Sullivan, this volume). These are immediate and urgent environmental threats where community livelihoods and environmental health are at stake. Scholars and activists refer to these sites of struggle as "frontline communities" or "environmental justice communities." The extraction of raw materials and commodities acts as the first phase of the Global Treadmill of Production (Gould et al., 2008; Schnaiberg, 1980; see chapter by Gould & Lewis, this volume). Extraction activities provide the energy and materials for subsequent metabolic conversion into manufactured goods until they become waste—leaving side-polluting products throughout the whole cycle (Bunker, 2003; Burkett & Foster, 2006; Givens et al., 2019). For example, the material foundations of high-tech economies require several raw materials to build and make its infrastructure function. Before their conversion—through industrial production—into building compounds, hardware instruments, highways, and so on, the inputs of this industry were raw materials, such as copper, lithium, and nickel, and fuels, such as oil, gas, and coal, that had to be extracted (Jorgenson, 2016). Antiextraction climate action increases when local communities are ecologically threatened from meeting their basic subsistence needs (Simmons, 2016).

From a long-term perspective, extractivism stands behind several anthropogenic changes over natural landscapes, with the consequent socioecological pressures and environmental threats (Freudenburg et al., 1995; Holleman, 2017). Extraction of raw materials is enmeshed in processes that drive GHG emissions and their socioenvironmental consequences. Structural features from extractivism shape the localized eco-territorial conflicts in opposition to it, including those of activities related to climate change and energy transitions. Once reservoirs become scarcer, extractivist activities move to other areas where reservoirs are less abundant or more costly to exploit.

The changes in the oil frontier are an example of this dynamic (Bebbington & Bury, 2014; Haluza-DeLay, 2012). The extraction of raw materials and commodities expands throughout "sacrifice areas": highly intervened landscapes of intense monoproduction, where negative social, economic, and environmental externalities are placed. In these sites, the disruption of everyday life, dispossession, and harms of extraction on workers become the norm (Kuletz, 1998; Mayer, 2018). The international division of labor and power imbalances explain this "unequal ecological exchange" in terms of the uneven distribution of the negative consequences of extractivism (see chapter by Jorgenson et al., this volume). These forms of environmental injustice act as a crucial component of the development model of nations in the Global South and marginalized communities in the Global North (Jorgenson, 2010; Rodney, 1981). Structural pressures from extractivism transform into concrete and localized environmental threats that mobilize subaltern populations. The struggles at times involve state violence against local communities opposing the extraction and transport of fossil fuels and minerals needed for a low carbon transition (e.g., nickel, copper, and lithium). In a global analysis of 649 extraction-type conflicts, 10% involved the assassination of eco-activists, with particularly heightened state and corporate violence in operations involving hydropower, biomass, pipelines, and coal extraction (Temper et al., 2020).

In the Maghreb region (especially Algeria, Tunisia, and Morocco), large-scale resource extraction, green energy development, and phosphate mining have polluted water bodies and arable land critical to human survival. The rural poor, unemployed, landless, rural workers, fisherfolks, and small-scale farmers who are the most affected by subsoil exploitation, land degradation, loss of livelihood, and severe weather conditions have responded with various forms of counterhegemonic resistances and mobilization framed in connection to environmental justice for environmental protection, public benefits from extractivism, and access to land (Hamouchene, 2019, 2022).

Large-scale extraction of raw material activities, such as an open-pit copper mine or shale fracking for oil and gas production, places pressure on the economic and ecological sustainability of local communities and their surrounding environment (see chapter by Arce & Sullivan, this volume). This is particularly felt on agricultural lands, forests, waterways, and mountains (Vasi et al., 2015; Widener, 2021). In the Global South, these spaces have been functionally defined in the treadmill of production as "resource reservoirs" and are often where peasant and Indigenous communities—sometimes originating from a previous colonial displacement—find their productive and reproductive niches (Bacon, 2019; Bartra Vergés, 2006). In contrast to the urban-based transnational climate movements outlined earlier, environmental threats from the expansion of extractivist activities are very tangible and immediate for the respective communities, attributable to specific actors and megaprojects. Nonetheless, mobilization does not instantaneously occur with the presence of concrete environmental threats from extractive activities; organizational resources and collective attributions also matter. Particularly in vulnerable populations, destruction of community ties and mutual trust—from long colonial histories and the form of implementation of the extractive megaproject—could undermine collective action (Auyero & Swistun, 2009; Reyes-García & Jofré, 2024).

There are three common types of environmental threats from extraction of commodities and raw materials related to fossil fuels and the green economy. The first one is the displacement of communities through land grabbing, rent, expropriation, and other mechanisms of capitalist accumulation, which disrupts preexisting "nature" and society-environment equilibria (Blanes et al., 2023; Scheidel et al., 2023). An example is the dispossession of communities by hydropower megaprojects, near the dam sites and the reservoirs (Khagram, 2005; Schroering, 2024). A second ecological threat from extractivism comes from the overexploitation of a particular raw material, which can cause the exhaustion of a critical component of a given natural environment. Such resource exhaustion could affect local economies or foster the creation of diseases by altering the food supply (Artiga Purcell, 2021; Longo et al., 2015). These processes are observed in agricultural extractivism to produce biofuels (e.g., soy and sugarcane), which can result in the loss of productivity of soils from overexploitation or in multiple disease outbreaks common in monocultural croplands (Leguizamón, 2019; Scheidel et al., 2023). A third noticeable environmental threat from extractivism derives from contamination. Pollutants could be released in the course of extracting raw materials or as side products from on-site processing. For instance, shale fracture procedures in natural gas and oil extraction in both the Global North and South may produce air contamination of the surrounding area and the risk of massive leaks of methane to the atmosphere—a greenhouse effect gas, more potent than CO_2 (Grosse, 2022). Moreover, as a growing share of extractivist activities, natural gas production affects water quality—by the contamination of rivers and the phreatic layer (Vasi et al., 2015). Although these environmental threats are inherent to extractivism, their harms are augmented by lack of proper regulation, narratives of expendability of sacrifice zones (and their inhabitants), and racial, class, and geopolitical inequalities (Hooks & Smith, 2004; La Duke, 2010; Taylor, 2014).

As depicted in the previous examples of hydrological-powered dams, biofuels, and shale fracture gas and oil, extractivism is well-entrenched in the conflicts and dilemmas of the energy transition. One of the most important sources of antiextraction mobilization relates to the opposition to the direct effects of the new extreme forms of hydrocarbon production: offshore, shale-fracking, open-cast coal mining, or extraction in protected areas (Chomsky & Stuffler, 2014; Delina, 2021; Grosse, 2022). Despite the efforts for pushing for renewable energies and more efficiency in energy use, the absolute production of hydrocarbons has kept increasing under favorable prices and increasing demand (Fernández Durán & González Reyes, 2018), particularly for oil as atmospheric concentrations of CO_2 continue to increase. As a result, the petroleum frontier has expanded, pushing for exploration and production projects in remote places, and through more extreme forms of extraction (as mentioned before with shale fracture) or in more difficult refining reserves (Ferrari et al., 2024). The uptick in mobilization by Indigenous communities in the Amazon—especially in the Ecuadoran, Colombian, Venezuelan, and Peruvian regions—against new oil projects illustrates the contentious reaction to this dynamic (Hecht, 2005; Lewis, 2016; Vasquez, 2014). These communities and their allies contest the socioeconomic dispossession and ecosystem disruption of petroleum exploitation in an area historically called the "Lungs of the Earth" (see chapter by Fenelon & Figueroa Helland, this volume).

Other community-based antiextraction mobilization against hydrocarbons is motivated by environmental health concerns of related industrial activities in particular localities. An example is the struggles of Louisiana communities in the "Cancer Alley" region against the oil and petrochemical industry near the Gulf of Mexico (Bullard & Wright, 2012; Human Rights Watch, 2024). Nevertheless, not all collective action against these projects takes the form of public opposition or through environmental concern: While some of them do, others link to the unequal distribution of harms, rents, or side effects from these energy megaprojects and the economic dependency of nearby communities (Arce, 2014), as well as everyday forms of resistance. Moreover, protests motivated by the threats of nonrenewable energy sources tend to be more successful than protests for renewable energy (Temper et al., 2020; Uba & Engeman, 2025).

Besides climate action against the expansion of hydrocarbons, mobilization in opposition to threats from extractivist megaprojects of *renewable energy* has also expanded, particularly in the second decade of the 21st century. After the 2008 great economic recession, a "decarbonization" paradigm became partially institutionalized in public and private energy investments. This has encompassed heavily state-subsidized megaprojects of "clean energy" production, which have been erected as an emerging economic growth and development strategy, including during the Biden administration in the United States, the European Union, and China (Bringel & Svampa, 2023; Méndez, 2020; Spash, 2010). The financial sector (public and private) has allocated substantial investments in these types of energy projects, increasing their share in the electricity and transportation sectors. A global wave of projects has occurred in wind farms, photovoltaic parks, hydrodams, biofuels, or mining for critical materials (i.e., lithium, nickel, cobalt, copper, and rare earth elements) for the ever-growing demand for electric vehicles and decarbonization initiatives (Andreucci et al., 2023; Kingsbury, 2022; Sellwood & Valdivia, 2018). A large portion of these projects follow an extractivist logic with similar consequences to other energy megaprojects, and therefore on several occasions they have become the target of antiextractive mobilization.

An example comes from the "Lithium Triangle" in South America (Bolivia, Chile, and Argentina). The area contains about half of the world's lithium reserves (Milewski, 2023). The massive extraction of this chemical for electric batteries (for smartphones and electric cars) has led to territorial displacement of Indigenous communities inhabiting these remote borderland areas (Argento & Puente, 2021; Kingsbury, 2022). Since lithium extraction concentrates in the salt flats and surrounding areas through major territorial interventions, it disrupts the main productive activities of local communities in salt production, local tourism, and agriculture. Moreover, there is the intense use of water in lithium extraction and the contamination of water sources in the desert ecosystem, which is vital for humans, animals, and plants (Jerez et al., 2021). Affected communities—mostly, Indigenous groups—and networks of national and international environmental movements resist the consequences of large-scale lithium extraction through public protest actions (Bringel & Svampa, 2023; Pragier, 2019). Besides opposing this form of extractivism, these movements underscore community-based solutions to climate change and point to the unsustainability of a mere energetic substitution. The movements denounce what scientists and scholars have pointed out before: Given the current and future energy demands, there are not enough lithium reserves to replace hydrocarbons. Even the attempt to reach such ambitious goals would likely have substantial ecological consequences, particularly in the extraction areas (Argento & Puente, 2021). The communities of the "Lithium Triangle" form part of a new constellation of forces that mobilize against the concrete and local environmental threats from the green economy. They underscore the contradictions, limitations, and environmental injustices of the prevalent decarbonization paradigm, without denying the urgency of climate action (Almeida et al., 2024).

Conclusion

This chapter emphasized the complex relationship between the environmental threat of planetary warming and some of the dominant forms of climate action. The present volume offers panoramic coverage of the forms of climate action discussed in this overview chapter. Scholarly experts weigh in and review major forms of institutional and noninstitutional climate action as well as geographical representation from every major world region. They provide state-of-the-art knowledge of the strengths, weaknesses, and viability of addressing the climate crisis through a variety of strategies, practices, and technologies. Taken together, the contributions of the Handbook also offer a coherent landscape to the vast and interdisciplinary literature on climate action. Readers passing through the following chapters will have a better understanding of the diverse uses and meanings of the term "climate action" by exploring its multiple dimensions and contexts where institutions and actors employ the concept.

Part I consists of chapters that cover the core environmental theories related to climate action. These chapters provide important conceptual and contextual roadmaps on how different approaches to climate action are applied, from the institutional reforms-oriented ecological modernization theory to perspectives that are concerned with current economic arrangements and equity, such as the treadmill of production, unequal exchange, intersectionality, and environmental justice. These approaches facilitate understandings of the varying emphases of climate action strategies from technological fixes to deep concerns with exploitive relations with nature and the unequal distribution of climate damages and harms on human populations.

The second section of the book (Part II) centers on noninstitutional forms of climate action. These chapters largely encompass the field of climate-based social movements around the globe. Contributions discuss the origins of climate activism as well as the most important achievements in terms of raising public awareness and education, increasing participation, and implementing climate-friendly policies. Experts in this subfield also demonstrate why nonconventional forms of collective action are necessary and effective when existing institutions fail to address global warming at an appropriate pace. Descriptions of the most influential climate movements are also provided.

Parts III through V of the Handbook focus on institutionalized forms of climate action. The entries offer extensive coverage of climate action planning, implementing alternative energy systems, and sustainable economic transitions, including Just Transitions. The final section, Part VI, emphasizes both institutional and noninstitutionalized climate actions in different world regions covering most of the globe. The volume concludes with a chapter on "The Pathway Forward for Climate Action," summarizing some of the major achievements of climate action to date as well as a heavy focus on the substantial obstacles to expanding action and civic engagement in the race against time.

References

The African Climate Foundation. (2025). "Advancing Senegal's just energy transition: A vision for a sustainable future." The African Climate Foundation. https://africanclimatefoundation.org/news_and_analysis/advancing-senegals-just-energy-transition-a-vision-for-a-sustainable-future/

African Union. (2022). *African Union climate change and resilient development strategy and action plan (2022–2032)*. African Union.

Almeida, P. (2019a). The role of threat in collective action. In D. A. Snow, S. A. Soule, & H. Kriesi (Eds.), *The Wiley Blackwell companion to social movements* (pp. 43–62). Wiley. https://onlinelibrary.wiley.com/doi/book/10.1002/9781119168577

Almeida, P. (2019b). Climate justice and sustained transnational mobilization. *Globalizations, 16*(7), 973–979.

Almeida, P. (2019c). *Social movements: The structure of collective mobilization*. University of California Press.

Almeida, P., González Márquez, L. R., & Fonsah, E. (2024). The forms of climate action. *Sociology Compass, 18*(2), e13177.

Andreucci, D., García López, G., Radhuber, I. M., Conde, M., Voskoboynik, D. M., Farrugia, J. D., & Zografos, C. (2023). The coloniality of green extractivism: Unearthing decarbonisation by dispossession through the case of nickel. *Political Geography, 107*, 102997. https://doi.org/10.1016/j.polgeo.2023.102997

Arce, M. (2014). *Resource extraction and protest in Peru*. University of Pittsburgh Press.

Argento, M., & Puente, F. (2021). 7 hipótesis sobre las dinámicas territoriales y el litio en Argentina. In *Salares Andinos. Ecología de saberes por la protección de nuestros salares y humedales* (pp. 1–17). Observatorio plurinacional de salares andinos.

Artiga Purcell, J. A. (2021). *Contesting extractivism: Gold, water and power in El Salvador* [Tesis de doctorado]. University of California, Santa Cruz.

Associated Press. (2023, August 21). Ecuadorians reject oil drilling in the Amazon, ending operations in a protected area. AP News. https://apnews.com/article/amazon-ecuador-oil-drilling-referendum-yasuni-5c72a325755976c47a3ec138bdab8537

Auyero, J. (2001). Glocal Riots. *International Sociology, 16*(1), 33–53.

Auyero, J., & Swistun, D. A. (2009). *Flammable: Environmental suffering in an Argentine shantytown*. Oxford University Press.

Babiker, M., Reilly, J. M., & Jacoby, H. D. (2000). The Kyoto Protocol and developing countries. *Energy Policy, 28*(8), 525–536. https://doi.org/10.1016/S0301-4215(00)00033-1

Bacon, J. M. (2019). Settler colonialism as eco-social structure and the production of colonial ecological violence. *Environmental Sociology, 5*(1), 59–69.

Bartra, V. A. (2006). *El capital en su laberinto: De la renta de la tierra a la renta de la vida* (1st ed.). Universidad Autónoma de la Ciudad de México.

Bebbington, A., & Bury, J. (Eds.). (2014). *Subterranean struggles: New dynamics of mining, oil, and gas in Latin America*. University of Texas Press.

Bezerra, J. C., Walker, T. R., Clayton, C. A., & Adam, I. (2021). Single-use plastic bag policies in the Southern African development community. *Environmental Challenges, 3*, 100029. https://doi.org/10.1016/j.envc.2021.100029

Blanes, R. L., Rodrigues, A. C., & Goncalves, E. (2023). The multiple paths of extraction, dispossession, and conflict in Mozambique: From Tete's coal mines to Cabo Delgado's LNG projects. *The Journal of Social Encounters, 7*(1), 4–25.

Bohringer, C. (2003). The Kyoto protocol: A review and perspectives. *Oxford Review of Economic Policy, 19*(3), 451–466. https://doi.org/10.1093/oxrep/19.3.451

Bringel, B., & Svampa, M. (2023, August). Del «Consenso de los commodities» al «Consenso de la descarbonización». *Nueva Sociedad, 306*, 51–70.

British Broadcasting Corporation. (2019, July 29). *Ethiopia "breaks" tree-planting record to tackle climate change.* BBC. https://www.bbc.com/news/world-africa-49151523

British Broadcasting Corporation. (2023, June 19). *Switzerland referendum: Voters back carbon cuts as glaciers melt.* BBC. https://www.bbc.com/news/world-europe-65946888

Buck, H. J. (2019). *After geoengineering: Climate tragedy, repair, and restoration.* Verso Books.

Bullard, R. D., & Wright, B. (2012). *The wrong complexion for protection: How the government response to disaster endangers African American communities.* New York University Press.

Bunker, S. (2003). Matter, space, energy, and political economy: The Amazon in the world-system. *Journal of World-Systems Research, 9*(2), 219–258. https://doi.org/10.5195/jwsr.2003.241

Burkett, P., & Foster, J. B. (2006). Metabolism, energy, and entropy in Marx's critique of political economy: Beyond the Podolinsky myth. *Theory and Society, 35*(1), 109–156. https://doi.org/10.1007/s11186-006-6781-2

Cha, J. M. (2024). *A just transition for all: Workers and communities for a carbon-free future.* MIT Press.

Chase-Dunn, C., & Almeida, P. (2020). *Global struggles and social change.* Johns Hopkins University Press.

Chomsky, A., & Stuffler, S. (2014). Empire, labor, and environment: Coal mining and anticapitalism environmentalism in the America. *International Labor and Working-Class History, 85*, 194–200.

Christoff, P. (2016). The promissory note: COP 21 and the Paris Climate Agreement. *Environmental Politics, 25*(5), 765–787.

Ciplet, D., & Harrison, J. L. (2020). Transition tensions: Mapping conflicts in movements for a just and sustainable transition. *Environmental Politics, 29*(3), 435–456. https://doi.org/10.1080/09644016.2019.1595883

Climate Emergency Declaration. (2024, September 22). Climate emergency declarations in 2,364 jurisdictions and local governments cover 1 billion citizens. *Climate Emergency Declaration.* https://climateemergencydeclaration.org/climate-emergency-declarations-cover-15-million-citizens/

DeBoom, M. J. (2025). Hedging energy transition: Green hydrogen, oil, and low-carbon resilience as state strategy in Namibia. *Geoforum, 161*, 104267.

Delina, L. L. (2021). Topographies of coal mining dissent: Power, politics, and protests in southern Philippines. *World Development, 137*, 105194. https://doi.org/10.1016/j.worlddev.2020.105194

EBAFOSA. (2022). Analysis of the impact of environmental Solutions in addressing the triple crisis for food and livelihood security. EBAFOSA. https://ebafosa.org/index.php/mediaa/item/918-analysis-report

Einwohner, R., & Maher, T. (2011). Threat assessment and collective-action emergence: Death-camp and ghetto resistance during the holocaust. *Mobilization: An International Quarterly, 16*(2), 127–146.

EU-UNEP Africa LEDS Project. (2017a). EU-UNEP Africa LEDS AMCEN breakfast session. https://www.africaleds.org/index.php/meddia/84-news/190-eu-unep-africa-leds-amcen-breakfast

EU-UNEP Africa LEDS Project. (2017b). AfricaLEDS at AMCEN: Africa LED countries. https://www.africaleds.org

Fanning, A. L., & Hickel, J. (2023). Compensation for atmospheric appropriation. *Nature Sustainability, 6*(9), 1077–1086.

Fernández Durán, R., & González Reyes, L. (2018). *En la espiral de la energía. Vol. II, Colapso del capitalismo global y civilizatorio* (2nd ed.). Libros en Acción; Baladre.

Ferrari, L., Flores Hernández, J. R., & Hernández Martínez, D. (2024). A 20 años del pico del petróleo en México: Análisis del sector hidrocarburos e implicaciones para el futuro energético nacional. *Revista Mexicana de Ciencias Geológicas, 41*(1), 66–86.

Fisher, D. R. (2024). *Saving ourselves: From climate shocks to climate action*. Columbia University Press.

Fonjong, L., Matose, F., & Sonnenfeld, D. A. (Eds.). (2024). Special issue on "Climate change adaptation and policy response in Africa." *Global Environmental Change*. https://www.sciencedirect.com/special-issue/10oRXDMQ797

Frankyln, L. (2009). Overview: The current climate change situation in Africa. In H. Besada & N. Sewankambo (Eds.), *Climate change in Africa: Adaptation, mitigation and governance challenges* (pp. 8–16). The Centre for International Governance Innovation.

Freudenburg, W. R., Frickel, S., & Gramling, R. (1995). Beyond the nature/society divide: Learning to think about a mountain. *Sociological Forum, 10*(3), 361–392.

Gakou-Kakeu, J., Di Gregorio, M., Paavola, J., & Sonwa, D. J. (2022). REDD+ policy implementation and institutional interplay: Evidence from three pilot projects in Cameroon. *Forest Policy and Economics, 135*, 102642. https://doi.org/10.1016/j.forpol.2021.102642

Givens, J. E., Huang, X., & Jorgenson, A. K. (2019). Ecologically unequal exchange: A theory of global environmental injustice. *Sociology Compass, 13*(5), e12693.

Gomez-Echeverri, L. (2018). Climate and development: Enhancing impact through stronger linkages in the implementation of the Paris Agreement and the Sustainable Development Goals (SDGs). *Philosophical Transactions of the Royal Society: Mathematical, Physical and Engineering Sciences, 376*(2119), 20160444. https://doi.org/10.1098/rsta.2016.0444

Gould, K. A., & Lewis, T. L. (2021). Resilience gentrification: Environmental privilege in an age of coastal climate disasters. *Frontiers in Sustainable Cities, 3*, 687670. https://doi.org/10.3389/frsc.2021.687670

Gould, K. A., & Lewis, T. L. (Forthcoming). The consequences of public investments following coastal climate disasters: An assessment of four strategies post-Superstorm Sandy. *Human Ecology Review*.

Gould, K. A., Pellow, D. N., & Schnaiberg, A. (2008). *The treadmill of production: Injustice and unsustainability in the global economy*. Paradigm Publishers.

Greenpeace Africa. (2020, May 19). *34 plastic bans in Africa: A reality check*. Greenpeace Africa. https://www.greenpeace.org/africa/en/blogs/11156/34-plastic-bans-in-africa/

Grosse, C. (2022). *Working across lines: Resisting extreme energy extraction*. University of California Press.

Gudynas, E. (2020). *Extractivisms: Politics, economy and ecology*. Fernwood Publishing; Practical Action Publishing.

Hadden, J. (2014). Explaining variation in transnational climate change activism: The role of inter-movement spillover. *Global Environmental Politics, 14*(2), 7–25.

Hadden, J. (2015). *Networks in contention*. Cambridge University Press.

Haluza-DeLay, R. (2012). Giving consent in the petrostate: Hegemony and Alberta oil sands. *Journal for Activism in Science & Technology Education, 4*(1), 1–6.

Hamouchene, H. (2019). *Extractivism and resistance in North Africa*. Transnational institute.

Hamouchene, H. (2022). The political ecology of extractivism in North Africa. In H. Veltmeyer & P. Bowles (Eds.), *The essential guide to critical development studies* (2nd ed., pp. 327–334). Routledge.

Hansen, J., Johnson, D., Lacis, A., Lebedeff, S., Lee, P., Rind, D., & Russell, G. (1981). Climate impact of increasing atmospheric carbon dioxide. *Science, 213*(4511), 957–966. https://doi.org/10.1126/science.213.4511.957

Hecht, S. (2005). Extraction, gender, and neoliberalism in the Western Amazon. In P. S. Ciccantell, D. A. Smith, & G. Seidman (Eds.), *Nature, raw materials, and political economy* (pp. 1–22). JAI Press. http://site.ebrary.com/id/10139533

Holleman, H. (2017). De-naturalizing ecological disaster: Colonialism, racism and the global Dust Bowl of the 1930s. *The Journal of Peasant Studies, 44*(1), 234–260. https://doi.org/10.1080/03066150.2016.1195375

Hooks, G., & Smith, C. L. (2004). The treadmill of destruction: National sacrifice areas and native Americans. *American Sociological Review, 69*(4), 558–575. https://doi.org/10.1177/000312240406900405

Human Rights Watch. (2024). *Summary and recommendations. "We're dying here." The fight for life in a Louisiana fossil fuel sacrifice zone.* Human Rights Watch.

Isyaku, U. (2021). What motivates communities to participate in forest conservation? A study of REDD+ pilot sites in Cross River, Nigeria. *Forest Policy and Economics, 133*, 102598.

Jenkins, K. E. H., Sovacool, B. K., Błachowicz, A., & Lauer, A. (2020). Politicising the Just transition: Linking global climate policy, nationally determined contributions and targeted research agendas. *Geoforum, 115*, 138–142. https://doi.org/10.1016/j.geoforum.2020.05.012

Jerez, B., Garcés, I., & Torres, R. (2021). Lithium extractivism and water injustices in the Salar de Atacama, Chile: The colonial shadow of green electromobility. *Political Geography, 87*, 102382.

Johnson, E. W., & Frickel, S. (2011). Ecological threat and the founding of U.S. national environmental movement organizations, 1962–1998. *Social Problems, 58*(3), 305–329.

Jorgenson, A. K. (2010). World-economic integration, supply depots, and environmental degradation: A study of ecologically unequal exchange, foreign investment dependence, and deforestation in less developed countries. *Critical Sociology, 36*(3), 453–477.

Jorgenson, A. K. (2016). Environment, development, and ecologically unequal exchange. *Sustainability, 8*(3), 227.

Kahneman, D., & Tversky, A. (2013). Prospect theory: An analysis of decision under risk. In L. C. MacLean & W. T. Ziemba, *World scientific handbook in financial economics series* (Vol. 4, pp. 99–127). World Scientific.

Khagram, S. (2005). *Dams and development: Transnational struggles for water and power.* Oxford University Press.

Kingsbury, D. V. (2022). Energy transitions in the shadow of a dictator: Decarbonizing neoliberalism and lithium extraction in Chile. *The Anthropocene Review, 10*(2), 556–572. https://doi.org/10.1177/20530196221087790

Kuletz, V. (1998). *The tainted desert: Environmental ruin in the American West.* Routledge.

La Duke, W. (2010). Red land and uranium mining: How the search for energy is endangering Indian tribal lands. In L. Nader (Ed.), *The energy reader* (pp. 105–109). Wiley-Blackwell.

Leffel, B. (2022). Toward global urban climate mitigation: Linking national and polycentric systems of environmental change. *Sociology of Development, 8*(1), 111–137.

Leguizamón, A. (2019). The gendered dimensions of resource extractivism in Argentina's soy boom. *Latin American Perspectives, 46*(2), 199–216.

Lewis, T. L. (2016). *Ecuador's environmental revolutions: Ecoimperialists, ecodependents, and ecoresisters.* MIT Press.

Longo, S., Clausen, R., & Clark, B. (2015). *The tragedy of the commodity: Oceans, fisheries, and aquaculture.* Rutgers University Press.

Mayer, A. (2018). A just transition for coal miners? Community identity and support from local policy actors. *Environmental Innovation and Societal Transitions, 28*, 1–13.

McAdam, D., & Tarrow, S. (2019). The political context of social movements. In D. A. Snow, S. A. Soule, H. Kriesi, & H. J. McCammon (Eds.), *The Wiley Blackwell companion to social movements* (pp. 17–42). John Wiley & Sons. https://doi.org/10.1002/9781119168577.ch1

Méndez, M. (2020). *Climate change from the streets: How conflict and collaboration strengthen the environmental justice movement.* Yale University Press.

Milewski, A. (2023, October 19). *Latin America's "lithium triangle": Opportunities and risks for US investors—The Oregon Group—Investment Insights.* The Oregon Group. https://theoregongroup.com/commodities/lithium/latin-americas-lithium-triangle-opportunities-and-risks-for-us-investors/

Mlaba, K. (2021, September 16). *Rwanda to plant over 43 million trees to fight climate change.* Global Citizen. https://www.globalcitizen.org/en/content/rwanda-plant-43-million-trees-climate-change/

Noman, M. A., Miao, L., Macreadie, P. I., & Adyel, T. M. (2025). Time to count plastics in climate action. *Science, 387*(6738), 1048.

Pett-Ridge, J., Ammar, H., & Aui, A. (2023). *Roads to removal: Options for carbon dioxide removal in the United States.* Lawrence Livermore National Laboratory.

Pragier, D. (2019). Comunidades indígenas frente a la explotación de litio en sus territorios: Contextos similares, respuestas distintas. *Polis, 18*(52), 1–20. https://doi.org/10.32735/S0718-6568/2019-N52-1368

Reyes-García, P., & Jofré, D. (2024). Acciones climáticas frente a la sequía severa: El caso de comunidades rurales costeras del Chile central. *Íconos - Revista de Ciencias Sociales*, *79*, 185–205. https://doi.org/10.17141/iconos.79.2024.5962

Rilling, B., & Tosun, J. 2021. Policy and political consequences of mandatory climate impact assessments: an explorative study of German cities and municipalities. *Policy and Society*, *40*(1), 99–115. https://doi.org/10.1080/14494035.2021.1907901

Rodney, W. (1981). *How Europe underdeveloped Africa*. Howard University Press.

Sabin Center for Climate Change Law. (2025a). Global climate change litigation. *Climate Change Litigation*. https://climatecasechart.com/non-us-climate-change-litigation/

Sabin Center for Climate Change Law. (2025b). U.S. climate change litigation. *Climate Change Litigation*. Sabin Center. https://climatecasechart.com/us-climate-change-litigation/

Scheidel, A., Sorman, A. H., Avila, S., Del Bene, D., & Ott, J. (2023). Renewable grabbing: Land and resource appropriations in the global energy transition. In A. Neef, C. Ngin, T. M. Shegro, & S. Mollett (Eds.), *Routledge handbook of global land and resource grabbing* (1st ed., pp. 189–204). Routledge, Taylor & Francis Group.

Schnaiberg, A. (1980). *The environment, from surplus to scarcity*. Oxford University Press.

Schroering, C. (2024). *Global solidarities against water grabbing: Without water, we have nothing*. Manchester University Press.

Sellwood, S. A., & Valdivia, G. (2018). Interrupting green capital on the frontiers of wind power in Southern Mexico. *Latin American Perspectives*, *45*(5), 204–221. https://doi.org/10.1177/0094582X17719040

Simmons, E. S. (2016). *Meaningful resistance: Market reforms and the roots of social protest in Latin America*. Cambridge University Press.

Snow, D., Cress, D., Downey, L., & Jones, A. (1998). Disrupting the "quotidian": Reconceptualizing the relationship between breakdown and the emergence of collective action. *Mobilization: An International Quarterly*, *3*(1), 1–22. https://doi.org/10.17813/maiq.3.1.n41nv8m267572r30

Sofaer, H. R., Barsugli, J. J., Jarnevich, C. S., Abatzoglou, J. T., Talbert, M. K., Miller, B. W., & Morisette, J. T. (2017). Designing ecological climate change impact assessments to reflect key climatic drivers. *Global Change Biology*, *23*(7), 2537–2553.

Spash, C. L. (2010). The brave new world of carbon trading. *New Political Economy*, *15*(2), 169–195. https://doi.org/10.1080/13563460903556049

Svampa, M. (2019). *Neo-extractivism in Latin America: Socio-environmental conflicts, the territorial turn, and new political narratives*. Cambridge University Press. https://doi.org/10.1017/9781108752589

Taylor, D. E. (2014). *Toxic communities: Environmental racism, industrial pollution, and residential mobility*. New York University Press.

Temper, L., Avila, S., Del Bene, D., Gobby, J., Kosoy, N., Le Billon, P., Martinez-Alier, J., Perkins, P., Roy, B., Scheidel, A., & Walter, M. (2020). Movements shaping climate futures: A systematic mapping of protests against fossil fuel and low-carbon energy projects. *Environmental Research Letters*, *15*(12), 123004.

Thomeczek, J. P. (2024). Die (geringe) Beteiligung am Berliner Klimaentscheid 2023: Empirische Erklärungsversuche. *Zeitschrift für Vergleichende Politikwissenschaft*, *18*, 303–322.

Tilly, C. (1978). *From mobilization to revolution*. Addison-Wesley.

Tosun, J., & Schoenefeld, J. J. (2017). Collective climate action and networked climate governance. *Wiley Interdisciplinary Reviews: Climate Change*, *8*(1), e440.

Uba, K., & Engeman, C. (2025). Defenders of the status quo: Energy protests and policy (in) action in Sweden. *Social Forces*, *103*(4), 1393–1415. https://doi.org/10.1093/sf/soae166

Vasi, I. B., Walker, E. T., Johnson, J. S., & Tan, H. F. (2015). "No fracking way!" Documentary film, discursive opportunity, and local opposition against hydraulic fracturing in the United States, 2010 to 2013. *American Sociological Review*, *80*(5), 934–959. https://doi.org/10.1177/0003122415598534

Vasquez, P. I. (2014). *Oil sparks in the Amazon: Local conflicts, indigenous populations, and natural resources*. University of Georgia Press.

Wanja, M. (2022, November 21). *Kenya marks five years since single use plastic bag ban*. CGTN. https://web.archive.org/web/20230205070739/https://africa.cgtn.com/2022/11/21/kenya-marks-five-years-since-single-use-plastic-bag-ban

Weller, M. P., & Tran, M. L. (2022). Climate litigation against companies. *Climate Action* 1(14), 1–17. https://doi.org/10.1007/s44168-022-00013-6.

Widener, P. (2021). *Toxic and intoxicating oil: Discovery, resistance, and justice in Aotearoa New Zealand*. Rutgers University Press.

Zepeda-Millán, C. (2017). *Latino mass mobilization: Immigration, racialization, and activism*. Cambridge University Press.

PART I

Theories of Climate Action

CHAPTER 2

Ecologically Unequal Exchange as a Catalyst for Climate Action

ANDREW JORGENSON, JENNIFER GIVENS, XIAORUI HUANG, AND HASSAN EL TINAY

Introduction

ECOLOGICALLY unequal exchange (EUE) theory's relevance for climate action is in drawing attention to the roles of trade, global inequality, and the larger political-economic structure in shaping the climate crisis. By accurately identifying multiple drivers of the problem and related barriers to change, research in this area informs our understanding of the social forces climate action must address in order to be more effective and equitable, and it provides empirical evidence to counter dominant paradigms.

EUE theory originates from a combination of political–economic and environmental social science approaches. The theory draws attention to global structures and relationships that shape the unequal distribution of environmental harms and human development. EUE suggests the existence of structural relationships that facilitate unequal material flows in which wealthier and more powerful Global North countries have greater access to both natural resources and the sink capacity for waste in countries within the Global South. In other words, Global South nations are structurally positioned as both a tap for resources and a sink for waste within the world economic system of extraction, production, consumption, and disposal. With this theory, many scholars aim to empirically examine the linked forms of environmental and social stratification that exist internationally and globally. The perspective also allows for analyses from the global to the local of environmental and social harms and the underlying mechanisms that drive these outcomes. Much research using an EUE perspective focuses on outcomes and processes tied directly or indirectly to the climate crisis. Scholars within this tradition note that global inequality, and specifically EUE, is "an underlying source of most of the environmental distribution conflicts in our time" (Hornborg & Martinez-Alier, 2016, p. 328). Thus, EUE has profound implications for climate justice and climate action.

In this chapter, which draws significantly from Givens and colleagues (2019), we review the theory of EUE and discuss its relevance for climate action. We first summarize the theory of EUE and describe its origins. We then briefly discuss the importance of a global approach and detail how research that utilizes EUE theory examines the production and unequal distribution of multiple environmental harms relevant for the climate crisis and human well-being. Next, we highlight research on the forces counteracting the harms and injustices of EUE, and then we explicitly connect EUE to climate justice and action. We conclude by briefly applying Almeida et al.'s recent scheme of climate action (Almeida et al., 2024), where we note how forms of action, both institutional and noninstitutional, may effectively challenge and restrict the climate-related impacts of EUE.

EUE Theory and Its Origins

A long-standing debate in environmental sociology focuses on the relationship between economic growth and the environment, with increasing attention given to the climate crisis. This debate has largely, but not entirely, occurred between treadmill of production theory (Gould et al., 1996, 2004, 2008) and ecological modernization theory (Mol, 1997; Mol et al., 2014; Spaargaren & Mol, 1992). In economics, a similar debate has occurred between critical analytical perspectives within ecological economics and more optimistic perspectives within environmental economics (Hornborg, 2003; Muradian & Martinez-Alier, 2001b). Treadmill of production theory and ecological economic perspectives emphasize the harmful, ever-intensifying environmental impacts of economic growth, such as enhanced land degradation, increased energy consumption, and growth in carbon emissions (see chapter by Gould and Lewis in this volume). In contrast, ecological modernization theory and environmental economic perspectives posit that an unproblematic relationship between economic growth and the environment is possible, suggesting that the environmental impacts of economic growth tend to diminish, especially after a certain level of economic prosperity is reached and when certain institutional, technological, and cultural changes occur within societies (see chapter by Patel and Shorette in this volume). EUE theory adds a useful global, distributive perspective to this debate (Givens et al., 2016; Jorgenson & Clark, 2012).

EUE theory argues that largely through the structure of international trade and global production networks, wealthier, more powerful countries within the Global North have disproportionate access to natural resources and sink capacity—the ability of the environment to absorb waste products—within Global South nations. The natural resources and waste are linked to all stages along global commodity chains, including extraction, production, consumption, and disposal.[1] These material flows result in both the unequal distribution of environmental harms and suppressed human well-being of populations within Global South nations (Gellert et al., 2017; Hornborg, 1998b, 2009; Jorgenson, 2006; Jorgenson & Clark, 2009a, 2009b; Rice, 2007a, 2007b). EUE theory draws attention to the displacement of some environmental harms spatially, to other locations across the planet, and temporally, to future generations; these spatial and temporal dynamics are commonly referred to as environmental load displacement (Hornborg, 2006, 2009; Muradian & Martinez-Alier, 2001a; Muradian et al., 2002). In addition to unequal trade relationships, other structural mechanisms, such as dependence on foreign direct investment within the Global South and the coercive, military power of Global North nations, also contribute to these global socio-environmental inequities (Jorgenson, 2016; Jorgenson et al., 2023).

EUE theory has roots in critical development perspectives (Amin, 1976; Frank, 1979; Prebisch, 1950), world-systems analysis (Wallerstein, 1974), and the work of Bunker (1984, 1985), which bridges them together. Building on critical development perspectives, the world-systems approach emphasizes historical context and the hierarchical global system's influence on nation-states' levels and paths of development (Wallerstein, 1974). Inequalities between nation-states and circumstances within nations are perceived to be shaped by both internal dynamics and global, external forces that are linked to a nation's historical context and position in the global hierarchy (Chase-Dunn, 1998; Chase-Dunn & Grimes, 1995). While helpful in understanding global inequality, these approaches do not focus explicitly on environmental issues, especially human-caused environmental degradation in general and the climate crisis in particular.[2]

Bunker (1985, see also 1984) in his groundbreaking work *Underdeveloping the Amazon: Extraction, Unequal Exchange, and the Failure of the Modern State* situates ecological factors as important aspects of national and regional development in the context of an unequal world system. Combining Marx's labor theory of value, Emmanuel's (1972) notion of wage differentials as the basis of unequal exchange, and a natural theory of value, Bunker's work is foundational to the formation of EUE theory (Bunker, 1985; Hornborg, 2007). Bunker (1985) finds environmental degradation and underdevelopment in the Amazon are co-produced by "modes of extraction organized in response to world-system exchange opportunities focused on the single natural products for which there was greatest global demand" (Bunker, 1985, p. 250). This degradation and underdevelopment were facilitated by local elites who, while benefiting from the extractive economy in the short term, "ultimately impoverished the resource base on which their own wealth and profits depended" (Bunker, 1985, p. 238). Resistance at the local level is necessary to overcome such dynamics; however, the local social, political, and economic capacity to do so is undermined by the extractive relationships linking the global to the local (Bunker, 1985, p. 247). Bunker's work, despite its regional focus on the Amazon, highlights the global, unequal forces at play in shaping local socio-environmental conditions and the unfolding climate crisis.

Hornborg (1998a) carefully articulates an ecological theory of unequal exchange to assist in conceptualizing the mechanisms that generate unequal levels of ecological destruction. In addition to building upon Bunker's work, Hornborg draws from scholarship on national ecological footprint accounting, which takes consumption and environmental load displacement into consideration by initially using the ecological footprint, a measure capturing the amount of land and water ecosystems required to produce the resources and process the waste of a designated population at a certain level of consumption. Hornborg (1998b) argues that ecology is an integral part of global political economic analysis and that there is a need to expand Emmanuel's (1972) concept of unequal exchange to include not only labor but also natural factors. Hornborg demonstrates that while raw materials have greater productive potential, they are priced lower than processed goods, a distorted valuation that drives global ecological degradation and inequalities. Hornborg's work was highly influential for later theoretical and empirical scholarship on EUE, including work focusing explicitly on carbon emissions and climate change (e.g., El Tinay, 2024; Huang, 2018; Jorgenson, 2012).

A Global Approach

EUE and earlier influential work, such as Bunker's (1984, 1985), highlights the need to account for issues of environmental equity and climate justice on a global scale. EUE theory and the related concept of environmental load displacement allow for the bridging of the local to the global in historical and contemporary contexts. EUE theory helps scholars pursue theoretically grounded research questions about how global contexts affect efforts toward and obstacles to climate mitigation and other sustainability concerns at various scales. More specifically, EUE directs attention to how the structure of international trade and material flows in an unequal world economic system affects both environmental and human well-being outcomes for countries in different structural and temporal contexts. Along these lines, next, we turn to a discussion of the creation, global distribution, and perpetuation of several environmental problems of relevance for the climate crisis, the analyses of which consistently provide support for the theory of EUE.

The Unequal Distribution of Environmental Harms

Much of the empirical research on environmental outcomes examining EUE has largely consisted of hypothesis testing with cross-national statistical analyses (Corsi et al., 2024; Givens & Huang, 2021; Givens et al., 2019). This, to some extent, has shaped the type of environmental harms studied because it necessitates such outcomes for which there are comparative cross-national data available. Fortunately, such data exist for multiple key environmental problems, many of which are highly relevant for climate change. For example, research on EUE has focused on deforestation, CO_2 emissions, biodiversity loss, water pollution and scarcity, and the ecological footprint. Other recent contributions using qualitative methods provide a deeper understanding of the underlying mechanisms that both structure and perpetuate EUE (e.g., Austin, 2017).

Deforestation is the environmental harm analyzed in an early study within this tradition (see chapter by Alvarez in this volume). To test EUE theory cross-nationally, Jorgenson (2006) analyzes the relationship between deforestation and weighted export flows, a measure he created that quantifies the relative extent to which the exports of a nation are sent "vertically" to more economically developed nations. This early measure included all primary and secondary sector exports and enabled the systemic examination of both trading partner concentration and commodity concentration, in addition to the percent of GDP made up of exports, that is, export intensity. The results of this initial study support the theory: Deforestation rates are substantially higher in Global South nations that send more of their exports to relatively wealthier nations within the Global North, net of how much they export in total. Later, studies improve on this early research in multiple ways: examining the effect of the vertical flow of exports of particular commodities or from certain economic sectors, utilizing longitudinal methods that are able to detect increasingly ecologically unequal relationships through time, and focusing on how both global and local contexts can facilitate global EUE relationships (e.g., Austin, 2010, 2012; Clark et al., 2019; El Tinay, 2024; Fitzgerald & Auerbach, 2016; Jorgenson, 2010, 2016; Jorgenson et al., 2009; Noble, 2017; Shandra et al., 2009; Shandra et al., 2009; Shandra et al., 2011, 2019).

A growing number of studies of anthropogenic CO_2 emissions, a primary cause of the climate crisis, also provide support for EUE theory (El Tinay, 2024; Huang, 2018; Jorgenson, 2011, 2012; Lawrence, 2009; Roberts & Parks, 2007). For example, Jorgenson (2012) finds that for Global South nations, a larger percentage of exports sent to high-income nations is associated with higher emissions, and the magnitude of this relationship has increased through time. Likewise, Roberts and Parks (2007, p. 193) find that "emissions are increasing sharply in developing countries as wealthy nations 'offshore' the energy- and natural resource-intensive stages of production." Jorgenson (2011) finds a heightening EUE effect on displaced emissions offloaded onto Central and Eastern European (CEE) nations. Hao (2020) has affirmed these earlier findings, identifying worsening trends of carbon externalization on middle- and low-income countries, while economists have begun incorporating EUE's broader impact on greenhouse emissions into forecasting models (Althouse et al., 2020). Vesia and colleagues (2023) find that countries at the periphery of the international natural resource production and trade network become more dependent on natural resource extraction as a result of EUE dynamics, resulting in a stronger coupling between economic growth and CO_2 emissions in these countries. Baker (2018) underscores the importance of considering energy and emissions embodied in trade, in order to account for EUE when evaluating the low-carbon transition in high-income countries. Indeed, Huang (2024) finds that in high-income countries, while economic growth is decoupled from the carbon emissions generated by their domestic industries serving domestic end users, it remains strongly coupled with emissions embodied in imports—a pattern indicative of EUE dynamics.

EUE also has human well-being implications. Empirical research shows the vertical flow of material value (i.e., unprocessed and processed natural resources exported to Global North countries) suppresses resource consumption within populations of Global South countries (Jorgenson & Rice, 2005; Rice, 2007b, 2008). The underconsumption in Global South nations, and the various aforementioned environmental harms, can lead to negative health outcomes such as increased maternal mortality (Rice, 2008) and infant mortality (Jorgenson, 2009).

The carbon intensity of well-being (CIWB) is a highly climate-relevant concept that combines environmental impact and human well-being into one indicator, which is becoming an increasingly popular measure of sustainability that is comparable cross-nationally and at multiple scales. CIWB is typically measured as a ratio between per capita carbon emissions (numerator) and average life expectancy at birth (denominator). In an analysis of nations' CIWB, Givens (2018) finds support for EUE: the vertical flow of exports to higher income nations in the Global North increases the CIWB for Global South nations. In other words, EUE makes it more carbon intensive for Global South nations to create well-being for their populations. In addition, this relationship is intensifying through time. Engaging in global trade, particularly trade with more developed nations, is often promoted as a strategy for development in Global South nations. However, Givens (2018) demonstrates that this strategy leads to increasing carbon emission per unit of human well-being, which is problematic for social and ecological sustainability. Givens (2017, 2018) also finds that trade in conjunction with other aspects of global integration is a mechanism by which higher income countries can displace the impacts of their consumption to lower income countries.

Forces Counteracting the Harms and Injustices of EUE

EUE dynamics and structural conditions unjustly distribute carbon emissions, other environmental harms, and related well-being outcomes across humans and other species, today and for future generations. This unjust distribution has wide-ranging implications for addressing the climate crisis. EUE displaces emissions driven by consumption in the Global North to the Global South, distorting the global distribution of the responsibility for climate mitigation among nations. EUE also locks nations in the Global South into a structural dependence on resource extraction and polluting industries, hindering their ability to pursue sustainable, low-carbon development. Furthermore, EUE weakens the resilience of populations, communities, and ecosystems in the South, limiting their capacity to adapt to the changing climate. Addressing EUE, therefore, is essential for advancing just and effective climate action.

Scholars have asked what might counteract these globally unjust dynamics of EUE (Hornborg & Martinez-Alier, 2016). Research centers on several possibilities, including slowing consumption or economic growth, and the potential of global governance, world society organizations, and especially environmental justice and climate justice movements, to address global environmental inequalities, each of which we discuss next.

Degrowth perspectives, dating back to Andre Gorz's introduction of the term in 1972, have taken the linkages between economic growth and environmental degradation as evidence of the need to advance environmental justice through "de-linking" the pursuit of socio-environmental goods away from the pursuit of growing GDP (D'Alisa et al., 2014). Ecologically unequal exchange scholarship has complemented these views by emphasizing the manner in which the environmental load of consumption in wealthy countries has often been displaced to the Global South. Huang (2018) finds that the 2008 economic recession alleviated the effect of the vertical flows of exports from Global South countries to the United States on increasing CO_2 emissions in the exporting countries. This finding points to the need for changes in consumption patterns in Global North countries that reduce the overall volume of consumption and shift toward less carbon-intensive goods, which is in line with Bunker's (1985, p. 253) observation that countering EUE dynamics requires "slow[ing] the flow of energy to the world center." However, the recession's mediating effect was short-lived; the economic recovery following the recession reversed the changes in consumption patterns that occurred during the recession and intensified the EUE dynamics post-recession to a level higher than before the recession (Huang, 2018). This indicates that conscious consumption reduction in the Global North is necessary to alleviate EUE dynamics and their climate impacts. Moreover, for this alleviating effect to last, the consumption reduction must be sustained beyond recession periods. These findings, in line with degrowth perspectives (Kallis, 2011; Kallis et al., 2018), suggest climate action should engage with the issue of economic growth that has long been studied in the literature.

Other than slowing economic growth and reducing consumption via a global recession, what other avenues might address the unequal distribution of global environmental harms? Multiple studies explore whether there is evidence that various institutions and/or global civil society organizations can mediate or moderate the processes of EUE and resulting environmental load displacement. Some research finds that global governance institutions could perpetuate rather than remedy EUE dynamics of global environmental harms and inequalities (Gareau & Lucier, 2018; Kill, 2016; Oulu, 2016; Roberts & Parks, 2007; Swamy, 2017). For example, Kill (2016) finds voluntary certification programs for wood pulp—intended to encourage socially and environmentally sustainable forest management—have enabled further expansion of environmentally damaging forestry industries. Researchers also highlight the increasing dominance of neoliberalism (Almeida & Pérez Martín, 2022; Gareau & Lucier, 2018) and pro-growth (i Manzano et al., 2016) ideologies in global governance institutions that obfuscate global environmental inequalities and amplify the climate crisis.

Others, engaging both EUE and the similarly global in scope world society theory, which emphasizes the importance of global institutions and global cultural norms, find support for both EUE and a modest protective effect of world society integration (see chapter by Patel and Shorette in this volume). World society integration, in this case connection to the global environmental regime, is often measured by the relative or absolute presence of environmental international nongovernmental organizations (INGOs) within countries. Shandra, Leckband, McKinney, and London (2009) find support for EUE in the context of biodiversity loss; they also find support for world society theory that, in poorer nations, a greater environmental INGO presence is associated with fewer endangered mammal species. Givens and Jorgenson (2014) find Global South countries' integration into the environmental world society modestly alleviates the impacts of the vertical flow of exports on increasing carbon dioxide emissions. Henderson and Shorette (2017), in a study of fifteen palm oil-producing nations in the Global South, find a similar effect of world society connection. However, up until now, empirical support for the ability of world society to mitigate the environmental harms from EUE dynamics is far less consistent than the evidence supporting the negative impacts of EUE, and when found to be beneficial, the beneficial effects of world society integration are modest at best.

Burns et al. (2003, 2006) introduced the concept of "recursive exploitation," as a conceptual approach to the way developing nations of the "semi-periphery" may end up reproducing EUE relationships toward more peripheral trading partners. Building on this approach, El Tinay (2024) highlights the role that different political-economic formations, which differentiate the nations of the semi-periphery, play in both amplifying and repressing their reproduction of EUE, which shapes the carbon emissions of nations. These political-economic formations include factors like large sovereign territorial holdings or high levels of market extroversion, which shape the developmental pathways charted out by countries and the subsequent ecological repercussions that are either internalized within or externalized to outside the country. Recent scholarship has also echoed a push to understanding the growing, potentially contradictory roles of the semi-periphery and developing states in EUE (Tausch & Althouse, 2024).

Connecting EUE to Climate Justice and Action

Climate justice "is the concern that the causes and consequences of climate change and the impacts of efforts to reduce the magnitude of climate change and adapt to it, are inequitably distributed" (Dietz et al., 2020, p. 144). Climate justice builds on insights from environmental justice but is also informed by perspectives such as EUE that draws attention to systemic and structural drivers and impacts of the climate crisis (Mohai et al., 2009). These perspectives link local and global institutional and noninstitutional efforts and offer important implications for effective climate action by drawing attention to the relationships between global inequalities and climate change. Climate action must address the inequalities that will stem from climate inaction and address the inequalities that will stem from policies to address climate change, including the unequal distribution of costs and benefits (Dolsak & Prakash, 2022; see chapter by Alvarez and Norton-Smith in this volume).

The occurrence and consequences of EUE can increase awareness and motivate action on environmental and climate injustices at both local and global scales. For example, Jorgenson and Givens (2014) find in a multilevel study that ecological unequal relationships in cross-national contexts are associated with increased environmental concern at the individual level. This suggests that experiencing climate change or environmental degradation from EUE may increase environmental concern in those places that experience such disproportionate environmental damage, with implications for mobilizing support for climate action (Fisher, 2024). For example, widespread experiences of EUE could act as facilitating conditions for a protest wave, such as by raising concern as a unifying frame and/or by being perceived as an economic threat or threat to well-being (Almeida, 2023). Furthermore, by drawing attention to the harmful aspects of trade and the need to address policies on environmental, trade, and labor protection, EUE research offers those engaged in climate research and action narratives to counter more dominant paradigms such as green growth or the uncritical neoliberal belief in the universal benefits of free trade (Almeida & Pérez Martín, 2022; Corsi et al., 2024; Hornborg, 2023).

More broadly, a global climate and environmental justice framework based on EUE theory and its ample empirical evidence empowers scholars, activists, and the affected communities in their call for more equitable climate action (Bullard, 2023). Researchers and environmental justice groups have increasingly argued that citizens of the Global South are owed an "ecological debt" (Hornborg & Martinez-Alier, 2016; Martinez-Alier, 2003; Martinez-Alier et al., 2016; Roberts & Parks, 2007; Warlenius, 2016; Warlenius et al., 2015). Roberts and Parks (2009) demonstrate how, in international climate negotiations, Global South countries claim payment on the ecological debt by calling on Global North countries to not only mitigate their own greenhouse gas emissions but also help Global South countries improve economic well-being and transition to a less carbon-intensive developmental pathway (see chapter by Khan and Khan in this volume).

Activists, scholars, and community members have a long history of confronting environmental and social harms brought by global EUE relationships, with various degrees of success (Martinez-Alier et al., 2016; Smith & Patterson, 2019). Some researchers in this area argue that environmental justice and climate justice groups have the potential to create transformative change (Temper et al., 2018). These movements are multidimensional and intersectional, and they address issues concerning both humans and other species, such as human health outcomes and biodiversity loss tied to the climate crisis (see chapter by Martinez-Alier and Limacher in this volume). They also address multiple interconnected dimensions of inequality, such as race, indigeneity, class, gender, and structural position in the global hierarchy (Malin & Ryder, 2018; Pellow, 2018). Temper, Walter, et al. (2018) argue this makes climate justice and environmental justice activism well situated to return agency to those affected by taking approaches that combine social and environmental concerns, confront both global and local power relations, and that have awareness of impacts across space and time.

Martinez-Alier et al. (2016) discuss whether there is a global environmental justice movement and detail efforts to catalog ecological distribution conflicts around the world in the Atlas of Environmental Justice (EJAtlas; www.ejatlas.org; Hornborg & Martinez-Alier, 2016; Martinez-Alier et al., 2016; Scheidel et al., 2018; Temper, Demaria, et al., 2018; Temper, Walter, et al., 2018). This cataloging is an empirical effort that allows for systematic analysis of various aspects of such conflicts and efforts to address them from the local to the global (see chapter by Martinez-Alier and Limacher in this volume). It also demonstrates the connected nature of such conflicts to larger systemic dynamics that EUE theory captures. Cataloging multiple global grievances from around the globe in one database demonstrates how such local events "belong to classes of conflicts that appear regularly" (Martinez-Alier et al., 2016, p. 17). EJAtlas is an ongoing, collective endeavor. Notably, at the time of this writing, it has documented over 800 cases worldwide under the category of "Fossil Fuel and Climate/Energy Justice," amounting to nearly 20% of all cases. Cases in other categories such as "Biomass and Land Conflicts" and "Industrial and Utilities Conflicts" also have significant direct and indirect implications for climate justice, mitigation, and adaptation.

Temper, Demaria, et al. (2018, p. 574) note that the EJAtlas database allows for asking questions such as "What are the reasons for the vast numbers of 'ecological distribution conflicts'? What are these contentious movements... against and for? And what can they tell the world about our (un)sustainably crisis?" In this way, scholars argue that these environmental distribution conflicts, and research on such conflicts at multiple scales that are all ultimately linked to global processes, can actually become a force for minimizing/overcoming EUE while facilitating more equitable pathways to sustainability (Scheidel et al., 2018).

Conclusion: EUE and Almeida et al.'s Classification Scheme of Climate Action

Almeida and colleagues (2024) offer a classification scheme of climate action, delineated by institutional and noninstitutional characteristics. Using their scheme, we conclude this chapter by briefly noting how a global climate and environmental justice framework based on EUE is positioned to contribute to multiple forms of institutional and noninstitutional climate action.

In terms of institutional actions, such a global climate justice framework helps better evaluate the levels of coverage and equity in climate action planning, underscoring that the transnational displacement of carbon emissions and other linked environmental harms must be accounted for in climate mitigation planning at the national level. By incorporating EUE into climate action planning, policymakers can also better identify vulnerable communities and understand the global forces driving their vulnerability, leading to more targeted and equitable interventions. Policies promoting green and renewable energy must also account for EUE to prevent the unintended displacement of environmental harms. While developing renewable energy is a key strategy for climate mitigation, the extraction of raw materials needed for renewable infrastructure—often sourced from the Global South—can perpetuate exploitation and environmental degradation in these regions (e.g., Kingsbury, 2023).

When examining initiatives related to carbon markets, EUE provides a greater understanding of the social and structural inequality inherent in the current political economic system, which may need to be more critically addressed. Research on EUE provides insights into structural barriers, including sectors of the economy that benefit from the status quo, developed pathways of technology that are difficult to change, and cultural assumptions and existing political structures (Roberts et al., 2020). Climate action and research informed by an understanding of EUE and that tackles these barriers is not only needed, but there are already signs that the EUE paradigm has influenced the framing of climate justice in the Global South. In 2023, Colombian President Gustavo Petro spoke at length on the issue during the COP28 conference, declaring that "[t]he consumption of the richest part of humanity on the planet, based on carbon, is a consumption based on the death of others" (Petro, 2023).

The recognition of EUE is similarly vital for noninstitutional climate action, particularly in linking global dynamics with local environmental struggles. Understanding how EUE connects transnational drivers of resource extraction to local environmental degradation can help unify transnational climate movements and local anti-extraction efforts across multiple locales and scales. This shared understanding facilitates coalition building, enabling local and transnational movements to collaborate in their fight for effective and equitable climate action. This dovetails with fossil fuel divestment campaigns and other efforts that focus on corporations as actors in the system which climate action movements target at multiple scales, potentially creating pressures that can influence climate action planning to take into account environmental and labor protections. Such coalitions can be difficult to form and sustain, but they may be vital to just transitions and tackling the climate crisis, and they can be advocated for by movement campaigns (Ciplet, 2022; LeQuesne, 2019).

Notes

1. The "extractivism" literature makes similar arguments (e.g., Gudynas, 2021; Svampa, 2019).
2. See Chase-Dunn and Almeida (2020) for an exception.

References

Almeida, P. (2023). Protest waves in Latin America: Facilitating conditions and outcomes. In F. M. Rossi (Ed.), *Oxford handbook of Latin American social movements* (pp. 143–161). Oxford University Press.

Almeida, P., González Márquez, L. R., & Fonsah, E. (2024). The forms of climate action. *Sociology Compass, 18*(2), e13177.

Almeida, P., & Martín, A. P. (2022). *Collective resistance to neoliberalism*. Cambridge University Press.

Althouse, J., Guarini, G., & Porcile, J. G. (2020). Ecological macroeconomics in the open economy: Sustainability, unequal exchange and policy coordination in a center-periphery model. *Ecological Economics, 172*, 106628.

Amin, S. (1976). *Unequal development: An essay on the social formations of peripheral capitalism*. Monthly Review Press.

Austin, K. (2010). The "hamburger connection" as ecologically unequal exchange: A cross-national investigation of beef exports and deforestation in less-developed countries. *Rural Sociology, 75*(2), 270–299.

Austin, K. (2012). Coffee exports as ecological, social, and physical unequal exchange: A cross-national investigation of the java trade. *International Journal of Comparative Sociology, 53*(3), 155–180.

Austin, K. F. (2017). Brewing unequal exchanges in coffee: A qualitative investigation into the consequences of the java trade in rural Uganda. *Journal of World-Systems Research, 23*(2), 326–352.

Baker, L. (2018). Of embodied emissions and inequality: Rethinking energy consumption. *Energy Research & Social Science, 36*, 52–60.

Bullard, R. D. (2023). The quest for environmental, climate and energy justice in the United States. https://stateofblackamerica.org/index.php/authors-essays/quest-environmental-climate-and-energy-justice-united-states

Bunker, S. G. (1984). Modes of extraction, unequal exchange, and the progressive underdevelopment of an extreme periphery: The Brazilian Amazon, 1600–1980. *American Journal of Sociology, 89*(5), 1017–1064.

Bunker, S. G. (1985). *Underdeveloping the Amazon: Extraction, unequal exchange, and the failure of the modern state*. University of Chicago Press.

Burns, T. J., Kick, E. L., & Davis, B. L. (2003). Theorizing and rethinking linkages between the natural environment and the modern world-system: Deforestation in the late 20th century. *Journal of World-Systems Research, 9*(2), 357–390.

Burns, T. J., Kick, E., & Davis, B. L. (2006). A quantitative cross-national study of deforestation in the late 20th century: A case of recursive exploitation. In A. K. Jorgenson & E. L. Kick (Eds.), *Globalization and the environment* (pp. 37–60). Brill.

Chase-Dunn, C. K. (1998). *Global formation: Structures of the world-economy*. Rowman & Littlefield.

Chase-Dunn, C. K., & Almeida, P. (2020). *Global struggles and social change: From prehistory to world revolution in the twenty-first century*. Johns Hopkins University Press.

Chase-Dunn, C. K., & Grimes, P. (1995). World-systems analysis. *Annual Review of Sociology, 21*(1), 387–417.

Ciplet, D. (2022). Transition coalitions: Toward a theory of transformative just transitions. *Environmental Sociology, 8*(3), 315–330.

Clark, B., Longo, S. B., Clausen, R., & Auerbach, D. (2019). From sea slaves to slime lines: Commodification and unequal ecological exchange in global marine fisheries. In R. S. Frey, P. K. Gellert, & H. F. Dahms (Eds.), *Ecologically unequal exchange: Environmental injustice in comparative and historical perspective* (pp. 195–219). Springer.

Corsi, G., Guarino, R., Muñoz-Ulecia, E., Sapio, A., & Franzese, P. P. (2024). Uneven development and core-periphery dynamics: A journey into the perspective of ecologically unequal exchange. *Environmental Science & Policy, 157*, 103778.

D'Alisa, G., Demaria, F., & Kallis, G. (Eds.). (2014). *Degrowth: A vocabulary for a new era*. Routledge.

Dietz, T., Shwom, R. L., & Whitley, C. T. (2020). Climate change and society. *Annual Review of Sociology, 46*(1), 135–158.

Dolšak, N., & Prakash, A. (2022). Three faces of climate justice. *Annual Review of Political Science, 25*(1), 283–301.

El Tinay, H. (2024). The semi-periphery and ecologically unequal exchange: carbon emissions and recursive exploitation. *Environmental Sociology, 10*(2), 253–265.

Emmanuel, A. (1972). *Unequal exchange: A study of the imperialism of trade*. Monthly Review Press.

Fisher, D. R. (2024). *Saving ourselves: From climate shocks to climate action*. Columbia University Press.

Fitzgerald, J. B., & Auerbach, D. (2016). The political economy of the water footprint: A cross-national analysis of ecologically unequal exchange. *Sustainability*, *8*(12), 1263.

Frank, A. G. (1979). *Dependent accumulation and underdevelopment*. Monthly Review Press.

Gareau, B. J., & Lucier, C. A. (2018). Neoliberal restructuring of the world polity: The weakening of the Montreal Protocol and Basel Convention in historical perspective. *Environmental Sociology*, *4*(3), 325–342.

Gellert, P. K., Frey, R. S., & Dahms, H. F. (2017). Introduction to ecologically unequal exchange in comparative perspective. *Journal of World-Systems Research*, *23*(2), 226–235.

Givens, J. E. (2017). World society, world polity, and the carbon intensity of well-being, 1990–2011. *Sociology of Development*, *3*(4), 403–435.

Givens, J. E. (2018). Ecologically unequal exchange and the carbon intensity of well-being, 1990–2011. *Environmental Sociology*, *4*(3), 311–324.

Givens, J., Clark, B., & Jorgenson, A. (2016). Strengthening the ties between environmental sociology and sociology of development. In G. Hooks, P. Almeida, D. Brown, S. Cohn, S. Curran, R. Emigh, et al. (Eds.), *The sociology of development handbook* (pp. 69–94). University of California Press.

Givens, J. E., & Huang, X. (2021). Ecologically unequal exchange and environmental load displacement: Global perspectives on structural inequalities and the environment. In B. S. Caniglia, A. Jorgenson, S. A. Malin, L. Peek, D. N. Pellow, & X. Huang (Eds.), *Handbook of environmental sociology* (pp. 53–70). Springer.

Givens, J. E., Huang, X., & Jorgenson, A. K. (2019). Ecologically unequal exchange: A theory of global environmental injustice. *Sociology Compass*, *13*(5), e12693.

Gould, K. A., Pellow, D. N., & Schnaiberg, A. (2004). Interrogating the treadmill of production: Everything you wanted to know about the treadmill but were afraid to ask. *Organization & Environment*, *17*(3), 296–316.

Gould, K. A., Pellow, D. N., & Schnaiberg, A. (2008). *The treadmill of production: Injustice and unsustainability in the global economy*. Paradigm.

Gould, K. A., Schnaiberg, A., & Weinberg, A. S. (1996). *Local environmental struggles: Citizen activism in the treadmill of production*. Cambridge University Press.

Gudynas, E. (2021). *Extractivisms: politics, economy, and ecology*. Fernwood.

Hao, F. (2020). A study of ecologically unequal exchange for 89 countries between 1990 and 2015. *Social Science Journal*, *57*(2), 245–257.

Henderson, K., & Shorette, K. (2017). Environmentalism in the periphery: Institutional embeddedness and deforestation among fifteen palm oil producers, 1990–2012. *Journal of World-Systems Research*, *23*(2), 269–297.

Hornborg, A. (1998a). Ecosystems and world systems: Accumulation as an ecological process. *Journal of World-Systems Research*, *4*, 169–177.

Hornborg, A. (1998b). Towards an ecological theory of unequal exchange: Articulating world system theory and ecological economics. *Ecological Economics*, *25*(1), 127–136.

Hornborg, A. (2003). Cornucopia or zero-sum game? The epistemology of sustainability. *Journal of World-Systems Research*, *9*(2), 205–216.

Hornborg, A. (2006). Footprints in the cotton fields: The Industrial Revolution as time–space appropriation and environmental load displacement. *Ecological Economics*, *59*(1), 74–81.

Hornborg, A. (2007). Environmental history as political ecology. In A. Hornborg, J. R. McNeill, & J. Martinez-Alier (Eds.), *Rethinking environmental history: World-system history and global environmental change* (pp. 1–24). Rowman and Littlefield/Altamira Press.

Hornborg, A. (2009). Zero-sum world: Challenges in conceptualizing environmental load displacement and ecologically unequal exchange in the world-system. *International Journal of Comparative Sociology*, *50*(3–4), 237–262.

Hornborg, A. (2023). Identifying ecologically unequal exchange in the world-system: Implications for development. In E. S. Reinert & I. H. Kvangraven (Eds.), *A modern guide to uneven economic development* (pp. 367–388). Edward Elgar.

Hornborg, A. & Martinez-Alier, J. (2016). Ecologically unequal exchange and ecological debt. *Journal of Political Ecology*, *23*(1), 328–333.

Huang, X. (2018). Ecologically unequal exchange, recessions, and climate change: A longitudinal study. *Social Science Research*, *73*, 1–12.
Huang, X. (2024). Not all emissions are created equal: Multidimensionality in nations' greenhouse gas emissions and the affluence/emissions nexus. *Socius*, *10*. https://doi.org/10.1177/2378023124 1238962
i Manzano, J. J., Cardesa-Salzmann, A., Pigrau, A., & Borras, S. (2016). Measuring environmental injustice: How ecological debt defines a radical change in the international legal system. *Journal of Political Ecology*, *23*(1), 381–393.
Jorgenson, A. K. (2006). Unequal ecological exchange and environmental degradation: A theoretical proposition and cross-national study of deforestation, 1990–2000. *Rural Sociology*, *71*(4), 685–712.
Jorgenson, A. K. (2009). Political-economic integration, industrial pollution and human health: A panel study of less-developed countries, 1980–2000. *International Sociology*, *24*(1), 115–143.
Jorgenson, A. K. (2010). World-economic integration, supply depots, and environmental degradation: A study of ecologically unequal exchange, foreign investment dependence, and deforestation in less developed countries. *Critical Sociology*, *36*(3), 453–477.
Jorgenson, A. K. (2011). Carbon dioxide emissions in Central and Eastern European Nations, 1992–2005: A test of ecologically unequal exchange theory. *Human Ecology Review*, *18*(2), 105–114.
Jorgenson, A. K. (2012). The sociology of ecologically unequal exchange and carbon dioxide emissions, 1960–2005. *Social Science Research*, *41*(2), 242–252.
Jorgenson, A. K. (2016). The sociology of ecologically unequal exchange, foreign investment dependence and environmental load displacement: Summary of the literature and implications for sustainability. *Journal of Political Ecology*, *23*(1), 334–349.
Jorgenson, A. K., Austin, K., & Dick, C. (2009). Ecologically unequal exchange and the resource consumption/environmental degradation paradox: A panel study of less-developed countries, 1970—2000. *International Journal of Comparative Sociology*, *50*(3–4), 263–284.
Jorgenson, A. K., & Clark, B. (2009a). Ecologically unequal exchange in comparative perspective: A brief introduction. *International Journal of Comparative Sociology*, *50*(3–4), 211–214.
Jorgenson, A. K., & Clark, B. (2009b). The economy, military, and ecologically unequal exchange relationships in comparative perspective: A panel study of the ecological footprints of nations, 1975–2000. *Social Problems*, *56*(4), 621–646.
Jorgenson, A. K., & Clark, B. (2012). Are the economy and the environment decoupling? A comparative international study, 1960–2005. *American Journal of Sociology*, *118*(1), 1–44.
Jorgenson, A. K., Clark, B., Thombs, R. P., Kentor, J., Givens, J. E., Huang, X., El Tinay, H., Auerbach, D., & Mahutga, M. C. (2023). Guns versus climate: How militarization amplifies the effect of economic growth on carbon emissions. *American Sociological Review*, *88*(3), 418–453.
Jorgenson, A. K., & Givens, J. E. (2014). Economic globalization and environmental concern: A multilevel analysis of individuals within 37 nations. *Environment and Behavior*, *46*(7), 848–871.
Jorgenson, A. K., & Rice, J. (2005). Structural dynamics of international trade and material consumption: A cross-national study of the ecological footprints of less-developed countries. *Journal of World-Systems Research*, *11*(1), 57–77.
Kallis, G. (2011). In defence of degrowth. *Ecological Economics*, *70*(5), 873–880.
Kallis, G., Kostakis, V., Lange, S., Muraca, B., Paulson, S., & Schmelzer, M. (2018). Research on degrowth. *Annual Review of Environment and Resources*, *43*(1), 291–316.
Kill, J. (2016). The role of voluntary certification in maintaining the ecologically unequal exchange of wood pulp: The Forest Stewardship Council's certification of industrial tree plantations in Brazil. *Journal of Political Ecology*, *23*(1), 434–445.
Kingsbury, D. V. (2023). Energy transitions in the shadow of a dictator: Decarbonizing neoliberalism and lithium extraction in Chile. *The Anthropocene Review*, *10*(2), 556–572.
Lawrence, K. S. (2009). The thermodynamics of unequal exchange: energy use, CO2 emissions, and GDP in the world-system, 1975–2005. *International Journal of Comparative Sociology*, *50*(3-4), 335–359.

LeQuesne, T. (2019). Petro-hegemony and the matrix of resistance: What can Standing Rock's water protectors teach us about organizing for climate justice in the United States? *Environmental Sociology*, *5*(2), 188–206.

Malin, S. A., & Ryder, S. S. (2018). Developing deeply intersectional environmental justice scholarship. *Environmental Sociology*, *4*(1), 1–7.

Martinez-Alier, J. (2003). *The environmentalism of the poor: A study of ecological conflicts and valuation*. Edward Elgar.

Martinez-Alier, J., Temper, L., Del Bene, D., & Scheidel, A. (2016). Is there a global environmental justice movement? *Journal of Peasant Studies*, *43*(3), 731–755.

Mohai, P., Pellow, D., & Roberts, J. T. (2009). Environmental justice. *Annual Review of Environment and Resources*, *34*(1), 405–430.

Mol, A. P. (1997). Ecological modernization: Industrial transformations and environmental reform. In M. R. Redclift & G. Woodgate (Eds.), *The international handbook of environmental sociology* (pp. 138–149). Edward Elgar.

Mol, A. P., Spaargaren, G., & Sonnenfeld, D. A. (2014). Ecological modernization theory: Taking stock, moving forward. In S. Lockie, D. A. Sonnenfeld, & D. Fisher (Eds.), *Routledge international handbook of social and environmental change* (pp. 15–30). Routledge.

Muradian, R., & Martinez-Alier, J. (2001a). South-north materials flow: History and environmental repercussions. *Innovation: The European Journal of Social Science Research*, *14*(2), 171–187.

Muradian, R., & Martinez-Alier, J. (2001b). Trade and the environment: From a "Southern" perspective. *Ecological Economics*, *36*(2), 281–297.

Muradian, R., O'Connor, M., & Martinez-Alier, J. (2002). Embodied pollution in trade: Estimating the 'environmental load displacement' of industrialised countries. *Ecological Economics*, *41*(1), 51–67.

Noble, M. D. (2017). Chocolate and the consumption of forests: A cross-national examination of ecologically unequal exchange in cocoa exports. *Journal of World-Systems Research*, *23*(2), 236–268.

Oulu, M. (2016). Core tenets of the theory of ecologically unequal exchange. *Journal of Political Ecology*, *23*, 446–466. https://doi.org/10.2458/v23i1.20251

Pellow, D. N. (2018). *What is critical environmental justice?* Polity Press.

Petro, G. (2023). COP28 Speech. COP28. Dubai. https://peoplesdispatch.org/2023/12/03/in-cop28-speech-colombian-president-gustavo-petro-calls-for-a-free-palestine/

Prebisch, R. (1950). *Theoretical and practical problems of economic growth*. United Nations.

Rice, J. (2007a). Ecological unequal exchange: Consumption, equity, and unsustainable structural relationships within the global economy. *International Journal of Comparative Sociology*, *48*(1), 43–72.

Rice, J. (2007b). Ecological unequal exchange: International trade and uneven utilization of environmental space in the world system. *Social Forces*, *85*(3), 1369–1392.

Rice, J. (2008). Material consumption and social well-being within the periphery of the world economy: An ecological analysis of maternal mortality. *Social Science Research*, *37*(4), 1292–1309.

Roberts, J. T., & Parks, B. C. (2007). Fueling injustice: Globalization, ecologically unequal exchange and climate change. *Globalizations*, *4*(2), 193–210.

Roberts, J. T., & Parks, B. C. (2009). Ecologically unequal exchange, ecological debt, and climate justice: The history and implications of three related ideas for a new social movement. *International Journal of Comparative Sociology*, *50*(3–4), 385–409.

Roberts, J. T., Steinberger, J. K., Dietz, T., Lamb, W. F., York, R., Jorgenson, A. K., Givens, J. E., Baer, P., & Schor, J. B. (2020). Four agendas for research and policy on emissions mitigation and well-being. *Global Sustainability*, *3*, e3.

Scheidel, A., Temper, L., Demaria, F., & Martínez-Alier, J. (2018). Ecological distribution conflicts as forces for sustainability: An overview and conceptual framework. *Sustainability Science*, *13*, 585–598.

Shandra, J. M., Leckband, C., & London, B. (2009). Ecologically unequal exchange and deforestation: A cross-national analysis of forestry export flows. *Organization & Environment*, *22*(3), 293–310.

Shandra, J. M., Leckband, C., McKinney, L. A., & London, B. (2009). Ecologically unequal exchange, world polity, and biodiversity loss: A cross-national analysis of threatened mammals. *International Journal of Comparative Sociology, 50*(3–4), 285–310.

Shandra, J. M., Shircliff, E., & London, B. (2011). The International Monetary Fund, World Bank, and structural adjustment: A cross-national analysis of forest loss. *Social Science Research, 40*(1), 210–225.

Smith, J., & Patterson, J. (2019). Global climate justice activism: "The new protagonists" and their projects for a just transition. In S. R. Frey, P. K. Gellert, & H. F. Dahms (Eds.), *Ecologically unequal exchange: Environmental injustice in comparative and historical perspective* (pp. 245–272). Springer.

Sommer, J. M., Shandra, J. M., & Coburn, C. (2019). Mining exports flows, repression, and forest loss: A cross-national test of ecologically unequal exchange. In S. R. Frey, P. K. Gellert, & H. F. Dahms (Eds.), *Ecologically unequal exchange: Environmental injustice in comparative and historical perspective* (pp. 13–48). Springer.

Spaargaren, G., & Mol, A. P. (1992). Sociology, environment, and modernity: Ecological modernization as a theory of social change. *Society & Natural Resources, 5*(4), 323–344.

Svampa, M. (2019). *Neo-extractivism in Latin America*. Cambridge University Press.

Swamy, R. (2017). Humanitarianism and unequal exchange. *Journal of World-Systems Research, 23*(2), 353–371.

Tausch, L., & Althouse, J. (2024). Towards a theory of ecologically unequal exchange (EUE) as a multi-tiered hierarchy: Investigating the interdependence of global and domestic environmental inequalities to explain China's rise to power. *Working Papers hal-04555443*, HAL. Available at SSRN 4847554.

Temper, L., Demaria, F., Scheidel, A., Del Bene, D., & Martinez-Alier, J. (2018). The Global Environmental Justice Atlas (EJAtlas): ecological distribution conflicts as forces for sustainability. *Sustainability Science, 13*(3), 573–584.

Temper, L., Walter, M., Rodriguez, I., Kothari, A., & Turhan, E. (2018). A perspective on radical transformations to sustainability: resistances, movements and alternatives. *Sustainability Science, 13*, 747–764.

Vesia, D. J., Mahutga, M. C., & Bui, B. K. H. (2023). Flattening the curve? The structure of the natural resource exchange network and CO_2 emissions. *Social Networks, 75*, 118–136.

Wallerstein, I. (1974). *The modern world-system: Capitalist agriculture and the origins of the European world-economy in the sixteenth century*. Academic Press.

Warlenius, R. (2016). Linking ecological debt and ecologically unequal exchange: Stocks, flows, and unequal sink appropriation. *Journal of Political Ecology, 23*(1), 364–380.

Warlenius, R., Pierce, G., & Ramasar, V. (2015). Reversing the arrow of arrears: The concept of "ecological debt" and its value for environmental justice. *Global Environmental Change, 30*, 21–30.

CHAPTER 3

World Society Theory and Climate Action

KAJOL PATEL AND KRISTEN SHORETTE

Introduction

In November 2020, KlimaSeniorinnen, an organization of elderly women advocating for climate protection, challenged the Swiss government at the European Court of Human Rights (ECHR) with the financial support of Greenpeace Switzerland. They contended that Switzerland's inadequate actions on climate change compromised their health, particularly due to intensified heat waves. Their primary assertion was that Switzerland's failure to achieve its greenhouse gas reduction targets based on the Paris Agreement breached the European Convention on Human Rights, specifically infringing on the rights to life and respect for private and family life as held in Articles 2 and 8. The case argued that the Swiss government had not implemented the necessary legislative changes to fulfill its commitments to cap global temperature rises at 1.5°C above preindustrial levels.

In April 2024, the Court determined, by a 16-to-1 majority, that Article 8 had been violated, marking a historic judicial acknowledgment that environmental policy failures can infringe on human rights. Despite this landmark ruling, it remains uncertain whether the Swiss government will fully comply, as by January 2025, the Swiss parliament has voted to reject the ruling, claiming their existing climate action plan was sufficient to meet the Paris Agreement's standards. This case illustrates the challenges of translating legal victories into policy changes which sets a critical precedent: It demonstrates that civil society actors can hold governments accountable for environmental negligence under human rights frameworks, and this has potential implications far beyond the 45 member countries of the Council of Europe.

Global institutions are increasingly shaping the actions of both state and civil society actors as the expectation of environmental protection becomes salient in public opinion and political discourse. The ECHR's decision to hold Switzerland accountable marks not an isolated incident but a larger global pattern of institutions and non-state actors pushing for robust legislation to address environmental challenges. Unlike prevailing political economic theories of globalization, the neo-institutionalist paradigm can account for such increasing international unity around climate action. Just as greenhouse gas emissions travel across national boundaries, neo-institutionalists posit that global environmental norms spread across the world-system as international organizations and agreements empower and unite local activists and civil society groups like the KlimaSeniorinnen (Schofer et al., 2012).

In this chapter, we consider climate action through the lens of world society theory. We begin by introducing the theoretical perspective in general and as it relates specifically to environmental issues. We then distinguish between civil society and statist dimensions of the global institutional system. Next, we review prior research that examines the influence of world society and world polity on the adoption of state and corporate policies, on the presence of environmental activism in the news media, and on individual attitudes and behaviors. In the next section, we consider key critiques of the neo-institutional perspective, especially relating to decoupling between norms and practices and the role of power in shaping those dynamics. Finally, we conclude with suggestions for extending research in this growing field.

The World Society Perspective: Isomorphism, Legitimacy, and the Diffusion of World Culture

World society theory offers a unique lens to understand the enormous uptick we observe in efforts to address environmental problems.[1] It does not assume that climate action is a natural or inevitable response to the problem of climate change. Rather, this theory posits that global institutions propel macro-historical change, including the growing mobilization around environmental issues. This perspective provides a "top-down" understanding of cultural diffusion, where globally accepted scripts or paradigms have cascading effects, ultimately shaping motivations and outcomes at national, organizational, and individual levels (Amenta & Ramsey, 2010; Hironaka, 2014). From a world society perspective, agents are influenced and shaped by the global-institutional environment they find themselves within. Global institutions empower individuals and governments alike to address the problem of climate change.

World society theory first developed out of an observation that similar forms of state governance and mass-education systems were spreading across the globe at the turn of the 20th century (Boli et al., 1985; Meyer et al., 1992, 1997; Pizmony-Levy, 2011; Ramirez, 2012). These *isomorphisms*, or growing similarities in structures and practices, suggested a convergence toward common models of institutions such as standardized education across nations despite diverse cultural and geopolitical circumstances. Meyer and colleagues argued that this *diffusion*, or spread, of practices and models such as education emerged out of a larger pattern of global cultural homogenization as a universal model of nation-state sovereignty formed out of World War II (Krücken & Drori, 2009; Meyer et al., 1992). These early theorists argued that *mimetic isomorphism*, terminology drawn from classical institutional theorists DiMaggio and Powell (1983), instead drove nations to adopt taken-for-granted institutional forms from the world culture's set of tools (Schofer et al., 2012; Swidler, 1986). Through international channels, this adoption can be a strategy for nation-states to display *legitimacy* on the global stage.

Beyond educational forms alone, global interconnectedness has fundamentally changed the expectations, role, and function of state and social actors (Albrow, 1996). Now, not only are nations tasked with the role of educating their populations to be legitimate (Ramirez & Boli, 1987) but also with a variety of other responsibilities aligned with protecting the rights of people and nature within their jurisdiction (Frank et al., 2000). This shift reflects a core tenet of neo-institutionalist thinking: the pursuit of modernity through rational, universal models aimed at social progress (Boli & Thomas, 1997; Schofer et al., 2012) intertwined with Durkheimian insights on the social functions of these institutions (Durkheim, 1982). Durkheim emphasized that social facts—norms, values, and institutions—exist outside of individuals and exert a coercive influence, shaping the collective conscience of societies. This perspective complements the Weberian and modernization theories, where rationalism emerges from historical processes in which systematic ordering guides social institutions toward models emphasizing methods and objectives of logic and efficiency (Spaargaren & Mol, 1992; Weber, 1905).

At the institutional level, as social processes become increasingly organized, actors not only adapt to such scripts but also reflect their underlying collective values and norms and continuously react to mirror them. This ideal suggests that modernity is not just a set of technological or economic changes, as an ecological modernization perspective may suggest, but also encompasses cultural and institutional transformations that reflect the interwoven nature of our world society.

The Global Environmental Regime

Climate action and environmentalism overall are quintessential examples of scripts that have been widely embraced and implemented on a global scale (Shorette et al., 2017). In recent decades a *global environmental regime*—consisting of international treaties, organizations, and cultural understandings of the environment—has emerged (Hironaka, 2014). The cultural understanding that the natural environment is in need of preservation and protection permeates multiple levels of society. Building on Meyer's foundational work, scholars have observed a parallel expansion in environmental education (Žalėnienė & Pereira, 2021). Initially rooted in biology, this education now spans multiple disciplines across universities around the world (Frank et al., 2011).

Concurrently, there is an increasing trend of nation-states participating in international environmental agreements and incorporating a growing number of environmental provisions in their trade agreements (Morin et al., 2018). This escalation in environmentalism across various social institutions prompts the question: Why are actors increasingly engaging in the global environmental regime? For neo-institutionalists, this trend reflects a growing acknowledgment of nation-states' responsibilities toward environmental stewardship, driven by a collective commitment and push to sustain the natural world by international organizations (Frank et al., 2000). Understanding the complex sources of world culture, although challenging due to their dispersed nature, helps us analyze how global institutional processes and cultural diffusion occur (Hironaka, 2014). World society theory differentiates between state and civil society components of the global institutional system, where each plays a crucial role in the creation and implementation of world culture (Cole, 2017; Givens, 2017; Kim, 2020).

The World Polity

The state dimension of the global institutional system, or world polity, focuses on the interaction of states within formal political institutions such as intergovernmental organizations (IGOs) and international treaties (Cole, 2017). Throughout the formalization of the nation-state model, structures in world polity have grown increasingly dense and interconnected (Boli & Thomas, 1997). National governments are pushed to align with international norms and standards to maintain stability or legitimacy for their reputation at the global stage.

One source of environmental scripts in the world polity is global quasi-political institutions like the United Nations Framework Convention on Climate Change (UNFCCC). The UNFCCC, established by treaty in 1992, set forth a collective attempt to regulate the human impacts on climate. Annually, those signed onto the UNFCCC gather at the Convention of Parties (COP) to reassess the state of the climate emergency and establish the responsibilities of member states. The Paris Agreement in 2015 at COP21 marked the commitment and acknowledgment of nearly all countries to cutting global greenhouse gas emissions in an effort to limit climate change–related atmospheric temperature rises. The Paris Agreement created a framework of voluntary national pledges to climate action plans submitted by parties, or nationally determined contributions (NDCs).

This approach differs significantly from previous UNFCCC decisions such as the 1997 Kyoto Protocol, which instead conferred mandatory emission reduction targets and was exclusively applied to industrialized countries. While counterintuitive, the move to NDCs or "soft" commitments from "hard" ones was a calculated risk. This strategy, a blend of bottom-up flexibility and top-down monitoring, reporting, and review, was a response to the gridlock seen in global climate governance, where previous diplomatic efforts were hindered by diverse national interests and reliance on fossil fuels (Bäckstrand et al., 2017; Hale et. al 2015). In a world economy that has heavily relied on fossil fuels, binding agreements that attempted to create equitable targets and timelines for all nations were too contentious as "there was too much on the table; too many countries with diverse interests [who] were unable to agree" (Victor, 2011, p. 208). The pledge and review system of the Paris Agreement successfully brought all major emitters to the negotiation table (Hale, 2016). As of 2024, with 195 Parties committed to the Paris Agreement, it continues a global effort that encourages the adoption of similar climate policies tailored to local contexts (UNFCCC, 2023). The Paris Agreement exemplifies a continuous effort to establish global political institutions and interstate agreements to foster a consensus on environmental issues, despite diverse national interests.

The World Society

In contrast, world society encompasses the role of non-state actors and interpersonal linkages in global ontological and epistemological networks. International nongovernmental organizations (INGOs) have grown significantly in number in the past century (Boli & Thomas, 1997, 1999; Turner, 2010). As key actors in the world society, INGOs play a critical role as advocates, lobbyists, whistleblowers, and service and information providers (Dörfler & Heinzel, 2023; Finger, 1994a; Madon, 1999). Moreover, INGOs can bring new forms of world culture to the forefront, challenging traditional boundaries (Finger, 1994b; Tsutsui, 2004). Despite often lacking formal authority themselves, these non-state actors are instrumental in influencing and weaponizing interstate agendas. They have created new climate action frames of moral authority by linking traditional social issues to climate politics, as in the case of human rights and environmental outcomes seen in *KlimaSeniorinnen v. Switzerland* (Allan, 2018).

Increasingly, environmental INGOs have been conferring more recognition as authoritative voices in environmental policymaking. Many are supported or funded by governments, albeit to some frustration among activists who have splintered into their own more radical factions who prefer direct action such as Earth First! (Berny & Rootes, 2018). Article 6 and Decision 1 of the Paris Agreement have helped further institutionalize the role of non-state actors. Article 6 recognizes and promotes voluntary cooperation between parties, noting such connections can spread "higher ambitions" (UNFCCC, 2023).

Alongside cities, regions, companies, and other investors, INGOs have been identified as collaborators who may act in party, participant, or lead organization for cooperative climate initiatives (CCIs) (Hale, 2016). The UNFCC's CCI Tracking shows that of the current 95 active CCIs registered by the UNFCCC as of 2024, 55 include organizations as a participant (UNFCCC, 2024). While it is still unknown how this will affect organizations or the effectiveness of environmental initiatives worldwide, it is undeniable that conferring such authority has changed the political landscape of the global environmental regime (Bäckstrand et al., 2017; Hale et al., 2021).

Applications of Neo-Institutionalist Framework

The dynamic interaction between world polity and society is a key aspect of understanding global environmental governance. These forces, sometimes collaborating and at other times opposing each other, play a crucial role in shaping environmental policies and outcomes and are part of the same integrated global networks (Buzan, 2018; Tsutsui, 2004). Though notably, world society theory does not suggest a direct link to environmental outcomes. The "Bee Swarm" model, proposed by Hironaka (2014), describes global environmental change as not produced through any individual person, treaty, or organization but rather through their long-term cumulative actions focused around shared environmental goals. These structures are in constant interactions where they influence and amplify one another. World cultural forces have driven actors toward environmental practices by the emergence of "rules" or institutionalized paths. These scripts can be seen both in the policy space through the adoption and enforcement of policies, and in shaping individual values, norms, and behaviors.

Environmental Policy Adoption

Foundational work in the world society tradition demonstrated that a nation's involvement in domestic or international environmental initiatives is closely linked with its engagement in world society (Frank et al., 2000; Hironaka, 2002; Longhofer et al., 2016). For example, Frank et al. (2000) show that a nation's general embeddedness in world society, particularly via INGOs, is a stronger predictor of the adoption of national environmental protection activities than the presence of domestic associations focusing on ecology or natural sciences. They argue that INGOs exert pressure on nations to meet international expectations for authority and legitimacy, serving as both connectors to global networks and collaborators in supporting initiatives. In an extension of this work, Frank et al. (2007) and Longhofer et al. (2016) reveal the relative strength of international over domestic environmental organizational ties in predicting national environmental reforms. Their findings suggest that domestic nongovernmental organizations (NGOs) who were new, small, and poorly funded were themselves influenced by global institutions rather than producers themselves (Frank et al., 2007; Longhofer et al., 2016; Tasmim et al., 2020).

Others note that the agenda of political institutions are influenced by the presence of world society actors. For example, Dörfler and Heinzel (2023) find that the presence of environmental INGOs located in geographical proximity to the international organizations' headquarters significantly increases the likelihood of these international organizations adopting environmental issues into their policy agendas. This proximity allows environmental INGOs to leverage contacts to organization representatives and media outlets to increase pressure privately and publicly on organizations (Keck & Sikkink, 1998). Contrary to Longhofer et al. (2016), they argue that the influence of proximate INGOs is further enhanced when they collaborate with domestically focused NGOs within states by positioning themselves as authoritative environmental experts.

Further research shows that ties to environmental INGOs increase the chances of environmental treaty ratification as well as the speed at which they are ratified into law (Evans et al., 2020; Shim & Shin, 2020). For example, Shim and Shin (2020) find that ties to both international and domestic environmental NGOs tended to hasten national ratification of the 1992 UN Convention on Biological Diversity. These findings collectively suggest that the cultural influence stemming from integration into world society is a crucial factor in shaping the world polity terrain. It not only drives the formation of an environmentally conscious world polity but also can hasten the pace that environmental policies are adopted and implemented.

Enforcement: Monitoring and Shaming

International Organizations and the State

Global environmental norms are further reinforced through the interaction between civil society and state dimensions of the global institutional system. In particular, nongovernmental agents can promote environmental norms by holding governments and businesses to the standards they have signed onto.

One tactic is to gather and disseminate data that highlight gaps between environmental commitments and practices. The Paris Agreement relies heavily on "naming and shaming" countries who fail to accurately participate in transparent reporting of their emissions and commitments (Falkner, 2016). For example, Climate TRACE, a nonprofit coalition of over 100 collaborators from NGOs, universities, and technology firms, showed that countries and corporations are failing to report emissions accurately. In 2021, they revealed that greenhouse gas assessments reported by Russia at the annual COP excluded 1.5 billion tons of CO_2 emissions while the United States underreported 400 million tons of CO_2 emissions from oil and gas sources (Climate TRACE, 2023; Voosen, 2023). Without independent sources of accurate information, nations may knowingly or unknowingly skirt their responsibilities, leading to the *decoupling* of stated goals and real-world impacts, a concept we will return to in later addresses to critiques of neo-institutionalist theory.

The news media can also serve as a platform for shaming. A recent study focusing on climate-specific shaming events in the news shows that environmental INGOs tend to be more inclined to engage in climate-specific shaming activities against democratic countries that participate in fewer international climate agreements and have few climate laws (Koliev et al., 2022). In the case of climate, they argue that INGOs may choose these targets because they are more likely to respond to shaming. In a related study, Koliev et al. (2023) find that international news shaming events are particularly effective in influencing government action vis-à-vis climate laws for states with more trade dependencies and who are already committed to international climate agreements. In a complementary study on environmental INGO-led government shaming, Murdie and Urpelainen (2015) find that organizations tend to target states with inadequate state capacity rather than the worst environmental offenders.

Alongside reporting, scholars find that ties to international organizations can help amplify local environmental movements. For example, Evans et al. (2020) show that national ties to environmental INGOs increase the likelihood of environmental protests being reported in the global news media. They show that the effect of environmental INGOs on the presence of climate protests is especially strong in less democratic societies, suggesting that organizations may strategically focus efforts on repressive regimes. This strategic targeting can showcase both the boomerang effect and a concerted effort by INGOs to act as "strategic substitutes" themselves, stepping in to mitigate the gaps in environmental governance where institutional support is lacking. This may help bridge the gap between decoupling of stated policy goals and real-world effects and allows NGOs to allocate funding wisely while building their competencies.

For activists like the Ogoni in the Niger Delta where Shell Petroleum Development Company set up oil mines detrimental to the local environment (see chapter by Oviedo in this volume), condemnation by international audiences through Amnesty International and Greenpeace contributed to Shell's withdrawal from the region, ending a decade-long struggle of demands being ignored and protests being subdued violently by Nigerian police. INGOs such as Greenpeace, the Rainforest Action Network, and Friends of the Earth often establish links to numerous domestic groups and serve as a conduit through which domestic concerns and movements can use international pressure to create domestic-level policy or practice change (Bob, 2005). INGOs function as umbrella organizations that link local groups from a wide variety of countries. This dynamic can help facilitate a "boomerang effect," where local activists, faced with limited domestic avenues for change, strategically engage with international audiences to exert pressure back on their own governments (Tsutsui & Shin, 2008).

International Organizations and Corporations

In addition to putting pressure on states, international environmental organizations have been successful in affecting corporate policies. Historically, firms would distance themselves from responsibility for environmental controversies linked to their global supply chain, denying any responsibility for ecological or social impacts of their operations. The sportswear brand Nike learned firsthand in the 1990s how a blaisé response to anti-sweatshop activist and burgeoning college student calls to address poor working conditions in their global supply chain could severely damage their image and lead to their first negative quarterly earnings loss in 13 years and a 50% stock price dip in 1998 (Shaw, 1999). Global activism around labor standards shows how cultural expectations set by global civil society can pressure corporations to modify their practices (Barley & Child, 2014). The lessons from this experience were not lost on transnational corporations (TNCs), many of whom have established supplier code of conduct documents as the question of economic globalization's detrimental effects on developing nations grew (Lim & Tsutsui, 2012). In 2011, when Greenpeace International launched a boycotting campaign against global brands like Nike, Puma, and Adidas for using suppliers that dumped their toxic chemical waste into water systems, a number of companies, including Nike, were surprisingly quick to publish plans to go toxic-free by 2020 (Birch, 2012).

Further research shows the expansion of voluntary corporate social responsibility (CSR) pledges that include environmental issues (Lim & Tsutsui, 2012). Even if these efforts are sometimes performative, businesses continue to adopt such frameworks to minimize risk and maintain legitimacy and reputation with consumers, investors, and the larger global society, particularly in developing countries (Lim & Tsutsui, 2012). Environmental, social, and governance (ESG) monitoring organizations quantify CSR commitments to measure a firm's performance on the three dimensions. Here again, civil society organizations serve as policy enforcers by monitoring empirical data and supply chain relationships to reveal the extent to which commitments are maintained.

Still, arguments remain that these pledges do not go far enough and rely on balancing by controversial carbon credit offsets, which allow companies to buy their way into net-zero claims without fundamental changes in operations (Fankhauser et al., 2022). If left unchecked, global standards may be adopted in name but not practice. In a study on the Mexican automotive industry, Montiel et al. (2012) look at how local factors, such as corruption, affect the implementation of global standards. Their findings suggest that in environments with high policy-specific corruption, environmental certifications like ISO 14001 gain value as a symbol of compliance, but their significance may diminish in generally corrupt contexts. Future work in this field may be interested in the process through which formalized commitments in green trade markets through labeling practices and ESG ratings can alter markets and practices, potentially furthering calls for increased standardization and monitoring (Delgado-Ceballos et al., 2023; Shorette, 2021).

Environmental Attitudes, Beliefs, and Behaviors

Lastly, global institutional structures also play a role in fostering individual environmental attitudes directly. Education systems are pivotal in this transformation, charged with integrating environmental topics from primary through to higher education (Bromley et al., 2011; Frank et al., 2011; Pizmony-Levy, 2011) where more numerous ties to IGOs and INGOs are related to higher numbers of domestic environmental education organizations (EEOs). Pizmony-Levy (2011) shows that the immediate local environmental conditions—often assumed to be a primary driver for environmental education—do not substantially influence the number of EEOs. Instead, global cultural institutions and networks exert a more substantial influence. Along the same lines, Frank et al. (2011) find that a nation's integration into world society, particularly through memberships in environmental INGOs, plays a more decisive role in shaping environmental education than the actual levels of environmental degradation within the nation. Frank and colleagues further note that the topics covered in university curricula were unrelated to the magnitude of environmental challenges faced by individual countries. Bromley et al. (2011) likewise find that the environmental content in textbooks is more strongly influenced by global environmental movements than national circumstances or policies.

In addition, global cultural models emphasize the agentic role of individuals to impart change through their actions. However, evidence for the impact of such models on individual attitudes and behavior is mixed. On the one hand, Meyer (2010) finds that younger, more educated citizens are especially likely to take up such ideals. Givens and Jorgenson (2013) show that environmental INGOs and environmental ministries both have a positive relationship to individuals' environmental concern. They add that the influence of civil society is stronger than that of the state. Likewise, Hadler and Haller (2011) show that linkages to INGOs have a strong effect on public and private environmental behavior, whereas IGO ties only significantly influence private behavior. They suggest this is due to variation across local contexts and structures. Hadler (2016) expands on this work to find more limited support for the direct effects of world society ties and even less so for IGOs. Alternatively, Hadler (2016) proposes that world cultural effects are more indirect, pointing to a mediatory effect through nation-level institutionalism of environmentalism.

On the other hand, Knight and Messer (2012) fail to find evidence of world society's influence on environmental concern. This lack of impact is further supported by Hao (2016) and Hao and Wang (2018), who note that INGO ties are not robustly influential on environmental concern nor to individuals' willingness to pay higher taxes to protect the environment. Research in this area can further explore the role of world polity institutions as well as environmentally driven world society ties to elucidate how individuals are impacted by the institutionalization of environmentalism as prior literature has been largely mixed.

Critiques: Diffusion, Decoupling, and Power

The World-System

Critiques of the neo-institutional perspective include the frequent *decoupling* between policy adoption and practical change (Buttel, 2000). In response, a considerable part of contemporary work in the field examines when and under what circumstances ties to world society and/

or world polity are most impactful on real-world environmental outcomes (Cole & Perrier, 2019; Givens, 2017; Hironaka, 2014; Shorette & Philips, 2024). Scholars have found that countries with more ties to the global institutional system were associated with improved environmental outcomes for deforestation (Henderson & Shorette, 2017; Longhofer et al., 2016; Schofer & Hironaka, 2005; Shandra, 2007), water quality (Jorgenson et al., 2011; Shandra et al., 2009), agrochemical use (Shorette, 2012), solar energy generation (McCall, 2020), and a number of greenhouse gas emissions (Givens, 2017; Henderson, 2019; Jorgenson et al., 2011; Mejia, 2020).

Despite the abundance of work that shows a relationship between national-global ties and practical environmental change, a notable aspect of neo-institutional theory is the lack of expectation that specific policies or organizations will directly lead to their intended outcomes (Schofer et al., 2012; Shorette et al., 2017). In contrast to many prevailing perspectives, neo-institutional theory is a loosely coupled model of social change where numerous factors at multiple levels of analysis push in the same direction much like a bee swarm (Hironaka, 2014). The United Nations Environment Programme along with numerous multilateral treaties codify environmental norms which are taken up by environmental NGOs (both international and domestic), states, and individuals. Likewise, naming and shaming and other non–legally enforceable strategies nudge governments and businesses to improve their environmental practices. However, they do not always lead to fully favorable outcomes.

For example, in Brazil, the administrations under former Presidents Michel Temer and Jair Bolsonaro removed legislative environmental restrictions to protect the Amazon. This led to a sharp increase in deforestation in the region which prompted international backlash. Numerous international grocers and financial organizations threatened to boycott Brazilian products (Watts, 2021). Norway and Germany froze contributions going to the Amazon Fund (Boffey, 2019). Even so, Bolsonaro made few concessions during his term as commodity prices and demand remained high as nations like China sought resources like soybeans (Gaulier & Zignago, 2010).

While environmentalism is a strongly institutionalized norm, it is important to note that norms are not always theorized to diffuse to the national level (Strang & Meyer, 1993). Global cultural models may not effectively cross transnational boundaries if they fail to promote collective interests, demonstrate their compatibility with existing global institutions, overcome political-economic barriers, or spread sufficient international organization vectors (Schofer et al., 2012). As such, more recent work has examined not only when and how diffusion results in tangible social change but also when and how decoupling persists (Cole & Perrier, 2019; Shorette & Philips, 2024; Patel 2024).

Others are critical of research in the neo-institutional tradition for often overlooking the role of power dynamics within the world-system (Downey et al., 2020). Indeed, prior work finds that the extent to which norms and practices are decoupled is tied to world-systems positionality (CoatarPeter & Gareau, 2023; Mejia, 2020; Shorette, 2012; Henderson & Shorette, 2017). World-systems theory posits that a nation's position in the global political economy disproportionately impacts the environmental quality of peripheral and semiperipheral nations due to the extraction of their natural resources and exploitation of cheap labor, a consequence of their roles in the global economy and pressure from powerful nations (Bunker, 1988; Moore, 2003). In her study of agrochemical use, Shorette (2012) introduces an "integrated theory of global environmentalism" that considers the dual roles of global institutional and political economic forces. She finds that the extent of decoupling varies by world-system position and argues that this *differential decoupling* is a result of a combination of the pursuit of legitimacy in the global system and of state capacity to align with environmental norms in practice. Mejia (2020) extends this discussion, finding that decoupling varies by world-system position in the case of ambient air pollution levels as well.

In a qualitative application, Coatar and Gareau (2023) observe that as a semiperipheral nation-state, Chile, was effectively compelled to adopt international environmental norms—developing and implementing its national forestry strategy under the UNFCCC's Reducing Emissions from Deforestation and Forest Degradation (REDD+) framework to display outward alignment with global environmental norms. However, this adoption is strongly mediated by the country's domestic political and economic context, where powerful interests in the forestry sector, central to Chile's economy, exert considerable influence. They find that this decoupling results from "organized hypocrisy" on part of the Chilean government, where the outward alignment to global institutions masks a disconnection between world polity goals and the realities of domestic implementation. This decoupling exemplifies the challenges faced by semiperipheral nations, which must navigate their dual roles within both the core and periphery of the global system.

The Nation-State

States' resources or capacity to engage effectively in environmental initiatives is another considerable source of decoupling (Shandra et al., 2012; Shorette et al., 2017). In one study, Givens (2017) reveals a striking dichotomy: Integration into world society or polity significantly reduces production-based carbon intensity of well-being (CIWB) in high-income countries, while lower-income countries sometimes experience an *increase* in CIWB under similar conditions. This divergence, Givens suggests, may stem from global economic inequalities and the varied methods countries internalize global cultural ties. In lower-income countries, engagement with international organizations might be superficial, aimed more at gaining legitimacy rather than fostering sustainable development. It may also be that there is an inherent incapacity in these states or a deficiency in the organizations themselves to harness these global ties effectively. Conversely, high-income countries often see a negligible or even negative influence of world society or polity on consumption-based CIWB. This suggests that while they may reduce their production-based CIWB through global integration, their consumption patterns—benefiting from the outsourcing of polluting industries—do not mirror the same level of environmental sustainability.

Longhofer and Schofer's (2010) study of the growth of domestic environmental associations provides context here. They find that, compared to the United States and Western Europe, formal environmental associations in the developing world emerged later and were less numerous than in the West. However, environmental INGOs led to the formation of fewer domestic environmental associations in industrialized countries. They point to domestic factors such as democratic political institutions and environmental legislation as key drivers for association formation. The ability of industrialized nations to leverage environmental scripts effectively may be partly attributed to their robust domestic structures and resources that foster environmental associations. Conversely, in developing countries, global forces rather than domestic factors are found to be the primary catalyst for environmental organizing.

In addition, others have considered the indirect role of the global environmental regime in offsetting other drivers of environmental degradation (Henderson, 2019; Jorgenson et al., 2011; Longhofer & Jorgensen, 2017). A large body of literature has pointed out the differing role of state structure, development, capitalism, and other macro-social forces on environmental outcomes (Chang et al., 2018; Greiner, 2019; Jorgenson, 2012; Shandra et al., 2004; Tester, 2020). In this body of work, contradictory national and global contexts are tested to see if world cultural norms can diminish the negative effect traditionally documented.

In one such study, Longhofer and Jorgenson (2017) find that countries at the core of the environmental world society—those with strong ties and high integration within the network of environmental INGOs—experience a relative decoupling of economic growth from carbon emissions over time. This suggests that integration into pro-environmental global networks can play a pivotal role in mitigating the environmental impacts of economic development. However, peripheral countries witness an intensification in the correlation between economic development and carbon emissions. Jorgenson et al. (2011) find that world society integration lessens the harmful effect of foreign direct investment on deforestation, air, and water pollution. Similarly, Henderson (2019) finds that development aid's negative influence on carbon emissions is decreased for nations who are further integrated into the world society.

Some scholars also raise the issue that neo-institutionalists have not focused on critically analyzing whose interests are being served or represented by the global environmental regime, suggesting that decoupling is natural as world culture is created to benefit the few (Downey et al., 2020). Buttel (2000) points out the conflict-free portrayal of the transmission of environmental rationality from global society to nation-states ignores the role of coercion, contestation, and international conflict in such uptake. Similar to the concerns around Weber's rationalization thesis for being Eurocentric, and emphasizing "intellectualization and, above all, the 'disenchantment of the world'" (Weber, 1917), neo-institutionalism has been criticized for its lack of a detailed account of the specific actors shaping world culture themselves (Beckfield, 2010). Others also note it is important to look closer at global field dynamics to investigate why global institutions emerge in some cases, like the UN Global Compact's voluntary CSR framework, while for others, like calls for a legally binding code of conduct for TNCs, it does not (Lim, 2021). To address these gaps, more descriptive studies have sought to explore these dynamics in greater depth.

Utilizing comparative historical analysis, Gareau and Lucier (2018) investigate how economic structures, which environmental social scientists largely agree are barriers to environmental policy, transform the environmental world polity. In this paper, the authors studied how the Montreal Protocol and the Basel Convention, two environmental treaties, were affected by changes in the normative discourse in the world polity. Rather than seeing scientific expertise as pure, they argue that in these spaces, science is heavily influenced by neoliberalism and powerful state interests, giving an illusion of environmentalism's participation at the global stage. While both treaties still contend that the global ecosystem needs to be protected, they emphasize neoliberal, market-oriented policies which protect elites and provide them with new business opportunities. Through this, they argue that progress toward curtailing environmental harms has been stalled at the world polity stage. Echoing Weber's observation that rationalism is historically contingent (Weber, 1905), the study highlights the importance of looking closely at these norm-disseminating organizations.

Transnational Municipal Networks

Reflecting this need, a growing body of research within the neo-institutional field is now investigating the influence of global norms and contexts on actors at the city, national, and regional levels of analysis. Researchers have long argued that circular structures of decentralized learning are imperative to be responsive to local needs and capacities (Madon, 1999). Yazar et al. (2023) examine the diffusion of global climate policy dynamics at both national and local levels (see also chapter by Leffel in this volume). The study reveals a nuanced interplay: at the national level, there is a tendency to depoliticize climate change through incremental strategies and broader climate networks. These strategies often include energy efficiency and adaptation measures like water management and desertification risk actions, reflecting a cautious approach to climate policy implementation.

In stark contrast, local governments, especially at the district level, actively *re*politicize climate change, which manifests in Turkey through transformative climate actions such as aggressive solar energy projects, comprehensive emission plans, and advanced urban water adaptation strategies. These initiatives are frequently supported by transnational municipal networks (TMNs), universities, and other local actors, demonstrating a more engaged and diverse approach to climate policy compared to the national level. Local governments are directly faced with the immediate impacts of climate change, creating an urgency to adopt more proactive and tailored solutions. This reflects the expanding influence of cities, subnational governments, businesses, and other non-state actors in climate governance through the Paris Agreement moving beyond state-centric models (Hale, 2016).

Haupt et al. (2020) focus on the influence of TMNs on city-level climate governance. They highlight the role of isomorphic city-to-city learning (C2CL) in transmitting norms across cities facing similar challenges and aiming for increased climate resilience. This process involves not just the exchange of information but the application of knowledge in a meaningful way, often requiring adaptation to the specific context of each city. These networks are seen as key players in global climate governance, balancing the need for meaningful global climate action with addressing specific local challenges. Building upon this understanding, Leffel (2022) examines urban climate mitigation through the lens of national and polycentric systems. His study highlights the critical role of cities in mitigating greenhouse gas emissions and focuses on the function of polycentric systems, including environmental management consultancies and TMNs. Leffel underscores their importance in facilitating the transmission of expertise and norms from global sources directly to local urban settings, thereby contributing significantly to the reduction of urban emissions.

Continuing in this vein, Grant et al. (2023) extend the exploration of local climate action onto practical environmental outcomes. They show that TMNs have a significant role in decreasing CO_2 emissions from power plants in cities. This effect is more pronounced in less developed countries, underscoring the potential of TMNs to facilitate resource access and decarbonization strategies in contexts where such resources might otherwise be lacking. These studies all illustrate the critical role of global institutions and knowledge structures in global climate mitigation efforts, particularly by bolstering the capabilities of cities to engage in effective environmental practices that are locally constituted.

Conclusions: Directions for Future Work

The world society, or neo-institutional, field focuses on understanding how global cultural frameworks, norms, and institutions influence behavior and social structures. By examining the spread of global environmentalism as a top-down process, this perspective challenges conventional actor-centric approaches to understanding social and political-economic change. Through the influence of global political and civil society organizations, actors at the individual, national, and global levels have been guided toward policies and behaviors that align with increasingly standardized norms. In this chapter, we have described a number of pathways world polity and world society forces have driven the establishment of a global environmental regime. We suggest several avenues for future work to advance the world society perspective on climate action.

To advance our understanding of global environmental change, scholars should continue to examine the indirect effects of world society and world polity in mitigating the influence of other global and local factors. There is a pressing need for qualitative research to unpack the origins of the global environmental regime and to critically assess whose interests are represented within it. Additionally, the rising number of climate-related legal cases globally presents a new frontier for research (UNEP, 2023). We began this chapter by describing a case study, *KlimaSeniorinnen v. Switzerland*, as an example of environmental actors weaponizing international commitments to force governments to uphold their end of the deal. A report by the Sabin Center for Climate Change Law found a dramatic increase in climate litigation globally, revealing that the number of climate cases has more than doubled from 2017 to 2023, going from 884 cases to 2,180 (Burger & Tigre, 2023). For cases filed outside of the United States of America, the number of cases nearly tripled from 230 to 658 cases.

In particular, we recommend that scholars build on Negro and Longhofer's (2018) study, which scrutinizes the relationship among world society integration, legal formalism, and the execution of legal procedures, and provides a crucial groundwork for understanding how domestic legal structures might interact with and be influenced by broader, global legal norms and practices. Studying litigation spaces and laws, scholars can gain deeper insights into the mechanisms driving the environmentalization of legal systems and the increasing prevalence of cross-border legal challenges (García-Villegas, 2016).

Through legal systems, world society actors can utilize their legitimacy through an alternative avenue. For example, future scholars may look at the prevalence of mentions of international policies in court decisions and how this changes the narrative arguments used by plaintiffs and decisions argued by judges, which typically favor corporate or state defendants in climate cases. Such studies would not only enrich our understanding of legal formalism in the context of neo-institutionalism but also illuminate the ways in which legal systems globally are adapting to and being shaped by the environmental world polity.

Furthermore, future scholars can explore cases of *backlashes* to climate policy through a world society lens. World society theory has done well to explain the tremendous rise of environmental policies, practices, and attitudes across the world-system. However, populist and nationalist backlashes against well-established world cultural norms, including environmentalism, have gained momentum since the early 2000's (Cole et al., 2024; Lerch et al., 2022, 2024; Meyer, 2024; Velasco, 2020). These illiberal forces exploit cultural dissonance to sustain climate change skepticism. Current trends indicate that climate policies, including those adopted at local levels, are encountering organized resistance, leading in some cases to policy rollbacks and de-adoption. Hardline policies necessary for decarbonization often come hand in hand with the disruption of existing economic and social dynamics—resistance which goes beyond simple opposition to the policy but instead a backlash which fundamentally challenges the authority and legitimacy behind the policy and those enacting it (Patterson, 2023).

A prominent example is the United States under former President Donald Trump, who in 2017 announced a withdrawal from the Paris Agreement. Justifying the decision by citing it as an unfair economic burden on American workers, taxpayers, and industries, President Trump stated, "This [Paris] agreement is less about the climate and more about other countries gaining a financial advantage over the United States" (Trump, 2017). By framing the agreement as a threat to domestic sovereignty to benefit foreign nations by disconnected elites, this action effectively delegitimized the Paris Agreement as a genuine climate initiative—a sentiment echoed in various parts of the world (Vihma et al., 2021).

World society theory provides a useful framework to understand these backlash movements as conflicts arising during the institutionalization of global norms. As climate denialist movements challenge the validity of climate science and the authority of international institutions, they erode the taken-for-granted status of climate action. Initial work in this area finds that although embeddedness in world society often reduces climate skepticism at the national level—particularly in authoritarian regimes—it fails to meaningfully influence individual attitudes, especially among right-wing individuals (Kim et al., 2024). This suggests that while liberal global frameworks actively challenge illiberal narratives at the macro scale, they encounter significant resistance at the micro level, where entrenched ideological opposition remains largely impervious to such forces. For durable institutionalization, such legitimacy concerns need to be addressed. Future researchers may be interested in exploring how local perspectives on global climate norms affect the legitimacy of climate action, looking at how legitimacy is built—or lost—on this stage. By studying the local, scholars can help illustrate ways to design climate policies that resonate locally.

Ultimately, to create a robust climate framework worldwide, we need to look at where actors adopt or fail to adopt globally institutionalized environmental norms. For policymakers, a critical priority should be designing climate initiatives that are both ambitious and locally resonant, addressing the specific economic, cultural, and political contexts in which they operate while pushing to imagine a world with sustainable practices at the center of decision making. Strengthening mechanisms for monitoring, transparency, and enforcement will ensure that commitments translate into measurable progress. Furthermore, fostering public engagement and addressing resistance with inclusive, equity-centered approaches can help build the legitimacy and durability of climate policies. The journey toward a sustainable future is a collective endeavor, requiring persistent innovation, resilience, and a willingness to bridge divides. By deepening our understanding of the dynamics that drive both progress and resistance, we can chart a course toward a world where environmental stewardship is not just an aspiration but a shared reality.

Note

1. World society theory is alternatively referred to as neo-institutionalism, sociological institutionalism, and world polity theory.

References

Albrow, M. (1996). *The global age: State and society beyond modernity.* Stanford University Press.

Allan, J. I. (2018). Seeking entry: Discursive hooks and NGOs in global climate politics. *Global Policy,* 9(4), 560–569. https://doi.org/10.1111/1758-5899.12586

Amenta, E., & Ramsey, K. M. (2010). Institutional theory. In K. T. Leicht & J. C. Jenkins (Eds.), *Handbook of politics: State and society in global perspective* (pp. 15–39). Springer. https://doi.org/10.1007/978-0-387-68930-2_2

Bäckstrand, K., Kuyper, J. W., Linnér, B.-O., & Lövbrand, E. (2017). Non-state actors in global climate governance: From Copenhagen to Paris and beyond. *Environmental Politics, 26*(4), 561–579. https://doi.org/10.1080/09644016.2017.1327485

Bartley, T., & Child, C. (2014). Shaming the corporation: The social production of targets and the anti-sweatshop movement. *American Sociological Review, 79*(4), 653–679. https://doi.org/10.1177/0003122414540653

Beckfield, J. (2010). The social structure of the world polity. *American Journal of Sociology, 115*(4), 1018–1068. https://doi.org/10.1086/649577

Berny, N., & Rootes, C. (2018). Environmental NGOs at a crossroads? *Environmental Politics, 27*(6), 947–972. https://doi.org/10.1080/09644016.2018.1536293

Birch, S. (2012, July 6). How activism forced Nike to change its ethical game. *The Guardian*. https://www.theguardian.com/environment/green-living-blog/2012/jul/06/activism-nike
Bob, C. (2005). *The marketing of rebellion: Insurgents, media, and international activism*. Cambridge University Press.
Boffey, D. (2019, August 16). Norway halts Amazon fund donation in dispute with Brazil. *The Guardian*. https://www.theguardian.com/world/2019/aug/16/norway-halts-amazon-fund-donation-dispute-brazil-deforestation-jair-bolsonaro
Boli, J., Ramirez, F. O., & Meyer, J. W. (1985). Explaining the origins and expansion of mass education. *Comparative Education Review*, *29*(2), 145–170. https://doi.org/10.1086/446504
Boli, J., & Thomas, G. M. (1997). World culture in the world polity: A century of international non-governmental organization. *American Sociological Review*, *62*(2), 171. https://doi.org/10.2307/2657298
Boli, J., & Thomas, G. M. (1999). *Constructing world culture: International nongovernmental organizations since 1875*. Stanford University Press.
Bromley, P., Meyer, J. W., & Ramirez, F. O. (2011). The worldwide spread of environmental discourse in social studies, history, and civics textbooks, 1970–2008. *Comparative Education Review*, *55*(4), 517–545. https://doi.org/10.1086/660797
Bunker, S. G. (1988). *Underdeveloping the Amazon: Extraction, unequal exchange, and the failure of the modern state*. University of Chicago Press. https://press.uchicago.edu/ucp/books/book/chicago/U/bo3634235.html
Burger, M., & Tigre, M. (2023). Global climate litigation report: 2023 status review. *Sabin Center for Climate Change Law*. https://scholarship.law.columbia.edu/sabin_climate_change/202
Buttel, F. H. (2000). World society, the nation-state, and environmental protection: Comment on Frank, Hironaka, and Schofer. *American Sociological Review*, *65*(1), 117–121. https://doi.org/10.2307/2657292
Buzan, B. (2018). Revisiting world society. *International Politics*, *55*(1), 125–140. https://doi.org/10.1057/s41311-017-0065-5
Chang, C.-P., Wen, J., Dong, M., & Hao, Y. (2018). Does government ideology affect environmental pollutions? New evidence from instrumental variable quantile regression estimations. *Energy Policy*, *113*, 386–400. https://doi.org/10.1016/j.enpol.2017.11.021
Climate TRACE. (2023). *Climate TRACE emissions inventory*. https://climatetrace.org
Coatar, P., & Gareau, B. J. (2023). Combining world-system and world polity approaches to analyze international environmental governance: A case study of forest governance in Chile. *Environmental Sociology*, *9*(1), 67–79. https://doi.org/10.1080/23251042.2022.2115660
Cole, W. M. (2017). World polity or world society? Delineating the statist and societal dimensions of the global institutional system. *International Sociology*, *32*(1), 86–104. https://doi.org/10.1177/0268580916675526
Cole, W. M., & Perrier, G. (2019). Political equality for women and the poor: Assessing the effects and limits of world society, 1975–2010. *International Journal of Comparative Sociology*, *60*(3), 140–172. https://doi.org/10.1177/0020715219831422
Cole, W. M., Schofer, E., & Meyer, J. W. (2024). Global liberalism, emerging illiberalism, and human rights, 1980 to 2018. *Social Science Research*, *120*, 103001. https://doi.org/10.1016/j.ssresearch.2024.103001.
Delgado-Ceballos, J., Ortiz-De-Mandojana, N., Antolín-López, R., & Montiel, I. (2023). Connecting the sustainable development goals to firm-level sustainability and ESG factors: The need for double materiality. *BRQ Business Research Quarterly*, *26*(1), 2–10. https://doi.org/10.1177/23409444221140919
DiMaggio, P. J., & Powell, W. W. (1983). The iron cage revisited: Institutional isomorphism and collective rationality in organizational fields. *American Sociological Review*, *48*(2), 147–160. https://doi.org/10.2307/2095101
Dörfler, T., & Heinzel, M. (2023). Greening global governance: INGO secretariats and environmental mainstreaming of IOs, 1950 to 2017. *Review of International Organizations*, *18*(1), 117–143. https://doi.org/10.1007/s11558-022-09462-4

Downey, L., Lawrence, E., Pyles, M., & Lee, D. (2020). Power, hegemony, and world society theory: A critical evaluation. *Socius, 6*. https://doi.org/10.1177/2378023120920059

Durkheim, E. (1982). *The rules of sociological method* (S. Lukes, Ed.; W. D. Halls, Trans.; 1st American ed.). Free Press.

Evans, E. M., Schofer, E., & Hironaka, A. (2020). Globally visible environmental protest: A cross-national analysis, 1970–2010. *Sociological Perspectives, 63*(5), 786–808. https://doi.org/10.1177/0731121420908899

Falkner, R. (2016). The Paris Agreement and the new logic of international climate politics. *International Affairs, 92*(5), 1107–1125. https://doi.org/10.1111/1468-2346.12708

Fankhauser, S., Smith, S. M., Allen, M., Axelsson, K., Hale, T., Hepburn, C., Kendall, J. M., Khosla, R., Lezaun, J., Mitchell-Larson, E., Obersteiner, M., Rajamani, L., Rickaby, R., Seddon, N., & Wetzer, T. (2022). The meaning of net zero and how to get it right. *Nature Climate Change, 12*(1), 15–21. https://doi.org/10.1038/s41558-021-01245-w

Finger, M. (1994a). Environmental NGOs in the UNCED process. In M. Finger & T. Princen, *Environmental NGOs in world politics* (pp. 186–216). Routledge.

Finger, M. (1994b). NGOs and transformation: Beyond social movement theory. In M. Finger & T. Princen, *Environmental NGOs in world politics* (pp. 48–68). Routledge.

Frank, D. J., Hironaka, A., & Schofer, E. (2000). The nation-state and the natural environment over the twentieth century. *American Sociological Review, 65*(1), 96–116. https://doi.org/10.2307/2657291

Frank, D. J., Longhofer, W., & Schofer, E. (2007). World society, NGOs and environmental policy reform in Asia. *International Journal of Comparative Sociology, 48*(4), 275–295. https://doi.org/10.1177/0020715207079530

Frank, D. J., Robinson, K. J., & Olesen, J. (2011). The global expansion of environmental education in universities. *Comparative Education Review, 55*(4), 546–573. https://doi.org/10.1086/661253

García-Villegas, M. (2016). A comparison of sociopolitical legal studies. *Annual Review of Law and Social Science, 12*(1), 25–44. https://doi.org/10.1146/annurev-lawsocsci-110615-084613

Gareau, B. J., & Lucier, C. A. (2018). Neoliberal restructuring of the world polity: The weakening of the Montreal Protocol and Basel Convention in historical perspective. *Environmental Sociology, 4*(3), 325–342. https://doi.org/10.1080/23251042.2018.1436893

Gaulier, G., & Zignago, S. (2010). *BACI: International Trade Database at the product-level.* The 1994–2007 version [Working Paper]. CEPII Research Center. https://econpapers.repec.org/paper/ciicepidt/2010-23.htm

Givens, J. E. (2017). World society, world polity, and the carbon intensity of well-being, 1990–2011. *Sociology of Development, 3*(4), 403–435. https://doi.org/10.1525/sod.2017.3.4.403

Givens, J. E., & Jorgenson, A. K. (2013). Individual environmental concern in the world polity: A multilevel analysis. *Social Science Research, 42*(2), 418–431. https://doi.org/10.1016/j.ssresearch.2012.10.005

Grant, D., Leffel, B., & Johnson, E. (2023). Can transnational municipal networks mitigate the carbon pollution of the world's power plants? An empirical analysis. *NPJ Climate Action, 2*(1), 1–8. https://doi.org/10.1038/s44168-023-00070-5

Greiner, P. T. (2019). Time, power and environmental impact: A growth curve model of the relationship between temporal change and CO_2 emissions per capita. *Human Ecology Review, 25*(1), 43–68.

Hadler, M. (2016). Individual action, world society, and environmental change: 1993–2010. *European Journal of Cultural and Political Sociology, 3*(2–3), 341–374. https://doi.org/10.1080/23254823.2016.1209642

Hadler, M., & Haller, M. (2011). Global activism and nationally driven recycling: The influence of world society and national contexts on public and private environmental behavior. *International Sociology, 26*(3), 315–345. https://doi.org/10.1177/0268580910392258

Hale, T. (2016). "All hands on deck": The Paris Agreement and nonstate climate action. *Global Environmental Politics, 16*(3), 12–22. https://doi.org/10.1162/GLEP_a_00362

Hale, T., Held, D., & Young, K. (2015). Gridlock: Why global cooperation is failing when we need it most. *IPSHU, 51*, 20–26.

Hale, T. N., Chan, S., Hsu, A., Clapper, A., Elliott, C., Faria, P., Kuramochi, T., McDaniel, S., Morgado, M., Roelfsema, M., Santaella, M., Singh, N., Tout, I., Weber, C., Weinfurter, A., & Widerberg, O. (2021). Sub- and non-state climate action: A framework to assess progress, implementation and impact. *Climate Policy, 21*(3), 406–420. https://doi.org/10.1080/14693062.2020.1828796

Hao, F. (2016). A panel regression study on multiple predictors of environmental concern for 82 countries across seven years. *Social Science Quarterly, 97*(5), 991–1004. https://doi.org/10.1111/ssqu.12237

Hao, F., & Wang, Y. (2018). Acceptance of higher taxes for the environment: A cross-national and multilevel study based on seven years of data from ISSP and WVS surveys. *International Journal of Sociology, 48*(4), 340–365. https://doi.org/10.1080/00207659.2018.1515704

Haupt, W., Chelleri, L., Van Herk, S., & Zevenbergen, C. (2020). City-to-city learning within climate city networks: Definition, significance, and challenges from a global perspective. *International Journal of Urban Sustainable Development, 12*(2), 143–159. https://doi.org/10.1080/19463138.2019.1691007

Henderson, K. E. (2019). Contradictory governance norms within world society: Energy development strategies, environmental protection, and carbon dioxide emissions. *Environmental Sociology, 5*(4), 393–404. https://doi.org/10.1080/23251042.2019.1613031

Henderson, K., & Shorette, K. (2017). Environmentalism in the periphery: Institutional embeddedness and deforestation among fifteen palm oil producers, 1990–2012. *Journal of World-Systems Research, 23*(2), 269–297. https://doi.org/10.5195/jwsr.2017.699

Hironaka, A. (2002). The globalization of environmental protection: The case of environmental impact assessment. *International Journal of Comparative Sociology, 43*(1), 65–78.

Hironaka, A. (2014). *Greening the globe: World society and environmental change.* Cambridge University Press.

Jorgenson, A. K. (2012). The sociology of ecologically unequal exchange and carbon dioxide emissions, 1960–2005. *Social Science Research, 41*(2), 242–252. https://doi.org/10.1016/j.ssresearch.2011.11.011

Jorgenson, A. K., Dick, C., & Shandra, J. M. (2011). World economy, world society, and environmental harms in less-developed countries. *Sociological Inquiry, 81*(1), 53–87. https://doi.org/10.1111/j.1475-682X.2010.00354.x

Keck, M. E., & Sikkink, K. (1998). *Activists beyond borders: Advocacy networks in international politics.* Cornell University Press. https://www.jstor.org/stable/10.7591/j.ctt5hh13f

Kim, J. (2020). The diffusion of international women's rights norms to individual attitudes: The differential roles of world polity and world society. *Sociology of Development, 6*(4), 459–492. https://doi.org/10.1525/sod.2020.6.4.459

Kim, J., Soysal, Y. N., Boado, H. C., & Schimmöller, L. (2024). Inhibiting or contributing? How global liberal forces impact climate change skepticism. *International Journal of Sociology, 54*(5–6), 530–564. https://doi.org/10.1080/00207659.2024.2416279.

Knight, K. W., & Messer, B. L. (2012). Environmental concern in cross-national perspective: The effects of affluence, environmental degradation, and world society. *Social Science Quarterly, 93*(2), 521–537. https://doi.org/10.1111/j.1540-6237.2012.00846.x

Koliev, F., Page, D., & Tallberg, J. (2022). The domestic impact of international shaming: Evidence from climate change and human rights. *Public Opinion Quarterly, 86*(3), 748–761. https://doi.org/10.1093/poq/nfac026

Koliev, F., Park, B., & Duit, A. (2023). Climate shaming: Explaining environmental NGOs targeting practices. *Climate Policy, 23*(7), 845–858. https://doi.org/10.1080/14693062.2022.2143315

Krücken, G., & Drori, G. S. (2009). *World society: The writings of John W. Meyer.* Oxford University Press.

Leffel, B. (2022). Toward global urban climate mitigation: Linking national and polycentric systems of environmental change. *Sociology of Development, 8*(1), 111–137. https://doi.org/10.1525/sod.2021.0018

Lerch, J. C., Frank, D. J., & Schofer, E. (2024). The social foundations of academic freedom: Heterogeneous institutions in world society, 1960 to 2022. *American Sociological Review, 89*(1), 88–125. https://doi.org/10.1177/00031224231214000.

Lerch, J. C., Schofer, E., Frank, D. J., Longhofer, W., Ramirez, F. O., Wotipka, C. M., & Velasco, K. (2022). Women's participation and challenges to the liberal script: A global perspective. *International Sociology*, *37*(3), 305–329. https://doi.org/10.1177/02685809211060911.

Lim, A. (2021). Global fields, institutional emergence, and the regulation of transnational corporations. *Social Forces*, *99*(3), 1060–1085. https://doi.org/10.1093/sf/soaa038

Lim, A., & Tsutsui, K. (2012). Globalization and commitment in corporate social responsibility: Cross-national analyses of institutional and political-economy effects. *American Sociological Review*, *77*(1), 69–98. https://doi.org/10.1177/0003122411432701

Longhofer, W., & Jorgenson, A. (2017). Decoupling reconsidered: Does world society integration influence the relationship between the environment and economic development? *Social Science Research*, *65*, 17–29. https://doi.org/10.1016/j.ssresearch.2017.02.002

Longhofer, W., & Schofer, E. (2010). National and global origins of environmental association. *American Sociological Review*, *75*(4), 505–533. https://doi.org/10.1177/0003122410374084

Longhofer, W., Schofer, E., Miric, N., & Frank, D. J. (2016). NGOs, INGOs, and environmental policy reform, 1970–2010. *Social Forces*, *94*(4), 1743–1768. https://doi.org/10.1093/sf/sow031

Madon, S. (1999). International NGOs: Networking, information flows and learning. *Journal of Strategic Information Systems*, *8*(3), 251–261. https://doi.org/10.1016/S0963-8687(99)00029-3

McCall, J. (2020). Turning toward the sun: A cross-national analysis of solar energy generation. *International Journal of Sociology*, *50*(4), 310–324. https://doi.org/10.1080/00207659.2020.1730569

Mejia, S. A. (2020). Global environmentalism and the world-system: A cross-national analysis of air pollution. *Sociological Perspectives*, *63*(2), 276–291. https://doi.org/10.1177/0731121419857970

Meyer, J. M. (2024). "The people" and climate justice: Reconceptualizing populism and pluralism within climate politics. *Polity*, *56*(2), 252–274. https://doi.org/10.1086/729277

Meyer, J. W. (2010). World society, institutional theories, and the actor. *Annual Review of Sociology*, *36*(1), 1–20. https://doi.org/10.1146/annurev.soc.012809.102506

Meyer, J. W., Frank, D. J., Hironaka, A., Schofer, E., & Tuma, N. B. (1997). The structuring of a world environmental regime, 1870–1990. *International Organization*, *51*(4), 623–651.

Meyer, J. W., Ramirez, F. O., & Soysal, Y. N. (1992). World expansion of mass education, 1870–1980. *Sociology of Education*, *65*(2), 128–149. https://doi.org/10.2307/2112679

Montiel, I., Husted, B. W., & Christmann, P. (2012). Using private management standard certification to reduce information asymmetries in corrupt environments. *Strategic Management Journal*, *33*(9), 1103–1113. https://doi.org/10.1002/smj.1957

Moore, J. W. (2003). The modern world-systems environmental history? Ecology and the rise of capitalism. *Theory and Society*, *32*(3), 307–377. https://doi.org/10.1023/A:1024404620759

Morin, J.-F., Dür, A., & Lechner, L. (2018). Mapping the trade and environment nexus: Insights from a new data set. *Global Environmental Politics*, *18*(1), 122–139. https://doi.org/10.1162/GLEP_a_00447

Murdie, A., & Urpelainen, J. (2015). Why pick on us? Environmental INGOs and state shaming as a strategic substitute. *Political Studies*, *63*(2), 353–372. https://doi.org/10.1111/1467-9248.12101

Negro, G., & Longhofer, W. (2018). World society, legal formalism, and execution of legal procedures. *Social Forces*, *97*(2), 649–674. https://doi.org/10.1093/sf/soy052

Patel, K. (2024). Varying Vistas: Exploring the Uneven Impacts of World Society Integration on Cross-National Particulate Air Pollution Using Unconditional Quantile Regression. *International Journal of Sociology*, *54*(4), 286–307. https://doi.org/10.1080/00207659.2024.2361543

Patterson, J. J. (2023). Backlash to climate policy. *Global Environmental Politics*, *23*(1), 68–90. https://doi.org/10.1162/glep_a_00684

Pizmony-Levy, O. (2011). Bridging the global and local in understanding curricula scripts: The case of environmental education. *Comparative Education Review*, *55*(4), 600–633. https://doi.org/10.1086/661632

Ramirez, F. O. (2012). The world society perspective: Concepts, assumptions, and strategies. *Comparative Education*, *48*(4), 423–439. https://doi.org/10.1080/03050068.2012.693374

Ramirez, F. O., & Boli, J. (1987). The political construction of mass schooling: European origins and worldwide institutionalization. *Sociology of Education*, *60*(1), 2–17. https://doi.org/10.2307/2112615

Schofer, E., & Hironaka, A. (2005). The effects of world society on environmental protection outcomes. *Social Forces, 84*(1), 25–47. https://doi.org/10.1353/sof.2005.0127

Schofer, E., Hironaka, A., John, D., & Longhofer, W. (2012). Sociological institutionalism and world society. In *The Wiley-Blackwell companion to political sociology* (pp. 57–68). John Wiley & Sons. https://doi.org/10.1002/9781444355093.ch6

Shandra, J. M. (2007). The world polity and deforestation: A quantitative, cross-national analysis. *International Journal of Comparative Sociology, 48*(1), 5–27. https://doi.org/10.1177/0020715207072157

Shandra, J. M., Esparza, L. E., & London, B. (2012). Nongovernmental organizations, democracy, and deforestation: A cross-national analysis. *Society & Natural Resources, 25*(3), 251–269. https://doi.org/10.1080/08941920.2011.573841

Shandra, J. M., London, B., Whooley, O. P., & Williamson, J. B. (2004). International nongovernmental organizations and carbon dioxide emissions in the developing world: A quantitative, cross-national analysis. *Sociological Inquiry, 74*(4), 520–545. https://doi.org/10.1111/j.1475-682X.2004.00103.x

Shandra, J. M., Shor, E., & London, B. (2009). World polity, unequal ecological exchange, and organic water pollution: A cross-national analysis of developing nations. *Human Ecology Review, 16*(1), 53–63.

Shaw, R. (1999). Just don't buy it: Challenging Nike and the rules of the global economy. In R. Shaw, *Reclaiming America: Nike, clean air, and the new national activism* (pp. 13–96). University of California Press.

Shim, J.-M., & Shin, E. (2020). Drivers of ratification rates in global biodiversity governance: Local environmentalism, orientation toward global governance, and peer pressure. *Environmental Politics, 29*(5), 845–865. https://doi.org/10.1080/09644016.2019.1630070

Shorette, K. (2012). Outcomes of global environmentalism: Longitudinal and cross-national trends in chemical fertilizer and pesticide use. *Social Forces, 91*(1), 299–325. https://doi.org/10.1093/sf/sos053

Shorette, K. (2021). The rise of market approaches to social problems: The case of fair trade and its uneven expansion across the Global South. *International Journal of Sociology, 52*(1), 1–24. https://doi.org/10.1080/00207659.2021.2003996

Shorette, K., Henderson, K., Sommer, J. M., & Longhofer, W. (2017). World society and the natural environment. *Sociology Compass, 11*(10), e12511. https://doi.org/10.1111/soc4.12511

Shorette, K., & Phillips, N. E. (2024). (In)sufficient institutionalization? Norm articulation in the World Health Organization and infectious disease prevalence across the global South. *International Journal of Comparative Sociology, 65*(6), 752–781. https://doi.org/10.1177/00207152241226449

Spaargaren, G., & Mol, A. P. J. (1992). Sociology, environment, and modernity: Ecological modernization as a theory of social change. *Society & Natural Resources, 5*(4), 323–344. https://doi.org/10.1080/08941929209380797

Strang, D., & Meyer, J. W. (1993). Institutional conditions for diffusion. *Theory and Society, 22*(4), 487–511. https://doi.org/10.1007/BF00993595

Swidler, A. (1986). Culture in action: Symbols and strategies. *American Sociological Review, 51*(2), 273–286. https://doi.org/10.2307/2095521

Tasmim, S., Sommer, J. M., Shorette, K., & Shandra, J. M. (2020). Non-governmental organizations, boomerangs, and forest loss: A cross-national analysis. *Environmental Sociology, 6*(4), 416–432. https://doi.org/10.1080/23251042.2020.1796203

Tester, A. W. (2020). Deforestation in the Global South: Assessing uneven environmental improvements 1993–2013. *Sociological Perspectives, 63*(5), 764–785. https://doi.org/10.1177/0731121420908900

Trump, D. J. (2017, June 1). Statement by President Trump on the Paris Climate Accord. The White House Archives. https://trumpwhitehouse.archives.gov/briefings-statements/statement-president-trump-paris-climate-accord/

Tsutsui, K. (2004). Global civil society and ethnic social movements in the contemporary world. *Sociological Forum, 19*(1), 63–87. https://doi.org/10.1023/B:SOFO.0000019648.21481.91

Tsutsui, K., & Shin, H. J. (2008). Global norms, local activism, and social movement outcomes: Global human rights and resident Koreans in Japan. *Social Problems, 55*(3), 391–418. https://doi.org/10.1525/sp.2008.55.3.391

Turner, E. A. L. (2010). Why has the number of international non-governmental organizations exploded since 1960? *Cliodynamics, 1*(1), 81–91. https://doi.org/10.21237/C7clio11196

United Nations Environmental Programme (UNEP). (2023, July 27). *Global climate litigation report 2023: Status review.* UN Environment. http://www.unep.org/news-and-stories/press-release/climate-litigation-more-doubles-five-years-now-key-tool-delivering

United Nations Framework Convention on Climate Change (UNFCCC). (2023). Nationally determined contributions under the Paris Agreement (Synthesis Reports No. GE.23-22207(E)). UN Climate Change Conference. https://unfccc.int/documents/632334

United Nations Framework Convention on Climate Change (UNFCCC). (2024). Cooperative climate initiative tracking. Global Climate Action Portal. https://climateaction.unfccc.int/Initiatives

Velasco, K. (2020). Transnational backlash and the deinstitutionalization of liberal norms: LGBT+ rights in a contested world. *American Journal of Sociology, 128*(5), 1381–1429.

Victor, D. G. (2011). *Global warming gridlock: Creating more effective strategies for protecting the planet.* Cambridge University Press.

Vihma, A., Reischl, G., & Nonbo Andersen, A. (2021). A climate backlash: Comparing populist parties' climate policies in Denmark, Finland, and Sweden. *Journal of Environment & Development, 30*(3), 219–239. https://doi.org/10.1177/10704965211027748

Voosen, P. (2023, December 3). Al Gore's climate watchdog spots rogue emissions. *Science.* https://www.science.org/content/article/al-gore-s-climate-watchdog-spots-rogue-emissions

Watts, J. (2021, May 25). UK supermarkets will seek soy alternatives if Amazon protections weakened. *The Guardian.* https://www.theguardian.com/world/2021/may/25/uk-supermarkets-will-seek-soy-alternatives-if-amazon-protections-weakened

Weber, M. (1917). *Science as a vocation.* Translated from H. H. Gerth and C. Wright Mills (Eds.), *Max Weber: Essays in Sociology* (pp. 129–156). Oxford University Press, 1946.

Weber, M. (1905). *The Protestant ethic and the spirit of capitalism.* Translated by T. Parsons. Scribner's Press.

Yazar, M., Daloglu Cetinkaya, I., Baykal Fide, E., & Haarstad, H. (2023). Diffusion of global climate policy: National depoliticization, local repoliticization in Turkey. *Global Environmental Change, 81*, 102699. https://doi.org/10.1016/j.gloenvcha.2023.102699

Žalėnienė, I., & Pereira, P. (2021). Higher education for sustainability: A global perspective. *Geography and Sustainability, 2*(2), 99–106. https://doi.org/10.1016/j.geosus.2021.05.001

CHAPTER 4

Ecological Modernization Theory and Climate Action

KEIICHI SATOH

Introduction

In response to the accelerating climate crisis, scholars and practitioners often pose several key interrelated questions: What types of climate actions and policies are required to effectively address the crisis? What kind of society, one that is both less environmentally harmful and more prosperous for humanity, should we create? What socioeconomic structure characterizes our historical era? Ecological modernization theory (EMT) provides a framework for addressing these questions. EMT posits that environmentally adaptive socioeconomic reforms can serve as a source of future innovation and development in late modern societies by extending the theory of reflexive modernity (Beck et al., 1994). At the practical level, EMT frequently highlights policies and actions that exploit the innovation pressures inherent in competitive market economies to transform resource-intensive and environmentally intensive industrial modes. Examples of such climate action include greenhouse gas emissions trading, feed-in tariffs for renewable energy, and carbon taxation (Jänicke, 2020).

EMT was conceived not only as a social theory but also as a theory of environmental reform (Mol, 2010). Therefore, EMT emphasizes the importance of feasible policy measures to achieve progress in environmental reform. However, its reformist nature has attracted substantial criticism since its inception. Furthermore, it has become increasingly important, both theoretically and practically, to distinguish among the various strands of ecological modernization, as each strand implies distinct types of climate actions (Almeida et al., 2023). This distinction is particularly relevant today, where EMT or similar ideas such as "green growth" or the "green economy" have become embedded in mainstream policy discourse.

The objective of this chapter is to clarify the debate surrounding EMT as a whole and to distinguish among its various strands, thus allowing for a clearer conceptualization of the climate actions associated with each. To achieve this, the chapter begins by outlining the basic concepts of EMT and its theoretical development before examining key criticisms. This chapter then explores different forms of ecological modernization, including both classical and more recent varieties. In doing so, this chapter advocates for a more holistic conceptualization of EMT, considering broader observational units, geographical scope, and the range of environmental issues addressed—not limited to climate change.

Ecological Modernization Theory: Core Concepts and Assumptions

The concept of ecological modernization can be traced back to a policy proposal by Japan's Ministry of International Trade and Industry during the oil crisis of the 1970s, which linked efficient resource use with economic growth through technological innovation (Mol & Jänicke, 2009, pp. 17–18). This idea was further developed in Germany, the Netherlands, and the United Kingdom during the 1980s, influencing environmental policies such as those of Germany's Red-Green coalition government from 1998 to 2005 (Jänicke & Jörgens, 2023, p. 70). Beyond national policy, ecological modernization has become a major orientation for environmental policy in the European Union (Leipold, 2021; Machin, 2019). In recent years, ecological modernization and related terms such as "green growth" and "green development" gained prominence within the Organisation for Economic Co-operation and Development (OECD) and the United Nations (Jänicke & Jörgens, 2023). As a sociological and political science theory, EMT advanced the concept of ecological modernization and provided academic support for it. The core premise of EMT is that environmental reforms and economic growth are compatible (Hajer, 1995). EMT's basic strategy is to embed environmental rationality within social institutions so that environmental costs become integral to rational decision-making processes, such as industrial production. EMT scholars generally assume that environmentally oriented technological innovation is the key to ensuring this compatibility between economic development and environmental sustainability (Christoff, 1996; Huber, 1985; Humphrey et al., 2001).

It is important to note that the idea of internalizing environmental costs within social institutions, especially markets, is not exclusive to EMT. Similar concepts can be found in other disciplines, such as environmental economics. However, EMT is distinct in how it connects the internalization of environmental rationality with broader sociological theories, particularly the modernity theories of Ulrich Beck and Anthony Giddens (Spaargaren, 1997). According to EMT, environmental rationality emerges as an autonomous rationality in late modernity, emphasizing the negative impacts of modernization, especially industrialization. This, in turn, drives ecologically motivated transformations and reforms in conventional social practices and institutions (Mol, 1996; Spaargaren & Van Vliet, 2000).

The development of EMT as a social theory can be divided into three phases (Mol & Sonnenfeld, 2000a). In the first phase, during the early 1980s, EMT emphasized the role of technological innovation in environmental reform, particularly in industrial production. It also adopted a critical stance against excessive bureaucracy, as bureaucratic inflexibility in adapting to technological innovation often hindered the implementation of new production processes (Huber, 1991/2009). During this early phase, Western European scholars such as Joseph Huber, Marten Jänicke, Gert Spaargaren, and Arthur P. J. Mol, to name a few, made significant contributions to the theory's development. In the second phase, spanning from the late 1980s to the mid-1990s, EMT shifted focus, placing less emphasis on technological innovation and more emphasis on the role of markets and governments in driving ecological transformation (Jänicke, 1997; Weale, 1992). This period also saw EMT integrate environmental reforms with institutional and cultural dynamics (Hajer, 1995; Spaargaren & Mol, 1992). The third phase marked the geographical expansion of EMT to the global level, including its application in non-Western countries, particularly developing nations (Mol & Sonnenfeld, 2000b). Although standard EMT texts, such as Mol and Jänicke (2009), typically describe its theoretical evolution through these three phases, contemporary debates, within the current era of normalized discourse, can be seen as representing a fourth phase, which will be discussed later in this chapter.

How have these theoretical developments shaped EMT's fundamental concept of re-embedding environmental rationality into social institutions? EMT acknowledges that modernization is a process through which each social system becomes detached from traditional structures and begins to operate according to its own values and goals. This separation has led to the neglect of ecological concerns, contributing to environmental degradation. Therefore, the solution lies in re-embedding environmental interests so that social systems once again integrate ecological considerations into their operations (Mol, 1996; Mol & Spaargaren, 1993; Spaargaren, 1997). This process involves recognizing ecological rationality as an independent element within each social subsystem and incorporating it into systemic actions. Furthermore, EMT argues that this re-embedding process had already begun in certain subsystems by the 1970s, particularly in the political and sociocultural spheres of most Western societies (Mol, 2006). A concrete example of this is the establishment of government bodies dedicated to environmental issues (Frank et al., 2000) and the rise of Green parties (Carter, 2001), supported by the spread of green belief systems among citizens (Giddens, 1994). Consequently, the economic domain remains a key area yet to be addressed.

What is required for ecological considerations to be integrated into the economic domain? Early EMT scholars (e.g., Huber, 1991/2009) proposed two key concepts: the "ecologization of the economy" and the "economization of ecology" (Mol, 1995). Ecologization of the economy refers to transforming the economy to achieve more ecologically rational material and energy inputs and outputs in economic activities while monitoring these material flows. The economization of ecology, conversely, involves assigning economic value to the environment to internalize ecological consideration within market mechanisms. However, it is unrealistic to expect these processes to occur solely through an economy's autonomous functioning, as environmental issues arise precisely because the economic system does not consider ecological factors. Therefore, government intervention is necessary. However, if this intervention is purely bureaucratic, it risks creating the inefficiencies commonly associated with centralized governance. Instead of relying solely on centralized control, a flexible, consensual style of governance that includes non-state actors, such as businesses and nongovernmental organizations (NGOs)—reflecting the concept of political modernization—is needed alongside market reforms (Jänicke, 1993; Mol & Buttel, 2002). Notably, during the second phase of theoretical development, consumers and self-employed individuals gained prominence (Spaargaren & Van Vliet, 2000). Particularly, if consumers shift toward becoming "green consumers," they can play a crucial role in transforming the economic system. Additionally, the visualization and measurement of these environmental elements require support from science and technology.

In essence, EMT does not advocate abandoning modern institutions but rather achieving "hypermodernization." As such, it places high value on globalization, maintains strong trust in science and technology, and retains confidence in the capitalist system (Mol, 2001). This perspective stands in contrast to other theories, such as those in the political economy tradition (e.g., O'Connor, 1994; Schneiberg et al., 2008) and "de-modernization" perspectives (e.g., Commoner, 1971; Devall & Sessions, 1999), which primarily blame modern capitalism for environmental problems (Moore, 2015; York & Rosa, 2003).

The specific climate actions aligned with EMT are diverse. Examples of efforts to integrate ecological concerns into the business sector include environmental management systems, the establishment of environmental departments within companies, environmentally driven liability and insurance mechanisms, natural resource conservation, recycling initiatives, and consulting (Mol, 2006). This integration is further supported by ecologically motivated economic policies such as eco-taxes, carbon markets, and renewable energy policies from national and subnational governments. Additionally, governmental climate action plans, such as setting carbon reduction targets, play a significant role in shaping businesses' long-term strategies. Civil society actors and environmental movements are also increasingly involved in decision-making processes alongside state and market actors (Mol, 2000; Sonnenfeld, 2002; Toke, 2011). In some cases, coalitions between emerging industries and environmental movements replace older, environmentally costly industries (Toke, 2011).

Empirical Evidence and Controversy

Empirical studies have revealed that EMTs have been highly influential in shaping global environmental discourse. Concepts like EMT and green growth have become key frameworks for policymakers and business leaders globally (Jänicke & Jörgens, 2023). Countries and subnational governments that promote environmental technology and innovation are increasingly being viewed as leaders in sustainable development. Businesses are responding to consumers' concerns about environmental impacts by incorporating sustainability into their operations. Furthermore, businesses recognize that preventive measures to address environmental challenges are more cost-effective in the long run than addressing their consequences (Young, 2001). This shift is gradually rendering the traditional belief that the environment and the economy conflict obsolete.

EMT has proven effective in fostering coordination across various sectors. The theory posits that environmental reform creates a win-win scenario, enabling collaboration among governments, businesses, and environmentally conscious citizens (Jänicke, 2020). Numerous case studies (e.g., Mol, 1995; Sonnenfeld, 1998) and comparative studies (e.g., Andersen, 2002; Novotný et al., 2021; Satoh et al., 2022) have highlighted the positive relationship between the spread of ecological modernization discourse and progress in climate reform. For instance, Novotný et al. (2021) compared climate policy networks in Germany and Japan. Policy networks, which refer to organizational communication structures—such as collaboration and information-sharing networks—among government agencies, corporate actors, NGOs, and scientific institutions, play a significant role in policymaking processes (Broadbent, 2016; Kenis & Schneider, 1991; Ylä-Anttila et al., 2018). Although both Germany and Japan have been considered pioneers in ecological modernization and have the technological capacity to advance renewable energy adoption, Japan lags behind Germany in transforming from fossil fuels to renewable energy sources and in broader climate policy reform (Mez et al., 2019). By comparing key actors and their policy perspectives, Novotný et al. (2021) found that in Germany, proponents of ecological modernization hold more central and influential positions within collaboration networks compared to their Japanese counterparts. Similarly, Satoh et al. (2022) examined the influence of ecological modernization discourse within climate policy networks in both countries and found that scientific organizations—those collecting and analyzing social and natural science data—played a significant role in disseminating ecological modernization ideas. However, the primary distinction between Germany and Japan lies in how these organizations operate. In Germany, both social scientist-based and natural scientist-based organizations actively contribute to spreading ecological modernization discourse, which has had a more significant impact on policy networks than in Japan.

These empirical findings demonstrate that ecological modernization ideas are more widely accepted and influential in societies actively engaged in environmental reform. This supports the theoretical expectation that ecological modernization can drive meaningful environmental reform. However, EMT remains controversial regarding its effectiveness in addressing environmental problems. One of the most comprehensive criticisms of EMT is that of York and Rosa (2003), who outlined four key points. First, an ecological orientation of institutions is necessary, but it is not sufficient condition for environmental reform. Second, some examples of institutional reforms are not representative of broader trends in hypermodernization. Third, improvements in one sector can lead to increased ecological impacts in another sector. Fourth, EMT's focus on resource use efficiency does not necessarily reduce total environmental impacts, as greater efficiency can lead to increased overall resource consumption (Clark & Foster, 2001). This phenomenon is known as "the treadmill of production" at the political-economic systems level (Gould et al., 2008; see also Gould & Lewis in this volume).

York and Rosa further provided empirical evidence (York et al., 2003a, 2003b) to support their critique. They argue that highly modernized societies—measured by per capita GDP, urbanization, and economic structure—often have the highest environmental impacts. EU nations, typically considered leaders in ecological modernization, have some of the world's highest per capita environmental impacts, challenging the notion that ecological rationality correlates with environmental reform. Their data suggest that nations supporting environmental treaties tend to have disproportionately high impacts on the global environment, potentially reversing the causal mechanism assumed by the EMT.

Beyond this critique regarding the limitations of technological innovation and institutional reform, unequal exchange theory emphasizes the need to evaluate environmental burdens on a global scale rather than focusing on individual countries (Jorgenson, 2016; see also Jorgenson et al. in this volume). Unequal exchange theory argues that economically advanced, ecologically modernized countries often extract raw materials from and export environmental degradation to peripheral countries, thereby shaping their developmental trajectories (Huang, 2018; Jorgenson, 2012). These peripheral countries typically face greater environmental damage, such as increased carbon emissions, as they often have more capacity for initial economic growth (Alderson & Pandian, 2018). Consequently, environmental efficiency gains in developed nations may be offset by their resource exchange partners in developing countries. This suggests that modernization efforts must have a global scope, including meaningful commitments from developed countries, such as technology transfers and development aid (Vesia et al., 2023).

Variations in Ecological Modernization

Up to this point, this chapter has reviewed the theoretical development of EMT and associated controversies. The debate surrounding the significance of EMT in fostering environmental reform and its limitations in providing practical solutions remains a central issue in contemporary discourse. However, it is important to recognize that these debates are, at least partially, framed by EMT's role as a counter-discourse. In the 1980s and 1990s, when the prevailing assumption was that the economy and the environment were fundamentally at odds, EMT presented a societal model aimed at resolving this impasse, which has since influenced how its effectiveness is debated. Over time, the initial phase of overcoming these ideological standoffs has been largely completed, leading to discussions about the next steps for ecological modernization. However, this shift has also resulted in deviations from the original premise of EMT, creating new theoretical challenges. To contextualize this ongoing debate, the following sections will examine strong and weak ecological modernization, ecological modernization within compressed modernity, and the concept of Ecological Modernization 2.0.

Strong Versus Weak Ecological Modernization

Even in the relatively early stages of its theoretical development, it was recognized that variations exist within EMT. A frequently discussed topic is the distinction between analytical and practical EMT, and whether the focus should be on market-based regulation and technological innovation or on broader societal structures (Mol, 1996). Among these distinctions, the classical and comprehensive division is between strong and weak ecological modernization, first proposed by Christoff (1996) and later summarized by several scholars, including Dryzek (1997/2021) and Toke (2011).

In the case of weak ecological modernization, environmental considerations are ultimately reduced to market calculations. The efficiency of resource use, environmental management, and technological innovation is evaluated in monetary terms. Christoff further criticized weak ecological modernization for implicitly assuming a uniform evolutionary process of industrial transformation characterized by Western science, technology, and consumer culture.

Conversely, strong ecological modernization places a direct emphasis on ecological concerns. It considers various global political and social changes, inherently involving political upgrading, such as democratic decision-making processes and expanded participatory opportunities for diverse stakeholders. These global market-social-political transformations are envisioned as following diverse, rather than uniform, trajectories. Therefore, strong ecological modernization contrasts with the weak version by rejecting its perceived hegemonic and Western-centric outlook. Christoff (1996) does not argue for abandoning the technological and economic tools that are emphasized in weak ecological modernization. Instead, advocates for the normative principles of strong ecological modernization to guide these efforts serve as guiding efforts.

Ecological Modernization in Compressed Modernity

As EMT becomes a global topic, its application in non-Western countries presents new challenges to theoretical expectations. An especially interesting case is that of emerging economies in East Asia (Barrett, 2005; Mol, 2006; Satoh, 2023; Wong, 2012). For instance, Mol (2006) examined China's institutional reforms through the lens of EMT and concluded that Chinese economic reforms are influenced by ecological modernization ideas. However, China's developmental path diverges from the Western version of ecological modernization in three key ways. First, the degree to which environmental interests are institutionalized remains relatively limited in China. Environmental considerations are not fully embedded in production and consumption practices, at least as observed by Mol in 2006. Second, many economy-environment indicators differ significantly from those seen in European societies. Third, China's economic liberalization and market reforms have not been paralleled by political liberalization and democratic reforms.

At first glance, the characteristics of Chinese ecological modernization seem to align more with weak ecological modernization than with strong ecological modernization. However, the dynamics of China's modernization process do not fit neatly into this distinction because ecological modernization in China extends beyond the economic sphere, which contradicts the assumption of weak ecological modernization. It also encompasses political and social institutions. The key to understanding modernization in China—as well as in South Korea, Taiwan, and Japan—is the concept of "compressed modernity" (Han & Shim, 2010), which has shaped these societies' unique responses to environmental issues (Satoh, 2023). "Compressed modernity" refers to a process in which the first wave of modernization (industrialization, urbanization, national boundary formation, and modern rational identities) is swiftly followed by the second (reflexive responses to the first modernization, such as the rise of green ideology).

In contrast to the Western experience, environmental reforms in East Asian societies have been key drivers of both the first and second modernization for two main reasons (Satoh, 2023). First, efficient resource use and technological innovation are prerequisites for modernization in these regions. Sugihara (2020) highlighted the differences in resource availability between Western societies and East Asia. While industrialization in the West was driven by the intensive use of locally available coal, East Asia, apart from the interior regions of China, faced a scarcity of natural resources. As a result, East Asia's industrialization relied heavily on the efficient use of natural resources and the development of industrial cities in coastal areas where resources could be imported. This industrialization was led by a bureaucratic "developmental state" that planned and invested in these industrial cities, supported by a labor force migrating from rural areas through state-led education systems. The "developmental state" refers to a state structure capable of intervening in the economy through long-term planning, with coalitions of bureaucrats, dominant political parties, and large corporations (Woo, 2022). In East Asia modernization has coincided with the development of environmental policy (Mol, 2006). For example, pollution control efforts in China began in the early 1970s, and the National Environmental Protection Office was established in 1974 during economic reforms. Similarly, in Korea and Taiwan, environmental movements since the 1980s have been closely linked to their democratization processes (Ku, 2011).

This unique combination of economic and political modernization, coupled with the institutionalization of environmental policy, has had complex effects on the development of climate policy in East Asia. First, the tradition of the developmental state is associated with noninstitutional climate regulation led by bureaucracies in China (Mol, 2006) and Japan, where voluntary agreements between governments and industrial sectors prevail (Schreurs, 2002). Second, the success of developmental state-led modernization has conferred strong *output legitimacy* (Scharpf, 1997) on state bureaucracies, which contrasts with EMT's conventional assumption that environmental reform is driven by increased *input legitimacy*, such as through participation, transparency, and accountability. Third, the industrial success of East Asian countries has fostered strong domestic trust in technologies. A notable example is the high public support for nuclear energy, even after the 2011 Fukushima nuclear accident (Kim et al., 2013). This trend is reflected in ongoing nuclear power plant construction, with 30 new reactors currently under construction in East Asia as of 2024 (26 in China, 2 in Korea, and 2 in Japan). Indeed, 50% (30 of 60) of all new reactors worldwide are located in East Asia (WNA, 2024). Furthermore, both Korea and Japan, which had previously announced long-term nuclear phase-out plans following the Fukushima disaster, have since reversed these decisions—Japan in 2013 and Korea in 2022 in Korea—returning to the promotion of nuclear energy as part of their climate policies.

In the long run, the experience of compressed modernity in East Asian societies may not be an exception but could become more common compared to the experiences of Western societies. Therefore, it is crucial to understand how ecological modernization evolves in these contexts and to conceptualize future models tailored to their specific circumstances.

Typologies of Green Capitalism and EMT 2.0

EMT posits that capitalism will persist and seeks mechanisms for environmental reform. Given this theoretical stance, it becomes crucial to explore the various future directions that capitalism may take. Tienhaara (2014) offers a particularly insightful analysis, suggesting that even if one accepts the compatibility of the environment and economy, three distinct pathways of reform emerge: "Green New Deal," "Green Stimulus," and "Green Economy." It is worth noting that Tienhaara's terminology may lead to some confusion, as scholars and practitioners often use these terms differently and, at times, interchangeably. However, given the challenges of identifying a universal alternative, this chapter retains its terminology.

These three typologies of green capitalism differ significantly in their perspectives on future economic growth, the technologies they support, and the role they assign to the financial sector, as outlined in Table 4.1. Tienhaara developed this framework by analyzing influential reports from think tanks, NGOs, and international organization bodies published between 2008 and 2012. Notably, Tienhaara's objective is not to provide a comprehensive survey of the

TABLE 4.1 Varieties of Green Capitalism

	Green New Deal	Green Stimulus	Green Economy
Position on economic growth	Aiming for low/no growth in the long term Skeptical about the decoupling of environmental impact and economic growth However, holds a somewhat inconsistent position that economic recovery is necessary	Continuing economic growth in the future Implicitly assumes relative decoupling of environmental impact and economic growth	Continuing economic growth in the future Explicitly claims absolute decoupling of environmental impact and economic growth
Position on specific technologies	Investment in renewable energy and energy efficiency Also considers reduction in consumption	Investment in renewable energy, energy efficiency, nuclear power, and CCS	Investment in renewable energy, energy efficiency, nuclear power, CCS, and agricultural/waste/water management technologies
Position on the market	Assumes a certain role for carbon pricing and carbon trading	Assumes a major role for carbon pricing and carbon trading	Assumes not only carbon pricing but also pricing of natural capital
Position on finance	Aims to strengthen regulation and shrink the financial sector	(No clear position)	Aims to make conservation profitable through future deregulation of the financial sector
Representative proponents	Green New Deal Group	HSBC Bank	UNEP
Background leading to these positions	Proposed amid the financial recession	Some financial institutions are among the proponents. Emphasizes the protection of workers' employment in line with traditional New Deal policies.	Some financial institutions are among the proponents. Although UNEP officially denies it, there is a connection with sustainable development.

Note: Adapted by the author from the original table in Tienhaara (2014, p. 194).

CCS, carbon capture and storage; UNEP, United Nations Environment Program.

various forms of green capitalism but rather to highlight the logical differences among these proposals.

The first comprehensive English report advocating a *Green New Deal* was produced by the Green New Deal Group (GNDG) in response to the 2008 global financial crisis. The GNDG observed that the prevailing model of financial capitalism led to unsustainable energy and resource consumption, which contributed to the crisis. As a remedy, the GNDG proposed structural reforms to both national and international financial and taxation systems to address climate change. Specifically, the group recommended a program of public and private investment in energy conservation and renewable energy, funded by taxes on oil and gas companies. The plan also endorsed the imposition of government-led capital controls and the elimination of tax havens.

The second model of green capitalism, *Green Stimulus*, traces its roots back to the Great Depression of the 1930s. This approach involves increased public spending and reduced taxation to stimulate economic activity. The modern, "green" iteration of fiscal stimulus combines the economic measures with environmental objectives. A central argument for this approach is that renewable energy industries are more labor-intensive than fossil fuel industries, and thus, they create more jobs. Additionally, investments in renewable energy can generate economic savings for individuals and businesses by reducing fuel costs and health expenditures.

The third model, the *Green Economy*, is particularly championed by the United Nations Environment Program (UNEP). UNEP (2010) defines a green economy as one that improves human well-being and social equity while significantly reducing environmental risks and ecological scarcities. As this definition suggests, the green economy extends the concept of sustainable development to encompass the entire economic system.

A key distinction between these three models of green capitalism, as Tienhaara (2014) points out, is that while both *Green Stimulus* and the *Green Economy* implicitly or explicitly align with the principles of ecological modernization, the *Green New Deal* does not. The GNDG's *Green New Deal* proposal envisions the eventual need for low or zero economic growth, whereas *Green Stimulus* and the *Green Economy* assume that continued economic growth is achievable through ecological modernization. The notable difference between *Green Stimulus* and *Green Economy* is that the *Green Economy* places greater emphasis on financial mechanisms, seeking to integrate environmental policy with financial capitalism and include natural capital within market calculations.

The concept of a *Green Economy* closely parallels "Ecological Modernization 2.0" (hereafter EM 2.0), as discussed by Jänicke and Jörgens (2023) (Table 4.2). Martin Jänicke, one of the founders of ecological modernization, acknowledges the limitations of conventional ecological modernization. Traditional EMTs have focused on institutional

TABLE 4.2 Comparison Between Ecological Modernization and EM 2.0

Policy Paradigm	Ecological Modernization	Ecological Modernization 2.0
Interpretative framework	Compatibility of environmental protection and economic growth	Absolute decoupling of growth from emissions/resource consumption, climate engineering
Policy objectives	Increase eco-efficiency through technological innovation	Green transformation of branches and infrastructures, "green solutions"
Policy instruments	Eco-taxes, incentives for green/eco-efficient investment, front-runner programs, etc.	Long-term investment programs, consensual transition management, etc.
Institutional requirements	Institutional mechanisms for environmental policy integration	Green new deal–type long-term strategy

Note: This work was adapted from Jänicke and Jörgens (2023, p. 79).

reforms such as carbon taxes and emissions trading to address climate change. However, many conventional policies supported by EMTs fail to fundamentally transform existing production systems, allowing "dirty" industries to coexist alongside cleaner ones. In response, Jänicke and Jörgens (2023) called for *ecological restructuring*—the second phase of ecological modernization, or EM 2.0. Long-term financial investments and systematic integration of environmental indicators into market accounting are central to EM 2.0 policies. These long-term investments and policy programs, including compensation mechanisms, represent efforts not just to reduce "dirty" industries but to phase them out entirely.

Furthermore, EM 2.0 policies seek a shift from marginal increases in resource efficiency to absolute reductions. This shift represents the goals of absolute, rather than relative, decoupling of economic growth from environmental impact. In theory, this is feasible because economic growth is measured in economic value, not physical terms. If economic value can be decoupled from material consumption by transforming economic activities so that the dominant form of exchange becomes immaterial and environmental impact is minimized through technological innovation, economic growth can occur without environmental harm (Ferguson, 2015; Jackson & Victor, 2019).

Another notable difference between classical ecological modernization and EM 2.0 is the latter's support for high-risk technologies such as geoengineering. Examples of geoengineering include carbon capture and storage (CCS), which involves storing carbon in geological formations (Krüger, 2017), and solar radiation management, which aims to reduce the amount of sunlight absorbed by the Earth's atmosphere. Unlike conventional environmental policies that focus on reducing environmental burdens as their source, geoengineering seeks to break the causal link between emissions and harmful environmental changes. Moreover, geoengineering policies do not necessarily require international collaboration; in theory, a single country can implement them independently (Victor et al., 2009). In line with its endorsement of geoengineering, EM 2.0 also supports nuclear energy as tools for combating climate change.

In many respects, EM 2.0 is far more controversial than classical ecological modernization. Phase-out policies often provoke strong opposition from incumbent industries, significantly reducing their political feasibility. Moreover, geoengineering and nuclear energy, as high-risk technologies, face lower levels of public acceptance than traditional efficiency-based approaches. Beyond these surface-level differences, it is essential to explore how these new features of EM 2.0 have developed. First, as the term implies, EM 2.0 can be viewed as the next step in environmental reform following the first mainstream adoption of the original ecological modernization project. Second, EM 2.0 represents a concerted effort to address climate change challenges that conventional EMTs arguably cannot adequately address. However, this reorientation introduces a significant contradiction to the original EMT premises.

The original concept of ecological rationality in classical EMT was heavily influenced by the theory of reflexive modernity, which is grounded in concerns over environmental risks stemming from technological development. Although EM 2.0 acknowledges the risks posed by climate change, it appears to overlook other environmental risks that may emerge from climate-centric technologies. In other words, the rationality promoted by EM 2.0 is fundamentally different and operates independently from traditional environmental rationality. This tension between climate-specific and broader environmental rationality has a profound effect on different conceptualizations of climate action, as discussed earlier.

Discussion and Conclusion

The EMT is arguably one of the most influential sociopolitical theories in shaping contemporary society. At its core, EMT asserts that ecological modernization drives innovation, thus fueling future development. Moreover, EMT systematically offers a comprehensive analysis of the historical context of current environmental crises, a vision of society's future, and necessary actions for reform. According to EMT, the root cause of current catastrophic environmental degradation lies in modernity, where social systems operate in isolation, each pursuing its own values and goals without accounting for environmental factors. To address this need, social systems must be conceptualized to reintroduce environmental rationality into their core operations. For this systemic transformation to occur, reforms must extend beyond market mechanisms to encompass culture (e.g., the greening of consumer culture) and political dimensions (i.e., political modernization aimed at fostering more open, inclusive, and consensus-oriented governance). These reforms are mutually reinforcing. One significant theoretical and practical benefit of EMT is its ability to bridge structure and agency. When a theory adopts a structurally deterministic stance, it can leave actors without a clear direction, resulting in inaction. However, inaction, in this context, is effectively the "action" that perpetuates the current environmentally harmful system.

The concept of ecological modernization has gained global acceptance, with many policies and business practices now incorporating environmental considerations. At the discursive level, EMT has become the dominant framework in global society. Nevertheless, whether ecological modernization translates into real environmental solutions remains a subject of debate (York & Rosa, 2003). EMT excels at setting agendas and advancing technological solutions, such as renewable energy, which is a central component. However, despite its ambition to transform modern society, EMT has struggled in sectors dominated by politically powerful incumbents. Instead of transforming social structures, EMT often permits the coexistence of "old" industrial actors alongside "new" actors (Jänicke & Jörgens, 2023). Furthermore, EMT has predominantly focused on environmental reform within individual entities (e.g., a country or a company) rather than recognizing the interconnectedness of a broader social system.

Considering EMT's strengths and limitations, recent scholarship (e.g., Fisher & Jorgenson, 2019; Vesia et al., 2023) advocates for integrating multiple approaches rather than debating on the relative merits of EMT and other theories. Given the complexity of climate change, no single approach is sufficient to address these challenges in a timely manner. For instance, EMTs are particularly relevant for explaining and driving institutional reforms within entities (e.g., countries or businesses), whereas unequal exchange theory may offer better insights into relational dynamics between entities. Moreover, critiques of EMT can be incorporated into its evolution. As unequal exchange theory points out, environmental improvements in one entity (e.g., a company in a developed nation) are often achieved by externalizing costs to other entities (e.g., local businesses in developing countries). The EMT must develop mechanisms to "reinternalize" these externalized costs within the original entity's calculations. In this regard, the ongoing carbon accounting initiative, which includes indirect emissions from the supply chain (i.e., "scope 3") is an important step forward.

In addition to reviewing existing debates on conventional EMT, this chapter emphasizes the need to examine the varieties of ecological modernization more closely. During the 1980s and 1990s, EMT served as a strong counter-theory to the prevailing notion that environmental protection was incompatible with economic growth. This old fault line has weakened significantly, if not disappeared entirely. Today, ecological modernization has become the dominant discourse in global politics. However, this shift presents a serious challenge for EMTs, as new fault lines are emerging among those that support ecological modernization (see also Krüger, 2017).

In particular, this chapter highlights three varieties of ecological modernization identified in the literature. The first variety is the classical distinction between weak and strong ecological modernization. While weak ecological modernization subordinates environmental concerns to economic interests, strong ecological modernization advocates for the transformation of entire social institutions based on environmental rationality. The second variety is ecological modernization in the context of compressed modernity, which is particularly evident in East Asian societies but potentially observable globally. Understanding this form of ecological modernization requires an awareness of its unique context, where early and late modernity occurs simultaneously or in rapid succession. These cases pose challenges for EMT scholars, as the realization of ecological modernization in these societies differs significantly from that in Western societies, despite shared drivers of modernization. To fully understand the experiences and future trajectories of these societies in compressed modernity, detailed case studies and comparative research are necessary. It is also crucial, both empirically and theoretically, to investigate whether ecological modernization led by developmental states in these societies is accompanied by long-term political modernization, as predicted by EMT.

The third variety is EM 2.0. Unlike classical ecological modernization, EM 2.0 explicitly aims for absolute decoupling of environmental burdens from economic growth rather than merely achieving relative improvements. It also places greater emphasis on long-term financial investments and government planning as key drivers of reform for a green economy. Notably, EM 2.0 supports high-risk technologies, such as geoengineering, as part of a novel approach to environmental policy that seeks to break the causal link between pollution (e.g., CO_2 emissions) and catastrophic outcomes (e.g., climate change). This chapter specifically introduced EM 2.0 as conceptualized by Jänicke and Jörgens (2023), though it is important to note that alternative versions of "EM 2.0" may emerge.

Each of these varieties of ecological modernization invites widespread debate, differing from the context in which classical ecological modernization was originally discussed. At that time, despite vigorous debate over the merits of ecological modernization compared with other theories, the proponents of EMT generally shared basic principles. However, as ecological modernization becomes the dominant discourse, internal cohesion within the framework may weaken. The debate surrounding the different directions of ecological modernization is thus becoming more crucial, as various strands imply different sets of climate actions. For instance, weak ecological modernization may lead to market reforms and technological innovation, primarily for profit-making. A developmental state version might prioritize technological solutions, such as nuclear energy and voluntary agreements between states and businesses, while placing less emphasis on actions that promote *input legitimacy*, such as broader participation in policymaking and transparency in policy implementation. Furthermore, noninstitutional forms of climate action, such as social movements, may be seen as disruptive to effective ecological modernization rather than as integral parts of it. Similarly, EM 2.0 seeks to address climate change by combining phase-out policies, geoengineering, and nuclear energy while disregarding other environmental risks that these technologies may introduce.

By reviewing the existing strands and corresponding climate actions, this chapter concludes by calling for a holistic approach to EMT. Such an approach would involve expanding the observational units, extending its geographical scope, and broadening the range of issues targeted by ecologically modernized reforms. EMTs need to consider multiple entities, rather than a single representative one, to accurately evaluate the impact and trajectory of environmental reforms (e.g., York & Rosa, 2003). Additionally, EMTs need to conceptualize reforms that internalize environmental costs exported from developed to developing countries through unequal exchange (Jorgenson, 2012). EMTs must emphasize the need for more democratic and humane environmental reform, not merely technological solutions to climate change. EMT must (re-) broaden its scope of solutions to address environmental risks, rather than merely viewing them through a climate-centric lens. While these goals are ambitious, they are essential to driving theoretical innovation in an age when the discourse of ecological modernization has become normalized.

References

Alderson, A. S., & Pandian, R. K. (2018). What is really happening with global inequality? *Sociology of Development*, *4*(3), 261–281. https://doi.org/10.1525/sod.2018.4.3.261

Almeida, P., Márquez, L. R. G., & Fonsah, E. (2023). The forms of climate action. *Sociology Compass*, *2024*, e13177. https://doi.org/10.1111/soc4.13177

Andersen, M. S. (2002). Ecological modernization or subversion? The effect of Europeanization on Eastern Europe. *American Behavioral Scientist*, *45*(9), 1394–1416. https://doi.org/10.1177/00027642 02045009006

Barrett, B. F. D. (2005). *Ecological modernization and Japan*. Routledge.

Beck, U., Giddens, A., & Lash, S. (1994). *Reflexive modernization*. Stanford University Press.

Broadbent, J. (2016). Comparative climate change policy networks. In J. N. Victor, A. H. Montgomery, & M. Lubell (Eds.), *The Oxford handbook of political networks* (pp. 875–900). Oxford University Press.

Carter, N. (2001). *The politics of the environment: Ideas, activism, policy* (1st ed.). Cambridge University Press. https://doi.org/10.1017/CBO9781139163859

Christoff, P. (1996). Ecological modernisation, ecological modernities. *Environmental Politics*, *5*(3), 476–500. https://doi.org/10.1080/09644019608414283

Clark, B., & Foster, J. B. (2001). William Stanley Jevons and "the coal question": An introduction to Jevons's "Of the Economy of Fuel." *Organization & Environment*, *14*(1), 93–98.

Commoner, B. (1971). *The closing circle: Nature, man, and technology* (1st ed.). Knopf.

Devall, B., & Sessions, G. (1999). *Deep ecology*. Smith.

Dryzek, J. S. (1997/2021). *The politics of the earth: Environmental discourses* (4th ed.). Oxford University Press.

Ferguson, P. (2015). The green economy agenda: Business as usual or transformational discourse? *Environmental Politics*, *24*(1), 17–37. https://doi.org/10.1080/09644016.2014.919748

Fisher, D. R., & Jorgenson, A. K. (2019). Ending the stalemate: Toward a theory of anthro-shift. *Sociological Theory*, *37*(4), 342–362.

Frank, D. J., Hironaka, A., & Schoter, E. (2000). The nation-state and the natural environment over the twentieth century. *American Sociological Review*, *65*(1), 96–116. https://doi.org/10.1177/0003122 40006500106

Giddens, A. (1994). *Beyond left and right: The future of radical politics*. Stanford University Press.

Gould, K. A., Pellow, D. N., & Schnaiberg, A. (2008). *The treadmill of production: Injustice and unsustainability in the global economy*. Paradigm Publishers.

Hajer, M. A. (1995). *The politics of environmental discourse: Ecological modernization and the policy process*. Clarendon Press.

Han, S.-J., & Shim, Y.-H. (2010). Redefining second modernity for East Asia: A critical assessment. *The British Journal of Sociology*, *61*(3), 465–488. https://doi.org/10.1111/j.1468-4446.2010.01322.x

Huang, X. (2018). Ecologically unequal exchange, recessions, and climate change: A longitudinal study. *Social Science Research*, *73*, 1–12. https://doi.org/10.1016/j.ssresearch.2018.03.003

Huber, J. (1985). *Die Regenbogen-Gesellschaft: Ökologie und Sozialpolitik*. Fischer. https://cir.nii.ac.jp/crid/1130282269460140288

Huber, J. (1991/2009). Ecological modernization: Beyond scarcity and bureaucracy. In A. P. J. Mol, D. A. Sonnenfeld, & G. Spaargaren (Eds.), *The ecological modernisation reader: Environmental reform in theory and practice* (pp. 42–55). Routledge.

Humphrey, C. R., Lewis, T. L., & Buttel, F. H. (2001). *Environment, energy, and society: A new synthesis*. Wadsworth Thomson Learning.

Jackson, T., & Victor, P. A. (2019). Unraveling the claims for (and against) green growth. *Science, 366*(6468), 950–951. https://doi.org/10.1126/science.aay0749

Jänicke, M. (1993). Über ökologische und politische Modernisierungen. *Zeitschrift Für Umwelt-Politik Und Umweltrecht, 16*(2), 159–175.

Jänicke, M. (1997). The political system's capacity for environmental policy. In M. Jänicke, H. Jörgens, & H. Weidner (Eds.), *National environmental policies: A comparative study of capacity-building* (pp. 1–24). Springer. https://doi.org/10.1007/978-3-642-60507-9_1

Jänicke, M. (2020). Ecological modernization: A paradise of feasibility but no general solution. In L. Mez, L. Okamura, & H. Weidner (Eds.), *The ecological modernization capacity of Japan and Germany. Comparing nuclear energy, renewables, automobility and rare earth policy* (pp. 13–23). Springer VS.

Jänicke, M., & Jörgens, H. (2023). Ecological modernization and beyond. In H. Jörgens, C. Knill, & Y. Steinebach (Eds.), *Routledge handbook of environmental policy* (1st ed., pp. 68–87). Routledge. https://doi.org/10.4324/9781003043843-7

Jorgenson, A. (2012). The sociology of ecologically unequal exchange and carbon dioxide emissions, 1960–2005. *Social Science Research, 41*(2), 242–252. https://doi.org/10.1016/j.ssresearch.2011.11.011

Jorgenson, A. (2016). Environment, development, and ecologically unequal exchange. *Sustainability, 8*(3), 227. https://doi.org/10.3390/su8030227

Kenis, P., & Schneider, V. (1991). Policy networks and policy analysis. In B. Marin & R. Mayntz (Eds.), *Policy networks. Empirical evidence and theoretical considerations* (pp. 25–59). Campus Verlag.

Kim, Y., Kim, M., & Kim, W. (2013). Effect of the Fukushima nuclear disaster on global public acceptance of nuclear energy. *Energy Policy, 61*, 822–828. https://doi.org/10.1016/j.enpol.2013.06.107

Krüger, T. (2017). Conflicts over carbon capture and storage in international climate governance. *Energy Policy, 100*, 58–67.

Ku, D. (2011). The Korean environmental movement: Green politics through social movement. In J. Broadbent & V. Brockman (Eds.), *East Asian social movements: Power, protest, and change in a dynamic region* (pp. 205–229). Springer. https://doi.org/10.1007/978-0-387-09626-1_10

Leipold, S. (2021). Transforming ecological modernization "from within" or perpetuating it? The circular economy as EU environmental policy narrative. *Environmental Politics, 30*(6), 1045–1067. https://doi.org/10.1080/09644016.2020.1868863

Machin, A. (2019). Changing the story? The discourse of ecological modernisation in the European Union. *Environmental Politics, 28*(2), 208–227. https://doi.org/10.1080/09644016.2019.1549780

Mez, L., Okamura, L., & Weidner, H. (2019). *The ecological modernization capacity of Japan and Germany: Comparing nuclear energy, renewables, automobility and rare earth policy*. Springer VS.

Mol, A. P. J. (1995). *The refinement of production: Ecological modernization theory and the chemical industry*. International Books.

Mol, A. P. J. (1996). Ecological modernisation and institutional reflexivity: Environmental reform in the late modern age. *Environmental Politics, 5*(2), 302–323. https://doi.org/10.1080/09644019608414266

Mol, A. P. J. (2000). The environmental movement in an era of ecological modernisation. *Geoforum, 31*(1), 45–56. https://doi.org/10.1016/S0016-7185(99)00043-3

Mol, A. P. J. (2001). *Globalization and environmental reform: The ecological modernization of the global economy*. MIT Press.

Mol, A. P. J. (2006). Environment and modernity in transitional China: Frontiers of ecological modernization. *Development and Change, 37*(1), 29–56. https://doi.org/10.1111/j.0012-155X.2006.00468.x

Mol, A. P. J. (2010). Ecological modernization as a social theory of environmental reform. In M. R. Redclift & G. Woodgate (Eds.), *The international handbook of environmental sociology* (2nd ed., pp. 63–76). Edward Elgar.

Mol, A. P. J., & Buttel, F. H. (2002). The environmental state under pressure: An introduction. In *Research in social problems and public policy* (Vol. 10, pp. 1–11). Emerald. https://doi.org/10.1016/S0196-1152(02)80003-5

Mol, A. P. J., & Jänicke, M. (2009). The origins and theoretical foundations of ecological modernisation theory. In A. P. J. Mol, D. A. Sonnenfeld, & G. Spaargaren (Eds.), *The ecological modernisation reader: Environmental reform in theory and practice* (pp. 17–27). Routledge.

Mol, A. P. J., & Sonnenfeld, D. A. (2000a). Ecological modernisation around the world: An introduction. *Environmental Politics*, 9(1), 1–14. https://doi.org/10.1080/09644010008414510

Mol, A. P. J., & Sonnenfeld, D. A. (Eds.). (2000b). *Ecological modernisation around the world: Perspectives and critical debates*. Frank Cass.

Mol, A. P. J., & Spaargaren, G. (1993). Environment, modernity and the risk-society: The apocalyptic horizon of environmental reform. *International Sociology*, 8(4), 431–459. https://doi.org/10.1177/026858093008004003

Moore, J. W. (2015). *Capitalism in the web of life: Ecology and the accumulation of capital*. Verso.

Novotný, V., Satoh, K., & Nagel, M. (2021). Refining the multiple streams framework's integration concept. *Journal of Comparative Policy Analysis: Research and Practice*, 23(3), 291–309. https://doi.org/10.1080/13876988.2020.1770089

O'Connor, M. (Ed.). (1994). *Is capitalism sustainable? Political economy and the politics of ecology*. Guilford Press.

Satoh, K. (2023). East Asian compressed ecological modernization: Modus of developmental state and technological response to the environmental crisis. In *Handbook of post-Western sociology: From East Asia to Europe* (pp. 799–811). Brill. https://brill.com/display/title/63690

Satoh, K., Nagel, M., & Schneider, V. (2022). Organizational roles and network effects on ideational influence in science-policy interface: Climate policy networks in Germany and Japan. *Social Networks*, 75, 88–106. https://doi.org/10.1016/j.socnet.2022.01.014

Scharpf, F. W. (1997). Economic integration, democracy and the welfare state. *Journal of European Public Policy*, 4(1), 18–36. https://doi.org/10.1080/135017697344217

Schneiberg, M., King, M., & Smith, T. (2008). Social movements and organizational form. *American Sociological Review*, 73(4), 635–667.

Schreurs, M. A. (2002). *Environmental politics in Japan, Germany, and the United States*. Cambridge University Press.

Sonnenfeld, D. A. (1998). From brown to green?: Late industrialization, social conflict, and adoption of environmental technologies in Thailand's pulp industry. *Organization & Environment*, 11(1), 59–87. https://doi.org/10.1177/0921810698111003

Sonnenfeld, D. A. (2002). Social movements and ecological modernization: The transformation of pulp and paper manufacturing. *Development and Change*, 33(1), 1–27. https://doi.org/10.1111/1467-7660.00238

Spaargaren, G. (1997). *The ecological modernization of production and consumption: Essays in environmental sociology* [PhD thesis]. Wageningen Agricultural University.

Spaargaren, G., & Mol, A. P. J. (1992). Sociology, environment, and modernity: Ecological modernization as a theory of social change. *Society & Natural Resources*, 5(4), 323–344. https://doi.org/10.1080/08941929209380797

Spaargaren, G., & Van Vliet, B. (2000). Lifestyles, consumption and the environment: The ecological modernization of domestic consumption. *Environmental Politics*, 9(1), 50–76. https://doi.org/10.1080/09644010008414512

Sugihara, K. (2020). *Sekaishi no naka no Higashi Ajia no kiseki* [The East Asian miracle in global history] (Shohan). Nagoya Daigaku Shuppankai.

Tienhaara, K. (2014). Varieties of green capitalism: Economy and environment in the wake of the global financial crisis. *Environmental Politics*, 23(2), 187–204.

Toke, D. (2011). Ecological modernisation, social movements and renewable energy. *Environmental Politics, 20*(1), 60–77. https://doi.org/10.1080/09644016.2011.538166
UNEP. (2010). *Green economy: Developing country success stories*. United Nations.
Vesia, D. J., Mahutga, M. C., & Bui, B. K. H. (2023). Flattening the curve? The structure of the natural resource exchange network and CO_2 emissions. *Social Networks, 75*, 118–136. https://doi.org/10.1016/j.socnet.2021.07.004
Victor, D. G., Morgan, M. G., Apt, J., Steinbruner, J., & Ricke, K. (2009). The geoengineering option: A last resort against global warming? *Foreign Affairs, 88*(2), 64–76.
Weale, A. (1992). *The new politics of pollution*. Manchester University Press.
WNA (World Nuclear Association). (2024, April 30). *Plans for new reactors worldwide*. World Nuclear Association. https://world-nuclear.org/information-library/current-and-future-generation/plans-for-new-reactors-worldwide.aspx
Wong, C. M. L. (2012). The developmental state in ecological modernization and the politics of environmental framings: The case of Singapore and implications for East Asia. *Nature and Culture, 7*(1), 95–119. https://doi.org/10.3167/nc.2012.070106
Woo, M. (2022). The developmental state. In J. Han, R. P. Pardo, & Y. Cho (Eds.), *The Oxford handbook of South Korean politics* (pp. 104–119). Oxford University Press. https://doi.org/10.1093/oxfordhb/9780192894045.013.7
Ylä-Anttila, T., Gronow, A., Stoddart, M. C. J., Broadbent, J., Schneider, V., & Tindall, D. B. (2018). Climate change policy networks: Why and how to compare them across countries. *Energy Research & Social Sciences, 45*, 258–265.
York, R., & Rosa, E. A. (2003). Key challenges to ecological modernization theory: Institutional efficacy, case study evidence, units of analysis, and the pace of eco-efficiency. *Organization & Environment, 16*(3), 273–288. https://doi.org/10.1177/1086026603256299
York, R., Rosa, E. A., & Dietz, T. (2003a). Footprints on the Earth: The environmental consequences of modernity. *American Sociological Review, 68*(2), 279–300. https://doi.org/10.1177/000312240306800205
York, R., Rosa, E. A., & Dietz, T. (2003b). STIRPAT, IPAT and ImPACT: Analytic tools for unpacking the driving forces of environmental impacts. *Ecological Economics, 46*(3), 351–365. https://doi.org/10.1016/S0921-8009(03)00188-5
Young, M. S. C. (2001). *The emergence of ecological modernisation: Integrating the environment and the economy?* Routledge.

CHAPTER 5

The Treadmill of Production Theory and Climate Action

KENNETH A. GOULD AND TAMMY L. LEWIS

Introduction

THIS chapter provides an overview of the treadmill of production (ToP) theory and applies it to institutionalized climate action, including state-oriented strategies, market-oriented strategies (including technological strategies), and to noninstitutionalized strategies, notably climate social movements. It concludes with a discussion of approaches that could potentially slow or disrupt the treadmill.

The Treadmill of Production

The treadmill of production (ToP) theory is a foundational theory in environmental sociology that explains how and why industrial societies create environmental and social disruptions. Allan Schnaiberg (1980) elaborated the theory in *The Environment: From Surplus to Scarcity* to explain the acceleration and consequences of resource withdrawals (extraction) and additions (pollution) in the industrialized, petrochemical world in the period following World War II. Subsequent elaborations of the ToP theory (Gould et al., 2004, 2008; Lewis, 2016, 2018; Leguizamón, 2020; Schnaiberg & Gould, 1994) apply the concepts in empirical analyses and extend its scale to account for the process of globalization.

In short, Schnaiberg and his collaborators argue that major social actors (states, corporations, and labor) consistently promote economic growth and production expansion (the treadmill of production), despite their negative consequences of environmental degradation and social dislocation. Growth increases corporations' profits and governments' tax base and provides workers with jobs. The downside is that corporations' use their profits to replace labor with energy and chemicals in production in a process of technological displacement. This shift increases ecological harm (through additional natural resource withdrawals and resultant pollution) and decreases social benefits (unemployment). The state must step in to provide social welfare for the displaced. To maintain a constant level of social benefits, the treadmill must expand to absorb displaced workers. The response to the problem is the cause of the problem, and thus the cycle continues, deepening social and ecological crises. Despite the fact that this does not work, the promise of economic growth has become a powerful ideology (Leguizamón, 2023).

As an example, consider the trajectory of the coal industry. As the industry mechanized, private fossil fuel companies laid off workers and blamed the employment reductions on environmental regulation. Displaced citizen-workers sought relief from the government. The government responded by providing social welfare to the unemployed, thus legitimizing their role, while continuing to seek taxes (accumulation) from the coal industry. Industry sought more coal and more profits through more extraction using more technology. In this way, the need for growth of the coal industry prevailed as a logic at the same time it created social and ecological damage.

Globalization of the economy extended these processes. As governments sought regulations to protect workers and their environments, private capital moved abroad to where regulations were limited, they could expand their access to resources, and they could increase profits to shareholders. As communication and transportation technology facilitated global supply chains, capital was able to seek low regulation and low wage contexts worldwide. Competition among developing states to attract foreign direct investment generated a race to the bottom with governments seeking to make their economies most attractive to capital, increasing the ecological costs of and reducing the social benefits (wages) of those investments. As environmental impacts mounted, calls for environmental protection emerged among citizen-workers in the Global South (Gould et al., 2008; Schnaiberg & Gould, 1994).

Despite the power of the political economic system, Schnaiberg and others have suggested motivations for why key actors would want to slow the treadmill and ways that it could be done. For instance, states and labor (in later writings "citizen-workers"[1]), have a contradictory relationship to the treadmill of production. While the state benefits by accumulating during periods of growth, it also has the responsibility to provide protections for citizen-workers to legitimate its role, and thus must protect citizen-workers through environmental and labor regulations. In parallel, citizen-workers also benefit from growth in terms of employment and opportunity, yet they demand quality-of-life assurances from the state, including healthy environments and workplaces, which are also basic principles of the environmental justice movement. In this way, states and citizen-workers have the potential to form alliances for practices that both ensure the benefits of the treadmill of production while limiting its negative outcomes. In this logic, citizen-workers, through pressure on states, are a source of slowing of the treadmill to ensure environmental and social protection. Corporations, with their general short-term profit orientation, are less likely to share these interests.

Schnaiberg (1980, pp. 424–425) explains three paths that societies could take to resolve the tension between economic growth and socio-environmental protection: (1) economic synthesis, (2) managed scarcity synthesis, and (3) ecological synthesis. These vary regarding the extent to which the state limits access to the environment for production, and in effect, ecological impacts. Schnaiberg emphasizes the importance of states' choices on slowing or speeding up the treadmill of production. Capital seeks to externalize production costs, limit regulation, expand access to resources, and reduce labor costs.

The economic synthesis is unfettered capitalism. The state facilitates corporations' environmental withdrawals, production expands, and consequently environmental and social problems increase. State subsidies to fossil fuel corporations exemplify this. In the managed scarcity synthesis, the state uses its power to limit corporations' environmental withdrawals and additions, which limits production and protects the environment and citizen-workers. States' attempts to regulate greenhouse gas (GHG) emissions are such an approach. The ecological synthesis requires the most intervention from the state. It limits corporations' access to the environment based on scientifically determined biophysical limits of what is environmentally sustainable. The ecological synthesis best protects the environment, but it requires disrupting existing power relationships. The ecological syntheses "largely focus on ecosystem protection, especially on sustaining bio-diversity in all ecosystems" (Schnaiberg, 2010, p. 17). To do this "would entail the state's substantial control over ecosystems, without regard to issues of profitability and of wages/employment" (Gould et al., 2004, p. 305). An illustration of this approach is government efforts to block fossil fuel extraction by "keeping the oil in the soil" (see chapter by Martinez-Alier and Lechner in this volume).

To attain a more ecological synthesis, ToP theorists point to the importance of coalitions among social movements (Van Dyke & Amos, 2017). For example, labor movements and environmental movements that come together to form blue-green alliances have the potential to push the state toward generating the policies for managed scarcity and ecological syntheses (Gould et al., 2004; Obach, 2004). They do this by using tactics from civil disobedience to lobbying. Social movements arising from citizen-workers' concerns are a key catalyst for applying pressure on states, which can then change policy and impact production outcomes.

Schnaiberg suggested some modest policy changes to the treadmill of production that could "maximize social welfare . . . within the context of limited biospheric resources" (1980, p. 431). These include some relatively small changes that limit withdrawals and additions and encourage employment rather than investment in technology. For instance, he proposes "materials and energy taxes, to encourage conservation" and "incentives (tax relief, credits) for material recycling," which would reduce withdrawals. To reduce pollution, he suggests "high effluent charges." To encourage employment, Schnaiberg recommends "employment tax credits" and "labor training tax credits." To sway capital away from energy-intensive technological investments, he proposes "rapid depreciation allowances for preexisting fixed capital that was highly material- and energy-intensive" and "increased taxation of new fixed capital, to discourage greater capital intensity" (Schnaiberg, 1980, pp. 431–432). These reform-oriented changes could slow the treadmill, reducing both worker displacement and ecological harm. These strategies are also consistent with green workforce development initiatives and just labor transitions (see chapter by Kerrissey and Coon in this volume).

In sum, the treadmill of production theory is a political economic theory that explains the acceleration of environmental damages that have occurred since World War II. It lays out the mechanisms by which a capitalist system's growth imperative causes ecological disorganization. In the next section, we explain how climate change is both a symptom of the treadmill and a threat to it.

Climate Change as Symptom of, and Threat to, the Treadmill of Production

Effective climate action is hard to imagine within the existing political economy due to the growth imperative of capitalism. The treadmill of production requires constant expansion of production and markets, hence constant expansion of natural resource extraction, transformation, and disposal as waste. All of this must occur within the ecological constraints of the finite natural resource pool of our planet. Expansion of production within the treadmill takes a particular form in that the wealth generated by natural resources and labor is used to replace labor with energy and chemicals in the form of fixed capital investment in labor-saving technology (Schnaiberg, 1980). Each round of production expansion displaces labor (reducing the social benefits of production) and increases energy demand (increasing the ecological costs of production). Recent advancements in artificial intelligence (AI) technology exemplify this process. Economic growth, a goal which orients all major social institutions, is therefore predicated on constant increases in energy demand. Steadily increasing energy demand makes it nearly impossible for renewable energy expansion to effectively displace fossil fuels, as increases in total energy demand outpace the rate at which new non–fossil fuel sources can be brought online (York, 2012). New energy sources thereby tend to increase the capacity of the treadmill of production to convert natural resources into products and then waste without decreasing GHG emissions. GHG emissions represent the accumulated atmospheric externalization of production expansion costs to the global commons. The negative climate consequences of treadmill expansion have been externalized to the future through atmospheric dumping of carbon waste, forcing future generations to pay the costs of social and material benefits received by prior generations, thus creating intergenerational climate injustice.

It is common for us to consider the long-term costs to capital of atmospheric externalization of the energy demands of the treadmill in terms of supply chain disruptions, agricultural failures, and damage to infrastructure and other fixed capital investments. It is also important to consider the opportunities that climate change represents for treadmill expansion. As the cryosphere retreats, new land becomes available for mineral exploration and extraction, and new sea routes open to facilitate international trade. Coastal climate risks generate opportunities for new real estate development through resilience gentrification processes, densifying populations along precarious coasts at ever-increasing prices (Gould & Lewis, 2021). Climate change expands markets for air conditioning and refrigeration, as well as new genetically modified and patented climate-resilient food and textile crops. It also creates opportunities for investments in new carbon capture and storage technologies, often with the state subsidizing fossil fuel industries to develop and implement. Climate change, at least in the short to medium run, represents much more of a threat to living things than it does to capital, which explains in part the concern gap between vulnerable populations and treadmill elites who see economic opportunity where others see ecological catastrophe.

More threatening than the fact of climate change to the treadmill is the possibility of effective action to mitigate climate change. Effective mitigation of climate change will require the rapid phasing out of fossil fuel use with an overall net reduction in total energy available to the economy (Bell, 2021). That is, effective climate change action necessitates economic contraction, which is anathema to the treadmill and the social institutions that support it (Stuart et al., 2020). Modern industrial capitalism is best understood as carboniferous capitalism, as it is fundamentally dependent on fossil fuels and their increasing availability. It is the portability and energy density of fossil fuels that allowed this form of political economy to evolve and expand, and fossil fuels are largely nonfungible for this purpose. While it is fully possible to run a global civilization on renewable energy sources, it is not at all clear that it would be possible to run our current global civilization without fossil fuels. This makes the specter of climate action threatening to treadmill elites, whose extraordinary social power derives from their position in this particular political economy. This, combined with a deep ideological opposition to government regulation of capital in the form of emissions controls, GHG reduction targets, carbon taxes, restrictions on extraction and transport of fossil fuels, and energy efficiency standards, impels treadmill elites (and often those dependent on them) to strongly oppose effective climate action. Treadmill elites can enlist workers in particularly vulnerable economic sectors (e.g., the fossil fuels industries) to their cause of blocking climate mitigation action. They are also able to enlist those citizens who are dependent upon or enamored of particularly fossil fuel–dependent social formations (e.g., suburbanites). Both treadmill workers and consumers may be called to resist the "war on cars" and the "war on gas stoves," let alone the coming "war on suburbia."

Despite an overwhelming social dependency on fossil fuels, growing realization of the consequences of changing the chemical composition of Earth's atmosphere in ways that heighten the global greenhouse effect represents a significant political threat to the treadmill. As the public experiences the realities of more powerful and frequent extreme weather events, including heat waves, droughts, cloudburst rain events, fires, and sea-level rise, demand for states to address the climate crisis grows. States must both act to expand the treadmill to generate jobs under conditions of technological (energy) displacement and be seen to be acting to protect the public from the worst impacts of climate change. This dual pressure impels states to pursue nonsolutions that create the public illusion of being proactive about the climate crisis while not threatening economic expansion and the interests of treadmill elites upon whom the state depends. No longer able to appease the public through direct climate denialism, a series of market-friendly pseudo-solutions are currently being deployed to resolve the state's contradictory position between growth and legitimacy. These political strategies include carbon capture technologies, "natural" gas expansion, "clean" coal technology, carbon offsets, cap-and-trade schemes, coastal reengineering, and geoengineering. All these strategies are designed to primarily protect the treadmill from climate change rather than to protect the climate from the treadmill. In the following sections, we outline how institutionalized actors' and noninstitutionalized actors' main climate strategies (as elaborated by Almeida et al., 2024) fit into Schnaiberg's three syntheses.

Capital and Market Strategies Generating Economic Synthesis

In the face of both growing political pressure to address climate change and rising threats to capital investments due to climate change (real estate loss, supply chain disruption, etc.), treadmill elites must strategize how to protect their interests, power, and the structural arrangements that serve them. For decades their primary strategy was outright climate denial, orchestrating a complex campaign to generate and publicize denialist talking points, and attacking climate scientists (Brulle, 2021; Farrell, 2019). As they funded climate denial networks which contradicted their own internal analyses that indicated dire climate consequences of continued fossil fuel use, they simultaneously pursued the expansion of fossil fuel exploration, extraction, and infrastructure construction. As these denialist strategies became increasingly hard to defend in the face of mounting climate disasters, capital shifted to efforts to shape social responses in ways that benefit rather than threaten their interests. They are thus deploying a series of market-based strategies and technological fixes that may address aspects of climate mitigation and resilience from which they may profit. One such approach is the expansion of renewable energy and expanded electrification to decarbonize the economy. This includes solar, wind, and geothermal power generation and electric vehicle production.

Although decarbonization of the economy is essential to address climate change, it is not clear that these market-based technological fixes can achieve that goal within existing political economic arrangements (Huber, 2022). Renewable energy tends to add to existing energy sources rather than fully replacing them, thus increasing our capacity for ecological withdrawals (extraction) and additions (pollution). Unless paired with state intervention to sunset fossil fuel use, renewable energy may prove to be a net ecological loss. Similarly, a transition to electric vehicles is intended to preserve automobile culture, sales, and dependency, which is central to high energy demand lifestyles. Electric vehicles are the treadmill alternative to mass public transit, 15-minute cities, and a lower consumption culture. Focusing political conflict on renewable expansion and electric vehicles is a strategy for defining the parameters of climate debate within market-friendly terms. When we debate what kinds of cars to build our future around, treadmill elites have scored a political victory in that their interests are protected.

The creation of carbon markets is another approach to apply an economic synthesis to the climate crisis. Cap-and-trade schemes propose to marketize using our atmosphere as a carbon dump. The political project of carbon markets is to propose that only markets can solve the problems generated by markets. The economic project of carbon markets is to secure the right to pollute in a way that allows for market exchange, speculation, and profitability. The failure of carbon markets as an ecological project is that carbon dumping allowances are typically set too high, and the cost of dumping carbon in our atmosphere is typically set too low (Stuart et al., 2020). Additionally, for cap and trade to have any real ecological benefit, the cap must be low and rapidly decrease to near zero, and it must be global, something proponents of cap-and-trade schemes tend to reject. Commodification of the atmosphere as a global dump is an effort to protect the treadmill of production, not the atmosphere.

Perhaps the ultimate technological fix to address the climate crisis is geoengineering (Buck, 2019), manipulating Earth's reception of solar radiation through a range of aerosol applications, shields, and albedo modification strategies. Approaching climate change as an engineering problem rather than as a social (political-economic) problem is key to a treadmill economic synthesis. Geoengineering offers a wide range of possibilities for profitable investment and corporate technological dependence. It also offers a wide range of potentially catastrophic impacts as both weather and climate changes that will result are unknown prior to real-world experimentation. Geoengineering offers the possibility of a future where the habitability of the planet and all life on it is fully dependent on treadmill institutions, which represents the ultimate power grab of treadmill elites. Geoengineering could create a global climate protection racket.

State and Policy Strategies Generating Managed Scarcity (United States and Abroad)

As the private sector slowly falls back from outright public climate denial to reliance on markets to solve the climate crisis in an economic synthesis approach, government finds itself needing to intervene in markets to the extent necessary to maintain political legitimacy. While many conservative politicians continue to cling to denialism, growing demands on the state from both their citizens and global institutions to do something to reduce GHGs and to protect people from the impacts of climate change have necessitated the adoption of a managed scarcity approach. Largely beholden to private capital interests, states seek to do as little as politics will allow without significantly constraining corporate actors (Gould, 1994). This mild managed scarcity approach is evident in recent US legislation (both passed and proposed).

The Inflation Reduction Act (IRA), which the US Treasury Department notes is a pro-growth policy, incentivizes the expansion of renewable energy. In doing so, it aligns with a managed scarcity synthesis in that it makes reforms to the current political economic structure through government-funded incentive structures. Through the IRA, the state intervenes in the market to provide economic benefits for market actors who pursue the government's modest climate goals. With much of the IRA funding going toward highway construction and repair, it remains consistent with the interests of dominant market actors, while promoting the use of construction materials with less embodied carbon. The IRA also includes new regulations to reduce methane emissions from fossil fuel extraction operations, but it does not prohibit such operations. The United States intends to remain the world's leading fossil fuel producer, as evidenced by ongoing fossil fuel infrastructure expansion and extraction subsidies. A significant portion of IRA funds are intended to be invested in disadvantaged communities with environmental vulnerabilities.

Like the IRA, the Bipartisan Infrastructure Law (BIL) invests in existing transportation networks, focusing on bridges, roads, and airports. As a managed scarcity approach, it also invests in rail and public transit infrastructure, and incentivizes electric vehicle expansion and infrastructure. While funding for rail and public transit is essential to address the climate crisis, the BIL represents a minor tweak to existing transportation systems, albeit with a slight shift toward electrification. The BIL also includes funding for resilience, emphasizing engineering responses to structurally mitigate cities, ports, and other infrastructure. Such resilience funding is intended to defend key treadmill investments from the climate impacts of the treadmill. The areas targeted for these protections are those with the highest real estate values (Gould & Lewis, forthcoming). Ultimately, the BIL represents the kind of pork barrel politics compromise currently possible without disrupting the treadmill. Mild reformist legislation is the essence of managed scarcity and is wholly inadequate to address the climate crisis.

A more aggressive and potentially impactful managed scarcity approach is represented by the Green New Deal (GND). The GND is proposed legislation that seeks to transition from fossil fuels to renewable fuels while creating jobs and increasing economic equity. It seeks to create a post–fossil fuel economy charged by electrification and green job growth. Although the GND elaborates a path toward sunsetting fossil fuels, it remains committed to a growth-oriented economy that would decarbonize the treadmill with renewable energy. The proposal aims to address carbon emissions without addressing the ecological unsustainability of the existing political economy more broadly. Even so, the proposal threatens the fossil fuel industry, arguably the most powerful and central industry within the global treadmill of production, and is therefore considered unacceptable by treadmill elites.

The content and fate of these three pieces of legislation in the United States ostensibly connected to fighting the climate crisis indicate that some managed scarcity synthesis is still possible within the treadmill, but that a managed scarcity approach will continue to prioritize economic interests over environmental interests. The managed scarcity policies are intended to incentivize some ecological modernization of the treadmill.

Outside of the United States, similar pressures exist for states to continue the treadmill of production, with growth of the economic system being the driving force for environmental regulation. For instance, the European Union (EU), which plays an influential role in generating global environmental standards, promotes a strategy that is aligned with a managed scarcity synthesis in that it seeks efficiency and environmental protection *for the economy*. The EU's official website spells out this commitment: "Environmental protection and innovation help to create new business and employment opportunities, which stimulate further investment. Green growth is at the heart of EU policy to ensure that Europe's economic growth is environmentally sustainable" (European Union, n.d.). The European Commission adopted a European Green Deal that is described as "the new growth strategy for an EU economy that is sustainable, cleaner, safer and healthier" (European Union, 2022). Again, this formulation puts the economy first and the environment second, encouraging some ecological modernization. While the EU also has legislation to protect and restore biodiversity, it is a secondary goal. Environmental policy in the EU, as a rule, applies the precautionary principle (acting to protect the public and environment when there is scientific uncertainty), and there have been several legal cases in the past years applying the principle to climate action (Falcao, 2021).

Unsurprisingly, other nations have followed suit. For instance, many developing countries have argued for their right to put their economic development ahead of their environmental protection as the "developed" countries have historically done. Protecting the global commons, they argue, is not just their responsibility, but the responsibility of the nations that have emitted the most GHGs. In this way, middle- and low-income nations' rights to economic development is a form of climate justice (Roberts & Parks, 2007). This sentiment is especially true in fossil fuel resource-dependent countries like Ecuador, Nigeria, and Mexico, which depend on foreign exports of fossil fuels to literally fuel accumulation into government budgets that provide states with the capacity to legitimate their roles by delivering services, such as health care, education, and infrastructure, to their people. The quantity and rates of global emissions depend on numerous factors and some big players. The choices of nations like China and India, with large populations, land masses, and economies, can have huge consequences for global GHG emissions.

International governing bodies, and specifically, the United Nations, have long deployed a managed scarcity strategy. Highlighted by the Brundtland Commission's *Our Common Future* (1987) packaging of economic growth and environmental protection within the concept of "sustainable development," the organization has promoted economic growth and environmental protection in that order. The United Nations has been the convener of international conferences focused on the environment since 1972 in Stockholm. The United Nations Conference for Environment and Development in 1992 in Rio de Janeiro was critical for global environmental policy. At that time, nations signed the United Nations Framework Convention on Climate Change that is the foundation for later agreements, including the Paris Agreement signed in 2015 and adopted by 196 countries. The signers pledged to do their part to limit the rise in global temperatures below 2°C from preindustrial levels, with further goals to keep it below 1.5°C. Each country agreed to nationally determined contributions (NDCs), which include elements such as national climate plans, GHG emissions reduction targets, and dates by which the nation will reach net-zero emissions. In addition to national pledges, the International Panel on Climate Change (IPCC) regularly produces reports that summarize scientific research on climate change. The most recent report—the IPCC Sixth Assessment Report (2023)—paints a dire picture.

Despite international reports and agreements, the outcomes of the international process have been limited. A recent academic assessment concludes, "In the last 30 years the amount of human-emitted CO_2 has doubled.... However, the climate negotiations and the resultant changes in the global economy mean that predicted levels of global warming above 4°C this century are highly unlikely. If all the NDCs made ... are fulfilled then global warming would be kept to between 2.4°C and 2.8°C. Despite this progress we can still view the negotiations as a political failure, as climate change impacts are increasing and we are nowhere near keeping global warming to just 1.5°C above pre-industrial levels" (Maslin et al., 2023, p. 13). Global warming above 2°C threatens to push Earth systems beyond critical ecological global warming tipping points (IPCC, 2023).

Finally, of special note is the role of international conflict in generating and potentially lessening emissions. For instance, the Russia-Ukraine war disrupted fossil fuel flows from Russia to the EU and led to a faster than anticipated transition to clean energy in the EU. According to the International Energy Agency, "The biggest legacy of the global energy crisis triggered by Russia's invasion of Ukraine may be that it accelerates the end of the fossil fuel era—with IEA projections now showing that under today's policy settings, demand for oil, gas and coal is set to peak within the decade" (IEA, n.d.). This conflict also highlights the extreme vulnerability of nuclear power in an unstable world.

Citizen-Workers Generating Ecological Synthesis

There is no single ideal example of citizen-workers acting to create an ecological synthesis, at least not in the United States, but there are efforts being made that would lead the political economic system toward a path that prioritizes ecology over the economy. The concept and practice of "just transition" aims to bypass wedges that corporate elites have promoted regarding "jobs versus the environment" by centering the concerns of workers (see chapter by Kerrissey and Coon in this volume). The idea derives from labor union activists who formed coalitions with environmentalists, and especially environmental justice advocates ("blue-green alliances"), to promote transitions away from dirty fossil fuel jobs to clean jobs in renewable energies, such as wind and solar (Gould et al., 2004; Obach, 2004). Key to the concept is justice and the idea that the costs and benefits of transitions to low-carbon economies should be distributed fairly.

While the public theoretically would benefit from this transition, the costs also need to be addressed. Given its origins in labor, one of the key policy actions for just transition includes job retraining for workers in the fossil fuel sectors. There are numerous organizations and examples of such policies (Brostrom et al., 2024). The Climate Justice Alliances elaborates its understanding of the goal: "Just Transition is a vision-led, unifying and place-based set of principles, processes, and practices that build economic and political power to shift from an extractive economy to a regenerative economy. This means approaching production and consumption cycles holistically and waste-free. The transition itself must be just and equitable; redressing past harms and creating new relationships of power for the future through reparations. If the process of transition is not just, the outcome will never be" (https://climatejusticealliance.org/just-transition/). This work builds alliances between classes and movement sectors (Obach, 2004). In a just transition, labor is displaced for pro-ecological rather than anti-ecological reasons, and that displacement is addressed through retraining and social support rather than production expansion.

Degrowth—the idea that societies can downscale production and consumption to be within ecological limits while increasing peoples' quality of life, social equity, and improve ecological conditions—originated as an academic concept. Degrowth reframes well-being around terms such as sufficiency, conviviality, and public wealth. "Put simply, the degrowth hypothesis is that it is possible to organize a transition and live well under a different political-economic system that has a radically smaller resource throughput" (Kallis et al., 2018, p. 292). Its history can be traced to work in the 1970s that questioned economic growth ideologies. It is growing in visibility and as a movement (it was recently highlighted in the *New York Times*, 2024) alongside actual and potential policy applications. A systematic review of degrowth policies highlights the most popular proposals for degrowth, including "universal basic incomes, work-time reductions, job guarantees with a living wage, maximum income caps, declining caps on resource use and emissions, not-for-profit cooperatives, holding deliberative forums, reclaiming the commons, establishing ecovillages, and housing cooperatives" (Fitzpatrick et al., 2022, p. 8).

Degrowth is akin to other ideologies around the world, such as the concept of *buen vivir* (good living) in Ecuador, which for a while guided official development policy. Like degrowth, *buen vivir* does not privilege economic growth over environmental protection or societal well-being; rather, it privileges the rights of nature (Kothari et al., 2014; Lewis, 2016). Given that degrowth, and *buen vivir*, and other associated concepts threaten the growth ideology and require less economic inequality puts elites at odds with it. Academic proponents of the concept conclude that "Planned degrowth is politically unlikely, given established interests and power relations" (Kallis et al., 2018, p. 309). Nevertheless, a growing number of social movements include elements of degrowth among their demands.

Noninstitutionalized Actors Engaged in Climate Movements

Why are people motivated to act? Citizen-workers and their families want clean environments and healthy living conditions. Youth, who have been especially outspoken on this issue, want opportunities for clean and healthy futures. Progressive activists who are regularly focused on the unequal distribution of the benefits of economic growth, have connected that concern to the unequal distribution of environmental degradation. Countless others have been materially affected by floods, hurricanes, droughts, wildfires, mudslides, and changes to their livelihoods. These noninstitutionalized actors have come together to form thousands of organizations around the world focused on addressing climate change, with a wide range of goals that vary in relation to their embrace of the treadmill of production and an equally wide range of tactics.

On the one end of the spectrum are groups that work within the system to reform political and economic practices, reinforcing current power relationships, and an economic synthesis. These organizations often use institutionalized tactics, such as education and lobbying. For instance, Citizens' Climate Lobby (CCL) organizes to build bipartisan political support to change laws, such as reforming permitting processes so that a transition to green energy can happen more quickly. They also seek to incorporate climate change measures into market processes through putting a price on carbon, and thus using the economic system to reduce carbon emissions. They also encourage the use of technology to increase the efficiency of buildings and to electrify processes for waste and pollution reduction (https://citizensclimatelobby.org/). In short, groups like CCL seek to "green" the treadmill of production, not change it.

Other groups work to address the importance of power and inequality and seek changes to the system aligned with a managed scarcity approach. For example, 350.org's goals are "to dismantle the influence and infrastructure of the fossil fuel industry and to power up clean systems rooted in justice" (https://350.org/about/). Through lobbying and protest they use the system to punish polluters and reward investments and laws that will lower carbon emissions (see chapter by Bove, Bates, and Henn in this volume). They recognize the outsize role of corporations and have numerous campaigns that seek to limit the financing and expansion of fossil fuel projects, often targeting insurance companies. For example, "We're calling on the President and CEO of Marsh to stop working with French oil giant, TotalEnergies, and not to broker insurance for the East African Crude Oil Pipeline" (https://act.350.org/). They also focus on changing government legislation. In Canada, for example, they are pressuring the Trudeau administration to tax oil companies to fund a transition to clean energy and to create stronger regulations that would require 100% clean electricity by 2035. Using policy as a focal point is in line with Schnaiberg's managed scarcity approach, which calls on the use of the state to enact limits on polluters, in this case, the fossil fuel industry. Working with frontline communities, justice is a central tenet of 350.org's efforts, as they explicitly connect economic injustice to environmental injustice. Groups like 350.org aim to use the power of "the people" to empower governments' positive influence within the treadmill of production to limit fossil fuel expansion and to shine a light on fossil fuel companies' interests in "block[ing] every serious attempt to stop the climate crisis" (https://350.org/finance/).

On the other end of the climate movement are organizations that call for system change. These groups advocate for states to take even stronger actions to reprioritize sustaining biodiversity and ecosystems over corporate profit, which aligns with Schnaiberg's ecological synthesis. Many of these organizations, such as Extinction Rebellion (XR), Fridays for the Future, and Just Stop Oil, use disruptive, direct-action tactics to achieve their goals (see chapters by Uba and Cavalho/Gardner in this volume). For instance, XR employs civil disobedience to demand that governments declare "a climate and ecological emergency ... to halt biodiversity loss and reduce greenhouse gas emissions to net zero by 2025" and to abide by decisions made by Citizens' Assembl[ies]" (https://rebellion.global/). They envision a different social order—a "regenerative culture" that is "healthy, creative, resilient and adaptable." Their Citizens' Assemblies are modeled on deliberative democracy, in which citizens are informed, deliberate, and make recommendations to governments regarding policy, which is voted on by nations, fundamentally disrupting existing power relationships. A transnational climate justice movement has emerged from this work, built upon former progressive organizing and structures established by the global justice movement (Almeida, 2019).

XR's actions have raised important questions regarding which tactics are most effective at generating change. Their direct-action tactics have included disrupting performances and sporting events, blocking roads, chaining themselves to banks, and throwing paint and soup on famous works of art (behind glass) in major museums. These controversial acts have been met with mixed responses: on one hand, they have garnered media attention and brought awareness to the climate crisis; on the other hand, the nonactivist public has viewed these actions unfavorably. Nevertheless, as is often the case in social movements with a wide range of activists, goals, and tactics, XR serves as a "radical flank" within the climate movement, which in turn "increases support for a moderate faction within the same movement" (Simpson et al., 2022). In other words, because XR engages in radical activity, the public is more likely to support Citizens' Climate Lobby and 350.org, which by contrast, seem moderate and reasonable. In early 2023, XR decided to move away from disruptive tactics—prioritizing "attendance over arrest and relationships over roadblocks" (https://extinctionrebellion.uk/2022/12/31/we-quit/) though other groups are ramping up direct action, such as Ende Gelände (German for "here and no further"), which employs blockadia, focuses on shutting down coal plants and other fossil fuel production through civil disobedience (Malm, 2021).

The final classification of climate change activists who are critical to highlight are those associated with Indigenous groups, whose worldviews often fall outside of the dominant Western political economy (see chapter by Fenelon and Figueroa Helland in this volume). Indigenous activists are networked globally through continental organizations such as the Indigenous Environmental Network and beyond through coalitions like the International Indigenous Peoples Forum on Climate Change. Many Indigenous-led campaigns are infused with ideologies that demand a system that prioritizes nature and other living beings, and thus aligns with Schnaiberg's ecological synthesis. Additionally, in many parts of the world, Indigenous groups' rights have been routinely overlooked and their efforts also align with human rights and justice. For example, in North America, Standing Rock Sioux tribal members led the protest the Dakota Access Pipeline (DAPL). The pipeline was rerouted from a more populous area to near one of their sacred sites. The Sioux argued that the proposed pipeline would threaten their water quality, put sacred land at risk, and violate a treaty that guarantees their use of the land. The Indigenous activists engaged in civil disobedience by setting up camps and occupying the space; they identified themselves as "water protectors." Their direct action was met with violent repression by private security forces—including being shot at with water cannons—and after one and a half years at the site, police cleared them from the area and the pipeline was built. While these actions were part of the movement to reduce dependence on fossil fuels and thus considered a climate change action, the campaign also highlights the role of colonialism in clearing the way for fossil fuels. Commentators have noted that historically the violations of treaties enable land to be taken for extraction; this isn't just unjust, but in this case is a form of climate injustice—"the idea that it is ethically wrong for some groups of people to suffer the detrimental effects of climate change more than others" (Whyte, 2016).

Another example that brings home the intertwined nature of Indigenous rights and climate justice actions can be found in the eastern Amazonian region of Ecuador in South America, in Yasuní National Park, the home of two Indigenous tribes—the Tagaeri and Taromenane. In 2007, the president of Ecuador, an oil-producing nation, proposed to preserve the oil reserves in the park—in other words, to "keep the oil in the soil" if the international community would donate half of what the oil was worth. This government initiative failed because the international community failed to fund the proposal, and the government has exploited some of the lands (Lewis, 2016). However, climate change activists kept the campaign alive. A group in the capital called the Yasunidos worked for over a decade to use their power to put up the question of drilling for a national referendum. In 2023, the citizens of Ecuador voted to keep the oil in the ground—a historical precedent (Bernal & Holst 2023). If instituted by the government, keeping the oil in the soil will protect both Indigenous culture and limit fossil fuel extraction.

This final example also highlights the relationships between institutionalized and noninstitutionalized actors. In the Yasuní case, the original proposal was formally presented by the president—an institutionalized actor—in front of the United Nations—another institutionalized body. When those actions failed, noninstitutionalized actors (the Yasunidos) continued the fight. Noninstitutionalized actors often are seeking to influence institutionalized actors, to change policies. In addition, noninstitutionalized actors with outside-the-box ideas also demonstrate that "another world is possible" and our present system of functioning is a choice and other choices have been and are possible. In terms of the treadmill of production, dominant elites resist movement actors who threaten their power, profits, and ideology. They are more likely to support and legitimize groups like the Citizens' Climate Lobby than actors that threaten business as usual—which Standing Rock Sioux tribal members did when they put their bodies in the way of the pipeline.

Conclusion: What Is to Be Done?

A key obstacle to effective climate action is extreme and expanding economic inequality. Economic inequality undermines democratic governance by institutionalizing the overwhelming policy interest of private capital, providing fossil fuel and related interests with effective veto power over climate legislation. Additionally, extreme economic inequality insulates treadmill elites from the consequences of their climate action/inaction. From multiple luxury residences they can afford to lose to fire and sea, to private nature reserves in remote locations, to doomsday bunkers, the key global decision-makers feel secure that their vast wealth will protect them from climate consequences. The trade-off for treadmill elites between unprecedented wealth and power and facilitating the possible collapse of the biosphere is better for them than anyone else on Earth. In contrast, those frontline communities feeling the impacts of the climate crisis on their lives and future, who recognize the need for immediate and decisive action, have the least political power to effect policy change. Extreme economic inequality in effect severs the feedback loop between social systems and ecosystems which have always been the key to human survival. In view of that, the restoration of the socio-environmental feedback loop through redistributive domestic and global economic equity policies may be the most important climate action we could take.

The key to reducing GHG emissions is reducing the energy demand of the global social system. A process of rapid phase-out of fossil fuels, rapid expansion of renewable energy sources, and a rapid powering down of total energy demand is the minimum necessary social action required to mitigate climate change. Such a dramatic energy transition (both in terms of sources and volume) would still leave our planet facing 1.5°C+ of global warming, requiring extensive resilience efforts for generations. A simultaneous transformation of agriculture, phasing out most meat consumption along with substantial reforestation, will also be necessary. A shift from private to public transit and substantial economic relocalization would be necessitated by such energy and agricultural transformations. That none of these options is being seriously discussed by policymakers is a good indication of total global institutional failure on the climate crisis.

Energy transformation and reduction, agricultural transformation, and reforestation would also go a long way toward addressing the larger ecological crisis of which climate change is only a part. Additionally, many of the proposed policy changes advocated by proponents of degrowth—a significant shift from private to public (shared) resources, a ban on advertising, work time reduction, and basic minimum income—would reduce production and consumption while allowing for substantial progressive redistribution of wealth (a more equitable distribution of a smaller economic pie). Finally, demilitarization would reduce one of the greatest sources of GHG emissions and the greatest waste of resources and social production (Gould, 2007).

The private capital actor most likely to have a direct positive impact on climate policy is the insurance industry. As climate impacts on businesses and residences mount, the insurance industry is raising premiums and walking away from some precarious locales completely. As insurance becomes more expensive or impossible to attain, construction in climate-vulnerable locations is bound to decrease. As reinsurance companies decline to insure residential and business insurance providers, a shift away from development in flood- and fire-prone areas may generate a neoliberal version of managed retreat from fire-prone areas and precarious coasts.

It will be important for institutional actors and citizen-workers to be cautious of the tyranny of sunk costs in relation to coastal armoring schemes. The impulse to protect valuable real estate and infrastructure with ever more expensive massive coastal reconstruction projects like the Ike Dike in Texas or the proposed HATS seawall to protect New York harbor is strong, especially since a large percentage of people live in coastal regions. Similar projects from Jakarta to Lagos use tens of billions of public dollars to protect private real estate, encouraging further development in precarious flood zones. As sea levels rise, each coastal armament will need to be reconstructed at ever-greater cost, incentivized by the increasing capital investment and population rising behind the coastal defenses (Gould & Lewis, forthcoming). Massive structural mitigation threatens to counter some of the positive impacts that insurance industry withdrawal might otherwise generate.

There are some mechanisms through which climate change can be conceptualized as an ecosystem defense to slow the treadmill of production that is rapidly degrading it—in other words, the second contradiction of capitalism (O'Connor, 1988). As supply chains are disrupted by droughts, storms, fires, and sea-level rise, the global economy is likely to slow its pace of growth. Unreliable supply chains may incentivize some relocalization of production and distribution. Agricultural failures will similarly depress economies, albeit with devastating social costs. It remains to be seen if the economic opportunities created by climate change are less than the economic costs incurred by climate disruption.

Material changes are essential, but so are ideological changes. A degrowth/sufficiency mindset can be witnessed in many variants around the world, including Indigenous cosmologies of *buen vivir* and *sumak kawsay*. The generation of conscious constituents who appreciate these alternative ways of thinking will be essential for supporting policies. There are segments that can be tapped in unexpected places, such as religious groups who consider their roles as earth stewards (e.g., the creation care movement). Diverse groups need to find commonality to work together for these causes.

It is difficult to order which actions need to occur first because we are late in the game. The simplest way to express what needs to happen is "all of the above, and fast." That said, the key obstacles to climate action in the United States are the domination of the political system by concentrated private capital interests, and the domination of the information environment by concentrated private capital interests. In both cases the concentration of power generated by extreme economic inequality allows a handful of billionaires and corporations to define the parameters of climate policy action. Policies aimed at reducing economic elite control of politics and information flows would make all else that needs to happen more possible.

Institutionalized political actors at the national (US) level need to reduce economic inequality by ensuring social welfare and economic redistribution while decoupling the political system from the economic system. Specific strategies could include overturning Citizens United to lessen corporations' power in the political system; campaign finance reform; renewing the fairness doctrine so that corporate "news" cannot enrage citizens into inaction; and providing universal basic income. In Europe, the EU is already a leader in work-life policies, maintains lower levels of economic inequality, and has adopted policies like the European Green Deal that begins to rebalance the environment and economy. The EU could use its influence to push other nations to follow their lead. This includes petro-states, whose profits are not well distributed within their countries (see chapter by Al-Sarihi in this volume). With the coming end to oil, they will need to diversify economically. Nations of the Global South should continue to push for debt forgiveness so that their cycles of debt do not push them into cycles of resource extraction in the name of growth that don't benefit their populations. Recent United Nations agreements around debt forgiveness and loss and damage move this in the right direction, though it needs to happen faster (see chapter by Khan and Khan in this volume).

Institutionalized public actors at subnational levels have key roles to play in laying the infrastructure for electrification. In *Climate Change as Class War*, Matthew T. Huber makes the case that the organized working class would benefit substantially from an electrification-based decarbonization of the economy. Organized labor can use its power at the point of production to promote a green-jobs-oriented process of decarbonization that shifts the benefits of growth toward the working class. Increased labor power will be necessary to decrease economic inequality and open the political opportunity structure for the environmental, health, and social welfare demands of labor.

Noninstitutionalized actors have two huge jobs. The first is to keep up the pressure on institutionalized actors by raising the cost of business as usual. The second is to convince people that a better world is possible—to demonstrate that economic growth does not lift all boats and that there is more to life than markets. As Brechin and Fenner (2017, p. 408) point out in a discussion of Karl Polanyi's work in relationship to environmental sociology, "[E]conomic activities, exchanges, the market, arose to serve the needs of people and society, and not the other way around, with society and people serving the needs of the market, as under liberalism.... To do otherwise treats people and nature as simply objects for sale in the market, nothing more, leading inevitably to social and ecological dislocation." Polanyi's analysis and the treadmill of production theory both conclude that it will be the actions of social movements that force market regulation for the good of people and the planet.

Note

1. Here "citizen" is used to indicate a participant in a society, not formal legal citizenship. The term "citizen-worker" was intended to highlight the role of individuals as both political participants and economic participants. The term replaced "labor" in treadmill analyses, which reduced nonelite participants in a society to their economic roles as workers. Here the term "citizen-worker" includes those with formal legal citizenship as well as those participants and inhabitants of a society without legal citizenship. This is commonly the second definition of "citizen" as an inhabitant of a particular place.

References

Almeida, P. (2019). Climate justice and sustained transnational mobilization. *Globalizations, 16*(7), 973–979. https://doi.org/10.1080/14747731.2019.1651518

Almeida, P., González Márquez, L. R., & Fonsah, E. (2024). The forms of climate action. *Sociology Compass*, e13177. https://doi.org/10.1111/soc4.13177

Bell, S. (2021.) Energy, society, and the environment. In K. A. Gould & T. L. Lewis (Eds.), *Twenty lessons in environmental sociology* (3rd ed., pp. 157–196). Oxford University Press.

Bernal, A. M., & Holst, J. (2023). Ecuador votes to keep Yasuní oil in the ground in historic referendum. *NACLA*, August 23. https://nacla.org/ecuador-yasuni-oil-historic-referendum

Brechin, S. R., & Fenner, W. H. (2017). Karl Polanyi's environmental sociology: A primer. *Environmental Sociology*, 3(4), 404–413. https://doi.org/10.1080/23251042.2017.1355723

Brostrom, I. E., Flores, R., Alatriste-Diaz, K., Luna, E., Fonsah, K., Juarez, J., Guerrero, A., Arias Murillo, E., Arana, E., & Almeida, P. (2024). *Kern strategic workforce development report*. UC Merced Community and Labor Center.

Brulle, R. J. (2021). Networks of opposition: A structural analysis of US climate change countermovement coalitions 1989–2015. *Sociological Inquiry, 91*, 603–624.https://doi.org/10.1111/soin.12333

Buck, H. J. (2019). *After geoengineering: Climate tragedy, repair, and restoration*. Verso Books.

European Union. (n.d.). The environment. https://european-union.europa.eu/priorities-and-actions/actions-topic/environment_en

European Union. (2022). EU environment action programme to 2030. https://eur-lex.europa.eu/legal-content/EN/TXT/?uri=LEGISSUM:4591047

Falcao, T. (2021). EU Courts use precautionary principle to prevent climate change. *Tax Notes International*, p. 199. https://ssrn.com/abstract=4193058

Farrell, J. (2019). The growth of climate change misinformation in U.S. philanthropy: Evidence from natural language processing. *Environmental Research Letters, 14*(3), 1–10. DOI: 10.1088/1748-9326/aaf939

Fitzpatrick, N., Parrique, T., & Cosme, I. (2022). Exploring degrowth policy proposals: A systematic mapping with thematic synthesis. *Journal of Cleaner Production, 365*, 132764. https://doi.org/10.1016/j.jclepro.2022.132764.

Gould, K. A. (1994). Legitimacy and growth in the balance: The role of the state in environmental remediation. *Industrial and Environmental Crisis Quarterly, 8*(3), 237–256.

Gould, K. A. (2007). The ecological costs of militarization. *Peace Review, 19*(3), 331–334.

Gould, K. A., & Lewis, T. L. (2021). Resilience gentrification: Environmental privilege in an age of coastal climate disasters. *Frontiers in Sustainable Cities*, https://doi.org/10.3389/frsc.2021.687670

Gould, K. A., & Lewis, T. L. (Forthcoming). The consequences of public investments following coastal climate disasters: An assessment of four strategies post-Superstorm Sandy. *Human Ecology Review*.

Gould, K. A., Lewis, T. L., & Roberts, J. T. (2004). Blue-green coalitions: Constraints and possibilities in the post 9-11 political environment. *Journal of World Systems Research, 10*(1), 91–116. https://doi.org/10.5195/jwsr.2004.314

Gould, K. A., Pellow, D. N., & Schnaiberg, A. (2004). Interrogating the treadmill of production: Everything you wanted to know about the treadmill but were afraid to ask. *Organization and Environment, 17*(3), 296–316. DOI: 10.1177/1086026604268747

Gould, K. A., Pellow, D. N., & Schnaiberg, A. (2008). *The treadmill of production: Injustice and unsustainability in the global economy.* Paradigm.

Huber, M. T. (2022). *Climate change as class war.* Verso.

IEA. (n.d.). Russia's war on Ukraine. https://www.iea.org/topics/russias-war-on-ukraine

Intergovernmental Panel on Climate Change. (2023). *AR6 synthesis report: Climate change 2023.* Interlaken, Switzerland. https://www.ipcc.ch/report/ar6/syr/downloads/report/IPCC_AR6_SYR_LongerReport.pdf

Kallis, G., Kostakis, V., Lange, S., Muraca, B., Paulson, S., & Schmelzer, M. (2018). Research on degrowth. *Annual Review of Environment and Resources, 43*(1), 291–316.

Kothari, A., Demaria, F., & Acosta, A. (2014). Buen Vivir, degrowth and ecological Swaraj: Alternatives to sustainable development and the green economy. *Development*, 57(3–4), 362–375.

Leguizamoń, A. (2020). *Seeds of power: environmental injustice and genetically modified soybeans in Argentina.* Duke University Press.

Leguizamón, A. (2023). Treadmill of production. In M. Long, M. Lynch, & P. Stretesky (Eds.), *Handbook on inequality and the environment* (pp. 11–27). Elgar.

Lewis, T. L. (2016). *Ecuador's environmental revolutions: Ecoimperialists, ecodependents, and ecoresisters.* MIT Press.

Lewis, T. L. (2018). Globalizing the treadmill of production: A solutions-oriented application to Ecuador. *Environmental Sociology*, 5(3), 219–231. https://doi.org/10.1080/23251042.2018.1514942

Malm, A. (2021). *How to blow up a pipeline.* Verso.

Maslin, M. A., Lang, J., & Harvey, F. (2023). A short history of the successes and failures of the international climate change negotiations. *UCL Open Environment*, 5. https://doi.org/10.14324/111.444/ucloe.000059

Obach, B. (2004). *Labor and the environmental movement: The quest for common ground.* MIT Press.

O'Connor, J. (1988). Capitalism, nature, socialism a theoretical introduction. *Capitalism Nature Socialism, 1*(1), 11–38. https://doi.org/10.1080/10455758809358356

Roberts, J. T., & Parks, B. C. (2007). *A climate of injustice: global inequality, north-south politics, and climate policy.* MIT Press.

Schnaiberg, A. (1980). *The environment: From surplus to scarcity.* Oxford University Press.

Schnaiberg, A. (2010). The economy and the environment. In N. J. Smelser & R. Swedberg (Eds.), Handbook of Economic Sociology (pp. 703–726). Princeton University Press.

Schnaiberg, A., & Gould, K. A. (1994). *Environment and society: The enduring conflict.* St. Martin's Press.

Simpson, B., Willer, R., & Feinberg, M. (2022). Radical flanks of social movements can increase support for moderate factions. *PNAS Nexus, 1*(3), 1–11. https://doi.org/10.1093/pnasnexus/pgac110.

Stuart, D., Gunderson, R., & Petersen, B. (2020). *Climate change solutions: Beyond the capital-climate contradiction.* University of Michigan Press.

Van Dyke, N., & Amos, B. (2017). Social movement coalitions: Formation, longevity, and success. *Sociology Compass, 11*(7), e12489.

Whyte, K. (2016). Why the Native American pipeline resistance in North Dakota is about climate justice. *The Conversation*, September 16. https://theconversation.com/why-the-native-american-pipeline-resistance-in-north-dakota-is-about-climate-justice-64714

York, R. (2012). Do alternative energy sources displace fossil fuels? *Nature Climate Change*, 2, 441–443. https://doi.org/10.1038/nclimate1451

CHAPTER 6

The Logics of Antiblackness and Coloniality in the Push for Climate (In)Action

MARILYN GRELL-BRISK

Introduction

> The problem is that those who need to make the decisions are kicking the can down the road, and they believe that they can because *they're not seeing us*. They see themselves. And for them, they don't reach that period of peril probably for another 15, 20 years. I mean, you heard this morning, 4 degrees to Shanghai and Miami to go—to be eradicated. Well, it's 1.5 and 2 for us. So, there are a lot of us that are going to be affected before Shanghai and Miami.
>
> —Mia Mottley,[1] prime minister of Barbados

In its sixth Assessment Report (AR6) in 2023, the Intergovernmental Panel on Climate Change (IPCC) crystallized the significance of global warming and climate change to our current global sociopolitical economic crises. Climate change, the report writers note, is unequivocally anthropogenic and "has caused widespread adverse impacts and related losses and damages to nature and people that are unequally distributed across systems, regions and sectors; has caused substantial damages, and increasingly irreversible losses, in terrestrial, freshwater, cryospheric, and coastal and open ocean ecosystems; and some ecosystems are approaching irreversibility" (IPCC Core Writing Team et al., 2023, p. 5). From climate change deniers to climate action activists, everyone has a solution for the ever-apparent crises of global warming and climate change. And while it has become obvious that the impending catastrophe *is* all about inequality, as Timmons Roberts (2001) articulates in *Global Inequality and Climate Change*, I argue that any climate action—be it concerned with adaptation, resiliency, mitigation, or readiness—must come to terms with the sociocultural and political-economic impulses that bolster and structure the Western modernity project and therefore fuel the impending climate catastrophe.

Climate action must attend to the antiblackness[2] and coloniality that are constitutive of modernity, as well as the industrialization imperative of our times and the resulting scourge across the globe. This chapter examines the logics of antiblackness (Dumas, 2016; grell-brisk, 2022, 2023; Vargas, 2018; Vargas & Jung, 2021) and coloniality (Escobar, 2007; Maldonado-Torres, 2007; Quijano, 2007), and how they undermine veritable attempts at addressing climate change in a meaningful way. That is, when measures are taken to combat the deleterious effects of climate change, if they are taken at all, they overlook or downright ignore those most historically marginalized, such as Black and Indigenous people.[3] And through coloniality, they uphold the systems of oppression that maintain Blackness as abjected and relegated to the bottom of a hierarchy of being and power. This leads to ineffective climate change solutions, adaptation, and/or mitigation. Further, in much of the discourse on the climate crisis, antiblackness and coloniality are treated perfunctorily rather than as integral features of the analysis.

This failure to address the logics of antiblackness and coloniality in the discourse on climate change leads to inaction in the face of suffering and devastation as a result of human-driven global warming and/or misguided and even defective approaches to climate action. This is most obvious when we observe the response of countries like the United States or institutions like the International Monetary Fund (IMF) to the continued devastation of the Caribbean islands during the past 10 tropical cyclone seasons. In 2019, Category 5 Hurricane Dorian, perhaps the worst natural disaster the Bahamas ever experienced (The Guardian Staff, 2019), ravaged the islands. The United States' response to this disaster was contemptible. The Bahamas are only 50 miles off the coast of Florida. When survivors of Dorian attempted to temporarily relocate to the United States, they were initially told they could evacuate to the United States if they were in possession of a valid Bahamanian passport and a clean police record. While on the ferry, Bahamanian nationals were told to disembark (Mansoor, 2019) and only holders of US visas could continue on to safety. After suffering through two consecutive Category 5 hurricanes, Irma and Maria in 2017, and having just barely recovered from Tropical Storm Erica, the Caribbean Island of Dominica was forced to depend on its Citizenship by Investment program funds to help recover and fund climate change mitigation and adaptation development programs. In 2017, the IMF rightly cautioned against dependence on citizenship by investment due to the economic and political precarity of the program (International Monetary Fund, 2017). And yet, in 2023, the IMF recommended that the funds raised through Dominica's Citizenship by Investment program be earmarked to support posthurricane relief and climate change adaptation and resilience efforts (IMF Executive Board, 2023), tacitly approving the high-risk program undergirding the Dominican economy. This meant that the very program that was previously deemed highly unstable was now recommended as a possible fix for climate change and its deleterious effects on the country.

For both the United States (nation state) and IMF (international organization), the wholesale treatment of Caribbean nations and their people is one of indifference and negligence, a lack of care, and a lack of accountability and responsibility, particularly given North America's share in global carbon emissions, which are the number-one driver of global warming. The push for supposed "green technology"[4] takes on a pattern of coercive extractivism of natural resources in Africa and Latin America. Of note here is the extraction of lithium, cobalt, manganese, and other natural resources that support the production of electric vehicles (EVs), which are supposed to reduce dependence on oil and natural gas and reduce carbon emissions. EVs are predicted to make up about half of all new car sales in the United States by 2030 (Castelvecchi, 2021; Sisson, 2023), and in 2020, California's governor issued a mandate that all new cars and passenger trucks sold in California by 2035 be zero-emission vehicles. EVs are generally powered by lithium-ion batteries, as lithium is highly reactive, allowing batteries made with it to be very efficient. Lithium batteries are made up of cathodes containing metal oxides composed of nickel, manganese, and cobalt, with cobalt being particularly useful as it reduces overheating and extends the life of the battery (Castelvecchi, 2021; Searcey & Lipton, 2023; Sisson, 2023). With the rise of EVs came the increased demand for lithium carbonate and for cobalt. The Democratic Republic of the Congo, Angola, and Chile are major sources of cobalt and of lithium (Mniga, 2024; Sisson, 2023). But as Sisson and others have pointed out, there are significant environmental (mining, including that of the seabed in Chile, disrupts ecosystems), political (corruption within both large-scale and artisanal mining reverberating in the political system), and social costs (forced labor, including that of children).

More recently, in the western United States, we have also seen the destruction of life, habitat, and property from increasingly intense wildfires. Increasingly warmer temperatures and shifting precipitation patterns have transformed the landscape in the western United States, making it more susceptible to high-severity fire (Cunningham et al., 2024; National Aeronautics and Space Administration, 2025; Wasserman & Mueller, 2023). This impacts landscape processes and ecosystem functioning, leading to changes in vegetation structure and composition (Cunningham et al., 2024). The increasing frequency and magnitude of extreme fire events have been linked to climate change (Cunningham et al., 2024; National Aeronautics and Space Administration, 2025; Wasserman & Mueller, 2023). In January 2025, we witnessed the damage caused by climate change when coupled with the spatial arrangement of difference here in Southern California. As the wealthy Pacific Palisades burned, the small historically Black Altadena (70% of Black residents in Altadena are homeowners[5]) burned, too, but while the loss was extreme in both cases, Altadena's was a culmination of both climate change and institutionalized, structural, and systemic inequalities. A University of California, Los Angeles, report noted that 61% of Black households in Altadena were located within the fire's perimeter, compared to 50% of non-Black households, and nearly half of Black households were destroyed or sustained major damage, compared to 37% of non-Black households. Also, 57% of Black homeowners in Altadena are older than age 65 and may be especially vulnerable to incomplete or insufficient insurance coverage or predatory financial scams as they navigate the process of rebuilding or restoring their homes. Finally, rising property values and preexisting barriers to homeownership for Black buyers could prevent younger Black people from buying in the area and accelerate the decline of Altadena's Black community (Bonner, 2025). The result of this is a disruption and perhaps the demise of any intergenerational wealth production among Black families in that area (Bonner, 2025; Talabong, 2025). In neighborhoods and cities like Altadena, post disaster, you see an influx of land vultures feeding on the despair and loss of their residents.

In all the instances mentioned here, what we are observing is the practice of spatial Black abjection,[6] and the manifestation of the limit of care and consideration for the impact of the ongoing climate catastrophe on historically marginalized communities. This is coupled with the emergence of a vigorous market for climate change funding, from the Green Climate Fund[7] and its facsimiles to carbon offset programs, which seek to address some of these inequities but are limited at best and more likely exploitative. Embedded within the logic of antiblackness is Black disposability (grell-brisk, 2022, 2023) wherein Black selves and bodies are understood as readily accessible and available to be extracted from and easily discardable when they are no longer useful. If Black and Brown bodies are seen as disposable and are spatially differentiated based on a hierarchy of being that places them at the lowest points, then it is easier to dismiss or disregard their suffering. Through regimes of disposability, the law (e.g., visa requirements), trade regulations and agreements, and institutions (e.g., Washington Consensus,[8] IMF, and World Bank) or social policies (e.g., redlining), Black and Brown bodies are domesticated, used, ignored, and disposed of.

The Antiblack Geography of the Ongoing Climate Crisis

> We are scared here, and angry too! It was Europe, and America, and China, who caused this heat wave, not us! I know we have burned a lot of coal in the last few decades but it's nothing compared to the West. And yet we signed the agreement to do our part, which we have done. But no one else is fulfilling commitments. No one is paying the developing nations, and now we have this heat wave. And another one could happen next week. Everyone knows but no one acts. ... So, [India] had burned coal, yes, like everyone else had up until just a few years before, now India was being told not to burn coal when everyone else had finished burning enough of it to build up the capacity to afford to shift to cleaner sources of power. India had been told to get better, without any financial help to do so whatsoever. Told to tighten the belt and embrace austerity, and be the working class for the bourgeois of the developed world and suffer in silence until better times came; but better times could never come. That plan was shot. The deck had been stacked. The game was over and now 20 million people were dead.
>
> —Kim Stanley Robinson, the Ministry of the Future[9]

Coloniality of being and power, the "afterlives"[10] of colonialism, provides a critical and analytical lens for us to understand how the patterns of power and exploitation, and an operating hierarchy of being (Mignolo, 2011; Quijano, 1989, 2007; Wynter, 2003) established through the Western modernity project (Foucault, 1994) continue to compromise the work that climate activists do. Despite having eliminated formal colonial rule, the asymmetric power distribution that defined that period and informed our understanding of culture, labor relations, knowledge production, and sociality endures beyond formal colonization. This is buttressed by antiblackness, a constituent element of coloniality of being and transhistorical in scope, wherein through modernity, man—White, cis, heteronormative, male, and rich—is regarded as the inherent stand-in and exemplar for human and by which "the other" is measured. This other's human(ness) is assessed by this exemplar, with Black(ness) being the furthest from that. It is also what supports the modern capitalist world economy (Quijano, 2007; Quijano & Wallerstein, 1992).

Therefore, the incapacity to engage in meaningful climate action goes beyond environmental racism, which Dorceta Taylor rightly finds to be the discrimination or racial advantages that are the result of environmental decisions, actions, and policies based on the execution of racial prejudice and discrimination, coupled with power, privilege, and institutional support, through business, as well as the political, legal, penal, and education institutions (Taylor, 2018). And while environmental racism results in the increased likelihood of being exposed to environmental hazard, disproportionate negative impact of environmental policies and processes on Black and Brown people, and the deliberate targeting and siting of noxious facilities in their neighborhoods (Bullard, 2001; Taylor, 2014, 2018), as formulated earlier, it only begins to get at the structural roots of the problem. However, it does not sufficiently address the continued dehumanization and the disposability regimes levied against Black, Indigenous, and non-White people across the globe. Grassroots climate organizers are intuitively grasping that many climate actions are inadequate. Bronx Climate Justice North notes, "without a focus on correcting injustice, work on climate change addresses only symptoms, and not root causes" (Cohen et al., 2020, p. 3).

In contemplating environmental racism and the environmental justice[11] movement, Laura Pulido (2017) theorizes that environmental racism is intimately tied to the prevailing capitalist economic structure that underpins our everyday lives. This marks an important shift analytically toward the source of climate inaction or nonsustainable, impermanent changes that are supposed to address the climate crisis. For Pulido, environmental racism is constituent of racial capitalism (race as a structuring logic of capitalism). Drawing on the works of Black Studies scholars like Cedric Robinson (1983),[12] critical geographers like Ruth Wilson Gilmore (2002), and critical ethnic studies scholars like Jodi Melamed (2015), Pulido does two things: critiques the environment justice movement for what she identifies as the "environmental racism gap" or the "persistent[13] *inequality* between white and nonwhite communities" (2017, p. 525), and in the process, Pulido also operationalizes the connection between environmental racism and racial capitalism. For Pulido, when Sidney Howe, director of the Human Environment Center, coined the term "environmental justice" in the late 1970s to name the steps needed to correct the disproportionate burden of air, noise, and water pollution on the poor, what was missing in the formulation was the centrality of race. She contends that racial capitalism creates value through the production of social difference, specifically the devaluation of non-White bodies, hence the environmental racism gap.

Therefore, even within the environmental justice movement, and among those pushing for action on climate change, we continue to see the leveraging of disposability regimes against Black, Indigenous, and Brown people, missing the mark on effective climate action across the globe. For example, despite the enactment of the landmark National Environmental Policy Act (NEPA) in 1970, which requires federal agencies to assess the environmental effects of their proposed actions prior to making decisions, exclusionary urban planning and predatory banking practices mean that this policy largely failed to meet the needs of Black and Brown people (Adisa-Farrar, 2023). This is how Black residents of Altadena were significantly affected by the wildfires in January 20205. NEPA is supposed to ensure that the federal government uses all practicable means to create and maintain conditions under which humans and nature can exist in productive harmony while also protecting communities from dangerous and poorly planned projects. However, Black and Brown communities continue to be have disproportionate siting for toxic waste incinerators, dangerous oil pipelines, and petrochemical plants (such as Cancer Alley in Louisiana). Consistent with Pulido (2017), when organizing around environmental and climate issues, an antagonistic stance in relation to the state is absolutely necessary. Black and Brown bodies are not accounted for in the formula of human-nature productive harmony. In our organizing, we must center environmental and climate justice, because the state is deeply implicated in this devaluation process. Instead of leveraging its mandate on regulatory and legal power to curtail environmental racism, implement just environmental and climate policies, and advocate on behalf of its citizens most adversely impacted by climate change, it has engaged in "[performing] regulatory activity" (Pulido, 2017, p. 529). As such, we should regard the state as an integral part of the current apparatus that is nonproductive for Black and Brown people and therefore a site of contestation if we want forward momentum and concrete action on the issue of climate change and climate justice.

The climate justice movement, which sees climate change as "an issue of justice and human rights, a question of equity over time and space ... that dangerously intersects class, race, gender, caste, sexuality and bodily abilities, among others" (Michael et al., 2023, p. 4), moves closer to the framework necessary to get *to the source*, to borrow a phrase from Amilcar Cabral (1973), of the current climate catastrophe. Alesia Montgomery and Marcos Magaña (2025), for example, urge us to consider a power dimension when attempting to understand the diminished capacity of racialized communities to manage or effectively address increasingly severe heat waves that kill them. Still, it is the climate justice framework (Michael et al., 2023; Porter et al., 2020; Schlosberg & Collins, 2014; Sultana, 2022) that brings coloniality into the logic of the current contours of climate (in)action. It emphasizes questions of who benefitted the most from colonialism and, by extension, the economic development that accelerated climate change and the climate crisis. And rightly so. The world-systems perspective is also attentive to these logics, given that the current world order is sustained through unequal exchange, among other durable sociocultural and political structures in the global ordering of wealth and power inherited from the formal colonial period. The global hierarchy of wealth, based on continuous extraction from peripheral countries (mostly made up of countries who had been formerly colonized by European nations), remains resistant to radical change (grell-brisk & Chase-Dunn, 2019; Wallerstein, 1974, 1979). This hierarchy remains largely in place due to the Western colonial and neocolonial projects and the constant, endless accumulation of capital for capitalists (Wallerstein, 1999) driving climate collapse. Although climate justice scholars tend to focus on differential vulnerabilities (as do world-system scholars who insist that the unit of analysis should be the world-system itself and not nation states) between countries or the historical responsibilities at the global level, it is also productive to look at the local and regional (Fisher, 2015) dimensions of the effects of climate change and how it presents spatially.

For this analysis, it is essential that we understand that coloniality and antiblackness work as structuring logics of the modern world-system with their capitalist economies. When we do, we can begin to grasp the racialized cartography of climate (in)action, and we can make sense of why and how, for example, the United States has very little interest in taking structural and sustainable measures to curtail its anthropogenic emissions, join the Paris Agreements, or take a progressive stance on our dependence on fossil fuels. As the Biden administration claimed a need for "green" or "clean" energy, it also noted a commitment to investing even more in domestic oil production to reduce the United States' dependence on foreign oil (Daly, 2024). Despite incorporating several progressive funding initiatives for frontline communities, Biden's Inflation Reduction Act of 2022 handed a windfall to the fossil fuel industry (M. Huber, 2022b). In 2015, when the United States, under Donald Trump's first regime, abandoned its commitment to the Paris Agreements, it was one of a small handful of countries that failed to sign the agreement. This list included Syria, which was embroiled in war and turmoil and therefore unable to commit to climate action, and Nicaragua, which believed that given its capacity to provide low-carbon-emitting energy to its citizens and their own contribution to global carbon emissions in relation to other countries like the United States, it did not need to sign the international agreement (Rhodes, 2017).

According the Our World In Data, the United States's share of global cumulative CO_2 emissions as of 2023 was 23.83% compared to China's at 15.04%, Canada's at 1.94%, Syria's at 0.11%, Nicaragua at 0.01%, and the Caribbean islands of St. Lucia, St. Kits & Nevis, and Dominica at <0.01% respectively.[14] Caribbean leaders have repeatedly called out the asymmetry in pollution contribution (or ecologically unequal exchange[15]) and have unequivocally underscored the resulting disparity between those of us who are on the frontline and negatively affected by this pollution. They have also pointed to the lack of accountability on the part of these major polluters like the United States. In 2017, reeling from devastating loss exacted by Hurricane Maria and two years after enduring the destructive Tropical Storm Erica, Dominica's prime minister addressed the United Nations General Assembly appealing to countries like the United States to take responsibility for their contribution to global carbon emissions, noting that the Caribbean islands were on the frontline of the effects of climate change. He charged: "We in the Caribbean do not pollute or overfish our oceans. We have made no contribution to global warming that can move the needle. But yet, we are among the main victims ... on the frontline! I repeat—we are shouldering the consequences of the actions of others!" (Skerrit, 2017, pp. 4, 12). More recently, in an interview with Christiane Amanpour, Barbados Prime Minister Mia Mottley not only spoke to the question of disparities in carbon emissions and the asymmetric effect on the islands but also the underlying disposability that informs the decisions that world leaders are making vis-à-vis climate change:

MOTTLEY: And you have not heard me say floods or hurricane, and those are the other ones as well. This year, Barbados had its first hurricane in 66 years. And before that, we had a freak storm for 90 minutes that had 46,000 lightning strikes in 90 minutes.

AMANPOUR: Mia, do you find it difficult to convince the people who you need to convince that this is all due to climate change?

MOTTLEY: I think people are getting there. The problem is that those who need to make the decisions are kicking the can down the road, and they believe that they can because *they're not seeing us. They see themselves.*[16] And for them, they don't reach that period of peril probably for another 15, 20 years. I mean, you heard this morning, 4 degrees to Shanghai and Miami to go—to be eradicated. Well, it's 1.5 and 2 for us. So, there are a lot of us that are going to be affected before Shanghai and Miami.

(Mottley & Amanpour, 2021)

On January 20, 2025, at the beginning of Trump's second regime, the United States again initiated the process of withdrawing from the Paris Agreements (Neuberger et al., 2025). Then, like today, this is a clear abdication of the United States' responsibilities as a polluter and global leader. The discourse that is presently emerging on the United States's objection to taking responsibility for its share in global carbon emissions, and the adverse effects of anthropogenic climate change on the planet, as well as the implications of the United States's withdrawal from the Paris Agreements, is instructive. Concerns are centered not on the impacts of the lives of Black, Indigenous, and non-White people living in the Caribbean or around the globe, but on the voluntary carbon markets (VCMs), carbon offset programs (Neuberger et al., 2025), and the United States' influence and leadership in the clean energy race and market (McBride, 2017).

Our lives, our being(ness), are not centered or prioritized in the calculus even in the climate change mitigation and adaptation discourse. For many on the "global left," one way to motivate action around climate change is to create economic incentives. The United Nations (UN), one of our leading global governance institutions, through the Paris Agreement established the UN Carbon Offset Platform to incentivize companies and individuals to take steps toward net-zero emissions in support of climate mitigation efforts. The institution, via the platform, issues certified emission reductions (CERs) or carbon credits which individual, businesses, and organizations can purchase "to reach carbon neutrality by compensating their footprints, while they help project developers to continue to finance their green projects. The prices per unit are set by project developers, who also receive the full profits from the sale of the units directly" (United Nations Framework Convention on Climate Change, 2022). These credits are traded from various projects certified and verified by carbon standards such as Verra, Gold Standard, American Carbon Registry, and Climate Action Reserve (E. Huber et al., 2024; Jennifer, 2023). This voluntary carbon market allows organizations and companies to outsource their decarbonization, permitting them the opportunity to sidestep the complete transformation of their business away from fossil fuels and claim contribution to carbon neutrality (CarbonCredits.Com, 2021; Trencher et al., 2024; United Nations Climate Change, 2024; United Nations Framework Convention on Climate Change, 2022). This has generated new forms of speculative trading and is becoming part of the global finance sector. The real voluntary carbon market was valued at $2 billion as of 2022 (CarbonCredits.Com, 2021; Trencher et al., 2024) and continues to expand, with Bloomberg noting that by 2050, the carbon offset market could be valued at half a trillion dollars annually (BloombergNEF, 2023). Globally, climate change mitigation has been financialized, and it is the financialization that is prioritized in the conversation about the United States' rejection of the Paris Agreements.

At the same time, there has been considerable discussion as to the effectiveness of carbon offsets in addressing decarbonization (Álvarez, this volume; Pande, 2024; West et al., 2020, 2023). Fundamentally, if your company is largely dependent on fossil fuels and your only mitigation practice or efforts come from purchasing offsets from a "green project," you remain a polluter and an obstacle for global efforts to net-zero emissions. Danny Cullenward et al. argue that offsets could work only if most countries were trying to reduce emissions on the margin as envisioned under the Kyoto Protocol, but based on our current context, "planetary temperature stabilization under the Paris Agreement requires global CO_2 emissions to fall to near-net-zero levels. This goal cannot be achieved when governments and companies justify their own ongoing pollution by paying someone else not to pollute" (2023, p. 1085). For those who argue that both offsetting and active mitigation measures by companies are important in today's context, carbon credits do not work if the quality of the offset is poor or substandard. This means that if the supposed emissions avoided by the green project do not equate with the amount that is supposedly being offset, then the quality of the credit is poor. And while there are established standards regulating the offsetting projects and the associated carbon credits, there is concern about quality consistency (E. Huber et al., 2024), access to high numbers of low-quality offsets (Trencher et al., 2024), and particularly the overestimation of offset credits like Reducing Emissions from Deforestation and Forest Degradation (REDD+) "green projects" in the Brazilian Amazon (West et al., 2020, 2023).

All of this subverts climate mitigation through CERs. In March 2025, the World Federation of Exchanges claimed that the voluntary carbon markets were not as efficient as mainstream markets (WFE Research Team, 2025). In April 2024, the board of the Science Based Targets initiative (SBTi) decided to include voluntary carbon credits in scope 3,[17] which made it more challenging for companies to reach climate targets as scope 3 includes emissions up and down a company's value chain. Some were outraged, even within the SBTi itself with a staff member being quoted in the *Financial Times* as saying "this decision would not incentivize companied to reduce their emissions; it would be a perverse incentive to just buy cheaper carbon credits" (Pande, 2024, p. 2). Note that the economic (dis)incentive is prioritized here, as opposed to the actual reduction in carbon emissions. It becomes clear that we cannot rely on these schemes and must be radical in our approach to addressing climate change; we must, as Davis urges, "adopt a revolutionary stance ... we must get to the root" (Davis, 1989, p. 14).

While countries like the United States and large corporations shirk their responsibility and obligations around their own pollution and seek to profit from climate change mitigation, the costs—to the planet, humans, and the material conditions of life on Earth—continue to increase astronomically every year for nation states like those in the Caribbean. The warming of the Atlantic primarily due to anthropogenic carbon emissions has led to an increased number and intensity of tropical cyclones per year (Cook et al., 2016; Emanuel et al., 2008; Slangen et al., 2016; Strauss et al., 2015). Based on the data collected by the *Global Carbon Project*, the United States' and Canada's emissions in the region are exponentially higher than that of the other countries combined and yet the United States' decision to join the Paris Agreement on Climate Change and take action to reduce its emissions is dependent on the politics of its executive branch and their perspective on whose lives are important (Guerrero, 2001), whose lives matter. The Caribbean as an example takes on even deeper meaning if we consider the region as central not only to colonial resistance movements but the recasting of Blackness as a "banner for unity and mobilization around a common project of freedom and equality that defied racial and economic injustice worldwide" (Guerra, 2014, p. 1). The management of climate change and lack of regard for its effects on the region are an extension of modernity's construction of an antiblack world and a signal of foreclosure for the possibilities of Black futures. Still the Caribbean islands persist and are at the forefront of climate adaptability programs as we will discuss later in the chapter. Action on climate change must address coloniality and antiblackness. Antiblackness and coloniality is intimately linked to spatiality, with entire regions being written off in the calculus of worthiness and value.

The Economics of Climate Catastrophe

> Civilizations are shaped and defined by their priorities: by deciding what things and what relations are valuable. Their rules of reproducing power and wealth turn on these choices of what is—and what is not—valuable.
>
> The Anthropocene makes for an easy story. Easy, because it does not challenge the naturalized inequalities, alienation, and violence inscribed in modernity's strategic relations of power and production. It is an easy story to tell because it does not ask us to think about these relations *at all*.
>
> — Jason W. Moore, *Capitalism in the Web of Life*[18]

Coloniality and antiblackness are intertwined with capitalist accumulation, and scholars have found it easy enough to link climate change to the extractiveness of capitalism (M. Huber, 2022b; Satgar, 2018). While coloniality has found its place in critical theory spaces in the academe, scholars remain ambivalent to the idea that antiblackness is constitutive of the modern capitalist world-system. Eco-Marxists who would ostensibly draw out the relationship between the exploitation of humans and nature for capitalist production get bogged down in the details of what counts as an authentically Marxist understanding of the climate catastrophe and a way out, whether it is through "degrowth" (Hickel, 2019; Schneider et al., 2010) or the notion that capitalism's contradictions will eventually lead to its end (Bastani, 2019; Mason, 2015). For some, degrowth, which is not the opposite of economic growth but a reduction of material or biophysical throughput (Kallis, 2011, 2018; Kangal, 2024; Martinez-Alier, 2009; Schmelzer et al., 2022; Victor, 2008) and energy demands (Grublier et al., 2018; van Vuuren et al., 2018) in wealthy countries, sounds like austerity measures and a "politics of less" (M. Huber, 2022a), which is of course automatically rejected. For degrowth's detractors it is rejected because it subscribes to a notion of growth and wealth that is narrow and limited to that which benefits only the wealthy, and it is translated as increased GDP/economic activity. There is no space for wealth redistribution or for a more equitable theory of "socialist growth," and it does not account adequately, for "specific social processes within a social structure that allocates different amounts of power to individuals to pursue their material interests" (Brooks, 2023, p. 7).

The sticky part in the logic of degrowth is that within nation states, whether Global North or South, antiblackness reigns, and the effects of climate change or ecological catastrophes are disproportionately felt by those at the bottom of the hierarchy of being. These structural and systemic inequalities are nested within each part of the world-system, resulting in profound inequities across the globe. This highlights the unevenness with which capitalist-generated inequality is experienced throughout the planet and thus poses a particular type of challenge for climate activists motivated by the disparities of economic wealth and well-being. And while it is important to theorize climate change and possibilities for addressing them, the eco-Marxists fall into the trap of developing a dichotomized discourse of the impending catastrophe (Holgersen, 2023), missing essential aspects of the formula, such that they conclude: "We need ecosocialist transitional programs for planning, building, and organizing new hegemony and ecosocialist movement to bring it into fruition, for a world that gives priority to human needs within ecological limits" (Holgersen, 2023, p. 8). Thus, not accounting for the fact that our very systems of knowledge, power relations, and international law see large portions of the world's population as less than human, using dehumanizing language to reflect on their positions which itself has been defined by the power relation. This is even as many scholars and mainstream activists point to the asymmetric power distribution that drives the climate change movement.

Ståle Holgersen, writing for *Climate & Capitalism*, notes, "We cannot grasp global warming without understanding the dynamics of the profit motive, capital accumulation, metabolic rifts, class struggle and class fractions" (2025, pp. 3–4). For Holgersen, at the center of the climate crisis and the way forward is to articulate the movement against global warming as a class struggle. The inability to mobilize solely around class for two centuries is noted but seen as a tool of reactionary forces who attempt to divide the working class. Holgersen contends that we should not give up on the idea that at the heart of eco-socialism should be a uniting of labor and climate movements. We cannot and will not move forward without attending to the sociocultural fissures that hinder deep restructuring of our economic and political systems that preclude climate action that benefit the most marginalized on the planet. Holgersen rightly argues that the climate crisis can lead to fascist movements, but whether this happens through a coalition between fossil fuel interests and the far right, or the far right blaming the climate catastrophe on minoritized groups, we are ultimately talking about fundamental problems within our existing systems of power, wealth, and hierarchy. We are in essence talking about how we have organized ourselves in the modern world.

While the eco-Marxists discourse can get encumbered by dogmatic discussions as to whether Karl Marx was anthropocentric in his analysis because he did not view nonhuman species, beings, and things on the same ontological plane as human beings, or the Cartesian Nature/Society dualism discussion, or any of the growth-degrowth disputes on the Marxist left,[19] a robust analytical lens serves to sharpen focus, technique, and strategy to address material conditions generated by climate change. This is why John Bellamy Foster's perceptiveness on an eco-Marxist approach to the climate catastrophe is so important (2000; Foster & Keqing, 2023). It leaves room for an analysis of the climate crisis that incorporates the work of Marx and Frederick Engels to confront the ecological struggle beyond simple class/economic struggle. Foster reminds us in his interview with Jia Keqing, a research fellow at the Academy of Marxism of the Chinese Academy of Social Sciences, that Engels's *The Condition of the Working Class in England* was a groundbreaking epidemiological study of disease under capitalism, with Engels concluding that "bourgeois relations of production promoted 'social murder'" (Foster & Keqing, 2023, p. 4).

Foster's points are instructive. Engels was concerned with "the capitalist city, housing conditions, air and water pollution, the spread of disease and illnesses of all kinds, and the much higher mortality rate of the working class. … The struggles of the working class in the early nineteenth century were a product of their whole living conditions,[20] not just factory conditions" (2023, p. 4). So, yes, we must look at the working-class's entire material existence and, as Foster notes, the conditions of social reproduction. This is the juncture to articulate the connection to antiblackness because the very material conditions that Foster is pointing to depend on a structure that relies on racial differentiation and domination embedded within it. In referencing István Mészáros's[21] call for a total transformation of the system of social metabolic reproduction, Foster does acknowledge the necessity of deep structural change to address climate action effectively while also acknowledging the asymmetric benefits of the modern capitalist world-system, but antiblackness is nowhere in this conversation.

Roberts (2001), Hickel (2019, 2020), Mészáros (2001), and others point to economic inequality, inequality in carbon emissions and their impacts on marginalized people especially in the Global South. Still, while we are considering (re)organizing new hegemonies, we must remain attentive to the reality that Black, Indigenous, and Brown people around the globe are already dying, their lives upended by climate disaster, and their culture and way of life at peril and on the brink of annihilation. We are engaging in planetary destruction.

A Way Out?

> The most magnificent drama in the last thousand years of human history is the transportation of ten million human beings out of the dark beauty of their mother continent into the new-found Eldorado of the West. They descended into Hell; and in the third century they arose from the dead, in the finest effort to achieve democracy for the working millions which this world had ever seen. … [In] a day when the human mind aspired to a science of human action a history and psychology of the mighty effort of the mightiest century, we fell under the leadership of those who would compromise with truth in the past in order to make peace in the present and guide policy in the future.
>
> —W. E. B. Du Bois, *Black Reconstruction in America*[22]

Johan Rockström et al. (2009, 2024) frame the current climate crisis, not as a global one but a planetary one, triggering a shift away from management of the crisis from a shared resource perspective dependent on national boundaries and toward a planetary commons that articulates ecosystem stewardship. The global commons as it currently exists shifts governance to multiple states governing, not their individual nation states but some of the "large elements of the Earth system, namely parts of the geosphere (deep seabed), hydrosphere (high seas), cryosphere (antarctica), and the atmosphere (the climate system)" (Rockström et al., 2024, pp. 1–2). However, Rockström et al. (2024) argue that the global commons were negotiated by nation states primarily through the context of our current geological epoch. The authors write: "the planetary commons will require moving away from global commons as a means of governing resource use … to universal rules of how to collectively secure critical biophysical Earth system functions that regulate livability on Earth for everyone" (Rockström et al., 2024, p. 8). This is appealing—universal rules that strengthen planetary resilience, livability on Earth, and justice. And like broad coalitions along class lines in the struggle against climate catastrophe, if we do not contend with constituent parts of the modern world-system, antiblackness in particular, we cannot attain systemic and structural transformation.

Could *democratic* eco-socialism—"a radical practice and conception of democracy as people's power, mediated by an ethics to sustain life" (Satgar, 2018)—provide some version of a way out, at least ideologically? It could be a struggle against climate change that is fundamentally antisystemic, fueled by a justice-oriented grassroots struggle urged to move across geographic and national scales and that builds red-green alliances could have a desired effect of structural transformation. Vishwas Satgar (2018), in his introductory chapter in *The Climate Crisis*, outlines five moments and approaches that would allow us to do this very important work: a Marxist ecology that situates itself within the existing "socialism"; the greening of Marxism through ecological critiques of productivist[23] Marxism; unearthing the ecological dimension of Marxism; using a historical materialist ecology and theory of capitalist crisis to address our current climate crisis; and the rise of systemic alternatives informed by climate justice movements (2018, pp. 8–14). Satgar's emphasis on the necessity of organically created systemic alternatives "informed by indigenous cultures, cosmologies and rights-based discourses" (2018, p. 13) is spot on. Indeed, those of us on the margins, Black queer trans Indigenous peoples, bring an epistemology to antisystemic movements that is crucial to envisioning novel configurations and alternatives to our present socioeconomic and political arrangements.

The Caribbean region continues to confront the challenges of climate change and the demands of a highly unequal global economy. In 2019, Notre Dame's Global Adaptation Initiative (ND-GAIN) found several Caribbean islands ranked high in terms of being climate ready but low in terms of climate vulnerability.[24] In September 2023, the World Bank named Dominica the World's First Climate Resilient Country (World Bank Group, 2023) 6 years after his appeal at the United Nations General Assembly for mass polluters to take ownership and responsibility for climate change and its effect on regions like the Caribbean. Accessing the UN's Green Climate Fund[25] and World Bank funds directed at climate change resilience and mitigation, the country developed a hydromet network that records localized heavy rainfall events, providing data for weather forecasting that can inform disaster preparedness and planning. Dominica has constructed enough water tanks throughout the island carrying water to at least 15% of the population in a bid to minimize water shortages post hurricanes; has invested in cultivating and maintaining native plant populations that complement its slope stabilization efforts; has enhanced access to and for the Kalinago through infrastructural projects that increase resiliency to extreme weather events; and provided training to government employees and profession association members to assess the country's hazard exposure in real time to make data-driven informed decisions vis-à-vis risk reduction (World Bank Group, 2023).

The overwhelming amounts of sargassum on the beaches and waters of the Caribbean, in part due to climate change and increased fertilizer use, has had significant impacts on the ocean and sea life, as well as tourism and fishing industries. Researchers across the Caribbean have teamed up with those around the globe to grapple with this growing issue. For example, Clifford Louime of the University of Puerto Rico is collaborating with researchers at Princeton University, Rutgers University, University of California, Los Angeles, Marine Biological Laboratory at University of Chicago, and Woods Hole Oceanographic Institute (Poore, 2024). The sargassum problem that researchers are hoping to convert into productive material like biofuel (Coxworth, 2020; Jones et al., 2020) for humans and reprieve for ocean life demonstrates the possibility of remarkable innovation through collaboration (Schmidt Sciences, n.d.). Still, like the resiliency projects undertaken by the country of Dominica, inasmuch as we should celebrate them, they are approaching the challenges of climate change from the standpoint of individual countries or institutions (even if they are collaborative ones like the Princeton project on sargassum biorefinery).

April Karen Baptiste and Stacy-Ann Robinson's paper (2023) on the contours of the environmental justice in the Caribbean develop a striking combination of coloniality (framework for understanding exploitation), sovereignty (struggle for self-determination informs success or failure of environmental justice movements), and resistance for broadening the conversation on environmental justice theory in the region (2023, p. 559). Another provocative way of thinking through the climate crisis is with what Archana Ramanujam calls a Du Boisian approach, one that requires interpreting the phenomenon from a specific standpoint—that of the racialized and communities (2023). Ramanujam argues that Du Bois's concept of the racial state made space for race in labor–capital relations, and he identified a possible continuation of the colonial order in the postcolonial state which remained largely "dependent on the colonizing states for investment and military support" (2023, p. 285). Satgar calls for us to "draw on the rich and emancipatory traditions of socialism" for constructing a viable and sustainable world (2018, pp. 9, 10), and Ramanujam appropriately notes that any conception of democracy that neglects or excludes economic and political emancipation of colonized and racialized people is not acceptable.

Our approach to confronting the climate change catastrophe must be one that is antisystemic and pursues total transformation of our antiblack existing structures, our conceptions of the human and the social, and our relationship to the planet, and elevates novel sustainable, planet-affirming ways of being. This can only emerge at the grassroots level but cannot remain there. It should actively build emancipatory coalitions across geographies so that a new world can be possible.

Notes

1. Mottley and Amanpour (2021).
2. Antiblackness is understood here as "a specific, transhistorical, and structural position that Black people, Black being, and personhood are relegated to, that leads to Black suffering, oppression, disposability, abjection, and death" (grell-brisk, 2022, p. 244).
3. Please see James V. Fenelon and Leonardo Figueroa-Helland's (2025) chapter on Indigenous climate activism, in this volume. The authors offer that Indigenous ways of being and existing allow Indigenous peoples and nations activism to embody alternative social structures that resist the extractivism and exploitation of the hegemonic colonialist, capitalist state system that energizes climate collapse. Similarly, Joan Martínez-Alier (2025) asserts that Black and Indigenous peoples in Latin America and Africa have long understood the unjust nature and the ill effects of extractivist practices by large corporations from industrialized countries on their lived environments. They have a century-old history of resisting this, which gives credence to claims of ecological debt—the amount owed by the Global North to the Global South for the historical plundering of their natural resources and the resulting environmental damage and degradation.
4. Of particular note here is the extraction of lithium, cobalt, and other natural resources that support the production of electric vehicles, which are supposed to reduce dependence on oil and natural gas and reduce carbon emissions.
5. Note that according to the US Census Bureau, the 2020 percentage share of White home ownership in the United States is 73%, and Black home ownership nationwide is 7.9%. In the state of California, the figures are 51.7% for White home ownership, compared to 3.8% for Black home ownership (United States Census Bureau, 2023). So Altadena is a remarkable place, and yet systemic inequalities and redlining practices contributed to its vulnerability to wildfires and their impacts (Bonner, 2025).

6. Note that the geography of antiblackness and the spatial reality of the effects of climate change is not limited to a purely Global North-South divide. This does not negate the North-South difference.
7. The Green Climate Fund (GCF) was established during the Cancún, Mexico, talks of the United Nations Framework Convention on Climate Change (UNFCCC) in 2010. The GCF is one of the world's largest climate funds, accessible to developing countries (UN classification) for climate change mitigation and adaptation projects. It provides financing for these countries to pursue their nationally determined contributions (NDCs—commitment to reducing greenhouse gas emissions) and for developing climate-resilient pathways to net zero. It operates through a network of accredited entities that work directly with the countries toward designing and implementing their projects. These funds can be structured through various combinations of grants, private investments, and concessional loans.
8. For examples of how this plays out on the African continent, see grell-brisk (2019) and Mniga (2024).
9. K. S. Robinson (2020).
10. In *Loose Your Mother*, Saidiya Hartman explains how the conditions and configurations generated under slavery that measures the ranking of a man's life and worth have extended beyond that institution such that Black lives continue to be devalued. "[B]ecause black lives are still imperiled and devalued by a racial calculus and a political arithmetic that were entrenched centuries ago . . . [they experience] the *afterlife* of slavery—skewed life chances, limited access to health and education, premature death, incarceration, and impoverishment" (Hartman, 1997, p. 6, my emphasis). In much the same way, I use *afterlife* here to convey how colonial culture, labor, intersubjective relations, and knowledge production operate well beyond the strict limits of formal colonization.
11. Robert D. Bullard's scholarship provides a deep grounding on environmental *justice* as a framework to understand environmental and climate issues (Bullard, 2001, 2015). Bullard maintains that a justice mindset allows us to ask questions that expose the "underlying assumptions that may contribute to and produce unequal protection" (2001, p. 153) from environmental hazards while also centering community and bottom-up organizing efforts. The justice framework is expansive and preemptive in approach apropos environmental threats. It incorporates other social movements that seek to eliminate harmful practices in housing, land use, industrial planning, healthcare, and sanitation services, while articulating an explicit connection between issues of redlining, economic disinvestment, pollution, poverty, unemployment, and so on.
12. The works of the two foundational scholars of racial capitalism, Cedric Robinson and Stuart Hall, are often viewed as interchangeable, yet there are distinct differences. It is important to note that Pulido is leaning heavily on Cedric Robinson's (1983) understanding of racial capitalism wherein capitalism is always racial, as racial differentiation and domination are the primary ordering of society (Paret & Levenson, 2024, p. 4). This is in tension with Stuart Hall's articulation of the relationship between capitalism and racism, wherein the expression of racism in capitalist formations is historically specific such that racism is "secondary to the organization of society around the demands and interest of capital" (Paret & Levenson, 2024, p. 4). For excellent discussions on Robinson and Hall's scholarship on racial capitalism, see Marcel Paret and Zachary Levenson (2024) and Yusef Al-Bulushi (2020).
13. My emphasis.
14. At Our World In Data provides various ways to compare emissions. You can find comparative cumulative CO_2 emissions here: https://ourworldindata.org/co2-emissions.
15. See Andrew Jorgenson et al. (2025).
16. My emphasis.
17. The Greenhouse Gas Protocol (standard by which greenhouse gas emissions are accounted for) are categorized into three scopes. Scope 1: fuel combustion, company vehicles, fugitive emissions; Scope 2: purchased electricity, heat, and steam; Scope 3: purchased goods, business travel, employee commuting, waste disposal, use of sold products, transportation and distribution, investments, leased assets, and franchises.

18. Moore (2015, pp. 51, 170).
19. To be clear, I am not dismissing the work of activists who view water as bearing rights (like the Māori, who won a 2017 court ruling that recognized the Whanganui River as a living entity and gave it the status of legal personhood, or the Río Atrato being granted legal personhood in Colombia), or of scholars such as Montgomery and Magaña (2025), who urge us to consider the dynamic relationship between humans and the more-than-human, or Hans Asenbaum et al. (2023) or Timothy Morton and his philosophy of solidarity with the nonhuman (2019), or the work of Jason Moore on capitalism as a way of organizing nature (2003, 2011, 2015).
20. My emphasis.
21. Mészáros (2001).
22. Du Bois (1935).
23. Productivist Marxists claim that Marx advocated unlimited material growth, dubbed Prometheanism.
24. In 2019, the study (https://gain.nd.edu/) had created a country index using data over a 20-year period and 45 indicators ranking over 181 countries annually based on their level of vulnerability and their readiness to successfully implement adaptation solutions (a country's ND-GAIN score). It found that there was a significant correlation between ND-GAIN scores and GDP per capita. So the countries were reranked based on the distance of a country's measured ND-GAIN score and its expected value based on the regression of ND-GAIN and GDP. Based on this measurement, small-island developing states demonstrated a stronger readiness to deal with global climate change. The most recent rankings are much more nuanced and based on multiple variables and metrics.
25. For critical appraisals of the Green Climate Fund, see Stacy-Ann Robinson and Matthew Dornan's work that speak to the limits of the GCF for small-island developing states (2017), Sanjay Kumar's report in Nature (2015), and Jessica Omukuti et al.'s article on the delivery of finance at the local level (2022).

References

Adisa-Farrar, T. (2023, February 2). How 600 years of environmental violence is still harming Black communities. https://earthjustice.org/article/overlooked-connections-between-black-injustice-and-environmentalism

Al-Bulushi, Y. (2020). Thinking racial capitalism and black radicalism from Africa: An intellectual geography of Cedric Robinson's world-system. *Geoforum*, January. https://doi.org/10.1016/j.geoforum.2020.01.018

Alier, J. M. (2009). Socially sustainable economic de-growth. *Development and Change*, *40*(6), 1099–1119.

Asenbaum, H., Machin, A., Gagnon, J. P., Leong, D., Orlie, M., & Smith, J. L. (2023). The nonhuman condition: Radical democracy through new materialist lenses. *Contemporary Political Theory*, 22(4), 584–615. https://doi.org/10.1057/s41296-023-00635-3

Baptiste, A. K., & Robinson, S. (2023). The contours of environmental justice in the Caribbean. *Geographical Journal*, *189*(4), 554–561. https://doi.org/10.1111/geoj.12545

Bastani, A. (2019). *Fully automated luxury communism: A manifesto*. Verso.

BloombergNEF. (2023, July 18). Long-term carbon offsets outlook 2023. https://www.bloomberg.com/professional/insights/commodities/long-te

Bonner, K. (2025, January 28). Altadena's Black community disproportionately affected by Eaton Fire, report shows. https://newsroom.ucla.edu/releases/altadenas-black-community-disproportionately-affected-eaton-fire-report-shows

Brooks, C. (2023, June 7). The left did a no growth. https://damagemag.com/2023/06/07/the-left-did-a-no-growth/

Bullard, R. D. (2001). Environmental justice in the 21st century: Race still matters. *Phylon*, *49*(3/4), 151–171. https://publicseminar.org/essays/the-promise-of-black-lives-matter/

Cabral, A. (1973). *Return to the source: Selected speeches of Amílcar Cabral* (Africa Information Service, Ed.; Monthly Review Press). Africa Information Service and the African Party for the Independence of Guinea and the Cape Verde Islands.

CarbonCredits.Com. (2021, July). What is the voluntary carbon market? Investor's Education Center. https://carboncredits.com/what-is-the-voluntary-carbon-market/

Castelvecchi, D. (2021, August 19). Electric cars: The battery challenge. *Nature, 596*, 336–339. https://www.nature.com/articles/d41586-021-02222-1

Cohen, I., Nieves, E., Fahys, J., Lavelle, M., & Bruggers, J. (2020). As protests rage over George Floyd's death, climate activists embrace racial justice. https://insideclimatenews.org/news/03062020/george-floyd-racial-justice-police-brutality-environment-climate-activism/

Cook, J., Oreskes, N., Doran, P. T., Anderegg, W. R. L., Verheggen, B., Maibach, E. W., Stuart, C. J., Lewandowsky, S., Skuce, A. G., Green, S. A., Nuccitelli, D., Jacobs, P., Richardson, M., Winkler, B., Painting, R., & Rice, K. (2016). Consensus on consensus: A synthesis of consensus estimates on human-caused global warming. *Environmental Research Letters, 11*(4), 48002. http://stacks.iop.org/1748-9326/11/i=4/a=048002

Coxworth, B. (2020). Inexpensive process converts waste seaweed into biofuel, fertilizer and more. https://newatlas.com/good-thinking/waste-seaweed-biofuel-fertilizer/

Cullenward, D., Badgley, G., & Chay, F. (2023, September 15). Carbon offsets are incompatible with the Paris Agreement. *One Earth, 6*(9), 1085–1088. https://doi.org/10.1016/j.oneear.2023.08.014

Cunningham, C. X., Williamson, G. J., & Bowman, D. M. J. S. (2024). Increasing frequency and intensity of the most extreme wildfires on Earth. *Nature Ecology and Evolution, 8*(8), 1420–1425. https://doi.org/10.1038/s41559-024-02452-2

Daly, M. (2024, September 13). Why Harris is promoting domestic oil drilling at the same time as clean energy jobs. https://www.pbs.org/newshour/politics/why-harris-is-promoting-domestic-oil-drilling-at-the-same-time-as-clean-energy-jobs

Davis, A. (1989). *Women, culture & politics*. Random House.

Du Bois, W. E. B. (1935). *Black reconstruction in America, 1860–1880*. The Free Press.

Dumas, M. J. (2016). Against the dark: Antiblackness in education policy and discourse. *Theory into Practice, 55*(1), 11–19. https://doi.org/10.1080/00405841.2016.1116852

Emanuel, K., Sundararajan, R., & Williams, J. (2008). Hurricanes and global warming: Results from downscaling IPCC AR4 simulations. *Bulletin of the American Meteorological Society, 89*(3), 347–367. https://doi.org/10.1175/BAMS-89-3-347

Escobar, A. (2007). Worlds and knowledges otherwise: The Latin American modernity/ coloniality research program. *Cultural Studies, 21*(2–3), 179–210. https://doi.org/10.1080/09502380601162506

Fenelon, J. V., & Figueroa-Helland, L. (2025). Global indigenous struggles and climate change activism. In P. Almeida (Ed.), *The Oxford handbook of climate action*. Oxford University Press. https://doi.org/10.1093/oxfordhb/9780197762097.001.0001

Fisher, S. (2015). The emerging geographies of climate justice. *The Royal Geographical Society with IBG, 181*(1), 73–82.

Foster, J. B. (2000). *Marx's ecology: Materialism and nature*. Monthly Review Press.

Foster, J. B., & Keqing, J. (2023). Ecological Marxism. *Monthly Review, 75*(4), 1–10. https://monthlyreview.org/2023/09/01/ecological-marxism/

Foucault, M. (1994). *The order of things: An archaeology of the human sciences (A translation of Les Mots et les choses)*. Vintage Books.

Gilmore, R. W. (2002). Fatal couplings of power and difference: Notes on racism and geography. *Professional Geographer, 54*(1), 15–24. https://doi.org/10.1111/0033-0124.00310

grell-brisk, m. (2019). Arrested development? Sub-Saharan Africa in the stratified world economy 1965–2015. *Journal of Contemporary African Studies, 37*(1), 93–111. https://doi.org/10.1080/02589001.2019.1585525

grell-brisk, m. (2022). Critical world-systems analysis: Thoughts on organizing against antiblackness across global-local boundaries. *Journal of World-Systems Research, 28*(2), 242–266. https://doi.org/10.5195/jwsr.2022.1103

grell-brisk, m. (2023). Introduction to the symposium: Parasitism and the logics of anti-Indigeneity and antiblackness. *Journal of World-Systems Research*, 29(1), 4–24. https://doi.org/10.5195/JWSR.2023.1179

grell-brisk, m., & Chase-Dunn, C. (2019). Mind the gap! Clustered obstacles to mobility in core-periphery hierarchy. In I. Rossi (Ed.), *Frontiers of globalization research: Theoretical and methodological approaches* (2nd ed., pp. 331–355). Springer.

Grublier, A., Wilson, C., Bento, N., Boza-Kiss, B., Krey, V., McCollum, D. L., Rao, N. D., Riahi, K., Rogelj, J., de Stercke, S., Cullen, J., Frank, S., Fricko, O., Guo, F., Gidden, M., Havlík, P., Huppmann, D., Kiesewetter, G., Rafaj, P., … Valin, H. (2018). A low energy demand scenario for meeting the 1.5°C target and sustainable development goals without negative emissions technologies. *Nature Energy*, 3, 515–527.

The Guardian Staff. (2019, September 9). Search for survivors of Hurricane Dorian continues in Bahamas. *The Guardian*. https://www.theguardian.com/world/2019/sep/09/search-for-bodies-and-survivors-continues-after-hurricane-dorian-in-bahamas

Guerra, L. (2014). Why Caribbean history matters. *Perspectives on History (American Historical Association)*. https://www.historians.org/perspectives-article/why-caribbean-history-m

Guerrero, D. G. (2001). The limits of capitalist solutions to the climate crisis. In V. Satgar (Ed.), *The climate crisis: South African and global democratic eco-socialist alternatives* (pp. 30–46). Wits University Press.

Hartman, S. (1997). *Lose your mother: A journey along the Atlantic slave route*. Farrar, Straus, and Giroux.

Hickel, J. (2019). The contradiction of the sustainable development goals: Growth versus ecology on a finite planet. *Sustainable Development*, 27(5), 873–884. https://doi.org/10.1002/sd.1947

Hickel, J. (2020). Quantifying national responsibility for climate breakdown: An equality-based attribution approach for carbon dioxide emissions in excess of the planetary boundary. *The Lancet Planetary Health*, 4(1), e399–e404. https://www.thelancet.com/journals/lanplh/article/PIIS2542-5196(20)30196-0/fulltext

Holgersen, S. (2023, September 4). Neither productivism nor degrowth: Thoughts on eco-socialism. *Spectre Journal*, 1–9. https://spectrejournal.com/neither-productivism-nor-degrowth/

Holgersen, S. (2025, March 26). Growth or degrowth? Ecosocialism confronts a false dichotomy. *Climate & Capitalism*. https://climateandcapitalism.com/2025/03/25/growth-or-degrowth-ecosocialism-challenges-a-false-dichotomy/

Huber, E., Bach, V., & Finkbeiner, M. (2024). A qualitative meta-analysis of carbon offset quality criteria. *Journal of Environmental Management*, 352, 10. https://doi.org/10.1016/j.jenvman.2023.119983

Huber, M. (2022a). *Climate change is class war: Building socialism on a warming planet*. Verso.

Huber, M. (2022b, August 18). Mish-mash ecologism. *Sidecar*, 1–10. https://newleftreview.org/sidecar/posts/mish-mash-ecologism

IMF Executive Board. (2023). *Dominica: IMF Country Report No. 23/246*. http://www.imf.org

International Monetary Fund. (2017, April 27). Dominica: Selected issues. *IMF Country Report 17/392*, 42 pp.

IPCC Core Writing Team, Lee, H., & Romero, J. (2023). IPCC, 2023: Summary for policymakers. In Core Writing Team (Ed.), *Climate change 2023: Synthesis report. Contribution of Working Groups I, II and III to the Sixth Assessment Report of the Intergovernmental Panel on Climate Change* (pp. 1–34). IPCC. https://doi.org/10.59327/IPCC/AR6-9789291691647

Jennifer, L. (2023). Who issues carbon credits? (Everything You Need to Know). https://carboncredits.com/who-issues-carbon-credits-everything-you-ne …

Jones, E. S., Raikova, S., Ebrahim, S., Parsons, S., Allen, M. J., & Chuck, C. J. (2020). Saltwater based fractionation and valorisation of macroalgae. *Journal of Chemical Technology and Biotechnology*, 95(8), 2098–2109. https://doi.org/10.1002/jctb.6443

Jorgenson, A., Givens, J., Huang, X., & Tina, H. El. (2025). Ecologically unequal exchange as a catalyst for climate action. In P. Almeida (Ed.), *The Oxford handbook of climate action*. Oxford University Press. https://doi.org/10.1093/oxfordhb/9780197762097.001.0001

Kallis, G. (2011). In defense of degrowth. *Ecological Economics, 70*(5), 873–880. https://doi.org/10.1016/j.ecolecon.2010.12.007
Kallis, G. (2018). *Degrowth*. Agenda Publishing.
Kangal, K. (2024). In defense of the planet. *European Legacy, 29*(2),204–207. https://doi.org/10.1080/10848770.2023.2265712
Kumar, S. (2015). Green Climate Fund faces slew of criticism. *Nature, 527*, 419–420.
Maldonado-Torres, N. (2007). On the coloniality of being: Contributions to the development of a concept. *Cultural Studies, 21*(2–3), 240–270. https://doi.org/10.1080/09502380601162548
Mansoor, S. (2019, September 10). After Dorian survivors are kicked off rescue ship, U.S. Border Patrol says all Bahamians welcome—Except those who aren't. *Time Magazine*. https://time.com/5672378/bahamas-dorian-survivors-us-visa/
Martínez-Alier, J. (2025). Environmentalism of the poor and climate action. In P. Almeida (Ed.), *The Oxford handbook of climate action*. Oxford University Press. https://doi.org/10.1093/oxfordhb/9780197762097.001.0001
Mason, P. (2015). *Postcapitalism: A guide to our future*. Allen Lane.
McBride, J. (2017). The consequences of leaving the Paris Agreement. *Council on Foreign Relations*. https://www.cfr.org/backgrounder/consequences-leaving-paris-agreement
Melamed, J. (2015). Racial capitalism. *Critical Ethnic Studies, 1*(1), 76–85.
Mészáros, I. (2001). *Socialism or barbarism: From the "American century" to the crossroads*. Monthly Review Press.
Michael, K., Mikulewicz, M., & Crawford, N. J. W. (2023). Introduction: Climate justice beyond the minority world—Towards decolonial knowledges. In *Climate justice in the majority world: Vulnerability, resistance, and diverse knowledges* (pp. 1–17). Taylor and Francis. https://doi.org/10.4324/9781003214021-1
Mignolo, W. D. (2011). Coloniality: The darker side of western modernity. In *The darker side of western modernity: Global futures, decolonial options* (pp. 1–26). Duke University Press.
Mniga, M. (2024). *Recolonizing Africa: An ethnography of land acquisition, mining, and resource control*. Routledge.
Montgomery, A., & Magaña, M. (2025). "Overwhelming heat" in racialized communities. In P. Almeida (Ed.), *The Oxford handbook of climate action*. Oxford University Press. https://doi.org/10.1093/oxfordhb/9780197762097.001.0001
Moore, J. W. (2003). "The modern world-system" as environmental history? Ecology and the rise of capitalism. *Theory and Society, 32*(3), 307–377.
Moore, J. W. (2011). Ecology, capital, and the nature of our times: Accumulation & crisis in the capitalist world-economy. *Journal of World-Systems Research, XVII*(1), 107–146.
Moore, J. W. (2015). *Capitalism in the web of life: Ecology and the accumulation of capital*. Verso.
Morton, T. (2019). *Humankind: Solidarity with non-human people*. Verso.
Mottley, M., & Amanpour, C. (2021, November 1). Barbadian prime minister calls on rich nations for climate financing. https://www.pbs.org/wnet/peril-and-promise/2021/11/barbadian-prime-minister-calls-on-rich-nations-for-climate-
National Aeronautics and Space Administration. (2025). *Wildfires and climate change*. https://science.nasa.gov/wildfires-and-climate-change/
Neuberger, N., Fatula, G. D., & Segal, S. H. (2025). US withdrawal from the Paris Climate Accord and its impact on the voluntary carbon market introduction: A withdrawal with broad implications. https://natlawreview.com/article/us-withdrawal-paris-climate-accord-and-its-impact-voluntary-carbon-market
Omukuti, J., Barrett, S., White, P. C. L., Marchant, R., & Averchenkova, A. (2022). The green climate fund and its shortcomings in local delivery of adaptation finance. *Climate Policy, 22*(9–10), 1225–1240. https://doi.org/10.1080/14693062.2022.2093152
Pande, R. (2024). Can the market in voluntary carbon credits help reduce global emissions in line with Paris Agreement targets? Science, *384*(6696), 8. https://doi.org/10.1126/science.adp5223
Paret, M., & Levenson, Z. (2024). Two racial capitalisms: Marxism, domination, and resistance in Cedric Robinson and Stuart Hall. *Antipode, 56*(5), 1802–1829. https://doi.org/10.1111/anti.13054

Poore, C. (2024). Transforming troublesome seaweed into a feedstock of the future. https://www.schmidtsciences.org/viff/

Porter, L., Rickards, L., Verlie, B., Moloney, S., Lay, B., Latham, B., Anguelovski, I., Pellow, D., Porter, L., Bosomworth, K., Rickards, L., Verlie, B., Moloney, S., Lay, B., Latham, B., BCNUEJ, I. A., Pellow, D. N., Naarm/Birrarung-ga, Porter, L., … Lay, B. (2020). Climate justice in a climate changed world. *Planning Theory and Practice*, *21*(2), 293–321. https://doi.org/10.1080/14649357.2020.1748959

Pulido, L. (2017). Geographies of race and ethnicity II: Environmental racism, racial capitalism and state-sanctioned violence. *Progress in Human Geography*, *41*(4), 524–533. https://doi.org/10.1177/0309132516646495

Quijano, A. (1989). Paradoxes of modernity in Latin America. *International Journal of Politics, Culture, and Society*, *3*(2), 147–177.

Quijano, A. (2007). Coloniality and modernity/rationality. *Cultural Studies*, *21*(2–3), 168–178.

Quijano, A., & Wallerstein, I. (1992). Americanity as a concept. *International Social Science Journal*, *44*(4), 549–557.

Ramanujam, A. (2023). Climate scholarship needs Du Bois: Climate crisis through the lens of racial and colonial capitalism. *Sociological Inquiry*, *93*(2), 273–295. https://doi.org/10.1111/soin.12516

Rhodes, C. J. (2017). US withdrawal from the COP21 Paris climate change agreement, and its possible implications. *Science Progress*, 100(4), 411–419. https://doi.org/10.3184/003685017X15063357842600

Roberts, J. T. (2001). Global inequality and climate change. *Society and Natural Resources*, *14*(6), 501–509. https://doi.org/10.1080/08941920118490

Robinson, C. (1983). *Black Marxism: The making of a Black radical tradition*. Zed Press.

Robinson, K. S. (2020). *Ministry of the future*. Orbit Books (Hachette Book Group).

Robinson, S., & Dornan, M. (2017). International financing for climate change adaptation in small island developing states. *Regional Environmental Change*, *17*(4), 1103–1115.

Rockström, J., Kotzé, L., Milutinović, S., Biermann, F., Brovkin, V., Donges, J., Ebbesson, J., French, D., Gupta, J., Kim, R., Lenton, T., Lenzi, D., Nakicenovic, N., Neumann, B., Schuppert, F., Winkelmann, R., Bosselmann, K., Folke, C., Lucht, W., … Steffen, W. (2024). The planetary commons: A new paradigm for safeguarding Earth-regulating systems in the Anthropocene. *Proceedings of the National Academy of Sciences of the United States of America*, *121*(5), 10. https://doi.org/10.1073/pnas.2301531121

Rockström, J., Steffen, W., Noone, K., Persson, Å., Chapin, F. S., Lambin, E., Lenton, T. M., Scheffer, M., Folke, C., Schellnhuber, H. J., Nykvist, B., De Wit, C. A., Hughes, T., Van Der Leeuw, S., Rodhe, H., Sörlin, S., Snyder, P. K., Costanza, R., Svedin, U., … Walker, B. (2009). Planetary boundaries: Exploring the safe operating space for humanity. *Ecology and Society*, *14*(2). https://about.jstor.org/terms

Satgar, V. (2018). The climate crisis and systemic alternatives. In V. Satgar (Ed.), *The climate crisis: South African and global democratic eco-socialist alternatives* (pp. 1–28). Wits University Press.

Schlosberg, D., & Collins, L. B. (2014). From environmental to climate justice: Climate change and the discourse of environmental justice. *Wiley Interdisciplinary Reviews: Climate Change*,*5*(3), 359–374. https://doi.org/10.1002/wcc.275

Schmelzer, M., Vetter, A., & Vansintjan, A. (2022). *The future is degrowth: A guide to a world beyond capitalism*. Verso.

Schmidt Sciences. (n.d.). *Virtual Institute on Feedstocks of the Future (VIFF)—Sargassum Biorefinery (SaBRe)*. https://www.schmidtsciences.org/viff/

Schneider, F., Kallis, G., & Martinez-Alier, J. (2010). Crisis or opportunity? Economic degrowth for social equity and ecological sustainability. Introduction to this special issue. *Journal of Cleaner Production*, *18*(6), 511–518. https://doi.org/10.1016/j.jclepro.2010.01.014

Searcey, D., & Lipton, E. (2023, June 22). Hunt for the "blood diamond of batteries" impedes green energy push. *New York Times*. https://www.nytimes.com/2021/11/29/world/congo-cobalt-albert-yuma-mulimbi.html

Sisson, P. (2023, February 17). How does an EV battery actually work? Are lithium batteries sustainable enough to fulfill the dream of the electric-car revolution? *MIT Tech Review*. https://www.technologyreview.com/2023/02/17/1068037/how-do-ev-batteries-work/

Skerrit, R. (2017). United Nations General Assembly 72nd session statement. https://gadebate.un.org/sites/default/files/gastatements/72/dm_en.pdf

Slangen, A. B. A., Church, J. A., Agosta, C., Fettweis, X., Marzeion, B., & Richter, K. (2016). Anthropogenic forcing dominates global mean sea-level rise since 1970. *Nature Climate Change*, 6, 701–706. https://doi.org/10.1038/NCLIMATE2991

Strauss, B. H., Kulp, S., & Levermann, A. (2015). Carbon choices determine US cities committed to futures below sea level. *Proceedings of the National Academy of Sciences*, *112*(44), 13508–13513. https://doi.org/10.1073/pnas.1511186112

Sultana, F. (2022). The unbearable heaviness of climate coloniality. *Political Geography*, 99. 16. https://doi.org/10.1016/j.polgeo.2022.102638

Talabong, R. (2025, April 4). Black residents of Altadena struggle to hang on to their community after LA fires. *Inside Climate News*. https://insideclimatenews.org/news/04042025/black-residents-of-altadena-struggle-after-la-fires/

Taylor, D. (2014). *Toxic communities: Environmental racism, industrial pollution, and residential mobility*. New York University Press.

Taylor, D. (2018, *October* 2–4). Environmental racism. *Pollution Issues*. http://www.pollutionissues.com/Ec-Fi/Environmental-Racism.html

Trencher, G., Nick, S., Carlson, J., & Johnson, M. (2024). Demand for low-quality offsets by major companies undermines climate integrity of the voluntary carbon market. *Nature Communications*, *15*(1). https://doi.org/10.1038/s41467-024-51151-w

United Nations Climate Change. (2024). What is the United Nations carbon offset platform? https://unfccc.int/climate-action/united-nations-carbon-offset-platform

United Nations Framework Convention on Climate Change. (2022). What is the United Nations carbon offset platform? https://unfccc.int/climate-action/united-nations-carbon-offset-platform

United States Census Bureau. (2023, September 28). Homeownership by race and ethnicity of householder. https://www.census.gov/library/visualizations/interactive/homeownership-by-race-and-ethnicity-of-householder.html

van Vuuren, D. P., Stehfest, E., Gernaat, D. E. H. J., van den Berg, M., Bijl, D. L., de Boer, H. S., Daioglou, V., Doelman, J. C., Edelenbosch, O. Y., Harmsen, M., Hof, A. F., & van Sluisveld, M. A. E. (2018). Alternative pathways to the 1.5°C target reduce the need for negative emission technologies. *Nature Climate Change*, *8*, 391–397.

Vargas, J. H. C. (2018). *The denial of antiblackness: Multiracial redemption and Black suffering*. University of Minnesota Press.

Vargas, J. H. C., & Jung, M.-K. (2021). Antiblackness of the social and the human. In M.-K. Jung & J. H. C. Vargas (Eds.), *Antiblackness* (pp. 1–16). Duke University Press.

Victor, P. (2008). *Managing without growth: Slower by design, not disaster*. Edward Elgar.

Wallerstein, I. (1974). Dependence in an interdependent world: Limited possibilities of transformation within the capitalist world economy. *African Studies Review*, *17*, 1–26.

Wallerstein, I. (1979). *The capitalist world economy*. Cambridge University Press.

Wallerstein, I. (1999). Ecology and capitalist costs of production: No exit. In W. Goldfrank, D. Goodman, & A. Szasz (Eds.), *Ecology and the world-system* (pp. 76–86). Greenwood Press.

Wasserman, T. N., & Mueller, S. E. (2023). Climate influences on future fire severity: A synthesis of climate-fire interactions and impacts on fire regimes, high-severity fire, and forests in the western United States. *Fire Ecology 19*, Article 43. https://doi.org/10.1186/s42408-023-00200-8

West, T. A. P., Börner, J., Sills, E. O., & Kontoleon, A. (2020). Overstated carbon emission reductions from voluntary REDD+ projects in the Brazilian Amazon. *Proceedings of the National Academy of Sciences*, *117*, 24188–24194. https://www.pnas.org/doi/full/10.1073/pnas.2004334117

West, T. A. P., Wunder, S., Sills, E. O., Börner, J., Rifai, S. W., Neidermeier, A. N., Frey, G. P., & Kontoleon, A. (2023). Action needed to make carbon offsets from forest conservation work for climate change mitigation. *Science 381*(6660), 873–877. https://doi.org/10.1126/science.ade3535. https://www.science.org

WFE Research Team. (2025, March 5). Voluntary carbon markets found to be over 10 times less efficient than mainstream markets, WFE research reveals. *World Federation of Exchanges*. https://www.world-exchanges.org/news/articles/voluntary-carbon-markets-found-be-over-10-times-less-efficient-mainstream-markets-wfe-research-reveals

World Bank Group. (2023). Dominica's journey to become the world's first climate resilient country. https://www.worldbank.org/en/news/feature/2023/09/26/dominica-s-journey-to-become-the-world-s-first-climate-resilient-country.print

Wynter, S. (2003). Unsettling the coloniality of being/power/truth/freedom: Towards the human, after man, its overrepresentation—An argument. *The New Centennial Review*, 3(3), 257–337.

CHAPTER 7

Intersectional Climate Action

CAMILA ALVAREZ AND KATHRYN NORTON-SMITH

In 2023, the 28th meeting of the Conference of Parties (COP28) hosted by the United Nations Climate Change Conference (UNCCC) took place in Dubai, United Arab Emirates (UAE). It was lauded as the largest event of its kind, with 85,000 participants representing governments, Indigenous nations, international organizations, activists, and lobbyists. However, group representation varies; for instance, there were 300 Indigenous people in attendance and over 2,400 fossil fuel lobbyists (Cultural Survival, 2023). The COP28 exemplifies climate action, defined as "collective efforts to address global warming" (Almeida et al., 2024, p. 2). Demonstrations by human rights and climate change activists at the conference provide a powerful intersectional critique of global climate governance and emphasize the need for climate action to address social and political inequities. While demonstrations were permitted within the specifically designated zones of the conference, protesters from various groups and identities voiced concerns that regulations on size, venue, and type of speech were "killing the movement" (Batrawy, 2023). These protests were perceived as destabilizing and threatening to national security, challenging the structural purpose of the COP (Batrawy, 2023). A closer examination of COP28 with an intersectional lens reveals whose voices, experiences, interests, and epistemologies were included and excluded from both the physical space and climate action planning, reflecting important overlapping scales of power and processes.

This chapter echoes the calls from environmental justice scholars and activists who argue that we cannot fully understand environmental (in)justice or climate action without employing a critical intersectional perspective (Amorim-Maia et al., 2022; Daniel & Dolan, 2020; Ducre, 2013, 2018; Ergas et al., 2021; Kaijser & Kronsell, 2014; Malin & Ryder, 2018; Mikulewicz et al., 2023). Importantly, intersectional thought has deep roots in the activism of Black, Indigenous, and People of Color (BIPOC). At COP28, Indigenous leaders discussed how climate change impacts their peoples and communities through intersecting injustices. As Dane de Souza from Métis National Council stated, "Climate change affects every aspect of our existence . . . it impacts how we gather at the table as a family, it impacts how we pass on our knowledge and our memories to our children" (Prete, 2023, para. 1). Climate action must address the inequities faced by historically marginalized communities that are disproportionately affected by climate change, extreme weather events, and displacement. By bridging intersectionality as theory and praxis with an environmental justice framework, this chapter explains how climate injustices occur through various structural mechanisms, putting specific communities at risk while privileging others. As such, an intersectional approach to climate action must center on those communities that are seen as expendable to the needs of global capitalism. We conclude by offering caution to climate actions that do not use an intersectional framework.

We begin by elaborating on the early history of intersectional thought among activists and academics. Then, we examine why risk, vulnerability, and resilience to climate change are intersectional. Central to an intersectional perspective is examining levels of power on multiple scales, from institutions to individuals. Given the emphasis on praxis in intersectional activism and research, we conclude by highlighting climate action resistance. A central goal of climate justice research is to reveal and resist the disproportionate impacts of climate-related hazards on BIPOC, immigrants in global cities, rural areas, low-income communities, women, nonbinary and transgender individuals, the unhoused, people with disabilities, and populations in the Global South. In doing so, environmental and climate justice scholars center multiple axes of inequality to illuminate how race, class, gender, Indigeneity, ability, nationality, and other facets of difference intersect to shape resistance to injustice.

Intersectionality: Theory and Origin

Intersectionality refers to the idea that mutually constructing phenomena, such as race, class, gender, nation, sexuality, and ability, shape social inequality in complex ways (Collins, 2000, 2015; Collins & Bilge, 2016). The concept of intersectionality was used throughout the 1960s–1980s to highlight the interconnectedness of race, class, gender, and sexual oppression in the social and political movements of Black, Chicana, Indigenous, and Asian American women (Collins & Bilge, 2016, pp. 63–87). Rooted in epistemological resistance, intersectional thinking critiqued the power relations that reinforced privilege and erasure in both political and academic realms (Collins, 2019). Social movements organized around unitary issues often present essential and universal experiences based on the most privileged group members. For example, Black women's experiences were often erased in the political and legal dialogues of both feminist and civil rights movements. The women's liberation movement emphasized patriarchy as the single source of women's oppression, reflecting the experiences of White, middle-class women. This frustration is captured in bell hooks's (1982) *Ain't I a Woman?*:

> [The] force that allows white feminist authors to make no reference to racial identity in their books about "women" that are in actuality about white women is the same one that would compel any author writing exclusively on black women to refer explicitly to their racial identity. That force is racism . . . it is the dominant race that can make it seem that their experience is representational. (p. 138)

These experiences of sexism and racism in the mainstream feminist movement led Black women to form their own Black feminist groups and community organizations, such as the National Black Feminist Organization. One notable group was the Combahee River Collective, whose "Combahee River Collective Statement" emphasized the need for complex analyses that recognized that Black women's lives are shaped by race, sexuality, gender, and class and that "major systems of oppression are interlocking" (The Combahee River Collective, 1977). The goals of these Black feminist organizations were to create more effective political and social movements that could more adequately address the needs of women of color.

Legal scholar Kimberlé Crenshaw is often cited as coining the term "intersectionality" in academic and legal conversations; however, there is a significant history of intersectional thinking inside and outside the academy (Collins, 2019). In "Mapping the Margins," Crenshaw (1991) discusses the limitations of identity politics to transcend difference, arguing that it "ignores intergroup difference." Using violence against women of color as an example, Crenshaw demonstrates the inability of advocacy and social movements organizing around violence against women to capture the vulnerability of women of color (Crenshaw,

1991). Crenshaw (1991) argues that such movements construct identities in an either/or manner, marginalizing the experiences of women of color "to a location that resists telling." She mitigates this silence by exploring how race and gender intersect to shape violence against women of color through (1) structural structurally, (2) political intersectional, and (3) representational intersectionality (Crenshaw, 1991, p. 1241). While this work establishes connections between central themes in Black feminist thought, Crenshaw's work is often framed as the origin of intersectionality. This framing can depoliticize intersectionality, shifting it from a collective epistemological project of resistance by Black feminism and other women of color to one woman's ideas and can reinforce "Western narratives of colonialism and capitalism" (Collins, 2019, p. 123).

According to Patricia Hill Collins and Sirma Bilge (2016):

> Intersectionality is a way of understanding and analyzing the complexity in the social world, in people, and in human experiences. The events and conditions of social and political life and the self can seldom be understood as shaped by one factor. They are generally shaped by many factors in diverse and mutually influencing ways. When it comes to social inequality, people's lives and the organization of power in a given society are better understood as being shaped not by a single axis of social division, be it race or gender or class, but by many axes that work together and influence each other. Intersectionality as an analytic tool gives people a better access to the complexity of the world and themselves. (p. 2)

As a concept, intersectionality emphasizes the simultaneity of systems, the intersecting nature of hierarchies, and the tension between social structure and agency. It allows for complex analysis that avoids erasure and recognizes shifting context and power relations. Understanding intersectionality is crucial for addressing complex social issues, including climate action.

An Intersectional Perspective on Climate Risk, Vulnerability, and Resilience

It is widely recognized that the impacts of climate change will be experienced differently based on factors such as race, class, gender, and geography, which reflect positions of privilege and oppression within the "matrix of domination" (Collins, 2000). Intersectional analysis illuminates how those who experience multiple forms of oppression are also those who will experience the most severe consequences of climate-related disasters and severe weather events. Research consistently shows that poor and BIPOC individuals, people with disabilities, and formerly colonized and marginalized communities experience the most severe impacts of climate change and climate-related disasters (Amorim-Maia et al., 2022; Kaijser & Kronsell, 2014; Sultana, 2022a, 2022b). Intersectional analyses reveal shortcomings to singular frameworks that essentialize such experiences; for example, systems of patriarchy often render women more vulnerable to the impacts of climate change, yet not all women's experiences of vulnerability are universal. In the context of Bangladesh, gendered vulnerability to water-related hazards and disasters is not only about gender but also caste, class, religion, and age due to differential rights, responsibilities, and resources (Mikulewicz et al., 2023, p. 1276; Sultana, 2010). Highlighting the tension between structure and agency, scholars caution against portraying climate change–affected individuals and communities as mere victims and instead emphasize the need to critically examine conventional understandings of *vulnerability* and *resilience*. Vulnerability is diverse and intersectional, varying by social status. While these social positions reflect social identities, they are socially constructed through structural mechanisms. Vulnerability to climate change and environmental injustice is not inherent to communities. Instead, it results from the positionality of marginalized communities within systems of domination like capitalism, patriarchy, racism, colonialism, imperialism, nationalism, militarism, and others. As Mikulewicz et al. (2023) state, "vulnerability is a process mediated by structural forces that are both local and global" (p. 1278).

Climate change emerges from the "extractivism" of high-productivity economies and the creation of sacrifice zones where profit is prioritized over people and the planet (also see chapters by Jorgenson et al.; Arce and Sullivan in this volume) (Sultana, 2021b, p. 448). Therefore, there is an interconnection between issues of environmental degradation with larger systems of power such as capitalism, colonialism, and patriarchy. Social problems stem from "exploitative practices towards nature and feminized and reproductive labor.... The combined extraction of nature and the exploitation of cheap labor from poor women and colonized and racial groups," supporting neoliberal and market-driven economic growth (also see chapter by Grell-Brisk in this volume) (Resurrección, 2021, p. 3). For example, the vulnerability of Caribbean states to climate change cannot be detached from colonialism and imperialism, which have resulted in significant indebtedness and forced reliance on climate-sensitive sectors like tourism and agriculture, exacerbated by neoliberal development models imposed during debt crises (Sealey-Huggins, 2017; Smith & Rhiney, 2016). At the same time, we must also investigate how representations of climate vulnerability deny agency and complexity to communities, serving as a tool for further exploitation (Perez, 2021). For example, the Pacific Islands and Islanders are framed as highly vulnerable to climate change, characterized as disappearing or on the frontline of climate impacts. This visibility has been exploited by "green" and "blue" corporations for profit, which mask neoliberal and neocolonial practices with the facade of sustainability to "exploit, harm, and displace Pacific Islands, waters, animals" (Perez, 2021, p. 3). In this context, Perez (2021) argues for critically examining such representations of vulnerability to avoid reductionist views, instead advocating for those that highlight agency, diversity, complexity, and resistance, as climate victim narratives also limit climate action strategies.

Intersectional approaches reveal how overlapping social identities and power structures intersect to shape capacities for resistance and forms of climate action, while also highlighting the agency of individuals and communities. The historical, material, and epistemological contexts are essential when examining climate impacts and justice. As such, intersectionality is an analytic tool for understanding instances of climate injustice because it positions the individual—or community—within simultaneous, complex, and relational systems of power. For example, Kyle Whyte (2014) describes how some Indigenous women, motivated by a commitment to cultural responsibilities, take on leadership positions in climate change adaptation and mitigation efforts. In this way, it is crucial to emphasize differential climate impact and how marginalized communities are *resilient to climate change—engaging in climate action*.

Despite being the most negatively affected, women, racial minorities, and Indigenous peoples are underrepresented in climate change adaptation and decision-making processes. Intersectional thinking helps move beyond exclusion based on a single variable: While women are excluded from international climate policy, Indigenous women experience disproportionate exclusion (Perkins, 2019). Such a perspective exemplifies the importance of examining who is prioritized in climate change decision making and policy implementation. Scholars like Gonda (2019) emphasize the need to integrate power analysis into intersectional climate justice action, asserting, "[p]ower is a difficult concept to grasp, let alone integrate into climate justice action" (p. 87). Too often, developmental institutions address only the "practical" potential negative effects of climate change on existing inequalities without tackling the more profound social, political, and environmental transformations needed. An intersectional perspective calls for a comprehensive understanding of climate change as an ongoing phenomenon deeply rooted in structures of power.

Intersectionality, Climate Action, and Multiple Scales of Power

Climate change and climate action operate through various scales of power, from local to global. An intersectional framework draws attention to the multiple levels of power and their relation to each other. Intersectionality allows us to see that climate change is more than temperature degrees but rather a systemic outcome predicated on the intersections of racial capitalism, heteropatriarchy, settler colonialism, and other interlocking systems of oppression (Pellow, 2017). Whyte (2018) describes settler colonialism as undermining Indigenous "collective continuation" by disrupting relationships with the environment and creating a unique form of environmental injustice that affects Indigenous peoples' ability to manage and protect their lands, waters, and resources (see also chapter by Fenelon and Figueroa-Hetland in this volume). Such power relations happen at the micro, meso, and macro level, and an intersectional perspective encourages us to assess relationships within and between these levels of power. A multiscalar analysis is central to a critical environmental justice framework, as laid out by David Pellow, to connect local action to global solutions (Pellow, 2017). Using an intersectional perspective to examine the range of levels within climate action reveals the significance of bodily health, community action, governmental power, and global partnerships. Central to intersectionality is to link everyday identity troubles to larger structural patterns. Therefore, we use intersectionality to examine climate action as a way to assess power relations beyond identity, including systems of power of capitalism and settler colonialism.

Macro-Level Institutions and Climate Action

To examine power at the largest scale is to focus on institutions, their interactions with each other, and their power toward groups. Such institutions include governments, militaries, religion, nonprofits, governmental organizations, and corporations. Institutions can embody equitable or inequitable power through resource redistribution, cultural norms, and policy enforcement. This enforces inequalities through the power of who or what groups benefit and lose from certain activities. Generally, examples of systems of power are structural racism, patriarchy, ableism, imperialism, and settler colonialism. Scholars using a structural intersectionality framework will focus on the overlapping macro-level systems by understanding labor patterns, land-use policies, and residential segregation (Homan et al., 2021). Intersectional critiques of macro-level global climate governance emphasize the interconnectedness of climate change with a system of racism, capitalism, imperialism, and colonialism (Sealey-Huggins, 2017; Sultana, 2022a, 2022b; Táíwò, 2022).

Supranational global governance partnerships such as COP28 mentioned earlier are typical examples of the macro-level climate governance institutions that work to shape climate action on a global level. Activists draw attention to the influence of money and the preference toward wealthier nations within global climate action planning institutions. Scholars argue that global governance institutions are organized to benefit those in the Global North, with powerful nations like the United States exercising influence through bribery and blackmail, intelligence gathering, and backroom deals to shape climate agreements in their favor

(Sealey-Huggins, 2017). An intersectional perspective illuminates why we see resistance toward COP where activists from around the world dissent. Indeed, even establishing a degree threshold—or the tipping point after which climate change is unmanageable—frames climate change as an objective measure and assumes a universal human subject that does not exist (Sealey-Huggins, 2017). In doing so, thresholds prioritize the future conditions of those in the Global North at the expense of those in the Global South (Catney & Doyle, 2011; Sealey-Huggins, 2017). For example, the 2°Celsius limit agreed upon at the UNFCCC COP in Paris would have significant consequences for the Caribbean, as reflected in the slogan "1.5°C to Stay Alive!" adopted by the Caribbean Community Climate Change Centre (Sealey-Huggins, 2017). Similarly, the Pan African Climate Justice Alliance argues that a global degree threshold of 2°C is functionally 3.5°C for African countries, creating a "certain death for Africa" and pushes for "climate fascism" (Bonds, 2013, p. 207). Global climate action planning should include all major state institutions—for example, the Intergovernmental Panel on Climate Change has within their guidelines that emissions from overseas bunker fuels, military bases abroad, and actions associated with multilateral operations all be exempted (Crawford, 2022). This can result in an undercounting of greenhouse gas emissions and climate action accountability from military sources. Intersectional perspectives can critique much of the climate action proposed at the macro-institutional level by asking who is prioritized in climate-change decision making and who benefits from such arrangements.

We want to highlight states' roles in climate action because of their significant influence in climate action planning. States are important institutions for making (in)equalities because of the power they carry, the materials they distribute, and the policies they implement. Racial scholars demonstrate the state has historically made racial inequities through institutional policies such as slavery, redlining, and census (Lipsitz, 1995; Massey & Denton, 1993). An intersectional perspective must evaluate institutions and how state agencies reinforce who is worthy of being saved from climate change. For example, institutional programs like Reducing Emissions from Deforestation and Forest Degradation Program (REDD+), through which the United Nations and national governments implement forest preservation projects (see chapter by S. Alvarez in this volume), may harm Indigenous peoples and communities often not recognized as forest "owners" (Forest Peoples Programme, 2023). As such, forest-dependent communities increasingly question REDD + programs, reflecting concerns about inequitable benefits and the risk of a global green land grab, emphasizing the need to safeguard their full participation in the planning process (Forest Peoples Programme, 2023). As such, an intersectional perspective on state action envisions "an approach to politics which is non-institutional" and the importance of noninstitutional solutions within the community and outside of the state (Dupuis-Déri, 2016). This echoes an anarchist approach to social and political action that prioritizes community action and "commons (commonly owned resources and means of production through voluntary associations practicing mutual aid)" (Dupuis-Déri, 2016, p. 52). As many have raised concerns about climate action planning at COP, it is imperative to incorporate community-based solutions.

Meso Level—Community and Collective Action

The meso level of power relations includes smaller institutions, such as city governments. The number of people living in cities continues to grow (Davis, 2006). As most people reside in urban areas, cities can play an important role in climate action. City governments are "key laboratories of change and spaces of opportunity" (Sareen & Waagsaether, 2023, p. 1). The shortfalls of neoliberalism, such as constraining governmental funds and lack of social safety nets, have led grassroots organizations to become politically involved at the city level (Finley, 2017; Sareen & Waagsaether, 2023). Although some early literature may view cities as an issue, they are places for resisting and mobilizing climate action (Pellow, 2024). Throughout the United States, there are examples of municipal-level movements, including Cooperation Jackson, Seattle Neighborhood Action Councils, Portland Assembly Olympia Assembly, and Genese Grill's District City Councilor (Finley, 2017).

One example of meso-level climate action is the movement of new municipalism, which centers on transforming local government toward a democratic state and economy by focusing on urban sustainability and meeting low-carbon needs (Sareen & Waagsaether, 2023). New municipalism works on climate action in three ways: (1) building inclusion by expanding democratic governance with new forms of "experimental, flexible and open-ended governance processes"; (2) strengthening alignment by improving the implementation across [vertical and horizontal] levels of government; and (3) directing orientation by redirecting efforts from incumbent practices that reinforce fossil fuel industries (Sareen & Waagsaether, 2023, p. 6). An example of new municipalism can be seen in Spain's *Decidim Barcelona*, which worked to build inclusion and address social-environmental injustices by incorporating virtual platforms of municipal proposals, and through these mechanisms, urban sustainable legislation has proceeded (Sareen & Waagsaether, 2023). Other municipal-level efforts include developing a "Low Emissions Zone," which limits individual car use in the historical district (Sareen & Waagsaether, 2023). Such municipal-level actions address neoliberal climate action's pitfalls by embracing intersectional components, focusing on a polycentric approach, and breaking down hierarchies through community building.

Micro-Level Bodies and Climate Action

We now turn to ways to examine power at the micro level by focusing on bodies. Climate change affects environmental health, and our bodies are our direct connection to the environment. Public health scholar Nancy Krieger (1994) developed ecosocial theory to stress the structural and ecological conditions of health exposure over genetic explanations. The systemic powers of racism, patriarchy, and other forms of oppression become embodied through health outcomes. For example, in Los Angeles, immigrants of color face heightened health risks due to pollution from ports, highways, and railways (Kim, 2021). Climate legislation tends to overlook the effects of the fossil fuel system on human health. For example, Michael Méndez (2020) builds on Krieger's ecosocial theory to argue that greenhouse gases have many co-pollutants associated with air pollution.

Carbon embodiment is the "human embodiment of climate change," recognizing the cumulative impacts and includes "activities such as mining, driving trucks, running factories, and refining chemicals to produce building materials and products result in greenhouse gas emissions" (Méndez, 2020, p. 42). Carbon embodiment is an intersectional perspective because it relies on "the multiple knowledges and the human impacts of climate change, as well as how these impacts are facilitated through the diffuse infrastructural body (for example, oil refining, extraction of raw materials, transportation, construction of buildings)" (Méndez, 2020, p. 42). In *Climate Change From the Streets*, Méndez (2020) documents activists in Richmond, California, which is home to California's largest oil refinery, arguing for a more holistic perspective on climate change legislation that encompasses all the various pollutants, their sources, and communities who experience negative health consequences as a result of the refinery and all its associated industrial facilities. Carbon embodiment acknowledges power differentials and moves for a multiscalar approach to mitigate health exposure (Méndez, 2020). Climate legislation should be adaptable and flexible and recognize the history that made unevenness.

Farhana Sultana's (2022b) concept of climate coloniality further enriches this discussion by highlighting how colonial and imperial violence shape embodied experiences of climate change (Sultana, 2022b). Communities in formerly colonized regions face disproportionate climate risks not simply due to historical patterns but the West's overdevelopment while actively underdeveloping resource-rich countries. The impacts manifest in intersectional ways, including the colonization of women's bodies and knowledge systems, the imposition of Eurocentric gender norms that subjugate local practices, the sacrifice of racialized women's bodies for capitalism, and the use of overpopulation narratives to exert control over women's reproductive decision making (Sultana, 2022b, pp. 5–6). As such, the lived experiences of climate change—from health impacts to displacement—are shaped by and reinforce colonial power dynamics (Sultana, 2022b). Jade Sasser (2024) further illuminates the embodied experience of climate change through its implications on deeply personal reproductive decision making, showing how the threat of climate change and ecological anxiety shape family formation and decisions about if, when, and how to have children. While Sasser (2024) demonstrates how women of color are most likely to consider having fewer children due to concerns about climate change, for some, childrearing is an act of racial hope amid the climate crisis, reflecting multigenerational tools of resistance (Sasser, 2024). As the embodied impacts of climate change become increasingly tangible, from ecological disasters to health effects, Sasser's research demonstrates how the lived experiences and environmental anxieties around climate change shape reproductive decision making (Sasser, 2024). By taking an intersectional perspective, we can see the multiple levels of power based on histories of oppression, and we see concepts and actions regarding climate action to address this.

Intersectional Climate Action: Case Studies and Strategies

Intersectional climate action involves collective efforts to address systemic power imbalances contributing to climate change risk and impacts. Communities around the world from different backgrounds are organizing climate action. Activists and scholars engage with intersectional climate action from many diverse frameworks, including decolonization, feminist political ecology, intersectional feminist praxis of care, decolonial feminist political ecology, and non-Western epistemological perspectives (Gonda, 2019; Resurrección, 2021, 2024; Sultana, 2021a, 2021b, 2022b). Such climate action reframes dominant discussions of climate change to be inclusive and accountable, emphasizing the need for reparations, reciprocity, and redistribution of power. Here, we highlight forms of intersectional climate action, including resistance to fossil fuels, municipal movements, climate debt and reparations, epistemological justice, and youth mobilization. While these cases suggest a multitude of ways to organize potential sites of intersectional climate action, the innate critical approach to intersectionality should not be reduced to a "buzzword" (Davis, 2008).

Resistance to Fossil Fuel Extraction

Worldwide, resistance to fossil fuel extraction occurs using diverse tactics to defend natural resources and disrupt the fossil fuel industry. Scholarship in climate action documents numerous examples, from legal challenges to direct action to campaigns for divestment. These efforts aim to defend land, air, water, and forests by disrupting various states of fossil fuel use, including infrastructure development, extraction, transpiration, and burning of fossil fuels

(Parkin, 2020; Temper et al., 2020). Resistance recognizes that climate injustice is rooted in systems of oppression, including racism, colonialism, capitalism, heteropatriarchy, imperialism, and other intersecting axes of domination. These systems intersect to create "sacrifice zones," areas where nearby residents and environments are disproportionately burdened with environmental harms, allowing corporations and consumers to obtain inexpensive products and energy without acknowledging or paying the full environmental and social impacts of production (Bell, 2013; Bell & York, 2010; Juskus, 2023; Kim, 2021). As capitalism emerged alongside the widespread adoption of fossil fuels as the dominant energy source, there is a resulting ideology of limitless production, consumption, and wealth accumulation (Foster, 1999; Tokar & Gilbertson, 2020).

Direct action against fossil fuel extraction includes efforts to halt the construction and operation of fossil fuel infrastructure and disrupt transportation. Examples include resistance to mountaintop removal and protests of the Keystone XL and Dakota Access Pipeline in the United States. Such protests demonstrate the interconnectedness of tribal sovereignty and climate action (Parkin, 2020; Whyte, 2017). In Newcastle, Australia, a coalition of Indigenous groups, environmental activists, and farmers formed the "Lock the Gate" alliance to block the largest coal port (Temper et al., 2020). Such place-based mobilization often consists of Indigenous people, women of color, the working class, young people, and environmental groups. For example, in North America, Bell (2013) details the grassroots resistance of working-class women to coal mining in central Appalachia. Additionally, immigrant-led resistance in Los Angeles, as highlighted by Kim (2021), showcases Asian and Latina women organizing against air pollution due to heightened health risks from ports, highways, and railways, motivated by their unique social location as immigrant women and mothers within the context of neoliberal capitalism. Because the "matrix of oppression" reveals women from different positionalities have varied lived experiences, this can make building intersectional alliances difficult. However, research on intersectional praxis from a major women of color organization shows that activists build solidarity using logics of recognizing the similarities or acknowledging the differences while building coalitions across different groups (Luna, 2016). These examples show community-led forms of direct resistance to fossil fuels and the importance of solidarity and alliance building between members of different marginalized communities, recognizing the interconnected nature of oppression is an essential component of intersectional climate action, as inequality is mutually constituted and reinforcing (Mikulewicz et al., 2023).

Municipal and Local Climate Action

Municipal and local spaces shed light on intersectional climate action because they draw attention to the importance of community-led efforts to combat climate change. Climate activists have been pushing for climate action that has radical social changes and not just technical solutions (Pellow, 2024). The Occupy Climate Change framing builds on the movements of Occupy Wall Street to radically adapt ways of living and using resources that challenge capitalism and the state. Grassroot organizations have built solidarity to address issues after major natural disasters such as Hurricane Sandy and Katrina. These movements "wide[n] practices of reclaiming and managing spaces under noncapitalist logics" (Pellow, 2024, p. 27). Some community-based organizations (CBOs) make explicit use of intersectional frameworks to advocate for urban climate justice (Strange et al., 2024). For example, Strange et al. (2024) identify three strategies employed by CBOs in San Francisco grounded in intersectionality: (1) recognizing and rectifying historical and compounding vulnerabilities, (2) people-centric and place-based solutions, and (3) collaborative partnerships and collective action (Strange et al., 2024). Additionally, moving to the state level, in Chile there has been a lot of movement to reduce poverty, and a new constitution was recently introduced, which was pushed by intersectional efforts across environmental, Indigenous, and feminist groups (Visconti, 2024). The discussion of the new constitution brought up issues around "plurinationality, environmental justice, territorial sovereignty and protection of the commons pursuing a post-extractive transition" (Visconti, 2024, p. 233). One of the most critical challenges is to move "on the protection of the commons to overcome the utilitarian view of nature as a stockpile of commodities and strategic resources" (Visconti, 2024, p. 234). In this context, grassroots and locally driven climate action is significant in addressing the climate crisis.

Climate Debt and Reparations

A comprehensive framework centers ethics and justice to organize claims against those most responsible for climate change impacts (Burkett, 2009). Climate reparations emphasize the need for "accountability and regress" for the disproportionate impacts of climate change and involves compensating low-income countries for the loss and damage they have suffered as a result of the historical and ongoing emissions of high-income countries (Chapman & Ahmed, 2021, p. 81). This approach advocates for the establishment of mechanisms like the Global Climate Reparations Fund (GCRF) to compensate for damages inflicted by climate change on small island states and low-income countries (also see chapter by Khan and Khan in this volume; Chapman & Ahmed, 2021). Importantly, Olúfẹ́mi O. Táíwò (2022) asserts that rather than focusing on past wrongs or reconciliation, climate reparations should be future-orientated with the goal of building a more equitable world system:

> The racially advantaged, the Global North, and institutional repositories of plunder should bear more of the burdens of constructing the just world order, *not* because of the relationship that they, *as responsible* moral agents, hold to the injustices of the past. Those advantaged by global racial empire should bear more burdens because of the relationship that their *advantages* hold to that history.
>
> (Táíwò, 2022, p. 123, emphasis in original)

The concept of climate reparations can also be extended to address intranational climate injustice, including domestic climate reparations to address climate change's disproportionate impact on marginalized communities within a country. Both concepts exemplify intersectional climate action as they mitigate climate change by placing the climate crisis in the context of historical and ongoing structural violence to support just worldmaking (Chapman & Ahmed, 2021; Sultana, 2022b; Táíwò, 2022; Warlenius, 2018).

Epistemological Justice

Intersectional climate action includes centering the epistemologies of those most impacted by climate change, decolonizing knowledge production and methodologies, and recognizing the agency of communities' climate solutions. As "alternative epistemologies and cosmologies emerge from lived experiences that were/are devalued in Eurocentric modernity and climate coloniality," dominant framings of climate solutions often devalue local and Indigenous knowledge systems (Sultana, 2022b, p. 3). As such, epistemological justice necessitates challenging the hegemony of Eurocentric scientific knowledge in favor of approaches that respect non-Western and Indigenous knowledge systems, practices, and methodologies (Smith, 2021; Sultana, 2022b; Whyte et al., 2020). This is linked to recognizing and protecting Indigenous rights, including land rights and self-determination (Whyte et al., 2020). As such, climate politics and practices must give Indigenous communities the power to manage their lands and sustain their cultural practices (Whyte, 2018). Indigenous and other communities on the frontlines of climate action are advancing locally led solutions that integrate traditional ecological knowledge (TEK). For example, American Indian and Alaska Native Climate Change Working Groups, tribal colleges and universities, and others integrate Indigenous knowledge into geoscience education and research, reminding us that "the world is full of relatives, not resources" (Wildcat, 2009, p. 139). At COP28, the Indigenous Environmental Network submitted "an Ethical Protocol for the Protection and Use of Traditional Indigenous Knowledge" to guide climate policies and actions (Indigenous Environmental Network, 2023). Organizations like the NDN Collective are developing the infrastructure to implement Indigenous-led climate solutions that uphold Indigenous sovereignty and self-determination (see also chapter by Fenelon and Figueroa-Hetland in this volume) (NDN Collective, n.d.).

It is crucial to understand that Indigenous practices can only be adopted with full consideration of their cultural context and histories of resistance to settler colonialism (Whyte et al., 2020). Indigenous sustainability approaches are deeply rooted in specific cultural contexts, traditional knowledge systems, and histories of environmental adaptation. They should be understood as strategies for Indigenous collective continuation and self-determination in the face of ongoing settler colonialism (Whyte et al., 2020). Non-Indigenous communities should seek to learn from and support Indigenous-led sustainability efforts while respecting Indigenous sovereignty, land rights, and self-determination (Whyte et al., 2020). For example, Indigenous people often play a limited role in major renewable energy decisions despite the proximity of many energy sites to Indigenous territories (Whyte, 2023). This highlights how the setter state has curtailed Indigenous peoples' administrative and infrastructure decisions (Whyte, 2023). In addition, a review of Canada's environmental policy finds that TEK, especially the knowledge held by women, is underemphasized in environmental decision making, perpetuating gender blindness, failing to recognize the unique climate impacts on Indigenous women, and devaluing Indigenous women's TEK (Gricius & Martel, 2024).

Youth Mobilization

Youth mobilization is another significant example of intersectional climate action because age is an overlooked dimension of climate impact where youth have less social power than older people (see chapters by Uba and Karsgaard in this volume). Heejin Han and Sang Wuk Ahn (2020) draw attention to the rich history of youth activism within social movements, such as student protests, and continuing with the rise of climate action worldwide. Youth activism involves activists between the ages of 15 and 24, a demographic that comprises an estimated 1.2 billion people, or roughly 16% of the population (Han & Wuk Ahn, 2020). However, areas expected to be most affected by climate change also have a lot of youth, such as African countries (Han & Wuk Ahn, 2020). An important example within the youth climate action is the court case *Juliana v. U.S.*, where 21 youth plaintiffs from Oregon are suing the government for irresponsible management toward climate change, and other countries are following (Han & Wuk Ahn, 2020). Han and Wuk Ahn (2020) show the rise in youth movements gaining global popularity in 2018–2019, particularly with Swedish Greta Thunberg's Fridays For Futures to boycott schools as a climate strike. By March 2019, over a million young people participated in hundreds of cities in the climate strike during the Paris

Agreement. They filed a complaint with the UN Convention on the Rights of the Child at the UN Youth Climate Summit. The Fridays For Future campaign worked on improving the European Commission, such as reaching net zero by 2035. Han and Wuk Ahn (2020) conclude by discussing the cross-coalition support for youth movements with other activist organizations, such as laborers and teachers. Youth climate action has important insights into intersectional climate action in acknowledging the urgency of time and future generations.

Concluding Toward a Truly Intersectional Climate Action

In this chapter, we have reviewed how intersectionality theory allows us to see the disproportionate impacts of climate change between and within communities, examine multiple levels of power, and examine various case studies and strategies. Intersectional climate action provides ways for evaluating and posing the most effective solutions toward equitable climate action. Central to this approach is the illumination of the institutional and political roots of the climate crisis, providing the analytical framework for interpreting climate change adaptation and mitigation strategies. According to Mwenda and Bond, intersectionality is a tool to resist "reformist reforms," those that "accept and legitimize status quo institutional forms, endorse market mechanisms, and neglect to incorporate analysis highlighting class, race, gender, generation, and geographic power relations" (Mwenda & Bond, 2020, p. 6). This perspective echoes climate justice advocates who have problematized technological and economic "corporate-driven false solutions" to the climate crisis, or policies that rely on market-mechanisms like renewable energy, carbon offset, geoengineering, biofuel, and nuclear power (Tokar, 2014, p. 3). Institutional climate action often lacks depth, thus maintaining existing power structures. As a result, resistance is often co-opted and watered down to make only superficial change. It is important to keep the critical edge of intersectionality and avoid reducing it to a "buzzword" (Davis, 2008). This is common within reformist reforms that address concerns of racial and climate justice movements, thus avoiding meaningful transformation or the disruption of the status quo (Tokar, 2014). Building on the conceptualization by André Gorz, nonreformist reforms aim to challenge and transform existing power structures (Gorz, 1968; Tokar, 2014). Intersectional climate justice emphasizes the interconnections of social identities and power structures in climate action. To address the urgency of the climate crisis, we must stress the critical lens; therefore, we focus on intersectionality as a power analysis to illuminate whose interests, experiences, and epistemologies are centered.

A truly intersectional climate action must center marginalized voices and knowledge systems, addressing the epistemic violence often present in framing climate issues and knowledge production. Moreover, it must ground intersectional climate action in the lived experiences of impacted individuals and communities as "legitimate sites of geopolitical knowledge" that "counterbalance hegemonic framings and structural forces" (Sultana, 2022b, p. 11). Rather than engaging with climate change through abstractions, this perspective emphasizes the importance of "feeling with climate change" and exploring its "heaviness" through embodied personal experience (Sultana, 2022b, p. 3). This means climate action should be (1) rooted in the lived experiences and epistemologies of those most affected by climate change, (2) incorporating diverse perspectives in climate policy and climate organizations, and (3) prioritizing community-driven climate initiatives. Intersectional approaches also acknowledge the epistemic violence present in dominant framings of climate change, which frame the other as epistemologically and ontologically deficient and seek to create more inclusive and effective solutions to the climate crisis (Mignolo, 2017; Sultana,

2022b). A truly intersectional climate action is grounded in feminist, antiracist, decolonial, anticapitalist, and anti-imperial frameworks, and recognizes the need to break down these hierarchies while working toward a transformative, more equitable future.

References

Almeida, P., González Márquez, L. R., & Fonsah, E. (2024). The forms of climate action. *Sociology Compass, 18*(2), e13177. https://doi.org/10.1111/soc4.13177

Amorim-Maia, A. T., Anguelovski, I., Chu, E., & Connolly, J. (2022). Intersectional climate justice: A conceptual pathway for bridging adaptation planning, transformative action, and social equity. *Urban Climate, 41*, 101053. https://doi.org/10.1016/j.uclim.2021.101053

Batrawy, A. (2023, December 12). Climate activists struggle to be heard at this year's U.N. climate talks. NPR. https://www.npr.org/2023/12/12/1218675784/climate-activists-struggle-to-be-heard-at-this-years-u-n-climate-talks

Bell, S. E. (2013). *Our roots run deep as ironweed: Appalachian women and the fight for environmental justice*. University of Illinois Press. https://www.jstor.org/stable/10.5406/j.ctt3fh586

Bell, S. E., & York, R. (2010). Community economic identity: The coal industry and ideology construction in West Virginia. *Rural Sociology, 75*(1), 111–143. https://doi.org/10.1111/j.1549-0831.2009.00004.x

Bonds, P. (2013). Climate justice in, by, and for Africa. In M. Dietz & H. Garrelts (Eds.), *Routledge handbook of the climate change movement* (pp. 205–221). Routledge.

Burkett, M. (2009). Climate reparations. *Melbourne Journal of International Law, 10*(2), 509–542.

Catney, P., & Doyle, T. (2011). The welfare of now and the green (post) politics of the future. *Critical Social Policy, 31*(2), 174–193. https://doi.org/10.1177/0261018310395921

Chapman, A. R., & Ahmed, A. K. (2021). Climate justice, human rights, and the case for reparations. *Health and Human Rights, 23*(2), 81–94.

Collins, P. H. (2000). *Black feminist thought: Knowledge, consciousness, and the politics of empowerment* (2nd ed.). Routledge.

Collins, P. H. (2015). Intersectionality's definitional dilemmas. *Annual Review of Sociology, 41*, 1–20. https://doi.org/10.1146/annurev-soc-073014-112142

Collins, P. H. (2019). *Intersectionality as critical social theory*. Duke University Press.

Collins, P. H., & Bilge, S. (2016). Intersectionality: *Key concepts*. Polity Press.

Combahee River Collective, The. (1977, April). *The Combahee River Collective statement*. Boston, MA.

Crawford, N. C. (2022). *The Pentagon, climate change, and war: Charting the rise and fall of U.S. military emissions*. The MIT Press.

Crenshaw, K. (1991). Mapping the margins: Intersectionality, identity politics, and violence against women of color. *Stanford Law Review, 43*(6), 1241–1299. https://doi.org/10.2307/1229039

Cultural Survival. (2023, December 7). *Reporting on week one of COP28! Cultural survival*, Cultural Survival. https://www.culturalsurvival.org/news/reporting-week-one-cop28A

Daniel, T., & Dolan, M. (2020). Intersectionality and collective action: Visioning a Feminist Green New Deal in the US. *Gender & Development, 28*(3), 515–533. https://doi.org/10.1080/13552074.2020.1843829

Davis, K. (2008). Intersectionality as buzzword: A sociology of science perspective on what makes a feminist theory successful. *Feminist Theory, 9*(1), 67–85.

Davis, M. (2006). *Planet of slums*. Verso Books.

Ducre, K. A. (2013). *A place we call home: Gender, race, and justice in Syracuse*. Syracuse University Press.

Ducre, K. A. (2018). The Black feminist spatial imagination and an intersectional environmental justice. *Environmental Sociology, 4*(1), 22–35. https://www.tandfonline.com/doi/full/10.1080/23251042.2018.1426089

Dupuis-Déri, F. (2016). Is the state part of the matrix of domination and intersectionality? An anarchist inquiry. *Anarchist Studies*, 24, 52.

Ergas, C., McKinney, L., & Bell, S. E. (2021). Intersectionality and the environment. In B. S. Caniglia, A. Jorgenson, S. A. Malin, L. Peek, D. N. Pellow, & X. Huang (Eds.), *Handbook of environmental sociology* (pp. 15–34). Springer.

Finley, E. (2017). The new municipal movements. *ROAR*, *6*, 14–20.

Forest Peoples Programme (FPP). (2023, September 29). *REDD+ and related initiatives*. Forest Peoples Programme. https://www.forestpeoples.org/en/work-themes/global-finance-trade/redd-and-related-initiatives

Foster, J. B. (1999). *The vulnerable plant: A short economic history of the environment*. Monthly Review Press.

Gonda, N. (2019). Re-politicizing the gender and climate change debate: The potential of feminist political ecology to engage with power in action in adaptation policies and projects in Nicaragua. *Geoforum*, *106*, 87–96. https://doi.org/10.1016/j.geoforum.2019.07.020

Gorz, A. (1968). Reform and revolution. *Socialist Register*, *5*, 111–143.

Gricius, G., & Martel, A. (2024). Integrating Indigenous women's traditional knowledge for climate change in Canada. *AlterNative: An International Journal of Indigenous Peoples*, *20*(3), 377–387. https://doi.org/10.1177/11771801241241816

Han, H., & Ahn, S. W. (2020). Youth mobilization to stop global climate change: Narratives and impact. *Sustainability*, *12*, 4127. https://doi.org/10.3390/su12104127

Homan, P., Brown, T. H., & King, B. (2021). Structural intersectionality as a new direction for health disparities research. *Journal of Health and Social Behavior*, *62*, 350–370. https://doi.org/10.1177/00221465211032947

hooks, b. (1982). *Ain't I a woman: Black women and feminism*. Pluto Press. Reprinted 1986, 1990.

Indigenous Environmental Network. (2023, November 28). *IEN introduces ethical protocol at the UN LCIPP FWG10*. https://www.ienearth.org/ien-introduces-ethical-protocol-at-the-un-lcipp-fwg10/

Juskus, R. (2023). Sacrifice zones. *Environmental Humanities*, *15*(1), 3–24. https://doi.org/10.1215/22011919-10216129

Kaijser, A., & Kronsell, A. (2014). Climate change through the lens of intersectionality. *Environmental Politics*, *23*(3), 417–433. https://doi.org/10.1080/09644016.2013.835203

Kim, N. Y. (2021). *Refusing death: Immigrant women and the fight for environmental justice in LA*. Stanford University Press.

Krieger, N. (1994). Epidemiology and the web of causation: Has anyone seen the spider? *Social Science & Medicine*, *39*, 887–903. https://doi.org/10.1016/0277-9536(94)90202-X

Lipsitz, G. (1995). The possessive investment in whiteness: Racialized social democracy and the 'white' problem in American studies. *American Quarterly*, *47*(3), 369–387.

Luna, Z. (2016). "Truly a women of color organization": Negotiating sameness and difference in pursuit of intersectionality. *Gender & Society*, *30*(5), 769–790.

Malin, S. A., & Ryder, S. S. (2018). Developing deeply intersectional environmental justice scholarship. *Environmental Sociology*, *4*(1), 1–7. https://doi.org/10.1080/23251042.2018.1446711

Massey, D. S., & Denton, N. A. (1993). *American apartheid: Segregation and the making of the underclass*. Harvard University Press.

Méndez, M. (2020). *Climate change from the streets: How conflict and collaboration strengthen the environmental justice movement*. Yale University Press.

Mignolo, W. D. (2017). Coloniality is far from over, and so must be decoloniality. *Afterall: A Journal of Art, Context and Enquiry*, *43*, 38–45. https://doi.org/10.1086/692552

Mikulewicz, M., Caretta, M. A., Sultana, F., & Crawford, N. J. W. (2023). Intersectionality & climate justice: A call for synergy in climate change scholarship. *Environmental Politics*, *32*(7), 1275–1286. https://doi.org/10.1080/09644016.2023.2172869

Mwenda, M., & Bond, P. (2020). African climate justice: Articulations and activism. In B. Tokar & T. Gilbertson (Eds.), *Climate change resistance and renewal: Campaigns and strategies from the frontlines* (pp. 202–218). Routledge.

NDN Collective. (n.d.). *Protect the He Sapa*. https://ndncollective.org/protect-the-he-sapa/

Parkin, S. (2020). Resistance is fertile: Direct action vs. fossil fuels across North America. In *Climate justice and community renewal* (pp. 113–136). Routledge.

Pellow, D. (2017). *What is critical environmental justice?* Polity Press.
Pellow, D. (2024). Foreword: From Occupy Climate Change! to confronting loss and damage. In M. Armiero, E. Turhan, & S. P. De Rosa (Eds.), *Urban movements and climate change: Loss, damage and radical adaptation* (pp. 11–17). Amsterdam University Press.
Perez, C. S. (2021). Thinking (and feeling) with Anthropocene (Pacific) islands. *Dialogues in Human Geography, 11*(3), 429–433. https://doi.org/10.1177/20438206211017453
Perkins, P. E. (2019). Climate justice, commons, and degrowth. *Ecological Economics, 160*, 183–190. https://doi.org/10.1016/j.ecolecon.2019.02.005
Prete, G. (2023). Indigenous leaders take center stage at COP28. https://earth.org/indigenous-leaders-champion-a-stronger-role-in-addressing-climate-change-at-cop28/
Resurrección, B. P. (2021). Gender, climate change and disasters: Vulnerabilities, responses, and imagining a more caring and better world. Background paper prepared for the Expert Group Meeting, Commission on the Status of Women (CSW66), UN Women.
Resurrección, B. P. (2024). Colonial erasures in gender and climate change solutions. *WIREs Climate Change*, e890. https://doi.org/10.1002/wcc.890
Sareen, S., & Waagsaether, K. L. (2023). New municipalism and the governance of urban transitions to sustainability. *Urban Studies, 60*(11), 2271–2289. https://doi.org/10.1177/00420980221114968
Sasser, J. (2024). *Climate anxiety and the kid question: Deciding whether to have children in an uncertain world*. University of California Press.
Sealey-Huggins, L. (2017). "1.5°C to stay alive": Climate change, imperialism and justice for the Caribbean. *Third World Quarterly, 38*(11), 2444–2463. https://doi.org/10.1080/01436597.2017.1368013
Smith, L. T. (2021). *Decolonizing methodologies: Research and Indigenous peoples* (3rd ed.). Zed Books.
Smith, R. J., & Rhiney, K. (2016). Climate (in)justice, vulnerability and livelihoods in the Caribbean: The case of the indigenous Caribs in northeastern St. Vincent. *Geoforum, 73*, 22–31. https://doi.org/10.1016/j.geoforum.2015.11.008
Strange, K. F., Satorras, M., & March, H. (2024). Intersectional climate action: The role of community-based organisations in urban climate justice. *Local Environment, 29*(7), 865–885. https://doi.org/10.1080/13549839.2024.2315992
Sultana, F. (2010). Living in hazardous waterscapes: Gendered vulnerabilities and experiences of floods and disasters. *Environmental Hazards, 9*(1), 43–53. https://doi.org/10.3763/ehaz.2010.SI02
Sultana, F. (2021a). Political ecology 1: From margins to center. *Progress in Human Geography, 45*(1), 156–165. https://doi.org/10.1177/0309132520936751
Sultana, F. (2021b). Climate change, COVID-19, and the co-production of injustices: A feminist reading of overlapping crises. *Social & Cultural Geography, 22*(4), 447–460. https://doi.org/10.1080/14649365.2021.1910994
Sultana, F. (2022a). Critical climate justice. *The Geographical Journal, 188*(1), 118–124. https://doi.org/10.1111/geoj.12417
Sultana, F. (2022b). The unbearable heaviness of climate coloniality. *Political Geography, 99*, 102638. https://doi.org/10.1016/j.polgeo.2022.102638
Táíwò, O. O. (2022). *Reconsidering reparations*. Oxford University Press.
Temper, L., Avila, S., Del Bene, D., Gobby, J., Kosoy, N., Le Billon, P., Martinez-Alier, J., Perkins, P., Roy, B., & Scheidel, A. (2020). Movements shaping climate futures: A systematic mapping of protests against fossil fuel and low-carbon energy projects. *Environmental Research Letters, 15*(12), 123004. https://doi.org/10.1088/1748-9326/abc197
Tokar, B. (2014). Movements for climate justice in the US and worldwide. In *Routledge handbook of the climate change movement* (pp. 131–146). Routledge.
Tokar, B., & Gilbertson, T. (Eds.). (2020). *Climate justice and community renewal: Resistance and grassroots solutions*. Routledge.
Visconti, C. (2024). Practices of resilience: Questioning urban adaptation in the Chilean social upsurge. In M. Armiero, E. Turhan, & S. P. De Rosa (Eds.), *Urban movements and climate change: Loss, damage and radical adaptation* (pp. 233–234). Amsterdam University Press.

Warlenius, R. (2018). Decolonizing the atmosphere: The climate justice movement on climate debt. *Journal of Environment & Development*, 27(2), 131–155. https://doi.org/10.1177/1070496517744593

Whyte, K. P. (2014). Indigenous women, climate change impacts, and collective action. *Hypatia* 29(3), 599–616. https://doi.org/10.1111/hypa.12089

Whyte, K. (2017). The Dakota Access Pipeline, environmental injustice, and US colonialism. *Red Ink: An International Journal of Indigenous Literature, Arts, & Humanities*, 19(1), 154–169.

Whyte, K. (2018). Settler colonialism, ecology, and environmental injustice. *Environment and Society*, 9, 125–144. https://www.jstor.org/stable/26879582

Whyte, K. (2023). Indigenous environmental justice, renewable energy transition, and the infrastructure of sovereignty. In P. C. Rosier (Ed.), *Environmental justice in North America* (pp. 307–330). Routledge.

Whyte, K., Caldwell, C., & Schaefer, M. (2020). Indigenous lessons about sustainability are not just for "all humanity." In J. Sze (Ed.), *Sustainability* (pp. 149–179). New York University Press. https://doi.org/10.18574/nyu/9781479894567.003.0007

Wildcat, D. (2009). *Red alert!: Saving the planet with Indigenous knowledge*. Fulcrum.

CHAPTER 8

"Overwhelming Heat" in Racialized Communities

ALESIA MONTGOMERY AND MARCOS MAGAÑA

FROM the preindustrial era to the digital era (1850–2020), the burning of fossil fuels increased the global average surface temperature on Earth by roughly 1 degree Celsius (2 degrees Fahrenheit) (Lindsey & Dahlman, 2024). The change seems small, but it has lethal consequences. The World Health Organization (2024a, 2024b) calls for urgent action to reduce the risk of heat stroke, citing estimates that an average of 489,075 heat-related deaths occurred globally each year between 2000 and 2019.[1] As the planet warms, extreme heat events (EHEs) are becoming more frequent and severe. EHEs are periods of high temperatures and humidity that are unusual for particular locations (McAnneny, 2021). Public awareness of the threat has spread with news of mass deaths from EHEs in cities such as Chicago in 1995, Paris in 2003, Karachi in 2015, and Mecca in 2024. Beyond EHEs, the chronic stress of high temperature intervals that are longer in duration imperils lives, livelihoods, and the quality of life (Oppermann et al., 2021).

Racialized communities are disproportionately endangered by warming trends (Alizadeh et al., 2022; Herold et al., 2017). In the United States, communities of color are exposed to higher temperatures than predominately White communities (Benz & Burney, 2021; Johnson, 2022; Mitchell & Chakraborty, 2015). Latino workers are overrepresented in agricultural and construction jobs that are vulnerable to extreme heat (Castillo et al., 2021; Dong et al., 2019), and Black and Indigenous people nationwide (and in some states, Latinos) die at higher rates from heat than do non-Hispanic White people (California Department of Insurance, 2024; Guo et al., 2023; J. D. Sharpe & Wolkin, 2022; Vaidyanathan et al., 2020). Across the American tropics, warming trends in southern Mexico, Central America, and the Caribbean are projected to increase droughts, monster hurricanes, floods, and vector-borne diseases that may imperil lives and livelihoods and thus drive migration (Andrade-Velázquez et al., 2021; Kaenzig & Piguet, 2013). In parts of Asia, sub-Saharan Africa, and the Middle East, rising temperatures threaten habitability (Kunda et al., 2024; Saeed et al., 2021; Vecellio et al., 2023). Around the world, agricultural workers face severe heat-related health risks. In Central America, an epidemic of kidney disease among sugarcane and cotton plantation workers is attributed to the physical impacts of hard labor in a warming climate (Peraza et al., 2012; Wesseling, 2020).

Critical environmental justice (CEJ) scholars do not deny the threat, yet they voice concerns about the epistemological assumptions that underlie heat research, and they critique the priorities that shape whether and how heat interventions are made (Hamstead, 2023; Pellow, 2024; Pulido, 2018). In line with CEJ, in this chapter we review the interdisciplinary literature on deadly heat, reading with and against dominant schools of thought about the crisis. Rather than framing race as a mere "risk factor" in the crisis, we note that the rise of climate science contributed to the making of race, and that heat interventions are entangled in racial politics. Dominant conceptions of extreme heat are ahistorical and undersocialized.[2] Since the early modern era, scholarly discourses about dangerously hot climates have obscured power and culture. In the imagined "torrid zones" of Enlightenment thinkers, hot climates produced "savages." This understanding of climate and race spread as Europe colonized and enslaved people in the tropics (Safier, 2014).

Climate science has advanced over the centuries, and its language has become more benign. As the tropics expand latitudinally, high temperatures are framed as a universal concern (Yang et al., 2020). However, it is problematic that, in policy circles, climate science is elevated above inquiries about planetary warming in the humanities and social sciences. Heat research and interventions that focus solely on physical processes do little to address the political, economic, and ideological structures that make racialized communities vulnerable to extreme heat. It must also be acknowledged that the humanities and social sciences have their own problematic relations with these communities. Across disciplines, there is a history of specious reasoning about climate and race, and the pathologization of "savage" places lingers.

Myriad interventions, informed by research, combat extreme heat. In cities of the Global North and South, government agencies, public–private partnerships, and nongovernmental organizations (NGOs) organize emergency services that include heat wave alerts, informational campaigns, and cooling centers, as well as infrastructural changes such as albedo surfaces (e.g., white roofs and cool pavements), tree canopies, and shade covers (Errett et al., 2023; Kotharkar & Ghosh, 2022; Sheridan et al., 2024). Beyond cities, interventions such as cooling bandanas, night harvests, crop changes, and irrigation innovations are used to protect farm workers and subsistence farmers from the impacts of extreme heat on health and livelihoods (Chicas et al., 2021; Rosa, 2022; Tan & Bogage, 2023).

These interventions have the potential to save lives, yet we must deepen our understanding of and responses to lethal temperatures. To advance interdisciplinary research and action that align with the needs of racialized communities, we must address *overwhelming heat*—the experience of struggling with systemic forces that threaten to overpower the sociocultural aims and capacities of community members for thermal regulation. The lead author for this chapter derives the concept of overwhelming heat from her study of Latina and Black mothers in Los Angeles. The study explores how the mothers deal with heat, given their gendered duties to care for vulnerable family members—the very young and the very old. In a sense, the mothers serve as the thermoregulation system of their households.

The term "overwhelming heat" and its synonyms are used by the mothers to describe their struggles to manage their families' exposure to hot temperatures amid housing and neighborhood problems, which bring them into conflict with property owners and government agencies. Naming a problem in the words of those who experience it is a necessary step in developing agendas for research and action that are responsive to community concerns. The specific challenges of racialized mothers in a US city may not be generalizable to other places, but the experience of struggling with the overwhelming heat of systemic forces is widely shared.

The main focus of our review is the Americas, yet we also draw on relevant studies from other regions. We begin by framing the interactions among thermal regulation, care practices, and overwhelming heat. To do so, we bring interdisciplinary research on thermal regulation as well as gender studies on caregiving into conversation with the narratives of two immigrant mothers in Los Angeles: "Beatriz" from southern Mexico and "Esther" from Haiti.[3] We note that thermal regulation in racialized communities is at once a physiological and sociocultural struggle in which caregivers such as mothers play a critical role. The concepts of EHEs and chronic heat stress direct inquiries into physical processes and their societal impacts (Luber & McGeehin, 2008). Complementing and challenging these inquiries, the concept of overwhelming heat—which frames community members as thinking and feeling agents amid rising temperatures—raises questions about how systemic processes, structural positions, and sociocultural practices interact to influence heat dangers, responses, and outcomes.

Next, we provide a history of the systems of power that have limited the ability of racialized people to (physiologically, socially) do thermal regulation. To trace this history, we review scholarly works from previous centuries as well as recent historical critiques. We start our historical analysis in the tropical regions where Beatriz and Esther were born. As Europe colonized the Americas, Africans and Indigenous people were exposed to overwhelming heat on sugar plantations and in mining operations. As we expand our focus across time and place to early industrial cities and then to today's urban areas, we make the case that overwhelming heat and its legitimations began before climate change concerns, and that the impacts of early social and environmental degradations persist. The making of the race-class order, with its discursive construction of disposable people and its material construction of dangerous places, magnifies the perils of the Anthropocene. For centuries, alongside the environmental harm that drives climate change, the sociocultural capacities of racialized people to thrive have been undermined.

After this historical overview, we illustrate how overwhelming heat can be used as a lens to evaluate heat interventions. Specifically, we review work on climate action and resiliency strategies such as cooling centers, shade structures, and urban greening. We must ask how these interventions and their governance interact with the existing struggles of community members, including caregivers such as Beatriz and Esther, to do thermal regulation.

We conclude by discussing the need to move beyond the "torrid zones" imaginary in ways that deepen our relations with other humans and with the more-than-human in hot places. In pursuit of this aim, we offer a *loving critique* of heat research—including within CEJ studies—that implicates ourselves.[4] This critique draws in part on the research of the second author that focuses on deserts in southern California. Our separate projects converge on this realization: Framing hot places as solely sites of death—even when the aim is to shed light on injustices—limits efforts to imagine and build sustainable futures.

Framing Thermal Regulation, Care Practices, and "Overwhelming Heat"

Thermoregulation enables humans and other mammals to maintain homeostasis—a stable body temperature (Mota-Rojas et al., 2021). Researchers in science and engineering examine the interaction between the natural-and-built environment and thermoregulation, while scholars in the humanities and the social sciences study political, economic, and cultural influences on thermal regulation.

The bridge between disciplines is the concept of *thermoreception* or *thermoception*, which refers to the process of sensing temperature. The latter spelling appears more often in humanities and social science journals, where sensing temperature is described as a signal not simply of a bodily state but also of one's place in the world—one's social status, environmental relation, and/or cosmological orientation (H. L. Hsu, 2023; Vannini & Taggart, 2014). Much more is at stake in thermoception than bare survival, including one's dignity. For humans, heat is an interpreted experience, and its benefits and harms extend beyond bodily impacts.

Thermoreception is the first step in thermoregulation: The hypothalamus receives information from temperature receptors on the body and then sends signals to respond to heat or cold. Key mechanisms in the heat responses of humans include perspiration, which cools the body through evaporation, and vasodilation, in which blood vessels expand to increase blood flow to the skin and thus facilitate heat transfer to the air. While it is clear that high humidity can impair the evaporation of sweat, physiological and epidemiological studies disagree on the extent to which humidity impacts health (Baldwin et al., 2023).

Beyond autonomic responses to temperature, mammals also have individual and group behavioral responses, such as seeking shade for coolness, huddling for warmth, and digging burrows to avoid temperature extremes (Harshaw & Alberts, 2012). The influence of temperature on human "behaviors" is debated, yet there is a move away from biological and environmental determinism toward recognizing that humans respond to heat with sociocultural practices that are embedded in complex meanings and materialities that differ across time and place (Auliciems, 2009; J. D. Moore, 2012).

When autonomic and behavioral responses are overwhelmed, high temperatures can result in heat-related illness (HRI). The severe effects include mental impacts such as anxiety, depression, aggression, and reduced cognitive performance (Thompson et al., 2018). Ultimately, heat stress—the overheating of the body—can lead to deadly heat stroke. For humans, high temperatures are particularly dangerous for elders and small children, pregnant women, and people with chronic health conditions such as diabetes, heart disease, and respiratory disease (Kenny et al., 2010; Kuehn & McCormick, 2017). In the United States, people of color are at greater risk of suffering heat-related illness and death, and non-US citizens are particularly at risk (Berberian et al., 2022).

Thermoception—defined as a physiological and social signal—complicates responses to heat. Hsuan Hsu, reflecting on Christina Sharpe's (2016) observations about "the weather" as anti-Blackness, explores the experience and interpretations of urban heat, as expressed in Black literature, art, and film. He notes that "temperature disparities produce and sustain racial differences in U.S. cities ... [these] differences ... can precipitate dramatically different modes of embodied thought, feeling, and action" (H. L. Hsu, 2023, pp. 773–774). Citing Black writers, Hsu contrasts air-conditioned spaces of White privilege with the searing heat of low-income Black neighborhoods, describing the rage that erupts when unauthorized thermal regulation, such as opening a fire hydrant, is blocked by the police. On one level, this rage might be understood as a bodily reaction to heat stress; on another level, it might be read as a response to the stress of being disrespected and disempowered. Physiological and sociocultural responses to heat combine.

Disparities in the natural and built environment can worsen physical and mental health outcomes. Over the life course, racialized and low-income people tend to be more exposed to environmental harms such as air pollution, which increase their risks for chronic health conditions (Brook et al., 2010; Erqou et al., 2018), which in turn make them more susceptible to HRI. In addition, urban heat island (UHI) effects—intensifications of temperature caused by multiple factors that include the absorption of heat by buildings and roads—disproportionately impact racialized and low-income groups who are unhoused or poorly housed and who reside in areas with little vegetation to cool their streets (Gabbe et al., 2023; Harris & Albrecht, 2024; A. Hsu et al., 2021).

Studies indicate that shade and air conditioning decrease the dangers of heat waves, while inadequate access to water, sanitation, and hygiene (WaSH) increases risks (Gronlund, 2014; Sorensen & Balbus, 2021). These cooling resources are embedded in "sociotechnical systems" that shape heat vulnerabilities: the social infrastructure (e.g., regulations, codes, conventions) and technical infrastructure (e.g., buildings, roads, power plants) combine to create lock-in that makes it hard in terms of politics and costs to transform them (Pincetl et al., 2016). Beyond their direct physical impacts, infrastructure inequities contribute to disparate mortality rates during heat waves by influencing sociality (Klinenberg, 2003). Impoverished streetscapes emptied of shops and parks offer few opportunities not only for relief from hot residences but also for social contact and support.

There is much research on the effects of informal support on HRI and death rates during extreme heat (Mondal et al., 2024), and there is an emerging literature on the formal organization of support that constitutes *heat governance*, which includes actors such as chief heat officers who are responsible for mitigating heat hazards in cities (Keith et al., 2021). However, conceptions of the social nature of heat responses need better grounding, starting at the most basic level with the role of thermoception in fostering care. Sensing one's own heat discomfort may lead to more than self-care—it also may prompt touching a child's forehead, placing a water bowl under a dog's nose, checking the leaves on a plant, or knocking on the door of an elderly neighbor. If heat stress is detected, help may be provided. These practices of monitoring and support for human and more-than-human others contribute to thermal regulation in households and communities.

Existing research suggests that practices of self-care and care for others to regulate heat (which may at times be in conflict) differ based on intersecting identities such as gender, ethnicity, and class, which are expressed in diverse ways across cultures (Cupples et al., 2007; Shih et al., 2022). The targets of care may range in scale from the family to the entire planet, and care may range from personal caregiving to political action.

More research is needed to understand the strategies, actions, and needs of caregivers as escalating temperatures become the new normal around the world. The scope of needs may overwhelm and "burn out" caregivers, which may endanger themselves and those who depend on them. Adapting concepts from the literature on unpaid care work, much of it in gender and family studies, would benefit heat research. This transnational literature—centering on the experiences of the caregiver and the cultural and structural contexts that shape those experiences—increasingly focuses on the challenges of heat (especially in the Global South), but much more work must be done to integrate heat research with the study of caregiving (Blair-Loy et al., 2015; Butt et al., 2020; Kadio et al., 2024; Vinyeta et al., 2016). Centering caregivers in heat research would remind us that they also may be suffering the impacts of heat stress. In addition, they may have risk factors that endanger them during heat waves—for example, they may be pregnant or have preexisting physical and mental health conditions. The pressures of caring for the needs of others during heat waves may add to their everyday burdens.

To develop a coherent body of interdisciplinary knowledge in this area, there is a need to develop shared frames for theorizing the experience and meanings of heat (thermoception) during caregiving, and there is a need to model the multiscalar interactions that shape heat dangers as a physiological and sociocultural phenomenon.

The concept of *overwhelming heat*—derived from the lead author's research in Los Angeles—contributes to grounding this line of inquiry in the thermoceptions of caregivers. Brief highlights from this research are provided next to explain the concept.

Beatriz and Esther exemplify the heat struggles of caregivers in a racialized urban community within the Global North. The mothers live in South Los Angeles, a low-income area. Compared to Whiter, more affluent areas in the city, South Los Angeles has less vegetation, including fewer parks, to cool its streets (Wolch et al., 2014), and a greater percentage of its residents are unhoused or poorly housed. In Los Angeles County, older residential buildings with poor insulation and a lack of air conditioning tend to be in lower-income areas (Porse et al., 2016).

Beatriz, a Latina mother in her mid-30s, migrated to Los Angeles from a region of southern Mexico with tropical thunderstorms that basted her town to humid triple-digits. Yet she recalls her "really, really warm" home near a river—where neighbors would gather on hot nights to share barbecue and chilled fruit drinks—as "really, really nice" with its "warm water, warm rain." Thermometers do not register higher temperatures in Los Angeles than in her hometown, but she feels her barely habitable, rat-infested apartment in the United States is hotter—an "overwhelming heat" that leaves her stressed, angry, and exhausted. She says that the constant heat in the summer makes family members irritable and causes "problems" between them.

Beatriz appears to have the resources that heat studies quantify to calculate health risks: an air conditioner; a nearby park; and access to water, sanitation, and hygiene. However, her physiological experience of heat in the summer is entangled with the stench of the rodents in her family's broiling apartment, the mold that she worries will harm her little boy's lungs when he tries to cool off in the shower, the costly gusts from the air conditioner that escape through cracks in the walls, the sagging ceiling that fell and hurt her eye, and the landlord whom she and her husband fear will raise their rent and make them homeless if they complain. "It's overwhelming how hot it is.... It's scary living like this," she sighs. "Me being a mother ... I feel overwhelmed and stressed.... I worry about [my child] getting heat stroke." Beatriz lives only a few blocks from a leafy park, but the lack of shade along the route to the park makes her worry about walking her child there on hot days. At any rate, mentally ill and drug-dependent people—lacking access to affordable housing and social services—have built an encampment in the park, so she fears taking her son to enjoy the shade trees. She feels trapped as she struggles to distance her child from the heat and from the unhoused, for whom she feels a mix of anger and empathy. She says that city officials and property owners do not care about people like her—they cater to folks who can pay higher rents. Beatriz sees gentrification unfolding nearby, and she worries that her building will be next. She plans to attend a meeting of a group that fights for housing rights. At its meetings, attendees discuss their experiences and plan actions. They also reflect on strategies that have been used in other cities and during other historical periods to get decent, affordable housing.

A few miles from Beatriz, Esther—a Haitian American mother in her mid-30s—became alarmed when her little girl got a high fever that she feared was heat stroke. She put the 4-year-old in the shower and placed an ice pack on her skin. Esther says, "Her body was so hot, she welcomed the ice pack." She sighs that she and her husband told the apartment manager and a housing inspector that the building needed central air, but they were rude to them and did nothing. If she and her husband can save enough money, they might buy a cheap air conditioner, but they are not sure if any of the aging, cracked window frames could hold it.

Similar to Beatriz, Esther lists housing and neighborhood problems that make it hard for her to keep her family cool and comfortable in the summer: the bad-smelling water from the faucets; the apartment's poor insulation; the hot stench of trash on the street; the squatters downstairs, the street crime, and the lack of shade trees that make her afraid to let her children play outside. Esther recently had a baby, and she says that these problems have increased her postpartum depression. Despite her depression, she tries to keep her children healthy and happy: She uses water gun fights to cool them off and to get them to laugh and have fun. But she feels deeply stressed. She is trying to get her local councilperson to address her family's housing and neighborhood problems.

Drawing on insights gained from the mothers and from relevant literature, a frame for studying thermal regulation as a sociophysiological process is outlined in Figure 8.1. Overwhelming heat is an embodied and interpersonal experience. As exemplified by Beatriz and Esther, one's physical and emotional well-being may be imperiled by high temperatures

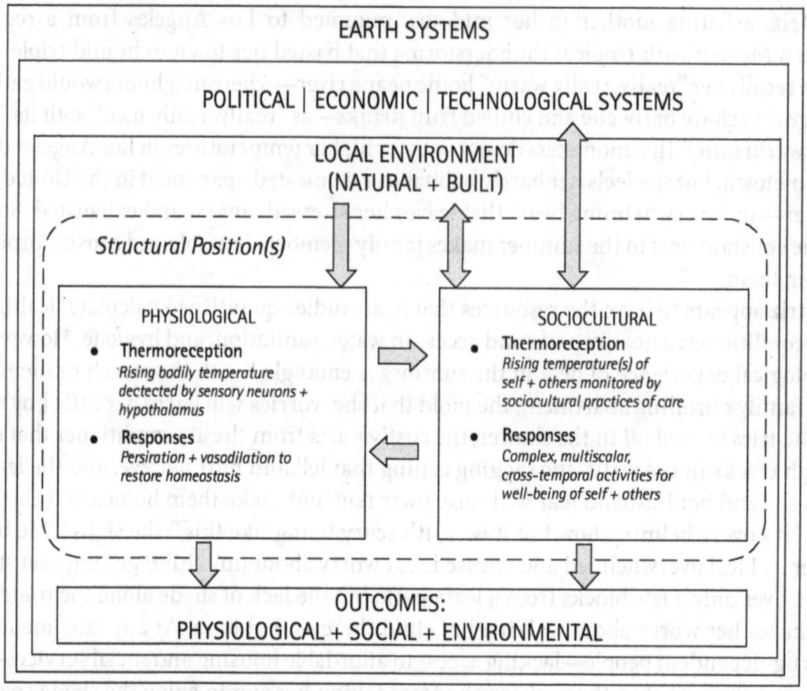

FIGURE 8.1. Processes and outcomes of overwhelming heat.

not simply because of its direct impacts on one's own body, but also because it threatens those for whom one is in relations of care (as a caregiver and/or as a recipient of care). The well-being of Beatriz and Esther is influenced by their children's conditions, and their children's conditions are related to their mothers' well-being.

As shown in the diagram, structural positions (e.g., race/class/gender) interact with local environments and ultimately with political, economic, and technological systems, which are embedded in Earth systems. The outcomes of concern in struggles for thermal regulation go beyond immediate physiological effects—communities need sustainable social and environmental outcomes that support their sociocultural aims and capacities to maintain thermal well-being.

Overwhelming heat (to use Beatriz's words) emerges from the interplay of forces. We define overwhelming heat as the experience of fighting systemic processes that threaten to overpower the physiological and sociocultural capacities of community members to regulate high temperatures. Sociocultural capacities for thermal regulation include family caregiving, community organizing, and collective understandings of nature. As community members attempt thermal regulation, their structural positions (e.g., race/class/gender) within systemic processes bring them into conflict with the state, the market, and at times each other (as evidenced by the hostility that Beatriz and Esther feel toward the unhoused and squatters, who are also attempting thermal regulation by their unauthorized use of spaces that provide shade and shelter).

Overwhelming heat is a complex, fluid experience. The heat narratives of Beatriz and Esther reveal at least three shades of meaning that it takes on across time and places, influenced but not determined by the "objective" temperature:

- Meaning 1: Overwhelming (*unbearable*) heat. The experience of unbearable heat is enmeshed with all the senses and other stressors and conditions—for example, postpartum depression (in the case of Esther) and odors that worsen in the summer heat (the stench of rodents in Beatriz's apartment and the smell of trash outside Esther's apartment). Reducing these personal and environmental problems might make the heat more bearable.
- Meaning 2: Overwhelming (*beating the*) heat. At times, the experience of high temperatures is transformed by individual and collective practices to overwhelm the heat—as exemplified by Esther's water gun fights with her children and both mothers' political actions to "beat the heat."
- Meaning 3: Overwhelming (*joy of*) heat. Within certain settings, heat overwhelms the senses with pleasure—as in Beatriz's remembrance of "really nice... warm water, warm rain" in her hometown that was warm in terms of its weather and its friendly gatherings.

Wherever there is overwhelming heat—in terms of feeling on the verge of being overpowered—there also may be moments of exhilaration in the fight against oppressive heat, and there may be community conditions under which high temperatures (within survivable parameters) bring happiness. This complexity suggests the need for heat researchers to go beyond framing vulnerable communities as passive victims who live in lethal places. There is a need to illuminate oppressive heat *and* the life-giving joys of "warm water, warm rain."

Sociogenesis of the Anthropocene in the Age of Man

The root causes of the heat struggles of Beatriz and Esther extend to the distant past of their birthplaces in Latin America and the Caribbean. The history of the American tropics reveals that the mechanisms that generate extreme heat and those that produce overwhelming heat overlap, but they are not one and the same.

Scientists use the term "Anthropocene" to refer to the current geological age in which human activity—most notably, the generation of greenhouse gases—has made climate change and other environmental degradations an existential threat (Steffen et al., 2011). But centuries before climate change began to melt Arctic ice, European expansion produced, on the one hand, apocalyptic spaces of overwhelming heat and, on the other, refuges of thermal comfort as colonists established sugar plantations and mining operations in the Americas. Attempts by privileged race-class groups to distance overwhelming heat were and continue to be fraught, challenged by environmental degradations, social protests, and violent rebellions. From the start, the race-class order was legitimated by ideologies, camouflaged as science, about the Earth and "Man."

Kyle Whyte (2020) warns that the "crisis epistemology" for understanding climate change—with its rhetoric about the urgent need to address an unprecedented threat—masks colonial power and misrepresents human needs as it obscures the long history of environmental injustices perpetuated against Indigenous people in the Americas. Sylvia Wynter (2003) critiques the conception of the human by Western bourgeois men, who have overrepresented themselves as Man in shifting ways over the centuries. She argues that current struggles around the world, including climate battles, have a basis in the *sociogenesis* of Man, who arises not simply as a creature of nature but also of society. The epistemologies of Man appear to address universal concerns, yet they have roots in hierarchical ways of thinking about which lives matter.

As sugar plantations spread across the Americas in the early 16th century, Europeans justified the enslavement of Africans by arguing that their biology and their equatorial origins made them capable of working in the hot, humid conditions (De las Casas, [1516] 1992; Graves, 2001). It was soon apparent that the heavy labor in hot temperatures, combined with malnutrition, infection, and abuse, was deadly for the Africans (Dubois, 2009). The slave regime forced overwhelming heat exertion—the tropical climate was blameless. In Saint-Domingue (present-day Haiti), the most profitable plantation colony in the Americas, many Africans died within a few years of stepping off the slave boats. In the decades before the bloody Haitian revolution (1791–1804), physicians in Saint-Domingue suggested interventions to reduce the death rate of Africans (Quinlan, 1996). They recommended reform rather than abolition; they described themselves as protectors of the slaveholders' property. Their prescriptions made little difference. The slaveholders regarded the Black laborers as disposable, calculating that it was cheaper to replace them with fresh imports from Africa than to improve their conditions.

The Africans had been brought to the cane fields to take the place of the Taínos, the original inhabitants of Haiti, who had been forced into genocidal labor on plantations and in gold mines after the arrival of Columbus. Throughout the Americas—at towns such as Potosí and Huancavelica (in present-day Bolivia and Peru, respectively) and Zacatecas and Guanajuato (in present-day Mexico)—the mining of valuable metals relied on the forced labor of Indigenous peoples and at times the work of enslaved Africans. Chiseling the metals from the earth exposed the miners to hot and cold temperature extremes, and the ore refining process required the laborers to tend ovens where they breathed in poisonous fumes (Brown, 2012, pp. 63–64).

In European intellectual circles, debates ensued about colonialism, slavery, and the worth of African and Indigenous lives (Flikschuh & Ypi, 2014). Many scholars portrayed the tropics as geographically, intellectually, and morally below the realm of Man. The purported backwardness of the "dark races" was attributed in part to their hot climates. Kant (1802/2012), referring to travel reports—the ethnographies of his day—declared in his lectures on physical geography that climate influenced the "form and temperament" of people; he argued that "in the torrid zones, humans mature more quickly ... than in the temperate zones, but they fail to reach the same ... perfection.... Humanity has its highest degree of perfection in the white race.... The Negroes are much lower, and lowest of all is part of the American races" (pp. 576–577).[5]

The *environmental determinism* of Kant and other Enlightenment scholars, building on the ideas of ancient philosophers such as Hippocrates and Aristotle about climate, constructed hierarchies of race and place (Davis, 2016). The temperate climates were posited to breed civilized, productive men of rational thought, in contrast to the torrid passions and indolence that were imagined to characterize the natives of jungles and deserts. While travel logs often described hot climes as places of brutality, there were accounts that romanticized them as serenely beautiful, yet their inhabitants were often still minoritized as childlike creatures who lacked the mental and material means to decide their futures. This "climatic imagination" (Livingstone, 2002), which pathologized the weather and the inhabitants of supposedly savage lands, enabled the Europeans who profited from torturous spaces to psychologically distance themselves from blame by imagining that the wretched condition of their victims was inherent in nature.

In the "New World," Europeans had power and privileges, but it was hard for them to reduce their exposure to lethal landscapes. On sugar plantations, slaveholders built elegant houses with wide verandas and shuttered windows to maximize airflow and shade (Ross, 1951), but they found it hard to escape the diseases that swept the cane fields (Senior, 2018). In mining towns, rich families could shut their beautiful doors on scenes of misery, but they could not avoid the poisoned air (Robins, 2011).

Far from the plantations and the mines, the rise of the industrial city during the 19th century created hot, polluted living conditions for all urban residents, yet the emergent race-class order shaped differences in exposure. During heat waves and disease outbreaks in summer months, wealthy families fled to their country houses and to seaside vacations. On hot days, the suffering of working-class residents was compounded by their sweltering factory jobs, their poorly ventilated housing, and the stench from their unsanitary streets (Hopkins, 1994; Schlichting & Kiechle, 2020). During the 1858 "Great Stink" heat wave in London, the members of Parliament—unable to bear the smells that invaded their meetings—approved a new sewage system (Halliday, 2001). After the deadly heat wave of June 1872 in New York, a charity organized trips for working-class children to the countryside so that they could escape the high temperatures, bad air, and immoral habits of the city, which the charity assumed to be interrelated (Plater, 2023). The project, patronizing and condescending to the impoverished European immigrants whom it served, largely excluded Black children. Across the United States, the development of de jure and de facto segregation in the late 19th century would, until the enforcement of the Civil Rights Act of 1964, limit the access of Black families to the public parks and pools at which White families found relief from the summer sun (Wolcott, 2012).

During the 19th and early 20th centuries, science and social science research—weaponized with environmental determinism and, at times, eugenics—became part of the intellectual armaments that served empire and its race-class order (Jablonski, 2021; Venkat, 2022). Francis Galton, a British polymath and explorer, is credited with making the first weather map; he also ranked human races and became the father of eugenics. Henry Blanford, a British meteorologist and paleontologist who served as the Imperial Meteorological Reporter in India, stated (as an aside, in a book thick with meticulous geographical observations) that race and climate interact to shape the health and "character" of a people (Blanford, 1890, p. 20). Henry Pratt Fairchild, who would serve as president of the American Sociological Society and the American Eugenics Society over his long career, speculated—in terms similar to Blanford—that "physical environments" and "blood … admixtures" interact to shape the "history and character of races of men" and that "the temperate zone is best fitted to develop a progressive people" (Fairchild, 1911, pp. 8–20). Robert DeCourcy Ward (1929), a prominent American climatologist and eugenicist, mused about combining military authority with improved sanitation and "artificially cooled" houses to colonize the "the moist hot tropics" in ways that enriched the "white race" and maintained its "physical, mental, and moral vigor" as it extracted the tropics' natural wealth (pp. 618–619, 623–624, 627).

In the early 20th century, engineers and architects developed a temperate zone of sorts indoors. Expectations about thermal comfort increased as air conditioning spread in Western cities (Cooper, 1998). In the postcolonial era, tropical countries sought air-conditioned, high-rise buildings to achieve "thermal modernity" and spur development, via paths that were contingent on local circumstances (Chang & Winter, 2015).

Ironically, the upper classes of the West, who reviled the heat of the tropics and the industrial city, set in motion the interplay of these zones in production chains and consumption practices that have driven the rise of global temperatures. Their artificial climate of air conditioners, powered by fossil fuel, at once reduced and contributed to the danger. Janelle Baker and her coauthors (2020), advocating a new conceptualization of the Anthropocene, reflect on the legacies of European "invasion, extraction, and exploitation" that have led to a warming planet and other ecological degradations. They suggest that planetary disasters force realization that there can be no more "deflections from the structural inequalities we manufactured to invent 'safety' for some and danger for others.... Now most everyone can taste the fumes of the Anthropocene."

In the United States, the history of settler-colonialism and racial segregation still impacts thermal inequities. For centuries, the US military forced Indigenous nations from their ancestral lands to areas that today make them more vulnerable to extreme weather; these areas lack adequate infrastructure (e.g., plumbing), natural resources (e.g., groundwater), and health services to deal with heat waves (Tanana, 2023). For much of the 20th century, redlining—racist disinvestment in Black and immigrant neighborhoods—led to wealth disparities, housing injustices, and infrastructure inequities (e.g., fewer parks and less vegetation on streets) that continue to shape heat vulnerabilities (A. Hsu et al., 2021).

In the humanities and social sciences, heat research increasingly engages with critical social theories and with cultures of resistance. A strand of critical approaches is inspired by CEJ, which emphasizes intersectional analyses, multiscalar studies, multispecies indispensability, and a distrust of state power and the market. Zoe Hamstead (2023) calls for a critical heat studies that investigates how the social production of heat by the state and the market, which she frames as a form of violence, impacts people with intersecting identities. David Pellow (2024), studying the influence of extreme heat in prisons on mental health, denounces the "eco-psychological violence" of the state.

The Age of Man set in motion severe environmental harms. Equally pernicious, ideas about the ability of non-White "races" to endure heat—and the self-interested inattention to their perilous conditions—endure (Derickson, 2019). Yet we should keep in mind that extreme heat and heat vulnerability are multicausal. Today the global order interlocks with systems of power enforced by elites of myriad countries, who legitimate their control and their heat policies with harmful ideologies in and beyond the West.

Heat Interventions

Cooling centers, shade structures, and urban greening are broad sets of (at times overlapping) interventions to reduce the effects of extreme heat on racialized and low-income areas such as Beatriz and Esther's community in Los Angeles. These interventions should be tailored to the needs of community members as they go about their everyday lives—for example, the need for shade on Esther's block so that her children can play outside, and the need for shade on the route from Beatriz's home to her nearby park. To identify the best locations for heat interventions—and to track and evaluate use of those that already exist—researchers use a mix of methods, including GIS mapping of daily rounds and heat risks in an area, and qualitative, community-engaged mapping projects (e.g., interactive walks) to understand local needs and experiences (Derakhshan et al., 2023; Foshag et al., 2024; Holec et al., 2021).

Various studies, including mobile phone tracking research (Derakhshan et al., 2023), suggest that officially designated cooling centers—air-conditioned spaces such as public libraries that provide emergency relief—are underutilized during extreme heat. It is unclear whether this is because they have limited appeal or inadequate publicity. Resilience hubs—community spaces organized by and for local residents that provide year-round services in addition to emergency relief during heat waves—might be more engaging (Aoun, 2022; Lou et al., 2020; NorCal Resilience Network, 2024; Tu, 2023; Urban Sustainability Directors Network, n.d.). Some resilience hubs offer health and mental health services as well as facilitate community organizing to address local concerns, which might be useful to mothers such as Esther and Beatriz who struggle with the complex stresses of overwhelming heat. Others provide workforce development (e.g., green jobs training), and still others strive to further community resilience by improving local understandings of and relationships with nature.

The various types of heat interventions have the same physiological aim—cooling the body—yet their actual use, features, and outcomes are influenced by an array of sociocultural, political, economic, and environmental processes, as exemplified by urban greening. The literature on urban greening indicates that tree canopies and other vegetation can reduce air and ground temperatures in neighborhoods (Knight et al., 2021), yet the type and mix of green infrastructure must be carefully matched with local settings to meet cooling needs (Zölch et al., 2016). It is also necessary to take into account how green interventions will be impacted over the long term by climate change and municipal budgets—for example, one must evaluate whether various tree species are sustainable (ecologically, economically), given hotter temperatures and escalating maintenance costs, and one must take into account the impacts of green infrastructure on water and other resources threatened by climate change (Jenerette et al., 2011).

Beyond environmental impacts, it is also important to consider the social impacts of urban greening: Studies show that adding green features such as tree-lined sidewalks and leafy parks to a low-income area can contribute to gentrification (Anguelovski et al., 2022). When attractive new green spaces are placed in a low-income area, the heightened market demand for residential and commercial properties in the area may increase rents, which may lead to the displacement of existing residents and small business owners. To reduce gentrification, some researchers argue that green interventions should be combined with policies that retain the existing housing at affordable rents, limit construction of new luxury housing, and help existing small business owners to remain (Gould & Lewis, 2016). In addition, green spaces must be crafted in inclusive, culturally sensitive ways. If racialized, low-income people feel that green spaces were designed for gentrifiers, they may not use these spaces and thus their potential health benefits will not be achieved (Jelks et al., 2021).

The governance of green spaces is not typically coordinated with heat governance. In US cities, *neoliberal austerity*—reductions in government spending and control (Almeida & Pérez Martín, 2022)—have led to a proliferation of entities that fund and manage green projects. These entities include business improvement districts (BIDs) and tax increment financing districts (TIFs). BIDs are publicly authorized, privately administered entities that collect fees from local businesses to fund maintenance and improvement projects (e.g., new green spaces) within a defined area. TIFs are public and private entities that use increases in property tax revenues within a defined area to fund redevelopment projects.

Some of the literature on this topic describes these entities as practical ways to "stack" funds for green space interventions, given that local governments have limited resources (Zemaitis, 2024). However, studies also indicate that the top-down decision making of these entities is sometimes opaque and unresponsive to community members, and that it may further gentrification (Montgomery, 2020). Longtime residents may enjoy these green interventions for a short while, but market forces may eventually displace them. There seems to be an inherent conflict in funding green interventions to benefit low-income people through entities that get their funds by increasing market demand and property values in an area—outcomes associated with gentrification. Yet it should be noted that these entities differ across places in their governance practices and in the nature of their green interventions. More research is needed on the long-term effects of BIDs and TIFs and their green space interventions on the well-being of low-income, racialized people. There is also a need for more study regarding whether and how these entities collaborate with government agencies and grassroots groups as they design and manage their green interventions.

Beyond concerns about gentrification, there is a need to consider how heat interventions fit within the larger social and environmental contexts of overwhelming heat. As illustrated by the narratives of Beatriz and Esther, the use of green space is complicated by street crime and by conflicts between housed and unhoused people. If history is a guide, it seems likely that climate change will intensify existing struggles over status and resources within communities, as exemplified by the Little Ice Age in the early modern era—a period of regional cooling that led to reduced harvests, desperate hunger, wild speculations, and violent scapegoating in parts of Europe (Behringer, 1999; Fagan, 2019). Socially isolated and stigmatized individuals in communities may be endangered not only from the direct impacts of more severe heat waves but also from their indirect influences on social cohesion and caregiving, mediated by political and economic systems. These complex challenges require multiscalar interventions and societal transformations.

Beyond the Torrid Zones Imaginary

Current studies about extreme heat seem a sharp break with the knowledge regime of empire and its race-class order. Gone is the reliance on the fanciful travel logs about climate and race that plagued the early modern era. Yet centuries-old ways of knowing in the academy still prime us to discover "savage" places. What would it mean if we disentangled oppressive heat from the heat of deserts and rain forests? How might it help us to reimagine and remake our futures on a warming planet if we appreciated "warm water, warm rain" (in Beatriz's words)?

In closing, the second author of this chapter suggests a path forward for heat studies beyond the torrid zones imaginary, drawing from his research on deserts in southern California. Critiquing the limitations of contemporary heat studies, he strives to develop what he calls a "desert epistemology" that understands heat in relation to the human and more-than-human life that thrives in these environments. Collective understandings of nature—informing family care and community practices—contribute to climate resilience.

The portrayal of deserts and other hot environments as hostile landscapes is entrenched in colonial and settler ideologies. Equally problematic are calculations of these places as unpopulated sites from which riches (e.g., oil, lithium) can be extracted without "cost," or their romanticization as serene refuges for travelers to enjoy nature and the "children of nature" (the original inhabitants). These representations have legitimated the exploitation and marginalization of human communities in these regions.

In reality, these environments have deep ecological and cultural richness and complexity. In profound ways, human and more-than-human entities have coexisted, adapted, and thrived in these places. Deserts are far from empty wastelands; they are dynamic landscapes teeming with life, knowledge, and history (Herbst, 2023). The deserts of southern California—including the Mojave that stretches across the mountains of four US states and the Colorado that extends across valleys which reach into Mexico—exemplify this richness. These deserts, which have a history of being used as mining sites by settlers, have become tourist destinations (Magaña, 2024).

The region has an older history. The Indigenous peoples of these regions, such as the Cahuilla who still live in the Coachella Valley within the Colorado desert, cultivated an intimate understanding of their environment, developing sustainable practices in tune with the desert ecosystem that enabled them to thrive under conditions that might otherwise be experienced as harsh and unlivable. This relationship with the desert is not merely one of survival but also of symbiosis, where the rhythms of human life are deeply entwined with the natural cycles of the desert ecosystem. The ability of these communities to live in harmony under such extreme conditions challenges the simplistic notion that heat is something to be feared or combated.

This relationship between humans and the more-than-human is not static but dynamic, characterized by ongoing processes of learning and adaptation. In desert environments, more-than-human entities—plants, animals, and even geological formations—play an active role in shaping human knowledge and practices (Voyles, 2021). For instance, the mesquite tree, a key species in the desert ecosystem, not only provided indispensable resources (including food) to both humans and more-than-humans but also taught the Cahuilla people about the ever-changing climatic cycles of the desert. The desert serves as a site of knowledge generation, where humans and more-than-humans interact in ways that are crucial for survival in extreme conditions. The desert is not a passive backdrop but an active participant in creating and transmitting knowledge that helps communities understand and respond to the challenges of heat.

Life has thrived in deserts and rain forests, within survivable parameters, for millennia. Cultures in these hot regions feel attached to their lands and have learned to live in them. Our challenge is to learn with them how to survive and thrive on a warming planet.

Notes

1. The World Health Organization cites Zhao et al. (2021), an international team of researchers, who state, "Globally, 5 083 173 deaths . . . were associated with non-optimal temperatures per year, consisting of 4 594 098 cold-related deaths . . . and 489 075 heat-related deaths."
2. The term "undersocialized" refers to models that frame humans as rational actors whose actions are driven by self-interest. These models ignore how culture and emotions influence care for others, and they inadequately consider how power relations shape choices of action.
3. "Beatriz" and "Esther" are pseudonyms.
4. Django Paris and H. Samy Alim describe a "loving critique" as the critical examination of a school of thought in which one is embedded. The school of thought is reflected on with respect and appreciation for its contributions, yet the need to move away from its problematic aspects is not shied away from.
5. Kant apparently changed some of his views about race toward the end of his life. See Flikschuh and Ypi (2014).

References

Alizadeh, M. R., Abatzoglou, J. T., Adamowski, J. F., Prestemon, J. P., Chittoori, B., Akbari Asanjan, A., & Sadegh, M. (2022). Increasing heat-stress inequality in a warming climate. *Earth's Future*, *10*(2), e2021EF002488. https://doi.org/10.1029/2021EF002488

Almeida, P., & Martín, A. P. (2022). *Collective resistance to neoliberalism*. Cambridge University Press.

Andrade-Velázquez, M., Medrano-Pérez, O. R., Montero-Martínez, M. J., & Alcudia-Aguilar, A. (2021). Regional climate change in Southeast Mexico-Yucatan Peninsula, Central America and the Caribbean. *Applied Sciences*, *11*(18), 1–17. https://doi.org/10.3390/app11188284

Anguelovski, I., Connolly, J. J., Cole, H., Garcia-Lamarca, M., Triguero-Mas, M., Baró, F., Martin, N., Conesa, D., Shokry, G., Del Pulgar, C. P., Ramos, L. A., Matheney, A., Gallez, E., Oscilowicz, E., Máñez, J. L., Sarzo, B., Beltrán, M. A., & Minaya, J. M. (2022). Green gentrification in European and North American cities. *Nature Communications*, *13*(1), 3816.

Aoun, G. (2022, September 23). "A living, breathing building": The rise of resilience centers amid extreme heat in the US. *The Guardian*. https://www.theguardian.com/us-news/2022/sep/23/california-extreme-heat-resilience-centers

Auliciems, A. (2009). Human adaptation within a paradigm of climatic determinism and change. In K. L. Ebi, I. Burton, & G. R. McGregor (Eds.), *Biometeorology for adaptation to climate variability and change* (Vol. 1, pp. 235–267). Springer. https://doi.org/10.1007/978-1-4020-8921-3_11

Baker, J., Ebron, P., Ficek, R., Ho, K., Ramirez, R., Todd, Z., Tsing, A., & Vaughn, S. E. (2020). The snarled lines of justice. *Orion Magazine*. https://orionmagazine.org/article/the-snarled-lines-of-justice/

Baldwin, J. W., Benmarhnia, T., Ebi, K. L., Jay, O., Lutsko, N. J., & Vanos, J. K. (2023). Humidity's role in heat-related health outcomes: A heated debate. *Environmental Health Perspectives*, *131*(5), 055001. https://doi.org/10.1289/EHP11807

Behringer, W. (1999). Climatic change and witch-hunting: The impact of the Little Ice Age on mentalities. *Climatic Change*, *43*(1), 335–351. https://doi.org/10.1023/A:1005554519604

Benz, S. A., & Burney, J. A. (2021). Widespread race and class disparities in surface urban heat extremes across the United States. *Earth's Future*, *9*(7). https://doi.org/10.1029/2021EF002016

Berberian, A. G., Gonzalez, D. J. X., & Cushing, L. J. (2022). Racial disparities in climate change-related health effects in the United States. *Current Environmental Health Reports*, *9*(3), 451–464. https://doi.org/10.1007/s40572-022-00360-w

Blair-Loy, M., Hochschild, A., Pugh, A. J., Williams, J. C., & Hartmann, H. (2015). Stability and transformation in gender, work, and family: Insights from the second shift for the next quarter century. *Community, Work & Family*, *18*(4), 435–454. https://doi.org/10.1080/13668803.2015.1080664

Blanford, H. F. (1890). *An elementary geography of India, Burma, and Ceylon*. Macmillan.

Brook, R. D., Rajagopalan, S., Pope, C. A., Brook, J. R., Bhatnagar, A., Diez-Roux, A. V., Holguin, F., Hong, Y., Luepker, R. V., Mittleman, M. A., Peters, A., Siscovick, D., Smith, S. C., Whitsel, L., & Kaufman, J. D. (2010). Particulate matter air pollution and cardiovascular disease. *Circulation*, *121*(21), 2331–2378. https://doi.org/10.1161/CIR.0b013e3181dbece1

Brown, K. W. (2012). *A history of mining in Latin America: From the colonial era to the present*. UNM Press.

Butt, M. N., Shah, S. K., & Yahya, F. A. (2020). Caregivers at the frontline of addressing the climate crisis. *Gender & Development*, *28*(3), 479–498. https://doi.org/10.1080/13552074.2020.1833482

California Department of Insurance. (2024). *Impacts of extreme heat to California's people, infrastructure, and economy*. California Department of Insurance.

Castillo, F., Mora, A. M., Kayser, G. L., Vanos, J., Hyland, C., Yang, A. R., & Eskenazi, B. (2021). Annual review of public health. *Annual Review of Public Health*, *42*, 257–276. https://doi.org/10.1146/annurev-publhealth-012420-105014

Chang, J.-H., & Winter, T. (2015). Thermal modernity and architecture. *Journal of Architecture*, *20*(1), 92–121. https://doi.org/10.1080/13602365.2015.1010095

Chicas, R., Xiuhtecutli, N., Dickman, N. E., Flocks, J., Scammell, M. K., Steenland, K., Hertzberg, V., & McCauley, L. (2021). Cooling interventions among agricultural workers: Qualitative field-based study. *Hispanic Health Care International*, *19*(3), 174–181. https://doi.org/10.1177/1540415321993429

Cooper, G. (1998). *Air-conditioning America: Engineers and the controlled environment, 1900–1960*. JHU Press.

Cupples, J., Guyatt, V., & Pearce, J. (2007). "Put on a jacket, you wuss": Cultural identities, home heating, and air pollution in Christchurch, New Zealand. *Environment and Planning A: Economy and Space*, *39*(12), 2883–2898. https://doi.org/10.1068/a3932

Davis, D. K. (2016). *The arid lands: History, power, knowledge*. MIT Press.

De las Casas, B. ([1516] 1992). Memorial de remedios para Las Indias. In *Obras Completas*, Vol. 13, 23–48. Alianza Editorial.

Derakhshan, S., Bautista, T. N., Bouwman, M., Huang, L., Lee, L., Tarczynski, J., Wahagheghe, I., Zeng, X., & Longcore, T. (2023). Smartphone locations reveal patterns of cooling center use as a heat mitigation strategy. *Applied Geography*, *150*, 102821. https://doi.org/10.1016/j.apgeog.2022.102821

Derickson, A. (2019). "A widespread superstition": The purported invulnerability of workers of color to occupational heat stress. *American Journal of Public Health*, *109*(10), 1329–1335. https://doi.org/10.2105/AJPH.2019.305246

Dong, X. S., West, G. H., Holloway-Beth, A., Wang, X., & Sokas, R. K. (2019). Heat-related deaths among construction workers in the United States. *American Journal of Industrial Medicine, 62*(12), 1047–1057. https://doi.org/10.1002/ajim.23024

Dubois, L. (2009). *Avengers of the New World: The story of the Haitian Revolution.* Harvard University Press.

Erqou, S., Clougherty, J. E., Olafiranye, O., Magnani, J. W., Aiyer, A., Tripathy, S., Kinnee, E., Kip, K. E., & Reis, S. E. (2018). Particulate matter air pollution and racial differences in cardiovascular disease risk. *Arteriosclerosis, Thrombosis, and Vascular Biology, 38*(4), 935–942. https://doi.org/10.1161/ATVBAHA.117.310305

Errett, N. A., Hartwell, C., Randazza, J. M., Nori-Sarma, A., Weinberger, K. R., Spangler, K. R., Sun, Y., Adams, Q. H., Wellenius, G. A., & Hess, J. J. (2023). Survey of extreme heat public health preparedness plans and response activities in the most populous jurisdictions in the United States. *BMC Public Health, 23*(1), 811. https://doi.org/10.1186/s12889-023-15757-x

Fagan, B. (2019). *The Little Ice Age: How climate made history, 1300–1850.* Basic Books.

Fairchild, H. P. (1911). *Greek immigration to the United States.* Yale University Press.

Flikschuh, K., & Ypi, L. (2014). *Kant and colonialism: Historical and critical perspectives.* Oxford University Press.

Foshag, K., Fürle, J., Ludwig, C., Fallmann, J., Lautenbach, S., Rupp, S., Burst, P., Betsch, M., Zipf, A., & Aeschbach, N. (2024). How to assess the needs of vulnerable population groups towards heat-sensitive routing? An evidence-based and practical approach to reducing urban heat stress. *Erdkunde, 78*(1), 1–34.

Gabbe, C. J., Mallen, E., & Varni, A. (2023). Housing and urban heat: Assessing risk disparities. *Housing Policy Debate, 33*(5), 1078–1099. https://doi.org/10.1080/10511482.2022.2093938

Gould, K., & Lewis, T. (2016). *Green gentrification: Urban sustainability and the struggle for environmental justice* (1st ed.). Routledge.

Graves, J. L. (2001). *The emperor's new clothes: Biological theories of race at the millennium.* Rutgers University Press.

Gronlund, C. J. (2014). Racial and socioeconomic disparities in heat-related health effects and their mechanisms: A review. *Current Epidemiology Reports, 1*(3), 165–173. https://doi.org/10.1007/s40471-014-0014-4

Guo, C., Lanza, K., Li, D., Zhou, Y., Aunan, K., Loo, B. P., Lee, J. K. W., Luo, B., Duan, X., Zhang, W., Zhang, Z., Lin, S., & Zhang, K. (2023). Impact of heat on all-cause and cause-specific mortality: A multi-city study in Texas. *Environmental Research, 224,* 115453.

Halliday, S. (2001). *The great stink of London: Sir Joseph Bazalgette and the cleansing of the Victorian metropolis.* The History Press.

Hamstead, Z. A. (2023). Critical heat studies: Deconstructing heat studies for climate justice. *Planning Theory & Practice, 24*(2), 153–172. https://doi.org/10.1080/14649357.2023.2201604

Harris, A. W., & Albrecht, J. (2024). Association between homelessness and temperature-related health emergencies in the United States. Weather, Climate, and Society, *16*(4), 633–638.

Harshaw, C., & Alberts, J. R. (2012). Group and individual regulation of physiology and behavior: A behavioral, thermographic, and acoustic study of mouse development. *Physiology & Behavior, 106*(5), 670–682. https://doi.org/10.1016/j.physbeh.2012.05.002

Herbst, M. (2023). Journey to the center of the world: Memory and the sacred in the Colorado Desert. *World History Connected, 20*(2), 1–33.

Herold, N., Alexander, L., Green, D., & Donat, M. (2017). Greater increases in temperature extremes in low versus high income countries. *Environmental Research Letters, 12*(3), 034007. https://doi.org/10.1088/1748-9326/aa5c43

Holec, J., Šveda, M., Szatmári, D., Feranec, J., Bobáľová, H., Kopecká, M., & Šťastný, P. (2021). Heat risk assessment based on mobile phone data: Case study of Bratislava, Slovakia. *Natural Hazards, 108*(3), 3099–3120. https://doi.org/10.1007/s11069-021-04816-4

Hopkins, E. (1994). *Childhood transformed: Working-class children in nineteenth-century England.* Manchester University Press.

Hsu, A., Sheriff, G., Chakraborty, T., & Manya, D. (2021). Disproportionate exposure to urban heat island intensity across major US cities. *Nature Communications, 12*(1), 2721. https://doi.org/10.1038/s41467-021-22799-5

Hsu, H. L. (2023). Race, urban heat, and the aesthetics of thermoception. *American Literary History, 35*(2), 769–794. https://doi.org/10.1093/alh/ajad001

Jablonski, N. G. (2021). Skin color and race. *American Journal of Physical Anthropology, 175*(2), 437–447. https://doi.org/10.1002/ajpa.24200

Jelks, N. O., Jennings, V., & Rigolon, A. (2021). Green gentrification and health: A scoping review. *International Journal of Environmental Research and Public Health, 18*(3), Article 3. https://doi.org/10.3390/ijerph18030907

Jenerette, G. D., Harlan, S. L., Stefanov, W. L., & Martin, C. A. (2011). Ecosystem services and urban heat riskscape moderation: Water, green spaces, and social inequality in Phoenix, USA. *Ecological Applications, 21*(7), 2637–2651. https://doi.org/10.1890/10-1493.1

Johnson, D. P. (2022). Population-based disparities in U.S. urban heat exposure from 2003 to 2018. *International Journal of Environmental Research and Public Health, 19*(19), Article 19. https://doi.org/10.3390/ijerph191912314

Kadio, K., Filippi, V., Congo, M., Scorgie, F., Roos, N., Lusambili, A., Nakstad, B., Kovats, S., & Kouanda, S. (2024). Extreme heat, pregnancy and women's well-being in Burkina Faso: An ethnographical study. *BMJ Global Health, 8*(Suppl. 3), e014230. https://doi.org/10.1136/bmjgh-2023-014230

Kant, I. (2012). Physical geography (1802). In E. Watkins (Ed.), *Kant: Natural science* (pp. 434–679). Cambridge University Press. https://doi.org/10.1017/CBO9781139014380.019

Keith, L., Meerow, S., Hondula, D. M., Turner, V. K., & Arnott, J. C. (2021). Deploy heat officers, policies and metrics. *Nature, 598*(7879), 29–31. https://doi.org/10.1038/d41586-021-02677-2

Kenny, G. P., Yardley, J., Brown, C., Sigal, R. J., & Jay, O. (2010). Heat stress in older individuals and patients with common chronic diseases. *CMAJ, 182*(10), 1053–1060. https://doi.org/10.1503/cmaj.081050

Klinenberg, E. (2003). *Heat wave: A social autopsy of disaster in Chicago.* University of Chicago Press.

Knight, T., Price, S., Bowler, D., Hookway, A., King, S., Konno, K., & Richter, R. L. (2021). How effective is "greening" of urban areas in reducing human exposure to ground-level ozone concentrations, UV exposure and the "urban heat island effect"? An updated systematic review. *Environmental Evidence, 10*, 1–38.

Kotharkar, R., & Ghosh, A. (2022). Progress in extreme heat management and warning systems: A systematic review of heat-health action plans (1995–2020). *Sustainable Cities and Society, 76*, 103487. https://doi.org/10.1016/j.scs.2021.103487

Kuehn, L., & McCormick, S. (2017). Heat exposure and maternal health in the face of climate change. *International Journal of Environmental Research and Public Health, 14*(8), Article 8. https://doi.org/10.3390/ijerph14080853

Kunda, J. J., Gosling, S. N., & Foody, G. M. (2024). The effects of extreme heat on human health in tropical Africa. *International Journal of Biometeorology, 68*(6), 1015–1033. https://doi.org/10.1007/s00484-024-02650-4

Lindsey, R., & Dahlman, L. (2024, January 18). Climate change: Global temperature. http://www.climate.gov/news-features/understanding-climate/climate-change-global-temperature

Livingstone, D. N. (2002). Tropical hermeneutics and the climatic imagination. *Geographische Zeitschrift, 90*(2), 65–88.

Lou, Z., Raval, A., Young, M., & Appel, S. (2020). Resilience before disaster: The need to build equitable, community-driven social infrastructure. https://apen4ej.org/wp-content/uploads/2020/10/Resilience-Before-Disaster-FINAL-UPDATED.pd

Luber, G., & McGeehin, M. (2008). Climate change and extreme heat events. *American Journal of Preventive Medicine, 35*(5), 429–435. https://doi.org/10.1016/j.amepre.2008.08.021

Magaña, M. R. (2024). Digital placemaking in the high desert: Airbnb's and perception of desert communities in Joshua Tree, CA [UCLA]. https://escholarship.org/uc/item/77t7c591

McAnneny, C. (2021). Extreme heat. In *Oxford Bibliographies Online in Geography.* Oxford University Press. https://doi:10.1093/obo/9780199874002-0234

Mitchell, B. C., & Chakraborty, J. (2015). Landscapes of thermal inequity: Disproportionate exposure to urban heat in the three largest US cities. *Environmental Research Letters, 10*(11), 115005. https://doi.org/10.1088/1748-9326/10/11/115005

Mondal, T., Sen, J., Goswami, R., & Nag, P. K. (2024). Community adaptation to heat stress—Social network analysis. *Climate Risk Management, 44*, 100606. https://doi.org/10.1016/j.crm.2024.100606

Montgomery, A. (2020). *Greening the Black urban regime: The culture and commerce of sustainability in Detroit.* Wayne State University Press.

Moore, J. D. (2012). *The prehistory of home.* University of California Press.

Mota-Rojas, D., Titto, C. G., Orihuela, A., Martínez-Burnes, J., Gómez-Prado, J., Torres-Bernal, F., Flores-Padilla, K., Carvajal-de la Fuente, V., & Wang, D. (2021). Physiological and behavioral mechanisms of thermoregulation in mammals. *Animals (Basel), 11*(6), 1733. https://doi.org/10.3390/ani11061733

NorCal Resilience Network. (2024). Our story. https://norcalresilience.org/

Oppermann, E., Kjellstrom, T., Lemke, B., Otto, M., & Lee, J. K. W. (2021). Establishing intensifying chronic exposure to extreme heat as a slow onset event with implications for health, wellbeing, productivity, society and economy. *Current Opinion in Environmental Sustainability, 50*, 225–235. https://doi.org/10.1016/j.cosust.2021.04.006

Pellow, D. N. (2024). Climate justice, public health, and mental health in prisons and jails. In R. J. Moore (Ed.), *Climate change and mental health equity* (pp. 269–295). Springer. https://doi.org/10.1007/978-3-031-56736-0_10

Peraza, S., Wesseling, C., Aragon, A., Leiva, R., García-Trabanino, R. A., Torres, C., Jakobsson, K., Elinder, C. G., & Hogstedt, C. (2012). Decreased kidney function among agricultural workers in El Salvador. *American Journal of Kidney Diseases, 59*(4), 531–540. https://doi.org/10.1053/j.ajkd.2011.11.039

Pincetl, S., Chester, M., & Eisenman, D. (2016). Urban heat stress vulnerability in the U.S. Southwest: The role of sociotechnical systems. *Sustainability, 8*(9), Article 9. https://doi.org/10.3390/su8090842

Plater, M. (2023). Tonic for body or soul: Fresh air for poor children in Progressive Era New York City. *Journal of Urban History, 49*(4), 865–890. https://doi.org/10.1177/00961442211040453

Porse, E., Derenski, J., Gustafson, H., Elizabeth, Z., & Pincetl, S. (2016). Structural, geographic, and social factors in urban building energy use: Analysis of aggregated account-level consumption data in a megacity. *Energy Policy, 96*, 179–192. https://doi.org/10.1016/j.enpol.2016.06.002

Pulido, L. (2018). Racism and the Anthropocene. In G. Mitman, M. Armiero, & R. Emmett (Eds.), *Future remains: A cabinet of curiosities for the Anthropocene* (pp. 116–128). University of Chicago Press. https://doi.org/10.7208/9780226508825-014

Quinlan, S. (1996). Colonial encounters. Colonial bodies, hygiene and abolitionist politics in eighteenth-century France. *History Workshop Journal, 42*(1), 107–126. https://doi.org/10.1093/hwj/1996.42.107

Robins, N. A. (2011). *Mercury, mining, and empire: The human and ecological cost of colonial silver mining in the Andes.* Indiana University Press.

Rosa, L. (2022). Adapting agriculture to climate change via sustainable irrigation: Biophysical potentials and feedbacks. *Environmental Research Letters, 17*(6), 063008. https://doi.org/10.1088/1748-9326/ac7408

Ross, M. D. (1951). Caribbean colonial architecture in Jamaica. *Journal of the Society of Architectural Historians, 10*(3), 22–27. https://doi.org/10.2307/987447

Saeed, F., Schleussner, C.-F., & Ashfaq, M. (2021). Deadly heat stress to become commonplace across South Asia already at 1.5°C of global warming. *Geophysical Research Letters, 48*(7), e2020GL091191. https://doi.org/10.1029/2020GL091191

Safier, N. (2014). The tenacious travels of the torrid zone and the global dimensions of geographical knowledge in the eighteenth century. https://doi.org/10.1163/15700658-12342388

Schlichting, K. M., & Kiechle, M. A. (2020). *Invisible inequalities: Persistent health threats in the urban built environment*. http://hdl.handle.net/10919/101613

Senior, E. (2018). *The Caribbean and the medical imagination, 1764–1834: Slavery, disease and colonial modernity*. Cambridge University Press.

Sharpe, C. (2016). *In the wake: On blackness and being*. Duke University Press.

Sharpe, J. D., & Wolkin, A. F. (2022). The epidemiology and geographic patterns of natural disaster and extreme weather mortality by race and ethnicity, United States, 1999–2018. *Public Health Reports®*, *137*(6), 1118–1125. https://doi.org/10.1177/00333549211047235

Sheridan, S., de Guzman, E. B., Eisenman, D. P., Sailor, D. J., Parfrey, J., & Kalkstein, L. S. (2024). Increasing tree cover and high-albedo surfaces reduces heat-related ER visits in Los Angeles, CA. *International Journal of Biometeorology*, *68*(8), 1603–1614. https://doi.org/10.1007/s00484-024-02688-4

Shih, W.-Y., Lung, S.-C. C., & Hu, S.-C. (2022). Perceived heat impacts and adaptive behaviours in different socio-demographic groups in the subtropics. *International Journal of Disaster Risk Reduction*, *71*, 102799.

Sorensen, C. J., & Balbus, J. (2021). Climate change and women's health: Risks and opportunities. In K. E. Pinkerton & W. N. Rom (Eds.), *Climate change and global public health* (pp. 403–426). Springer. https://doi.org/10.1007/978-3-030-54746-2_19

Steffen, W., Grinevald, J., Crutzen, P., & McNeill, J. (2011). The Anthropocene: Conceptual and historical perspectives. *Philosophical Transactions of the Royal Society A: Mathematical, Physical and Engineering Sciences*, *369*(1938), 842–867. https://doi.org/10.1098/rsta.2010.0327

Tan, E., & Bogage, J. (2023, September 9). Extreme heat is forcing America's farmers to go nocturnal. *Washington Post*. https://www.washingtonpost.com/business/2023/09/09/heat-night-harvesting-farmers/

Tanana, H. (2023). Protecting tribal health from climate change. *Northeastern University Law Review*, *15*, 89.

Thompson, R., Hornigold, R., Page, L., & Waite, T. (2018). Associations between high ambient temperatures and heat waves with mental health outcomes: A systematic review. *Public Health*, *161*, 171–191. https://doi.org/10.1016/j.puhe.2018.06.008

Tu, M. (2023, December 1). The Black churches lifting up climate resilience. *Next City*. https://nextcity.org/urbanist-news/the-black-churches-lifting-up-climate-resilience

Urban Sustainability Directors Network. (n.d.). Resilience hubs. https://resilience-hub.org/

Vaidyanathan, A., Malilay, J., Schramm, P., & Saha, S. (2020). Heat-related deaths—United States, 2004–2018. *Morbidity and Mortality Weekly Report*, *69*. https://doi.org/10.15585/mmwr.mm6924a1

Vannini, P., & Taggart, J. (2014). Making sense of domestic warmth: Affect, involvement, and thermoception in off-grid homes. *Body & Society*, *20*(1), 61–84. https://doi.org/10.1177/1357034X13499381

Vecellio, D. J., Kong, Q., Kenney, W. L., & Huber, M. (2023). Greatly enhanced risk to humans as a consequence of empirically determined lower moist heat stress tolerance. *Proceedings of the National Academy of Sciences*, *120*(42), e2305427120. https://doi.org/10.1073/pnas.2305427120

Venkat, B. J. (2022). Through a glass darkly: Race, thermal sensation and the nervous body in late colonial India. *BJHS Themes*, *7*, 117–138. https://doi.org/10.1017/bjt.2022.3

Vinyeta, K., Whyte, K., & Lynn, K. (2016). Climate change through an intersectional lens: Gendered vulnerability and resilience in Indigenous communities in the United States (SSRN Scholarly Paper 2770089). https://papers.ssrn.com/abstract=2770089

Voyles, T. B. (2021). *The settler sea: California's Salton Sea and the consequences of colonialism*. University of Nebraska Press.

Ward, R. D. (1929). The acclimatization of the white race in the tropics. *New England Journal of Medicine*, *201*(13), 617–627.

Wesseling, C., Glaser, J., Rodríguez-Guzmán, J., Weiss, I., Lucas, R., Peraza, S., Silva, A. S. da, Hansson, E., Johnson, R. J., Hogstedt, C., Wegman, D. H., & Jakobsson, K. (2020). Chronic kidney disease of non-traditional origin in Mesoamerica: A disease primarily driven by occupational heat stress. *Revista Panamericana de Salud Pública*, *44*, e15. https://doi.org/10.26633/RPSP.2020.15

Whyte, K. (2020). Against crisis epistemology. In B. Hokowhitu, A. Moreton-Robinson, L. Tuhiwai-Smith, C. Andersen, & S. Larkin (Eds.), *Routledge handbook of critical Indigenous studies* (pp. 52–64). Routledge.

Wolch, J. R., Byrne, J., & Newell, J. P. (2014). Urban green space, public health, and environmental justice: The challenge of making cities "just green enough." *Landscape and Urban Planning, 125*, 234–244. https://doi.org/10.1016/j.landurbplan.2014.01.017

Wolcott, V. W. (2012). *Race, riots, and roller coasters: The struggle over segregated recreation in America*. University of Pennsylvania Press. https://doi.org/10.9783/9780812207590

World Health Organization. (2024a). United Nations Secretary-General calls for global action on extreme heat. https://www.who.int/news/item/25-07-2024-united-nations-secretary-general-calls-for-global-action-on-extreme-heat

World Health Organization. (2024b). Heat and health. https://www.who.int/news-room/fact-sheets/detail/climate-change-heat-and-health

Wynter, S. (2003). Unsettling the coloniality of being/power/truth/freedom: Towards the human, after man, its overrepresentation—An argument. *CR: The New Centennial Review*, 3(3), 257–337.

Yang, H., Lohmann, G., Lu, J., Gowan, E. J., Shi, X., Liu, J., & Wang, Q. (2020). Tropical expansion driven by poleward advancing midlatitude meridional temperature gradients. *Journal of Geophysical Research: Atmospheres, 125*(16), e2020JD033158. https://doi.org/10.1029/2020JD033158

Zemaitis, L. (2024). *Unlocking community resilience: Innovative strategies to access climate adaptation funding*. Center for Climate and Energy Solutions.

Zhao, Q., Guo, Y., Ye, T., Gasparrini, A., Tong, S., Overcenco, A., Urban, A., Schneider, A., Entezari, A., Vicedo-Cabrera, A. M., Zanobetti, A., Analitis, A., Zeka, A., Tobias, A., Nunes, B., Alahmad, B., Armstrong, B., Forsberg, B., Pan, S.-C., … Li, S. (2021). Global, regional, and national burden of mortality associated with non-optimal ambient temperatures from 2000 to 2019: A three-stage modelling study. *The Lancet Planetary Health*, 5(7), e415–e425. https://doi.org/10.1016/S2542-5196(21)00081-4

Zölch, T., Maderspacher, J., Wamsler, C., & Pauleit, S. (2016). Using green infrastructure for urban climate-proofing: An evaluation of heat mitigation measures at the micro-scale. *Urban Forestry & Urban Greening, 20*, 305–316.

PART II

Noninstitutional Climate Action

CHAPTER 9

The Emergence and Implications of the Global Climate Justice Movement Ecosystem

FERNANDO TORMOS-APONTE AND JACKIE SMITH

Introduction

CLIMATE justice consists of a series of counterhegemonic principles, actors, and processes that seek to transform human and ecological relations to enable a transition away from fossil fuel dependence in pursuit of just futures. The principles that guide the struggle for climate justice emerged from radical and feminist challenges to capitalism, neoliberal globalization, racism, and colonialism. Collectively, actors involved in these struggles form a global environmental movement ecosystem: a field with actors, strategies, and conceptual and strategic frameworks that vary over time and across regions on the extent to which they coordinate action (Rodríguez-Garavito, 2014). This conceptualization of movements transcends modernist, compartmentalized thinking to capture the fluid and dynamic ways that diverse arrays of actors—both formal and informal—interact, learn, and respond strategically to changing environments. As Evans and Rodríguez-Garavito (2018, p. 11) argue:

> Seeing transnational advocacy as having "the structure and logic of an ecosystem" … encourages us to look at differentiation within the field of transnational advocacy and at the interconnections among its different elements. The overall prospects for transnational advocacy are enhanced if different parts of the ecosystem system build connections with one another, thus enhancing the prospects of "counter-hegemonic globalization" (Evans, 2005). As in any ecosystem, the field's robustness will depend on the collaboration and complementarity among different types of issues, frames, organizational structures, actors, and strategies.

As many activists (and others) now recognize, climate justice requires work to transform the capitalist order and its hierarchies while *operating within it*. This process resembles natural processes of adaptation and evolution, as activists help advance alternatives to the prevailing social order while simultaneously resisting capitalist hegemony, repression, and attempts at co-optation. Such work requires collaboration, experimentation, and complementarity among a wide array of diverse actors. While some of this may be coordinated among different actors in the ecosystem, such coordination may emerge organically through loosely shared visions for a better world. We argue that such shared visions and principles as well as models for how to build "unity in diversity" have been advanced through decades of struggles led by "climate justice protagonists," who include actors often obscured and marginalized in mainstream scholarship, movements, and public discourse—that is, Indigenous peoples, people of color communities, and peasants engaged in frontline struggles to protect livelihoods (see Smith & Patterson, 2018). We trace this history of struggle through three main threads: global Indigenous peoples organizing, the globalization of US people of color-led environmental justice advocacy, and the convergence of movements in the global justice movement following the "Battle in Seattle."

The contemporary movement ecosystem related to global climate struggles involves two prominent networks of actors (Hadden, 2015). The climate justice sector emerges from grassroots and community-based struggles against environmental harm and in pursuit of a harmonious relationship with the earth. Under the leadership of marginalized groups historically excluded from mainstream environmentalism, the climate justice tradition bridges actors fighting neoliberal globalization, colonialism, and racism and mobilizing for climate action, human rights, and economic justice. The global struggle for justice and democracy is rooted in various histories of mobilization in the Global South and among the marginalized in the Global North, many of which converged in the late 1990s to challenge the global neoliberal order. This history of translocal struggle has informed the climate justice sector's development of a structural understanding of the problem of climate change (Cochabamba, 2010).

Indigenous Struggle, Learning, and Leadership

While many critics point to the ways the United Nations (UN) system supports prevailing power relations, it is critical to recognize its role in helping support translocal counterhegemonic movements. The development of movement analyses of the links between global capitalism and local experiences of injustice has been supported by the meeting spaces provided by the various forums created under the UN umbrella to support international understandings of shared problems and efforts to address them. By convening diverse constituencies across national and other divisions, the UN fosters dialogue and exchanges that support more expansive understandings of systemic problems. These spaces also allow participants to develop friendships and long-term bases for cooperation that evade state-centric and nationalist frameworks and the compartmentalized thinking that characterizes national and global policy realms.

For instance, the creation of the UN Working Group on Indigenous Peoples in 1982 opened spaces for Indigenous peoples from around the world to meet and explore shared experiences and challenges. These led to the 1990 Indigenous Alliance of the Americas Continental Gathering, "500 Years of Indian Resistance," in Quito, Ecuador. Representatives from 120 Indian Nations and organizations met for several days to develop a declaration that called for Indigenous people to come together to defend their autonomy and control of their territories. The meeting declaration concluded that resistance to the global capitalist order

> is a principal task for Indian Peoples; however, through our struggles we have learned that our problems are not different, in many respects, from those of other popular sectors. We are convinced that we must march alongside the peasants, the workers, the marginalized sectors, together with intellectuals committed to our cause, in order to destroy the dominant system of oppression and construct a new society, pluralistic, democratic and humane, in which peace is guaranteed. (Indigenous Alliance of the Americas, 1990)

From that 1990 meeting emerged some of the core ideas that have and that continue to guide today's environmental justice movements, including the need for leadership to emerge from those most impacted by capitalism; the recognition that the key struggle is over fundamentally different ideas about how to organize society and humans' relationship with the earth;

the centrality of local autonomy and control over local territories; and the idea that changing the status quo requires work to build diverse alliances that support and amplify the leadership and voices from frontline communities.[1]

Translocal Indigenous organizing supported ongoing attention to Indigenous issues in the UN, producing two International Decades of the World's Indigenous Peoples (1995–2004 and 2005–2014), which helped Indigenous groups further cultivate and popularize a global identity around the experiences of diverse Indigenous communities (see chapter by Fenelon and Figueroa Helland in this volume). It helped create transnational solidarities across highly diverse and localized Indigenous cultures. These connections led to organizing that sustained global attention to the concerns of this globally marginalized and oppressed constituency and created new legal protections and a standing Permanent Forum on Indigenous Issues within the UN system. These achievements by Indigenous movements have helped proliferate new global spaces where the intersections of environment and human rights are routinely scrutinized, and where novel initiatives for both institutional and transgressive collective action can be developed. For instance, it is through these spaces where counterhegemonic ideas such as food sovereignty and rights of peasants progressed from movement ideas into national and global policy frameworks (Edelman & James, 2011; McKeon, 2015; Shattuck et al., 2015; Shawki, 2014).

This global Indigenous movement-building was happening as other global developments created more openings for critical social movements to develop transnational dialogues and movements. By the 1990s, the end of the Cold War had helped open new avenues for global dialogue through a series of global conferences (Friedman et al., 2005; Willetts, 2011). These encouraged the proliferation of transnational organizations and networks, which in turn helped consolidate critical ideas and thinking across diverse movements and peoples around the world (Carroll, 2016; Carroll & Sapinski, 2013; Smith et al., 2018). It is these networks that supported protests capable of closing down the opening meeting of the 3rd World Trade Organization Ministerial Meeting in Seattle in 1999, generating a further expansion and flourishing of transnational networking and protests known as the global justice movement (Almeida & Lichbach, 2003). A key argument in our chapter is that much of the discourse and organizational origins of today's environmental justice movement can be traced to this period in the late 1980s and early 1990s and the movement-based learning it unleashed (Smith et al., 2017).

In the early 1990s, US-based environmental justice groups organized a challenge against the domination of white-led, Global North–based, mainstream environmentalist groups. They worked to articulate a counterhegemonic vision to challenge conventional thinking among mainstream environmental groups, beginning with a 1990 letter to the largest environmental groups in the United States, known as the Group of Ten (including large long-standing groups such as the Sierra Club, Greenpeace, Friends of the Earth, and National Wildlife Federation),[2] and following with the 1991 Principles of Environmental Justice,[3] and later the 1996 Jemez Principles of Democratic Organizing (Jemez Principles for Democratic Organizing, 1996; Moore, 1990; The Principles of Environmental Justice, 1991).[4] The Southwest Organizing Project (SWOP) convened a growing network of radical environmental justice groups to develop the letter to the Group of Ten (Almeida, 1994). This letter challenged mainstream green groups' claims to represent the viewpoints of grassroots communities and communities of color. Letter writers and signatories also pointed out the lack of accountability among green groups and their complicity in the disruption of environmental justice communities.[5] They called for a people-oriented approach to addressing environmental issues that included, but were not limited to, conservation, which was articulated in the Environmental Justice Principles.[6] Few mainstream green groups answered the letter or adapted their activities in response to environmental justice critiques. Nevertheless, environmental justice groups continued their efforts to broaden and build an environmental justice movement that built on their grassroots campaigns to fight the siting of environmentally hazardous facilities and toxic dump sites in marginalized communities. The Jemez Principles for Democratic Organizing grew from work in this ecosystem to identify values and organizing strategies to facilitate movement building across diverse organizations and constituencies.

Despite falling on deaf ears within the United States, the ideas emerging with the Environmental Justice Principles found fertile ground as they were introduced in global spaces. At this time, social movements from across the Global North and South encountered new opportunities to come together through UN conferences of the early 1990s, including those addressing Indigenous peoples as noted earlier, and around the 1992 UN Conference on Environment and Development (UNCED). In fact, organizers from the National People of Color Environmental Leadership Summit attended the UNCED meeting within months of concluding their seminal environmental justice statement (Chavis, 1993). Thus, we see how this global movement ecosystem developed through both official UN conferences and movement-generated convenings.

Such global spaces and the networks they generated supported the spread of critical insights from Indigenous and US environmental justice activism globally. And here US activists gained important insights into the similarities between their struggles within the United States and those of Black, Indigenous, and other people of color communities around the world, deepening shared understandings of how global capitalism reinforces racial and gender oppressions in diverse contexts worldwide. Since the 1990s, environmental justice has thus grown from an emergent and insurgent idea and movement to becoming one of the dominant environmental frames globally. The persistent struggles of environmental justice activists to contest dominant conceptual frameworks and organizing strategies accomplished that. Over time, the global expansion of environmental justice activism has supported policy changes from local to global scales.

The development and popularization of justice principles have helped shape the strategies of the contemporary environmental justice movement. Subsequent efforts to develop new norms included the Bali Principles of Climate Justice (2002), which internationalized environmental justice principles and contextualized them in relation to the issue of climate change, and the Just Transition principles, which emerged from long-term work by US grassroots and labor coalitions to challenge the notion that thriving economies and environments were mutually exclusive aspirations (Indigenous Environmental Network, 2018; Just Transition Alliance, 2024). The global growth and success of climate justice organizing benefited from the emergence of a global justice movement that broke down issue siloes and supported the development of this global movement ecosystem by creating iterative, counterhegemonic spaces for activists to convene, exchange ideas, engage in collective actions, and develop long-term networks and relationships (Smith, 2008, 2020). This movement did not merely resist neoliberal globalization. It also articulated visions for alternative approaches to human and ecological relations, supporting the proliferation of new modes of thinking that radicalized translocal environmental organizing and have helped support a paradigm shift (Smith, 2014). In the sections that follow, we discuss the radicalization of environmental mobilization following the emergence of global protests against the world trade system, how radicals resisted domination within environmental movement networks, and the demands that radical climate justice groups popularized. Further, we offer some thoughts on the future of environmental mobilization and research on it.

Paradigm Shift and Radicalization

Especially important in the period following the Battle in Seattle was the series of global-to-local convenings of radical activists organized under the framework of the World Social Forum process, which was most active from its origins in 2001 through roughly the first decade of the 21st century (Sen & Waterman, 2012; Smith et al., 2011). The World Social Forums created a space for activists to develop shared analyses of climate change and its intersections with other concerns of global justice movements, contributing to the evolution and popularization of radical movement thinking initiatives of environmental justice leaders, including Tom Goldtooth, director of the Indigenous Environmental Network (IEN), to articulate formal climate-specific justice principles (Tokar, 2018). During the 2002 second People of Color Leadership Summit in Washington, DC, organizers drafted 10 Principles for Just Climate Change Policies in the United States. Later that year, 14 organizations, making up the International Climate Justice Network, met in Bali, Indonesia, and developed the Bali Principles of Climate Justice.

As a global movement for global justice emerged, Global North and White-led green groups that constituted the majority of environmental organizing resources formed an international network, the Climate Action Network (CAN), in the late 1990s. This network was initially diverse yet dominated by green group leaders from the Global North, who imposed their policy analysis and prescriptions on the rest of the network and engaged in the practice of speaking on behalf of all civil society during their UN Framework Convention on Climate Change (UNFCCC) interventions. These practices contradicted environmental justice principles and the Jemez Principles for Democratic Organizing, which called on letting people speak for themselves, being inclusive, emphasizing bottom-up organizing, working together in solidarity, and engaging in equitable resource distribution and decision-making.

In 2007, the exclusion of climate justice groups from CAN motivated the formation of Climate Justice Now! (CJN), which subsequently split due to the growing discontent with the UNFCCC process. Critiques of the UNFCCC had emerged long before in the context of the World Social Forums and related networks (Bond, 2012).

A notable breaking point followed the UN climate negotiations in Cancún in 2010, where delegates of various ALBA (Bolivarian Alliance for the Peoples of Our America) sought to block market-based approaches to funding climate action and called for emissions reduction targets (among other demands). These objections were sidelined by the Mexican presidency of the meeting. After the meeting chair, Mexican Minister of Foreign Affairs Patricia Espinosa, gaveled over the Bolivian delegate Pablo Solón's objections to the text, the Cancún meeting set the UNFCCC on a course with no return (Solon, 2012). The Cancún Agreements laid the groundwork for the Paris Agreement, a weak and voluntary emissions reduction commitment-based treaty that included market-based mechanisms to finance climate action and left human rights provisions out of the legally binding sections of the agreement.

Green groups' dominance of UN-based global environmental organizing and their support for market-based mechanisms for addressing climate change posed various challenges to climate justice. Green groups captured climate change expertise, portraying themselves as the main experts who knew the science and policies needed to address the crisis. However, the policy proposals that green groups often forwarded included harmful market-based approaches that involved putting a price on carbon and other natural resources and allowing the sale of the right to pollute the environment. Carbon trading is one of these harmful policies that green groups and environmental policy technocrats successfully pushed governments to rely on as measures to address climate change.

The failure of the 2009 negotiations in Copenhagen and the sidelining of radical challenges to market-based approaches to addressing climate change motivated radical and grassroots climate justice groups within the environmental movement ecosystem to articulate an analysis of the structural roots of the climate crisis. Opportunities for articulating these structural analyses included the 2010 World People's Conference on Climate Change and the Rights of Mother Earth in Cochabamba, Bolivia.

The suppression of dissent in UNFCCC negotiations and the failures of the UN process to meet the demands of climate justice groups motivated various groups to question the focus on influencing the UN process and place greater emphasis on local efforts to prevent fossil fuel extraction and human rights violations. Some climate justice groups kept a presence at the UNFCCC, reconstituting as the Campaign to Demand Climate Justice (DCJ), and continued to demand climate justice at the UNFCCC. In their view, keeping a presence in the UNFCCC space was a strategy aiming to block harmful policies from being enacted and implemented and to push for long-standing climate justice demands of non-market-based financing to address the climate crisis.

Various global climate justice movement meetings facilitated breaking issue siloes and enabled coordination between grassroots and community-based struggles and those taking their demands to international fora. The World Social Forum process, as well as two climate justice meetings in Cochabamba, Bolivia, were among those meetings that provided such spaces for building global solidarity across movements. First convening in 2010 after the failed 2009 UN meeting in Copenhagen, the World People's Conference on Climate Change and the Rights of Mother Earth convened around 10,000 activists and government representatives, generating a "People's Agreement" that reflects shared visions and analyses of global climate justice activists and fundamentally challenges capitalist, state-centric approaches to the climate emergency.[7] The agreement echoes language seen in other contexts of global justice organizing, demonstrating processes of movement knowledge-building. It begins, "We confront the terminal crisis of a civilizing model that is patriarchal and based on the submission and destruction of human beings and nature," and it presents a fundamental paradigm shift in its plan of action, which relies upon the principal agency of popular movements rather than states. It calls for a global popular referendum to recognize the rights of Mother Earth, reallocate resources to halt climate change, and create a Climate Justice Tribunal to "judge those who destroy Mother Earth" and promote accountability for environmental debts. Although "unrealistic" in the geopolitics of climate negotiations, the document puts forward ideas for a pathway to a new way of organizing social relations that is more "realistic" in the sense that it would enable the continuation of human society on Earth, while the "realist" approach guarantees an end to human civilization.

Documents like the People's Agreement of the World People's Conference on Climate Change demonstrate the paradigmatic shift in thinking by large and diverse numbers of people and movements who both recognize and confront the failures of the interstate system's inherently modernist, colonial project that drives climate change. The prevailing capitalist paradigm separates people from natural ecosystems and from each other while destroying the socio-ecological reproductive foundations of society. Yet it continues to determine the parameters of intergovernmental climate negotiations, thereby preventing any meaningful government action in the face of a deepening climate crisis (Salleh, 2025). In this leadership vacuum, climate justice movements are articulating ideas of global agency that can inspire and empower the growing numbers of people who must come to grips with the realities of climate change and recognizing that survival will require a break from the current system.

The growth of global and translocal activism over recent decades has helped popularize the idea that other worlds are possible as they foster new understandings and identities in activist communities, helping break down issue siloes to connect environmental concerns with critical analyses of the global economy. The growth of movements for social and economic rights has helped infuse climate activism with demands for human rights, decolonization, and racial justice—ideas now integral to notions of climate justice and "just transition." At the same time, global Indigenous activism that has been encouraged by processes surrounding the UN Permanent Forum on Indigenous Issues (established in 2000) has helped introduce new "epistemological resources" by popularizing ideas of buen vivir and rights of nature in global movement networks (Evans & Rodríguez-Garavito, 2018; see also Salleh, 2025; Smith, 2014).

Movement organizations emerging in the second half of 2010, including Fridays For Future, the Sunrise movement, and Extinction Rebellion, attracted increased visibility to climate mobilization globally. Their emergence within a crowded field of green groups and environmental and climate justice groups caused some conflict for various reasons. These tensions tend to emerge from tactical differences (Berglund & Schmidt, 2020), resource competition (Bush & Hadden, 2019), and the violation of principles like the Jemez Principles for Democratic Organizing. The Sunrise movement, for example, popularized a Green New Deal proposal that surprised labor and environmental justice groups when it became public (Adrar et al., 2019). Despite ongoing tensions, a climate of coexistence and collaboration has come to mark the interactions among emerging groups, green groups, and environmental justice groups in UNFCCC spaces, such as the Climate Camp that youth host in the sidelines of the yearly intersessional negotiations in Bonn, Germany. In this setting, diverse groups tend to gather to strategize, share meals, and support popular education about climate policies.

The Future of Climate Activism and Implications for Research

The focus of much social movement research, particularly that in the Global North, tends to privilege states and the interstate system that dominate formal politics. However, when we consider the relationships among the prevailing political order, the global capitalist system, and climate change, it is essential to look beyond the actors and institutions that are grounded in that system for ideas about what a more just and sustainable social order might look like and what actors and actions might lead to such a place. We argue that the movement knowledge emerging in a vast and increasingly networked ecosystem of environmental justice movements and activists is popularizing ideas that are necessary—if not sufficient—for generating the kind of paradigm change required to ensure the survival of humanity and a livable planet.

We are currently watching the unfolding of a climate emergency. The failure of social movement scholars to identify and illuminate the work of social movement actors that have most directly confronted the failures of governments to take meaningful action on climate change has contributed to their marginalization and slowed progress. Professional norms and incentive structures that encourage dispassionate science and claim "objectivity" contribute to our failure to take radical action on climate change and constitute—as leading climate scientist James Hansen, Jeffrey Sachs, and their collaborators called—"an act of extraordinary witting intergenerational injustice."[8] Conventions guiding social research encourage attention to actors and social settings deemed worthy of our attention. But few question how such "worthiness" is determined. For much social movement research, a movement's visibility in the public sphere or in the academic literature or its actual or potential impact on policy is a leading determinant of what is worth studying. Thus, those studying climate politics have tended to focus on the interstate arenas in which climate-related negotiations take place. As a result, we have extensive documentation of the monumental failure of states to act in the face of what is clearly an urgent, existential problem for human society. But there is relatively less scholarly attention on emerging alternatives to interstate politics, although our discussion herein shows that this is finally changing. Also, in much of the social sciences and even in social movement research, there is a strong state-centric bias. Methodological nationalism prevents scholars from seeing actions and actors operating outside of—and possibly prefiguring—alternatives to the state system for organizing human society. The protagonists of climate justice struggles are calling us to imagine and help give birth to a system that supports life on Earth for the long haul. It's time for us to tune into their messages and contribute our energies to the search for pathways to such a system.[9]

Notes

1. This account draws from the account in Smith and Patterson (2018).
2. https://www.ejnet.org/ej/swop.pdf
3. https://www.ejnet.org/ej/principles.html
4. https://www.ejnet.org/ej/jemez.pdf
5. The letter is available at https://www.scribd.com/doc/26270824/SWOP-Letter-to-Group-of-10-Environmental-Organizations-1990
6. https://www.ejnet.org/ej/principles.html
7. https://pwccc.wordpress.com/2010/04/24/peoples-agreement/
8. https://journals.plos.org/plosone/article?id=10.1371/journal.pone.0081648
9. This section draws from "Sociology's Nero Syndrome?," Jackie Smith, Mobilizing Ideas, May 4, 2015, https://mobilizingideas.wordpress.com/2015/05/04/sociologys-nero-syndrome/

References

Adrar, A., Burlingame, O., Rogers-White, A., & Tormos-Aponte, F. (2019, July 16). Green new deal policies should be fueled by frontline and grassroots power. *Public Administration Review*. https://www.publicadministrationreview.com/2019/07/16/gnd16/

Almeida, P. (1994). The network for environmental and economic justice in the Southwest: Interview with Richard Moore. *Capitalism, Nature, Socialism*, 5(1), 21–54.

Almeida, P., & Lichbach, M. (2003). To the Internet, from the Internet: Comparative media coverage of transnational protests. *Mobilization*, 8(3), 249–272.

Berglund, O., & Schmidt, D. (2020). *Extinction rebellion and climate change activism: Breaking the law to change the world*. Springer International.

Bond, P. (2012). *Politics of climate justice: Paralysis above, movement below*. South Africa University of Kwazulu-Natal Press.

Bush, S. S., & Hadden, J. (2019). Density and decline in the founding of international NGOs in the United States. *International Studies Quarterly*, 63(4), 1133–1146.

Carroll, W. (2016). *Expose, oppose, propose: Alternative policy groups and the struggle for global justice*. Zed.

Carroll, W., & Sapinski, J. P. (2013). Embedding post-capitalist alternatives? The global network of alternative knowledge production and mobilization. *Journal of World-Systems Research*, 19(2), 1–31.

Chavis, B. Jr. (1993). Foreward. In R. D. Bullard (Ed.), *Confronting environmental racism: Voices from the grassroots* (pp. 3–5). South End Press.

Cochabamba, T. (2010, May 6). *Final conclusions working group 1: Structural causes*. World People's Conference on Climate Change and the Rights of Mother Earth. https://pwccc.wordpress.com/2010/04/30/final-conclusions-working-group-1-structural-causes/#more-1775

Edelman, M., & James, C. (2011). Peasants' rights and the UN system: Quixotic struggle? Or emancipatory idea whose time has come? *Journal of Peasant Studies*, *38*(1), 81–108.

Evans, P. (2005). The challenges of the "institutional turn": New interdisciplinary opportunities in development theory. In V. Nee & R. Swedberg (Eds.), *The economic sociology of capitalism* (pp. 90–116). Princeton University Press.

Evans, P., & Rodríguez-Garavito, C. (2018). "Introduction: Building and sustaining the ecosystem of transnational advocacy." In P. B. Evans & C. Rodríguez-Garavito (Eds.), *Transnational networks advocacy: Twenty years of evolving theory and practice* (pp. 8–25). Dejusticia.

Friedman, E. J., Clark, A. M., & Hochstetler, K. (2005). *Sovereignty, democracy, and global civil society: State-society relations at the UN World Conferences*. State University of New York Press.

Hadden, J. (2015). *Networks in contention: The divisive politics of climate change*. Cambridge University Press.

Indigenous Alliance of the Americas on 500 Years of Resistance. (1990, July). *Declaration of Quito, Ecuador*. http://www.cumbrecontinentalindigena.org/quito90.php

Indigenous Environmental Network. (2018). *Indigenous principles of just transition*. https://www.ienearth.org/wp-content/uploads/2018/08/PrinciplesJustTransition-BW.pdf

Jemez Principles for Democratic Organizing. (1996, December). *Ejnet.Org*. https://www.ejnet.org/ej/jemez.pdf

Just Transition Alliance. (2024, October 3). *Just transition alliance principles*. https://jtalliance.org/what-is-just-transition/

McKeon, N. (2015). *Food security governance: Empowering communities, regulating corporations*. Routledge.

Moore, R. (1990). Southwest Organizing Project. https://www.ejnet.org/ej/swop.pdf

The Principles of Environmental Justice. (1991, October 27). *EJnet.Org*. https://www.ejnet.org/ej/principles.html

Rodríguez-Garavito, C. (2014). The future of human rights: from gatekeeping to symbiosis. *SUR–International Journal on Human Rights*, *11*(20), 499–509.

Salleh, A. (2025). *Decolonize ecomodernism!*. Bloomsbury Publishing Plc.

Sen, J., & Waterman, P. (2012). *The World Social Forum: Critical explorations*, Vol. 3. OpenWord.

Shattuck, A., Schiavoni, C., & Vangelder, Z. (2015). Translating the politics of food sovereignty: Digging into contradictions, uncovering new dimensions. *Globalizations*, *12*(4), 421–433. http://dx.doi.org/10.1080/14747731.2015.1041243

Shawki, N. (2014). New rights advocacy and the human rights of peasants: La Via Campesina and the evolution of new human rights norms. *Journal of Human Rights Practice*, *6*(2), 306–326.

Smith, J. (2008). *Social movements for global democracy*. Johns Hopkins University Press.

Smith, J. (2014). Counter-hegemonic networks and the transformation of global climate politics: Rethinking movement-state relations. *Global Discourse: An Interdisciplinary Journal of Current Affairs and Applied Contemporary Thought*, *4*(1), 1–19. http://dx.doi.org/10.1080/23269995.2013.874111

Smith, J. (2020). Making other worlds possible: The battle in Seattle in world-historical context. *Socialism and Democracy*, *34*(1), 114–137.

Smith, J., Byrd, S., Reese, E., & Smythe, E. (2011). *Handbook of World Social Forum activism*. Paradigm.

Smith, J., Gemici, B., Plummer, S., & Hughes, M. (2018). Transnational social movement organizations and counter-hegemonic struggles today. *Journal of World-Systems Research*, *24*(2), 372–403.

Smith, J., & Patterson, J. (2018). Global climate justice activism: "The new protagonists" and their projects for a just transition. In R. Scott Frey, Paul K. Gellert, & Harry F. Dahms (Eds.), *Ecologically unequal exchange: Environmental injustice in comparative and historical perspective* (pp. 245–272). Palgrave MacMillan.

Smith, J., Plummer, S., & Hughes, M. (2017). Transnational social movements and changing organizational fields in the late twentieth and early twenty-first centuries. *Global Networks, 17*(1), 3–22.
Solon, P. (2012). Why Bolivia stood alone in opposing the Cancun Climate Agreement. In *What next: Climate, development, and equity* (pp. 106–107). What Next Forum. https://whatnext.org/wp-content/uploads/2020/03/wnv3_web_single_72.pdf
Tokar, B. (2018). On the evolution and continuing development of the climate justice movement. In T. Jafry (Ed.), *Routledge handbook of climate justice* (pp. 13–25). Routledge.
Willetts, P. (2011). *Non-governmental organizations in world politics: The construction of global governance*. Routledge.

CHAPTER 10

The Drivers of Individual Participation in Climate Action

CLARE SAUNDERS AND EMILY RAINSFORD

Introduction

The prominence of climate protest actions in the media and in academic books and papers in recent years (Chen et al., 2023) might make us surprised to learn that in the United Kingdom only around 6% of the population participate in a public legal demonstration, on any issue, in any given 12 months (the figure was 6.5% in the 2023 European Social Survey). Climate change is only one of a variety of issues and causes about which citizens protest. This means that a smaller proportion, still, participates in protest actions specifically on climate change. Given they are such a small minority, it is therefore likely that there is something special or different about people who participate in climate protest. What is it that they do or have that is different? These different behavior and attitudinal traits are collectively known as drivers for protest participation.

There are three prominent, and overlapping, theoretical models underpinning the majority of research on the drivers for participation in protest. It is important to note that most of the literature in this area is about protest generally, rather than specifically about climate protest. In order of publication date, the first is Klandermans and Oegema's (1987) stepwise model, which involves consensus mobilization (sharing the need to address a cause) followed by action mobilization (turning out) (Klandermans, 1988). The second is Verba et al.'s (1995) resource-based model, which points out the need for the right set of resources to become a protest participant. The third is Schussman and Soule's (2005) summary of the drivers of protest, which we call the tripartite model due to its emphasis on three key sets of drivers: biographical availability (such as being young or free from a routine job), structural availability (being in the right time at the right place to hear about a protest), and psychological/political engagement (having the dispensation to engage in politics more generally).

This chapter summarizes these models and links them to other related bodies of literature, contextualizing them, where possible, in relation to climate protest. It then briefly introduces a range of climate protest tactics and explores protest survey data to illustrate the extent to which climate change protesters have the biographical, attitudinal (including environmental attitudes), structural, and psychological traits that mobilized protesters are expected to possess. The bodies of literature that we summarize tend to assume that a lack of the aforementioned traits is what leads to nonparticipation. However, there is also a smaller body of literature that explicitly seeks to understand nonparticipation in collective action. Klandermans and Van Stekelenburg (2014) produce a social-psychological model of nonparticipation that largely confirms the notion that nonparticipation is simply the flip side of participation. However, they also noted that people tend to fit in with their social milieu—those who protest have friends and family who approve of their protest; those who do not protest have friends and family who do not approve of protest. Relatedly, Feldman (2021) found that young people do not engage with Global Climate Strikes when they feel too busy, if their school or college does not support striking from school, if they generally disapprove of protest, or if they have difficulty traveling to the protest site.

In particular, this chapter examines the characteristics of protesters who turn out to routine climate marches, Extinction Rebellion's (XR's) rebellious activities, XR's less disobedient and legally sanctioned marches, and participants in the Global Climate Strike. It does this to allow us to discover which drivers matter for which types of protests. Examining two different climate marches also allows for a comparison over time. One of the challenges of understanding the extant literature on drivers of protest participation is that findings vary across studies, making it hard to generalize. The variation in findings is due to several factors, including the theoretical emphasis of different studies, as well as differences in the causes examined, the types of action under scrutiny, and the political and country context. Some of these reasons are interrelated. For example, a cause and set of protest demands deemed radical by the public and state institutions may result in civilly disobedient actions that are at best discouraged and at worst repressed by the state. The types of people who turn out to actions that vary in their personal consequences are likely to vary markedly from one another. Those participating in actions that come with a chance of arrest or injury might be thought of as a subgroup of activists with extra special characteristics who have surpassed additional barriers to protest participation (Hope et al., 2019). Rarely is the relevance of the same set of variables tested across different types of protest. Findings also vary because survey instruments designed for different studies do not always contain comparable question batteries. Indeed, our own use of illustrative data in this chapter is limited by data availability.

In the illustrative data we use in this chapter, the cause (climate change) is kept constant, the political context is *relatively* stable (2009–2023 in the United Kingdom), but the forms of action differ. There have been changes in the political party holding government during this time span, and the United Kingdom has introduced more restrictive regulations on protest deemed as a nuisance (Saunders et al., forthcoming), but this makes little difference to the practice of legally sanctioned demonstrations. Legally sanctioned protests have always been permitted in the United Kingdom, and our own observations of climate protests over the years (except for civilly disobedient ones) have been of facilitative policing. Let us now turn to a summary of some of the key drivers of protest participation.

The Stepwise Model

Klandermans and Oegema (1987) famously found that although 74% of the Dutch population sympathized with the demands of peace demonstrations in the Netherlands, only 59% were aware that the demonstrations were taking place, as few as 10% were motivated to attend, and just 4% actually turned out to the streets. We can see similar patterns in relation to climate change in the United Kingdom. Here, 80% of the adult population are estimated to be concerned about climate change (Burgueno Salas, 2023, statisca.com), and yet only around 30,000 of them participated in the peaceful The Big One demonstrations in April 2023. Why is this?

According to Klandermans and Oegema (1987), concern needs to be translated first into consensus mobilization and second into action. Consensus mobilization involves sharing framings of the causes and potential solutions to a political issue. Snow et al. (1986) argued that social movement organizations actively work on frame alignment through sets of "interactive and communicative processes" that encourage individuals' and an organization's interpretive orientations to become "congruent and complementary" (Snow et al., 1986, p. 464). Action mobilization involves the steps required to convert sympathizers into participants: They need to be targeted by mobilization campaigns—directly or indirectly—and they need to be willing to attend; only then can they jump the final hurdle of attending.

For Snow et al. (1986), consensus mobilization is achieved through networks and the mechanisms of bridging, amplifying, extending, and/or transforming. Frame bridging helps to link existing grievances into collective action via organized efforts; frame amplifying involves appealing to latent values (e.g., family, home, ethnicity) or beliefs (e.g., injustice, inequality); frame extending means elaborating the goals of a movement to gather up more support from cognate grievances; and frame transforming redefines the ways people conceive of issues to make once tolerated niggles become unbearable and disgraceful. Snow et al. (1986) assume a top-down set of framing processes—that is, from an organization to individuals. Klandermans and van Stekelenburg (2019) call these types of identity formation deductive. They are deductive in the sense that the identity of an organization is taken on board by individuals. Klandermans and van Stekelenburg (2019) encourage us to think beyond top-down framings by coining the term "inductive identify formation." Inductive identity formation happens when activists come together to shape their own identities in the absence of a formal organization, through horizontal rather than vertical discussions. Whether inductive or deductive, networking is key to recruitment success, even if that is sometimes only through being in receipt of movement communications (Tindall, 2002). Interestingly, it has been found that identifying as an environmentalist leads to people engaging in private environmental activism, whereas identifying with the environmental movement leads to membership in environmental organizations (McCright & Dunlap, 2015).

A prerequisite for frame alignment is concern for the issue. For frame alignment to kick in, that concern needs to resonate, at least to some extent, with the demands of a protest. But what do we know about the drivers of environmental concern? Feeling connected to nature, perhaps by spending a significant amount of time in it, is often thought to encourage environmental concern (see, e.g., Nisbet et al., 2008; Tam et al., 2013). Many studies show that women have higher levels of environmental concern (Blocker & Eckberg, 1989; Dunlap, 1983; Mohai, 1997), a fact often associated, perhaps wrongly (Blocker & Eckberg, 1997; MacGregor, 2017), with our supposed innate maternal ethic of care. Women are also found to more consistently engage in environmental activism (see Zelenzy et al., 2000), particularly in the private sphere (e.g., boycotting and buycotting; Saunders & Roth, 2019), compared to men. Hunter et al.'s (2004) study of gender differences in private and public environmental behaviors across 22 countries shows that gendered patterns of participation vary across countries. In most countries' studies, there were no significant differences in public environmental behaviors between men and women, but in three countries more men participated than women (Bulgaria, Poland, and Spain) and in three more women participated than men (Netherlands, Australia, and New Zealand). The authors suggest a correlation with the economic performance of countries, which resonates with the postmaterialist and new middle-class theories we introduce next.

Environmental concern is thought to translate into action through several pathways. One of them is "environmental cognition," that is, having the ability to imagine ways in which people could live in harmony with nature. Wright et al. (2022), for example, found a positive association between environmental cognition and the signing and sending of a pro-environmental letter to someone in a position of authority. In a landmark early study into the causes of environmentalism, Cotgrove and Duff (1980) found that environmentalists (i.e., by their measurement, those with membership in Friends of the Earth and the Conservation Society) have a specific set of values: They are skeptical of science and technology, opposed to public institutions, uncomfortable with hierarchical power structures, have core values cognizant with postmaterialism, and believe that humans should and can live in harmony with nature. Drawing on Inglehart (1977/2015), postmaterialism means thinking beyond the material goals of economic growth, strong stable military economies, and defense forces toward the higher-level goals of giving people more say in important decisions (at work and in government), being more humane, and prioritizing ideas over money. For Cotgrove and Duff (1981), environmentalists holding postmaterialist values comprise a "new middle class" who work predominantly in service, welfare, and creative industries. They were thought to have become disillusioned with politics as usual and therefore took to the barricades, instead. In a follow-up piece, Cotgrove and Duff (1981) argued that, although many environmentalists are from the middle classes, and affluence can help shape the development of postmaterialist ideas, postmaterialist attitudes are actually a value to which environmentalists aspire in their utopic vision, rather than a precondition of their actions.

Environmental concern can be absent when people interpret other threats as more existential—such as living in poverty or being homeless (Sloam, 2020). As the climate crisis becomes increasingly obvious to many, this will likely change. Saunders (forthcoming) recently argued that the increased entanglements between people and more-than-humans (a term that refers to the ways in which human lives are entangled in complex ecologies with other species), in what has become known as the Anthropocene, mean that postmaterialism as a driver is being replaced by a sense of prematerialism. That is, we need to secure a livable planet first as the ultimate resource before we can be concerned about other types of materialism (Saunders, 2024).

Chawla (1999) does not distinguish between factors that lead to consensus mobilization from the ones that lead to action mobilization. But she did find that important drivers for environmental activism include having the values of environmental protection and social justice, being linked through family and organizational ties to environmental concerns and/or time in nature (e.g., being in the Scouts movement), reading an influential book, spending time outside, and having negative environmental experiences (e.g., witnessing or suffering from pollution). These factors we pick up in more depth in our account of the resource-based and tripartite models.

In addition to spending time in nature, environmental concern can be spread through political socialization, through parenting, or through friendship groups (Pickard et al., 2020), hinting at the importance of networks for recruitment into environmental activism in relation to both consensus and action mobilization (Saunders, 2022). As Snow et al. (1980, p. 798) famously put it: "recruitment cannot occur without prior contact with a movement agent. The potential participant has to be informed about and [be] introduced into a particular movement." In contemporary times, the agent could be online resources as much as face-to-face contacts. Indeed, Boulianne and Ohme (2022) have found that political use of social media is one of the most important predictors of participation in environmental activism among 18- to 24-year-olds in the United Kingdom, the United States, Canada, and France. In the resource-based and tripartite models, most of the emphasis is on action mobilization rather than consensus mobilization. However, the importance of education, crucial to consensus mobilization, is also part of the resource model, as we shall now explore. Concern alone does not automatically translate into activism.

To summarize, the stepwise model proposes that you first must share the ideology of the movement and then be mobilized (and be available to be mobilized) for participation. The sharing of ideology can come from various channels such as socialization through friends, family, or organizations both online and offline, as well as personal values of identification with a movement or organization and appreciation for nature (in the case of environmental movements).

The Resource-Based Model

Verba et al.'s (1995) model resonates with Klanderman's and Oegma's (1987) stepwise approach. They argue that people participate in political actions because *they want to, they can*, and *they have been asked to*. For someone to want to attend, the threshold of consensus mobilization must be achieved. One of the most important resources in developing a *want* to attend a protest is education. Evidence is found, again and again, supporting the notion that protesters are very highly educated, although this is axiomatically less true of young protesters, such as Global Climate Strike participants, who are currently undertaking their education. Our own protest survey research finds that climate change protesters on marches and in XR's actions are very highly educated, with 80% or more having completed, or undertaking, university degrees (see Saunders et al., 2020, 2024). Liberal education theory posits that protesters are highly educated because universities are carriers of liberal values, especially on humanities and social science programs of study (Rootes, 1980, 1995). Moreover, groups like XR privilege science framings, and the professionalized and highly educated middle classes are heavily invested in rationality Marxian theories and concepts of social justice that have permeated many young people's lives and are thought to have underpinned the student radicalism of the 1970s (Rootes, 1980). Education continues to shape people's lives, including their proclivity to protest (Olcese et al., 2014). When they are drawn into protest, those who are highly educated are less like to resort to violence and, as a result, are more likely to be successful in helping to generate new and more democratic regimes (Dahlum, 2019).

Having both the agency (the "*I can*") and the network links to attend (the "*I have been asked to*") are necessary for action mobilization. People who feel they can protest have a sense of their political agency, and they tend to believe that their actions will be efficacious (Chan, 2016). In line with Klandermans and Oegema's (1987) model, Schussman and Soule (2005) find that certain factors encourage intention to protest (and this includes many aspects of consensus mobilization), but that being asked to attend is driven by other factors that encourage protest participation. Network links also provide people with civic skills that equip them to participate in protest (the sense of "*I can*") (Putnam, 2000). In the resource-based model, networks themselves are both resources and important carriers of resources.

The Tripartite Model

Verba et al.'s "*I can*" also maps onto aspects of Schusmman and Soule's (2005) tripartite model. Having a sense of agency and of efficacy (i.e., believing that my or our actions *can make a difference*) is one of their political engagement variables. *Being asked* is an element of structural availability, which means being in the right place at the right time to hear about a demonstration. This could mean being a university or college student attending a campus where posters and leaflets are displayed, being a member of organizations or being signed up to social media pages that mobilize participants to attend a protest, or being a part of a friendship network comprised of activists. Personal networks and organizational networks both seem to matter, although inductive networking has been found to be particularly important in countries that clamp down on activism that challenges the state and on overtly political environmental organizations (Keshavarsi et al., 2023). Driscoll (2018) also argues for the importance of networking links outside of the organizational context. He finds that sometimes people become persistent activists precisely because they have distanced themselves from organizations with which they have become disillusioned.

Networks are deemed especially important in shaping identities. Collective identity, which develops through networking (Larana et al., 1994; Saunders, 2022), has been found—alongside connection to nature—to help generate a moral obligation to act to protect the environment and to predict politicized environmental behaviors, including protest (Schmitt et al., 2018). Several scholars have found that identifying with other environmentalists can encourage actions favorable to that group (Hogg, 2018; van der Werff et al., 2014). Furlong and Vignoles (2021), for example, suggest that fostering identification as part of a group is an important step along a pathway toward generating moral outrage that leads to mobilization.

Such processes can involve developing a taste for tactics (Jaspers, 2006), perhaps one developed through intense solidarity with similar others (Saunders, 2008a). Yet this does not mean that activists necessarily always develop strong collective identities and solidarities that preclude participation in multiple tactics. Tripp (2018), for example, identified environmental movement participants who were politically active, privately active, *and* both. He refers to the process of engaging in multiple tactics as "crowding in" and the exclusive use of single tactics as "crowding out."

Biographical availability, part two in the tripartite model, evolved from McAdam's (1986) work on the "high-risk" civil rights actions in the Deep South of the United States known as Freedom Summer. Biographical availability refers to having the personal circumstances to provide freedom to participate in actions. Being young (without family) or retired and free from full-time work are often seen as markers of biographical availability. However, the evidence in support of biographic availability is weak, even in McAdam's (1986) own work (Roth and Saunders, forthcoming). Like McAdam (1986), Nepstad and Smith (1999) also found that people who are in full-time or inflexible jobs are actually *more* likely to protest, rather than less—perhaps by virtue of the resources their jobs provide. Note here the link with Verba et al.'s resource model.

Psychological availability relates to the literature on consensus mobilization. It primarily incorporates political interest, access to political information, perceived efficacy, and political attitudes (Schussman & Soule, 2005), but others have added emotions and motivations to this model (Goodwin & Jasper, 2006; Saunders et al., 2012; see also Poma and Gravante in this volume). Approach emotions like anger and frustration are thought to drive participation in ways that avoidance emotions like fear and worry do not. Approach emotions have been associated with the perception of joining a powerful group that can affect change; and so they interact with senses of efficacy. Avoidance emotions, however, tend to result in withdrawal via association with groups found less powerful. Young people join climate movements as a result of their anger, which is combined with their sense of agency. They feel that protesting on climate change, using their own preferred repertoires, gives them a voice that can shape better futures. They have more of a sense of efficacy than their nonprotesting counterparts (Feldman, 2021).

Believing that one's actions can make a positive difference to the world has been found to be a significant predictor of environmental activism in particular, although collective senses of efficacy (i.e., the notion that *together* with others we can make a difference) do not always lead to protest participation (Hamann & Reese, 2020). This might be because a combination of both individual and group efficacy is required in order to avoid the free-rider syndrome (Olson, 1971). Free-riderism is the puzzle of collective action expressed by the difficult-to-answer question of "Why should I bother acting when someone else can do it on my behalf?"

Those who engage in protest have, historically, been left-libertarian. However, this has been changing over time as right-wing protesters have come to increasingly engage in activism (van Aelst & Walgrave, 2001), especially in the United States (Caren, 2023). In relation to environmental action, we have already seen that those with a propensity to be concerned about the environment think differently about human–nature relations and can envisage ways of running human societies outside of the dominant paradigms of techno-centric economic growth (Cotgrove & Duff, 1981). In our own work, we have found very few right-wingers attending environmental protest and hardly any attendees who most closely identify with the Conservative Party (Saunders et al., 2012, 2020, 2024).

In summary, the tripartite model covers similar aspects as the resource and stepwise model by arguing that for someone to participate in protest action they must be *structurally available* to be mobilized, they need to be *biographically available* to be able to participate, and they need to be *psychologically willing* to participate by being interested in politics and having approach emotions of anger and frustration. As it is the most comprehensive of the three models, we provide illustrative data on the tripartite model later in the chapter.

Varieties of Climate Action

In April 2023, around 30,000 people descended on the streets of London to participate in The Big One. This was a protest staged by XR, with limited support from a range of other nongovernmental organizations (NGOs; Saunders et al., 2024). It was a 4-day event that incorporated a range of protests, including pickets outside government departments, marches, and a die-in. This marked a change in the action repertoire of XR, which had previously used forms of civil disobedience to shut down cities and to clog up the judicial system with the processing of arrests, hoping that this would force the government's hand to seriously address climate change (Hayes et al., 2024; XR, 2019). In early 2023, XR announced that "we quit," signaling its shift away from civil disobedience to mass movement building (Saunders et al., 2024).

XR became a big household name in the United Kingdom from 2018 onward when it began its rebellious shutdowns of London city center (Doherty & Hayes, 2023). It was part of a wave of pre-COVID-19 climate action that was also marked by the unprecedented participation of young people in climate protest (Pickard, 2019). Fridays For Future is a youth-led and youth-centered movement that encouraged school pupils and college students to strike from their education to reclaim their future. Motivated by Swedish teenager Greta Thunberg, who began individual school strikes outside the Swedish parliament in 2018, the strikes at their peak mobilized around 7.6 million participants in over 180 countries (de Moor et al., 2020; see chapter by Uba in this volume). These actions build on a long history of climate protest in the United Kingdom. From 2005 onward, UK environmental NGOs, most prominently involving the Campaign Against Climate Change and Stop Climate Chaos coalition, have organized significant marches almost once a year—often timed to coincide with the negotiations of the Conference of Parties to the United Nations Convention on Climate Change (known as COPs) (Saunders, 2008b). What do the characteristics of people present at these different events tell us about the relative importance of different drivers for protest in encouraging participation in environmental protest?

A Note About the Data

The data used to illustrate the importance of drivers into climate activism come from protest surveys that have been conducted using systematized data collection protocols of the Caught in the Act of Protest project (Klandermans et al., 2009). These protocols take every effort to randomize distribution of surveys and to measure response rate biases. For more information on these protocols, see van Stekelenburg et al. (2012). There are two limitations of using this protest survey data. The first is that we do not have data on nonparticipants. The second is that we do not have the exact same questions in the protest survey questionnaires over time. This is because of evolution of our interests, as well as our work in partnership with XR for the 2023 survey. We present data on participants in the climate marches of 2009 ($n = 299$) and 2015 ($n = 296$), the London-based rebellious XR actions in 2019 ($n = 232$) and nonrebellious Big One demonstrations in 2023 ($n = 611$), and the Global Climate Strike protests in 2019 in Manchester and Truro, Cornwall ($n = 116$). Note also that the data are not weighted to account for response rate bias (we know that women, the well-educated, and those more politically interested have greater proclivity to respond to the surveys). It is therefore included for illustrative purposes only. The response biases are likely to be similar across all surveys; therefore, we can still compare across the different types of protesters.

Attitudes

Some of our protest survey questionnaires asked respondents to provide a qualitative response to a question asking them who or what is to blame for climate change, and another on what should be done to fix it. This question was asked at climate marches of 2009 and 2015, and the 2019 Global Climate Strike and XR rebel actions. The answers are very similar over time across different types of actions. Commonly blamed are polluting technologies, limitless economic growth, human greed, and profit-driven fossil fuel companies. Each time we do a survey, several people blame "everyone" for our contribution to the problem through our greenhouse-gas-intensive lifestyles.

Answers on how we should fix the problem range from policy initiatives to radically rethinking the way we live so that we can be more in tune with the planet. At the Global Climate Strike (GCS) and XR 2019 actions, we asked respondents the extent to which they agreed that modern science, governments, and companies can be relied on to solve the climate crisis. Climate protesters are considerably more optimistic about the role of science, with 55.5% of GCS and 36.9% of XR 2019 protesters agreeing or strongly agreeing that we can rely on it to solve climate change. In comparison, few place faith in the ability of governments to solve the problem (13.9% of GCS and 4.9% of XR participants), and even fewer believe companies can solve the problems that they created (9.0% GCS, 2.7% XR). Certainly we see evidence that climate protesters have a similar value profile to the environmentalists surveyed by Cotgrove and Duff back in the late 1970s.

Table 10.1 illustrates the structural availability of climate change protesters, exploring their network positioning and the embeddedness of individuals. From the literature, we know that protest participation happens when someone who agrees with the way an environmental organization frames an issue is also asked to participate, and in the stepwise model literature there was emphasis on the social milieu of friends and family approving of engagement or not influencing protest participation. We can see that friends and family are more likely to share information about the demonstration for the XR protests than for marches, for XR 15–16% heard about the actions from family/partner while only 8% of climate marchers did. For XR demonstrators 31% and 41% heard about it from friends/acquaintances while only 24% for the marchers. We can also see that an organization was a common channel for information for the 2009 march at 62.7%, but this was much lower for the XR demonstrations at under 35% and 25%. We can also see that XR and GCS demonstrators to a larger extent, at 81.9% and 75.9% respectively, were not asked by anyone to participate compared to the marchers at 29% and 35.8%. This difference could be a result of the significant press coverage of XR as well as its substantial social media presence.

TABLE 10.1 Structural Availability of Different Types of Climate Change Protesters

Variable	March 2009 (%)	March 2015 (%)	XR 2019 (%)	XR 2023 (%)	GCS 2019 (%)
Attended alone	13.7	21.5	21.0	—	14.9
Heard about demo from:	8.4	8.2	16.4	15.1	—
Partner/family	24.1	24.9	41.8	31.9	—
Friends/acquaintances	5.7	5.8	3.4	3.2	—
Work/school contacts	30.4	33.8	44.4	34.5	—
Organizational members	62.7	—	35.8	25.4	—
An organization					
Asked by no one	29.0	35.8	81.9	—	75.9*
Identifies with participants	80.6	76.7	79.9	—	80.6

Notes: Asterisks indicate that a different question was asked at the XR 2019 rebellious actions. In that survey round we instead asked, "To what extent do you agree that you participated because someone asked you to?" The number in the cell is the percent of respondents who listed that the disagreed, strongly disagreed, or neither agreed nor disagreed. Dashes indicate that we do not ask the question in that survey round.

Across all demonstrators we can see a high level of identification with the other participants, but we also see that there is a significant proportion of people across all demonstrations who have attended the demonstration alone, ranging from 13% for the 2009 march to approximately 21% for the 2015 march and 2019 XR demonstration. Overall, we can see that, of the structural availability variables, the social milieu is important for protest participation in terms of being aware of the demonstration, and possibly supporting the issue, but it is less important in terms of actual mobilization of demonstrators.

Table 10.2 illustrates the biographical availability of a range of environmental protesters. The patterns in the data support theories in the literature that women, middle-class, and highly educated people are more likely to engage in environmental protest. These patterns hold across all types of demonstrations. We see a relatively consistent proportion of people working part-time participating in all types of demonstrations (from 14.8% to 19.5%). However, we see a lower proportion of XR attendees in full-time employment compared to marchers, at 18.5% and 15.1% for XR while 35.4% and 38.6% for marchers. Some clue to interpreting this pattern might be seen in the age distribution of attendees; a larger proportion of XR attendees were under 25 at approximately 12% at each demonstration versus 8% and nearly 7% for the marchers. Younger attendees are more likely to be students than being in full-time work.

Table 10.3 explores the psychological and political engagement variables for protest engagement across different environmental demonstrations. We can see that all demonstrators express high levels of all feelings; thus, there does not seem to be support for the approach and avoidant dichotomy of feelings where the former encourage, and the latter dissuade, protest participation. We do see a slight decrease in interest in politics in the 2019 demonstrations (XR and GCS) compared to the 2009 and 2015 marches, but it is still well

TABLE 10.2 Biographical Availability of Different Types of Climate Change Protesters

Variable	March 2009 (%)	March 2015 (%)	XR 2019 (%)	XR 2023 (%)	GCS 2019 (%)
Gender (female)	60.3	54.5	60.2	56.4	62.1
Education (university educated)	72.3	80.1	85.8	84.2	73.0
Identifies as middle class or higher	70.7	71.3	67.0	—	75.0
Employed	38.6	35.4	18.5	15.1	—
Full-time	19.5	14.8	16.8	19.1	—
Part-time					
Biographically	8.1	6.9	11.9	12.1	60.7
available age	12.8	20.1	15.0	13.6	3.7
<25 years					
65+ years					

TABLE 10.3 Psychological and Political Engagement of Different Types of Climate Change Protesters

Variable	March 2009 (%)	March 2015 (%)	XR 2019 (%)	GCS 2019 (%)
Emotions (quite or a lot)	58.2	72.8	79.9	87.1
Angry	78.8	81.0	80.4	76.0
Worried	39.7	60.4	66.9	67.8
Fearful	83.6	90.8	86.9	90.3
Frustrated				
Interested in politics (quite or a lot)	92.5	91.4	84.9	87.8
Talks about politics (quite or a lot)	55.9	62.0	—	66.7

above 80%. Interestingly we see the pattern reversed with regard to talking about politics, and the proportion of people saying they discuss politics is generally lower than those who say they are interested (55–66%). This disconnect among those who actively engage in politics and who are being surveyed while they are engaging in a collective political act is puzzling and is an area for further research.

Conclusion

This chapter has introduced three different models that seek to understand the drivers of protest—the stepwise model, the resource model, and the tripartite model—and tailored them to environmental activism. We structured our illustrative use of data around the tripartite model as it covers many of the same variable as the other two, even though the internal logics of the models differ.

We used protest survey data to illustrate the importance of drivers into environmental activism across different types of environmental protests. The illustrative data confirm the importance of structural availability, as many survey respondents reported that they had been asked by friends and or family to attend, but interestingly they did not attend with them. School and workplace are not a mobilizing network, and in the XR demonstrations organizations seem to be less important than individual people in organizations or in close family or friendship circles. This is largely because XR relies heavily on social media for sharing news of its mobilizations. Generally, those who participate in environmental protest identify with the other participants, indicative of the importance of movement identification in driving protest engagement.

Second, biographical availability is illustrated as an important driver, with highly middle-class women being well represented across all demonstrations. Third, elements of psychological availability are illustrated to be important. People need to have a concern for the issue and have some interest in politics; certainly the environmentalists we survey have a worldview cognizant with that identified by Cotgrove and Duff (1980) back in the 1970s. However, the approach and avoidant feelings present a more complex pattern.

We also note that the presence of known drivers for participation is differentially distributed across our protesters. The biographical profiles of people attending different environmental protests vary, as do the mobilization patterns. XR and GCS participants were much more likely to have not been asked by anyone to participate. But they are more likely to have heard about the demonstration from their networks, so we can surmise their participation is "approved" by their social milieu.

To conclude, the theories explored in this chapter have some overlap, and the data further suggest that each theory has some support, but it differs across different protests and over time. More analysis is needed to explore the effects of type of protest participation and whether national context plays a part as well. We anticipate different drivers mattering in contrasting political contexts, as well as across new and emergent forms of environmental activism. Anantharaman et al. (2019) have begun this work by comparing who participates in three very different environmental community initiatives across the world: zero waste management in India, a local food movement in Canada, and a local sustainability group in Congleton, United Kingdom. They similarly found that the middle classes dominate these scenes. And yet other forms of environmentalism are purposefully environmentalisms *for* the poor *by* the poor (Martinez-Alier, 2014).

References

Anantharaman, M., Huddart Kennedy, E., Middlemiss, L., & Bradbury, S. (2019). Who participates in community-based sustainable consumption projects and why does it matter? A constructively critical approach. In C. Isenhour, M. Martiskainen, & L. Middlemiss (Eds.), *Power and politics in sustainable consumption research and practice*, Chapter 9. Routledge.

Blocker, T. J., & Eckberg, D. L. (1989). Environmental issues as women's issues: General concerns and local hazards. *Social Science Quarterly, 70*(3), 586–593.
Blocker, T. J., & Eckberg, D. L. (1997). Gender and environmentalism: Results from the 1993 general social survey. *Social Science Quarterly, 78*(4), 841–858.
Boulianne, S., & Ohme, J. (2022). Pathways to environmental activism in four countries: Social media, environmental concern, and political efficacy. *Journal of Youth Studies, 25*(6), 771–792.
Burgueño Salas, E. (2023). Level of concern about climate change in the United Kingdom. https://www.statista.com/aboutus/our-research-commitment/1898/erick-burgueno-salas
Caren, N. (2023). Right-wing protest in the United States, 2017 to 2022. *Socius, 9*. https://doi.org/10.1177/23780231231181900.
Chan, M. (2016). Psychological antecedents and motivational models of collective action: Examining the role of perceived effectiveness in political protest participation. *Social Movement Studies, 15*(3), 305–321.
Chawla, L. (1999). Life paths into effective environmental action. *Journal of Environmental Education, 31*(1), 15–26.
Chen, K., Molder, A. L., Duan, Z., Boulianne, S., Eckart, C., Mallari, P., & Yang, D. (2023). How climate movement actors and news media frame climate change and strike: Evidence from analyzing twitter and news media discourse from 2018 to 2021. *International Journal of Press/Politics, 28*(2), 384–413.
Cotgrove, S., & Duff, A. (1980). Environmentalism, middle-class radicalism and politics. *Sociological Review, 28*(2), 333–351.
Cotgrove, S., & Duff, A. (1981). Environmentalism, values, and social change. *British Journal of Sociology, 28*(2), 92–110.
Dahlum, S. (2019). Students in the streets: Education and nonviolent protest. *Comparative Political Studies, 52*(2), 277–309.
De Moor, J., Uba, K., Wahlström, M., Wennerhag, M., & De Vydt, M. (2020). Protest for a future II: Composition, mobilization and motives of the participants in Fridays for Future climate protests on 20–27 September, 2019, in 19 cities around the world. https://www.diva-portal.org/smash/get/diva2:1397070/FULLTEXT01.pdf
Doherty, B., & Hayes, G. (2023). Extinction Rebellion. In D. A. Snow, D. della Porta, B. Klandermans, & D. McAdam (Eds.), *The Wiley Blackwell encyclopedia of social & political movements* (2nd ed., 264–269). Wiley Blackwell.
Driscoll, D. (2018). Beyond organizational ties: Foundations of persistent commitment in environmental activism. *Social Movement Studies, 17*(6), 697–715.
Dunlap, R. E. (1983). Male-female differences in concern for environmental quality. *International Journal of Women's Studies, 6*(4), 291–301.
Feldman, H. R. (2021). Motivators of participation and non-participation in youth environmental protests. *Frontiers in Political Science, 3*, 662687.
Furlong, C., & Vignoles, V. L. (2021). Social identification in collective climate activism: Predicting participation in the environmental movement, Extinction Rebellion. *Identity, 21*(1), 20–35.
Hamann, K. R., & Reese, G. (2020). My influence on the world (of others): Goal efficacy beliefs and efficacy affect predict private, public, and activist pro-environmental behavior. *Journal of Social Issues, 76*(1), 35–53.
Hayes, G.., Doherty, B., Cammiss, S., & Saunders, C. (2024). Extinction Rebellion's disobedient citizenism. *Environmental Politics*, 1–20. https://doi.org/10.1080/09644016.2024.2406183.
Hogg, M. A. (2018). Social identity, self-categorization, and the small group. In E. White & J. H. Davis (Eds.), *Understanding group behavior: Small group processes and interpersonal relations* (Vol. 2, pp. 227–253). Psychology Press.
Hope, E. C., Pender, K. N., & Riddick, K. N. (2019). Development and validation of the Black community activism orientation scale. *Journal of Black Psychology, 45*(3), 185–214.
Hunter, L. M., Hatch, A., & Johnson, A. (2004). Cross-national gender variation in environmental behaviors. *Social Science Quarterly, 85*(3), 677–694.

Inglehart, R. (1977/2015). *The silent revolution: Changing values and political styles among Western publics*. Princeton University Press.
Jaspers, J. (2006). *Getting your way: Strategic dilemmas in the real world*. University of Chicago Press.
Jasper, J. M., & Goodwin, J. (2006). Emotions and social movements. In J. E. Stets & J. H. Turner (Eds.), *Handbook of the sociology of emotions* (pp. 611–635). Springer.
Kesharvarzi, S., Saunders, C., & Karimi, M. (2023). Persistent anti-littering activism in a non-western context: The case of the Nature Cleaners Movement in Iran. *Society & Natural Resources*, *36*(10), 1181–1199.
Klandermans, B. (1988). The formation and mobilization of consensus. In B. Klandermans, H. Kriesi, & S. Tarrow (Eds.), *International social movement research. Vol.1, From structure to action: Comparing movement participation across cultures* (pp. 173–197). JAI Press.
Klandermans, B., Giugni, M., Peterson, A., Sabucedo, J. M., Saunders, C., & Walgrave, S. (2009). Caught in the act of protest: Contextualising contestation survey project. ECRP Eurocores Project Funded by the ESF.
Klandermans, B., & Oegema, D. (1987). Potentials, networks, motivations, and barriers: Steps towards participation in social movements. *American Sociological Review*, *52*(4), 519–531.
Klandermans, B., & van Steklenburg, J. (2014). Why people don't participate in collective action, *Journal of Civil Society*, *4*, 341–352.
Klandermans, B., & van Stekelenburg, J. (2019). Identity formation in street demonstrations. In J. E. Stets and R. T. Sterpe (Eds.), *Identities in everyday life* (pp. 309–327). Oxford University Press.
Laraña, E., Johnston, H., & Gusfield, J. R. (Eds.). (1994). *New social movements: From ideology to identity*. Temple University Press.
MacGregor, S. (2017). Moving beyond impacts: More answers to the "gender and climate change" question. In S. Buckingham & V. Le Masson (Eds.), *Understanding climate change through gender relations* (pp. 15–30). Routledge.
Martinez-Alier, J. (2014). The environmentalism of the poor. *Geoforum*, *54*, 239–241.
McAdam, D. (1986). Recruitment to high-risk activism: The case of Freedom Summer. *American Journal of Sociology*, *92*(1), 1–256.
McCright, A. M., & Dunlap, R. E. (2015). Comparing two measures of social movement identity: The environmental movement as an example. *Social Science Quarterly*, *96*(2), 400–416.
Mohai, P. (1997). Gender differences in the perception of most important environmental problems. *Race, Gender and Class*, *5*(1), 153–169.
Nepstad, S., & Smith, C. (1999). Rethinking recruitment to high-risk/cost activism: The case of Nicaragua exchange. *Mobilization: An International Quarterly*, *4*(1), 25–40.
Nisbet, E. K., & Gick, M. L. (2008). Can health psychology help the planet? Applying theory and models of health behaviour to environmental actions. *Canadian Psychology/Psychologie Canadienne*, *49*(4), 296–303.
Olcese, C., Saunders, C., & Tzavidis, N. (2014). In the streets with a degree: How political generations, educational attainment and student status affect engagement in protest politics. *International Sociology*, *29*(6), 525–545.
Olson Jr, M. (1971). *The logic of collective action: Public goods and the theory of groups, with a new preface and appendix*, Vol. 124. Harvard University Press.
Pickard, S. (2019). Young environmental activists are doing it themselves. *Political Insight*, *10*(4), 4–7.
Pickard, S., Bowman, B., & Arya, D. (2020). "We are radical in our kindness": The political socialisation, motivations, demands and protest actions of young environmental activists in Britain. *Youth and Globalization*, *2*(2), 251–280.
Putnam, R. D. (2000). *Bowling alone: The collapse and revival of American community*. Simon and Schuster.
Rootes, C. A. (1980). Student radicalism: Politics of moral protest and legitimation problems of the modern capitalist state. *Theory and Society*, *9*(3), 473–502.
Rootes, C. (1995). A new class? The higher educated and the new politics. In L. Maheu (Ed.), *Social movements and social classes* (pp. 220–235). SAGE.

Saunders, C. (2008a). Double-edged swords? Collective identity and solidarity in the environment movement, *British Journal of Sociology*, *59*(2), 227–253.

Saunders, C. (2008b). The Stop Climate Chaos Coalition: Climate change as a development issue. *Third World Quarterly*, *29*(8), 1509–1526.

Saunders, C. (2022). Social networks and recruitment. In M. Guigni & M. Grasso (Eds.), *Routledge handbook of environmental movements* (pp. 390–404). Routledge.

Saunders, C. (forthcoming). Environmental movements and environmental political theory in the Anthropocene. In M. Wissenburg & A. Machin (Eds.), *Edward Elgar handbook of environmental political theory in the Anthropocene*. Edward Elgar.

Saunders, C., Doherty, B., & Hayes, G. (2020). A new climate movement? Extinction Rebellion's activists in profile. https://cusp.ac.uk/themes/p/xr-study

Saunders, C., Grasso, M., Olcese, C., Rainsford, E., & Rootes, C. (2012). Explaining differential protest participation: Novices, returners, repeaters, and stalwarts. *Mobilization: An International Quarterly*, *17*(3), 263–280.

Saunders, C., Hayes, G., Doherty, B., Cammiss, S., & Walker, N. (2024). Who participated in the Big One? A profile of participants in Extinction Rebellion's 2024 action. HaSS Cornwall Working Paper 1. https://hass-cornwall.exeter.ac.uk/v8media/facultysites/hass/hasscornwall/XR_Report_-_Final_Draft_(PDF)_(4)-compressed.pdf

Saunders, C., & Roth, S. (2019). NGOs and social movement theory. In T. Davies (Ed.), *Routledge handbook of NGOs and international relations* (pp. 138–152). Routledge.

Saunders, C., Walley, B., & Lloyd, R. (forthcoming). Environmental protest and civil obedience. In J. Connelly, R. Izci, & J. Mouton (Eds.), *Edward Elgar handbook of environmentalism*. Edward Elgar.

Schmitt, M. T., Aknin, L. B., Axsen, J., & Shwom, R. L. (2018). Unpacking the relationships between pro-environmental behavior, life satisfaction, and perceived ecological threat. *Ecological Economics*, *143*, 130–140.

Schussman, A., & Soule, S. A. (2005). Process and protest: Accounting for individual protest participation. *Social Forces*, *84*(2), 1083–1108.

Sloam, J. (2020). Young Londoners, sustainability and everyday politics: The framing of environmental issues in a global city. *Sustainable Earth*, *3*, 1–7.

Snow, D., Zurcher, L. Jr., & Ekland-Olson, E. (1980). Social networks and social movements: A microstructural approach to differential recruitment. *American Sociological Review*, *45*, 787–801.

Snow, D. A., Rochford Jr, E. B., Worden, S. K., & Benford, R. D. (1986). Frame alignment processes, micromobilization, and movement participation. *American Sociological Review*, *51*(4), 464–481.

Tam, K. P., Lee, S. L., & Chao, M. M. (2013). Saving Mr. Nature: Anthropomorphism enhances connectedness to and protectiveness toward nature. *Journal of Experimental Social Psychology*, *49*(3), 514–521.

Tindall, D. B. (2002). Social networks, identification and participation in an environmental movement: Low-medium cost activism within the British Columbia wilderness preservation movement. *Canadian Review of Sociology/Revue Canadienne de Sociologie*, *39*(4), 413–452.

Tripp, W. B. (2018). Being green: Patterns of participation in the environmental movement. *Current Sociology*, *66*(5), 788–809.

Van Aelst, P., & Walgrave, S. (2001). Who is that (wo) man in the street? From the normalisation of protest to the normalisation. *European Journal of Political Research*, *39*(4), 461–486.

van der Werff, E., Steg, L., & Keiser, K. (2014). Follow the signal: When past pro-environmental actions signal who you are. *Journal of Environmental Psychology*, *40*, 273–282.

van Stekelenburg, J., Walgrave, S., Klandermans, B., & Verhulst, J. (2012). Contextualising contestation: Framework, design, and data. *Mobilization: An International Journal*, *17*(3), 249–262.

Verba, S., Schlozman, K. L., & Brady, H. E. (1995). *Voice and equality: Civic voluntarism in American politics*. Harvard University Press.

Wright, J. D., Schmitt, M. T., & Mackay C. M. L. (2022). Access to environmental cognitive alternatives predicts pro-environmental activist behaviour. *Environment and Behaviour*, *54*(3), 712–742.

Zelezny, L. C., Chua, P. P., & Aldrich, C. (2000). New ways of thinking about environmentalism: Elaborating on gender differences in environmentalism. *Journal of Social Issues*, *56*(3), 443–457.

CHAPTER 11

The Role of Emotions in Climate Activism

ALICE POMA AND TOMMASO GRAVANTE

Introduction

RESEARCH on emotions and climate change has increased in recent years, providing valuable knowledge to understand how climate change affects us on a psychological level—framing it as a mental health problem (Clayton, 2024; Clayton et al., 2014, 2017; Innocenti, 2022; Ojala et al., 2021)—but also finding the barriers that hinder a collective and forceful response to the crisis (Corraliza, 2022; Norgaard, 2011; Stoknes, 2015),[1] and showing how we can use emotions to improve the communication of this problem (Harth, 2021; Huntley, 2020; Moser, 2007; Salama & Aboukoura, 2017). In addition, there are studies that have focused on understanding the role of emotions in climate activism (Kleres & Wettergren, 2017; Poma, 2024; Poma & Gravante, 2021, 2024; Wettergren, 2024). Like everyone else, activists are people with emotions, but the emotions that characterize their participation in social movements are different from those we feel every day, because they are constructed in a context of protesting and questioning the status quo. For this reason, when studying the emotional dimension of climate activism, we deal with climate emotions (e.g., eco-anxiety, climate fear, and sorrow) that are constructed in a context of crisis but also with a wide range of emotions that depend on the protest experience. From a sociocultural approach to the study of emotions in protest, the aim of this chapter is to show the relevance of emotions in climate activism and not on their effects on the nonactivist population.

This approach considers emotions as sociocultural constructions, in a sense that they depend not only on human physiology and psychology but also on the social and cultural context where we live (Hochschild, 1979, 1983, 2016). Applied to the study of social movements, this approach has allowed us to know which emotions activists feel and how these emotions interact with each other, how they are constructed, how they are managed, and the effects that they have on the protest (Flam & King, 2005; Goodwin et al., 2001; Gould, 2009; Jasper, 1997, 2018). It is important to consider a questioning of the universality of emotions that characterizes this approach and leads to the study of specific types of emotions and how they are constructed in each context. This is highly relevant in the case of climate activism, mostly studied in the Global North. The rationale behind studying the emotional dimension of a social movement is that, from a micro level of individual social movement participation, it offers a deep understanding of activists' values, beliefs, priorities, degree of commitment, motivation, and lifestyle; thus, emotions allow us to know how activists understand the world around them, giving a meaning to their experience, and acting consequently. At the same time, emotions can mobilize and demobilize, energize and deenergize, so, from a meso level of analysis, emotions help to explain the organizational and strategic dimensions of social movements and the construction of a collective identity.

The chapter presents the theoretical background for the study of the emotional dimension in social movements, as well as the current literature on emotions and climate activism from different analytical approaches. The second part explores some of the emotions that stand out in climate activism. We will show how emotions such as concern, powerlessness, fear, guilt, anxiety, and sorrow influence climate activism. In the final section, we will discuss some of the emotions that characterize activism, focusing on anger, hope, trust, and love.

A Way to Study Emotions: The Sociocultural Approach in Social Movements

There are different approaches to understand emotions, so we start by presenting the theoretical background that supports our analysis. Since the 1990s, a line of research has been developed in the study of social movements focused on understanding the role of emotions in protests (Collins, 2019; Flam, 2015; Flam & King, 2005; Jasper, 2011, 2018; Jasper et al., 2001; Poma & Gravante, 2017, 2024). The sociocultural and constructivist approach to emotions, although interdisciplinary (see Feldman-Barrett, 2017), has been developed in the study of social movements, thanks to the contributions of Hochschild (1979, 1983) in the sociology of emotions, and to the work of authors such as James M. Jasper, Helena Flam, and Verta Taylor, who were pioneers in this field of study. Starting in the 2000s, young researchers with case studies in different countries joined the challenge of understanding the emotional dimension in protest and activism (e.g., Flam & King, 2005; Gould, 2009; Jasper et al., 2001). This literature has had a lot of resonance, not only in the Global North but also in Global South countries, such as those of Central and Latin America (Gravante & Poma, 2023; Massal, 2021).

From a sociological point of view, emotions are complex experiences, and Jasper comes to see them as *feeling-thinking processes*, since according to the author, "our feelings are mostly ways of processing information, orienting ourselves to the world, signaling to others, and preparing ourselves for purposive action" (2018, p. 9). Thus, analyzing the emotional dimension of activism from a sociocultural approach means understanding how emotions are constructed by activists throughout their own experience. From this perspective, the complexity of the emotional dimension is untangled by identifying the different emotions based on some of their characteristic properties, such as their duration, intensity, or direction, as well as their degree of cognitive processing. That is why the typology of emotions proposed by Jasper (2018)—which includes reflex emotions, moods, moral emotions, and affective commitments—has become a core tool for studying the emotional dimension of protest, since it allows systematizing data and overcoming the limits of the "emotion labels" with which we name emotions (Poma & Gravante, 2024). In this way, we can distinguish a primary or reflex anger (or rage), which is a quick stimulus to an event, from a moral anger, which is the result of perceiving something as unjust, such as the dignified anger of the Zapatistas in Mexico.[2] Likewise, we can distinguish between a hopeful mood and a moral hope. The first one may be constructed by activists participating in a successful protest event, and it can provide them with emotional energy for a few days. In contrast, moral hope is what activists feel about collective action efficacy, and it differs from the former in that it depends on their values and is more lasting. Jasper (2011) also suggests focusing on moral batteries, that is, the opposite emotions (e.g., joy and sadness; pride and shame) that can help us to better understand a movement's emotion culture (Taylor, 1995, 2010). Therefore, in the following discussion, some moral batteries will be addressed, showing how to focus on opposite emotions allows for a better understanding of their effects in climate activism.

Other concepts developed by Jasper (1997) are reciprocal and shared emotions, which he considers as collective emotions. Reciprocal emotions are those activists feel for each other, while shared emotions are those emotions that can be shared by different actors, uniting them, coping with loneliness, and in some cases strengthening collective identity (Poma & Gravante, 2016).

Finally, some papers are analyzing how climate activists cope with climate emotions. While psychology applies the concept of "emotion regulation," in sociology we rely on Hochschild's concepts (1975, 1979) of "emotion work or management" and "feeling rules." Hochschild shows two typologies of emotion management: evocation and suppression, which can be achieved through superficial and deep acting through cognitive, bodily, and expressive techniques. We can add the concept of "channeling emotions" used by authors in the field of social movement studies like Taylor and Whittier (1995), Jasper (1998), Reger (2004), and Gould (2009). This concept means the ability to move from a potentially paralyzing emotion, such as fear or grief, to a mobilizing or energizing emotion, such as anger. In the social movements literature, there are several examples, such as the experience of channeling grief into anger for a missing son/daughter (Bayard de Volo, 2006) and for the victims of AIDS (Gould, 2009), as well as fear of repression into anger (Gravante & Poma, 2018), or eco-grief in anger in Extinction Rebellion (XR) activists (Giæver & Russell, 2024). Linked to emotion management, Hochschild also developed the concept of *feeling rules*, which refers to those emotions that are socially expected in every social and cultural context. Feeling rules regulate which emotions are appropriate in different situations, as well as how to express them, when, to whom and with what intensity. Feeling rules are linked to culture but also to the ideology and political beliefs, that is, to the framing rules through which we understand the world, just as they can follow gender patterns (Hochschild, 1975, 1983). Feeling rules are spread and strengthened by institutions, states, and corporations, until they constitute the emotional culture of a country, ideology, or religion. The feeling rules of the dominant system (e.g., capitalism) can be challenged individually and collectively, just as social movements do. In addition, social movements, like the feminist movement, can create their own emotion culture (Gravante, 2020a; Taylor, 1995). This is a relevant element in the creation of strategy and practices, and in the process of empowerment and self-affirmation of groups. To conclude, applying the sociocultural approach to understand climate activism means identifying the emotions that are most relevant in the experience of activists, the interaction between these emotions and their possible effects on the protest. Emotions, as a part of culture, can also become signs of cultural change, when the dominant feeling rules are challenged by activists and social movements.

Emotions, Climate Change, and Activism

In recent years, research on the emotional impact of climate crisis has intensified. Psychology has focused on the effects of mental health crises, advancing knowledge on eco-anxiety, depression, climate sorrow or grief, and the sense of hopelessness, among others (Clayton, 2024; Innocenti, 2022; Ojala et al., 2021); added to this are the studies on trauma, shock, and posttraumatic and pretraumatic stress, which can be experienced by people directly affected by some extreme meteorological events, such as a hurricane or a long-lasting drought (Clayton et al., 2014, 2017; Neria & Schultz, 2012).The emotional response to the climate crisis differs depending on its impacts, with different emotions observed in the face of disasters than in the face of the gradual impacts of climate change, as shown by Neria and Schultz (2012). After analyzing the emotional impact of Hurricane Sandy in the United States in 2012, the stress caused by the effects of extreme atmospheric events can be intertwined with concern and anxiety about future disasters, and feelings of vulnerability, such as, helplessness, depression, grief, and despair. In addition, depending on culture, beliefs, and values, there may be different emotional responses even to a direct impact, which in turn can lead to different perceptions and responses to climate change. As evidenced in Clayton et al.: "Although people's understanding and knowledge of climate change can increase by experiencing the effects directly, perception, politics, and uncertainty can complicate this link" (2017, p. 6).

Most studies on the emotional impact of climate change have been conducted on nonactivist populations, and especially on students (Calixto Flores & Terrón Amigón, 2018; Ojala, 2015, 2020; Ramírez et al., 2023; Verlie, 2022), with the new generations being those who are expressing the most intense emotional responses to the crisis—although there are studies on other categories, such as climate researchers (Head & Harada, 2017). In this area, research from psychology and pedagogy stands out, with some relevant contributions from sociology, such as the work of Norgaard (2011) that shows the role of emotions in a passive denial of the problem—when people act showing indifference—in a Norwegian community at the end of the 20th century. More recently, Davidson and Kecinski (2022) from affective neurosciences propose a hypothetical emotion–cognition model of climate change response, but in their model culture context is missing, while Davidson's last book (2024) offers an insight of her exploration on the role of some emotions (e.g., pride, shame, guilt, anxiety) in different responses to climate change. Moreover, Neckel and Hasenfratz discuss the emotions that are characteristic of the ecological crisis, presenting a map of climate emotions but also questioning the impacts, not always positive, of the emotionalization of the ecological crisis and highlighting that "little attention has been paid to the high degree of reflexivity with which climate emotions are dealt with in public and in science" (2021, p. 255). It should be noted that most of this literature has been developed with populations from industrialized countries and that, with some exceptions, knowledge about emotional responses to climate change in the Global South is still lacking (Calixto Flores & Terrón Amigón, 2018; Iniguez-Gallardo et al., 2021; Poma, 2018, 2019; Ramírez et al., 2023; Rodríguez Gudiño & Poma, 2023; Wilf et al., 2024).

Despite the contribution of all the research conducted to understand the emotional impact of the climate crisis on nonactivist people, in the study of climate activism we are dealing with empowered subjects, whose perception of the problem has led them to organize. Therefore, it is worth paying attention to how activists are constructing their emotions, both from their perception of the problem and in the practice of their activism. Particularly, it is worth both understanding how activists feel and manage climate emotions such as eco-anxiety or climate sorrow and comprehending the impact of the emotions generated during activism on their lifestyle and political strategies, because this knowledge could shed light to suggest strategies to help to overcome eco-paralysis on a broader level, especially in nonactivist people.

Although the study of the emotional dimension of climate activism is much less extensive and more recent than the general literature on emotions and climate change, in recent years, research studies on this topic have increased; for example, Kleres and Wettergren (2017) adopted the sociology of emotions approach to analyze the role and interactions among different emotions linked to climate change (fear, hope, guilt, and anger) that can influence activism depending on their management and combination. These researchers analyze the role of hope—which is emerging as a fundamental emotion in countering fear and encouraging climate action—while noting some differences between the Global North and South (e.g., the prevalence of guilt in countries of the North and anger in the South). In the field of studies on social movements, Lorenzini and Rosset (2023) analyze the emotions of fear and anger in the face of climate change and their role to motivate protests in the 2019 climate strikes in some cities of the Global North, showing how adults and elders feel less fear and more anger about the problem, compared to younger generations. Using communication studies, Molder and colleagues analyze hope, anger, and fear in Thunberg's Instagram posts, highlighting that hope is used "as a practice of empowerment channeled toward a better future" (2022, p. 682), while Belotti and colleagues (2022) examine the online emotional contagion and engagement that Fridays For Future (FFF) activists achieve through an emotional narrative shared on social media, which works through empathy. Soler-i-Martí and colleagues (2022) investigate the social media discourse of FFF Barcelona, highlighting the use of hope in a context of despair.

Applying the affect theory to the study of climate mobilizations, Knops (2021) examines the indignation of young climate activists in Belgium, which emerges when linked to other emotions (such as anger, disgust, hope, fear, and compassion), and intervenes in the production of different imaginaries that compete and are integrated into the activists' narratives. Knops (2023) discusses the emotional dimension of the tipping points in FFF activists, and this has led her to identify affective tipping points that influence the mobilization of climate activists in general. Pickard (2021) briefly analyzes the emotional dimension of the experiences of British XR activists, showing how pleasure and support help to overcome the feelings of isolation and cope with the feelings of helplessness or hopelessness. Although Neckel and Hasenfratz (2021) report some examples of the use of emotions by XR activists, showing how anxiety can stimulate participation in activism, and mourning can become a part of political performances. Stuart (2020) discusses the radical hope that XR activists feel, even when they are pessimistic, which differs from false and passive hope, and Wilf and colleagues (2024), studying Indian youths' climate Activism, highlight that the sense of community among youths, generations commitment (versus adults' failure), and hope in their actions are supporting factors for activism; meanwhile, climate anxiety had mixed effects on both sustaining and detracting from youth climate activism.

Finally, there is an emerging literature on emotion management or regulation among climate activists. For instance, Holthaus (2023) claims that climate activists become emotion entrepreneurs who promote climate-related emotions like eco-grief, aiming to catalyze cultural change and sustainable practices in Western societies. Climate activists face apathy, challenging the prevailing feeling rules within the climate movement and in the broad Western culture, and countering the dismissal or rejection of climate-related emotions to foster a more engaged and proactive approach to address environmental challenges. Giæver and Russell (2024) show that emotion regulation is useful not only to cultivate emotional resilience but also to facilitate a bonding with nature and individuals, acting as a mechanism for socio-ecological advancement.

The aforementioned literature is based on the experience of climate activists in the Global North, with research in the Global South being scarce. To overcome this lacuna, Poma and Gravante (2021, 2024) show the most relevant emotions in climate activism in Mexico, and Poma (2024) emphasizes the emotional dimension of Mexican climate activism from a radical and intersectional approach. These authors show, for example, that emotions such as concern, fear, and anxiety felt by climate activists in Mexico are linked more to repression from the state than to the effects of climate change. Their work highlights that, although similar emotions can be observed in different sociocultural contexts, it is important to study how activists construct these emotions (identifying the object, direction, intensity, and duration of each emotion), because it is in the construction process where most differences can be observed between the Global North and South.

A final observation on the methodology of the researches cited earlier is that the analysis of the emotions publicly expressed through social media, speeches, texts, videos, and so on—which are connected to the strategic dimension of a movement, in terms of communicating, at least, certain emotions—fails to encompass the complexity of the emotional dimensions of climate activism. To do that, we also need to know the emotions that activists feel but do not publicly express or share, which requires research techniques that allow a more in-depth approach, such as interviews, diaries, or focus groups (Poma, 2023). Furthermore, while quantitative studies that analyze emotions are common in psychological and communication studies, a qualitative and cultural approach is prevalent in social movements studies, as we will see later, discussing the most relevant emotions in climate activism.

Emotions That Characterize Climate Activism

Analyzing the emotional dimension of activism requires facing the complexity of many emotions. Although in empirical work it is more common to be guided by research questions that focus on certain dynamics of collective action (recruitment, mobilization, burnout, etc.), in this chapter we have opted to discuss the research results on those emotions that stand out in the literature on activism and the climate crisis. We will show the dynamics related to concern, powerlessness, fear, guilt, anxiety, and pain felt by climate activists, and then we will discuss some of the emotions that characterize activism, such as anger, hope, trust, and love, as we show in Tables 11.1 and 11.2, at the beginning of the next two sections.

Climate Emotions

Climate emotions, that is, those felt after experiencing the consequences of climate change, are studied as responses to the crisis by different subjects, for instance, people and communities that have been directly affected and/or have suffered some trauma due to any effect of climate change, younger generations (generally more sensitive to the problem), climate researchers, climate change deniers, and those people suffering a mental health problem that may be negatively affected by climate change. The guiding question here is, How do these emotions influence climate activism? Let us start with *concern*, which would be a necessary starting point to face the climate crisis. In relation to the construction of this emotion, climate activists are concerned, in general, for the climate change effects and for the lack of response to the problem. Huntley (2020) shows that in Australia, the youngest are the most concerned, and daughters express more concern than their parents and brothers. She considers that this gender pattern, observed in different countries, can be associated with patriarchal views of being concerned for the environment and can be seen as weak and womanish, while dominance over the environment and accepting the possible risk of climate damage were associated with masculine strength (Huntley, 2020, p. 138).

Thinking in terms of moral batteries (Jasper, 2011), the counterpart of concern is indifference, which is the antithesis of any action. Concern for climate change effects is a moral emotion (Jasper, 2018) constructed by activists from their understanding of the problem, and it is shared among them. Moreover, we suggest that this concern has become one of the movement's feeling rules, because climate activists consider it wrong not to feel concern to the climate crisis, to the point that indifferent people generate anger and frustration among activists. Poma and Gravante (2024) show that activists in Mexico use empathy in their attempt to understand indifferent people, mostly if they are known individuals, trying to raise their awareness instead of increasing the distance between them. Nevertheless, the concern for the climate crisis is not a feeling rule that characterizes the economic and cultural

TABLE 11.1 Climate Emotions

Emotion	What Generates Them or to Whom They Are Aimed in Climate Activism	Outcomes in Climate Activism
Concern	The effects of climate change The lack of response The future For internal conflicts	Sustaining activism if felt with moral hope to collective action Detracting factors if accompanied with fear and anxiety, or caused by internal dynamics
Fear	The repression against activists The future The effects of climate change	Detracting factors if it is not managed
(Eco) Anxiety	The effects of climate change The future	A certain level can mobilize if accompanied with hope Detracting factors if accompanied with powerlessness or resignation
Powerlessness	In the face of climate change as a global problem The lack of institutional response and participation	Detracting factors if it is not managed
Guilt	When activists cannot act in favor of the environment	Sustaining activism when it is linked to a collective responsibility Detracting factors if it is managed individually as a unpleasant emotion
Pain and grief	The effects of climate change and climate injustice Ecosystem degradation (solastalgia) The suffering of nonhuman animals Social injustices The future awaiting the new generations Not being able to change things Accusations of acting out of self-interest Not being taken seriously Feeling alone in the fight against climate change	Sustaining activism when they are channeled into anger

TABLE 11.2 Activism Emotions

Emotion	What Generates Them or to Whom or Where They Are Aimed	Outcomes in Climate Activism
Anger	Toward people who are accountable for climate change and their lack of response	Sustaining activism
	Toward indifferent people	It can be channeled in order to avoid the *Naughty or Nice* dilemma, or repression
Hope	For and during collective action	Sustaining activism, when it is a constructive and collective hope
		Help to manage some unpleasant emotions
Trust	Toward activists	Sustaining activism
	Toward other actors	
	Toward climate science	
Love	Toward the planet	Sustaining activism
	Toward other activists and people	

capitalist system (Hochschild, 2023) that is distinguished by the pursuit of individualist happiness (Cabanas & Illouz, 2019), which requires not being affected by issues considered external to the individual.

Thus, the concern about the effects of climate change can be considered in climate activism as a mobilizing emotion that activists share and evoke. However, concern alone is not enough to mobilize people, so it must be accompanied by other emotions like hope in collective action (see next section). It is necessary to clarify that very intense concern (e.g., worry) could also generate anxiety and anguish, emotions that can become mental health problems; hence, Huntley notes, "what constitutes an appropriate level of worry remains in question" (2020, p. 155). To study the role of concern in different cultural contexts of climate activism, it is important to explore, on the one hand, what concerns activists and, on the other, what other emotions they feel along with concern. For example, Poma and Gravante (2024) show that in Mexico concern is accompanied by fear and anxiety and is more felt due to repression from the state and paramilitary groups and the lack of response to the problem than due to the effects of climate change. Nevertheless, the authors also show that concern emerges as a consequence of internal conflicts and strategic decisions, which can generate internal dilemmas such as those of "naughty or nice?" or radicalism (Jasper, 2006). While climate concern can unite and mobilize people, other concerns may require rethinking the strategies or dynamics internal to the group or movement, as happens when discussions arise in an organization about whether to claim values like anticapitalism or antispeciesism.

Fear can emerge as a primary emotion when it is a rapid and intense response to a perceived threat or risk, but it also can be constructed from values and experiences as something more complex. The most common fear in climate activism is fear of the future, mainly among the youngest people. However, Poma and Gravante (2024) and Poma (2024) show that, in Mexico, fear of repression is more common than the fear of the future, due to the violence against environmental activists but also to the normalization of uncertainty and emergencies (Global Witness, 2022).[3] Huntley (2020) questions whether fear can be useful in communicating climate change, concluding that it cannot be ignored, but we must be aware that fear may undermine trust in the messenger and produce unwanted effects such as denial. In this sense, Norgaard (2011) shows that a type of passive denial—which involves acting with indifference before a problem and evading it—could be the response to the emotional management of the fear, guilt, and powerlessness that climate change generates in some people. Neckel and Hasenfratz (2021) highlight the role of fear in perceiving the dangers of climate change, as well as the association between a climate of fear and responses such as apathy or denial.

Related to the management of fear emanated from the effects of climate change, Kleres and Wettergren (2017) claim that it is possible to counteract the fear with hope, and Poma and Gravante (2021, 2024) add to this that climate activists learn to live with fear, thanks to self-training and the consolidation of support networks, which also allow them to share their fear among activists and close people. Fear is one of the emotions that climate activists are not used to expressing publicly, so much that Molder et al. (2022) show that Thunberg's posts used fear only 3.6% of the time, and "when she did use anger or fear framing, the posts were often targeted to people directly responsible for climate inaction, such as specific politicians and 'older generations'" (2022, p. 683). To study what type of fear climate activists feel in different cultural contexts, it is necessary to know their direct experience using techniques such as interviews, focus groups, or diaries.

Anxiety is linked to the concern of the effects of climate change and fear of the future. As explained by Innocenti (2022), "the emotional impact [to climate change] will be excessive, the worry will become real anxiety, generating a dysfunctional behavioral response, which will increase in direct proportion to the increase in anxiety" (2022, p. 41). As in the case of concern, a certain level of anxiety can mobilize (Jasper, 2018), particularly when it is connected to hope (Neckel & Hasenfratz, 2021). But if it is too intense, or related to emotions like powerlessness or resignation, anxiety can manifest itself as panic attacks, eating and sleeping disorders, irritability, and weakness, among other conditions leading to nonaction. Eco-anxiety and climate anxiety, which are emotional responses to environmental degradation and climate instability, could add to the anxiety already felt due to other aspects of life, and it is observed primarily in young people from different cultural contexts (Clayton 2020; Huntley, 2020; Innocenti, 2022; Ojala et al., 2021; Ramírez et al., 2023). Nevertheless, there are not sufficient studies to demonstrate a problem of climate anxiety in activists, and this could be because activism allows them to organize and to cope with the situation and its related emotions, increasing their feelings of effectiveness and empowerment. In addition to facilitating anxiety management (Giæver & Russell, 2024; Poma & Gravante, 2021), activism also helps to prevent them from feeling powerlessness, another climate emotion that can lead to denial (Norgaard, 2011), and burnout.

Huntley (2020) associates powerlessness with despair, thinking of an emotional response when facing an overwhelming situation that does not improve. Powerlessness in the face of climate change derives from the fact that it is a global, large-scale problem, against which we feel incapable of doing anything. In addition, a lack of results from climate policies and diplomacy, reflected on the persistent degradation of ecosystems and the already visible effects of climate change, strengthens this emotion. In this sense, it is possible to assert that the unwillingness to face anthropogenic causes of climate change, resulting from industrial production, is a mechanism designed to spread powerlessness and despair, just as the neoliberal model has been consolidated spreading the idea that There Is No Alternative (TINA). Pickard shows that "protesting was a way to counteract or cope with feelings of helplessness or hopelessness" (2021, p. 11), as it has been observed with other climate emotions such as anxiety.

Another emotion that has been studied in relation to climate change is guilt, often associated with shame. These moral emotions (Jasper, 2018) are different, because shame is built from the fear of losing the respect of people who are relevant to us or to our identity, while "guilt is the feeling of being responsible" (Neckel, 2020, p. 44). In this regard, Huntley (2020) highlights that, contrary to shame, guilt could be constructive when it is linked to a collective responsibility. Although feeling guilty when acting against our values could influence a behavioral change, Neckel and Hasenfratz (2021) show that shaming someone, or making them feel guilt for their actions, such as traveling by plane, eating meat, polluting, and so on, could create an emotional resistance that generates anger in those who feel judged. This process has been described by Hochschild (2016), when she shows that imposing a feeling rule, such as compassion to someone, can generate resentment in those who are not targeting the same emotion to the same beings. Thus, in the climate movement it is common to avoid the use of guilt to involve more people in the struggle, considering also the observations that many audiences already are very sensitive to charges of being somehow guilty or "bad people" (Stoknes, 2015). Poma and Gravante (2021, 2024) show that, although climate activists could feel guilt, because they cannot reduce their ecological footprint or dedicate more time to activism, this guilt is compensated by their commitment and participation. Poma (2018) also shows that Mexican working-class socio-environmental activists manage their guilt by comparing their lifestyle and ecological footprint with people with more purchasing power. The guilt related to the climate crisis has been studied mainly in populations from industrialized countries or upper middle classes, who may feel guilty when they learn that their lifestyle is impacting vulnerable populations that are being affected by the pollution that they are not generating (Huntley, 2020; Neckel & Hasenfratz, 2021; Norgaard, 2011). However, we know little about its construction in other contexts.

This section ends with one last emotion associated with the climate crisis: the pain of loss, also studied as climate grief and sorrow. As Orbach writes, "to come into knowing is to come into sorrow. A sorrow that arrives as a thud, deadening and fearful. Sorrow is hard to bear. With sorrow comes grief and loss" (2019, p. 77). Stoknes talks about the Great Grief referring to pain that "was not triggered only by climate change itself, but by the extended loss of beauty, diversity, and habitat" (2015, p. 274). This emotion is shared by climate activists from different geographies and is present in the repertoire of protest of the climate movement in the form of die-in protests or other acts of mourning. Poma and Gravante (2024) show that in Mexico and in Latin America, grief is an emotion that characterizes social movements due to the systemic violence in these countries; therefore, it is not something characteristic only of the climate movement. However, Neckel and Hasenfratz highlight that "grief for the loss of the ecosystem seems to have no ending"; hence, it becomes an "interminable emotional burden" (2021, p. 261). Gould (2009), who studied grief in the ACT UP movement during the AIDS pandemic in the 1990s, shows that the organization had to channel their grief into anger to be able to mobilize, which is also the case in Latin American movements for victims (Bayard de Volo, 2006; Gravante, 2020b). Embracing the tradition of the feminist movement, the first one to consider emotions as a part of the political struggle, Orbach invites climate activists to collectively manage their emotions, exhorting them "to create spaces in which we can share how difficult this hurt is and how to deal with our despair and rage" (2019, p. 69). Similarly, Stoknes asserts, from the eco-psychology approach, that grief is an emotion that "rarely appears in conversations or the popular media" (2015, p. 278), although it is important to find spaces to share it, because "maybe there is a community to be found among like-hearted people, among those who also can admit they've been inside the Great Grief feeling the earth's sorrow, each in their own way" (2015, p. 279). To conclude, climate activists also feel climate emotions, but their effects are different from those studied in the population that is not mobilized, because in general, instead of becoming mental health problems, or leading to the denial of the problem, they can strengthen the collective identity when they are shared and mobilize when they are associated with emotions such as anger and hope.

Activism Emotions

We also need to consider emotions that are key to understanding activism that include two moral emotions, anger and hope, as well as two affective commitments, trust and love. Although these are not the only emotions that characterize activism, we are guided by those that stand out in the literature on the climate crisis, as in the previous section. Anger is a relevant emotion in the study of social movements because it has a mobilizing effect. While a classical approach to emotions associates anger with aggressive behaviors, being considered a primary and negative emotion, in the study of emotions in protest and from a sociocultural approach, we analyze complex and moral forms of anger (Jasper, 2018), which are constructed by people from their idea of (in)justice.

Concerning the climate movement, the most observed anger is that directed to those people accountable for the crisis and their lack of response. In this way, Stoknes highlights that "anger can be most appropriate when reacting to outrageous conditions. Feeling a mix of grief and anger about the state of the climate, rivers, and forest is only appropriate" (2015, p. 285). Huntley (2020) reports that Indigenous activists she has interviewed have declared that they feel anger and frustration, given the lack of action by the Australian government on climate change, as well as resentment and anger because their situation does not matter to the world. Kleres and Wettergren (2017) show that while climate activists from the Global South express their anger, Europeans tend to suppress it because it is considered a suspicious emotion in the liberal democracies of the old continent. Poma and Gravante (2024) show that Mexican radical activists, with an anticolonial, anticapitalist, and antipatriarchal intersectional vision, express more anger against politicians and corporations than activists with a more moderate vision. As Huntley stated, anger, frustration, and rage "are now an integral part of the new emotional regime of the climate age, and we have to find a way to live with them" (2020, p. 127), particularly taking into account that "emotions such as hatred are as essential to political action and reaction as they are to one's sense of worth and identity" (Cabanas & Illouz, 2019, p. 164), and "funneling anger into action can make for enduring, effective motivation" (Stoknes, 2015, p. 285). As noted in the book by XR, the climate movement "need[s] to harness the energy in people's anger with the status quo into a movement that gathers in classrooms, living rooms, churches and village halls across the country" (Lucas, 2019, p. 72).

Hope, for its part, is one of the most studied emotions in relation to the crisis and climate activism. This has fueled the debate about its effects on the response to climate change and the generation of specific knowledge about different types of hope. In terms of moral battery, the counterpart of hope is hopelessness, associated with powerlessness and despair. Wettergren also suggests that hope is entwined with the "fear that the bad present may prevail or worsen" (2024, p. 6).

One of the reasons of the attention paid to hope is its potential for mobilizing and transforming paralyzing emotions such as fear or sorrow. However, it is important to understand this emotion in depth to know all its possible effects, since it does not always mobilize or promote a response to the problem. Applying Jasper's typology (2018), we could observe different types of hope: hopeful moods, including optimism, which expands "people's sense of the possible" (Jasper, 2018, p. 82), and another form of moral hope, also defined as constructive (Ojala, 2012) or active, in response to climate change. Moral hope is the emotion felt from collective action, and it emerges when we study climate activism. In this sense, Kleres and Wettergren show that hope inspires action, helps to manage fear, and "propels action even though prospects are negative" (2017, p. 509). Soler-i-Marti et al., who analyzed climate activists' social media discourses, see hope "as a mechanism that operates as a form of agency in future thinking for youth at an individual level, and, at the same time, it is also fundamental to understand how the future at collective level is perceived" (2022, p. 5). For their part, Molder et al. show that "Thunberg's Instagram communicates hope as a practice of empowerment channeled toward a better future" (2022, p. 682), and Wettergren shows that "collective hope, unlike individual hope, is always responsive" (2024, p. 16).

Poma and Gravante (2024) discuss the importance of moral hope in climate activism in Mexico, which exists even when activists are pessimistic about the future. In light of this finding, the authors claim that hope in collective action has become a feeling rule in Mexico's activism, as a way out of resignation. This idea is shared by authors from other geographies such as Huntley, who from Australia states that "Hope is an ethical imperative. Action generates hope. And hope attracts others to your cause" (2020, p. 193). Stuart (2020) has analyzed radical hope (Thompson, 2010) in British XR activists who declare that they do not feel hope about the future, about stopping climate change and species extinction, or about this world. In this sense, radical hope is different from a false and passive hope, because it represents a new hope "for saving what can still be saved" (Stuart, 2020, p. 503). Huntley (2020) talks about radical hope referring to that hope we feel even without having any idea about the outcome of the protest. In these cases, we refer to a moral hope, because people continue fighting, for they believe it is the fair thing to do, and because they could not keep living without trying. But, at the same time, they are aware that their fight is unlikely to solve the problem.

Finally, we refer to false hope in the field of climate change when it is associated with technological optimism, that is, with the hope that someone will invent a technology that will allow us to mitigate and adapt to climate change without changing the status quo and the economic model. As an academic and XR activist in Argentina, writes:

> The false hope of returning to a known and comfortable situation is not a good thing to maintain. It turns us fearful, installs a fear to explore the boundaries of the possible; when we abandon the hope that a way of life is going to continue, it opens a pluriverse for alternative hopes. (Broffoni, 2020, p. 86)

As Huntley (2020) argues, this optimism reduces the perception of the threat and makes people more complacent to climate change. Another false hope that has been discussed in the field of climate activism is the hope placed in younger generations. While it is common that the mobilization of young climate activists around the world generates hope in concerned adults and in the more experienced activists (Huntley, 2020; Poma & Gravante, 2021), in the climate movement, the question arises as to whether the hope placed in new generations could be a mechanism to avoid facing the crisis in the present.

In summary, in collective action, hope is necessary to avoid surrendering and giving up. Nevertheless, in the study of climate activism, it is important to understand which type of hope people and activists feel, and which hope is promoting the communication of climate change. When Greta Thunberg, at the World Economic Forum in Davos (Switzerland), declared, "I don't want your hope. I don't want you to be hopeful. I want you to panic," she was evidencing the false hopes that go with the lack of response to climate change, helping to spread the idea that hope is not always a mobilizing or positive emotion, because it can also fuel denial and inaction.

An emotional battery that is also very important in climate activism is trust and distrust. As Ojala writes, "trust is important for humans at different levels, from concrete interpersonal trust, via trust in various institutions, to a more abstract, existential, belief in humanity, and the possibility of existence itself" (2020, p. 3). Following Jasper's (2018) typology, we are dealing with affective commitments that could be felt among activists—in this case, we are talking about reciprocal emotions (Jasper, 1997)—or that could be felt toward other actors. For instance, Doerr and Porsild Hansen (2024) discuss how Danish and German climate activists tried to build trust over local green transition processes, in different political and cultural contexts, through horizontal and, more confrontational, vertical repertoires of practices of trust building.

Trust and distrust are also critical for the target audience of climate activists, because humans are prone to believing information coming from sources we trust. Trust and distrust can also be outcomes of climate activism (Poma & Gravante, 2024) since the experience of activists can lead to building or losing trust in other actors. In the discussion of the communication of climate change, Huntley (2020) states that the use of fear can undermine trust in the messenger and have unwanted effects such as denial.

In addition, Ojala (2020) analyzes young people's trust in climate science, an element that has been substantial in the climate movement, although questioned from a decolonial perspective due to the influence it receives from Western culture and cosmovision. Consequently, as Ojala claims, "it is vital to see both young people's trust and their distrust as a form of active engagement with climate change" (2020, p. 5). Having trust in oneself and in collective action is also a feeling of empowerment necessary to generate constructive hope (Huntley, 2020; Kleres & Wettergren, 2017). As Jasper states, "trust, for instance, arises out of the interaction between expectations and experience with groups and individuals. These positive loyalties—along with negative ones toward outsiders, enemies, and other threats—motivate or allow much political action" (2018, p. 89).

Similar to moral emotions, such as the aforementioned forms of anger and hope, affective bonds like trust are also durable. However, they can change, for example due to a betrayal, with negative impacts on the personal and political experience of activists. Because of this, organizations stress the importance of building trust, as a basic component of social movements (Almeida 2019), since "while information can perhaps inspire rebellion, it cannot build the mutual trust and solidarity which enables people to sustain that rebellion [. . .] personal trust-building is crucial to mobilizing committed and sustainable action" (Burns & Reimann, 2019, p. 53). Confirming this idea, Poma and Gravante (2024) show that, in the framework of climate activism in Mexico, a feeling rule of trust toward other activists has spread, to overcome the general climate of distrust that some activists associate with the repressive context (Poma, 2024).

We conclude this discussion by talking about love, "a deep and abiding affection, steady attachment, and unshakable commitment to a person, place or thing" (Huntley, 2020, p. 278). Unlike anger and hope, which are emotions that climate activists communicate and publicly express on social media or in official documents—and they are therefore more studied—affective commitments, such as trust and love, characterize the internal experience of activism, and can only be known through a direct relationship with activists. However, organizations such as XR have decided to turn an emotion such as love into a feeling rule for their activists, spreading slogans like "Rage and Love," or writings such as the following:

> You may end up feeling sad, or empty, or guilty, or angry, or frightened, or numb. But, ultimately, this is a book about love. About the love we should have for humanity, and the love we should have for the planet. About the love we are currently lacking. And the love that we desperately need. (Knights, 2019, p. 11).

As can be read from the quote, the type of love referred to by XR activists is the love for humanity and the planet, but in the book, we also can find references to love between activists (as a reciprocal emotion) and toward their neighbors. Thus, love becomes an emotion to recruit more people, take care of each other, care for the planet, and manage despair as well, because "there seems to be a way through despair. It is to love" (Bendell, 2019, p. 41).

The mobilizing capacity of love lies in the fact that it is the loss of something or someone we love that influences our beliefs and actions, and that can mobilize us. For instance, Huntley (2020) states that all the people she has interviewed have joined climate activism because climate change threats to destroy what they love. For this reason, the author asserts that to raise people's awareness about climate change, rather than talking to them about the problem, or about the science explaining it, it is necessary to understand what they are concerned about and what they love. Solidarity and mutual support reside in altruistic, nonpossessive love. In the struggles to defend the territory in Mexico, one of the slogans that has gone viral is "Land/water is not for sale, it is loved and protected." Love is at the basis of care (for people, places, or species) and of life itself, and it is being vindicated both in the Global South and North, as can be seen in this statement: "Anger, love and joy may sound like odd bedfellows, but these are the seeds of a future that will offer life—not success, but life" (Williams, 2019, p. 92).

Conclusions

Emotions guide our actions: They show who we are, and what our values and identity are. That is why they are so relevant to understanding activism, including climate activism (Gravante & Poma, 2023), which, as we face the crisis of civilization, has an intense emotional charge. Studying the emotional dimension of climate activism requires focusing on the emotions that emerge from the perception of the problem but also on those that are constructed through the practice of activism. It is important to keep in mind that climate emotions do not have the same effects on activists as on the general population, because the contentious collective action allows them to create spaces and strategies to collectively manage these emotions. In this sense, we have observed in different organizations and countries that climate activism could, in some way, be considered a therapeutic way to cope with climate emotions, and it would be interesting to confirm whether people who are facing alone the climate crisis are more vulnerable to mental health problems.

Finally, as evidenced by Jasper (1997, 2006, 2011, 2018), emotions are also a central dimension in the strategy of social movements, and this is prominent in climate activism, both in studies of publicly expressed emotions and in research that interviews climate activists.

Due to the relevance of the emotional dimension of protest and the spreading of climate activism in different geographies and cultures, it is necessary to extend research focusing on emotions in climate activism, in order to understand the distinctive strategies and repertoires that are emerging in different contexts, as well as the emotion management strategies to face climate emotions, which can contribute to strengthening the climate movement and promoting a broader collective response facing the climate crisis.

Notes

1. In this text, we use "climate change" and "climate crisis" as synonyms, although we understand that their use varies depending on cultural and social contexts, because "change" can have a more positive meaning and "crisis" implies a shorter period of time. For the same reason, in some contexts it is preferred to use "climate emergency" or "climate collapse" to emphasize the severity and urgency of the problem, although both have been criticized for having alarmist and pessimistic connotations.

2. The Zapatista Army of National Liberation (Spanish: *Ejército Zapatista de Liberación Nacional*) or EZLN, often referred to as the Zapatistas, is an Indigenous armed organization in Chiapas, the southernmost state of Mexico. EZLN declared war on the Mexican government in 1994 against neoliberal politics, demanding work, land, housing, food, health, education, independence, liberty, democracy, justice, and peace. ELZN has a huge solidarity from around the world.
3. For the third year in a row, in 2022, Global Witness documented an increase in lethal attacks in Mexico—54 defenders were killed in 2021, almost half of them Indigenous people. Forced disappearances are common—Global Witness recorded 19 in 2021. Impunity remains rife, with over 94% of crimes not reported and only 0.9% resolved.

References

Almeida, P. (2019). *Social movements: The structure of collective mobilization.* University of California Press.

Bayard de Volo, L. (2006). The dynamics of emotion and activism: Grief, gender, and collective identity in revolutionary Nicaragua. *Mobilization: An International Quarterly, 11*(4), 461–474. https://doi.org/10.17813/maiq.11.4.q21r34325611217

Belotti, F., Donato, S., Bussoletti, A., & Comunello, F. (2022). Youth activism for climate on and beyond social media: Insights from Fridays For Future–Rome. *The International Journal of Press/Politics, 27*(3), 718–737. https://doi.org/10.1177/19401612211072776

Bendell, J. (2019). Doom and bloom: Adapting to collapse. In AA. VV., *This is not a drill: The extinction rebellion handbook* (pp. 38–41). Penguin.

Broffoni, F. (2020). *Extinción. La supervivencia de la humanidad en juego. ¿Qué estás haciendo para evitar el colapso?* Editorial Sudamericana.

Burns, D., & Reimann, C. (2019). Movement building. In AA. VV., *This is not a drill: The extinction rebellion handbook* (pp. 191–197). Penguin.

Cabanas, E., & Illouz, E. (2019). *Manufacturing happy citizens: How the science and industry of happiness control our lives.* Polity Books.

Calixto Flores, R., & Terrón Amigón, E. (2018). Las emociones en las representaciones sociales del cambio climático. *Educar em Revista, 34*(68), 217–233. https://doi.org/10.1590/0104-4060.55684

Clayton, S. (2020). Climate anxiety: Psychological responses to climate change. *Journal of Anxiety Disorders, 74.* http://dx.doi.org/10.1016/j.janxdis.2020.102263.

Clayton, S. (2024). A social psychology of climate change: Progress and promise. *British Journal of Social Psychology, 63*(4), 1535–1546. https://doi.org/10.1111/bjso.12749

Clayton, S., Manning, C. M., & Hodge, C. (2014). *Beyond storms & droughts: The psychological impacts of climate change.* APA/ecoAmerica.

Clayton, S., Manning, C. M., Krygsman, K., & Speiser, M. (2017). *Mental health and our changing climate: Impacts, implications, and guidance.* APA/ecoAmerica.

Collins, R. (2019). Emotional micro bases of social inequality: Emotional energy, emotional domination and charismatic solidarity. *Emotions and Society, 1*(1), 45–50. https://doi.org/10.1332/263168919X15580836411823

Corraliza, J. A. (2022). Emociones y emergencia climática. Trampas y barreras psicológicas a la proambientalidad. In T. Gravante & A. Poma (Eds.), *Emociones y medio ambiente* (pp. 29–41). CEIICH-UNAM.

Davidson, D. J. (2024). *Feeling climate change: How emotions govern our responses to the climate emergency.* Routledge.

Davidson, D. J., & Kecinski, M. (2022). Emotional pathways to climate change responses. *Wiley Interdisciplinary Reviews: Climate Change, 13*(2), e751. https://doi.org/10.1002/wcc.751

Doerr, N., & Porsild Hansen, J. (2024). "Climate translators" building trust and local democratic cooperation on green transition: Denmark and Germany. *International Journal of Comparative Sociology, 65*(4), 479–498. https://doi.org/10.1177/00207152231219489

Feldman Barrett, L. (2017). *How emotions are made: The secret life of the brain.* Houghton Mifflin Harcourt.

Flam, H. (2015). Micromobilization and emotions. In D. della Porta & M. Diani (Eds.), *The Oxford handbook of social movements* (pp. 1–15). Oxford University Press. https://doi.org/10.1093/oxfordhb/9780199678402.013.31

Flam, H., & King, D. (2005). *Emotions and social movements.* Routledge.

Giæver, F., & Russell, S. (2024). Courage without the certainty of a happy ending: The emotion regulation of environmental activists. *Social Movement Studies.* https://doi.org/10.1080/14742837.2024.2345177

Global Witness. (2022). Decade of defiance: Ten years of reporting land and environmental activism worldwide environmental activism worldwide, September 2022. *Global Witness.* https://www.globalwitness.org/en/about-us/annual-report-2022-rising-challenge-world-crisis/

Goodwin, J., Jasper, J. M., & Polletta, F. (2001). *Passionate politics: Emotions and social movements.* University of Chicago Press.

Gould, D. (2009). *Moving politics: Emotion and ACT UP's fight against AIDS.* University of Chicago Press.

Gravante, T. (2020a). Emociones y reglas del sentir como impactos culturales de los movimientos sociales. *Interdiscipina, 8*(22), 157–179. https://doi.org/10.22201/ceiich.24485705e.2020.22.76423

Gravante, T. (2020b). Forced disappearance as a collective cultural trauma in the Ayotzinapa movement. *Latin American Perspectives, 47*(6), 87–102. https://doi.org/10.1177/0094582X20951773

Gravante, T., & Poma, A. (2018). Manejo emocional y acción colectiva: las emociones en la arena de la lucha política. *Estudio Sociológico, 36*(108), 593–616. https://doi.org/10.24201/es.2018v36n108.1612

Gravante, T., & Poma, A. (2023). *Emociones y activismos de base.* CEIICH-UNAM.

Harth N. S. (2021). Affect, (group-based) emotions, and climate change action. *Current Opinion in Psychology, 42*, 140–144. https://doi.org/10.1016/j.copsyc.2021.07.018

Head, L., & Harada, K. T. (2017). The heart a long way from the brain: The emotional labour of climate scientists. *Emotion, Space and Society, 24*, 34–41. https://doi.org/10.1016/j.emospa.2017.07.005

Hochschild, A. R. (1975). The sociology of feeling and emotion: Selected possibilities. In M. Millman & R. M. Kanter (Eds.), *Another voice* (pp.280–307). Anchor.

Hochschild, A. R. (1979). Emotion work, feeling rules, and social structure. *American Journal of Sociology, 85*(3), 551–575.

Hochschild, A. R. (1983). *The managed heart: The commercialization of human feeling.* University of California Press.

Hochschild, A. R. (2016). *Stranger in their own land: Anger and mourning on the American right.* New Press.

Hochschild, A. R. (2023, September 21). The emotional culture of capitalism [Conference]. https://www.lacab.org.mx/index.php/video-conferences/145-the-emotional-culture-of-capitalism-arlie-hochschild

Holthaus, L. (2023). Feelings of (eco-)grief and sorrow: Climate activists as emotion entrepreneurs. *European Journal of International Relations, 29*(2), 352–373. https://doi.org/10.1177/13540661221136677

Huntley, R. (2020). *How to talk about climate change in a way that makes a difference.* Murdoch Books.

Iniguez-Gallardo, V., Lenti Boero, D., & Tzanopoulos, J. (2021). Climate change and emotions: Analysis of people's emotional states in southern Ecuador. *Frontier in Psychology, 12*, 644240. https://doi.org/10.3389/fpsyg.2021.644240

Innocenti, M. (2022). *Eco ansia, i cambiamenti climatici tra attivismo e paura.* Erickson.

Jasper, J. M. (1997). *The art mural of protest: Culture, biography, and creativity in social movements.* University of Chicago Press.

Jasper, J. M. (1998). The emotions of protest: Affective and reactive emotions in and around social movements. *Sociological Forum, 13*(3), 397–421. https://doi.org/10.1023/A:1022175308081

Jasper, J. M. (2006). *Getting your way. Strategic dilemmas in the real world.* University Chicago Press.

Jasper, J. M. (2011). Emotions and social movements: Twenty years of theory and research. *Annual Review of Sociology, 37*, 285–303. https://doi.org/10.1146/annurevsoc-081309-150015

Jasper, J. M. (2018). *The emotions of protest.* University Chicago Press.

Kleres, J., & Wettergren, Å. (2017). Fear, hope, anger, and guilt in climate activism. *Social Movement Studies, 16*(5), 507–519. https://doi.org/10.1080/14742837.2017.1344546

Knights, S. (2019). Introduction: The story so far. In AA. VV., *This is not a drill: The extinction rebellion handbook* (pp. 9–11). Penguin.

Knops, L. (2021). Stuck between the modern and the terrestrial: The indignation of the youth for climate movement. *Political Research Exchange*, *3*(1). https://doi.org/10.1080/2474736X.2020.1868946

Knops, L. (2023). The fear we feel everyday: Affective temporalities in Fridays for Future. *South Atlantic Quarterly*, *122*(1), 203–214. https://doi.org/10.1215/00382876-10242784

Lorenzini, J., & Rosset, J. (2023). Emotions and climate strike participation among young and old demonstrators. *Social Movement Studies*, *23*(1), 39–55. https://doi.org/10.1080/14742837.2023.2178406

Lucas, C. M. P. (2019). A political view. In AA. VV., *This is not a drill: The extinction rebellion handbook* (pp. 70–72). Penguin.

Massal, J. (2021). Las emociones en la movilización social: la agenda investigativa en América latina en la década del 2010. *Ciencia Política*, *16*(31), 73–115. https://doi.org/10.15446/cp.v16n31.96573

Molder, A. L., Lakind, A., Clemmons, Z. E., & Chen, K. (2022). Framing the global youth climate movement: A qualitative content analysis of Greta Thunberg's moral, hopeful, and motivational framing on Instagram. *International Journal of Press/Politics*, *27*(3), 668–695. https://doi.org/10.1177/19401612211055691

Moser, S. C. (2007). More bad news: The risk of neglecting emotional responses to climate change information. In S. C. Moser & L. Dilling (Eds.), *Creating a climate for change: Communicating climate change and facilitating social change* (pp. 64–80). Cambridge University Press.

Neckel, S. (2020). Sociology of shame: Basic Theoretical Considerations. In L. Frost, V. Magyar-Haas, H. Schoneville, & A. Sicora (Eds.), *Shame and social work: Theory, reflexivity and practice* (pp. 39–54). Policy Press.

Neckel, S., & Hasenfratz, M. (2021). Climate emotions and emotional climates: The emotional map of ecological crises and the blind spots on our sociological landscapes. *Social Science Information*, *60*(2), 253–271. https://doi.org/10.1177/05390184211996264

Neria, P., & Schultz, J. M. (2012). Mental Health Effects of Hurricane Sandy Characteristics, Potential Aftermath, and Response. *Journal of the American Medical Association*, *308*(24), 2571–2572.

Norgaard, K. M. (2011). *Living in denial: Climate change, emotions, and everyday life*. MIT Press.

Ojala, M. (2012). Hope and climate change: The importance of hope for environmental engagement among young people. *Environmental Education Research*, *18*(5), 625–642. https://doi.org/10.1080/13504622.2011.637157

Ojala, M. (2015). Hope in the face of climate change: Associations with environmental engagement and student perceptions of teachers' emotion communication style and future orientation. *Journal of Environmental Education*, *46*, 133–148. https://doi.org/10.1080/00958964.2015.1021662

Ojala, M. (2020). To trust or not to trust? Young people's trust in climate change science and implications for climate change engagement. *Children's Geographies*, *19*(3), 284–290. https://doi.org/10.1080/14733285.2020.1822516

Ojala, M., Cunsolo, A., Ogunbode, C. A., & Middleton, J. (2021). Anxiety, worry, and grief in a time of environmental and climate crisis: A narrative review. *Annual Review of Environmental Resource*, *46*, 35–58. https://doi.org//10.1146/annurev-environ-012220-022716

Orbach, S. (2019). Climate sorrow. In AA. VV., *This is not a drill: The extinction rebellion handbook* (pp. 66–67). Penguin.

Pickard, S. (2021). You are stealing our future in front of our very eyes. The representation of climate change, emotions and the mobilisation of young environmental activists in Britain. *E-Rea*, *18*(2). https://doi.org/10.4000/erea.11774

Poma, A. (2018). El papel de las emociones en la respuesta al cambio climático. *Interdisciplina* *6*(15), 191–214. https://doi.org/10.22201/ceiich.24485705e.2018.15.63843

Poma, A. (2019). Los impactos emocionales del reporte especial del IPCC. In J. C. Rueda Abad (Ed.), *¿Aún estamos a tiempo para el 1.5°C? Voces y Visiones sobre el Reporte Especial del IPCC* (pp. 177–195). PINCC-UNAM.

Poma, A. (2023). Emociones y activismo climático. In T. Gravante & A. Poma (Eds.), *Emociones y activismos de base* (pp. 25–43). CEIICH-UNAM.

Poma, A. (2024). You can't let yourself be overcome by fear: The experience of Mexican climate activists. *JEP—Austrian Journal of Development Studies, 4.*

Poma, A., & Gravante, T. (2016). Environmental self-organized activism: Emotion, organization and collective identity in Mexico. *International Journal of Sociology and Social Policy, 36*(9/10), 647–661. https://doi.org/10.1108/IJSSP-11-2015-0128

Poma, A., & Gravante, T. (2017). Emociones, protesta y acción colectiva: estado del arte y avances. *Aposta. Revista de Ciencias Sociales, 74*, 32–62.

Poma, A., & Gravante, T. (2021). Entre frustración y esperanza: emociones en el activismo climático en México. *Ciencia Política, 16*(31), 117–156. https://doi.org/10.15446/cp.v16n31.97635

Poma, A., & Gravante, T. (2024). The emotional dimension of Mexican climate activism: A sociocultural approach. *Mobilization: An International Quarterly, 29*(1), 103–122. https://doi.org/10.17813/1086-671X-29-1-103

Ramírez López, A., Rosetti, M., & Poma, A. (2023). Gender, exposure to news, knowledge about climate change and prosociality predict climate anxiety scores in Mexican students. *Ecopsychology, 15*(2), 184–192. https://doi.org/10.1089/eco.2022.0049

Reger, J. (2004). Organizational "emotion work" through consciousness-raising: An analysis of a feminist organization. *Qualitative Sociology, 27*(2), 205–222. https://doi.org/10.1023/B:QUAS.0000020693.93609.6c

Rodríguez Gudiño, A., & Poma, A. (2023). Emociones y emergencia climática: algunas claves para comprender la ecoparálisis. *Ambiente & Sociedad, 26*, 1–21. https://doi.org/10.31840/sya.vi26.2713

Salama, S., & Aboukoura, K. (2017). Role of emotions in climate change communication. In W. Leal Filho, E. Manolas, A. Azul, U. Azeiteiro, & H. McGhie (Eds.), *Handbook of climate change communication: Vol. 1. Climate change management.* Springer. https://doi.org/10.1007/978-3-319-69838-0_9

Soler-i-Martí, R., Fernández-Planells, A., & Pérez-Altable, L. (2022). Bringing the future into the present: The notion of emergency in the youth climate movement. *Social Movement Studies, 23*(19), 1–20. https://doi.org/10.1080/14742837.2022.2123312

Stoknes, P. E. (2015). *What we think about when we try not to think about global warming: Toward a new psychology of climate action.* Chelsea Green.

Stuart, D. (2020). Radical hope: Truth, virtue, and hope for what is left in extinction rebellion. *Journal of Agricultural and Environmental Ethics, 33*, 487–504. https://doi.org/10.1007/s10806-020-09835

Taylor, V. (1995). Watching for vibes: Bringing emotions into the study of feminist organizations. In M. M. Ferree & P. Y. Martin (Eds.), *Feminist organizations: Harvest of the new women's movement* (pp. 223–233). Temple University Press.

Taylor, V. (2010). Culture, identity, and emotions: Studying social movements as if people really matter. *Mobilization: An International Quarterly, 15*(2), 113–134. https://doi.org/10.17813/maiq.15.2.a610145145k43057

Taylor, V., & Whittier, N. (1995). Analytical approaches to social movement culture: The culture of the women's movement. In H. Johnston & B. Klandermans (Eds.), *Social movements and culture* (pp. 163–187). University of Minnesota Press.

Thompson, A. (2010). Radical hope for living well in a warmer world. *Journal of Agricultural and Environmental Ethics, 23*, 43–59. https://doi.org/10.1007/s10806-009-9185-2

Verlie, B. (2022). *Learning to live with climate change: From anxiety to transformation.* Routledge.

Wettergren, Å. (2024). Emotionalising hope in times of climate change. *Emotions and Society.* https://doi.org/10.1332/26316897Y2024D000000021

Wilf, S., Rudra, A., & Wray-Lake, L. (2024). "I will still fight for it till the end": Factors that sustain and detract from Indian youths' climate activism. *Youth, 4*(3), 1238–1259. https://doi.org/10.3390/youth4030078

Williams, R. (2019). Afterword. In AA. VV., *This is not a drill: The extinction rebellion handbook* (pp. 91–92). Penguin.

CHAPTER 12

Fridays For Future

A New Wave of Global Climate Activism

KATRIN UBA

Introduction

ENVIRONMENTAL movements are among the most important and influential social movements (Rootes, 2004). Many of these movements are well institutionalized and can be termed "social movement organizations." Others have evolved into political parties that, in some countries, have even attained governmental positions (van der Heijden, 1997). However, most of these movements have retained their distinct character and function primarily at the community level. The initially youth-led climate movement, Fridays For Future (FFF), belongs to this category of environmental movements. It has played a crucial role in mobilizing the latest wave of climate activism, which began in the autumn of 2018. Starting as a small group of activists in Sweden, it quickly grew into a substantial transnational movement. By September 2019, it had mobilized approximately 7.6 million people from 185 countries to participate in "school strikes for climate" (de Moor et al., 2020). Since then, a growing body of research has analyzed FFF's claims (Fernández-Zubieta et al., 2023; Svensson & Wahlström, 2023), its mobilization and communication strategies (Fisher & Nasrin, 2021a; Flörchinger et al., 2024; Martiskainen et al., 2020), the backgrounds of its activists (della Porta & Portos, 2021; Noth & Tonzer, 2022; Wahlström et al., 2019), as well as its challenges and impact on public opinion and politics (Berker & Pollex, 2023; Fisher, 2019; Fritz et al., 2023). Often, the scholarly focus has been on the youth climate movement in more general terms than just the FFF, and many studies also highlight the role of Greta Thunberg in this movement (e.g., Wahlström & Uba, 2024).

This chapter aims to provide an overview of the recent research on the FFF in social science outlets, with a caveat that one cannot cover all the 20,000 publications on the subject within a few pages. Additional information about youth climate activism can be found in various handbooks or companions (e.g., Almeida, 2023; Buzogány & Scherhaufer, 2023; Grasso & Giugni, 2022; Knops, 2024; Parks et al., 2023).

History and the Development of the FFF

The origins of the FFF can be traced back to September 2018, just after Greta Thunberg started the school strikes for the climate in front of the Swedish parliament in Stockholm in late August 2018. Shortly after that, she was joined by three other teenagers—Mina, Morrigan, and Edit (Ernman et al., 2020). At the People's Climate March in Stockholm on the September 8 (a Saturday), these young people announced that the school strikes would continue every coming Friday. From then on, the movement had the name "Fridays for

Future." In the background of this youth climate activism, organizing and coordinating activities were helped by several adults from the established environmental movement organizations such as Greenpeace, Nature and Youth Sweden (Fältbiologorna), and the Swedish Nature Conservation Society (Natuskyddsföreningen), and smaller networks like "We do not have time" or "Artists for the environment" (Ernman et al., 2020).

While the movement consisted first of a few schoolchildren in Sweden, it inspired many young people worldwide to take similar actions, primarily thanks to the social media and regular media coverage of the "school strikes." On March 15, 2019, when FFF organized its first Global Climate Strike, it attracted over 1 million participants in demonstrations across the world. Numerous local city-level branches were started, and these mobilized and continue to mobilize smaller school strikes or demonstrations every Friday. These events catalyzed a new wave of global climate activism, significantly leveraging the infrastructure of existing transnational climate change movements that included formal organizations, various informal groups, and grassroots movements (Nulman, 2022). Furthermore, climate movements were very much related to existing environmental movements. Previous waves of global climate activism can be traced back to pivotal developments such as the formation of the Climate Action Network in 1989 and the series of Climate Justice summits that began in the year 2000, which collectively marked significant milestones in the international environmental agenda (Almeida, 2019; della Porta & Parks, 2014). The FFF movement's rapid and extensive transnational mobilization significantly reinvigorated the global climate movement, which, according to Cassegard et al. (2017), was experiencing a sense of disillusionment stemming from a setback in Copenhagen in 2009 (COP15) and was partly disunited since the COP meeting in Paris in 2015.

The participation in the FFF's actions peaked with the global week of climate action in September 2019 when the UN Climate Summit took place in New York, and the movement's spokesperson, Greta Thunberg, participated in it. According to some estimations, the September Global Climate strike(s) gathered 7.6 million participants in 6,000 events in 185 countries (Chase-Dunn & Almeida, 2020). However, due to the eruption of the global COVID-19 pandemic in winter–spring 2020 and the accompanying restrictions on public gatherings, the number of FFF events and participants dwindled drastically and were partly replaced with online events. The movement adapted its actions by shifting to virtual campaigns and webinars and continuing social media activism via digital platforms such as Facebook, Twitter/X, YouTube, Instagram, and TikTok (Haßler et al., 2021; Sorce, 2023). After the pandemic, the movement faced new challenges as its prominent activists grew older, finished schools, started to participate in the actions of other movements (e.g., Extinction Rebellion [XR]) or joined political parties. FFF is still active but on a much smaller scale than in 2019.

Social media certainly aided the diffusion of the movement, but one should not forget that in many countries, it also coincides with the time of election campaigns (in Sweden in 2018; in Belgium in 2019). Although the "school strikes" strategy became widely used, the FFF has been predominantly Europe based. The existing research has focused on the developments in Belgium (e.g., Crouzé et al., 2024; Kenis, 2021; Knops, 2021; Knops & de Vydt, 2024), Finland (Huttunen, 2021; Piispa & Kiilakoski, 2022), Germany (e.g., Pollex & Soßdorf, 2023; Schürmann, 2024; Wallis & Loy, 2021), Italy (e.g., Belotti et al., 2022; Cugnata et al., 2024), Portugal (Campos & Martins, 2024), Spain (Soler-i-Martí et al., 2022), Sweden (Emilsson et al., 2020; Schack, 2024); Switzerland (e.g., Cologna et al., 2021; Fritz et al., 2023); or the United Kingdom (e.g., Doherty & Saunders, 2021). The movement also mobilized beyond Western Europe—to Eastern Europe (Císař, 2022) but also to Latin America (Poma & Gravante, 2024) and Africa—Uganda and Nigeria (Nakabuye et al., 2020), Gambia (Gardner et al., 2024), and Kenya and South Africa (Damoah et al., 2023). In other regions, regardless of the local sections of the FFF, there are other climate movements as well. For example, in the United States, the major youth climate movement is called the Sunrise movement (Fisher & Nasrin, 2021b); in New Zealand, the School Strike 4 Climate NZ (Nairn et al., 2021); in Australia, scholars mainly focus on the Australian Youth Climate Coalition (AYCC) (Hilder & Collin, 2022).

The Movement of Young Girls?

The literature on youth activism often distinguishes the movements by and for young people. While the FFF should be described as a youth-led movement, it is not only for young people. Initially, however, FFF was predominantly a movement of schoolchildren, and for many activists, it was important to emphasize their image as innocent children suffering from intergenerational injustice (Schack, 2024). While typically "first-timers" constitute only a small portion of protest event participants (Saunders et al., 2012), FFF was an exception. A unique protest survey, conducted by an international collaborative network of scholars, revealed that participants under 19 were the majority in the global school strikes for climate in March 2019 (Wahlström et al., 2019). Many had not participated in any protests before (Wahlström et al., 2019). Remarkably, regardless of the age group of participants in the global climate strikes of March or September 2019, most activists identified themselves as female (de Moor et al., 2020). It is unusual, as the typical demonstration participant used to be a young man. Furthermore, the major spokespersons for the movement were young female university or high school students: Greta Thunberg (Sweden), Anuna De Wever Van der Heyden (Belgium), Luisa Neubauer (Germany), Hilda Nakabuye (Uganda), Adenike Titilope Oladosu (Nigeria), Adenike Titilope Oladosu, and Sophie Handford (New Zealand). Most scholarly attention has been directed toward Greta Thunberg (e.g., Mede & Schroeder, 2024; Sorce, 2022). Wahlström and Uba (2024) have even shown a certain "Greta effect," as many young women were inspired and mobilized by Thunberg as a role model, and others saw her as a political icon.

The proportion of very young activists participating in the FFF events declined from March to September 2019, and this trend has continued (de Moor et al., 2020). This pattern suggests that not many new participants joined the movement during the pandemic, a phenomenon expected in social movements as the activists' average age increases over time. Additionally, scholars have studied the socioeconomic backgrounds of participants in global climate strikes and found a clear overrepresentation of highly educated individuals (de Moor et al., 2020). However, half of the activists participating in the FFF protests in Europe self-identified as working or from the lower-middle class (della Porta & Portos, 2023). In addition to their participation in climate protests, these activists generally report more climate-friendly behaviors, such as consuming less meat, flying less, and taking steps to compensate for their CO_2 emissions (Cologna et al., 2021). However, these trends also have notable cross-national variations (Martiskainen et al., 2020).

When examining the reasons for joining FFF actions, scholars have noted that activism is often driven by the perception that friends are also participating and a strong identification with other climate activists (Wallis & Loy, 2021). Similar to other social movements, the mobilization within FFF is linked to emotional drivers such as hope, fear, anger, perceptions of threats, and feelings of urgency (see chapter by Poma and Gravante in this volume). Consequently, researchers have placed significant emphasis on the emotional dimensions of climate activism. For instance, Lorenzini and Rosset (2024) have used data primarily from European protest surveys to demonstrate that fear and anger are primarily related to mobilization in FFF, with younger protesters more likely to experience fear, while older ones tend to feel more anger (cf. Rainsford & Saunders, 2021). In their study focusing on Mexico, Poma and Gravante (2024) explore the interactive nature of emotions in climate activism, examining how activists frame the problem and their perceptions of threats and outcomes from a sociocultural perspective.

Claims of the FFF

According to the activists, the main reasons for FFF mobilization were the perceived discrepancy between the scientific knowledge provided at schools (climate change is a threat) and the political response that does not address the emerging urgent crisis due to global warming. Therefore, the FFF called for getting world governments in line with the Paris Agreement on climate change (cutting global CO_2 emissions in half by 2030 and just below zero by 2050) and used school strikes as the first protest strategy (De Wever Van der Heyden et al., 2020). It has also been shown that—at least in Germany the participants of FFF events overwhelmingly derive their goals from scientific knowledge (Soßdorf & Burgi, 2022) and often relate their claim-making to the fact that the older/past generations have caused the occurring crisis (Pavenstädt & Rödder, 2024; Schack, 2024). The call "Listen to the Science" is strongly related to the idea that FFF bases its arguments on scientific evidence, but this also means that the movement had less to say about how to solve the crisis (Evensen, 2019). Although one can find the claims for intergenerational equity and climate justice in the FFF mobilizations, the last is arguably less present among the FFF claims than in the case of other climate movements, such as the XR (Pavenstädt & Rödder, 2024).

When analyzing the prognostic framing (Snow et al. 2018), more precisely, the proposed actions that should be taken to solve the identified problem of climate change, Svensson and Wahlström (2023) find that European participants of the FFF's global strikes in 2019 adhere to a "top-down" change within the existing societal system. Still, many activists also emphasize the importance of individual lifestyle changes (cf. Zamponi et al., 2022), and only left-leaning or older activists argue for the importance of a bottom-up change in society to address global warming. This reflects that, similarly to other social movements, FFF is heterogeneous and not fully united over its major claims. For example, internal tensions and friction over different ideas of social life and political order have been found in the FFF in Germany (Marquardt, 2020).

The competition for different future visions is nothing unique to the FFF or climate movement, although one could easily separate the different techno-centric, apolitical, and market-driven ideas here. FFF protesters believe that climate change should be primarily fought through the adoption of changes in lifestyle choices (Zamponi et al., 2022). In the Swedish case, Emilsson et al. (2020) found that a substantial majority of respondents to the global climate strike protest survey described above strongly agreed with the statement, "Protecting the environment should be given priority, even if it causes slower economic growth and some loss of jobs," and only a few disagreed with this statement. Still, when comparing the framing of the energy future by the FFF and XR in Germany, Buzogány and Scherhauf (2022) found that FFF followed the traditional institutional narratives somewhat more than the XR. On the same line, Della Porta et al. (2024) have found that FFF also considers supranational institutions such as the European Union to be important actors, criticizing their lack of actions for mitigating climate change. These perspectives characterize the certain dualism that is present among the FFF members. On the one hand, there is widespread distrust toward political institutions (Knops & De Vydt, 2024; Zamponi et al., 2022)—probably best described by Thunberg's "blah, blah, blah" comment during COP26. On the other hand, when FFF activists talk about "system change," the reference is not made so much to the political system but to the (capitalist) economic system. FFF has also advocated and participated in different citizens' deliberative climate assemblies, but this is still in the frame of a "green governmentality" approach (Svensson & Wahlström, 2023).

FFF has also broadened its claims over the years, adapting to specific environmental problems in specific countries (e.g., mines, wind farms), but broadening the political scope toward the issues of war, peace, and conflicts (Russia-Ukraine, Israel-Palestine) has come with several internal frictions. Further research on these trends would be fundamental.

Repertoires and Strategies

The primary protest repertoire that characterizes FFF is a school strike—skipping school as a protest against the lack of political action to solve the climate crisis. These "strikes" were quickly combined with demonstrations and marches, and since the pandemic, the focus is more on demonstrations than school strikes. The more disruptive repertoires are combined with more institutional ones, as the movement's spokespersons were invited to meet politicians in many countries in the initial phase of the movement (e.g., Greta Thunberg at the UN Climate Summit in 2019). During the global COVID pandemic, when freedom of assembly was restricted in most countries, a significant trend of innovation of protest strategies and the hybridization of protest actions took form. Hunger and Hutter (2020) report that in Germany, FFF opted for digital climate strikes to different degrees—mainly in urban areas with highly educated populations. Inevitably, there was also a decline in FFF activism during the pandemic (Haßler et al., 2021). While in the digital sphere, statistics show a declining number of tweets even in the spring of 2020. The lockdowns turned public attention to other topics, and there was also increasing social media discussion over the legitimacy of the FFF (Haßler et al., 2021). Arguably, the pandemic affected the intensity of the FFF mobilization and the framing of their arguments—the activists opted for broadening their claims (Sorce & Dumitica, 2023).

More recently, there have been reports of FFF activists joining other movements that adopt more disruptive forms of action—acts of civil disobedience in the form of blocking roads. The international movement's homepage regularly updates the number of strikes carried out, the number of participating activists, and future strikes (see www.fridaysforfuture.org). Only a few academic studies have systematically covered the spatial and temporal developments of the FFF protest events. There are, however, some country-specific studies. For example, Zamponi et al. (2023) show that in Italy since 2021, FFF initiated broad cooperation with many different social movements (e.g., feminist movement) and other climate groups (e.g., new group Last Generation/Ultima Generazione). The cooperation also correlates with the increasing use of tactics like civil disobedience (Zamponi et al., 2023).

Significant research has been focused on the FFF's social media presence and communication strategies (Sorce, 2023). In a review of 138 peer-reviewed articles about environmental and climate issues, Baran and Stoltenberg (2023) show that most of FFF's social media presence research focuses on X (formerly Twitter) and Instagram. As an innovative strategy, many FFF activists use memes in their mobilization (Johann et al., 2023). Belotti et al. (2022), in their long-term analysis of the Rome branch of the FFF, argue that even though these young activists perceive social media as a necessary "battlefield," they swiftly combine online and offline mobilization strategies. The increasingly important field of study in the context of the FFF's online mobilization is the visual analysis of their mobilization (e.g., Hayes & O'Neil, 2021; Herrmann et al., 2023), and there, we will likely see more future works.

Consequences of the FFF

The rapid proliferation of school strikes for climate illustrates the significant impact of FFF on mobilizing public action on climate issues. Research indicates that traditional strategies such as climate strikes (demonstrations) and civil disobedience have bolstered support for climate activism among the general populace in the United States, though with different effects on supporters of the Democratic and Republican parties (Bugden, 2020). Furthermore, the "Greta effect" has notably heightened political activism among young women (Wahlström & Uba, 2024). Although these short-term effects of social movement mobilization are readily observable, assessing the long-term outcomes—whether biographical, cultural, or political—requires a more extended perspective (Bosi et al., 2016). Additionally, given the concurrent activity of multiple climate movements such as XR, Just Stop Oil, and Last Generation, isolating the distinct societal impacts attributable solely to FFF becomes challenging.

Still, some studies have already estimated the effects of FFF mobilization on political agenda setting, public opinion, citizens' voting patterns, and politicians' reactions in parliament. In particular, Fritz et al. (2023) show that one third of the Swiss public has a positive attitude toward FFF's and Greta Thunberg's climate activism, and their mobilization arguably increased citizens' environmental awareness and climate-friendly behavior. Some studies also argue for the effects of climate strikes on voting patterns, demonstrating that, at least in Germany, the exposure to FFF activism increased the vote share of the Green Party (Valentim, 2023). In the case of the United Kingdom, Barrie et al. (2024) show that FFF's protests are correlated to politicians' changed social media activism. More specifically, more climate-friendly posts engage the United Kingdom's members of parliament. Similar patterns have been noted in the German case by Schürmann (2024). The lack of effect on politicians' offline actions, namely the parliamentary speech, is not surprising, considering the generally low expectancy for the political impact of environmental movements (Rootes & Nulman 2015). For instance, Berker and Pollex (2023) find that in Austria, Germany, and Sweden, the reactions to FFF by political parties vary substantially. While left parties uniformly support FFF, populist and radical right parties fiercely oppose the movement, and there is some variation among center-right parties. Despite initially relatively moderate strategies and claims, the parties' reactions were not accommodating.

Another reaction to the movements' mobilization is counteractivism. While there certainly was a countermovement to climate movement even before the rise of FFF in 2018 (see more in Staggenborg & Meyer, 2022), the first school strikes awakened this heterogeneous group. Much of this countermobilization emerged in social media rather than on the streets. It has also been shown that populist radical right parties such as Swedish Democrats or Alternative for Germany only became aware of climate change as a political issue since the FFF's mass protests in 2019 (Schwörer & Fernández-García, 2024). Both parties oppose the claims of FFF and have also won votes thanks to this position in recent elections.

Outlook for the Future

When FFF mobilized its first global climate strike in March 2019, it gained broad support from the old established environmental movements and the scholarly community (Hagedorn et al., 2019). By 2024, this youth-led movement had not achieved its primary objectives. There were still only a few signs of policies capable of keeping the rise in global temperatures below 1.5°C compared to preindustrial levels or advancing climate justice and equity. Nonetheless, the issue of climate change has not disappeared from the public agenda even though other topics, such as the global pandemic, wars, or political and economic crises, have certainly offered competition. Partly, it is related to the urgency of the matter—the climate crisis is emerging, but there has also been a sustained mobilization of climate activists (of FFF and beyond). As a result, public discourse has shifted, and more people are aware of the climate crisis and related threats. There is also increasing evidence of institutional-level climate action planning focusing on mitigation, adaption, and resiliency (Almeida et al., 2024) and changes in countries' renewable energy policies (Fleck & Anatolitis, 2023). In the same line, until 2023, over 2,000 cities/municipalities had declared a climate emergency (Henman et al., 2023), even though these actions have more of a symbolic value and are probably a combined effect of both "old" and "new" waves of global climate activism. Nevertheless, these changes have not been enough for the youth-led FFF and other climate movement activists, and large global strikes and the small every-Friday gatherings in cities and rural centers worldwide continue.

References

Almeida, P. D. (2019). Climate justice and sustained transnational mobilization. *Globalisations, 16*(7), 973–979.

Almeida, P. D. (2023). Climate change and climate justice movements. In D. A. Snow, D. della Porta, B. Klandermans, & D. McAdam (Eds.), *The Wiley-Blackwell encyclopedia of social and political movements* (pp. 1–3). Blackwell Publishing.

Almeida, P. D., González Márquez, L. R., & Fonsah, E. (2024). The forms of climate action. *Sociology Compass, 18*(2), e13177.

Baran, Z., & Stoltenberg, D. (2023). Tracing the emergent field of digital environmental and climate activism research: A mixed-methods systematic literature review. *Environmental Communication*, 17(5), 453–468.

Barrie, C., Fleming, T. G., & Rowan, S. S. (2024). Does protest influence political speech? Evidence from UK climate protest, 2017–2019. *British Journal of Political Science, 54*(2), 456–473.

Belotti, F., Donato, S., Bussoletti, A., & Comunello, F. (2022). Youth activism for climate on and beyond social media: Insights from Fridays For Future–Rome. *International Journal of Press/Politics, 27*(3), 718–737.

Berker, L. E., & Pollex, J. (2023). Explaining differences in party reactions to the Fridays for Future Movement—A qualitative comparative analysis (QCA) of parties in three European countries. *Environmental Politics, 32*(5), 755–792.

Bosi, L., Giugni, M., & Uba, K. (Eds.). (2016). *The consequences of social movements.* Cambridge University Press.

Bugden, D. (2020). Does climate protest work? Partisanship, protest, and sentiment pools. *Socius, 6*, 1–13.

Buzogány, A., & Scherhaufer, P. (2022). Framing different energy futures? Comparing Fridays for Future and Extinction Rebellion in Germany. *Futures, 137*, 102904.

Buzogány, A., & Scherhaufer, P. (2023). The new climate movement: Organization, strategy, and consequences. In H. Jörgens, C. Knill, & Y. Steinebach (Eds.), *Routledge handbook of environmental policy* (pp. 358–380). Routledge.

Campos, R., & Martins, J. C. (2024). Ontological insecurity and urgency as a political value: Discourses of youth climate activists in Portugal. *Globalisations, 21*(7), 1180–1196.

Cassegard, C., Soneryd, L. Thorn, H. & Wettergren, A. (Eds.). (2017). *Climate action in a globalizing world: Comparative perspectives on environmental movements in the Global North.* Vol. 27. Routledge.

Chase-Dunn, C., & Almeida, P. (2020). *Global struggles and social change: From prehistory to world revolution in the twenty-first century.* Johns Hopkins University Press.

Císař, O. (2022). Rhapsody in green: Environmental movements in East Central Europe. In M. Grasso & M. Giugni (Eds.), *The Routledge handbook of environmental movements* (pp. 32–44). Routledge.

Cologna, V., Hoogendoorn, G., & Brick, C. (2021). To strike or not to strike? An investigation of the determinants of strike participation at the Fridays for Future climate strikes in Switzerland. *PLoS One, 16*(10), e0257296. https://doi.org/10.1371/journal.pone.0257296.

Crouzé, R., Godard, L., & Meurs, P. (2024). Learning democracy through activism: The global climate strike movement and Belgian youth's democratic experience in times of environmental emergency. *Social Movement Studies, 23*(1), 56–71.

Cugnata, G., Cuel, J., Bertuzzi, N., Parks, L., & Zamponi, L. (2024). Convergence in action: Framing the climate crisis at the 2021 Pre-COP Counter-Summit in Milan. *European Societies, 26*(5), 1545–1571. https://doi.org/10.1080/14616696.2024.2362151

Damoah, B., Keengwe, S., Owusu, S., Yeboah, C., & Kekessie, F. (2023). The global climate and environmental protest: Student environmental activism a transformative defiance. *International Journal of Environmental, Sustainability, and Social Science, 4*(4), 1180–1198.

de Moor, J., Uba, K., Wahlström, M., Wennerhag, M., & De Vydt, M. (Eds.). (2020). Protest for a Future II: Composition, mobilization and motives of the participants in Fridays For Future climate protests on 20–27 September, 2019, in 19 cities around the world. https://osf.io/asruw/

De Wever Van der Heyden, A., Neubauer, L., & van der Heyden, K. (2020). Fridays for Future—FFF Europe and beyond. In C. Henry, J. Rockström, & N. Stern (Eds.), *Standing up for a sustainable world: Voices of change* (pp. 197–211). Edward Elgar.

Della Porta, D., & Parks, L. (2014). Framing processes in the climate movement: From climate change to climate justice. In *Routledge handbook of the climate change movement* (pp. 19–30). Routledge.

Della Porta, D., Parks, L. R., & Portos, M. (2024). Social movements and Europeanisation: Framing "responsibility" and "responsiveness" in times of multiple crises. *Journal of European Public Policy, 31*(4), 1100–1125.

Della Porta, D., & Portos, M. (2023). Rich kids of Europe? Social basis and strategic choices in the climate activism of Fridays for Future. *Italian Political Science Review*, 53(1), 24–49.

Doherty, B. J., & Saunders, C. (2021). Global climate strike protests and media coverage of the protests in Truro and Manchester. In A. M. Mesinas, S. Pickard, & J. Bessant (Eds.), *When students protest* (pp. 251–268). London: Rowman and Littlefield.

Emilsson, K., Johansson, H., & Wennerhag, M. (2020). Frame disputes or frame consensus? "Environment" or "welfare" first amongst climate strike protesters. *Sustainability*, 12(3), 882.

Ernman, B., Ernman, M., Thunberg, G., & Thunberg, S. (2020). *Our house is on fire: Scenes of a family and a planet in crisis.* Translated by P. Norlen & S. Vogel. London: Penguin. (Originally published 2019)

Evensen, D. (2019). The rhetorical limitations of the #FridaysForFuture movement. *Nature Climate Change, 9*, 428–430.

Fernández-Zubieta, A., Guevara, J. A., Caballero Roldan, R., & Robles, J. M. (2023). Digital activism masked—The Fridays for Future movement and the "Global Day of Climate Action": Testing social function and framing typologies of claims on Twitter." *Social Sciences, 12*(12), 676.

Fisher, D. R. (2019). The broader importance of #FridaysForFuture. *Nature Climate Change, 9*(6), 430–431.

Fisher, D. R., & Nasrin, S. (2021a). Climate activism and its effects. *Wiley Interdisciplinary Reviews: Climate Change*, 12(1), e683.

Fisher, D. R., & Nasrin, S. (2021b). Shifting coalitions within the youth climate movement in the US. *Politics and Governance*, 9(2), 112–123.

Fleck, A. K., & Anatolitis, V. (2023). Achieving the objectives of renewable energy policy—Insights from renewable energy auction design in Europe. *Energy Policy, 17*, 113357.

Flörchinger, D., Frondel, M., Sommer, S., & Andor, M. (2024). Pro-environmental behavior and environmentalist movements: Evidence from the identification with Fridays for Future. USAEE Working Paper No. 24-620. https://ssrn.com/abstract=4785583orhttp://dx.doi.org/10.2139/ssrn.4785583

Fritz, L., Hansmann, R., Dalimier, B., & Binder, C. R. (2023). Perceived impacts of the Fridays for Future climate movement on environmental concern and behaviour in Switzerland. *Sustainability Science, 18*(5), 2219–2244.

Gardner, P., O'Brien, T., Carvalho, T., & Adekola, O. (2024). PROFILE: Extinction rebellion in the Gambia. *Social Movement Studies, 23*(1), 122–128.

Grasso, M., & Giugni, M. (Eds.). (2022). *The Routledge handbook of environmental movements*. Routledge.

Hagedorn, G., Kalmus, P., Mann, M., Vicca, S., Van den Berge, J., van Ypersele, J. P., Bourg, D., Rotmans, J., Kaaronen, R., Rahmstorf, S., Kromp-Kolb, H., Kirchengast, G., Knutti, R., Seneviratne, S. I., Thalmann, P., Cretney, R., Green, A., Anderson, K., Hedberg, M., . . . Hayhoe, K. (2019). Concerns of young protesters are justified. *Science, 364*(6436), 139–140.

Haßler, J., Wurst, A.-K., Jungblut, M., & Schlosser, K. (2021). Influence of the pandemic lockdown on Fridays for Future's hashtag activism. *New Media & Society, 25*(8), 1991–2013. https://doi.org/10.1177/14614448211026575

Hayes, S., & O'Neill, S. (2021). The Greta effect: Visualizing climate protest in UK media and the Getty images collections. *Global Environmental Change, 71*, 102392.

Henman, J., Shabb, K., & McCormick, K. (2023). Slow emergency but urgent action? Exploring the impact of municipal climate emergency statements in Sweden. *Urban Climate, 49*, 101575.

Herrmann, C., Rhein, S., & Dorsch, I. (2023). # fridaysforfuture–What does Instagram tell us about a social movement? *Journal of Information Science, 49*(6), 1570–1586.

Hilder, C., & Collin, P. (2022). The role of youth-led activist organizations for contemporary climate activism: The case of the Australian Youth Climate Coalition. *Journal of Youth Studies, 25*(6), 793–811.

Hunger, S., & Hutter, S. (2020). Online strikes with the usual suspects: How Fridays for future has coped with the COVID-19 pandemic. LSE European Politics and Policy (EUROPP) blog. https://eprints.lse.ac.uk/105403/

Huttunen, J. (2021). Young rebels who do not want a revolution: The non-participatory preferences of Fridays For Future activists in Finland. *Frontiers in Political Science, 3*, 672362. DOI: 10.3389/fpos.2021.672362

Johann, M., Höhnle, L., & Dombrowski, J. (2023). Fridays for Future and Mondays for memes: How climate crisis memes mobilize social media users. *Media and Communication, 11*(3), 226–237.

Kenis, A. (2021). Clashing tactics, clashing generations: The politics of the school strikes for climate in Belgium. *Political Governance, 9*, 135–145. DOI: 10.17645/pag.v9i2.3869

Knops, L. (2021). Stuck between the modern and the terrestrial: The indignation of the youth for climate movement. *Political Research Exchange, 3*(1), 1868946.

Knops, L. (2024). Beyond the indignation of young climate activists: The political potential of climate-emotions. In J. Bessant, P. Collin, & P. O'Keeffe (Eds.), *Research handbook on the sociology of youth* (pp. 53–68). Edward Elgar.

Knops, L., & De Vydt, M. (2024). Capitalism at the roots: The crisis of representative democracy through the eyes of Belgian climate activists. *Environmental Politics, 33*(1), 161–189.

Lorenzini, J., & Rosset, J. (2024). Emotions and climate strike participation among young and old demonstrators. *Social Movement Studies, 23*(1), 39–55.

Marquardt, J. (2020). Fridays for Future's disruptive potential: An inconvenient youth between moderate and radical ideas. *Frontiers in Communication, 5*, 48.

Martiskainen, M., Axon, S., Sovacool, B. K., Sareen, S., Del Rio, D. F., & Axon, K. (2020). Contextualizing climate justice activism: Knowledge, emotions, motivations, and actions among climate strikers in six cities. *Global Environmental Change, 65*, 102–180.

Mede, N. G., & Schroeder, R. (2024). The "Greta effect" on social media: A systematic review of research on Thunberg's impact on digital climate change communication. *Environmental Communication, 18*(6), 801–818. https://doi.org/10.1080/17524032.2024.2314028.

Nakabuye, H. F., Nirere, S., & Oladosu, A. T. (2020). The Fridays For Future movement in Uganda and Nigeria. In C. Henry, J. Rockström, & N. Stern (Eds.), *Standing up for a sustainable world* (pp. 212–218). Edward Elgar.

Nairn, K., Kidman, J., Matthews, K.-R., Showden, C.-R., & Parker, A. (2021) Living in and out of time: Youth-led activism in Aotearoa New Zealand. *Time & Society, 30*(2), 247–269.

Noth, F., & Tonzer, L. (2022). Understanding climate activism: Who participates in climate marches such as "Fridays for Future" and what can we learn from it?. *Energy Research & Social Science*, *84*, 102360.

Nulman, E. (2022). Climate change movements in the Global North. In M. Grasso & M. Giugni (Eds.), *The Routledge handbook of environmental movements* (pp. 185–198). Routledge.

Parks, L., Della Porta, D., & Portos, M. (2023). Environmental and climate activism and advocacy in the EU. In T. Rayner, K. Szulecki, A. J. Jordan, & S. Oberthür (Eds.), *Handbook on European Union climate change policy and politics* (pp. 98–112). Edward Elgar.

Pavenstädt, C. N., & Rödder, S. (2024). Between evidence first and political fight—Understanding dynamics of (de-)politicization in US climate movements' future narratives. *Environmental Politics*, *33*(6), 1065–1086.

Piispa, M., & Kiilakoski, T. (2022). Towards climate justice? Young climate activists in Finland on fairness and moderation. *Journal of Youth Studies*, *25*(7), 897–912.

Pollex, J., & Soßdorf, A. (Eds.). (2023). *Fridays for Future: Einordnung, Rezeption und Wirkung der neuen Klimabewegung*. Springer.

Poma, A., & Gravante, T. (2024). The emotional dimension of Mexican climate activism: a sociocultural approach. *Mobilization: An International Quarterly*, *29*(1), 103–122.

Rainsford, E., & Saunders, C. (2021). Young climate protesters' mobilization availability: Climate marches and school strikes compared. *Frontiers in Political Science*, *3*, 713340.

Rootes, C. (2004). Environmental movements. In D. Snow, S. A. Soule, & H. Kriesi (Eds.), *Blackwell companion to social movements* (pp. 608–640). Blackwell.

Rootes, C., & Nulman, C. (2015). 'The impacts of environmental movements." In D. della Porta & M. Diani (Eds.), *The Oxford handbook of social movements* (pp. 729–742). https://doi-org.ezproxy.its.uu.se/10.1093/oxfordhb/9780199678402.013.55

Saunders, C., Grasso, M., Olcese, C., Rainsford, E., & Rootes, C. (2012). Explaining differential protest participation: Novices, returners, repeaters, and stalwarts. *Mobilization: An International Quarterly*, *17*(3), 263–280.

Schack, L. (2024). Climate subjectivity: youth, innocence and willfulness in the Swedish climate movement. *Mobilization: An International Quarterly*, *29*(2), 229–244.

Schürmann, L. (2024). The impact of local protests on political elite communication: evidence from Fridays for Future in Germany. *Journal of Elections, Public Opinion and Parties*, *34*(3), 510–530. https://doi-org.ezproxy.its.uu.se/10.1080/17457289.2023.2189729

Schwörer, J., & Fernández-García, B. (2024). Climate sceptics or climate nationalists? Understanding and explaining populist radical right parties' positions towards climate change (1990–2022). *Political Studies*, *72*(3), 1178–1202.

Snow, D. A., Vliegenthart, R., & Ketelaars, P. (2018). The framing perspective on social movements: Its conceptual roots and architecture. In D. A. Snow, S. A. Soule, and H. Kriesi (Eds.), *The Wiley Blackwell companion to social movements* (pp. 392–410). Blackwell.

Soler-i-Martí, R., Fernández-Planells, A., & Pérez-Altable, L. (2022). Bringing the future into the present: The notion of emergency in the youth climate movement. *Social Movement Studies*, *23*(19), 1–20.

Sorce, G. (2022). The "Greta effect": Networked mobilization and leader identification among Fridays for Future protesters. *Media and Communication*, *10*(2), 18–28.

Sorce, G. (2023). Stuck with the algorithm: Algorithmic consciousness and repertoire in Fridays for Future's data contention. *Media and Communication*, *11*(3), 214–225.

Sorce, G., & Dumitrica, D. (2023). # fighteverycrisis: Pandemic shifts in Fridays for future's protest communication frames. *Environmental Communication*, *17*(3), 263–275.

Soßdorf, A., & Burgi, V. (2022). "Listen to the science!"—The role of scientific knowledge for the Fridays for Future movement. *Frontiers in Communication*, *7*, 983929.

Staggenborg, S., & Meyer, D. S. (2022). Understanding counter-movements. In D. Tindall, M. C. J. Stoddart, & R. E. Dunlap (Eds.), *Handbook of anti-environmentalism* (pp. 23–42). Edward Elgar.

Svensson, A., & Wahlström, M. (2023). Climate change or what? Prognostic framing by Fridays for Future protesters. *Social Movement Studies*, *22*(1), 1–22.

Valentim, A. (2023). Repeated exposure and protest outcomes: How Fridays for Future protests influenced voters. https://osf.io/preprints/socarxiv/m6dpg[07092024]

van der Heijden, H. A. (1997). Political opportunity structure and the institutionalisation of the environmental movement. *Environmental Politics*, 6(4), 25–50.

Wahlström, M., & Uba, K. (2024). Political icon and role model: Dimensions of the perceived "Greta effect": Among climate activists as aspects of contemporary social movement leadership. *Acta Sociologica*, *67*(3), 301–316.

Wahlström, M., Kocyba, P., De Vydt, M., & de Moor, J. (Eds). (2019). Protest for a future: Composition, mobilization and motives of the participants in Fridays for future climate protests on 15 March, 2019 in 13 European cities. https://osf.io/xcnzh/

Wallis, H., & Loy, L. S. (2021). What drives pro-environmental activism of young people? A survey study on the Fridays For Future movement. *Journal of Environmental Psychology*, *74*, 101581. DOI: 10.1016/j.jenvp.2021.101581

Zamponi, L., Baukloh, A. C., Bertuzzi, N., Chironi, D., della Porta, D., & Portos, M. (2022). (Water) bottles and (street) barricades: The politicization of lifestyle-centred action in youth climate strike participation. *Journal of Youth Studies*, 25(6), 854–875.

Zamponi, L., Ferro, A., & Cugnata, G. (2023). Strikes, assemblies and blockades: The dynamics of repertoire change in grassroots climate action in Italy (2018–2023). *Italian Political Science*, *18*(3), 257–293.

CHAPTER 13

Beyond the Number

350.org Origins and Lessons

WILL BATES, MAY BOEVE, AND JAMIE HENN

STANDING at the edge of Central Park, the crowd of demonstrators extended as far as the eye could see. There were busloads of college students from across the country, some of whom had traveled over 24 hours to get there; thousands of union members, from the transit workers to local International Brotherhood of Electrical Workers; dozens of different faith groups, many of whom saw their participation in the day's events as a spiritual act; Indigenous leaders from across Turtle Island, many of whom were engaged in fights against major fossil fuel projects that threatened their ancestral homes; and tens of thousands of New Yorkers, including a number who had lost their homes to Hurricane Sandy just a couple of years before. To us, it looked like the movement we had always dreamed of.

At the time, each of us was in a leadership role at 350.org, the international climate advocacy organization: May as our Executive Director, Will as Global Organizing Director, and Jamie as Global Communications Director. A few months before, we and our allies had put out the call for a "People's Climate March" the weekend before the United Nations General Assembly, when world leaders would be coming together to build momentum for the Paris Climate Talks that were scheduled for the following year. Up until that point, the largest climate demonstration in US history had been about 50,000 people, and we optimistically hoped to double that.[1] So when over 400,000 people showed up that Saturday, September 21, 2014, we were absolutely blown away.

The march in New York was just one of hundreds of similar demonstrations around the world that day, each as vibrant and diverse as the crowd that snaked its way through the Big Apple. Two days later, when President Obama spoke at the United Nations, he pointed to the demonstrations as evidence that the time had come to finally agree to a global climate treaty, saying, "Our citizens are marching. We cannot pretend we do not hear them."[2] When the Paris Climate Agreement was adopted the following year, many, including UN Climate Secretary Christiana Figueres, pointed to the outpouring of public action as a decisive factor in closing the deal.

Now that a decade has passed since the Paris climate talks, arguably the pinnacle of global climate diplomacy, it's worth looking at how the climate movement got to that point and where it has gone since. As climate impacts intensify, and as right-wing governments and a desperate fossil fuel industry seek to roll back hard-fought progress, understanding where we have come from may help us chart a course for where we need to go. With a movement as global and diverse as the climate movement, it's impossible to do justice to all the work that has brought us to where we are today. Like many social movements, what we now refer to as a distinct "climate" or "climate justice" movement has its roots in many older struggles, from environmentalism to civil rights. The progress we have made has been born out of thousands of local fights and individual acts of bravery, most of which have gone undocumented or faded into memory.

But perhaps we can offer some useful snapshots from the journey that brought us to this point. The three of us, each a cofounder of 350.org, have spent the last two decades growing up inside the climate movement. Together with our friends and colleagues, we have coordinated numerous global days of action that have mobilized millions of people in the name of climate progress; organized mass marches, such as the aforementioned Peoples Climate March and the Global Climate Strikes; engaged in civil disobedience, including planning the White House sit-ins against the Keystone XL pipeline and supporting fossil fuel protests from the Ende Gelande movement against open pit coal mines in Germany to the Standing Rock encampment against the Dakota Access Pipeline in the United States; launched major campaigns, such as the fossil fuel divestment movement, which has now moved over $40 trillion away from coal, oil, and gas investments; attended over a decade's worth of UN climate negotiations; assisted frontline communities and small island nations to make their voices heard; brought together thousands of young people from around the world together for activist training camps; and sent many, many emails along the way.

Through it all, we've learned some lessons about what we think works (and does not work) when it comes to mobilizing the public for climate action. In this chapter, we will share some of those lessons through the stories of our successes, failures, and the many discoveries along the way. Before we begin, an important caveat: having had the privilege to organize in different contexts around the world, and work with hundreds of incredible global colleagues at 350.org, we are acutely aware of the limited perspective we bring as three White people from the United States who all attended the same small liberal arts college in Vermont (Middlebury College). Over the last couple of decades in this movement, we have repeatedly stumbled over our biases, grappled with our privilege, and been humbled by our blind spots. We offer these pages not as a definitive account, or a set of rules for how to organize a movement, but just as our thoughts, reflections, and hopes for the future. As we often say in the climate movement, "To change everything, it takes everyone." Our only hope is that our offering here inspires readers to participate in climate action. Maybe together, we can change the world.

The Beginning

"What Works: Building the New Climate Movement": The course description didn't read like much else in the college catalog. It was January of 2005 and the start of "J-Term" at Middlebury College, a 1-month semester in the middle of the cold Vermont winter, when students are given the opportunity to step outside of their usual course of study to explore something new. Our group of friends had each been engaged in activism in our own way on campus—volunteering at the organic garden, pushing for fair trade coffee at the cafeterias, organizing trips to Washington, DC, to protest the war in Iraq—but none of us had ever engaged in "climate activism" before. In truth, at that point in 2005, we knew hardly anyone who considered themselves a "climate activist." It's not that global warming was not an issue. It had been nearly 20 years since Dr. James Hansen had testified in the US Senate about the dangers of the greenhouse effect. But at that point, climate change was still seen as just another concern for the traditional environmental movement to address, something akin to acid rain or the hole in the ozone layer. A threat, yes, but not something to organize your life around.

Well, not something *most* people would organize their lives around. Already, there were a small but growing number of people in the United States who saw themselves not only as traditional environmentalists but as something new. They used different terms: climate activist, climate justice organizer, green jobs advocate, clean energy enthusiast. But whatever the label, each was homing in on the climate crisis as both a civilizational scale challenge *and* the opportunity to rethink nearly every aspect of our society. A growing group of people were waking up to the existential threat of climate change.

One of those people was Middlebury professor Jon Isham, an environmental economist by training, who had become obsessed with both global warming and the work of social movements. The syllabus he put together for that J-Term class was an eclectic mix of readings from both disciplines, *The Heat Is On*, a survey of the climate crisis by journalist Ross Gelbspan[3] but also *Doing Democracy* by Bill Moyer,[4] which laid out a "movement action plan" for how to build successful social movements.

We went to that first week of class not really knowing what to expect. We left totally fired up. Whereas a lot of people's first exposure to the topic of global warming leaves them feeling anxious or depressed, because we were simultaneously studying examples of successful social change, we instead felt inspired and optimistic. In fact, it was the very enormity of the climate crisis that made us so positive that a social movement could be built to address it. Here was an issue that touched every aspect of our lives, from the food we ate, to the air we breathed, to the way we made our energy, buildings, and the very clothes we wore to class. No matter your issue—racial justice, economic equality, global health—there was a connection to the climate crisis.

Our conversations in the classroom quickly spilled over to the dining hall and our dorm rooms. Since there seemed to be enough interest, we organized a meeting on a Sunday night, mostly because that was when no other student clubs were meeting, and we invited all the campus activists we knew from our various different groups. Middlebury has a "green" reputation, and there were at least seven different student groups just focused on the environment. For a school with only 2,400 undergraduates, that represented a lot of students! We invited them all to meet together, the idea being that climate change connected everything we all cared about, yet made room for our specific concerns.

Throughout January, and even after our initial J-Term class had wrapped up, we continued to meet each Sunday to discuss how we could address the climate crisis both on campus and off. As word got out about the meetings (and the free beer we brought along to each one), the group continued to grow. Soon, the question came up about what we were going to call ourselves. Something like the "campus climate club" seemed obvious, but we were dogmatic about this idea that climate was not supposed to be about just one issue, but about everything, and so we went with a name that was flexible enough to encompass all sorts of organizing: the Sunday Night Group. We liked the element of mystery embedded in the name.

We didn't know it at the time, but Sunday Night Group (SNG) set the template for a lot of our work with 350.org over the coming decade. Organizing on a small college campus in Vermont would seem the polar opposite of organizing protests on a global scale, but as we found out, a lot of the same rules applied. At SNG, we operated on the principle that there was a role for everyone: Anyone could propose a project they wanted to take on, and anyone else could volunteer to help them out. We encouraged a variety of tactics: from "Do It in the Dark" parties to encourage energy-saving measures, to serious conversations with facilities staff about decarbonizing the fuel mix of our campus boiler system. We prioritized having fun, singing songs at meetings, organizing mass bike rides, and incorporating art and creativity into all of our actions. And we thought at a grand scale: Why simply cut carbon, when we could go carbon neutral? Why just organize on our campus when we could organize with other schools across the state, and maybe even around the country?

Back in January, as part of Prof. Isham's class, we connected with another young activist, Billy Parish, who had recently dropped out of Yale to organize something called the Campus Climate Challenge, a nationwide effort to get colleges to reduce their carbon footprint. As part of the effort, Billy had started pulling together existing environmental youth networks, like the student arm of the Sierra Club, into a new coalition, the Energy Action Coalition, which was going to be exclusively focused on climate. As our organizing at Middlebury picked up steam, we kept in touch with Billy and began plotting a summer campaign that could help grow Energy Action, as well as try and make progress on an important climate front: transportation emissions. Another early influence, Liz Cunningham, had built a network connecting all Vermont's universities to focus on environmental issues. With her and Billy's help, we began to see a wider context beyond Middlebury.

Besides our youthful enthusiasm, we had another asset that we brought to the table: a 1990s school bus that some other friends at Middlebury had converted to run on vegetable oil and biodiesel for what they called Project BioBus. Our plan for the summer of 2005 was to drive the BioBus across the country, organizing local events and gathering signatures on a "Clean Car Pledge" that we would take to Detroit and deliver to the Big Three automakers, with the request that they develop a new line of fuel-efficient cars for the next generation of consumers (at that point, we couldn't even fathom that fleets of all-electric cars would be on the table). Billy leapt at the idea, both because it was ambitious and provided a way to organize more people but also because it sounded like a hell of a lot of fun.

"Road to Detroit," as we called our summer road trip, was our first nationwide campaign and introduced us to the broader environmental movement beyond our campus. As another 350.org cofounder, Jeremy Osborn, drove the bus around the country with a few other college students, Jamie and May headed out to San Francisco to run mission control out of the Greenpeace offices there. Then they made their way to Detroit, where they house-sat for the legendary civil rights organizer Grace Lee Boggs and worked out of a local Sierra Club office run by environmental justice advocate Rhonda Anderson. Rainforest Action Network helped us with our strategy and sent Jennifer Krill, Nile Malloy, and Sarah Connolly to help us formulate our demands. In August, we all met up in Detroit, delivering our petitions to General Motors, Ford, and Chrysler, alongside the United Auto Workers, and hosting a conference for a ragtag group of other youth activists from around the country.

Seven of those ragtag kids—the four of us, plus our friends and fellow organizers Phil Aroneanu, Kelly Blynn, Jon Warnow—were the students who went on to form 350.org together. We had sat in the SNG meetings, but our trip to Detroit planted the seeds for a collaboration that went on for years and had an impact far beyond what we could have imagined.

We returned to campus that fall, excited about the potential for a growing nationwide movement on climate. Over the next 6 months, we continued to connect with more activists who increasingly saw themselves as part of a movement distinct from traditional environmentalism. This was 2005 to 2006, still the early days of social media, so much of our online community revolved around a blog called It's Getting Hot in Here (yes, a reference to that Nelly song), where activists would post dispatches from the fight against mountaintop removal coal mining in Appalachia or early efforts to pass state-wide climate legislation. In December, many of us had the opportunity to meet up in person for the first time at the UN Climate Talks in Montreal, sleeping on the floor of a freezing warehouse and marching through the streets of the city, before returning to campus on Sunday night to make it to classes the next day.

These early connections with students from across North America were fuel to our passion. If our whole idea was that a social movement was needed to fight climate change, we were starting to meet dozens and dozens of others who shared that vision for change, and we were an emerging community with a set of common values: joy, ambition, inclusivity, and impact.

As our organizing intensified on campus and off, it caught the eye of someone who would go on to play a central role in the story of 350.org, and that of the broader climate movement, over the decades to come: the author and environmentalist, Bill McKibben. Back in 1989, when he was in his late 20s, and working as a staff writer for the *New Yorker* magazine, Bill had published the first book about global warming for a mass audience, *The End of Nature* (McKibben, 1989). Since then, he'd become a leading author on the subject, publishing hundreds of news articles and a number of follow-up books to his first bestseller. He would also engage sporadically in activism, giving speeches for environmental groups, attending protests, and connecting with the growing number of people across the globe who were increasingly concerned about the worsening climate crisis. But back in 2006, Bill still thought of himself primarily as a writer and a professor. A number of years earlier, he accepted a role as a "scholar in residence" at Middlebury, where he taught the occasional class, advised students on their thesis projects, and served as the faculty advisor for the Nordic ski team (one of his true joys in life).

We had known of Bill as a sort of distant celebrity on campus, someone whom we admired from afar and would invite to speak at our events, but didn't dare consider a colleague or co-conspirator. But just as we were beginning to deepen our commitment to climate activism, Bill was going on a similar journey of his own. As he told a fellow journalist years later, "At some point it became clear to me that we'd won the argument but we were losing the fight."[5] By 2006, Bill was ready to take on that fight; he just needed some compatriots to work with; and as he looked up from his keyboard at the growing activism on the campus where he taught, there we were.

That spring, a few of us began talking with Bill about what we could do in our small corner of the world to try and spark more of a movement around climate change. We quickly settled on perhaps the most tried and true tactic of social movements: a march. But this being Vermont, and Bill being a writer, we started planning a march with a certain literary flare. We would start high up in the Green Mountains outside Middlebury, at Robert Frost's cabin, and then spend the next 5 days walking north on country roads up to Burlington, the state's largest city. Will had conveniently just graduated, so he did not have those pesky classes to attend, and he put his hand up to be the coordinator of the effort. Over the next few months, we designed the posters and banners for the march, talked with farmers who were willing to host us along the way, and enlisted allies to help, especially Greenpeace, who sent full-time staff Rebecca Sobel, Chris Miller, and Gabe Wiesnieski. The Greenpeace staff actually knew what they were doing, and they were key to pulling off the march.

That August, Bill kicked off the march with a dispatch for the environmental website *Grist*, which has continued through today playing a central role in telling the story of the grassroots climate movement. "Why would anyone spend their Labor Day weekend wandering the shoulder of a highway? It's possible no one will—but if they do, it may signal the next wave in the global-warming fight. And not a moment too soon," Bill started his piece.[6] "By now, after almost 20 years, there's an amazing array of people working on global warming," he continued. "What there haven't been, oddly, are any people in the streets. That's about to change."

And so, unwilling to wait for someone more experienced or better prepared to launch the first big climate march, we laced up our hiking boots and set off from Robert Frost's cabin, down a twisting road out of the Green Mountains, and took off for Burlington. Over the next 5 days, our numbers grew, and by the time we marched into the city our march was over 1,000 people strong, which in little Vermont, felt like the revolution.[7] Enough of a revolution at least to convince every candidate running for Congress that fall to show up at our rally and sign a pledge to "cut carbon 80% by 2050," what scientists at the time were saying was the right target for reducing emissions (both climate scientists and policymakers now argue for much faster and stronger action).

Our featured speaker, Senator Bernie Sanders, had not yet risen to national prominence, but he was then and remains deeply committed to climate justice. With a giant Sharpie handed to him by our youngest walker, Schuyler Klien, Bernie signed our pledge to cut carbon 80% by 2050. If 1,000 people showing up in Burlington could convince our members of Congress to sign onto this demand, and it only took a summer to organize, could we do this in more places, we wondered?

This question planted the seed for a second effort, organizing marches across Vermont that coincided with the November 2006 election. Each march carried the same demand—80 by 50—but was different in that multiple organizers in multiple towns were involved, facilitated by digital tools. This model is what we now call "distributed organizing," and it is a hallmark of political and issue campaigns the world over, but back then we were helping to build some genuinely new technological tools. Anyone anywhere could find our website, register an event, and we could broadcast it to a wider audience who could find it. This enabled connections to be made among people who had never met, and it increased the size and power of our movement, similar to how activists mobilized online for the global economic justice movement in the preceding years (Almeida & Lichbach, 2003).

That same summer, we began to imagine what we might do when the rest of us graduated from Middlebury the next year. At first, we strategized plans to move to Montana to contribute to local organizing against coal expansion. We started to talk to some of our Energy Action Coalition friends, most notably Shadia Fayne Wood, about our ideas. She gently suggested that there might already be work happening on this coal boom, particularly from the Indigenous communities already located in coal country, and we might ask them what they were already doing and if they wanted our help.

This sort of question was one we would encounter many times in the course of our organizing. But another force was moving, too. Bill McKibben had been so inspired by the Vermont Walk and saw real potential in more projects like it. He approached us at a potluck and asked us if we were seriously planning this move to Billings, Montana. He wanted to know if we would we be interested in continuing to work together on more demonstrations together. We had to make a choice. One night we all sat together in Jamie's dorm room at the Weybridge House, Middlebury's environmental coop, where many of us lived. Was it going to be Bill or Billings? We weighed the pros and cons and ultimately decided on Bill. No knocks on Billings—but each of us can say that for the many things we have to thank Bill for, one of the first was helping us avoid moving there to an uncertain future.

Once we had made that decision, the next step was clear: a nationwide version of the Vermont actions. We called it Step It Up, and the message was directed at every Member of Congress and US Senator: you need to Step It Up and cut carbon 80% by 2050. In early 2007, we put out a call—again, on Grist[8]—and over the course of a winter and spring, thousands of people put their hands up to organize marches in every US state—1,400 in total. There was a "sea of people" dressed in blue, flooding lower Manhattan; skiers placing banners on California's threatened glaciers; a traditional "Second Line" in New Orleans featuring then presidential hopeful John Edwards; and so many more. All the big environmental groups, including the Sierra Club, the Natural Resources Defense Council, and the League of Conservation Voters, helped us out and emailed their memberships inviting them to a local event. We were later told by veteran climate policymakers that Step It Up swept the movement up with energy at a time it was sorely needed—right around the demise of the first attempt at federal climate legislation (the Waxman Markey bill). We brought 14 of the top organizers to the Blue Mountain Center, a writers' and activists retreat center run by movement fairy godmother Harriet Barlow, and wrote a book based on the experience.[9]

The remainder of 2007 continued as busily as it had begun. We co-organized a project with the Sierra Student Coalition ahead of the New Hampshire presidential primaries that we dubbed "Climate Summer." We organized a second round of Step It Up actions in the autumn also geared toward raising the climate agenda ahead of national elections. We participated in the legendary PowerShift conference organized by Energy Action Coalition and played a role in the formation of 1Sky, a nationwide climate policy platform (which we would merge with in 2011). But a part of our attention—particularly Bill's—had started to shift beyond the United States as well.

Gradually, governments around the world were starting to gear up for what was intended to be a momentous climate summit in late 2009 in Copenhagen. Yet the international climate discourse felt lacking in urgency for the crisis at hand. We began to wonder what we could do to help change that. With the last remaining budget we had from Step It Up, we traveled to the 2007 climate negotiations in Bali, Indonesia (COP 13). There we largely spent time learning about the United Nations climate processes and tried to nudge and push governments where we could. Most importantly, we met other youth and groups from around the world with similar ambitions to our own. We shared the same passion as people who would be living with the long-term consequences of the negotiations.

In Bali we received a message from Bill, who was back home doing his own research on how we could try and make a difference. He had just heard from James Hansen about an upcoming paper identifying 350 parts per million (ppm) as the safe upper limit of carbon in the atmosphere.[10] Until that point, 450 ppm was widely being discussed as the basis for setting ambitious, science-based targets for climate action, and at that time the level was 384. The weight of that new data point was an emotional blow that perhaps only activists deeply immersed in a cause might feel. We understood the significance and how far governments were from doing enough. Thankfully our determination outmatched our despair.

When we regrouped back in the United States, we concluded that a global version of what we attempted with Step It Up was at least worth a try. Bill had the creative insight. He asked, why not use the number itself, 350, as both the message and the name of the effort? He queried a friend at an ad agency to draw up a mock logo with a 350 and an arrow to its left, indicating that 350 was the upper limit.

By June of 2008 we launched 350.org with the claim that 350 "is the most important number in the world" and embarked on a kind of grassroots public relations campaign to make it the most well-known number as well. We released a 90-second animated video with no words as our translation-free way of introducing the science and the plan to the world—and it went viral by the standards of its time.[11]

In late 2008 it was time for another round of UN climate talks in Poznan, Poland (COP 14). There we teamed up with many of the same international youth we had met in Bali, this time with a clearer agenda. The camaraderie and connection we experienced with those other young people felt akin to our SNG days, and much of the 350's early progress was a testament to how much we aligned with other youth and how our efforts amplified one another. We still felt like we were pushing up against enormous odds with our 350 goal. But on the final day of the talks, when Al Gore got on stage and gave a speech mentioning 350,[12] we stood up and cheered (and tweeted!), feeling like momentum was on our side.

Following Poznan, we launched the call for a global day of action for October 24, 2009, 6 weeks ahead of the summit in Copenhagen. We had given ourselves more lead time (we were trying to organize the globe), and we spent the majority of that year traveling to meet partners, running workshops with youth interested in joining the effort, attending UN intersessional meetings, and staging 350 stunts. One of the real keys to this period was the continuation of our college-organizing approach—anyone could be a leader, and the more creative, the better. That ethos, combined with the simple fact that until this point there had just been very few efforts to mobilize people for climate action beyond individual lifestyle choices, meant that there was an untapped audience of people eager to participate, or even take a chance at local organizing.

We also recognized, in part inspired by Paul Hawken's insights from *Blessed Unrest* (Hawken, 2007), which described the enormous emergent movement dedicated to saving our shared planet, that the world was full of people taking all kinds of action related to the climate. The vast majority, however, was largely invisible. Part of our aspiration for the 350 international day of climate action was in part for it to be a coming-out party for those existing efforts that we admired—making the invisible visible. We had seen through our experience in the United States how movement building was possible, and it required *movement*—actually getting into public spaces, moving our bodies, joining with others, and capturing images to share the story of our efforts.

The simplicity of our plan combined with both the urgency of the climate crisis and the joy of taking action resonated with our targeted audiences. Gradually many major environmental and other international campaigning organizations joined the effort, including Avaaz, Friends of the Earth, Greenpeace, the World Council of Churches, and contemporary climate campaigns like tcktcktck. When the big day finally rolled around, there were 5,400 events in 188 countries making news headlines all around the world.

There were actions of all shapes and sizes. One of the largest took place in Addis Ababa, Ethiopia, led by two young women who had attended a 350 training. Then President Mohammed Nasheed of the Maldives organized an underwater cabinet meeting, complete with SCUBA gear, pointing out the risks posed to his island. One action that really moved us all was led by a young woman named Ola, standing by herself with a 350 sign in Iraq, during the height of the US-led war in the country. These examples evoke how that day felt: iconic, heartfelt, creative, courageous, contagious. We projected images on the TV screens in Times Square, where we held a press conference about the impact of the day with its historic scale, and how we believed it raised the stakes for Copenhagen.

We arrived in Copenhagen with our entire 350 team at that time—dozens of us from around the world, sleeping in tiny bunk beds but excited to be part of history in the making. Except that the history that was made wasn't what we hoped for—instead, Copenhagen will be remembered as an epic failure of geopolitics and diplomacy. Despite being dubbed "Hopenhagen," global governments could not agree on binding emissions reductions targets nor the financing required for developing economies to meet them. The rallying cry of "climate justice now" penetrated the movement, because its absence was deeply felt.

Our crew didn't have any plans following Copenhagen—we had expected success! After all, that's what our studies of social movements made change look like. It was sobering and also important for us to learn that you can work hard, have thousands of people with you, and still fail. It was humbling, to say the least, since every project we had tackled up to that point led to continued momentum. The entire climate movement felt deflated following the losses in Copenhagen.

Post-Copenhagen

We each took a bit of a break, and then we regrouped in Vermont for a mid-winter retreat. There was a lot to reflect on after the last few years of whirlwind activity, and big questions looming about what was next. Perhaps it was the Vermont setting, or perhaps it was being back in one another's company after being spread across the globe the year before, but during those days we reconnected with our movement-building visions and roots.

We were able to process the collapse of Copenhagen from the perspective of Bill Moyer's "perception of failure" phase that most social movements experience. We understood then that the movement would need to carry on, and the loose global network of people that had emerged from our 350 efforts thus far was too valuable to walk away from. It was then, for the first time, that we began considering how to take 350.org forward (though the idea that it would still be going more than a decade later still hadn't crossed our minds) and how as a movement we would be able to build sufficient power to address the political impasse.

Witnessing the failure of governments to meet the challenge in Copenhagen and the rise in despair across the movement that followed, our next plan was to double-down on the hopeful, creative leadership of the movement itself. We devised yet another global day of action, this time a party, celebrating the power of community solutions. We called it the Global Work Party and held it on October 10, 2010 (10/10/10). After that we tried planetary-scale art, making aerial images photographed by satellites. It was at once fun and also challenging and uncertain. How would any of this add up to the impact we needed? And how were we to keep this operation going after some years now of improvised, ad hoc organization?

The answers came in three parallel parts—and each took considerable time with countless learnings along the way: sharpening our strategy, building political power for impact, and building an institution.

Sharpening Strategy

Sharpening our strategy took place shortly after the Vermont retreat and the Global Work Party. The political context on climate had just shifted with the election of Barack Obama, and, around that same time, the failure of cap-and-trade legislation in the US Congress. Our experience in Copenhagen made the (negative) role of the United States in the international sphere painfully obvious—and so we set out to think about how to revive movement energy closer to home, alongside our efforts to continue strengthening the global network we had helped build.

It was at that time that the campaign we are perhaps best known for came into being—the fight over the Keystone XL pipeline. This pipeline was intended to unlock the potential of tar sands oil in Canada—some of the most polluting oil on the planet—to be shipped to international markets. There had been opposition from the start, by Indigenous peoples in the United States and Canada and farmers and ranchers along the route in the United States. They had been steadily building a case and wider support to end this project. We took note of the powerful alliance (which would later be dubbed "The Cowboy and Indian Alliance," or CIA, if you can believe it), the climate impacts of the project, and crucially, the fact that it required approval from then Secretary of State Hillary Clinton. Our "theory of change" was that if such a project required executive approval, we could reach President Obama and force him to make a choice in favor of the climate and against Big Oil.

Again, history was our guide. We began to learn about civil disobedience, knowing how vital it was to movements throughout history (Morris, 1984), but relatively less common in the climate movement. We put out a call for the "Tar Sands Action" and invited people to come to the White House for 2 weeks of sit-ins to protest the pipeline. Like in the civil rights movement, everyone was asked to "wear their Sunday best," and their Obama campaign buttons—trying to impress upon people the idea that this was a friendly protest aimed at a leader we had all voted for.

All of the 350 cofounders, alongside hundreds of other people, were arrested in this action in 2011, and it was a defining time in all of our lives. We garnered the national-level attention for the climate issue that had always been our goal. Surprised by the success of the first action, we quickly organized a follow up—where thousands of people encircled the White House a few months later. By that time we were actually meeting with members of the Obama team, and beginning to see that we might ultimately succeed through our contacts with government officials. But just as soon as we would get one piece of good news—there would be a setback. Each time a court ruled against the pipeline, Transcanada (now TC Partners) would apply for a new permit. We were reeling from the sheer power of the fossil fuel industry. This led us to our next major campaign: fossil fuel divestment, active from 2012 to this day. We knew we had to find a way to tackle the industry's power—its money and social license—head on. Even though the Keystone battle wouldn't be over until President Biden canceled it, on his first day in office in 2021, we knew we had to go bigger.

The fossil fuel divestment campaign has been written about more than any of our other projects—and it succeeded beyond our wildest dreams and continues to this day.[13] Part of its success was its simplicity: getting across the idea that if it's wrong to cause climate change, it's wrong to profit from it. With that idea in mind—a distributed campaigning model which we staffed helped to touch thousands of institutions all at once to divest their endowments from fossil fuels. Trillions of assets under management are now "fossil-free" as a result. Not only that, it helped unleash a wave of campaigns on climate finance around the world, and through it trained hundreds of activists (some of whom went on to run for elected office, start the Sunrise Movement, and other major victories). It's a campaigner's dream, because in many ways the campaign now runs itself.

Building Power

The last major juncture in our evolution was about learning how to build political power. We had learned how to grab headlines, raise money, and build coalitions—but how did we know if the work we were doing was creating power for the climate movement? Through that line of questioning we were able to embrace our roots in student organizing and focus on identifying, supporting, and connecting climate leaders all over the world. The first great attempt at this was called Global Power Shift (GPS), which took place in Turkey against the backdrop of the Gezi Park uprising in Istanbul in 2013. We brought together 500 young people from 135 countries for a week of training in climate policy, "artivism" (arts-based activism), mobilizing, storytelling, and more. For many participants it was the first time they were face to face with people who, like them, were organizing locally to fight the climate crisis. They were able to experience that thrill we'd first had in Montreal in 2006—that feeling that your efforts add up because it is more than just you and your group—it's everywhere. GPS led to the creation of over 500 volunteer groups that continue to this day and which form the backbone of 350's global organizing network.

Between 2014 and 2015, we began to combine all these threads helping lead some of the first truly large-scale mass climate mobilizations: the People's Climate March and the Global Climate March. During these moments, we could showcase the true scale (and potential voters!) that put the climate movement's power on display. These protests led up to the Paris Climate negotiations, which proved far more momentous than Copenhagen. The Paris Agreement kept alive the promise of limiting warming to 1.5°C—among other things—and it also cemented the idea that the climate movement could help deliver ambitious policy change. This time around we were more prepared with efforts to sustain the movement beyond Paris. In early 2016, we mobilized again with distributed mass actions directly targeting fossil fuel infrastructure with "Break Free." And by this stage, it was more than clear that the climate justice movement had grown far beyond the efforts of any individual group or coalition. Increasingly our internal conversations shifted toward how we could best contribute to that greater whole, rather than solely designing "the next big moment." And yet just a few years later those conversations converged, and we got to team up with Fridays For Future for the biggest moment to date—the Global Climate Strikes of 2019, which saw 7.6 million people in the streets demanding climate justice.

Building an Institution

Our story would be incomplete without describing the platform upon which all this action took place—350.org itself. By that we mostly mean the staff, which as of this writing is at 150 people, working in 32 countries, and grew from the first 7 cofounders back in 2007. We have numerous sister 350 entities in other countries and a global governing Board of Directors. The organization is the backbone which helps make sure people get paid, have a laptop to attend Zoom calls, a number to call if they're arrested at an action, and a team of people to strategize with. The unglamorous work of running an organization is nevertheless critical to all of our efforts, and hundreds of brilliant people have walked through our virtual doors and made 350 what it is today.

COVID-19 and Recent Campaigns

When the world ground to a halt during the COVID-19 pandemic, we were impacted alongside everyone else, but to a lesser extent. This has a great deal to do with having been a distributed network-based organization from day one. As the whole world turned to remote work and Zoom, we ran virtual trainings about the systems we'd relied on for years. [14] In some ways the COVID-19 pandemic brought our team closer together, because everyone around the world was experiencing it together at the same time. During that same period, historic mobilizations for racial justice following the murder of George Floyd once again underscored the necessity for our organizing base to understand the link between climate justice and racial justice.

Conclusion

There has been a lot written about campus activism, much of it characterized by the same clichés. The late-night debates in smoke-filled dorm rooms, the ranting students with no respect for decency or decorum, the pointless protests over issues too esoteric to really matter. In our experience, both as former student activists ourselves and having continued to work with young people to this day, there's something far more serious going on. Activism has always required a certain level of naivete. That's because to change the world, you first have to be enough of a dreamer to try. At their best, campuses provide a venue where students can make that attempt, gather others with them, and continually learn from both losses and wins.

If it worked across Middlebury, might it work across Vermont? If across Vermont, might it work across the country? If we translated all the materials, might it work across the world? The animating impulse of those early days was always about scale. That proved to be one of our greatest strengths and what we were most known for externally—our big ideas and willingness to try everything to pull them off. The inspiration and vision that has marked our work is a vital chapter in the story of climate action. More than anything, we believe we contributed to the now undeniable sense that climate change is a crisis demanding action, and that there are countless solutions available, if they are enacted in time and with attention to care and to justice.

It also led us down quite a lot of paths where we had to be shown, sometimes willingly, sometimes not, that quantity and quality aren't the same; that in some cases our recruitment of new activists undermined vital work that was already taking place; that sometimes going more slowly can create lasting change; and that our blind spots led us to privileging other activists who looked just like us: White, straight, and upper middle class.

Over the years we began to meet and build relationships with a much wider movement; much more diverse, with different theories of change, different tools to bring to the table, and long histories and lineages of organizing that were new to us. These interactions were often conflictual: It can be hard to be confronted with your own shortcomings by people who share your cause! But some of our best teachers helped us to understand a true synthesis—they welcomed our ambition, drive to fight climate change, and ability to bring vast numbers of people together. That, combined with a worldview oriented more toward equity, political power building, and the lessons of history, was a truly powerful combination. It was those forces coming together which enabled the power of that first Peoples Climate March; it will be those forces reinventing themselves in an age of rising authoritarianism that will determine if our dream of a just world, safe from the worst of climate change is possible. Our experiences have taught us that it is.

Notes

1. https://en.wikipedia.org/wiki/Forward_on_Climate
2. https://obamawhitehouse.archives.gov/the-press-office/2014/09/23/remarks-president-un-climate-change-summit
3. Gelbspan (1997).
4. Moyer (1987).
5. https://www.washingtonpost.com/lifestyle/magazine/it-was-not-arguing-it-was-fighting-bill-mckibben-on-efforts-to-save-the-planet/2019/02/08/fadad63e-1849-11e9-88fe-f9f77a3bcb6c_story.html
6. https://grist.org/climate-energy/mckibben2/
7. https://www.sevendaysvt.com/news/walk-of-ages-how-a-vermont-march-helped-launch-a-climate-movement-3633748
8. https://grist.org/climate-energy/stepitup1/
9. McKibben & the Step It Up Team (2007).
10. Hansen et al. (2008).
11. https://www.youtube.com/watch?v=LFlN1QTV47I
12. https://grist.org/politics/gore-to-un-350-or-bust/
13. https://www.vox.com/2015/11/8/9690654/keystone-climate-activism
14. https://trainings.350.org/

References

Almeida, P., & Lichbach, M. (2003). To the Internet, from the Internet: Comparative media coverage of transnational protests. *Mobilization: An International Quarterly, 8*(3), 249–272.

Gelbspan, R. (1997). *The heat is on.* Hachette.

Hansen, J., Sato, M., Kharecha, P., Beerling, D., Berner, R., Masson-Delmotte, V., Pagani, M., Raymo, M., Royer, D. L., & Zachos, J. C. (2008). Target atmospheric CO_2: Where should humanity aim? *Open Atmospheric Science Journal*, 2(1), 217–231.

Hawken, P. (2007). *Blessed unrest: How the largest social movement in history is restoring grace, justice, and beauty to the world.* Penguin.
McKibben, B. (1989). *The end of nature.* Random House. Trade Paperbacks.
McKibben, B., & the Step It Up Team. (2007). *Fight global warming now.* Holt.
Morris, A. D. (1984). *The origins of the civil rights movement.* Simon and Schuster.
Moyer, B. (1987). *Doing democracy.* https://activisthandbook.org/resources/doing-democracy

CHAPTER 14

Extinction Rebellion

The Emergence, Evolution, and Character of a Transnational Environmental Movement

TIAGO CARVALHO AND PETER GARDNER

Introduction

Active in more than 80 countries across all seven continents and with more than 1,000 local-level chapters at the time of writing (Gardner et al., 2022; XR, 2024a), Extinction Rebellion (XR) represents one of the most important transnational social movements of recent years. Since its emergence in 2018, it has had a profound impact across the world, engaging a wide variety of governments, corporations, media, and publics in questions relating to climate change and related ecological crises. Although it is best known for its use of dramatic and disruptive tactics, XR has utilized the full gamut of nonviolent contentious activities and repertoires. In the wake of XR's activities, climate emergencies have been declared by various local and national governments (including the United Kingdom, Ireland, France, Aotearoa New Zealand, and Canada), supranational confederations (such as the European Union), universities (e.g., the universities of Bristol, Bath, Plymouth, and Cardiff), and institutions (from wide-ranging businesses to the Catholic Church).

XR is part of a new wave of environmental movements that build on past movements while bringing new aspects to these mobilizations (Almeida, 2019; de Moor et al., 2021). From the start, the movement adopted a well-defined strategy which involved the development of an organizational brand that could be easily diffused across the world and drawing on past nonviolent repertoires to call on the state—and others—to take urgent action on the climate and ecological emergency. At the same time, their particular approach has been a marked departure from the kinds of strategies employed by mainstream environmental movements since the 1990s (de Moor et al., 2021) through their deployment of notably disruptive tactics. The movement has also acted as a stimulus for the formation of new, similar groups centered on addressing climate change. Overall, while XR can be viewed as a part of a longer history of nonviolent civil disobedience groups—environmental and otherwise—it also constitutes a "step change" in the arena of action (Almeida, 2019; Chase-Dunn & Almeida, 2020; Farrer et al., 2023).

XR began in the United Kingdom with three demands. The first, "Tell the Truth," calls on governments and the media to accurately communicate to the public the "extreme cascading risks humanity now faces" as a result of climate change, as well as "the injustice this represents, its historic roots, and the urgent need for rapid political, social and economic change" (XR, n.d.). The second, "Act Now," demands that net zero be achieved by 2025. To do so, XR saw a need for the "whole of society" to "move into a new precautionary paradigm, where life is sacred and all are in service to ensuring its future." The third demand, "Decide Together," petitioned the government to establish a citizens' assembly, allowing key decision making on climate change–related policy to be made through direct rather than representative democracy. Meanwhile, the movement in the United States added a fourth demand, calling for a transition that prioritized racial and Indigenous justice—a demand that has been adopted by a range of local chapters of the movement in the United Kingdom and elsewhere.

This chapter provides a comprehensive overview of XR from 2018 to 2024. Drawing on existing social movement studies research and our own research,[1] we cover XR's history; strategy; goals; discourses; organizational structure; transnational diffusion; issues of race, class, gender, and age; and a brief overview of movement spin-offs. In doing so, we critically assess, in particular, what kind of a social movement organization XR represents in relation to its decentralizing and centralizing tendencies. Alongside its prefigurative, horizontalist, and decentralized features, the literature also speaks of XR as entailing vanguardist elements (Weaver, 2024) with a hierarchical organizational structure that occasionally prioritizes efficiency over democracy (Berglund & Schmidt, 2020). Overall, we contend that while XR exhibits certain centralizing characteristics, such depictions need to be tempered by recognition of its polycentric and consensus-based aspects, especially in light of its transnational character.

The Emergence of XR

In 2016, a small set of activists connected online, coming together to form a movement called "Rising Up!". Comprising a range of seasoned activists, including Roger Hallam, Gail Bradbrook, and Simon Bramwell, among others, Rising Up! engaged in a variety of campaigns on climate change, most of which failed to garner traction or media attention (Melia, 2021; XRUK, 2024c). However, between 2016 and 2018, a calm period for the international climate movement after the 2015 Paris agreement, the group successfully built small-scale coalitions with key activist networks and established a set of core guiding principles and values, both of which subsequently emerged as fundamental to XR's operational framework (Melia, 2021). Rather than being established as a movement in its own right, "Extinction Rebellion" was initiated as another Rising Up! campaign. As Melia (2021, pp. 168–169) explains:

> In December 2017 Hallam wrote a short strategy paper entitled "Pivoting to the Real Issue," drawing on Rising Up!'s recent experience and on his reading of research evidence. It defined climate change as "the greatest mobilization story of all time." Unlike other threats this one was existential—it threatened the elites as well as the people, but campaigners, politicians and the wider public were stuck in a "death spiral of pragmatism." The conventional wisdom said that "if you tell people the truth about climate change, they will switch off," which might be true for most people, but "1 to 5 per cent will be shitting themselves over it and this will drive them to action." … For this new campaign Rising Up! should ask people to commit to break the law and, if necessary, go to prison.... Hallam's paper was discussed for the first time at a meeting in January 2018. There was much support for the plan, but Bradbrook and some others felt more preparatory work was needed before embarking on anything so ambitious.... Despite the misgivings of some members, and an argument over the name, they decided to go ahead with XR as a campaign—not an organisation.

Hence, XR rose to prominence as Rising Up!'s most successful campaign. Once it began to gain exponential momentum from late 2018, the campaign became the movement on the fly, incorporating the groundwork undertaken over the preceding 2 years.

Although both Rising Up! and the XR campaign began in England, its genesis transcended national borders, with pivotal aspects of its framing and objectives being experimented with beyond the UK context. Through the 2010s an international network of activists and psychologists—based mainly in Australia and the United States—had been advocating action to persuade governments to declare a climate emergency. This loose assortment of actors, networks, and movements had begun to challenge the approach taken at the time, as outlined by one Australian environmental activist (and later cofounder of XR Australia)[2]:

> [During his time as US President, Obama's messaging had been,] "Don't worry about the science, just jobs and growth: jobs, jobs, jobs, growing jobs. Don't mention the science, don't do anything to make people afraid," right? So that spread through the whole climate movement in America and then next thing, we had caught the flu over here [in Australia] as well. Even more than before, all the NGOs were saying "fear doesn't work," "you've just got to talk about renewables," "lots of renewables, lots of jobs." It was driving me crazy. So, in 2017, ... I spoke at this festival ... called *Don't Mention the Emergency*. Sometime after that, I got a call from Gail Bradbrook and she said, "I'm setting up a climate group in the UK, can we use your talk?" I said, "Sure, sure." I was thinking it was going to be like one of those grassroots groups. Anyway, we had some chat afterwards and, of course, it was turning into Extinction Rebellion.

As this quote illustrates, Rising Up!—and subsequently XR—was not merely influenced by the broad array of antecedent environmental and other movements across the world but also was consciously engaged with and actively incorporated such influences.

In early 2018, Rising Up! began to undertake a series of talks throughout the United Kingdom to mobilize for the XR campaign, initially only managing to garner support from "a few anarchists, vegan activists or anti-capitalists," with limited recruitment of local nonactivists (Melia, 2021, p. 170). In mid-October, the campaign was mentioned in an article by George Monbiot in the British newspaper *The Guardian* and shared on Twitter the following day by former US presidential candidate Bernie Sanders, affording the campaign considerable publicity and mobilizing "over a thousand people" for the first public protest event (Melia, 2021, p. 170). At the end of October 2018, around 1,000 people congregated outside the UK Houses of Parliament in London. A "Declaration of Rebellion" was read aloud. It stated that the "ecological crises that are impacting upon this nation, and indeed this planet and its wildlife can no longer be ignored, denied nor go unanswered by any beings of sound rational thought, ethical conscience, moral concern, or spiritual belief," that climate change will, "unless immediately addressed, ... catapult us further into the destruction of all we hold dear," and that they "refuse to bequeath a dying planet to future generations by failing to act now" (XR, 2018). After declaring themselves "in rebellion against our Government and the corrupted, inept institutions that threaten our future" (XR, 2018), they blockaded a major road in the city, ending in 15 arrests (Taylor, 2018).

In April 2019, XR undertook its first Rebellion, occupying core areas of central London for 2 weeks and employing a variety of disruptive, nonviolent actions to shut down the city, including the blockading of several of the bridges that form vital thoroughfares through the city (Berglund & Schmidt, 2020). This time, in addition to thousands more participating in the streets of London, the event was transnational, with simultaneous Rebellions taking place in at least 34 other countries across the globe (Gardner et al., 2022). A significant portion of participants in the protest were new to activism, with "nearly three-quarters" of the cohort in London previously "having participated in ten demonstrations or fewer" (Hayes et al., 2020, p. 3). Due to its sheer scale, widespread media attention, and transnational character, the Rebellion in April 2019 marked a pivotal moment in the formation of the movement. And yet the feat was repeated and extended in October 2019, drawing even larger numbers to the streets and taking place in more than 60 cities worldwide (XR, 2019b).

Around the same time that the first XR protest event was being planned in the United Kingdom, Fridays For Future (FFF) was emerging in Sweden, with Greta Thunberg beginning to engage in school strikes in the latter half of 2018. In early 2019, FFF gained momentum, with Friday schools strikes spreading across the world. Research has fruitfully compared the discourses, strategies, and approaches of the two groups, demonstrating overlaps and divergences between them (see Buzogány & Scherhaufer, 2022; Clot-Garrell, 2024; Rödder & Pavenstädt, 2023; Rowe & Ormond, 2023). However, the relationship between the two movements remains underexamined. From our own research, it would seem that this differs considerably according to context, ranging from being entirely distinct and unrelated movements through forms of collaboration and mutual support to being effectively a single entity (for an example of the latter, see the case of XR Gambia: Gardner et al., 2024). Although the overlaps in goals and demands between the two movements would render their coalition potentially productive, evidence of the precise role each played in the others' development through 2018 and 2019 is scant across the literature.

Sailing Boats on City Roads: XR's Strategy and Protest Methods

Looking across the many national contexts in which XR is active, one can see that the movement has used a plethora of strategies, repertoires, and goals. Best known for its disruptive and dramatic protest events, there is a tendency in the literature to reduce the movement to the form it has taken in the United Kingdom, Western Europe, Aotearoa New Zealand, Australia, and North America, and in so doing overlooking the rich diversity and plurality of XR as a global movement. In reality, its approaches, targets, and aims have included the full spectrum from massive city blockades calling for system change to leafleting projects aiming to educate publics (see Gardner et al., 2023). Nevertheless, it was XR UK's highly dramatic approach in 2018–2019 that drove public awareness of the movement, as well as its meteoric rise and diffusion (Gardner et al., 2022).

XR UK has broadly framed their action in terms of "nonviolent direct action" (NVDA). Drawing on Gandhi and Luther-King Jr's theory and practice, as well as in other past movements, they prioritized "mass civil disobedience, directed primarily at national governments and located mainly in national capitals," as the initial objective was to move beyond "small-scale direction action to environmentally damaging projects" to reach beyond left-wing environmentalism (Doherty & Hayes, 2022). When compared with previous environmental movements, an important characteristic of the new wave that emerged since 2018 "appears to be the use of a more politically 'neutral' framing of climate change that is directed more strongly at state than non-state actors" (De Moor et al., 2021, pp. 619–620).

XR's strategy of NVDA should be understood as a strategic choice aligned with its overall goals. This approach contrasts with less disruptive repertoires (e.g., protest marches, petitions, and letter writing) and more subversive approaches such as sabotage (Berglund, 2023; Malm, 2021; Scheuerman, 2022). In the XR mobilization toolbox, NVDA refers to occupations, blockages, lock-ons, banner drops, unauthorized demonstrations, and other actions that can consciously result in the mass arrest of the participants and, in so doing, applying pressure to the state (alongside, at times, other organizations and institutions).

Despite being branded as antidemocratic by some media commentators, XR's strategy has been explicitly within the confines of liberal democracy, calling for reform of—rather than overthrow or even fundamental restructuring of—the state (Berglund & Schmidt, 2020). Even in contexts where its activists engage in lawbreaking while undertaking NVDA, the movement has been remarkably dedicated to the ethic of nonviolence. Given XR's justification of such actions as acting to prevent a future societal collapse—including a crumbling of the rule of law and the replacement of democracy for authoritarianism—these more disruptive repertoires ought to be understood as nonetheless grounded in a "fidelity to law" in a broader sense (Berglund, 2023; Berglund & Schmidt, 2020; Extinction Rebellion, 2019). As such, "[c]ivil disobedience challenges legitimacy by being a communicative act," as it clearly takes advantage of the law to communicate that they are willing to accept the "punishment and using this punishment as a communicative platform" (Berglund & Schmidt, 2020, p. 30). Berglund (2023) posits such an approach as a form of prefigurative legitimacy, with the movement drawing on both liberal and anarchist traditions in their action. By putting themselves on the line, their objective is to gain legitimacy and to be seen to stand for and communicate what they believe.

Such an approach differs from those of the Global Justice movement and the Zapatista model (Berglund & Schmidt, 2020; De Moor et al., 2021) in relation to the transformation of the state, with these movements situating much of their action outside of the state, pointing to a bottom-up and grassroots alternative. Through its use of NVDA, XR has engaged directly with the state in ways that recognize and legitimize its authority. A paradox thus emerges, as their rejection of electoral politics conflicts with the need to work with current political forces to speak the truth, take immediate action, create Citizens' Assemblies (CAs), and to implement their recommendations (Falanga & Carvalho, forthcoming).

XR's action, strategy, and goals for NVDA are grounded in an interpretation of conclusions presented in the work of Chenoweth and Stephan (2011), specifically the claim that at a mobilization of 3.5% or more of the population in a given polity inevitably leads to political change. However, the movement has also been critiqued for its oversimplified application of Chenoweth and Stephan's work, particularly for the conceptualization of social change as obeying natural science–like laws and for the transplanting of analysis undertaken in nondemocratic settings into democratic ones (Berglund & Schmidt, 2020; Matthews, 2020; Scheuerman, 2022).

XR's repertoires have tended to be highly performative and theatrical (Coombs, 2020; Grenfell, 2023; Johnston & Holland, 2023), involving "vivid graphics and branding, and ... performative protests [that] have blocked traffic, enraged onlookers, and given a new sense [to] the power of movement, corporeality, and the colour red" (Coombs, 2020, p. 125). The desire to create a "fun atmosphere" is underpinned by an understanding that "most people respond to what is cultural and celebratory rather than political and solemn" (Hallam, 2019, p. 4). As Johnston and Holland (2023) demonstrate, XR crafted a public approach geared toward eye-catching displays that disrupt daily life, using attention-grabbing interventions to raise public awareness, prompt quick government action on environmental concerns, and foster a temporary feeling of solidarity among those involved. Overall, the approach has involved protest method innovations that challenge established norms and call for urgent measures to address climate change.

XR regularly uses symbolic objects in its protest actions, such as multicolored flags carrying the group's logo, coffins, giant octopuses, and "zombie koalas" (Sharp-Wiggins, 2022), and the much-documented use of sailing boats in their inner-city blockades (XR, 2020b). These objects serve multifaceted roles, as they are capable of making declarations and conveying narratives (Gardner & Abrams, 2023), raising awareness, mobilizing activists, expressing affect, and mitigating against activist exhaustion (Stammen & Meissner, 2024). Moreover, visual bodily practices such as the Red Brigade and public die-ins render the body itself as a symbolic object in protest, allowing it to transcend conventional argumentation and engage in "non-verbal levels of political action and argumentation" (Malafaia et al., 2024, p. 11).

The use of "disruptive" protest tactics have been broadly critiqued in the media, with the group routinely accused of precipitating a "backfire" effect, with the public turned against climate policy as a result (Scheuch et al., 2024). However, there is evidence to suggest that disruptive forms of climate protest have not alienated the public from pro-environmental behaviors, beliefs, or policy preferences: to the contrary, they appear to be associated with reductions in rates of public opposition to such measures (Kountouris & Williams, 2023). Furthermore, as Young and Thomas-Walters (2024) have cogently argued, garnering public support is at best a weak metric with which to analyze a social movement's success. "More important than whether or not a tactic wins over the majority of the public is whether it imposes disruption on elite decisionmakers" (Young & Thomas-Walters, 2024, p. 2). Nevertheless, further research is needed to investigate which kinds of disruptive protest have yielded the most productive impacts on policymaker decision making.

As indicated at the start of this section, XR operates in diverse contexts globally. However, the academic literature has been slow to analyze XR in the Global South. Taking a single line of latitude, little is known of the repertoires used by the movement in countries such as Costa Rica, Venezuela, Sierra Leone, Burkina Faso, Ghana, Sudan, Sri Lanka, Thailand, or the Philippines. Nevertheless, across the many contexts in which XR is active, it would appear that a wide variety of approaches have been adopted, with activists in nondemocratic settings and regions where protest is generally less common tending to avoid the use of dramatic antigovernment discourses or the tactics of mass arrest and large-scale city blockades (Gardner et al., 2023, 2024).

"People gonna rise like the water": XR's Framing and Discourse

Having outlined XR's approach to protest, we now turn to the movement's framing and discourse: the outlook and language used by the movement to describe the world as they see it, what they believe needs to change, and how they plan to change it. Scholarship has sought to analyze XR's framing and discourse using various methods, including interviews with activists in a local XR chapter (Smiles & Edwards, 2021), social media accounts, press releases, and online material produced by the movement (Buzogany & Scherhaufer, 2022), songs and chants used in protest events (Pieri, 2024), and XR UK's podcast (Gardner & Carvalho, 2023). Overall, with a heavy emphasis on evidence from natural science, XR posits a future in which ecological and societal collapse is inevitable without immediate and radical action by policymakers and others in power. Despite the broader framing of being "beyond politics," the discourses of its activists range from calls for sweeping changes to the current economic and political system (Gardner & Carvalho, 2023) to those more "reminiscent of mainstream environmental policy makers" (Smiles & Edwards, 2021, p. 1445).

Rohden (2021, p. 8) reports that the movement has "almost 900 links to pages related with climate change and biodiversity loss, pointing to scientific evidence." As such, they act as amplifiers of such results to justify and legitimize their action, create awareness, and mobilize further.

In our analysis of the XR UK podcast (Gardner & Carvalho, 2023), one of their core ideas, and in a way a motto for action, centers on environmental catastrophe—both present and future. Differently from a mere crisis, catastrophe corresponds to a situation where there is no possibility of returning to the pre-catastrophe status quo. Following the scientific consensus, the movement sees the future climate collapse as the result of the intersection, crosscut, and reinforcing temporal, psychological, democratic, representational, media, and political economy crises. In XR's discourse, there is a clear depiction of the state of the world in the hands of an economic and political elite. Nonetheless, and to counter such a future, XR emphasizes that radical action offers an opportunity for societal renewal and transformation.

XR's Rebellion Songbook—a collection of songs and chants to be used in the movement's protest events—also articulates their diagnosis of the situation and potential solutions to the climate crisis (Pieri, 2024). These songs serve a dual purpose: They clearly express the movement's grievances and proposed solutions, such as the need for adherence to scientific recommendations and the establishment of climate assemblies. The music underscores the urgent message that unless climate change is halted, we will face an ecosystem collapse. However, by incorporating rebellion as a form of action within their lyrics, the songs evoke strong emotional responses, mobilizing support and fostering a sense of solidarity and urgency among listeners. This musical strategy reinforces XR's broad message and enhances the movement's impact by engaging audiences on both an intellectual and emotional level.

Giæver and Russell (2024), through interviews with XR members across different country contexts, investigated how activists in the movement perceive the future as foretold by science. Activist discourses were found to be fraught with negative emotions, such as anxiety, depression, and guilt, while also expressing a feeling of being "grateful to be alive." In terms of their activism, two main emotions emerge: (1) burnout as a result of the weight of responsibility this outlook holds and the challenges of balancing multiple roles in the movement, and (2) a feeling that they have no choice but to act, while describing their activism as providing energy and hope for change as a result of their actions. As such, XR members develop two emotional regulatory actions to counter it: healing to sustain action and taking action to heal. The first corresponds to self-care practices, within regenerative culture, to heal from the negative experiences of climate change activism. In the second XR, members point out that besides pressuring the government, taking action also helps them to "process and make sense of negative emotions around environmental issues" (Giæver & Russell, 2024, p. 9).

Aware of the potential emotional burden of climate activism from the start, XR developed various community-based strategies for emotional and cultural healing. To foster resilience for coping with the heavy emotional and psychological burden of the future, prevent activist burnout, and countering feelings of hopelessness (Stuart, 2020), XR developed a "regenerative culture" (or "regen"). Based on self-care, people care, and planet care, regen acts as a form of prefigurative practice that in turn informs the ideology, organization, and strategy of the movement as a whole (Berglund & Schmidt, 2020; Westwell & Bunting, 2020). Such activities involve "the creation of time and space for sharing and processing emotional experiences, and the building of community, e.g., through debriefing and the creation of ceremonies to hold difficult emotions" (Giæver & Russell, 2024, p. 5). Although not ubiquitously applied throughout the movement, regen has been a core element of XR's practices and discourses from its inception, offering both a clear framing of the climate crisis as an emotionally challenging and transformative event and a potential solution to the issue for its activists and the local communities in which they are embedded.

XR's discourses and framing are also reflected in its imagery. Their hourglass logo, for instance, reflects quite clearly this double aspect of the impending doom, with time fast running out, while at the same time infusing action with hope: It is only by engaging in direct action that catastrophic environmental damage can be prevented. As we mentioned earlier, such declarations of the nature of the threat faced are also present in the symbolic objects used in XR's protest events, with sailing boats in urban centers speaking of the destructive power of rising seas, funeral coffins raising the specter of death, and the skeletons of humans and other animals acting as "a stark reminder of our mortality and through humor and humility bring us closer to ourselves and the loss of life on Earth" (XR, 2019a). In this way, XR visually argues for actions that can avert climate catastrophe, especially those that disrupt existing power structures, and presents disruption through collective action as a necessary step toward addressing the climate emergency.

Organizational Structure: A Decentralized Movement With a Vanguardist Core?

XR uses an organizational structure that it calls the "Self-Organising System (SOS)," generally considered to be sociocratic, holacratic, or a combination of both (Berglund & Schmidt, 2020; Rowe & Ormond, 2023; Zantvoort, 2023).[3] According to XRUK's website, this system aims to be fully nonhierarchical, mitigating power as much as possible such that "no single person has power over another person or the system itself" (XRUK, 2024). Working groups, or "circles," focus on particular tasks (such as media and messaging, actions and logistics, or data analysis circles) and operate autonomously within the remit of their responsibilities. Decision making is built by consensus, with facilitators tasked with ensuring that all voices are heard while ensuring that meetings are relatively efficient. Attempting to be both democratic and agile, the bar for consensus is whether it is "safe enough to try" rather than the ideal outcome for all present.

The "chapter" is a cornerstone of XR's decentralized organizational structure. Chapters serve as local hubs, fostering mobilization, participation, training, and community building at the grassroots level and allowing for the emergence of highly localized strategies and repertoires (Fotaki & Foroughi, 2022). Being largely autonomous, XR chapters have responded to a diverse range of climate and ecological issues from the provincial to the global (Smiles & Edwards, 2021), integrated into the broader movement through alignment with its core demands, principles, and values and dedication to nonviolent ecological protest (Burgess & Read, 2020; Smyth & Walters, 2020). In addition to their organizational role, chapters cultivate a sense of imagined community, aiding in the formation of an "idea of Extinction Rebellion as a whole, despite individual activists having never met most others in the movement and despite actual experiences and local realities of being associated with the movement being rather diverse in practice" (Gardner et al., 2022, p. 427; James & Mack, 2020; Morgan, 2021). In this way, chapters of the movement across the world have been able to remain both globally connected—whether merely symbolically or in more concrete ways—and locally rooted (Berglund & Schmidt, 2020; Gardner et al., 2022).

In addition to local chapters, and inspired by anarchist organizing strategies, the movement encourages the formation of "affinity groups." Like chapters, affinity groups are autonomous groups within the movement, required only to agree to XR's principles and values. However, unlike chapters, affinity groups are rooted in a specific area of focus, protest event, or common interest. Two broad types of affinity groups exist within XR. First, there are those that represent the "smallest organizational unit of XR" (Westwell & Bunting, 2020, p. 548), containing around "eight to twelve people," united around "a shared vision of what they want to achieve" and focused on engaging in nonviolent direct action as a team (Berglund & Schmidt, 2020, p. 43). However, a second type of affinity group has emerged that is far larger and more diffuse, such as XR Buddhists (XR, 2020a) or Cycling Rebellion (Beckett, 2023).

In the United Kingdom, the movement is organized on local, regional/national, and UK-wide levels. At the local level, XRUK is made up of a large number of "chapters" that are spread extensively throughout the country. By June 2021, 420 local chapters had been formed across the United Kingdom (Gardner et al., 2022), but this figure fell to 230 by May 2024 (XR, 2024b). The regional level aims to foster connections among local chapters, organized into 11 groups: London, Midlands, North, North East and Cumbria, East of England, South East, South West, Bristol, Wales, Scotland, and Northern Ireland. A national-level circle ("The Hive") connects all circles and chapters of the movement within the United Kingdom and coordinates country-level actions ("rebellions") in the capital.

There is some notable diversity in the organizational structure of the movement across the world. Similar approaches to the United Kingdom exist in the Netherlands (Rowe & Ormond, 2023; Zantvoort, 2023), France (Brunon, 2021), and Poland (Smółka, 2023), albeit without the regional level of organization seen in the United Kingdom. However, in many contexts, either a single (national-level) chapter (e.g., Bulgaria, Sierra Leone, Zimbabwe, and Thailand) or a small number of chapters (e.g., Slovakia, Brazil, Pakistan, and Ethiopia) have been established (Gardner et al., 2022; XR, 2024a). In The Gambia, a single national chapter exists, coordinating various activities across the country from Banjul to the rural interior (Gardner et al., 2024). In Australia, the movement has tended to organize around the regional level, connecting local groups within specific states and territories, with a national level proper only emerging during the COVID-19 pandemic (Gardner et al., working paper). In the United States, factional divisions led to a collapse at the national level, resulting in an array of isolated local chapters and emergent regional groupings (such as the North East Region, centered on New York, Boston, and Washington, DC) becoming established as core supra-local levels of organization (Gardner & Müller, working paper).

The movement also has a supra-national level of organization. Founded in 2018 as the "International Support Team" and renamed "XR Global Support" in 2020, this circle was established to support the worldwide movement, supplying "resources, training, media communications, tech infrastructure, and funding" to local- and national-level groups across the countries where XR is active (Smółka, 2023; XR, 2024c). However, we were unable to find a single research study analyzing this feature of the movement.

XR's organizational structure has been critiqued for being both too democratic and less democratic than it claims. On the former, while being highly decentralized allows XR to engage in a "high frequency of actions," it nonetheless renders it "difficult to avoid controversial activities" (Buzogány & Scherhaufer, 2023, p. 365). A case in point is the widely condemned Canning Town underground railway station blockade that took place during the 2019 Rebellion in London (Fotaki & Foroughi, 2022; Mansfield, 2020; Seagrave, 2023).

On the latter, however, the "dispersion of authority to smaller groups has made XR['s] ... internal decision-making more opaque" (Berglund & Schmidt, 2020, p. 54). For Fotaki and Foroughi (2022, p. 238), the claim to leaderlessness and a fully nonhierarchical organizing structure is inherently flawed, allowing for "power [to be] practiced and simultaneously disavowed." The fantasy of leaderlessness at play in the movement has obscured the forms of power at work in the movement, preventing the development of "more egalitarian and participatory ways of sharing and distributing roles and responsibilities" (Fotaki & Foroughi, 2022, p. 241).

In a provocative profile piece in *Social Movement Studies*, Duncan Weaver (2024) argued that XR ought to be considered a green vanguardist movement. Weaver's argument centers on the affinities between vanguardism and forms of civil resistance, XR's framing of climate change as an all-encompassing existential threat, and the movement's rhetoric opposing imperialism (perceived as the highest form of capitalism) for its inherently ecocidal character. Nevertheless, vanguardism involves—in Weaver's (2024, pp. 113, 118, 114) words, "a cadre of revolutionaries leading from the front to mobilize mass-constituencies" and "a catalyst creating conditions for revolution," wherein a "cadre, the vanguard, musters around universal(izable) grievances" such that the "regime can only be overthrown if the masses in totality revolt." As outlined earlier, the emergence of XR in 2018 as a particularly ambitious campaign of Rising Up! certainly has vanguardist qualities, with internal contestation over Hallam's "Pivoting to the Real Issue" paper somewhat reminiscent of Bolshevik disagreement over Lenin's *April Thesis*, wherein heated debate over the extent to which Lenin's proposal for immediate revolutionary action was premature and inchoate took place among the party's leaders. Overall, however, we view the vanguardist characterization of XR as too much of an oversimplification of what is, in reality, a rather broad, multifaceted, and diffused movement containing significant democratic and decentralized qualities.

Building a Global Movement

In a context of increasing debate, XR, alongside other movements, gave a new impetus to transnational environmental mobilizations. Given the global character of the climate crisis, responses to it demand endeavors that go beyond the local or even national level. Not that this is new: Past environmental movements have also engaged in building global networks of action, with environmental advocacy seeing increasing transnationalization since the 1970s (Smith et al., 2018). In fact, the new environmentalism that emerged in the 1970s shows "very little evidence of global environmental protest action or of groups working effectively across borders" (Doherty & Doyle, 2006, p. 698), as organizations remained highly centralized, institutionalized, and based in Global North, in what has been described as a "limited transnationalisation" (Rootes, 2005).

The year of 2018 has been considered a turning point, bringing about a new phase of environmental action (Almeida, 2019) with an enlargement and reconfiguration of the environmental arena in line with a renewed internationalist impetus. Continuing this global path of action, XR explicitly strategized its development toward a transnational reach that would spread beyond the United Kingdom in a coordinated fashion. To do so, they adopted a simple, flexible, and open strategy for diffusion, minimizing barriers to for activists wishing to create their own XR groups in other countries and making stencils for the movement's branding publicly available for use elsewhere (Gardner et al., 2022). To establish a new chapter and become officially included in XR, activists needed only to accept the movement's three demands and ten principles and values. Such an approach was key in the movement's transnational diffusion.

Elsewhere, we have argued that XR's protest events may also have precipitated its transnationalization (Gardner et al., 2022). XR's diffusion was, we contend, significantly propelled by pivotal protest events, notably the Declaration of Rebellion in Autumn 2018 and the first International Rebellion in Spring 2019, catalyzing the establishment of new chapters worldwide and effectively broadening its geographical and operational scope. These eventful protests served as a beacon, rallying support and inspiring action across the globe and facilitating rapid expansion. While this finding was initially interpreted from quantitative data, this has also been borne out in subsequent interviews we have conducted with XR activists across the world, many of whom referred described these events as having inspired their involvement in the movement or the creation of local chapters. In the wake of the Declaration of Rebellion in Autumn 2018, XR transnationalized remarkably rapidly, establishing 1,165 chapters across 79 countries and six continents by June 2021 (Gardner et al., 2022).[4] Despite its widespread reach, a predominant concentration of XR chapters is observed in Western Europe and the Anglosphere, with 70.2% of chapters located in Europe, 12.5% in North America, 8.5% in Oceania, 3.3% in Asia, 3.1% in Africa, and 2.5% in South America. This distribution highlights a denser organizational spread in these regions, particularly noting the United Kingdom as a crucial nexus, where one third of all XR chapters are based, emphasizing the movement's origins and initial growth momentum within this area. However, it is essential to underscore that XR's influence is not confined to the Global North; of the 79 countries hosting XR chapters, 41 are situated outside of these traditionally wealthier regions, showcasing XR's truly international scope and its efforts to localize climate activism across diverse global contexts.

The growth of XR up to the summer of 2021 unfolded in five key surges, each linked to significant increases in new chapters: starting from its foundation until January 2019, then in Spring and Summer 2019, followed by a period of slower growth, and finally, during the COVID-19 pandemic. The initial phase saw the creation of 277 chapters, sparked by the Declaration of Rebellion and further boosted by media attention, setting the stage for rapid expansion, especially in Spring 2019 with the first International Rebellion. This period demonstrated XR's global appeal and capacity for mobilization. However, growth varied in the later phases, with the pandemic notably slowing down new chapter formations, highlighting both the challenges and the resilience of XR in the face of global crises. This journey reflects XR's ability to adapt and navigate through the sociopolitical landscape, emphasizing its ongoing commitment to climate action.

XR's global diffusion occurred in waves, marked by a small set of key protest events (Gardner et al., 2022). Although Europe and Anglophone countries were prominent in all phases of its transnational spread, countries in the Global South were nonetheless also ever-present. However, sustained growth of the movement across the Global South was delayed in its expansion. Asia saw its apogee in new chapter creation in mid- to late 2019, with 15 local groups created, predominantly in India. An equivalent expansion took place in Africa between October 2019 and April 2020 with the creation of 13 groups spanning the continent from Ethiopia to Zambia and from Sierra Leone to Kenya. Despite some productive work analyzing the movement in North America, Western Europe, and Oceania, the dynamics and challenges of the movement's adoption and adaptation remain relatively underexplored for most places in which the movement is active. Work on XR in Canada and Aotearoa New Zealand report issues in adapting the movement to settler-colonial spaces as a result of the lack of consideration given to the roles played by Indigenous groups and organizations in these societies (James & Mack 2020; Morgan 2021). Meanwhile, our own work on the movement in western and southern Africa describes how local activists have adjusted their strategies, repertoires, and targets of protest to suit the local political and cultural context (Gardner et al., 2023, 2024). In other contexts, the XR model appears to have been adopted without scholars or activists explicitly reporting significant deviations from the original model (Zantvoort, 2023). Nevertheless, more research is needed to establish the different forms the movement has taken across the world.

Too White and Middle-Class? Demographics and Representation

The literature suggests that XR—at least in Western Europe, the Unites States, Canada, Australia, and Aotearoa New Zealand—has been predominantly, but far from exclusively, White, middle-class, and educated (Hayes et al., 2020). Statistics on the racial and ethnic background of those involved are scant, but within the movement there exists a prevailing perception that its demographic composition skews disproportionately White (Bell & Bevan, 2021; James & Mack, 2020; Morgan, 2021). Among the cohort of 1,032 activists arrested during the Rebellion in April 2019, for example, 90.3% were White, with only 1% Asian (N = 11), 0.4% Black (N = 5), 2.7% identified as mixed-race (N = 31), and the remaining 4.7% abstaining from specifying their ethnic background (Hayes et al., 2020, p. 12). However, some caution is necessary here due to the potential for self-selection bias, with the decision to designate oneself as "arrestable" having a potential racial dimension.

Nevertheless, beyond questions relating to the group's demographics, XR is often critiqued for being White, middle-class, and educated in its outlook, framing, and approach (Bell & Bevan, 2021; Berglund & Schmidt, 2020; Smiles & Edwards, 2021). The strategy of mass arrest employed while often explicitly expressing support or even solidarity with the police has been widely criticized both within and beyond the movement as an exclusionary approach (Berglund & Schmidt, 2020). Similarly, the claim of acting "beyond politics" has come into conflict with environmental and racial justice frames (James & Mack, 2020; Pavenstädt & Rödder, 2024). In many contexts, the movement has failed to recognize and act in solidarity with local Indigenous rights movements that preceded it, leading to distrust of XR among BIPOC activists (James & Mack, 2020). Meanwhile, XR has tended to view the state as a neutral actor or even potential ally in the fight against climate change, contrasting sharply with the perceptions of the state in antiracist activism and research that highlight its role in the construction of lethal and degrading conditions for racialized bodies (Gardner & Müller, working paper; James & Mack, 2020; Pulido, 2017). Roger Hallam has also been personally criticized for his approach to racial (in)justice in the environmental movement and flippant use of language when discussing the Holocaust (Connolly, 2019; Levene, 2022). In addition to these and other problematic aspects of the movement, the media has also "fixated on a critique of XR as being primarily middle-class, White and not representing the views of society as a whole" (Bell, 2021, p. 63) and, in the context of a broadly oppositional approach adopted by the press (Scheuch et al., 2024), shown scant interest in reporting on diversity in the movement or antiracist initiatives. The combination of problematic discourses and strategies used by the movement and negative reportage in the press on these issues has led to a situation in which Black, Asian, and minority ethnic groups tend to express alienation from the movement (Bell & Bevan, 2021).

Social class has also been an issue for the movement. Bell (2021) identified XR's tendency to adopt middle-class discourses (including the fetishization of poverty as an eco-lifestyle choice and emphasis on individual behavior modification), tactics (failing to consider the needs and concerns of working-class communities in protest planning), behaviors (such as classist microaggressions and paternalism toward working-class activists in meetings), passive rather than active approaches to inclusion (through an "open door" policy rather than engaging working-class mobilization), and a failure to recognize and support working-class environmental struggles, past and present.

A range of affinity groups exist within the movement that speak to a degree of internal diversity, such as XR Disabled Rebels, Extinction Rebellion Internationalist Solidarity Network (XRISN), and the faith-based blocs XR Muslims, XR Jews, and XR Hindus. However, to date, scholarly attention directed toward these groups remains lacking; in our review of the literature, no research endeavors undertaken to investigate their dynamics were found. In many local and national chapters in Western Europe, North America, Australia, and Aotearoa New Zealand, the movement has grappled with its approach to racial and class-based in/justices and striven to improve in these domains,[5] albeit with mixed results (Hiraide & Evans, 2023; Matthews, 2023; Morgan, 2021).

Considerably less is known about the movement in the Global South, where around half of all countries involved in the movement are located. Additionally, there is likely considerable diversity in the composition of the movement over both space and time, not just cross-nationally but also within states. In terms of local leadership, older women play a key role in this regard across the Global North (Gardner et al., 2024); however, it appears that the demographics of these roles in the Global South are quite different, with younger men tending to adopt such positions in countries such as The Gambia (Gardner et al., 2023). Nevertheless, the precise demographic composition of the movement across its various national contexts—and its relationship to the movement's framing, strategy, and identity—remains unknown.

Spin-offs, Offshoots, and Shooting for the Moon

Ostensibly, XR has been the progenitor of an array of new social movements since 2018. In many contexts, this has taken the form of radical flanks, such as Blockade Australia, and the "A22 Network" that includes Just Stop Oil in the United Kingdom, Declare Emergency in the United States, Riposte Alimentaire in France, Restore Passenger Rail in Aotearoa New Zealand, Nødbremsen in Denmark, Återställ Våtmarker in Sweden, Renovate Switzerland, and Last Generation (active in Italy, Germany, Canada, and Austria) (A22, 2024). Other spin-off movements include Animal Rising, Insulate Britain, Burning Pink, Ocean Rebellion, and People Demand Democracy in the United Kingdom; Rave Revolution and Climate Defiance in the United States; and Les Soulèvements de la Terre in France.

The precise links between XR and subsequent environmental movements are debated. There are often overlaps in personnel, tactics, and perspective between XR and post-2018 social movements potentially associated with it, despite often being explicit about their independence, divergence from, or even opposition to XR (Hinks & Rödder, 2023). Nevertheless, the precise connection between XR and emergent social movements is more challenging to identify. As a global movement, its potential impacts on subsequent activism are vast and diverse. In some places, XR represented the first climate change–related movement to emerge, radically altering the local social movement scene. In contrast, for XR in The Gambia, aligning with XR was more of an additive strategy than the development of a wholly new movement, and so the precise link between XR and subsequent movements is further blurred (Gardner et al., 2024). There are also divergences in perspective, outlook, strategy, and organizational structure between XR and subsequent environmental movements, with connections between them even rejected by the activists themselves. Overall, the literature on XR's (potential) spin-off movements is currently thin on the ground. Further research is needed to establish the precise connections (and disjunctures) between these movements, as well as the longer-term impact of XR on the varied contexts in which it has been active.

Although not uniformly, spin-off movements tend to retain XR's original use of dramatic, attention-grabbing, and disruptive forms of "politically motivated lawbreaking" (Scheuerman, 2022, p. 791), but not necessarily its organizational structure. In contrast to XR, for example, Just Stop Oil has adopted a much more centralized, top-down, and vanguardist approach. As one of our interviewees put it: "XR is meant to be non-hierarchical, right? And that's sort of, like, the democratic principle. I don't think it actually is non-hierarchical, but … Just Stop Oil isn't pretending to be non-hierarchical."[6] Just Stop Oil activists also tend to understand themselves—often explicitly—to be acting as the radical flank of the climate movement, largely drawing its membership from among those within the broader movement most dedicated to undertaking disruptive action. In this way, spin-off movements such as Just Stop Oil have attempted to maintain a sense of energy and urgency in their demands for the government to radically change course to avert environmental catastrophe.

Conclusion: Where Do We Go From Here?

In this chapter, we have provided a broad overview of XR's characteristics and development since its inception. Drawing on the social scientific literature, we outlined its core characteristics of the movement, including its goals, tactics, discourses, organizational structure, transnationalities, demographics, and spin-off movements. In doing so, we consider an overarching question about the nature of XR: namely, the extent to which it is best described as a fully decentralized and horizontal organization or a rather more centralized vanguardist one. Overall, as we have aimed to demonstrate through the sections herein, XR is a movement with multiple competing tendencies, incorporating important elements of both decentralization and vanguardism. We also contend that XR is also a transnational movement whose globally diverse character is often not considered. Although there is a lack of research on the character of the movement in many of the national contexts in which it operates, it seems likely that the organizational structure looks rather different at global, regional, national, and local scales.

Looking to the future, how the movement develops amid the evolving landscape of environmental activism in the United Kingdom and elsewhere remains uncertain. COVID-19 lockdowns in 2020 worked to effectively stall the momentum the movement had built throughout 2019, leading to a significant setback in its development (Gardner et al., 2022; Zantvoort, 2023). In December 2022, the movement made a strategic shift, announcing plans to cease engaging in forms of NVDA that disrupted the public, stating a desire to "prioritise attendance over arrest and relationships over roadblocks" (XR, 2022). However, newspaper content analysis has revealed that this strategic pivot did not elicit more favorable coverage of the group in the UK press (Scheuch et al., 2024). At the same time, the movement now finds itself in a much more crowded social movement space than before, with numerous other environmental groups vying for public and media attention. Despite its attempts to rebuild and adapt, the future trajectory of the movement remains uncertain.

Notes

1. Including a dataset of all XR local chapters (taken from rebellion.global in 2022), a protest event dataset of major XR protests worldwide, a series of semistructured interviews with 96 activists across 17 countries worldwide (conducted 2020–2023), and a dataset of 10 interviews conducted with activists in XR, Animal Rising, Insulate Britain, and Just Stop Oil (conducted by Gardner in 2023–2024).
2. Interview with the authors, 2022.
3. Sociocracy is a "semi-horizontal method of governance defined by the organisation of people into small semi-autonomous 'circles' to work on specific tasks and make decisions on issues outlined in their mandate; the use of elections to appoint people to circle coordinator and representative positions; and the use of consent-based decision-making, whereby all individuals in a circle must express consent for a decision to be actioned" (Rowe & Ormond, 2023, p. 146). The key features of holacracy that XR has adopted involve the use of "energizing roles" (with authority delegated to individuals who occupy clearly defined roles associated with task-focused mandates) and a "circle structure" (multiple self-organizing groups connected to the wider movement by individuals acting as links to other circles) (Berglund & Schmidt, 2020, pp. 42–43).

4. As of the first half of 2024, their website indicates there are 988 local groups across 89 countries on every continent, including Antarctica. This suggests that since our initial research, inactive groups have been "cleaned up" and removed from the listing.
5. This also came up repeatedly in the authors' own interviews with XR activists in these countries.
6. Interview with XR and Just Stop oil activist in 2023.

References

A22. (2024). *Who we are?* A22 network. https://a22network.org

Almeida, P. (2019). Climate justice and sustained transnational mobilization. *Globalizations, 16*(7), 973–979.

Beckett, A. (2023). Introducing cycling rebellion, Extinction Rebellion's two-wheeled cousin. *Cycling Weekly*. https://www.cyclingweekly.com/news/introducing-cycling-rebellion-extinction-rebellions-two-wheeled-cousin

Bell, K. (2021). Working-class people, Extinction Rebellion and the environmental movements of the global north. In K. Bell (Ed.), *Diversity and inclusion in environmentalism* (pp. 63–81). Routledge.

Bell, K., & Bevan, G. (2021). Beyond inclusion? Perceptions of the extent to which Extinction Rebellion speaks to, and for, Black, Asian and minority ethnic (BAME) and working-class communities. *Local Environment, 26*(10), 1205–1220.

Berglund, O. (2023). Disruptive protest, civil disobedience and direct action. *Politics*, https://doi.org/10.1177/02633957231176999.

Berglund, O., & Schmidt, D. (2020). *Extinction Rebellion and climate change activism: Breaking the law to change the world*. Palgrave Macmillan.

Brunon, H. (2021). Extinction Rebellion: Pratiques et motivations de l'activisme écologiste radical. *Métropolitiques*.

Burgess, C., & Read, R. (2020). Extinction Rebellion and environmental activism—The XR interviews. *Journal of Human Rights and the Environment, 11*(3), 171–180. https://doi.org/10.4337/jhre.2020.03.08

Buzogány, A., & Scherhaufer, P. (2022). Framing different energy futures? Comparing Fridays for Future and Extinction Rebellion in Germany. *Futures, 137*, 1–10.

Buzogány, A., & Scherhaufer, P. (2023). The new climate movement: Organization, strategy, and consequences. In H. Jörgens, C. Knill, & Y. Steinebach (Eds.), *Routledge handbook of environmental policy* (pp. 358–380). Routledge.

Chase-Dunn, C., & Almeida, P. (2020). *Global struggles and social change: From prehistory to world revolution in the twenty-first century*. Johns Hopkins University Press.

Chenoweth, E., & Stephan, M. (2011). *Why civil resistance works: The strategic logic of nonviolent conflict*. Columbia University Press.

Clot-Garrell, A. (2024). Voices of emergency: Imagined climate futures and forms of collective action. *Current Sociology, 72*(5), 967–985.

Connolly, K. (2019). Extinction Rebellion founder's Holocaust remarks spark fury. *The Guardian*. https://www.theguardian.com/environment/2019/nov/20/extinction-rebellion-founders-holocaust-remarks-spark-fury

Coombs, G. (2020). It's (red) hot outside! The aesthetics of climate change activists Extinction Rebellion. *Journal of Public Space, 5*(4), 123–136.

De Moor, J., De Vydt, M., Uba, K., & Wahlström, M. (2021). New kids on the block: Taking stock of the recent cycle of climate activism. *Social Movement Studies, 20*(5), 619–625.

Doherty, B., & Doyle, T. (2006). Beyond borders: Transnational politics, social movements and modern environmentalisms. *Environmental Politics, 15*(5), 697–712.

Doherty, B., & Hayes, G. (2022). Extinction Rebellion. In D. A. Snow, D. della Porta, D. McAdam, & B. Klandermans (Eds.), *The Wiley Blackwell encyclopedia of social and political movements* (2nd ed., pp. 1–5). Wiley.

Extinction Rebellion. (2019). *This is not a drill: An Extinction Rebellion handbook*. Penguin.

Falanga, R., & Carvalho, T. (forthcoming). Beyond (what) politics and towards (what) change? A look into the Extinction Rebellion's call for Climate Assemblies. *Environmental Science & Policy.*

Farrer, B., Doyle, L., & Smith, S. (2023). Goals, strategies, and tactics: Continuity and change in Extinction Rebellion in the United Kingdom. *Contention, 11*(2), 29–56.

Fotaki, M., & Foroughi, H. (2022). Extinction Rebellion: Green activism and the fantasy of leaderlessness in a decentralized movement. *Leadership, 18*(2), 224–246. https://doi.org/10.1177/17427150211005578

Gardner, P., & Abrams, B. (2023). Contentious politics and symbolic objects. In B. Abrams & P. Gardner (Eds.), *Symbolic objects in contentious politics* (pp. 13–39). University of Michigan Press.

Gardner, P., Adekola, O., Carvalho, T., & O'Brien, T. (2023). Confronting the climate crisis in Africa: Just transitions and Extinction Rebellion in Nigeria and South Africa. *Review of African Political Economy, 50*(177–178), 475–490.

Gardner, P., O'Brien, T., Carvalho, T., & Adekola, O. (2024). Profile: Extinction Rebellion in the Gambia. *Social Movement Studies, 23*(1), 122–128.

Gardner, P., Williams, S., & Macdonald, A. (2024). Glued on for the grandkids: The gendered politics of care in the global environmental movement. *Sociology Compass, 18*(1), e13148.

Gardner, P. R., & Carvalho, T. (2023). Intersecting planes of crisis: Geometrics of crises and catastrophe in Extinction Rebellion rhetoric. In K. Heidemann (Ed.), *Combating crises from below: Social responses to polycrisis in Europe* (pp. 223–242). Maastricht University Press.

Gardner, P. R., Carvalho, T., Newman-Grigg, E., & O'Brien, T. (Working paper). 1,000 km of just nothingness: Geography, networks and space in the organisation of Extinction Rebellion Australia.

Gardner, P. R., Carvalho, T., & Valenstain, M. (2022). Spreading rebellion? The rise of Extinction Rebellion chapters across the world. *Environmental Sociology, 8*(4), 424–435.

Gardner, P. R., & Müller, T. (Working paper). A movement rebirthed: Race, multiple boundary spanning, and the rise of Extinction Rebellion in the US.

Giæver, F., & Russell, S. (2024). "Courage without the certainty of a happy ending": the emotion regulation of environmental activists. *Social Movement Studies*, 1–16. https://doi.org/10.1080/14742837.2024.2345177

Grenfell, R. (2023). Platform capitalism and place relations in social movements: Environmentalism and Extinction Rebellion in Western Australia. *Media International Australia*, https://doi.org/10.1177/1329878X2312022

Hallam, R. (2019). *Common sense for the 21st century: Only nonviolent rebellion can now stop climate breakdown and social collapse.* Chelsea Green.

Hayes, G., Doherty, B., & Saunders, C. (2020). A new climate movement? Extinction Rebellion's activists in profile. *Environmental Politics, 29*(3), 546–551.

Hinks, E., & Rödder, S. (2023). The role of scientific knowledge in Extinction Rebellion's communication of climate futures. *Frontiers in Communication, 8*, 1007543.

Hiraide, L. A., & Evans, E. (2023). Intersectionality and social movements: A comparison of environmentalist and disability rights movements. *Social Movement Studies*. https://doi.org/10.1080/14742837.2023.2234828

James, D., & Mack, T. (2020). Toward an ethics of decolonizing allyship in climate organizing: Reflections on Extinction Rebellion Vancouver. In B. J. Richardson (Ed.), *From student strikes to the Extinction Rebellion: New protest movements shaping our future* (pp. 32–53). Edward Elgar.

Johnston, C., & Holland Bonnett, A. (2023). Picturing these days of love and rage: Extinction Rebellion's "Impossible Rebellion." *Cultural Geographies, 30*(2), 315–323.

Kountouris, Y., & Williams, E. (2023). Do protests influence environmental attitudes? Evidence from Extinction Rebellion. *Environmental Research Communications, 5*(1), 1–15.

Levene, M. (2022). The Holocaust paradigm as paradoxical imperative in the century of anthropogenic omnicide. *Genocide Studies and Prevention: An International Journal, 16*(1), 76–100.

Malafaia, C., Kettunen, J., & Luhtakallio, E. (2024). Visual bodies, ritualised performances: An offline-online analysis of Extinction Rebellion's protests in Finland and Portugal. *Visual Studies*, 1–14. https://doi.org/10.1080/1472586X.2023.2292620

Malm, A. (2021). *How to blow up a pipeline*. Verso.

Mansfield, F. (2020). Rebel for life: Extinction Rebellion's approach to the climate crisis. *Medicine, Conflict and Survival*, *36*(4), 375–382.

Matthews, K. R. (2020). Social movements and the (mis) use of research: Extinction Rebellion and the 3.5% rule. *Interface*, *12*(1), 591–615.

Matthews, K. R. (2023). Climate activist knowledge-practices and radicalism in Extinction Rebellion Aotearoa New Zealand (Doctoral dissertation, University of Otago). Melia, S. (2021). *Roads, runways and resistance: From the Newbury Bypass to Extinction Rebellion*. Pluto Press.

Morgan, H. (2021). Extinction Rebellion in Aotearoa: The possibilities and pitfalls of importing a social movement organisation. *Interface*, *13*(1), 157–173.

Pavenstädt, C. N., & Rödder, S. (2024). Between evidence first and political fight–understanding dynamics of (de-) politicization in US climate movements' future narratives. *Environmental Politics*, *33*(6), 1065–1086.

Pieri, Z. P. (2024). Framing the climate emergency: The role of music in extinction rebellion. *Social Movement Studies*, 1–16. https://doi.org/10.1080/14742837.2024.2304300

Pulido, L. (2017). Geographies of race and ethnicity II: Environmental racism, racial capitalism and state-sanctioned violence. *Progress in Human Geography*, *41*(4), 524–533.

Rödder, S., & Pavenstädt, C. N. (2023). Unite behind the science! Climate movements' use of scientific evidence in narratives on socio-ecological futures. *Science and Public Policy*, *50*(1), 30–41.

Rohden, F. G. (2021). Experts, influencers, and amplifiers—Exploring climate movements' hyperlinking practices. *Journal of Science Communication*, *20*(7), 1–20.

Rootes, C. (2005). A limited transnationalization? The British environmental movement. In D. Della Porta & S. Tarrow (Eds.), *Transnational protest and global activism* (pp. 21–43). Rowman and Littlefield.

Rowe, T., & Ormond, M. (2023). Holding space for climate justice? Urgency and "regenerative cultures" in Extinction Rebellion Netherlands. *Geoforum*, *146*, 1–10.

Scheuch, E. G., Ortiz, M., Shreedhar, G., & Thomas-Walters, L. (2024). The power of protest in the media: Examining portrayals of climate activism in UK news. *Humanities and Social Sciences Communications*, *11*(1), 1–12.

Scheuerman, W. E. (2022). Political disobedience and the climate emergency. *Philosophy & Social Criticism*, *48*(6), 791–812.

Seagrave, J. (2023). Protest in the face of catastrophe: Extinction Rebellion and the anti-politics of grief. *Suomen Antropologi: Journal of the Finnish Anthropological Society*, *47*(22), 49–73.

Sharp-Wiggins, B. (2022). "Zombie koala" greets parliamentarians in Canberra—video. *The Guardian*. https://www.theguardian.com/australia-news/video/2022/jul/27/zombie-koala-greets-parliamentarians-in-canberra-video

Smiles, T., & Edwards, G. A. (2021). How does Extinction Rebellion engage with climate justice? A case study of XR Norwich. *Local Environment*, *26*(12), 1445–1460.

Smith, J., Gemici, B., Plummer, S., & Hughes, M. M. (2018). Transnational social movement organizations and counter-hegemonic struggles today. *Journal of World-Systems Research*, *24*(2), 372–403.

Smółka, P. (2023). Aktywizm klimatyczny w mediach. Studium przypadku Extinction Rebellion Kraków. In M. Pałasz (Ed.), *Do serca: Zarządzanie zaangażowaniem społecznym przez media, muzykę i marketing* (pp. 57–94). Biblioteka Jagiellońska.

Smyth, I., & Walters, L. (2020). The seas are rising and so are we!—A conversation between two women in Extinction Rebellion. *Gender and Development*, *28*(3), 617–635.

Stammen, L., & Meissner, M. (2024). Social movements' transformative climate change communication: Extinction Rebellion's artivism. *Social Movement Studies*, *23*(1), 19–38.

Stuart, D. (2020). Radical hope: Truth, virtue, and hope for what is left in Extinction Rebellion. *Journal of Agricultural & Environmental Ethics*, *33*(3–6), 487–504.

Taylor, M. (2018). 15 environmental protesters arrested at civil disobedience campaign in London. *The Guardian*. https://www.theguardian.com/environment/2018/oct/31/15-environmental-protesters-arrested-at-civil-disobedience-campaign-in-london

Weaver, D. (2024). PROFILE: Extinction rebellion: Greening vanguardism? *Social Movement Studies*, *23*(1), 113–121.
Westwell, E., & Bunting, J. (2020). The regenerative culture of Extinction Rebellion: Self-care, people care, planet care. *Environmental Politics*, *29*(3), 546–551.
XR. (n.d.). About us. Extinction Rebellion UK. https://extinctionrebellion.uk/the-truth/about-us/
XR. (2018). Declaration of rebellion. Extinction Rebellion. https://extinctionrebellion.uk/declaration/
XR. (2019a). Creative rebellion: Arts and culture highlights from international rebellion London. Extinction Rebellion UK. https://extinctionrebellion.uk/2019/10/21/creative-rebellion-arts-and-culture-highlights-from-international-rebellion-london/
XR. (2019b). This Monday—The international rebellion continues in more than 60 cities around the world. Extinction Rebellion. https://extinctionrebellion.uk/2019/10/04/this-monday-the-international-rebellion-continues-in-more-than-60-cities-around-the-world/
XR. (2020a). Rebel stories: Grappling with hope and making connections. XR Buddhists. https://xrbuddhists.com/category/history/
XR. (2020b). We were the boat: The inside story of an April icon. Extinction Rebellion. https://rebellion.earth/2020/04/15/we-were-the-boat-the-inside-story-of-an-april-icon/
XR. (2022). We quit. Extinction Rebellion UK. https://extinctionrebellion.uk/2022/12/31/we-quit/
XR. (2024a). XR in your area. Extinction Rebellion Global. https://rebellion.global/groups/#countries
XR. (2024b). XR United Kingdom. Extinction Rebellion. https://rebellion.global/groups/gb-united-kingdom/#groups
XR. (2024c). Extinction Rebellion global support. XR Global Support. https://chuffed.org
Young, K. A., & Thomas-Walters, L. (2024). What the climate movement's debate about disruption gets wrong. *Humanities and Social Sciences Communications*, *11*(1), 1–7.
Zantvoort, F. (2023). Movement pedagogies in pandemic times: Extinction Rebellion Netherlands and (un)learning from the margins. *Globalizations*, *20*(2), 278–291.

CHAPTER 15

Environmentalism of the Poor and Climate Action

JOAN MARTINEZ-ALIER

Introduction

CONFLICTS with local populations sparked by coal, oil, and gas extraction and burning have led environmental justice organizations to campaign to leave fossil fuels in the ground. Here we retrace the history of grassroots movements opposing fossil fuels extraction and explore how the climate justice movement might influence climate policy negotiations by promoting such actions and by putting forward claims for an ecological debt from North to South.

The most wide-ranging concept encompassing most collective efforts to address global warming is referred to as "climate action." The required reduction of greenhouse gases (GHGs) to avoid further increases in concentrations in the atmosphere is of the order of at least half the present emissions. In this chapter we *do not focus* on the following: negotiations at the Conference of the Parties (COPs) establishing desirable or obligatory reductions in GHG emissions; policies such as taxes on GHG emissions; international agreements on resource caps; or a potential international treaty against the proliferation of extraction of fossil fuels. Whether demand side or supply side, these policies are top-down. They are initiated or require government action. This chapter is about grassroots climate actions, combining local opposition to extractions of fossil fuels with awareness of climate change. This can be deemed to have begun in the mid-1990s, when activists against Shell in Nigeria met activists against Chevron-Texaco from Ecuador, and with representatives of other countries founded Oilwatch in Lago Agrio and Quito (Ecuador) in early 1996 (Oilwatch, n.d.). The future "climate actions" that Environmental Rights Action (ERA) in Nigeria and Acción Ecológica in Ecuador strategized with other organizations at that meeting were local actions, such as marches, blockades, and street protests against oil spills. They were met with repression, particularly in the Niger Delta, where Ogoni people and others were killed by the military in demonstrations or given the death sentence in court cases (as Ken Saro-Wiva in 1995, with his companions). In response, Oilwatch launched the slogan "leave oil in the soil" and, somewhat later, "leave coal in the hole." Grassroots leaders of this movement brought together local complaints with awareness of climate change. They were Nnimmo Bassey, Esperanza Martinez (both now hold honorary doctorates at European universities), and others who later became well-known. Nnimmo Bassey was president of Friends of the Earth International from 2008 to 2018. The Ogoni in Nigeria managed to get Shell out of their small territory (though they failed to stop the carnage elsewhere in the Niger Delta, such as the oil spills and the gas flaring). Acción Ecológica failed in the long run to get Chevron-Texaco to effectively pay for the damage done in the northern Amazon part of Ecuador, though it managed to score a partial success with the Yasuni ITT proposal of 2007–2017. As it is well-known, this grassroots proposal, stopping the extraction of oil in the ITT fields of the southeastern part of the Amazon of Ecuador, was officially adopted by the government of the country with Rafael Correa as president and Alberto Acosta as minister for energy and mines. After a few years, Rafael Correa gave up on the idea, Alberto Acosta was no longer minister, and drilling began after 2013, and oil extraction is taking place today. The Yasuni ITT proposal with its calculations of the CO_2 emissions that would be avoided if the oil was kept underground linked very directly the defense of local biodiversity and the survival of some Indigenous groups to the issue of world climate change that by 2007 could no longer be denied. The Yasuni ITT proposal had worldwide influence (Temper et al., 2013). After Ecuador's executive and judicial power denied the possibility to hold a referendum on the Yasuni ITT proposal, on August 20, 2023, Ecuador finally voted to halt all future oil drilling in this sensitive protected area in the country's eastern Amazon. However, this was too late.

Blockadia and LFFU

The grassroots slogans, and occasional grassroots practices of "leaving the oil in the soil, leaving the coal in the hole," because of local and global initiatives that Oilwatch launched, were later brought together within the terminology of Blockadia that Naomi Klein (2105) used in her book *This Changes Everything*, and in the acronym LFFU that some academics use (myself included, Martinez-Alier, 2021), meaning "leave the fossil fuels underground." Naomi Klein's term of "Blockadia" has been successful among militants; she learned the word in some places where blockades were established against oil and gas pipelines. Naomi Klein also acknowledges that the idea of an ecological debt, or more precisely a "carbon debt," from North to South, was suggested to her by Angelica Navarro, a young Bolivian diplomat, brought up in Latin American activist circles. In an interview years later, Klein explained that in the book she quotes Angelica Navarro, Bolivia's then ambassador to the World Trade Organization, but who was also part of Bolivia's climate negotiation team at the 2009 United Nations climate talks in Copenhagen. "Exactly 10 years ago," she said, "Navarro called for a Marshall Plan for planet earth. She was building on a concept of ecological debt that comes from social movements in Ecuador, in Nigeria, in front line African-American and Latino communities in the United States. So that's the origin of this idea" (Klein, 2019, p. 3). We go back to the Latin American origins of the claim for the "carbon debt" in a later section.

One famous blockade was against the Dakota Access Pipeline (DAPL). The DAPL, with the famous Standing Rock protest, is an 1,886-km oil pipeline that would transport nearly 500,000 barrels of crude oil a day from North Dakota to Illinois. There is a growing awareness of the ways in which climate change and fossil fuel extraction are a continuation of colonial exploitation (see chapter by Fenelon and Figueroa in this volume). In this historic movement against the Dakota Access Pipeline (DAPL), the native Sioux mobilized Indigenous peoples and diverse supporters in 2015 (EJAtlas, 2022). To block construction of the oil pipeline, "spiritual and water protection camps" were founded. Broad-based support was received from all over the United States and beyond, including a visit from Sarayaku, Ecuador, a community that managed to stop oil exploration (EJAtlas, 2021). During protests, more than 750 people were arrested and others faced pepper spray attacks and rubber bullets. The Sioux argued that "water is sacred"; therefore, economic compensation was not the issue.

Given that local complaints against damages from coal mining and the pollution from coal-fired power plants (CFPPs) are many decades old already, and that complaints against oil spills became widespread 100 years ago, one can wonder why it took so long to bring together, in the mid-1990s and with Oilwatch, the local complaints with the alarm at climate change. The link was that relatively poor people were displaced and polluted by the coal and oil industries, and in parallel, relatively poor people were going to be damaged by global climate change in the future. Solidarity among the poor, and a sense of contemporary and intergenerational justice, demanded that the coal be left in the hole and the oil in the soil. Already by the 1990s, awareness of climate change became widespread.

Climate scientists and the early ecological economists were perfectly aware of climate change by the 1970s and 1980s. However, one main worry was "peak oil." The Hubbert curve was better known than the Keeling curve.[1] Nevertheless, there were early warnings as witnessed by Georgescu-Roegen in an article published in 1975 when he wrote: "The continuous accumulation of carbon dioxide in the atmosphere has a greenhouse effect which should aggravate the heating of the globe" (Georgescu-Roegen, 1975, p. 358). He already suspected that "thermal pollution could prove to be a more crucial obstacle to growth than the finiteness of accessible resources" (Georgescu-Roegen, 1975, p. 358). Nowadays emissions of CO_2 by humans reach about 40 billion tons per year, and they are still increasing slightly. Their distribution is very unequal (Figure 15.1). We humans all breathe in and out more or less the same, and it would be impracticable to reduce CO_2 emissions by slow respiration. There are livelihood emissions, and there are luxury emissions. One characteristic feature of human ecology is the extreme difference in the exosomatic use of energy. The Kyoto agreement in 1997 granted rights to the United States, Europe, and Japan equal to their 1990 emissions, on the promise of a reduction of 5.2% for the year 2010. This was unjust.

Concentration of CO_2 in the atmosphere as measured in the Keeling curve has reached 427 ppm (in May 2024) when it was 360 ppm in 1992. Every COP has meant 2 ppm extra. There is no doubt that it will reach 450 ppm before 2050 because the extraction of coal, oil, and gas around the world is still increasing. It is assumed the fossil fuels extraction will stop growing soon, and the emissions will reach a peak. Then, they must decrease emissions at least by half in order to reach a "net-zero" position. This means that emissions will be captured by the oceans and new vegetation and their concentration in the atmosphere will stop increasing. "Net-zero" means going down to 20 billion tons of CO_2 per year or less from the present 40 billion. It therefore requires reducing by half the extraction and burning of fossil fuels, starting mainly with coal. Net-zero does not mean stopping fossil fuel extraction, but it does involve drastically reducing GHG emissions, leaving the fossil fuel companies (and countries rich in fossil fuels) with so-called stranded assets. They will perhaps ask for monetary

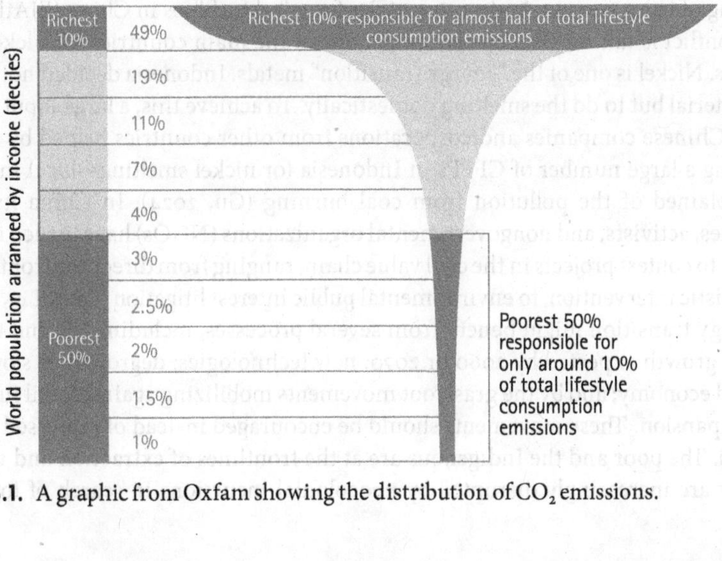

FIGURE 15.1. A graphic from Oxfam showing the distribution of CO_2 emissions.

compensation instead of facing the "ecological debt" they have created. The conflicts at the extraction frontiers for coal, oil, and gas will be less acute but will continue. The economy is not circular; it is entropic—coal, gas, and oil burned this year are no longer available, and every year there will be a search for new sources. The use of the atmosphere as an open-access reservoir is still increasing. The sinks (oceans, soils, new vegetation) are used on a first-come, first-serve basis, without payment. The warming climate has recognized consequences, including ocean acidification, mass forest fires, and melting of the permafrost with increased methane releases. President Trump's policies (2025–2028) will continue such trends by failing to decrease oil and gas extraction.

The environmental and social costs of this necessary energy transition will be large and, again, unequally distributed. To substitute current solar energy for the stock of photosynthesis of the past from fossil fuels would require a tremendous investment in wind turbines and solar panels and in the metals needed such as lithium for batteries but also copper, nickel, aluminum, and cobalt. This implies a large energy input in itself (for mining, for smelting metals, for building the solar panels or windmills), the invasion of new commodity frontiers, and an increase in water use and displacement of many people from their customary lands. Thus, it has been argued in detail that constructing, operating, and maintaining a low-carbon energy system will itself require energy, with much of it derived from fossil fuels (Slameršak et al., 2022).

The transition toward halving fossil fuel usage while maintaining economic growth will be difficult if not impossible. What we see so far is that India is fast increasing solar energy while simultaneously completing a transition toward coal (Roy & Schaffartzik, 2021), probably doubling annual coal production from 1 billion to 2 billion tons in 10 years (that will still be half the amount of China, which is 4 billion tons of coal per year). We see that there is change at the world level with a rapid increase in solar energy, but we are at a stage of "energy addition" rather than witnessing a substitution in absolute terms. In the 20th century at the global level, coal increased seven times, and oil and gas much more. In absolute terms, energy sources were maintained (biomass) or grew at different rates so that percentages varied (Fressoz, 2024; York & Wells, 2019). This is perhaps happening again; fossil fuel will be maintained or decrease very slowly in absolute terms, while solar energy (wind and photovoltaic) increases at a rapid pace.

Optimism regarding a quick energy transition is not reasonable. One must take into account the energy and material inputs of procuring solar energy. An unexpected conflict is the ravaging of balsa trees in Andean countries for wind turbines in China (EJAtlas, 2024). Another conflict is taking place in Indonesia, one of the main countries for nickel mining and exports. Nickel is one of the "energy transition" metals. Indonesia decided not to export the raw material but to do the smelting domestically. To achieve this, a large input of energy is needed. Chinese companies and corporations from other countries helped by financing and building a large number of CFPPs in Indonesia for nickel smelting—local inhabitants often complained of the pollution from coal burning (Gu, 2024). In China itself, local communities, activists, and nongovernmental organizations (NGOs) have turned to various approaches to contest projects in the coal value chain, ranging from direct confrontation and creative artistic intervention, to environmental public interest litigation (Gu & Liu, 2024).

The energy transition might benefit from several processes, including the end to human population growth expected by 2060 or 2070; new technologies; degrowth or slow growth of the world economy; and by the grassroot movements mobilizing against fossil fuel extraction and expansion. These movements should be encouraged instead of repressed (Scheidel et al., 2020). The poor and the Indigenous are at the frontlines of extraction and waste disposal. They are increasingly disrupted by neocolonial incursions in search of fossil fuels, metals, and biomass. The politically marginalized at the commodity extraction frontiers are also the main actors of a just socioecological transition, unless repression and fear hinder action. In many instances, mobilizations have led to successful outcomes. A mapping of 649 cases from the EJAtlas of resistance movements against both fossil fuels and low carbon-energy projects shows that over a quarter of such projects had been canceled, suspended, or delayed—demonstrating the success of place-based movements (Temper et al., 2020). Poor and Indigenous people often object to fossil fuels extraction, transport, and burning, adding arguments of climate change to local grievances (see chapter by Fenelon and Figueroa in this volume). They also mobilize, as one would expect, against land and water grabbing, and open-pit mining for the new green energy systems. Next, we give a small sample of cases in which local movements organize against fossil fuels, combining local grievances with climate change implications.

Glocal Movements Against Fossil Fuels

A few years ago, in 2021, the Goldman Prize was awarded to Kimiko Hirata, the leader of the Kiko Network in Japan. After the Fukushima disaster in 2011, many nuclear power plants were stopped in Japan at least for some years. The government planned to build many CFPPs instead; some of them were never built because of public protests. Over several years, Kimiko Hirata's grassroots campaign led to the cancellation of 13 coal power plants (7,030 MW). These coal plants would have released 1.6 billion tons of CO_2 over their lifetimes. One can think this opposition was a luxury of the rich Japanese who could afford to be environmentalists (Thiri & Borsi, 2024). But what about Gloria Capitan, from Mariveles near Manila in the Philippines? Gloria Capitan was one of the leaders of the Coal-Free Bataan movement and the president of United Citizens of Lucanin Association, a community-based organization that was peacefully opposing the operation and expansion of coal plants and open storage facilities in the Mariveles neighborhood, which would have had harmful consequences for the local population. As a part of her human rights work, Gloria Capitan organized campaigns, filed complaints with the courts, collected signatures for petitions, and initiated other public actions, calling for a permanent closure of a coal project undertaken in the region. In July 2016, she was shot dead by two unidentified men in Lucanin village, Mariveles municipality, Bataan province. Since the start of her work opposing a coal mining and storage project in 2015, Gloria Capitan (a grandmother) had faced intimidation and threats. She was also a campaigner for the Philippine Movement for Climate Justice, totally aware of the link between coal burning and climate change (Martinez-Alier, 2023, pp. 63–69, 89–90, 708).

In another case of coal mining repression, Fikile Ntshangase, a South African environmental activist, was assassinated on October 22, 2020. She was a leading member of the Mfolozi Community Environmental Justice Organisation (MCEJO), which was taking legal action against the proposed expansion of an open-pit coal mine operated by Tendele Coal Mining (Pty) Ltd, near Somkhele, situated adjacent to Hluhluwe—iMfolozi Park, the oldest nature reserve in Africa. Fikele was also a grandmother, and she resided in the territory where the mine operated. In the mourning ceremonies, there were banners with the slogan "keep the coal in the hole" (Martinez-Alier, 2023, p. 334).

In India, the people's struggle against coal extraction is active in several regions. In 2009, the Chhattisgarh state environment ministry designated Hasdeo Arand forest as a "no-go" area for mining for its rich, unfragmented forest cover. Soon after, in 2011, coal mines were opened up in Hasdeo Arand because the no-go policy never became implemented and enforced. In 2019, the Ministry for Environment gave environmental clearance for open-pit coal mining in Parsa to be operated by Rajasthan Collieries, a unit of Adani Enterprises Limited. In the Hasdeo-Arand region, two coal mines are operational apart from the Parsa mine, with a capacity of 5 million metric tons per annum. All three mines in the Hasdeo forests have been awarded to Adani Enterprises Limited as the mine developer and operator. The 90 million Adivasis are among India's poorest. Among them, the Gonds and other Adivasis who live outside and within parts of the Hasdeo Arand forest will become a displaced people. They are poor in financial terms, and now economic development deprives them of their land and water as the coal industry benefits. They would be removed from their land, lose their fields, lose their culture, and, perhaps, join the migration to the slums of Indian cities. For a decade, Adivasi communities have battled to preserve the Hasdeo Arand forest, even though successive governments have illegally bypassed local village councils to award coal-mining contracts to India's powerful Adani conglomerate. The Hasdeo Arand Bachao Sangharsh Samiti is a village-level organization set up to protect the forest. On October 2, 2021, marking Gandhi's birthday, hundreds of tribal villagers began a *padayatra*, a long protest march, against government plans for a major coal mining expansion. "This land is our land!" chanted men and women in Hindi as they navigated forest tracts, village paths, and highways on a 300-km trek to the state capital. The local Adivasi communities agitated to save their *Jal Jangal Zameen*, water, forest, and land, a slogan used in a long history of rebellions in central India. But they were also waging a much larger battle that goes beyond self-interest. They were trying to protect this entire forested landscape, the existence of wildlife emphasizing the interconnectedness of all these issues. To this, the organizers of the march added the argument of climate change to which coal burning in CFPP makes an important contribution. "The challenge of climate change and our role in it poses fundamental questions about the survival of the planet. In such a scenario, dense forests like Hasdeo need to be saved from destruction," said Alok Shukla, an activist in the 300-km march (Martinez-Alier, 2023, p. 573). By April 2024, Alok Shukla, convenor of the Chhattisgarh Bachao Andolan and founding member of the Hasdeo Aranya Bachao Sangharsh Samiti, was awarded the Goldman Prize from Asia (Shukla, 2024).

There is also extensive resistance against gas extraction. At times the struggle is against liquified natural gas (LNG), a fuel which is fast growing. Perhaps the gravest clash today is found in Cabo Delgado, Mozambique, where Total and other transnational firms are keen in acquiring the gas while displaced peoples and their allies are resisting. Many people have died already. Competition for oil and gas has caused wars (see chapter by Oviedo in this volume). A brutal neocolonial military encounter is making progress around the very large LNG project in Cabo Delgado. Total, the French fossil fuel firm, was forced in 2021 to suspend work at the Afunge airport and gas terminal because of an Islamic insurgency. The tragic events occurred at Palma, a town on the coast of Mozambique immediately south of the border with Tanzania, because this is the base of LNG operations for the growing offshore natural gas industry. Although at the beginning, community relocation seemed successful, the gas project led to major local changes and violence, displacing farming and fishing communities who lost their livelihoods. Over half the population of Cabo Delgado is Muslim. Total planned to secure with Mozambican and foreign troops about 7,000 hectares. A terrestrial and naval "counterinsurgent" army with Rwandan, South Africa and other forces was being organized to secure gas extraction and export of 50 million tons of LNG per year. Mozambique's Tete province is already suffering from coal extractivism by foreign companies. Three multinationals control the gas in Cabo Delgado: ENI (Italy), ExxonMobil (United States), and Total (France, after taking it over from Anadarko). The Total CEO told President Filipe Nyusi that Total would only continue if Mozambique guaranteed security in a 25-km cordon in the coast for the gas project. I have called this "bunker extractivism"

(Martinez-Alier, 2023, p. 342). It is still doubtful whether a moratorium will be established for some decades instead of resorting to military violence (Llavero-Pasquina et al., 2024).

In Mexico and Argentina, to give two cases out of many, gas and oil fracking has been resisted with different outcomes (Martinez-Alier, 2023, pp. 335–336). In Mexico, CORASON is a coordinating body born in June 2015; activists meeting at the ejido Emiliano Zapata, in Papantla, set up the Coordinadora Regional de Acción Solidaria en Defensa del Territorio Huasteca-Totonacapan. With an estimated 600 trillion cubic feet of recoverable shale gas and 13 billion barrels of recoverable shale oil, Mexico is reported to be one of six countries that account for almost two thirds of recoverable shale gas in the world. Pemex began the exploration process for shale gas production in Chihuahua and elsewhere in 2010, under the constitutional reforms in Mexico that ended Pemex's 75-year monopoly, allowing foreign companies to invest in gas and oil exploration and production. In February 2014, Pemex drilled 29 wells in Chihuahua, and officials claimed that more than 18 trillion cubic meters were found. Cesar Duarte, governor of Chihuahua, supported fracking and claimed that it would boost economic and social development. Duarte met and engaged with European investors, businesspersons, and Pemex in his attempts to bring in more investment in Chihuahua. However, Martha Beatriz Cordova, a Chihuahuan Congress woman, openly opposed fracking. Popular resistance also took place. Hundreds of Chihuahuan small farmers listed fracking in 2014 as one of their main grievances during a blockade of the Bridge of Americas linking Ciudad Juarez with El Paso. Then, the Chihuahua Citizen Alliance against Fracking formed in Chihuahua City. There is growing concern surrounding fracking due to the high levels of overlap with water-stressed regions, threat of toxic waste, threats to natural protected areas, and high levels of seismic activity. The Mexican Alliance Against Fracking formed in 2013, and it brought together a broad range of groups. In November 2014 in Saltillo, members of "Coahuilenses contra el Fracking" (Coahuilians Against Fracking) protested in front of the Autonomous University of Coahuila, where the first Energy Expo forum was taking place and where federal and state authorities were present. Those from San Buenaventura denounced the dispossession of their land by GPA Energy for their exploratory gas projects. The protestors blocked part of federal highway 57 (México-Saltillo). Many antifracking forces pushed Congress to pass a moratorium or ban on fracking in the country. By 2021, this ban was still in operation in Mexico.

In Argentina the outcome was different. In the provinces of Río Negro, Neuquén, and Mendoza, there has been resistance against gas and oil fracking or drilling. Significant deposits of shale oil and gas were discovered in Neuquén province in the Loma de la Lata field in 2010, adding to Argentina's overall shale reserves, estimated to be the second largest in the world. Argentina's Vaca Muerta shale formation is estimated to hold a vast amount of oil and gas. Over the past 20 years in Loma de la Lata and Loma Campana, there have been community-based conflicts over fossil fuel extraction. They first occurred with YPF (the national oil company), Repsol and Petrobras, and later with Chevron, on oil and gas extraction and fracking. The resistance involves some Indigenous people—such as the Confederación Mapuche de Neuquén. There have been blockades against Vaca Muerta. Allen, Río Negro, is a fruit growing zone settled by Italian immigrants decades ago. It belongs to the Vaca Muerta formation, connected from its principal area in Añelo. A government extraction agreement with Chevron was signed in 2013. This is both a "commodity widening" and "commodity deepening" process. There is some local resistance: Maristella Svampa, a leading Latin American political ecologist, was born in Allen, and in 2019 she wrote a sad book about the events (Svampa, 2019).

In the Argentinian dilemma between "civilization or barbarism" that Sarmiento announced in 1845 after Independence, Argentina has opted sometimes for both—that is, the barbarism of Western civilization as in the so-called Conquest of the Desert in Patagonia against Indigenous peoples in 1870 or in the barbarous military dictatorship of 1976–1983. At present, the Argentinian economy is based on extractivism of fossil fuels, metals, and soybeans as never before.

In addition to local grievances against oil and gas extraction, the issue of climate change is often incorporated. This was clearly the case in the Lofoten Islands in Norway, a country whose economy is largely based on oil extraction. The Lofoten Islands became an exception, at least for the time being. Success was achieved in the attempts to "leave oil in the ground," or rather under the sea, believed to hold about 1.3 billion barrels of oil and gas. But the area is also one of abundant fisheries and natural beauty. For many years there was a debate on whether to extract oil from the Lofoten Islands. By 2019, the question was temporarily settled. After years of civil society efforts, by 2019 the alarm at climate change meant that the Labour Party, the country's largest force in Parliament and a long-time backer of the oil industry, decided to stop pushing for oil exploration in the Lofoten Islands, but the Norwegian state continues to explore extraction elsewhere (Martinez-Alier, 2023, p. 339).[2] Other major campaigns in Europe include Ende Gelände in Germany and the Soulèvements de la Terre in France.

The Carbon Debt

Examples could be multiplied from the Global South and North of both successful and unsuccessful local movements to keep fossil fuels underground. The local movements in countries of the Global South are reinforced by the campaigns claiming an ecological debt from the North to the South. They put forward the issue of climate justice. This means that compensation should be provided for the damages that the extractive industries cause by abusing the environment. It arises also from the disproportionate use of the environment as a sink for waste (see chapter by Jorgenson et al. in this volume). This includes the disproportionate use of the atmosphere to dispose of GHGs. The rich countries of the world, and upper-class strata, should be accountable for the past and current inequality of emissions per capita. They should not be allowed to attend international meetings and deny liability for climate change.

The profits that corporations make are in part attributable to the use of the environment as a source for resources or as a pollution sink without any compensation. These two aspects of the ecological debt appear in paragraphs 51 and 52 of the *Encyclical Laudato si* (2015) in simple language quoted here and taken by Pope Francis from publications by Latin American academics and activists since 1992:

> Inequity affects not only individuals but entire countries; it compels us to consider an ethics of international relations. A true "ecological debt" exists, particularly between the global north and south, connected to commercial imbalances with effects on the environment, and the disproportionate use of natural resources by certain countries over long periods of time.... There is a pressing need to calculate the use of environmental space throughout the world for depositing gas residues which have been accumulating for two centuries and have created a situation which currently affects all the countries of the world. There is also the damage caused by the export of solid waste and toxic liquids to developing countries, and by the pollution produced by companies which operate in less developed countries in ways they could never do at home.... The foreign debt of poor countries has become a way of controlling them, yet this is not the case where ecological debt is concerned.

Pope Francis's argument in paragraphs 51 and 52 of *Laudato si* is that the ecological debt is not compensated for, while the external debt is a mechanism of control of economies forcing them into an extractivist pattern.

The words "ecological debt" go back to the early 1990s, coined by environmental justice organizations in Chile and Colombia, against ecologically unequal trade and in complaints against the damage to the ozone layer and also climate change because of fossil fuels burning (Borrero, 1994; Robleto & Marcelo, 1992). There was an alternative treaty in Rio de Janeiro in 1992 (outside the official Earth Summit) on the ecological debt from North to South. In India, Anila Agarwal and Sunita Narain (1991) had published a booklet convincingly and influentially showing the world injustices in the emissions of GHGs. Countries like Bangladesh which suffer most for climate change have no responsibility in the increase in CO_2 concentration in the atmosphere, because its poverty implies historically low per capita emissions, which continues to the present.

Acción Ecológica of Ecuador took up the term "ecological debt" in 1997, with several publications which included a definition. The Jubilee campaign for the abolition of the external financial debt of poor countries coalesced in 2000 with the campaign claiming an ecological debt from North to South because of the disproportionate use by rich countries of nature's services such as the free use of CO_2 sinks and reservoirs (Simms et al., 1999).

There are several methods for calculating the ecological debt from climate change. Basically, how much does the Global North save itself (over many decades) by not making the reduction in emission of GHGs that should be made to avoid climate change (Warlenius, 2015, 2016)? The position of the rich countries in all the COPs has been to deny liability for climate change. The US Climate Envoy repeated in Paris in 2015 what he had said in Copenhagen in 2009: The United States had long acknowledged there was a need to support countries acutely vulnerable to climate change impacts, but there was "one thing that we do not accept and will not accept in this agreement and that is the notion that there should be liability.... That is a line that we can't cross" (http://www.ejolt.org/2015/12/todd-stern-dont-acknowledge-ecological-debt/). The US Climate Envoy was supported by the EU climate commissioner, who said that loss and damage provisions would be included in the Paris Agreement, provided they did not expose wealthy countries to claims for compensation. Therefore, the Paris Agreement refused recognition of an ecological debt for climate change, although a discretionary charity in the form of payment for so-called loss and damage was contemplated. (Bhagwati [2010] suggested to apply "strict liability" internationally as in the US CERCLA legislation.)

The ecological debt seems in principle an economic concept. In the Earth Summit in Rio de Janeiro in 1992, Latin American activists persuaded Fidel Castro to use this concept in his own speech at the official conference. "Do not pay for the external debt, claim instead the ecological debt" —he said (Castro, 1992). Virgilio Barco, the president of Colombia at the time, had already used the expression in Colombia in 1989 (Barco, 1989), and shortly afterward in his speech on June 4, 1990, at an MIT commencement ceremony. Almost one decade later, at its annual general assembly of 1999, Friends of the Earth made ecological debt one of its campaigns for the following years. The notion of an ecological debt is not particularly radical. It can be compared to environmental liabilities incurred by firms (for instance, under the United States CERCLA or Superfund legislation) and in the environmental engineering field called "restoration ecology." However, rich countries have denied that acknowledging losses and damages caused by adverse climate impacts may be construed as an admission of legal liability that might trigger litigation and compensation claims on a major scale. They are afraid of this. They prefer "loss and damage" charity to the acknowledgment of liability. Rich countries, led by the United States and the European Union, included language in the Paris Agreement of 2015 that would prevent them "from being legally on the hook to provide compensation" (World Resources Institute, 2024). The "loss and damage" discussions at COP27 and COP28 concluded that "funding arrangements, including a fund, for responding to loss and damage are based on cooperation and facilitation and do not involve liability or compensation" (World Resources Institute, 2024).

The growth and changes of social metabolism (i.e., the flows of energy and materials that constitute the industrial economy) necessarily incorporate new spaces of extraction of commodities in a process that is called "commodity widening" and "commodity deepening." For instance, new mining in a territory is "commodity widening," while substituting open-pit mining for underground mining is "commodity deepening," increasing extraction with a lower grade in the same location. Fossil fuels, as sources of energy, are burnt and cannot be recycled. Materials such as metals are recycled only in a small proportion, and therefore new commodity extraction frontiers are opened up all the time. Copper, for instance, is recycled only about 30%. There is no compensation for social and environmental damages. Spatial relations are modified, and temporal relations are altered because the extraction of energy and materials in the newly incorporated territories can no longer be governed by the time of reproduction of nature. Capitalism needs new territories and accelerates the extraction times. The antagonism (noticed long ago by Frederick Soddy, 1926) between economic time, which proceeds according to the fast-paced rhythm imposed by capital circulation and the profit rate, and biogeochemical time controlled by the rhythms of nature, is expressed in the irreparable destruction of nature and of local cultures which valued their environment differently. Nature is an open system, and some of its organisms grow sustainably at very rapid rates, but this is not usually the case of the raw materials and products exported by the Global South. Overexploitation of natural resources is intensified when terms of trade worsen for the extractive economies which face payments of the external debt and must finance necessary imports (see chapter by Jorgenson et al. in this volume). Mine tailings and other production of waste increase. The sink capacity of the atmosphere, the oceans, and the new vegetation should not be appropriated unilaterally, and they cannot be expanded without limit. In fact, there is already far too much CO_2 in the atmosphere.

Calculations of the Carbon Debt

How can we determine the limit or target for emissions of GHGs? How much is enough? The carbon debt from the Global North to the South justifies the demands for contraction, convergence of emissions, and compensation. Attempts at using cost-benefit analysis of the increased greenhouse effect (as in some early reports of the Intergovernmental Panel on Climate Change [IPCC]) are not convincing because of the arbitrariness of the discount rate and because many items are not easily measured in physical terms, much less easily valued in monetary terms. Moreover, the very pattern of prices in the economy would be different without the free access to carbon sinks. In the IPCC process in 1995, when it was suggested that greenhouse policy should be guided by a calculus of the economic costs of climate change, including an estimate of the economic value of human lives to be lost in some poor countries, there were loud complaints (Pearce, 1955). Some said that the prices of human lives could not be so different around the world, and moreover should not be discounted. There was a discrepancy in the economic values themselves, but here it would seem that the economists were right in the sense that "the poor are cheap." Now, however, why should the poor stay poor, and will Bangladesh stay poor in 50 years? There was also another more substantial difference of opinion, whether economics should hold the key to an integrated assessment. It does not. A cost-benefit analysis goes against future generations (because of discounting) and also against the poor, whose willingness to pay is necessarily limited— hence, the plausibility of the appeal to noneconomic values. For instance, it can be stated that, while humans have different economic values, they all have the same value in the scale of human dignity. Limits to CO_2 emissions should not be established by economic reasoning of costs and benefits.

Going back to the economic language of the ecological debt, one can consider the case of the environmental service provided by the permanent carbon sink (oceans, new vegetation, soils) and by the atmosphere as a temporary deposit or reservoir where the CO_2 accumulates while waiting for a permanent sink. Notice also that the oceans are acidifying, and therefore they are becoming a doubtful sink for CO_2. The ecological debt can be calculated by the cost avoided by rich countries by evading their moral obligation to reduce emissions of GHGs to the necessary level to avoid climate change (Warlenius et al., 2015, 2016). This fits into K. W. Kapp's (1950) description of externalities not as market failures but as systematic cost-shifting successes. Such debt cannot always be quantified in monetary terms. In any case, the quantities claimed in money should be seen as a fine, not a compensatory payment that would allow a repeat of the same actions increasing the environmental liabilities even more.

Therefore, one possible objection to the notion of the ecological debt is that it implies monetization of Nature's services. The ecological debt must be expressed in terms of social and physical indicators. Another objection to the notion of an ecological debt is that debts are recognized obligations arising from contracts, such as a sale or a mortgage. A nonrecognized debt does not exist, according to this view. However, there are cases in which debts have arisen without a contract—for example, the obligation for Germany to pay reparations after the Second World War to the families of victims of the Holocaust (with the agreement of most citizens of the country).

What would be the cost of reduction of carbon emissions or the price of the extra absorption? When the commitment to reduce emissions is weak, as at present, then, in principle, the price of a ton of carbon in absorption projects will be low because the demand for sinks will be negligible. The price will also be low when the supply of projects in the South is large (whether as additional sinks, especially when conservation of threatened primary forests is also accepted) or when changes in techniques that diminish carbon emissions or capture CO_2 industrially are cheap. In other words, the stronger and quicker the commitment to reduce, the higher the marginal cost of the reduction. If the owners of carbon sinks are poor, the local selling price of carbon absorption will still be low—then intermediaries come into play.

Instead, if there is no reduction, this implies the persistent and disproportionate use of the sinks and the atmosphere as de facto property of the wealthy, and therefore a continuous increase year after year of the ecological debt. The ecological debt rises on this count because, by not doing the necessary reduction, the rich countries save themselves substantial wealth. Such calculations were already made by the mid-1990s (Parikh, 1995). If we took the human-made emissions of CO_2, industrialized countries produced three fourths of these emissions, instead of the one fourth that would correspond to them based on population. The difference is 50% of total emissions. Here the increasing marginal cost of reduction means that the first tons could be reduced at a low cost but then the cost increases rapidly a total of hundreds of billions of dollars if the needed reduction is half the present emissions (Parikh, 1995).

Such calculations were taken up again around the year 2000 and elaborated upon by NGOs concerned with the social and environmental burdens imposed on poor countries by the service and repayment of the external debt. Thus, Christian Aid (1999) made available a document on climate change, debt, equity, and survival (with the title "Who owes who?" and pictures of poor children with water up to their necks). They argued that to mitigate the effects of climate change "we will all have to live within our environmental budget." The atmosphere can only absorb a certain amount of GHGs before disruption begins. So, emissions need controlling. As industrialized countries delay action on the CO_2 emissions cuts that are needed, they go over budget and are running up an environmental or "carbon" debt. Christian Aid's calculation of the "carbon debt" was done in this way: The carbon intensity of gross national product (GNP) is taken as constant, a required reduction of carbon emissions in rich countries of 60% was assumed, and the corresponding decrease in GNP is calculated. The enormous decrease in GNP does not occur because the reduction in emissions does not take place; this is the avoided cost (i.e., the debt). Christian Aid's figures are too high because small reductions of carbon emissions can be achieved with small marginal costs, the marginal cost increasing with the volume and urgency of the reductions. One has to allow for changes in technologies and in the composition of output. But the answer was clear: The costs avoided by rich countries (and people) by not reducing CO_2 emissions to the necessary level are enormous.

Other Christian groups, such as the Canadian Ecumenical Council for Economic Justice, in 2000 estimated the "carbon debt" in the context of the increasing discussion on the ecological debt (www.ecej.org). In 2001, the World Council of Churches adopted this line of thought. There are many uncertainties as to how the future energy systems will develop. Photovoltaic energy becomes cheaper. The number of windmills and solar panels is increasing. If we look at the past century, we see that new energy systems are added on top of the existing ones, without substituting for them. Meanwhile, the carbon debt accumulates.

To sum up, countries and their governments that are in a creditor position in the ecological debt could and should give a sense of urgency to the negotiations on climate change by claiming the ecological debt, which is admittedly hard to quantify in monetary terms. The issue is not the money but the physical and social damage now and in the future by not reducing emissions. The claim of the ecological debt should become a crucial topic in the international political agenda, and this would strengthen the hundreds of grassroot movements organizing against, boycotting, and blockading the oil wells, the coal mines and CFPPs, and the gas and oil pipelines. The external debt from South to North has already been paid on account of the ecological debt the North owes to the South, and we must stop the ecological debt from increasing any further.

Notes

1. See https://keelingcurve.ucsd.edu/.
2. https://financialpost.com/commodities/energy/norway-is-walking-away-from-billions-of-barrels-of-oil.

References

Agarwal, A., & Narain, S. (1991). *Global warming in an unequal world: A case of environmental colonialism*. Centre for Science and Environment, Delhi.

Barco, V. (1989). Los Países Industrializados tienen una Deuda Ecológica con la Humanidad. *Colombia Internacional 6*, 29–31. https://journals.openedition.org/colombiaint/26513?lang=en

Bhagwati, J. (2010, February 22). A new approach to tackling climate change. *Financial Times*. https://on.ft.com/3OxZnuD

Borrero, J. M. (1994). *La Deuda Ecológica. Testimonio de una reflexión*. FIPMA.

Castro, F. (1992). Discurso de Fidel Castro en Conferencia ONU sobre Medio Ambiente y Desarrollo, 1992. Cuba Debate. http://www.cubadebate.cu/opinion/1992/06/12/discurso-de-fidel-castro-en-conferencia-onu-sobre-medio-ambiente-y-desarrollo-1992/

EJAtlas. (2021). Sarayaku—Oil extraction in Block 23, Ecuador. Environmental Justice Atlas (EJAtlas). https://ejatlas.org/conflict/sarayacu-oil-extraction-in-block-23-ecuador

EJAtlas. (2022). Native American's Sioux against the Dakota Access Pipeline (DAPL), USA. Environmental Justice Atlas (EJAtlas). https://ejatlas.org/conflict/dakota-access-pipeline?translate=en

EJAtlas. (2024). Balsa wood extraction, Ecuador. Environmental Justice Atlas (EJAtlas). https://ejatlas.org/conflict/balsa-wood-extraction-ecuador.

Fressoz, J. B. (2024). *Sans transition. Une nouvelle histoire de l'énergie*. Seuil.

Georgescu-Roegen, N. (1975). Energy and economic myths. *Southern Economic Journal*, *41*(3), 347–381.

Gu, B. (2024). Black gold and green BRI—A grounded analysis of Chinese investment in coal-fired power plants in Indonesia. *The Extractive Industries and Society*, *17*, 101411.

Gu, B., & Liu, J. (2024). Unburnable coal and unlabelled climate activism in China (Chapter 22). In S. von Mering, T. E. Bell, A. da Silva, F. Wendy Steele, A. Ward, & M. Arjona Soberón (Eds.), *The Routledge handbook of grassroots climate activism*. Routledge.

Kapp, K. W. (1950). *The social costs of private enterprise*. Harvard University Press.

Klein, N. (2019, September 17). Interview with Jon Queally. *Common Dreams*. https://www.commondreams.org/news/2019/09/17/goddamn-terrifying-time-be-alive-naomi-klein-explains-why-global-green-new-deal

Llavero-Pasquina, M., Navas, G., Cantoni, R., & Martínez-Alier, J. (2024). The political ecology of oil and gas corporations: TotalEnergies and post-colonial exploitation to concentrate energy in industrial economies. *Energy Research and Social Science*, *109*, 103434. https://doi.org/10.1016/j.erss.2024.103434

Martinez-Alier, J. (2021). Circularity, entropy, ecological conflicts, and LFFU. *Local Environment*, *27*, 1182–1207. https://www.tandfonline.com/doi/full/10.1080/13549839.2021.1983795

Martinez-Alier, J. (2023). *Land, water, air and freedom: The making of world movements for environmental justice*. Edward Elgar.

Oilwatch. (n.d.). About us: Oilwatch is a network of resistance to the impacts of fossil fuels (oil, gas and coal) industries on people's and their environments. https://www.oilwatch.org/about-us/

Parikh, J. K. (1995). Joint implementation and the North and South cooperation for climate change. *International Environmental Affairs*, *7*(1), 22–41.

Pearce, F. (1955). Global row over value of human life. *New Scientist*. 19 August. http://www.newscientist.com/article/mg14719910.900-global-row-over-value-of-huma

Robleto, M. L., & Marcelo, W. (1992). *Deuda ecológica*. Instituto de Ecologia Politica.

Roy, B., & Schaffartzik, A. (2021). Talk renewables, walk coal: The paradox of India's energy transition. *Ecological Economics*, *180*, 106871.

Scheidel, A., Del Bene, D., Liu, J., Navas, G., Mingorría, S., Demaria, F., Avila, S., Roy, B., Ertör, I., Temper, L., & Martinez-Alier, J. (2020). Environmental conflicts and defenders: A global overview. *Global Environmental Change*, *63*, 102104. https://doi.org/10.1016/j.gloenvcha.2020.102104

Shukla, A. (2024). Goldam prize awarded to Alok Shukla. https://www.goldmanprize.org/recipient/alok-shukla/

Simms, A., Meyer, A., & Robins, N. (1999). *Who owes who? Climate change, debt, equity and survival*. Christian Aid.

Slameršak, A., Kallis, G., & O'Neill, D. (2022). Energy requirements and carbon emissions for a low-carbon energy transition. *Nature Communications*, *13*, 6932.

Soddy, F. (1926). *Wealth, virtual wealth, and debt: The solution of the economic paradox*. G. Allen and Unwin.

Svampa, M. (2019). *Chacra 51, Regreso a la Patagonia del fracking*. Editorial Sudamericana.

Temper, L., Yánez, I., Sharife, K., Ojo, G., Martinez-Alier, J., CANA, Combes, M., Cornelissen, K., Lerkelund, H., Louw, M., Martínez, E., Minnaar, J., Molina, P., Murcia, D., Oriola, T., Osuoka, A., Pérez, M. M., Roa Avendaño, T., Urkidi, L., Valdés, M., Wadzah, N., & Wykes, S. (2013). *Towards a Post-Oil Civilization: Yasunization and other initiatives to leave fossil fuels in the soil*. EJOLT Report No. 6, 204 p.

Temper, L., Avila, S., Del Bene, D., Gobby, J., Kosoy, N., Le Billon, P., Martinez-Alier, J., Perkins, P., Brototi, R., Scheidel, A., & Walter, M. (2020). Movements shaping climate futures: A systematic mapping of protests against fossil fuel and low-carbon energy projects. *Environmental Research Letters, 15*(2), 123004.

Thiri, M. A., & Borsi, M. T. (2024) From Fukushima to fossil fuels: Carbon emissions, climate narratives, and grassroots movements in Japan's energy transition. *Energy Research and Social Sciences, 112*, 103520.

Warlenius, R. (2016). Linking ecological debt and ecologically unequal exchange: Stocks, flows, and unequal sink appropriation. *Journal of Political Ecology, 23*(1), 364–380.

Warlenius, R., Pierce, G., & Ramasar, V. (2015). Reversing the arrow of arrears: The concept of "ecological debt" and its value for environmental justice. *Global Environmental Change, 30*, 21–30.

World Resources Institute. (2024). What is "loss and damage" from climate change? 8 key questions, answered. https://www.wri.org/insights/loss-damage-climate-change

York, R., & Bell, S. E. (2019). Energy transitions or additions?: Why a transition from fossil fuels requires more than the growth of renewable energy. *Energy Research & Social Science, 51*, 40–43. https://doi.org/10.1016/j.erss.2019.01.008

CHAPTER 16

Fossil Fuel Extraction Conflicts as Climate Action

MOISES ARCE AND ZOË SULLIVAN

Introduction

THE world stands at a critical juncture in mitigating the threats of climate change, and extractive industries are facing increasing scrutiny and escalating resistance from grassroots social movements (Dolšak & Prakash, 2022; Temper et al., 2020). The fossil fuels economy, a primary driver of greenhouse gas emissions, has become a flashpoint for the broader struggle for climate justice. Climate justice is a framework that seeks equitable treatment and rights for all individuals and communities affected by climate change, with a focus on addressing the disproportionate impacts faced by marginalized and vulnerable populations. Its key principles include distributive justice (e.g., fair allocation of resources and responsibilities); procedural justice (e.g., meaningful participation of all stakeholders in decision-making processes); and recognitional justice (e.g., respect for the identities and rights of impacted communities) (Hasan, 2018; Schlosberg & Collins, 2014). These principles aim to redress historical injustices and promote inclusive policies that both mitigate climate change and enhance social equity.

In resource-abundant Latin America, states and fossil fuel companies play an outsized role in the development of the resource sector. Petrobras in Brazil, YPF in Argentina, YPFB in Bolivia, and their foreign partners wield considerable power in the decision-making and planning processes for oil extraction across the Global South (Dolšak & Prakash, 2022; Orta-Martínez & Finer, 2010; Wainwright & Mann, 2015). The larger global shifts toward renewable energy are also expanding new extractive industries in the region. This includes bridge fuels like natural gas in Peru and Bolivia, and the mining of lithium in Argentina, Bolivia, and Chile, which are essential for battery production and renewable technologies. The recent swing toward the political right has increased climate denialism (e.g., Argentina's Javier Milei in 2024 and Brazil's Jair Bolsonaro in 2019), undermining the scientific consensus on climate change and its impacts.

Indigenous communities often bear the disproportionate brunt of the environmental impacts of resource extraction. They are mobilizing to protect their lands, livelihoods, and cultural identities from the impacts of oil extraction, and many of their mobilizations are finding larger connections to the global struggle for climate justice (Muttitt & Kartha, 2020; Temper et al., 2020). The tension between oil extraction and Indigenous well-being is creating conflicts worldwide (see chapter by Fenelon and Figueroa Hellan in this volume). Recent data indicate that over 807 conflicts regarding oil extraction and climate justice are ongoing worldwide, with the highest concentrations of oil conflicts in Africa, Central and South America, and South and Southeast Asia (Temper et al., 2018).

South America exemplifies this broader struggle, with multiple and prolonged conflicts surrounding the resource sector. Today, oil extraction remains a significant economic activity in both Peru and Ecuador, yet it faces ongoing scrutiny and opposition (Arce et al., 2020; Eisenstadt & West, 2019; Orta-Martínez et al., 2018). Growing awareness of climate justice and the pursuit of a global climate governance agenda puts governments and oil companies under increasing pressure to adopt more sustainable practices and redress the historical wrongs (Orta-Martínez & Finer, 2010; Schlosberg & Collins, 2014). However, the expansion of fossil fuel economies continues to encroach on Indigenous territories, leading to displacement, loss of livelihoods, and cultural disruption (Orta-Martínez et al., 2018; Orta-Martínez & Finer, 2010). Environmental degradation has had and continues to have severe impacts on the Amazon's fragile ecosystems (Finer et al., 2008; Johnston et al., 2019). The increasing trend of purchasing forest lands from Indigenous communities for carbon credits also presents challenges for Indigenous peoples as they often bypass traditional knowledge and governance systems. The expansion of extractive activities, including the global shift toward renewable energy, presents pathways for utilizing lessons learned from the struggles over oil extraction to seek more inclusion of Indigenous voices in the planning and execution of these newer projects (Gustafsson & Schilling-Vacaflor, 2022; Gustafsson & Scurrah, 2023).

We are witnessing a marked change in how Indigenous communities are mobilizing against the impacts of extraction. These communities have brought their struggles to the forefront of international attention in an unprecedented manner and reveal new opportunities for climate justice. However, the demands and strategies of mobilization among these communities are varied, creating a diverse landscape of resistance that is different from one country to the next. This chapter examines the evolving landscape of oil conflicts in the South American context, highlighting the diverse demands and multifaceted resistance of Indigenous groups in the region. The chapter also explores the ways in which these struggles connect to larger themes of climate justice and what this could mean for the future of resource and climate governance as the world seeks a shift away from fossil fuels.

Oil Extraction in the Amazon Rainforest

Oil extraction in the Amazon regions of Ecuador and Peru has a long and complex history, marked by tensions between economic development and socio-environmental conflicts. The region saw oil exploration begin in the early 20th century, driven by growing global demand. However, significant discoveries of oil reserves and large-scale extraction operations did not commence until the late 1960s and early 1970s (Arce et al., 2024). In 1967, Texaco discovered substantial oil reserves in the northeastern Amazon region of Ecuador, known as the Oriente, which was and remains home to many Indigenous communities (Eisenstadt & West, 2019). This led to the construction of large-scale oil infrastructure and pipelines, and the start of large-scale extraction that would come to dominate Ecuador's economy.

Similarly, in Peru, large-scale oil operations began in the 1970s as the country, eager to develop its hydrocarbon resources to boost its economy and reduce dependency on foreign oil, welcomed companies like Occidental Petroleum (Oxy) (Orta-Martínez et al., 2018). Occidental Petroleum acquired concessions for oil exploration and production in Peru's northern Amazon region. Most notable were the concessions granted to Oxy for Block 1AB, located in the Loreto region, which became the center of large-scale oil production in the nation from the 1970s through the 1980s (Orta-Martínez et al., 2007; Orta-Martínez & Finer, 2010). Additionally, the discovery of the Camisea gas fields in the 1980s marked a significant shift toward increased hydrocarbon exploration and extraction activities in the Peruvian Amazon (Acosta, 2013; Arce et al., 2024).

Both governments were complicit in the expansion of the resource sector and have aggressively opened up the Amazon rainforest to large-scale oil activities. Both governments have traditionally framed oil extraction as a key driver of national development, promising jobs, infrastructure improvements, revenue generation, and economic stability. In Ecuador, revenues from oil exports have historically constituted a significant portion of the national budget, funding infrastructure projects, social programs, and public-sector salaries (Cueva & Diaz, 2018). In Peru, the government has portrayed oil revenues as crucial for funding public services, infrastructure, and social programs, especially in economically underdeveloped regions like Loreto that are heavily reliant on the oil industry (Orta-Martínez et al., 2018). While the Peruvian government markets oil extraction as a positive source of jobs, royalties, and development projects for local communities, the impacts of extraction often outweigh the potential benefits for the communities close to extraction sites.

Uneven Benefits, Lasting Costs

While oil companies and national governments profited greatly from the expansion of the fossil fuels economy, the Indigenous communities living closest to the extraction sites have rarely benefited from its presence. Quite the contrary, they experience persistent poverty and some of the lowest levels of human development. For example, in the Loreto region of Peru, where oil production is concentrated, Human Development Index (HDI) indicators reveal large disparities in education, health, and overall state density compared to national averages (Arce et al., 2024). The environmental impacts of oil extraction are also felt most sharply by Indigenous peoples. These impacts are well-documented and include illegal dumping, oil spills, and other related pollution problems that often cause severe effects on the health of Indigenous communities (Johnston et al., 2019). Complicit national governments have rarely placed the interests of Indigenous peoples over those of oil companies. Government oversight has been consistently weak and far from meeting international environmental standards.

Yet Indigenous peoples were not passive spectators to the growth of the fossil fuels economy within Ecuador and Peru. They have voiced their concerns about the environmental and socioeconomic impacts of oil extraction since the early days of operations in the region. Two major high-profile environmental disasters and the ensuing legal battles reflect early political action against oil companies and the attention of global audiences. But remediation is and was always late, reflecting the power asymmetries among governments, oil companies, and Indigenous communities.

In Ecuador, Texaco's operations led to one of the worst modern environmental disasters often referred to colloquially as the "Amazonian Chernobyl" (San Sebastián & Hurtig, 2004). From 1964 to 1992, over 16.8 million gallons of crude were spilled in the Amazon basin by the Texaco operation, in addition to substantial contamination caused by the dumping of toxic wastewater into unlined pits and rivers near the Lago Agrio oil fields (Pellegrini et al., 2020). This contamination polluted the soil and water of the area, leading to disastrous environmental, health, social, and economic effects for local communities comprised of mainly Indigenous peoples and small-scale farmers (San Sebastián & Hurtig, 2004). A class-action lawsuit against Texaco (purchased later by Chevron) on behalf of the native communities was filed in 1993 in the United States, where Texaco operations were based—becoming one of the first legal challenges to oil extraction and its consequences (Acosta, 2011; Pellegrini et al., 2020). The case was dismissed in the United States in 2001, but it was later brought to Ecuadorian courts, where in 2011 Chevron was found guilty of extensive pollution and ordered to pay billions in compensation (Acosta, 2011). Despite this victory, the Ecuadorian government—eager to maintain oil revenues and attract foreign investment—has continued to be criticized for failing to adequately clean up contaminated sites or hold the oil companies accountable for the damage caused (Acosta, 2011; Pellegrini et al., 2020).

In Peru, Oxy's operations, particularly those in Block 1AB, also led to significant environmental degradation and conflicts. Their extractive activities also created widespread contamination of soil, groundwater, and rivers with toxic wastewater, crude oil, and heavy metals. This led to similar catastrophic consequences for the health, environment, and livelihoods of the Indigenous communities in the areas of the Achuar, Quechua, and Urarina populations (Fraser, 2018; Orta-Martínez et al., 2007). The affected Indigenous communities began to organize and demand accountability from Oxy and the Peruvian government throughout the 1990s and 2000s. Increased mobilization and support from local and international nongovernmental organizations brought the struggle of the local people to the attention of global audiences. Mounting pressure from environmental and Indigenous campaigns led Oxy to sell its stake in Block 1AB to Pluspetrol in 2000 (Orta-Martínez et al., 2018). In 2007, a group of Achuar plaintiffs filed a lawsuit against Occidental Petroleum seeking compensation for the environmental and health impacts of their operations in Block 1AB, arguing that Oxy had violated multiple US and international environmental standards during its operations. In 2015, Oxy reached a confidential settlement with the Achuar plaintiffs (Orta-Martínez et al., 2018).

The Occidental and Chevron rulings served not only to establish significant legal precedents for holding oil companies accountable but also to bring the negative effects of oil extraction on the Amazon rainforest and the Indigenous peoples of the region to the attention of the global community (Orta-Martínez et al., 2007). They were key in the creation and recognition of the global frameworks for Indigenous consultation on extractive activities that are critical to the present era of Indigenous mobilization surrounding both oil extraction and climate justice—particularly the principle of procedural justice. They have galvanized collective action among Indigenous peoples, but this increased political action has often yielded limited concrete change, as the international and national frameworks that have emerged from these initial efforts often prove toothless in the face of powerful economic interests and entrenched systems of inequality.

Weak State Oversight

Weak state oversight of the oil industry leaves Indigenous communities vulnerable to environmental risks and other harms. While existing on paper, the international and national regulatory frameworks for environmental protection and Indigenous rights that have come to prominence since the 1990s have frequently been poorly enforced in practice. Environmental impact assessments (EIAs), for instance, are systematic evaluations of the environmental, social, and economic impacts of proposed extractive activities (Morrison-Saunders & Retief, 2012). EIAs typically include considerations such as air and water quality, wildlife, public health, and socioeconomic factors. The purpose of an EIA is to provide decision-makers with the necessary information to evaluate project proposals and to mitigate any adverse impacts before they occur (Morrison-Saunders & Retief, 2012). In both Peru and Ecuador, EIAs are conducted poorly, and they lack thoroughness and accuracy in assessing potential environmental and social impacts (Flemmer & Schilling-Vacaflor, 2016; Merino, 2018; Schilling-Vacaflor & Flemmer, 2015). Public participation and consultation in the EIA process are also limited, undermining the principles of environmental justice, particularly the principle of procedural justice (Flemmer & Schilling-Vacaflor, 2016; Merino, 2018). Weak regulatory frameworks and a lack of independent review mechanisms also contribute to the ineffectiveness of EIA enforcement and oversight (Flemmer & Schilling-Vacaflor, 2016; Merino, 2018).

Free, prior, and informed consent (FPIC) and prior consultation (*consulta previa*) are two other mechanisms of participation and oversight at play in the region. These are principles aimed at ensuring meaningful participation and decision making by Indigenous communities and local populations in matters that may affect their rights and interests (Flemmer & Schilling-Vacaflor, 2016). FPIC is an international standard set by international human rights instruments that oblige states and industries to obtain consent from Indigenous peoples before undertaking projects that could impact their lands, resources, or livelihoods.[1] Grounded on similar international human rights frameworks and building off this idea of FPIC, prior consultation is a more specific mechanism used to engage Indigenous communities in decisions that impact their rights and territories (Flemmer & Schilling-Vacaflor, 2016; Schilling-Vacaflor & Flemmer, 2015). It is meant to facilitate dialogue between Indigenous communities and governmental or corporate entities, allowing for the expression of community concerns, preferences, and rights, and is found in legal and policy frameworks of several Latin American countries (Becerra Valdivia, 2022).

These mechanisms of participation are seen as major achievements in championing Indigenous rights, but numerous problems have arisen in their implementation. Capacity and the responsibility of the state to engage with Indigenous communities under the bounds of the prior consultation in good faith have been a major issue. It seems that in multiple cases government officials see prior consultation as a box to check rather than a sincere driver of engagement—with the government not taking the needed steps to ensure that Indigenous communities can meaningfully participate in project deliberations, for example, presenting the information in culturally and linguistically appropriate ways, providing extra training to improve the capacity of Indigenous peoples to participate in prior consultation, failure of government officials to show up to planned meetings, and so on (Schilling-Vacaflor & Flemmer, 2015; Zaremberg & Torres Wong, 2018).

Moreover, even when consultation occurs, it appears that the government often does not create space for Indigenous peoples to have meaningful decision-making capability. There is limited effectiveness of joint and binding decisions between the state and the consulted group on the ground—particularly concerning extractive activities—as many claims and proposals from Indigenous communities have been excluded from the decisions ultimately adopted (Vargas & Torres, 2019). Genuine compromises and mutual agreements on solutions that represent the interests of all parties involved are rare. Decisions do not address unresolved conflicts, potentially escalating tensions, and the postponement of resolutions of protracted disputes (Vargas & Torres, 2019). Additionally, numerous critical violations of agreements reached during the consultation process occur. Many agreements are altered post consultation, sometimes very egregiously to the extent that they have the opposite meaning to what was agreed upon (Merino, 2018). Finally, a lack of specific mechanisms for monitoring the implementation of agreements and sanctioning of violations of said agreements leads to very limited effectiveness of the prior consultation process (Merino, 2018). This creates a system that speaks toward Indigenous empowerment and more equitable extraction—but lacks real effectiveness in much more than name. This process has been seen across the region as other nations have adopted similar frameworks and policies to similar effects.

In many ways, the creation of these participatory mechanisms and their subsequent challenges and failures became the catalyst for major Indigenous mobilization in the region (Jaskoski, 2020). Persistent weak state oversight incentivized greater opportunities for mobilization by Indigenous peoples, leading to further resistance and open conflict on the ground, more strategic litigation tactics, and increased capacity to connect regionally and internationally with other communities, raising local struggles to the global stage (Arce et al., 2020; Schilling-Vacaflor & Flemmer, 2015). However, it would be an oversimplification to characterize the demands of Indigenous populations in the Amazon region as uniform or homogeneous.

Diverse Demands

The demands of Indigenous peoples are diverse and include advocating for environmental remediation, equitable benefit-sharing of resource rents, recognition of communal property rights, and co-management of compensation funds in the extraction areas. The emphasis on advocating for improved environmental governance and fair distribution of benefits highlights a nuanced approach to addressing the impacts of extraction, beyond that of only stopping extraction, which appears insurmountable (Arce et al., 2024; Eisenstadt & West, 2019; Orta-Martínez et al., 2018).

In older oil fields, some Indigenous communities prioritize remediation and compensation for environmental damage, seeking to hold oil companies and the state accountable for cleaning up pollution and restoring affected ecosystems. For example, the Achuar peoples in Peru sought primarily remediation and compensation for the environmental damage caused by oil operations, as well as holding the state and oil companies responsible for cleaning up pollution (Orta-Martínez et al., 2018). In newer oil fields, Indigenous communities tend to focus on securing a more equitable distribution of oil wealth, demanding greater benefits, social investments, and decision-making power over resource development in their territories. The Kukama peoples, also from the Loreto region of Peru, primarily sought equitable benefit-sharing of resource rents, recognition of land rights, co-management of funds in the oil circuit, and the government distribution of public goods, with a focus on seeking greater social investments and decision-making power in the resource development of the region (Arce et al., 2024).

Competing demands reflect divisions within and among Indigenous peoples in the Amazon region. This presents obstacles to the effective representation of these communities in consultations and decision-making processes. For example, in some cases, government officials will choose to meet only with Indigenous communities who regularly receive compensation from oil companies to fulfill their obligation for FPIC, rather than other communities who stand in opposition to extraction (Gustafsson & Schilling-Vacaflor, 2022; Watson & Arce, 2024). Competing demands may hamper mobilization strategies in opposition to the fossil fuels economy but can also present opportunities for different mobilizations to connect to larger struggles for climate justice in different ways. For example, the mobilizations focused on seeking greater rights in the decision-making process on projects, self-determination, and equity may align themselves with broader movements seeking procedural justice while the mobilizations focused on more equitable sharing of resource rents may frame their demands in the larger context of distributional justice (Muttitt & Kartha, 2020; Watson & Arce, 2024).

Multifaceted Resistance

Indigenous communities have employed a range of strategies, from direct action such as occupying oil fields and legal challenges to media activism and international advocacy, to pressure for change, whether redressing environmental concerns or otherwise. In this section, we explore some examples of different mobilization strategies against oil extraction in the Amazon region embraced by Indigenous communities to varying degrees of effectiveness.

Mobilization

While open conflict is undesirable and dialogue is the preferred outcome in political struggles, research examining Indigenous mobilization in the Amazon indicates that conflict can sometimes be a productive solution. Orta-Martinez refers to the concept of a "conflict imperative," which posits that certain circumstances, such as overcoming environmental injustice or asserting Indigenous rights, necessitate various forms of direct action, including conflict with external actors like oil companies or the state (Orta-Martínez et al., 2018).

The "conflict imperative" challenges the notion that dialogue and negotiation alone are sufficient to address deeply rooted environmental and societal injustices (Orta-Martínez et al., 2018). Actions such as occupying oil facilities, blockading operations, and other methods to disrupt oil activities are not only a typical mobilization strategy but, in some cases, the best strategic move for Indigenous communities facing an uneven playing field. These actions are often the byproduct of broken commitments or poorly enforced regulations, leading to opposition as a last-resort strategy for Indigenous peoples (Orta-Martínez et al., 2018). The strategies of nonconventional protest as more efficacious for excluded groups are also consistent with the social movement literature (Almeida, 2019).

The Achuar people, who inhabit the Loreto region of the Northern Peruvian Amazon, provide a compelling example of the importance of direct action as a mobilization strategy. This region hosts the country's longest running and largest oil production block, known as Block 1AB. In October 2006, over 800 Achuar community members occupied the facilities at the Block 1AB oil field (Orta-Martínez et al., 2007, 2018). This occupation came just a week after the Achuar had released a statement expressing their frustration: "For many years we have tried to dialogue with companies and governments, and nevertheless nothing has changed, and contamination continues" (Orta-Martínez et al., 2018, p. 6). This occupation marked the culmination of over three decades of the Achuar's protracted struggle against oil activities, during which they had primarily employed dialogue and negotiation tactics.

During the occupation, the Achuar's key demand was that the government and the oil company, Pluspetrol, should reinject the toxic by-products of oil extraction back into the oil field, rather than allowing them to contaminate the surrounding lands. This was not a radical request, but rather a call for the Peruvian government to hold Pluspetrol accountable to existing laws. As Orta-Martinez notes, the Achuar decided to take matters into their own hands after decades of talks and negotiations failed, seeking to challenge "not only the willful negligence of the oil company" but also the "inability or unwillingness of the Peruvian state to enforce its own laws or to protect the wellbeing of its indigenous communities" (Orta-Martínez et al., 2018, p. 7). The Achuar claimed that the government would summon them to travel to the distant city of Lima for talks; this ultimately led to no concrete action and was seemingly a ploy to wear them down (Orta-Martínez et al., 2007).

In January 2005, the Achuar community engaged in their first direct action against the oil company Pluspetrol, blocking the oil road, in response to the company's failure to fulfil a community–company agreement to provide electricity and drill water wells. This initial mobilization was temporarily halted when a new commitment to work toward change and improved conditions was signed (Orta-Martínez et al., 2018). However, after a deadline passed in June 2005, the community once again blockaded the road for 3 days, prompting Pluspetrol to sign another written agreement (Orta-Martínez et al., 2018).

In May 2006, a health study revealed alarmingly high levels of cadmium and lead in the Achuar population, drawing greater attention to their ongoing struggle (Orta-Martínez et al., 2018). While a Health Action Plan was designed in June 2006, its implementation was severely lacking. This was followed by a series of disappointing negotiations between the government and Indigenous leaders, where the leaders were repeatedly required to travel to distant locations, government officials failed to attend, and officials met with an Indigenous organization known to receive compensation from oil companies. These failed negotiations ultimately led to the occupation of the oil facility, which was instrumental in implementing international environmental standards and national legislation that the Peruvian state had been unwilling to enforce.

The Dorissa Accords resulted in Pluspetrol agreeing to reinject all produced wastewater into the oil blocks within Achuar territories. Pluspetrol committed approximately US$ 10 million to implement the previously designed Health Action Plan (Orta-Martínez et al., 2018). The agreement also allocated funds for a development plan to provide food and infrastructure aid to the Achuar. However, the original agreements only benefited the Achuar people, without any changes or aid offered to the Kichwa and Quechua peoples in the same region, who were also affected. This led to another 5-year period of negotiations, breakdown of negotiations, and protests in the region (Orta-Martínez et al., 2018).

The Achuar's mobilization represented a turning point that showed Indigenous peoples sometimes had no choice but to take direct action to have their needs met. This would eventually lead to further legislation and agreements within the region and across the Amazon. Their actions drew international attention to the environmental and social injustices faced by Indigenous communities due to oil extraction, and the need for greater environmental oversight in the Amazon. Furthermore, the Achuar's struggle influenced public discourse on responsible resource extraction and prompted discussions on the rights of Indigenous peoples and the importance of preserving their cultural and ecological integrity.

Strategic Litigation

The use of strategic litigation has been an important avenue for Indigenous communities seeking to challenge the impacts of oil extraction in the Amazon. Strategic litigation refers to the deliberate and calculated use of legal action to bring about broader social or policy change, beyond just the immediate outcome of a particular case (Buckel et al., 2024). This approach involves identifying specific legal cases that have the potential to set precedents, challenge existing laws or practices, and advance human rights and/or social justice causes. Strategic litigation often targets systemic issues and aims to effect change at a societal level rather than an individual level. Strategic litigation can be seen as a form of institutionalized climate action and potentially an important source for climate reparation (Almeida et al., 2023).

The Amazon region has been a leader in using strategic litigation against oil companies. Multiple landmark cases, such as the Texaco-Chevron decision, have been among the first to use strategic litigation to combat oil extractive industries and create precedents that influence climate governance globally (Tigre et al., 2023). Moreover, the unique way that strategic litigation cases are framed by Indigenous peoples in the region—not only as struggles for the protection of biodiversity or direct environmental impacts but also as cases seeking the protection of Indigenous rights, human rights, and Indigenous right to self-determination—has created new avenues to seek climate justice at both national and international levels by invoking the language and principles of distributional, procedural, and recognitional justice (Auz, 2022; Rodríguez-Garavito, 2020; Tigre et al., 2023).

The Sarayaku litigation case is a prominent example of strategic legal action undertaken by Indigenous communities in the Amazon region to challenge oil extraction activities.

The case centers on the Sarayaku community in Ecuador, who took legal action against the government for granting oil exploration and extraction concessions within their territories without obtaining their free, prior, and informed consent, as required under international law and the Ecuadorian constitution (Rodríguez-Garavito & Baquero-Díaz, 2022). The government not only failed to consult the Sarayaku but also disregarded their opposition to these activities, which the community viewed as a violation of their land rights, cultural integrity, and environmental well-being.

In response, the Sarayaku, with the support of legal and advocacy organizations, pursued a strategic litigation approach. They took their case to the Inter-American Court of Human Rights, arguing that the government's actions had violated their rights to territory, self-determination, and a healthy environment, thereby jeopardizing their social, political, economic, environmental, and cultural survival. The Sarayaku framed their legal battle within a broader global context of FPIC, human and Indigenous rights, environmental rights, and self-determination, which not only garnered international recognition and support but also set a precedent for other Indigenous peoples seeking to use strategic litigation in the fight for climate justice (Rodríguez-Garavito & Baquero-Díaz, 2022).

By invoking the principles of FPIC and international human rights law, the Sarayaku demonstrated the flaws in its implementation and how this can lead to human rights violations—calling for stronger mechanisms to ensure procedural justice and demonstrating that legal action is one method to seek recourse for violations of FPIC. They also emphasized their rights to self-determination and cultural survival, portraying the issue as a matter of Indigenous sovereignty and autonomy, and highlighting the need to preserve their unique way of life and traditions and the importance of recognitional justice in climate governance. Furthermore, they framed their case within the broader context of the right to a healthy environment, arguing that the proposed oil exploration would not only undermine their cultural survival but also pose grave environmental risks to their territory, thereby underscoring the interconnectedness of their cultural survival and the health of the environment, as well as the significance of their land as a crucial aspect of their cultural identity and spiritual practices (Rodríguez-Garavito & Baquero-Díaz, 2022).

Additionally, the Sarayaku community strategically engaged with international human rights bodies and legal experts, effectively mobilizing global support and raising awareness about their plight through online advocacy (Rodríguez-Garavito & Baquero-Díaz, 2022). The strategic framing of their struggle enabled them to garner solidarity from human rights organizations, environmental groups, Indigenous communities, and environmentally conscious individuals worldwide, amplifying the visibility and significance of their cause. This also allowed them to forge connections between their fight for Indigenous rights and the larger global movements for climate action in the face of climate change by tapping into the core tenets of procedural and recognitional justice (Natcher et al., 2022; Rodríguez-Garavito & Baquero-Díaz, 2022).

In 2012, the Inter-American Court of Human Rights ruled in favor of the Sarayaku community, finding that the Ecuadorian government had violated their rights to life, physical integrity, consultation, community ownership, and cultural identity (Vargas & Torres, 2019). The court ordered the Ecuadorian government to undertake a meaningful free, prior, and informed consent process for any future large-scale projects on Indigenous lands. Furthermore, the court directed the government to provide compensation for material and nonmaterial damages to the Sarayaku, train public employees working with Indigenous peoples on their rights, and organize a public recognition of responsibility on Sarayaku lands (Vargas & Torres, 2019).

The Sarayaku litigation case stands as a landmark victory for Indigenous rights and environmental protection. The ruling not only secured the immediate rights of the Sarayaku community but also established vital legal precedents in Ecuador and beyond. It underscored the crucial importance of meaningful and clearly legislated FPIC for Indigenous communities facing resource extraction on their lands, and it highlighted the shortcomings of current FPIC recognition and implementation, emphasizing the need for meaningful procedural justice within the larger movement for climate governance (Auz, 2022; Rodríguez-Garavito, 2020). Beyond its legal impact, the case empowered the Sarayaku people, amplifying their voices and fostering a sense of agency within the community. By strategically leveraging international human rights mechanisms on a global stage, the Sarayaku demonstrated the power of strategic litigation to challenge government actions, protect Indigenous lands, and contribute to a growing body of jurisprudence on Indigenous rights and environmental justice.[2]

Media Activism

The expansion of Internet access in Ecuador and Peru has bolstered climate media activism, enabling Indigenous and environmental groups to leverage digital platforms for advocacy and mobilization. As Internet connectivity has surged across Latin America, increased access to social media has facilitated information dissemination and the organization of grassroots movements (Lupien et al., 2024). This digital shift has empowered various groups to share their experiences, raise climate awareness, and connect with broader networks, thereby amplifying their impact on climate justice efforts. Platforms like Facebook, Twitter, Whatsapp, and Instagram have enabled these groups to engage national and international audiences, fostering solidarity and support for their causes (Lupien et al., 2024).

Indigenous communities have effectively utilized media activism to articulate the environmental and socioeconomic repercussions of oil extraction. Social media, online petitions, and digital storytelling have empowered these communities to amplify their concerns, share narratives, and mobilize national and international solidarity (Watson & Arce, 2024). Digital platforms serve as critical tools for advocacy and accountability, allowing Indigenous communities to document environmental degradation, rights violations, and instances of corporate and governmental negligence (Arce et al., 2024; Orta-Martínez et al., 2018).

Consequently, the convergence of increased Internet accessibility and climate activism has created a dynamic space for Indigenous voices to challenge environmental injustices and advocate for sustainable practices. Media activism has effectively bridged the gap between the global community and the Amazon, providing visibility to local struggles that were once marginalized. This connection has strengthened alliances between grassroots mobilizations in the Amazon and the larger climate justice movement, with Indigenous communities framing their local efforts within the broader climate justice framework and global efforts gaining insight into the practical application of its principles in real-world contexts.

Political Consequences of Political Activity

The increased political mobilization by Indigenous peoples during the expansion of the fossil fuels economy has resonated globally, prompting a reassessment of power dynamics. Indigenous communities are no longer passive in the face of extractive activities; instead, they are seizing the political stage, demanding recognition of their rights, and articulating alternative visions of ecological development (Acosta, 2012). As highlighted previously, Indigenous peoples, particularly in the Amazon region, are leveraging diverse avenues of joint political action to connect with the growing global attention centered on climate action and justice. Media activism has enabled these communities to bring their struggles to the international stage and forge alliances with national, regional, and transnational movements. Moreover, as exemplified by the Sarayaku case, Indigenous peoples are strategically framing their struggles within the broader fight for climate justice, strengthening their legal battles by grounding them in international legal frameworks that recognize climate change as a human rights issue, thus helping to set legal precedents supporting procedural and recognitional justice frameworks at a global scale (Rodríguez-Garavito, 2020; Tigre et al., 2023). The movements and mobilizations of Amazonian Indigenous groups against oil extraction are taking on new forms and demanding a larger voice not just within their nation but as a part of a larger global vision for a different future rooted in climate justice.

Furthermore, Indigenous peoples in South America are actively shaping the global discourse on climate justice—articulating new approaches to climate governance and ecology grounded on their traditional beliefs and cosmovisions. Concepts like "Buen Vivir" or "Sumak Kawsay" in the Andean region, which emphasize living in harmony with nature and prioritize collective well-being, are gaining traction within both transnational Indigenous and peasant movements and transnational climate justice movements (Acosta, 2012; Eisenstadt & West, 2019). By promoting their unique cosmovision as an alternative framework for ecological sustainability and understanding, Indigenous peoples challenge dominant paradigms of development, advocating for holistic approaches that prioritize environmental integrity, social equity, and cultural diversity. Their efforts enrich the conversation on recognitional justice and alternative visions for the future of climate governance, demonstrating the value of Indigenous knowledge systems and offering pathways toward a more just and sustainable future (Acosta, 2012).

Moreover, this increased political voice of Indigenous groups is concretely manifested in the demands for meaningful implementation of participatory mechanisms, in ways that significantly empower Indigenous communities in more than name only. As discussed previously, on paper these institutions appear to signal a significant shift toward inclusive decision making, granting Indigenous communities a formal platform to voice their concerns and potentially influence the trajectory of extractive activities; however, in practice these mechanisms have often proved to do little good and in many cases actively stoke tensions between Indigenous peoples and the government and oil companies.

Critics argue that these institutions, often weakened by limited scope, resources, and enforcement mechanisms, are easily co-opted by powerful political and economic interests. Instead of genuine dialogue and consent, these consultations often become performative exercises, carefully orchestrated to legitimize predetermined outcomes and ultimately reproduce the very model of extractive activities they were ostensibly designed to challenge (Merino, 2018). However, while these mechanisms have proved thus far to be relatively weak, the increased political activity of Indigenous peoples can help to close "implementation gaps" and potentially make those mechanisms stronger or suitable to other ends (Arce & Jaskoski, 2024). The examples discussed herein highlight how mobilizations, litigation, and media activism have illuminated the shortcomings and valuable lessons from prior consultation mechanisms utilized in Latin American oil extraction contexts. This insight allows these mechanisms to serve as models for developing and implementing more robust frameworks for procedural justice within global climate governance. Furthermore, these models can be adapted to inform the creation of similar frameworks for prior consultation and procedural justice, which can be extended to various aspects of the broader struggle for a better climate future, extending beyond the specific domain of oil extraction.

Climate Change and Climate Action

The growth of the fossil fuels economy in the Amazon region lays bare the intricate and often fraught intersections of environmental justice, climate change, and Indigenous rights. While completely halting oil extraction is not a realistic option for the economies in the Global South, Indigenous communities are adopting a pragmatic approach by demanding remediation for past and present harms, equitable distribution of oil wealth, and stringent environmental safeguards for future activities. This does not, however, imply that the Indigenous struggles for distributional, procedural, and recognitional justice, which constitute their pursuit of climate justice, are confined to the issue of oil extraction. Rather, the growing Indigenous resistance against oil projects over the past decades serves as a starting point for a broader struggle to secure a better climate future and to gain recognition for the rights of Indigenous peoples, both regionally and globally, as the international community strives to chart a path toward a more equitable and better climate governance framework.

The global push to transition toward less carbon-intensive economies has fueled a surge in demand for critical minerals and elements such as lithium, often found in the very lands inhabited by Indigenous and rural communities (Dolšak & Prakash, 2022; Seefeldt, 2020). Patterns of mobilizations surrounding the impacts of lithium extraction, which are similar to the negative impacts of hydrocarbon and oil extraction, are already emerging in Chile and Argentina and will only continue to cause friction as global demand for the mineral increases (Petavratzi et al., 2022; Seefeldt, 2020, 2022). This parallel demand for extraction, albeit for a different set of resources, presents a new set of challenges and underscores the enduring vulnerabilities of these communities to extractive industries (Gustafsson & Schilling-Vacaflor, 2022; Gustafsson & Scurrah, 2023). Furthermore, the global reliance on bridge fuels like natural gas, exemplified by projects like the Camisea gas project in Peru further exacerbates the pressures on Indigenous communities in the region (Watson & Arce, 2024; Watson & Davidsen, 2021). The Camisea case, for example, highlights how the pursuit of seemingly cleaner energy sources can still perpetuate the same environmental injustices and territorial displacement as seen with oil extraction.

Moreover, as international climate governance frameworks continue to expand the inclusion of Indigenous peoples beyond just the FPIC and EIAs to issues such as necessary Indigenous inclusion in the reducing emissions from deforestation and forest degradation (REDD+) framework, the problems seen with prior consultation and extractive industries in the Amazon region are only expanding and reproducing on a larger scale (Lozano Flores et al., 2023; Osborne et al., 2024; Paredes et al., 2024). This presents new barriers for meaningful Indigenous participation that reproduce environmental injustices in resource governance on an international scale, such as the lack of influence of Indigenous peoples over the institutional design of governance instruments, the exclusion of Indigenous peoples in the domestication of global instruments, policy incoherences constraining the scope of decision making, and weak cross-scale linkages between Indigenous spaces (Gustafsson & Schilling-Vacaflor, 2022). While the world is recognizing a need for more Indigenous inclusion in climate governance decisions, their meaningful participation is still challenging; however, the legal precedents, lessons learned, and pathways to global connection formed by the early mobilizations against oil extraction in the Amazon can help to shape new approaches to participation and consultation and the improvement of existing frameworks. This confluence of factors underscores the persistent tension between the urgent imperatives of global climate action and the localized, often devastating, impacts of extraction on Indigenous communities.

New Realities

In response to oil extraction, Indigenous peoples in the Amazon have mobilized in diverse ways, aligning their struggles with the broader movement seeking distributional, procedural, and recognitional climate justice. This heightened awareness is crucial as the world seeks to shift away from fossil fuels dependence. However, this transition also presents new challenges, particularly in navigating resource governance issues related to critical minerals necessary for this transition, as similar problems continue to persist across various extractive industries. Embracing best practices toward global climate action and solutions to the threats of climate change is a complex endeavor that requires carefully balancing the need for less dependent fossil fuel economies with the well-being and self-determination of Indigenous communities.

The increased political mobilization by Indigenous peoples, coupled with new participatory mechanisms, holds the potential to mitigate resource governance issues related to critical minerals required for the global transition away from oil. When effectively implemented, these mechanisms may also be applied to other pressing climate challenges, such as deforestation or mineral extraction in the Amazon region. The need for Indigenous participation in initiatives like the REDD+ program mirrors the principles of FPIC, highlighting the potential for synergies but also showing similar challenges in execution (see chapter by Alvarez in this volume). This necessitates a shift in the way the global environmental governance community understands how to translate programs and frameworks calling for Indigenous inclusion into enforceable frameworks for meaningful participation on the ground (Gustafsson & Schilling-Vacaflor, 2022; Gustafsson & Scurrah, 2023; Osborne et al., 2024).

Furthermore, improving industry–community engagement, alongside greater oversight by governments and the enforcement of international and national environmental and human rights standards, is crucial (Dolšak & Prakash, 2022). The power dynamics between industry and local actors also need to be addressed, with more equitable distribution of economic benefits and environmental costs, resulting in greater benefits for Indigenous communities and a fairer allocation of costs for the oil companies (Arce et al., 2024).

As the world grapples with the urgency of climate change and the imperative to transition away from fossil fuels, the active involvement of Indigenous peoples and the utilization of participatory mechanisms may offer a pathway toward more inclusive and sustainable resource governance. This is crucial for addressing the need for critical minerals for transition economies, as well as other climate challenges beyond oil and hydrocarbon extraction, such as deforestation. However, it is important to acknowledge and overcome the inherent barriers and complexities of the effective utilization of these mechanisms in the pursuit of a better climate future.

Notes

1. This concept is rooted in well-known human rights instruments, particularly the United Nations Declaration on the Rights of Indigenous Peoples (UNDRIP) and the International Labour Organization (ILO) Convention 169 (Merino, 2018).
2. The lawsuits against Occidental Petroleum (Oxy) in Peru and Texaco (now Chevron) in Ecuador summarized in this section do not represent an exhaustive list of court cases. Strategic litigation as a source of climate reparation is ongoing in Peru. There, Indigenous plaintiffs filed a lawsuit against Germany's largest electricity producer and one of the world's biggest greenhouse gas emitters—RWE—seeking to remedy its responsibility for climate change (Frank et al., 2019).

References

Acosta, A. (2011). The Chevron-Texaco judgment: A triumph of humanity. *NACLA Report on the Americas*, *44*(2), 5–6. https://doi.org/10.1080/10714839.2011.11722159

Acosta, A. (2012). *Buen Vivir Sumak Kawsay: Una Oportunidad Para Imaginar Otros Mundos* (1st ed.). Editorial Universitaria Abya Yala. https://muse.jhu.edu/book/119685.

Acosta, A. (2013). Extractivism and neoextractivism: Two sides of the same curse. *Beyond Development: Alternative Visions from Latin America*, *1*, 61–86.

Almeida, P. (2019). *Social movements: The structure of collective mobilization* (1st ed.). University of California Press. https://doi.org/10.2307/j.ctvd1c7d7

Almeida, P., González Márquez, L. R., & Fonsah, E. (2023). The forms of climate action. *Sociology Compass*, *18*(2), e13177. https://doi.org/10.1111/soc4.13177

Arce, M., Franco, O. A., & Merino, R. (2024). Subnational oil resource governance after the commodity boom: The making and limitations of Peru's closing development gaps plan. *Studies in Comparative International Development*, *59*(2), 212–237. https://doi.org/10.1007/s12116-023-09418-8

Arce, M., & Jaskoski, M. (2024). Development elites, impacted communities, and environmental governance in Latin America. *Studies in Comparative International Development*, *59*, 199–211.

Arce, M., Polizzi, M. S., & Reeder, B. W. (2020). Willingness to protest over resource extraction in Latin America. *The Extractive Industries and Society*, *7*(2), 716–728. https://doi.org/10.1016/j.exis.2020.03.005

Auz, J. (2022). Human rights-based climate litigation: A Latin American cartography. *Journal of Human Rights and the Environment*, *13*(1), 114–136. https://doi.org/10.4337/jhre.2022.01.05

Becerra Valdivia, K. (2022). *Indigenous collective rights in Latin America: The role of coalitions, constitutions, and party systems*. Lexington Books.

Buckel, S., Pichl, M., & Vestena, C. A. (2024). Legal struggles: A social theory perspective on strategic litigation and legal mobilisation. *Social & Legal Studies*, *33*(1), 21–41. https://doi.org/10.1177/09646639231153783

Cueva, S., & Diaz, J. P. (2018). *The fiscal and monetary history of Ecuador: 1960–2017* (Working Paper No. 2018-65). University of Chicago, Becker Friedman Institute for Economics. https://doi.org/10.2139/ssrn.3238175

Dolšak, N., & Prakash, A. (2022). Three faces of climate justice. *Annual Review of Political Science*, *25*(1), 283–301. https://doi.org/10.1146/annurev-polisci-051120-125514

Eisenstadt, T., & West, K. J. (2019). *Who speaks for nature? Indigenous movements, public opinion, and the petro-state in Ecuador*. Oxford University Press.

Finer, M., Jenkins, C. N., Pimm, S. L., Keane, B., & Ross, C. (2008). Oil and gas projects in the western Amazon: Threats to wilderness, biodiversity, and Indigenous peoples. *PLOS ONE*, *3*(8), e2932. https://doi.org/10.1371/journal.pone.0002932

Flemmer, R., & Schilling-Vacaflor, A. (2016). Unfulfilled promises of the consultation approach: The limits to effective Indigenous participation in Bolivia's and Peru's extractive industries. *Third World Quarterly*, *37*(1), 172–188. https://doi.org/10.1080/01436597.2015.1092867

Frank, W., Bals, C., & Grimm, J. (2019). The case of Huaraz: First climate lawsuit on loss and damage against an energy company before German courts. In R. Mechler, L. M. Bouwer, T. Schinko, S. Surminski, & J. Linnerooth-Bayer (Eds.), *Loss and damage from climate change* (pp. 475–482). Springer International. https://doi.org/10.1007/978-3-319-72026-5_20

Fraser, B. (2018). Peru's oldest and largest Amazonian oil field poised for clean-up. *Nature*, *562*(7725), 18–34. https://doi.org/10.1038/d41586-018-06886-0

Gustafsson, M.-T., & Schilling-Vacaflor, A. (2022). Indigenous peoples and multiscalar environmental governance: The opening and closure of participatory spaces. *Global Environmental Politics*, *22*(2), 70–94. https://doi.org/10.1162/glep_a_00642

Gustafsson, M.-T., & Scurrah, M. (2023). Subnational governance strategies at the extractive frontier: Collaboration and conflict in Peru. *Territory, Politics, Governance*, *11*(1), 1–18. https://doi.org/10.1080/21622671.2020.1840425

Hasan, M. (2018). Climate justice and democracy: A normative approach. *International Journal of Political Theory*, 3(1), 1–17. https://doi.org/10.22609/1.3.1.5

Jaskoski, M. (2020). Participatory institutions as a focal point for mobilizing: Prior consultation and Indigenous conflict in Colombia's extractive industries. *Comparative Politics*, 52(4), 537–556. https://www.jstor.org/stable/26976025

Johnston, J. E., Lim, E., & Roh, H. (2019). Impact of upstream oil extraction and environmental public health: A review of the evidence. *Science of the Total Environment*, 657, 187–199. https://doi.org/10.1016/j.scitotenv.2018.11.483

Lozano Flores, L., Delgado Pugley, D., Casas Luna, S., Van Den Broeck, P., & Parra, C. (2023). Challenging state authority and hierarchical power: A case study of the engagement of Peru's Amazonian Indigenous peoples' organizations in the governance of REDD+. *Environmental Policy and Governance*, 34(2), 137–151. https://doi.org/10.1002/eet.2067

Lupien, P., Rincón, A., Lalama Vargas, A., Machaca, S., & Chiriboga, G. (2024). #Resistencia: Indigenous movements, social media, and mobilization in Latin America. *Latin American Research Review*, 59(4), 803–822. doi:10.1017/lar.2024.4

Merino, R. (2018). Re-politicizing participation or reframing environmental governance? Beyond Indigenous' prior consultation and citizen participation. *World Development*, 111, 75–83. https://doi.org/10.1016/j.worlddev.2018.06.025

Morrison-Saunders, A., & Retief, F. (2012). Walking the sustainability assessment talk—Progressing the practice of environmental impact assessment (EIA). *Environmental Impact Assessment Review*, 36, 34–41. https://doi.org/10.1016/j.eiar.2012.04.001

Muttitt, G., & Kartha, S. (2020). Equity, climate justice and fossil fuel extraction: Principles for a managed phase out. *Climate Policy*, 20(8), 1024–1042. https://doi.org/10.1080/14693062.2020.1763900

Natcher, D., Owens-Beek, N., Bogdan, A. M., Lu, X., Li, M., Ingram, S., McKay, R., & Rice, A. (2022). Scenario planning tools for mitigating industrial impacts on First Nations subsistence economies in British Columbia, Canada. *Sustainability Science*, 17, 469–484. https://doi.org/10.1007/s11625-021-00969-0

Orta-Martínez, M., & Finer, M. (2010). Oil frontiers and Indigenous resistance in the Peruvian Amazon. *Ecological Economics*, 70(2), 207–218. https://doi.org/10.1016/j.ecolecon.2010.04.022

Orta-Martínez, M., Napolitano, D. A., MacLennan, G. J., O'Callaghan, C., Ciborowski, S., & Fabregas, X. (2007). Impacts of petroleum activities for the Achuar People of the Peruvian Amazon: Summary of existing evidence and research gaps. *Environmental Research Letters*, 2(4), 045006. https://doi.org/10.1088/1748-9326/2/4/045006

Orta-Martínez, M., Pellegrini, L., & Arsel, M. (2018). The squeaky wheel gets the grease? The conflict imperative and the slow fight against environmental injustice in northern Peruvian Amazon. *Ecology and Society*, 23(3),7. https://doi.org/10.5751/ES-10098-230307

Osborne, T., Cifuentes, S., Dev, L., Howard, S., Marchi, E., Withey, L., & Santos Rocha da Silva, M. (2024). Climate justice, forests, and Indigenous peoples: Toward an alternative to REDD+ for the Amazon. *Climatic Change*, 177(8), 128. https://doi.org/10.1007/s10584-024-03774-7

Paredes, M., Gianella, C., & Olivera, S. (2024). Corrupting climate change institutions from the inside systemic collusion in the Peruvian forest governance. *Political Geography*, 108, 102991. https://doi.org/10.1016/j.polgeo.2023.102991.

Pellegrini, L., Arsel, M., Orta-Martínez, M., & Mena, C. F. (2020). International investment agreements, human rights, and environmental justice: The Texaco/Chevron case from the Ecuadorian Amazon. *Journal of International Economic Law*, 23(2), 455–468. https://doi.org/10.1093/jiel/jgaa016

Petavratzi, E., Sanchez-Lopez, M. D., Hughes, A., Stacey, J., Ford, J. R., & Butcher, A. S. (2022). The impacts of environmental, social and governance (ESG) issues in achieving sustainable lithium supply in the lithium triangle. *Mineral Economics*, 35(3–4), 673–699. https://doi.org/10.1007/s13563-022-00332-4

Rodríguez-Garavito, C. (2020). Human rights: The Global South's route to climate litigation. *AJIL Unbound*, 114, 40–44. https://doi.org/10.1017/aju.2020.4

Rodríguez-Garavito, C., & Baquero-Díaz, C. A. (2022). Reframing Indigenous rights: The right to consultation and the rights of nature and future generations in the Sarayaku legal mobilization. In G. De Búrca (Ed.), *Legal mobilization for human rights* (1st ed., pp. 73–88). Oxford University Press. https://doi.org/10.1093/oso/9780192866578.003.0005

San Sebastián, M., & Hurtig, A. K. (2004). Oil exploitation in the Amazon basin of Ecuador: A public health emergency. *Revista Panamericana de Salud Pública*, *15*(3), 205–211. https://doi.org/10.1590/S1020-49892004000300014

Schilling-Vacaflor, A., & Flemmer, R. (2015). Conflict transformation through prior consultation? Lessons from Peru. *Journal of Latin American Studies*, *47*(4), 811–839. https://doi.org/10.1017/S0022216X15000826

Schlosberg, D., & Collins, L. B. (2014). From environmental to climate justice: Climate change and the discourse of environmental justice. *WIREs Climate Change*, *5*(3), 359–374. https://doi.org/10.1002/wcc.275

Seefeldt, J. L. (2020). Lessons from the lithium triangle: Considering policy explanations for the variation in lithium industry development in the "lithium triangle" countries of Chile, Argentina, and Bolivia. *Politics & Policy*, *48*(4), 727–765. https://doi.org/10.1111/polp.12365

Seefeldt, J. L. (2022). Water as property: Contention between Indigenous communities and the lithium industry for water rights in Chile. *Latin American Policy*, *13*(2), 328–353. https://doi.org/10.1111/lamp.12265

Temper, L., Avila, S., Bene, D. D., Gobby, J., Kosoy, N., Billon, P. L., Martinez-Alier, J., Perkins, P., Roy, B., Scheidel, A., & Walter, M. (2020). Movements shaping climate futures: A systematic mapping of protests against fossil fuel and low-carbon energy projects. *Environmental Research Letters*, *15*(12), 123004. https://doi.org/10.1088/1748-9326/abc197

Temper, L., Demaria, F., Scheidel, A., Del Bene, D., & Martinez-Alier, J. (2018). The Global Environmental Justice Atlas (EJAtlas): Ecological distribution conflicts as forces for sustainability. *Sustainability Science*, *13*(3), 573–584. https://doi.org/10.1007/s11625-018-0563-4

Tigre, M. A., Urzola, N., & Goodman, A. (2023). Climate litigation in Latin America: Is the region quietly leading a revolution? *Journal of Human Rights and the Environment*, *14*(1), 67–93. ttps://doi.org/10.4337/jhre.2023.01.04

Vargas, S. C. F., & Torres, M. H. O. (2019). El Derecho a La Consulta Previa a Los Pueblos Indígenas en El Sistema Interamericano De Derechos Humanos: Casos De Estudio: Colombia Y Ecuador. *Civilizar*, *19*(36), 59–76. https://doi.org/10.22518/usergioa/jour/ccsh/2019.1/a04

Wainwright, J., & Mann, G. (2015). Climate change and the adaptation of the political. *Annals of the Association of American Geographers*, *105*(2), 313–321. https://doi.org/10.1080/00045608.2014.973807

Watson, A., & Arce, M. (2024). Indigenous mobilization and territorial ordering in the Amazon. *Political Geography*, *108*, 103013. https://doi.org/10.1016/j.polgeo.2023.103013

Watson, A., & Davidsen, C. (2021). Pandemic state failure, hydrocarbon control, and Indigenous territorial counteraction in the Peruvian Amazon. *Frontiers in Human Dynamics*, *3*, 654311. https://doi.org/10.3389/fhumd.2021.654311

Zaremberg, G., & Wong, M. T. (2018). Participation on the edge: Prior consultation and extractivism in Latin America. *Journal of Politics in Latin America*, *10*(3), 29–58. https://doi.org/10.1177/1866802X1801000302

CHAPTER 17

Resistance to Fossil Fuel Extraction Projects in Africa as Climate Action

JESSE SALAH OVADIA

Introduction

Although no new fossil fuel projects are needed in energy transition scenarios that achieve net-zero energy by 2050 (NZE2050), Africa has entered a new phase of an oil and gas boom that has been going strong since the early 2000s. Russia's invasion of Ukraine in 2022 has generated new interest in African oil and gas, especially in Europe, and new confidence that new fossil fuel extraction will be profitable despite global commitments through the Paris Agreement to limit global warming to "well below 2°C" (WB2C) and to pursue a target of 1.5°C above preindustrial levels.

Achieving Paris commitments generally requires NZE2050, yet current policies are projected to lead to 2.7°C of global warming.[1] This disconnect explains why so many African governments are moving ahead enthusiastically with new oil and gas extraction. The most recent African oil boom began in the Gulf of Guinea as Nigeria, Angola, and other producers began moving into deeper offshore oil extraction pushed by high oil prices and the George W. Bush administration's desire to diversify away from its reliance on Middle Eastern oil (Klare & Volman, 2006; Ovadia, 2008). African governments in the Gulf of Guinea, Congo basin, East Africa, and elsewhere rode waves of popularity associated with "Africa rising" promises of oil bringing economic development and structural transformation (Graham & Ovadia, 2019, pp. 593–594; Ovadia, 2016a). Across the continent, new natural gas discoveries and production generated further optimism due to the proliferation of potential markets across the globe (Ayaburi et al., 2021). It is estimated Africa will almost double its share of global gas from 6% to 11% by 2050 as it witnesses growth in gas supply second only to the Middle East (Orucho, 2023). While some of this gas will help feed rising domestic demand, most of it will be exported to Europe and Asia. The investment and optimism about fossil energy continued even as US demand dropped due to its domestic boom in shale oil, global energy prices collapsed post 2014, and the Paris Agreement in 2015 signaled the need to transition to renewable energy.

Optimistic promises about fantastic oil wealth run counter to the history of oil and gas extraction in Africa, which has generally been associated with numerous negative social outcomes, including poverty, poor economic performance, authoritarianism, and conflict. The extractive industries are also dirty, leading to environmental degradation and loss of livelihoods, especially for agriculture, fisheries, and related sectors (Arce et al., 2022; Givens et al., 2019). In the context of Latin America, the conceptual links between resource extraction, negative social outcomes, and the politics of resistance and redistribution are brought together through the framework of "neo-extractivism" (Bebbington & Bebbington, 2011; Gudynas, 2010; Svampa, 2015). In Africa, the theoretical framework of "resource curse" dominates discussions of negative developmental outcomes. Such outcomes have been seen most clearly in Africa's two largest oil and gas producers, Angola and Nigeria, who have also become prime examples of the "resource curse" hypothesis.

Angola and Nigeria, along with emerging oil and gas producers, have also ignored the risks associated with new oil and gas developments in the context of green energy transitions. New fossil fuel projects, even those far offshore, have brought significant opposition from civil society and especially from host communities. In most cases, resistance to fossil fuel extraction has been noninstitutional and rooted in concern over who benefits rather than concern over climate change and global warming. However, the environmental and social impacts from these projects have already impacted citizens in their everyday lives. Therefore, African organizations and activists have begun to sound alarm bells about climate change as well as the transition risks (stranded assets and fossil fuel lock-in) associated with global climate action.

The Resource Curse

The well-known resource curse hypothesis is perhaps better understood as three separate but related hypotheses: that resource wealth can be linked to poor economic performance, that it can be linked to poor political and institutional outcomes, and that it can be linked to conflict (Ovadia & Graham, 2022, p. 61). While there is significant anecdotal evidence and the resource curse hypothesis (or hypotheses) "feel" correct, there are numerous critiques of the resource curse perspective as well—particularly more deterministic formulations.

The first formulation of the resource curse argument was that dependence on natural resources extraction is related to poor economic performance. Scholars such as Gelb and associates (1988) and Karl (1997) noted numerous challenges for economics dependent on oil. The most common assertion was that economies with a high ratio of natural resources export to GDP leant toward low economic growth rates (Auty, 1993, 2001; Sachs & Warner, 2001). Resource dependency was and is also linked to high levels of poverty (Ross, 2001). Finally, these arguments are also associated with arguments about "Dutch Disease," which refers to the appreciation of a country's currency due to revenues from petroleum exports, as occurred in Holland in the 1960s and 1970s (Ovadia, 2020).

While there is strong empirical support for the concept of Dutch Disease, Rosser (2006) points out that studies on the resources curse do not conclusively illustrate the direction of causality from natural resource wealth to lower economic growth and other development outcomes, nor do they establish the nonexistence of an influencing third variable. Furthermore, some scholars have pointed out that the hypothesis linking resources to poor economic performance ignores reverse causality and adopts inappropriate proxies to measure resource endowment. When this is considered, the evidence of causality and or even correlation between resource endowment and poor economic outcomes is less clear (Brunnschweiler, 2008).

With regard to the second hypothesis, Ross (2001) and Jensen and Wantchekon (2004) argue that an abundance of natural resources leads to low levels of democracy. Poor management of the revenues from resources may be explained, as per Auty (2007) by political competition over resource rents and the role of those rents in fueling patronage. Mehlum and colleagues (2006) also suggest that weak institutions facilitate the resource curse, while numerous studies associate resource wealth, particularly oil wealth, with corruption (Ahmadov, 2014; Kolstad & Søreide, 2009; Petermann et al., 2007; Wright et al., 2015). Finally, other studies focus on how resource wealth can weaken pro-development institutions to fuel rent-seeking, patronage, and corruption (Bulte et al., 2005; Isham et al., 2005; Leite & Weidmann, 1999). Findings that suggest resource dependency creates authoritarian forms of government have been challenged as well on the grounds that they are methodologically flawed (Haber & Menaldo, 2011, p. 2). Additionally, to the extent it exists, the causes of the resource curse cannot be explained by governance issues alone as there are numerous actors in the global political economy that impact both the political and socioeconomic outcomes in resource-rich countries (Cramer, 2002; Rosser, 2006).

Some more critical approaches to the resource curse note that natural resource abundance makes a developing country a target for forced incorporation into the global capitalist system. This, in turn, impairs a country's ability to pursue autonomous programs of economic development (Davis & Tilton, 2005; Idemudia, 2012; Mehlum et al., 2006). While it is clear that many of the world's oil producers, especially in Africa, have experienced all the negative impacts of capitalist development without any of the benefits (Ovadia, 2016a), it is also clear that any negative outcomes associated or correlated with oil or resource wealth are not inevitable outcomes, but rather failures of policy (within specific political-economic local, national, and global contexts). As argued by Saad-Filho and Weeks, "the essential flaw in the resource curse hypothesis is that the 'disease' and the 'curse' are outcomes rather than causes.... It follows that what writers call diseases and curse are the result of failures to implement effective macro-management policies" (Saad-Filho & Weeks, 2013, pp. 1–2). As Papyrakis says, the "curse" is actually context specific and depends "on the type of resources, socio-political institutions and linkages with the rest of the economy" (Papyrakis, 2017, p. 176).

It is worth noting that the potential exists for a variety of different developmental outcomes, including "positive oil exceptionalism" emerging from new limits of the possible created by local content policies and other strategies of state intervention to encourage linkages between the extractive industry and domestic non-oil sectors (Ovadia, 2016a). However, as I argue elsewhere, the window for such outcomes may be closing or already have closed due to the imperatives of climate action (Ovadia, 2019).

In the third hypothesis associating resource wealth with conflict, research has focused on the link between resources and the onset, duration, and intensity of civil war (Collier & Hoeffler, 1998, 2005; Welsch, 2008) more than on community-based mobilization against resource extraction. Angola and Nigeria are prime examples of the link between oil and civil war (Collier & Hoeffler, 1998; Graham & Ovadia, 2019; Le Billon, 2008; Obi, 2010a; Ovadia, 2016b; Ross, 2001). However, while large-scale conflicts can and do occur, at different scales, the likelihood of conflict may be even greater. Numerous scholars have shown that different development trajectories are common within states between mineral-rich and mineral-poor areas (Papyrakis & Gerlagh, 2004; Angrist & Kugler, 2008; Buccellato & Mickiewicz, 2009; Papyrakis & Raveh, 2014; Shao & Qi, 2009; Zhang et al., 2008).

At the micro (community) level, the development impact of resource extraction is bound to be more negative for the same reasons African countries experience the negative impacts of capitalist development without the positive benefits. More peripheral areas that are immediate to extractive projects are unlikely to see job gains or increased economic activities, while they suffer the worst impacts of the extraction. Therefore, studies of the resource curse at the micro level look at the development outcomes and impacts as well as the questions of individual agency, community relationships, and cultural characteristics that explain difference in outcomes and impacts (Gilberthorpe & Papyrakis, 2015).

At the micro level, outcomes, impacts, and resistance to fossil fuel extraction are more clearly and convincingly observed. Scholars have shown that resource extraction exacerbates poverty for nearby communities (Ayelazuno, 2014; Hilson, 2010, 2012) and leads to conflict or tensions between the state or extractive firms and Indigenous communities in mineral-rich areas (Arellano-Yanguas, 2011; Obi, 2010b; Watts, 2001). This can be seen in the cases on Nigeria, Angola, and Ghana.

Not all conflict involves resistance to extraction. The large-scale conflicts most commonly discussed in connection with the third resource curse hypothesis such as the civil wars in Angola and Nigeria were not about resistance but rather about the control of resource rents. However, the more recent and ongoing low-intensity conflict in the Niger Delta is both a conflict over control of the oil industry and the revenue it generates and a conflict over the continuation of oil extraction and the social, economic, and environmental costs associated with it.

Part of what makes Nigeria such a relevant example of resource curse outcomes is that its oil and gas industry has generated hundreds of billions of dollars of revenue, yet has very little to show beyond a legacy of underdevelopment and conflict. Since 1970, it is estimated that 85% of oil revenues accrued to 1% of the population (Watts, 2006, pp. 11–12). Meanwhile, poverty and underdevelopment are the norm for the vast majority of Nigerians. The mass mobilizations of up to 2 million people in Lagos alone that occurred in response to the removal of a fuel subsidy can be connected to the fact that Nigerians do not see tangible benefits from their nation's oil wealth and feel cheated that they do not see a benefit from the wealth the oil industry creates (Ovadia, 2013, pp. 316–317). Such protests are regular occurrences whenever the government has attempted to remove the fuel subsidy, as it did finally under Bola Tinubu despite massive protests and a new #Endbadgovernance movement (Ezeamalu, 2024).

At the end of the Angolan civil war, poverty in general was very high—with particularly negative gendered outcomes in terms of maternal mortality, illiteracy, and high levels of internal displacement (Kibble, 2006). In the oil-producing areas of Angola, especially around Soyo in Zaire province and in Cabinda Province, poverty was made worse by threats to traditional livelihoods, especially artisanal fishing. People in these communities faced loss of livelihoods, appropriation of land without compensation, significant and harmful environmental degradation, threats to customary land claims, and a weak regulatory environment (Reed, 2009, p. 10).

Reed argues that resentment in Cabinda (a region with an ongoing low-intensity successionist conflict) is fueled by the contrast between the wealth visible in the oil enclaves protected by barbed-wire fences and land mines and the poverty outside (Reed, 2009, p. 106). In my own work, I quote an Angolan oil worker who said: "I am from Cabinda, but you can't see the benefit of petroleum. The roads are still very bad in Cabinda and Zaire and people are not benefitting from the oil" (Ovadia, 2016b, pp. 173–174).

Although Ghana is a much newer oil producer with oil extraction occurring offshore, communities along the western coast have been profoundly impacted by the new fossil fuel extraction. In the regional capital of Sekondi-Takoradi (STMA), oil production has transformed the city, producing a variety of negative social impacts, impacts on fisheries, impacts on local government and land use planning, and impacts on enterprise development—all without significant new benefits for citizens (Ovadia & Graham, 2022). Impacts on the STMA and other coastal communities will be explored further later. In summary, as Ayelazuno and Ovadia (2022) explain, fossil fuel extraction brought communities dispossession from livelihood assets on both land and ocean and dislocation from their socioecological environment (Ayelazuno & Ovadia, 2022, p. 13; Obeng-Odoom, 2014).

These examples demonstrate qualitatively that new conflicts and negative developmental outcomes result from fossil fuel extraction regardless of whether a causal link can be shown to lower economic growth or to large-scale violent conflict. They can also result from fossil fuel infrastructure such as oil pipelines or transportation corridors (Barlow, 2023; Enns et al., 2022). In fact, many of these negative outcomes emerge well before actual oil and gas production takes place due to the anticipation of benefit and rising expectations in what some have called a "presource curse" (Cust & Mihalyi, 2017; Frynas et al., 2017). In addition to the negative outcomes that occur at the micro level (and give rise to various local forms of resistance), there are also emerging macro-level transition risks associated with fossil fuel extraction. These will be discussed in the following section.

Transition Risks for New Fossil Fuel Extraction

Africa's oil and gas boom has also coincided with growing global recognition of the urgent threat posed by climate change. It has now been incontrovertibly established that anthropogenic emissions of CO_2 and other gases with global warming potential have led and are leading to rapid increases in global temperatures (IPCC, 2021). With the increasing shift to cleaner technologies, natural gas has been positioned by some as a transition fuel that will pave the way for greater renewable energy deployment. European leaders have also pushed for new gas extraction and liquid natural gas (LNG) export projects to deal with the fallout of Russia's invasion of Ukraine. As a result, some African policymakers continue to see the exploitation of gas, especially through LNG exports, as an opportunity to address energy access, garner export revenues, and foster economic development. For all countries, but especially for African fossil fuel exporters, transitioning away from fossil fuel dependence and bolstering economic diversification in the longer term is necessary to minimize economic vulnerability.

The global energy transition involves significant risk of unburnable carbon and stranded assets. The main factors for determining the extent of these outcomes is the speed of transitions and how they unfold geographically. Fortunately, there is strong evidence that rapid energy transitions are possible. This is good news for Africans in general, but not necessarily for African oil and gas (Blondeel et al., 2021; Grubler et al., 2016; Sovacool, 2016; Sovacool & Geels, 2016). While there are many unknowns, fossil fuels are increasingly being understood as a liability as the banking and insurance sectors hedge against what are now called "transition risks" (Eicke & Goldthau, 2021; Fattouh et al., 2018; van de Graaf & Bradshaw, 2018).

In the IEA's *Net Zero by 2050* report, the feasibility of an energy transition that meets the 1.5°C warming target is demonstrated through a scenario in which unabated coal demand falls by 98%, oil by 75%, and gas by 55% by 2050. The scenario requires no new oil and gas fields approved for development by 2021. Although we are not currently meeting this scenario, there are many pathways to meet the WB2C Paris Agreement target, and there are still pathways to NZE2050.

In a report on how Africa can navigate just transitions and stranded assets, Denton et al. write:

> Stranded assets provide a clear warning to Africa that it must diversify and use its hydrocarbon resources, especially in the face of climate change and increased disinvestment away from carbon-intensive industries. Fossil-fuel exploitation has not contributed significantly to Africa's prosperity story. The continent can seize the opportunity to direct its development away from exogenous shocks and short-term development prospects without value addition and the trappings of the primary product export pull. (Denton et al., 2021, p. 6)

In a report published by the African Climate Foundation (ACF), Ovadia et al. (2024) focus on the risks associated with natural gas. Their analysis begins categorizing African countries looking to exploit natural gas into three categories: mature producers (Nigeria, Egypt, Algeria, Angola, and Equatorial Guinea), emerging producers (Mauritania, Senegal,

Ethiopia, Tanzania, and Mozambique), and countries exploring for gas (South Africa and Namibia). They write:

> Exploiting Africa's gas reserves involves multiple trade-offs, uncertainty and significant transition risk, particularly as the adoption of renewable energy (RE) and the low-carbon transition gains speed. Using models of two Paris-compliant scenarios developed for the African Climate Foundation (ACF), we argue that limiting global warming to well below 2°C (WB2C) or achieving net zero emissions by 2050 (NZE2050), would lead to investments in African gas being ultimately value destructive We find that current and aspiring African petro-states should be wary of accepting new risk associated with gas and gas infrastructure and are incentivised, whether or not they pursue gas exploitation, to embrace rapidly scaling up renewables in their energy technology mix. (Ovadia et al., 2024, p. 1)

Both existing and committed fossil fuel infrastructure will have to be phased out before the end of its expected economic lifetime—if NZE2050 or WB2C is to be achieved—simply because it will become uneconomic. There is a finite amount of carbon that the world can emit into the atmosphere and achieve either of these goals. The gap between planned fossil fuel production and global production levels consistent with limiting climate change is called the "production gap" (SEI, 2021). The annual Production Gap Report, launched in 2019, shows that the gap remains large. However, it can be closed through a managed transition led by governments and their climate commitments or through market forces. This latter possibility is also one African fossil fuel producers should be concerned about.

The idea of market forces creating a transition risk was raised by Bank of England Governor Mark Carney in a 2015 speech.[2] The rapid exit of international capital from fossil fuels could provoke a sudden price collapse, or "Minsky Moment" once the scale of potentially stranded assets dawns on investors and other financial stakeholders. Blondeel et al. suggest this would likely cause spillover effects into the wider global economy, leading to US$1–4 trillion worth of fossil-fuel-related assets disappearing and as much as US$30 trillion in assets facing a risk of devaluation (Blondeel et al., 2021, pp. 6–7).

African governments remain deeply exposed to a variety of transition risks. Not only is there financial risk since many fossil fuel projects are backed by African governments, there is also an additional political risk for governments that comes with such outcomes. Goldthau and colleagues describe this risk as a risk of "stranded nations"—an evocative phrase that highlights how governments with less legitimacy, or governments that maintain power through autocratic rule, may lose domestic support as they face deteriorating terms of trade (Goldthau et al., 2020, p. 327, citing Manley et al., 2017).

Ovadia et al.'s research for ACF concludes:

> In Paris Agreement and Net Zero 2050 compliant scenarios, many such investments are value destructive. African governments could face billions of dollars in losses and set back their own transitions to green technologies. In total, liquefied natural gas (LNG) imports will fall 60 percent between 2020 and 2050 in the NZE2050 scenario. Impacts are more severe for emerging gas producers (Mozambique, Tanzania, Mauritania and Senegal) and countries currently exploring for gas (South Africa and Namibia). (Ovadia et al., 2024, p. 1)

The NZE2050 outcome in which the majority of current and future fossil fuel projects are stranded is extreme; however, the WB2C scenario also makes most new natural gas and almost all oil projects uneconomical—even if they are not outright value destructive, they will not provide an acceptable internal rate of return. The impact of this would be extremely negative, particularly for the finances and budgets of emerging producers.

Mozambique is the emerging producer that fares best in the NZE and WB2C scenarios. However, even in Mozambique, the WB2C scenario would eliminate 70% of total value even as most projects would still be economic. In NZE2050 all gas projects but one would be value destructive and 92% of the expected value of gas revenues overall would be lost. However, even in WB2C, peak revenues in 2041 would account for just over a third of possible revenues (Ovadia et al., 2024, p. 9). Given such possible outcomes and inherent risks with not only new but also current fossil fuel projects, it is reasonable to expect some resistance in the years to come and significant political upheaval if expected benefits on a national level fail to materialize due to lowered global demand for fossil fuel caused by climate-friendly energy policies.

Resistance to Fossil Fuel Extraction

As mentioned earlier, most resistance to fossil fuel extraction in Africa is not currently related to climate action. Nevertheless, such resistance does play out, particularly at the micro level. This section reviews such resistance in greater detail with reference to Nigeria and Ghana. The Niger Delta is the best known example of fossil fuel extraction leading to conflict, insurgency, loss of livelihoods, and environmental degradation (see Ikelegbe, 2001; Okonta, 2008; Omeje, 2006). Revenue sharing, special government initiatives to develop the Niger Delta, and corporate social responsibility (CSR) investments by international oil companies (IOCs) have all failed to mitigate negative impacts (Frynas, 2005; Idemudia, 2007).

As Ike Okonta and Oronto Douglas write, "Historically, the people of the Niger Delta have always been at the mercy of greedy outsiders who plunder their natural resources without giving them anything in return, from the days of slavery to the present day" (Okonta & Douglas, 2003, p. 19). By the 1990s, the people of the Niger Delta had already witnessed three decades of poverty and environmental destruction due to oil extraction.

In 1990, the Movement for the Survival of the Ogoni People (MOSOP) presented the federal government with the Ogoni Bill of Rights, in which they demanded local autonomy, control of resources, and compensation for pollution's effect on loss of livelihood. When MOSOP's leader, Ken Saro-Wiwa, and eight other Ogoni activists were hung by the Abacha regime in November 1995, the Ogoni struggle against Abacha and Shell came to the whole world's attention. The Ogoni example was later followed by the Ijaw, who issued the Kaiama Declaration, an assertion of local ownership of land and natural resources similar to the Ogoni Bill of Rights.

As described by Obi and Rustad (2011), communal conflicts, local resistance, and the struggle for resource control evolved into a full insurgency by 2006. In the meantime, the residents of the Niger Delta suffer the ill effects of oil extraction. A 2005 report by Environmental Rights Action and the Climate Justice Programme estimated that in Bayelsa State alone, gas flaring is statistically likely to cause 49 premature deaths, 5,000 respiratory illnesses among children, 120,000 asthma attacks, and 8 additional cases of cancer (ERA & CJP, 2005, cited in Rowell et al., 2005, p. 68). A recent report by the Stakeholder Democracy Network compares the environmental performance of 43 oil companies operating in the Niger Delta. The report also found that overall Nigeria's oil and gas industry compares poorly to international averages in terms of oil spills (7 times the global average) and gas flaring (6 times the global average). They conclude that Nigeria's oil and gas industry "is far dirtier than any other region in the world, when assessed on an emissions per barrel of production basis" (SDN, 2024, p. 41).

According to Ikelegbe (2005), escalating violence in the late 1990s and early 2000s was due to the federal government's forceful approach, which radicalized Niger Delta youth on issues of resource control and local autonomy. In the context of increasing poverty, loss of livelihoods, and insecurity, gangs and cults of school dropouts and unemployed youth also began to engage in various criminal activities. These activities included hostage taking, protection rackets, extortion, armed robbery, political thuggery, drug peddling, theft of crude oil (bunkering), piracy, and weapons proliferation (Ako, 2011; Ikelegbe, 2011; Ukiwo, 2011). Over time, these criminal activities became more commonplace in the Delta and were also adopted by militant groups, making it difficult to distinguish criminal gangs from insurgents. Protest and violence increasingly used because force seen as only language the state and IOCs would listen to (Obi & Rustad, 2011, p. 3).

By 2006, a broad range of youth groups and militants began carrying out attacks against oil facilities in the Niger Delta in a loose alliance known as the Movement for the Emancipation of the Niger Delta (MEND). Several different individuals have appeared to speak for MEND at various times; however, the group has consistently called for greater resource control for the people of the Niger Delta and a greater share of revenues from petroleum resources. On occasion, it has called on the IOCs to leave the Niger Delta. The inability of the Nigerian armed forces to stop the dozens of bombings, acts of sabotage, kidnappings, and attacks on oil facilities led to the shutdown of one third of Nigeria's oil production, which in turn led to the amnesty program announced by President Yar'Adua in 2009 (Ukiwo, 2011).

Even Nigerian IOC executives feel disillusioned despite their own personal benefit. In fact, when asked if the benefits provided by his company outweighed the negative aspects of oil production, a Nigerian general manager of the French oil company Total responded empathetically that in his community people have lost land and been forced out of their livelihoods with inadequate compensation. However, he asserted that Total is a business operating in Nigeria and that it is the responsibility of the government, which gets most of the revenue, to arrange compensation (Ovadia, 2013, p. 331).

Former petroleum minister Edmund Daukoru, who was later chosen as the Amayanabo, or traditional ruler, of the Nembe Kingdom in Bayelsa state,[3] now acknowledges the devastation oil production has caused to communities and to livelihood activities, noting "One day we will have to deal with this—the pollution and exposure to the negative aspects of the industry. There are quite a lot of negatives to go with the positives we are enjoying!" (Ovadia, 2013, p. 332).

In recent years, conflict in the Niger Delta has often been related to contracts that the IOCs provide to local businesses. Such businesses are generally set up by local elites, and the contracts may be for goods and services directly related to oil production, but more often are related to activities, such as road building or construction of schools and clinics, carried out under the terms of the memoranda of understanding (MOUs) that the IOCs sign with host communities and land owners. Contracts are also awarded to private companies in the host communities for security and surveillance. This is often a way to pay off militant groups.

Prominent political leaders and traditional chiefs have their own companies and are most likely to be awarded contracts. This can lead, however, to disputes among local elites over contracts, chiefs complaining about the level of support and threatening violence, and IOCs directly and indirectly supporting local organizations and militias associated with particular politicians and ruling parties (Ovadia, 2013, p. 333). Employees, including one I spoke to who worked for a surveillance and security contractor for Shell, expressed anger that his company is owned by a wealthy Nigerian who is part of his village's elite, yet the for the vast majority, oil and gas have destroyed farming livelihoods and produced no benefit (Ovadia, 2016b, p. 181).

> What is true in areas where oil production is carried out by one IOC is also true in others. The Italian IOC ENI (formerly known as AGIP) operates the Nigerian AGIP Oil Company (NAOC). Summarizing the dynamics of conflict in the Niger Delta, an employee told me NAOC prefers to work with local elites in host communities and that they are the ones who are supposed to recommend which companies to for. This community has experienced decades of non-stop gas flaring, acid rain, new diseases and rashes, and loss of livestock and fisheries from oil spills. (Ovadia, 2016b, pp. 187–188)

While the situation in the Niger Delta has boiled over to the point that violence is an accepted fact of life and militant groups can only be pacified with payoffs, the situation in Ghana has not yet become so bad. Oil production in Ghana only began in 2010. Yet Ovadia and colleagues (2020) make a direct comparison to the Niger Delta in Nigeria in their longitudinal study in the six districts along Ghana's western coastline most affected by oil and gas production based on field research conducted in 2010 and 2015.

Literature on community impacts of resource extraction describes how firms often use the rhetoric of "sustainability and corporate social responsibility" to legitimize activities that cause environmental degradation and social disruptions (Benson & Kirsch, 2010; Gilberthorpe & Banks, 2012)—even as such rhetoric and CSR projects cause tensions, such as those described earlier, between the government and private sector on one side and Indigenous groups on the other (Arellano-Yanguas, 2011; Watts, 2001). Ovadia et al. found few identifiable benefits beyond CSR projects that are "often disconnected from local development priorities, communities are growing angrier at their loss of livelihoods, increased social ills and dispossession from land and ocean." In these communities, the assumption that others must be benefiting from the petroleum resources being extracted near their communities has led to growing frustration, real and perceived grievances, and increasing social fragmentation that threaten to lead to conflict and underdevelopment (Ovadia et al., 2020, pp. 397–398).

In a telling example, they describe when their research assistants were greeted with hostility in one community in particular. As they describe;

> Eventually, they were told the community would only do the focus group because the assistants were students. If they were researchers it would mean they are benefiting financially somehow from the research, so they wouldn't do the interview. If they were from a nongovernmental organisation (NGO), they told the assistants, they would have "locked them up" and kept them until the IOCs came to pay for their release. (Ovadia et al., 2020, p. 403)

Anger in Ghana's coastal communities centers on the decline of fisheries (and fish processing) and a belief they would receive more benefits in terms of infrastructure, financial support, job creation, and social services than have materialized. While it is not clear what role oil production has actually played in the decline of fish catches given alternative explanations such as overfishing by locals and foreign trawlers or climate change, communities blame IOCs for negative impacts on their livelihoods because they know the industry produces great wealth for some and that they are largely excluded.

Anger has also built due to increased crime, prostitution, higher costs of living, higher costs of land and land rents, and various other social impacts. "While these were felt most severely in the STMA, they were also felt elsewhere" (Ovadia et al., 2020, p. 415). The anger has increased social fragmentation due to questions about who has benefited from oil and increased suspicion both of outsiders and also local elites and political leaders. Perhaps most significantly, a "war on the sea" has been building with high numbers of canoe incursions into the offshore exclusion zone where oil production occurs (Ovadia et al., 2020, p. 420). Ayelazuno and Ovadia argue that these clashes occur because locals feel they have customary collective property rights of use and control over ocean space. The no-fishing zone can thus be understood as a case of enclosure of common property—one that is enforced by the Ghana Navy with multiple reports of violence against local fishers (Ayelazuno & Ovadia, 2022, p. 17).

Communities in Ghana also have environmental concerns around gas flaring and its health impacts, oil spills and oil spill preparedness, and whale deaths. Local district and municipal environmental officials feel ill-prepared to address any of these concerns (Ovadia et al., 2020, pp. 418–419). As Otchere-Darko and Ovadia explain, the conflict between local communities, on the one hand, and both government and IOCs, on the other, can be understood as being related to incommensurable values in land. While local and non-state actors express value in land in terms of social market investments, gainful employment, food security, and protection from expropriation and pollution, state actors and IOCs emphasize efficiency, competitiveness, and pursuit of the "highest and best use of land and petroleum resources" (Otchere-Darko & Ovadia, 2020, p. 69).

Ackah-Baidoo explains that IOCs operating offshore often fail in their development efforts because they do not adequately understand local dynamics. She calls this a failure of "offshore CSR" (Ackah-Baidoo, 2012). Overall, the ways that communities seek to increase benefit and reduce risk reflect their different values, backgrounds, and positions of power. Thus, their approaches to value in land, which include customary stewardship, communal benefit, and addressing proximal risk (in terms of metabolic and calorific value as well as environmental costs) are incommensurable with the approach of the state and oil and gas industry—as are the preferred forms of state and oil industry intervention, namely voluntary CSR and carbon-offset projects (Otchere-Darko & Ovadia, 2020, pp. 72–74). This leads to deepening conflict and resistance.

Conclusion

Although resistance to fossil fuel extraction is already significant and growing in oil-producing areas due to health impacts, environmental harms, loss of biodiversity, loss of livelihoods, and so on, there is reason to expect growing resistance driven by climate action in the years to come. Although it has contributed very little to climate change historically and in terms of current emissions, Africa is the most vulnerable continent to the impacts of climate change (Adenle et al., 2017; Carabine & Lemma, 2014; Ovadia et al., 2024). A recent World Bank study has suggested that as many as 86 million Africans will be forced to migrate within their own countries by 2050 because of changing weather patterns (World Bank, 2022). A study by the AfDB, UN Environment Programme, and UN Economic Commission for Africa found that western and eastern Africa could lose up to about 15% of their gross domestic product (GDP) by 2050 due to climate inaction (Baarsch & Schaeffer, 2019).

From a just energy transitions perspective, current and historical responsibility needs to be taken into account when considering which countries should be allowed to continue exploiting fossil fuel (Chancel, 2021; Ramachandran, 2021). However, given the remaining emissions that can be emitted before we reach global warming of 1.5°C, there should be no new fossil fuel extraction anywhere in the world and significant curtailment of existing fossil fuel production. Fortunately, African and global energy needs can be met with renewable energy (Ackah & Graham, 2021; Adeniran & Onyekwena, 2020; Irowarisima, 2020; IRENA & AfDB, 2022; Quitzow et al., 2016) with sufficient finance (Gilder & Rumble, 2021; Power Shift

Africa, 2021; Rumble & Benkenstein, 2021). Should warming limits be breached, African energy needs be left unmet, climate mitigation and adaptation in Africa be ignored, or the frustrations of communities that have hosted fossil extraction without sufficient benefit be left unaddressed, anger, conflict, and resistance will multiply in the years to come.

Notes

1. Temperatures. *Climate Action Tracker.* https://climateactiontracker.org/global/temperatures/
2. Carney, M. (2005, September 29). Breaking the tragedy of the horizon: Climate change and financial stability. https://www.bis.org/review/r151009a.pdf
3. The Nembe Kingdom and the city of Brass (now home to a major LNG terminal) were featured in the opening of Okonta and Douglas's book *Where Vultures Feast.* Brass, chief city of the Ijaw, was razed to the ground in 1895 to secure a monopoly for the Royal Niger Company over the Indigenous traders of the Nembe Kingdom. Over 2,000 people, mostly women and children, were massacred in the attack (Okonta & Douglas, 2003, p. 1).

References

Ackah, I., & Graham, E. (2021). Meeting the targets of the Paris Agreement: An analysis of renewable energy (RE) governance systems in West Africa (WA). *Clean Technologies and Environmental Policy, 23*, 501–507.

Ackah-Baidoo, A. (2012). Enclave development and "offshore corporate social responsibility": Implications for oil-rich sub-Saharan Africa. *Resources Policy, 37*(2), 152–159.

Adeniran, A., & Onyekwena, C. (2020). *Accelerating green energy transition in Africa through regional integration.* South African Institute of International Affairs.

Adenle, A. A., Ford, J. D., Morton, J. Twomlow, S., Alverson, K., Cattaneo, A., Cervigni, R., Kurukulasuriya, P., Huq, S., Helfgott, A., & Ebinger. J. O. (2017). Managing climate change risks in Africa—A global perspective. *Ecological Economics, 131*, 190–201.

Ahmadov, A. K. (2014). Oil, democracy, and context: A meta-analysis. *Comparative Political Studies, 47*(9), 1238–1267.

Ako, R. (2011). The struggle for resource control and violence in the Niger Delta. In C. I. Obi & S. A. Rustad (Eds.), *Oil and insurgency in the Niger Delta: Managing the complex politics of petro-violence* (pp. 42–54). Zed Books.

Angrist, J. D., & Kugler, A. D. (2008). Rural windfall or a new resource curse? Coca, income, and civil conflict in Colombia. *Review of Economics and Statistics, 90*(2), 191–215.

Arce, M., Hendricks, M. S., & Polizzi, M. S. (2022). *The roots of engagement: Understanding opposition and support for resource extraction.* Oxford University Press.

Arellano-Yanguas, J. (2011). Aggravating the resource curse: Decentralisation, mining and conflict in Peru. *Journal of Development Studies, 47*(4), 617–638.

Auty, R. M. (1993). *Sustaining development in mineral economies: The resource curse thesis.* Routledge.

Auty, R. M. (2001). *Resource abundance and economic development.* Oxford University Press.

Auty, R. M. (2007). Patterns of rent extraction and deployment in developing countries: Implications for governance, economic policy and performance." In G. Mavrotas & A. Shorrocks (Eds.), *Advancing development: Core themes in global economics* (pp. 555–577). Palgrave Macmillan.

Ayaburi, J., Sharma, S., & Bazilian, M. (2021). Comparative analysis of selected African natural gas markets and related policies. *Energy for Sustainable Development, 63*, 67–77.

Ayelazuno, J. A. (2014). Oil wealth and the well-being of the subaltern classes in sub-Saharan Africa: A critical analysis of the resource curse in Ghana. *Resources Policy, 40*, 66–73.

Ayelazuno, J. A., & Ovadia, J. S. (2022). Ocean and land grabbing in Ghana's offshore petroleum industry: From the agrarian question to the question of industrialization. *Journal of Agrarian Change, 22*(4), 673–702.

Baarsch, F., & Schaeffer, M. (2019). *Climate change impacts on Africa's economic growth*. African Development Bank.

Barlow, A. (2023). Piping away development: The material evolution of resource nationalism in Mtwara, Tanzania. *Journal of Southern African Studies, 49*(3), 377–395.

Bebbington, A., & Bebbington, D. H. (2011). An Andean avatar: Post-neoliberal and neoliberal strategies for securing the unobtainable. *New Political Economy, 16*(1), 131–145.

Benson, P., & Kirsch, S. (2010). Capitalism and the politics of resignation. *Current Anthropology, 51*(4), 459–486.

Blondeel, M., Bradshaw, M. J., Bridge, G., & Kuzemko, C. (2021). The geopolitics of energy system transformation: A review. *Geography Compass, 15*(7), e12580.

Brunnschweiler, C. N. (2008). Cursing the blessings? Natural resource abundance, institutions, and economic growth. *World Development, 36*, 399–419.

Buccellato, T., & Mickiewicz, T. (2009). Oil and gas: A blessing for the few. hydrocarbons and inequality within regions in Russia. *Europe-Asia Studies, 61*(3), 385–407.

Bulte, E. H., Damania, R., & Deacon, R. T. (2005). Resource intensity, institutions, and development. *World Development, 33*(7), 1029–1044.

Carabine, E., & Lemma, A. (2014). *The IPCC's fifth assessment report: What's in it for Africa?* Climate and Development Knowledge Network.

Chancel, L. (2021). *Climate change and the global inequality of carbon emissions (1990–2020)*. World Inequality Lab, Paris School of Economics.

Collier, P., & Hoeffler, A. (1998). On economic causes of civil war. *Oxford Economic Papers, 50*, 563–573.

Collier, P., & Hoeffler, A. (2005). Resource rents, governance, and conflict. *Journal of Conflict Resolution, 49*(4), 625–633.

Cramer, C. (2002). Homo economicus goes to war: Methodological individualism, rational choice and the political economy of war. *World Development, 30*, 1845–1864.

Cust, J. F., & Mihalyi, D. (2017). *Evidence for a resource curse? Oil discoveries, elevated expectations, and growth disappointments*. The World Bank.

Davis, G. A., & Tilton, J. E. (2005). The resource curse. *Natural Resources Forum, 29*, 233–242.

Denton, F., Arko, T., Forkuor, G., Tekie, B., Anderson, S., Yeboah, S., Moyo, Q., Bombande, M., Ampomah, G., & Abebe, S. K. (2021). *Blind alleys and bright prospects: Africa navigating stranded assets and just transitions*. United Nations University Institute for Natural Resources in Africa.

Eicke, L., & Goldthau, A. (2021). Are we at risk of an uneven low-carbon transition? Assessing evidence from a mixed-method elite study. *Environmental Science & Policy, 124*, 370–379.

Enns, C., Bersaglio, B., & Awiti, A. (2022). The promises and pitfalls or pursuing inclusive, sustainable development through resource corridors in Africa. In N. Andrews, J. A. Grant, & J. S. Ovadia (Eds.), *Natural resource-based development in Africa: Panacea or Pandora's box?* (pp. 221–238). University of Toronto Press.

ERA & CJP. (2005). *Gas flaring in Nigeria: A human rights, environmental and economic monstrosity*. Environmental Rights Action and Climate Justice Programme.

Ezeamalu, B. (2024, August 5). Tinubu calls for protests to end; organizers say it will continue. *The Africa Report*. https://www.theafricareport.com/357453/tinubu-calls-for-protest-to-end-organisers-say-it-will-continue/

Fattouh, B., Poudineh, R., & West, R. (2018). *The rise of renewables and energy transition: What adaptation strategy for oil companies and oil-exporting countries?* Oxford Institute for Energy Studies.

Frynas, J. G. (2005). The false developmental promise of corporate social responsibility: Evidence from multinational oil companies. *International Affairs, 81*(3), 581–598.

Frynas J. G., Wood, G., & Hinks, T. (2017). The resource curse without natural resources: Expectations of resource booms and their impact. *African Affairs, 116*(463), 233–260.

Gelb, A., & Associates. (1988). *Oil windfalls: Blessing or curse?* The World Bank.

Gilberthorpe, E., & Banks, G. (2012). Development on whose terms? CSR discourse and social realities in Papua New Guinea's extractive industries sector. *Resources Policy, 37*, 185–193.

Gilberthorpe, E., & Papyrakis, E. (2015). The extractive industries and development: The resource curse at the micro, meso and macro levels. *The Extractive Industries and Society*, 2(2), 381–390.
Gilder, A., & Rumble, O. (2021). *Implications of COVID-19 for climate finance in Africa*. South African Institute of International Affairs.
Givens, J. E., Huang, X., & Jorgenson, A. K. (2019). Ecologically unequal exchange: A theory of global environmental injustice. *Sociology Compass*, 13(5), e12693.
Goldthau, A., Eicke, L., & Weko, S. (2020). The global energy transition and the Global South. In M. Hafner & S. Tagliapietra (Eds.), *The geopolitics of the global energy transition* (pp. 319–339). Springer.
Graham, E., & Ovadia, J. S. (2019). Oil exploration and production in Sub-Saharan Africa, 1990–present: Trends and developments. *The Extractive Industries and Society*, 6, 593–609.
Grubler, A., Wilson, C., & Nemet, G. (2016). Apples, oranges and consistent comparisons of the temporal dynamics of energy transitions. *Energy Research & Social Science*, 22, 18–25.
Gudynas, E. (2010). *The new extractivism of the 21st century: Ten urgent theses about extractivism in relation to current South American progressivism*. Americas Policy Programme.
Haber, S., & Menaldo, V. (2011). Do natural resources fuel authoritarianism? A reappraisal of the resource curse. *American Political Science Review*, 105(1), 1–26.
Hilson, G. (2010). "Once a miner, always a miner": Poverty and livelihood diversification in Akwatia, Ghana. *Journal of Rural Studies*, 26(3), 296–307.
Hilson, G. (2012). Poverty traps in small-scale mining communities: The case of sub-Saharan Africa. *Canadian Journal of Development Studies*, 33(2), 180–197.
Idemudia, U. (2007). Community perceptions and expectations: Reinventing the wheels of corporate social responsibility practices in the Nigerian oil industry. *Business and Society Review*, 112(3), 369–405.
Idemudia, U. (2012). The resource curse and the decentralization of oil revenue: The case of Nigeria. *Journal of Cleaner Production*, 35, 183–193.
IEA. (2021). *Net zero by 2050: A roadmap for the global energy sector*. International Energy Agency.
Ikelegbe, A. (2001). Civil society, oil and conflict in the Niger Delta region of Nigeria: Ramifications of civil society for a regional resource struggle. *Journal of Modern African Studies*, 39(3), 437–469.
Ikelegbe, A. (2005). Encounters of insurgent youth associations with the state in the oil rich Niger Delta region of Nigeria. *Journal of Third World Studies*, 14(2), 208–234.
Ikelegbe, A. (2011). Popular and criminal violence as instruments of struggle in the Niger Delta region. In C. I. Obi & S. A. Rustad (Eds.), *Oil and insurgency in the Niger Delta: Managing the complex politics of petro-violence* (pp. 125–135). Zed Books.
IPCC. (2021). *Climate change 2021: The physical science basis sixth assessment report*. Intergovernmental Panel on Climate Change.
IRENA & AfDB. (2022). *Renewable energy market analysis: Africa and its regions*. International Renewable Energy Agency and African Development Bank.
Irowarisima, M. (2020). African energy challenges in the transition era: The role of regional cooperation. In V. R. Nalule (Ed.), *Energy transitions and the future of the African energy sector: Law, policy and governance* (pp. 37–72). Palgrave Macmillan.
Isham, J., Woolcock, M., Pritchett, L., & Busby, G. (2005). The varieties of resource experience: Natural resource export structures and the political economy of economic growth. *World Bank Economic Review*, 19(2), 141–174.
Jensen, N., & Wantchekon, L. (2004). Resource wealth and political regimes in Africa. *Comparative Political Studies*, 37(7), 816–841.
Karl, T. L. (1997). *The paradox of plenty*. University of California Press.
Kibble, S. (2006). Can the politics of disorder become the politics of democratisation and development? *Review of African Political Economy*, 33(109), 525–542.
Klare, M., & Volman, D. (2006). The African "oil rush" and US national security. *Third World Quarterly*, 27(4), 609–628.
Kolstad, I., & Søreide, T. (2009). Corruption in natural resource management: Implications for policy makers. *Resources Policy*, 34(4), 214–226.

Le Billon, P. (2008). Diamond wars? Conflict diamonds and geographies of resource wars. *Annals of the Association of American Geographers*, *98*, 345–372.

Leite, C. A., & Weidmann, J. (1999). *Does mother nature corrupt? Natural resources, corruption, and economic growth*. International Monetary Fund.

Manley, D., Cust, J. F., & Cecchinato, G. (2017). Stranded nations? The climate policy implications for fossil fuel-rich developing countries. OxCarre Policy Paper 34. https://ssrn.com/abstract=3264765

Mehlum, H., Moene, K. O., & Torvik, R. (2006). Cursed by resources or institutions? *The World Economy*, *29*, 1117–1131.

Obeng-Odoom, F. (2014). *Oiling the urban economy: Land, labour, capital, and the state in Sekondi-Takoradi, Ghana*. Routledge.

Obi, C. I. (2010a). Oil as the "curse" of conflict in Africa: Peering through the smoke and mirrors. *Review of African Political Economy*, *37*, 483–495.

Obi, C. I. (2010b). Oil extraction, dispossession, resistance, and conflict in Nigeria's oil-rich Niger Delta. *Canadian Journal of Development Studies*, *30*, 219–236.

Obi, C. I., & Rustad, S. A. (2011). *Oil and insurgency in the Niger Delta: Managing the complex politics of petro-violence*. Zed Books.

Okonta, I. (2008). *When citizens revolt: Nigerian elites, big oil and the Ogoni struggle for self-determination*. Africa World Press.

Okonta, I., & Douglas, O. (2003). *Where vultures feast: Shell, human rights, and oil*. Verso.

Omeje, K. C. (2006). *High stakes and stakeholders: Oil conflict and security in Nigeria*. Ashgate.

Orucho, B. (2023, February 13). Africa's share of global gas supply will almost double by 2050. *Quartz Africa*. https://qz.com/africas-share-of-global-gas-supply-will-almost-double-b-1850106651

Otchere-Darko, W., & Ovadia, J. S. (2020). Incommensurable languages of value and petro-geographies: Land-use, decision-making and conflict in south-western Ghana. *Geoforum*, *113*, 69–80.

Ovadia, J. S. (2008). Development and security in Africa's 'American lake': The political economy of oil and exploitation. In M. Ayyash & C. Hendershot (Eds.), *Violent interventions' conference proceedings* (pp. 39–56). York Centre for International and Security Studies.

Ovadia, J. S. (2013). The Nigerian "one percent" and the management of national oil wealth through Nigerian content. *Science & Society*, *77*(3), 315–341.

Ovadia, J. S. (2016a). Oil-backed capitalist development in the Global South: A case of positive oil exceptionalism? In T. Di Muzio & J. S. Ovadia (Eds.), *Energy, capitalism and world order: Toward a new agenda in international political economy* (pp. 79–93). Palgrave Macmillan.

Ovadia, J. S. (2016b). *The petro-developmental state in Africa: Making oil work in Angola, Nigeria and the Gulf of Guinea*. Hurst.

Ovadia, J. S. (2019, October 16). A window of opportunity for African fossil fuels. *Project Syndicate*. https://www.project-syndicate.org/commentary/africa-oil-industry-window-of-opportunity-by-jesse-salah-ovadia-2019-10

Ovadia, J. S. (2020). Natural resources and African economies: Asset or liability? In T. Falola and & S. O. Oloruntoba (Eds.), *Palgrave handbook of African political economy* (pp. 667–678). Palgrave MacMillan.

Ovadia, J. S., Andreoni, A., Roberts, S., Monjane, C. M., Davies, E., & Fakir, S. (2024). *Sustainable energy transitions for African petroleum producers*. African Climate Foundation.

Ovadia, J. S., Ayelazuno, J. A., & van Alstine, J. (2020). Ghana's petroleum industry: Expectations, frustrations and anger in coastal communities. *Journal of Modern African Studies*, *58*(3), 397–424.

Ovadia, J. S., & Graham, E. (2022). The resource curse and limits of petro-development in Ghana's "oil city": How oil production has impacted Sekondi-Takoradi. In N. Andrews, J. A. Grant, & J. S. Ovadia (Eds.), *Natural resource-based development in Africa: Panacea or Pandora's box?* (pp. 59–78). University of Toronto Press.

Papyrakis, E. (2017). The resource curse: What have we learned from two decades of intensive research: Introduction to the special issue. *Journal of Development Studies*, *53*, 175–185.

Papyrakis, E., & Gerlagh, R. (2004). The resource curse hypothesis and its transmission channels. *Journal of Comparative Economics*, 32, 181–193.

Papyrakis, E., & Raveh, O. (2014). An empirical analysis of a regional Dutch disease: The case of Canada. *Environmental and Resource Economics*, 58, 179–198.

Petermann, A., Guzmán, J. I., & Tilton, J. E. (2007). Mining and corruption. *Resources Policy*, 32(3), 91–103.

Power Shift Africa. (2021). *COP26 delivering the Paris Agreement: A five-point plan for solidarity, fairness, and prosperity.* Power Shift Africa.

Quitzow, R., Roehrkasten, S., Jacobs, D., Bayer, B., Jamea, E. M., Waweru, Y., & Matschoss, P. (2016). *The future of Africa's energy supply: Potentials and development options for renewable energy.* Institute for Advanced Sustainability Studies.

Ramachandran, V. (2021, November 3). Rich countries' climate policies are colonialism in green. *Foreign Policy.* https://foreignpolicy.com/2021/11/03/cop26-climate-colonialism-africa-norway-world-bank-oil-gas/

Reed, K. (2009). *Crude existence: Environment and the politics of oil in northern Angola.* University of California Press.

Ross, M. L. (2001). Does oil hinder democracy? *World Politics*, 53, 325–361.

Rosser, A. (2006). Escaping the resource curse. *New Political Economy*, 11, 557–570.

Rowell, A., Marriott, J., & Stockman, L. (2005). *The next gulf: London, Washington and oil conflict in Nigeria.* Constable & Robinson.

Rumble, O., & Benkenstein, A. (2021). *An African climate finance agenda for COP26.* South African Institute of International Affairs.

Saad-Filho, A., & Weeks, J. (2013). Curses, diseases and other resource confusions. *Third World Quarterly*, 34(1), 1–21.

Sachs, J., & Warner, A. M. (2001). The curse of natural resources. *European Economic Review*, 45(4–6), 827–838.

SDN. (2024). *2020 Nigerian oil industry environmental performance index.* Stakeholder Democracy Network. https://www.stakeholderdemocracy.org/wp-content/uploads/2024/05/EPI-Report-2020.pdf

SEI. (2021). *The production gap.* Stockholm Environment Institute.

Shao, S., & Qi, Z. (2009). Energy exploitation and economic growth in western China: An empirical analysis based on the resource curse hypothesis. *Frontiers of Economics in China*, 4(1), 125–152.

Sovacool, B. K. (2016). How long will it take? Conceptualizing the temporal dynamics of energy transitions. *Energy Research & Social Science*, 13, 202–215.

Sovacool, B. K., & Geels, F. W. (2016). Further reflections on the temporality of energy transitions: A response to critics. *Energy Research & Social Science*, 22, 232–237.

Svampa, M. (2015). Commodities consensus: Neoextractivism and enclosure of the commons in Latin America. *The South Atlantic Quarterly*, 114(1), 65–82.

Ukiwo, U. (2011). The Nigerian state, oil and the Niger Delta crisis. In C. I. Obi & S. A. Rustad (Eds.), *Oil and insurgency in the Niger Delta: Managing the complex politics of petro-violence* (pp. 17–27). Zed Books.

van de Graaf, T., & Bradshaw, M. (2018). Stranded wealth: Rethinking the politics of oil in an age of abundance. *International Affairs*, 94(6), 1309–1328.

Watts, M. (2001). Petro-violence: Community, extraction, and political ecology of a mythic commodity. In N. L. Peluso & M. Watts (Eds.), *Violent environments* (pp. 189–212). Cornell University Press.

Watts, M. (2006). Empire of oil: Capitalist dispossession and the scramble for Africa. *Monthly Review*, 58(4), 1–17.

Welsch, H. (2008). Resource abundance and internal armed conflict: Types of natural resources and the incidence of "new wars." *Ecological Economics*, 67(3), 503–513.

World Bank. (2022, January 5). Climate migration in Africa: How to turn the tide. https://www.worldbank.org/en/region/afr/publication/climate-migration-in-africa-how-to-turn-the-tide

Wright, J., Frantz, E., & Geddes, B. (2015). Oil and autocratic regime survival. *British Journal of Political Science*, *45*(2), 287–306.
Zhang, X., Xing, L., Fan, S., & Luo, X. (2008). Resource abundance and regional development in China. *Economics of Transition*, *16*(1), 7–29.

CHAPTER 18

Global Indigenous Struggles and Climate Change Activism

JAMES V. FENELON AND LEONARDO FIGUEROA HELLAND

Global Indigenous Struggles and Climate Change Activism

INDIGENOUS peoples and nations embody worldviews and biocultural societies that nurture healthier relationships with "Mother Earth" (Fenelon & Alford, 2020; Figueroa-Helland, 2022; FPP, 2020; Maffi & Woodley, 2010). Indigenous survivance and resistance—often considered "activism"—(re)assert sustainable Indigenous worlds versus a hegemonic colonial, capitalist, state system bent on accumulation, extraction, and exploitation, which drive climate change (Fenelon, 2015; Figueroa-Helland et al., 2021). This chapter provides a global overview and regional examples of Indigenous struggles, centering Indigenous contestations over "Anthropocene" crises. With Indigenous spirit, we proceed from a place-based story.

Grounding Story

Dakota elder Henry Swift Horse pointed to where buffalo would cross the wide but shallow Missouri River in days long ago, where his father would set up a small tipi, singing his *pte oyate* song, awaiting one to break from their crossing and be taken to feed their two-legged relatives. Just north was where the swirling rivers met, creating whirlpools producing great stone medicine balls, which the *wasicu* would call Cannonball. Many stories tell of this creation, along with peoples' origins, also of *mni*—water making life—and of *inyan*—stones used in ceremony—to honor Earth, *maka*, and the ancestors. We[1] had come to interview him with other elders—Reginald Bird Horse and Vernon Iron Cloud of Grand River—who lost their homes (and livelihoods) when the Army Corps of Engineers built the Oahe Dam just south near Mobridge, filling rich river valley forests of flora and fauna, with dam water,[2] destroying ecosystems upon which their culture had developed and depended on interaction with all life—animate and inanimate, which lived in a harmony of give and take, of life, death, and future generations. These Standing Rock elders speaking in 1987 forebode the past, having resisted dams and destruction over power and "ownership" in the 1950s and 1960s—we wrote a pamphlet, *The Taken Land*, which they took to the US Congress, only to be turned away—and foretold the present into the future, when their descendants and many Native nations stood at a confluence of the rivers, now a lake, to resist the Dakota Access Pipeline—2016 on to 2017. This militaristic destruction resembled "Indian wars" and continuing settler colonialism of the past 200 years since exploring invaders Lewis and Clark came to claim land, without negotiating, but only by governmental mandate over people they had never met nor made any "purchase" from. Elders had met on the same spot where the *Oceti Sakowin* camp of "water protectors" gathered to oppose ongoing takings for oil and gas, money and profits over Earth, people, spirituality, and place.

This event evokes Indigenous peoples' struggles against forces subjugating and desecrating their worlds (Scheidel et al., 2023). Upon colonial domination, capital accumulation, and state violence, these forces seek to eradicate ecosphere cultures/lifeways, risking systemic ecological-climate catastrophe (Fernández-Llamazares et al., 2021; Figueroa-Helland et al., 2021; Goldtooth, 2017). This eco-genocidal destruction underpins a Western civilization on a mad dash toward a world devoid of healthy environments, upon "progress" and "development" that sacrifice future generations and societies who care for all life, thereby limiting humanity (Fenelon & Hall, 2008; Whyte, 2020). Globally, for over 500 years, species, landscapes, ecosystems, biomes, and the biosphere have been devastated, and with that, many Indigenous nations—two intimately tied dystopias (Davis & Todd, 2017; Toledo, 2009; Whyte, 2018).

In Turtle Island/North America, for instance, buffalo would be nearly eradicated in a 200-year period of "development" and "progress" with most natural lifeforms suppressed, living near extinction. Indigenous peoples there, as elsewhere, were forced unto ever-smaller territories ("reservations"), evermore suppressed, while huge agricultural and mineral markets operating from highly stratified urban areas brought pollution and destruction to air, water, and land across regions, generating catastrophic climate change and poisonous toxicity on all life. Secretary Haaland recently stated:

> Of all the losses this continent has suffered, the destruction of the Great Plains is among the starkest. When the ... government began its conquest of Indigenous peoples, it surmised that the "Indian problem"—wherein Native communities lived and thrived on lands sought by colonial settlers and the ... government—could be solved by destroying Native food sources and livelihoods. And so, men with guns killed tens of millions of bison. This set off a chain reaction of ecosystem degradation that, now intertwined with the ... climate crisis, spelled catastrophe for ancestral homelands and ecosystems we all depend on.
>
> (Haaland, 2024, p. 38)

Global and Historical Framing

Colonial, capitalist, state exploitation causes "Anthropocene" crises: climate change, mass extinctions, species displacements, endemic biota replacements, degrading land-use changes, biosphere disruptions—eroding adaptability and resiliency (Bonneuil & Fressoz, 2016; Whyte, 2017). The prospect is a Hothouse Earth where nonlinear feedbacks too complex to comprehensively model/predict—let alone control—irreversibly trigger massive planetary upheavals at tipping points in Earth system boundaries/processes (Folke et al., 2021). By 2024–2028, the all-too-high 1.5°C Paris Agreement warming threshold—already sacrificing Indigenous, Global South, and coastal communities—is being exceeded; with 2023 and 2024 the warmest recorded years (EUC3S, 2024; WMO, 2024).

The founding violence and structural underbelly of "Anthropocene" catastrophe is the 500+ years of genocide-ecocide of Indigenous peoples and their biocultural territories, enabling the imposition of modern, Eurocentric, colonial, capitalist, patriarchal, and state-centric orders (Davis & Todd, 2017; Fenelon, 2015; Figueroa-Helland et al., 2021). European colonial invasions, first of Abya Yala-Turtle Island/Americas and then globally, alongside the enslavement of Indigenous Africans, for land, labor, and resource theft, tell the origin story (Davis & Todd, 2017; Fenelon & Norgaard, 2021; Figueroa-Helland, 2022). Beyond a historical event, colonialism is a continuing structure enforcing climate change, ecosystem destruction, and forced relocation on Indigenous nations (Whyte, 2018). Colonialism desacralizes, objectifies, and commodifies human and other-than-human nature, coercively reengineering living territories and biotas into exploitable resources (Figueroa-Helland et al., 2025a). Colonialism violently reduces a bioculturally diverse Earth into monocultures of dominion, dispossession, extraction, enclosure, accumulation, and waste, enacting ecocide upon racial capitalism (Gonzalez, 2024). A minority of colonially privileged humans disproportionately drives planetary crises through capitalist-state systems—not an undifferentiated "humanity" (Figueroa-Helland et al., 2021).

Contrastingly, Indigenous nations carry millennia of sustainable biocultures where peoples, biotas, and ecosystems, interwoven through webs of relational codependence and reciprocal care and nurturance, constitute *living territorial communities* whose governance and cultures must ensure integrated socioecological reproduction through generations (Fenelon & Hall, 2008; Figueroa-Helland, 2022; McGregor, 2020). Understandably, most biodiversity is in or overlaps with Indigenous territories, though assailed by actors whose worldviews and socioeconomic systems require objectification, desacralization, commodification, homogenization, domination, and exploitation of life and Earth (Fernández-Llamazares et al., 2021; FPP, 2020; Scheidel et al., 2023). Given their intimate territorial ties, Indigenous communities are distinctively sensitive to environmental destruction and climate injustice (Fenelon & Norgaard, 2021; Nelson & Shilling, 2018; Whyte, 2018). Territorial loss devastates knowledges, food/material sovereignty, waters, homelands, self-determination, and millennia-adapted lifeways (Fernández-Llamazares et al., 2021). While dominant actors finally notice Earth crises, Indigenous peoples always recognized and resisted their drivers, while surviving genocides/ecocides, colonialism, unsustainable terraforming, forced displacement/relocation, assimilation, and racism by many such actors (Davis & Todd, 2017; Whyte, 2018).

Indigenous territorial governance ensures rich biodiversity and carbon stocks; it must be defended where present and restituted where dispossessed (Figueroa-Hellan, 2022). Presently, Indigenous peoples dwell within or next to about 85% of protected areas, and Indigenous-tended forests show slower or no deforestation or degradation (Dooley, 2018; Gabay & Alam, 2016). Almost 38 billion tons of carbon are in community-managed forests globally, about 28% of countries' forests—mostly where Indigenous/forest peoples' customary governance sustains livelihoods by nurturing diverse biomass agroecologically, enhancing biodiversity (Figueroa-Helland et al., 2021; Gabay & Alam, 2016). Indigenous practices maintain, enrich, and expand biodiverse and carbon-rich habitats, and they can repair damaged habitats. Consider Indigenous forest gardening; it densifies carbon sinks, stably manages carbon cycles, sustains and enhances biodiversity, and ensures food/material sovereignty (Cardona-Calle, 2017; FPP, 2020; Figueroa-Helland et al., 2021; Gabay & Alam, 2016; GFC, 2018). Indigenous practices prove sustainable over millennia, *without* market mechanisms, exclusionary enclosures (fortress conservation, carbon offsetting enclosures), or risky techno-fixes (Figueroa-Helland, 2022; FOEI, 2022, 2023; GFC, 2018). Real solutions to Earth crises require restituting autonomous/sovereign Indigenous territories so biocultural diversity and climate resilience resurges (Figueroa-Helland et al., 2025a; FOEI, 2024).

Deep histories of nurturing and defending living territories mean Indigenous peoples have long been "climate/environmentally" active, well before "Anthropocene" crises became "threats" to the dominant Western society which predominantly caused them (Dhillon, 2021; Whyte, 2018). Since we discuss histories elsewhere (Fenelon, 2018, 2015, 2008; Fenelon & Norgaard, 2021; Figueroa-Helland, 2022; Figueroa-Helland et al., 2021), we discuss contemporary Indigenous climate politics/"activism"—internationally and then regionally.

International Indigenous Climate Politics/Activism

Though Indigenous "climate" politics/actors are complex and diverse, working across levels, regions, and subjects, we highlight key contemporary tensions with hegemonic climate governance, as noted by international bodies—the International Indigenous Peoples Forum on Climate Change (IIPFCC) and Indigenous Peoples Global Coordinating Committee (IPGCC), coalitions foregrounding diversity while advancing shared positions and demands. IPGCC, a global coordination and decision-making mechanism for the Indigenous Peoples Major Group for Sustainable Development, articulates Indigenous leaders and organizations from all seven sociocultural regions. Meanwhile, IIPFCC is an Indigenous representative body in the United Nations Framework Convention on Climate Change (UNFCCC). Partly rooted in the Indigenous Women's Biodiversity Network's activism at UNFCCC processes since 1998, by 2008 the IIPFCC constitutes the Caucus for Indigenous Peoples at Conferences of Parties (COPs) and intersessional meetings.

IIPFCC centers climate-related adaptation, mitigation, and funding, Indigenous rights, self-determination, and free, prior, and informed consent (FPIC), including regarding climate change–driving projects and policies (e.g., extractive) and mitigation projects and policies (e.g., forest, terrestrial, and ocean carbon, "green" energy, and carbon sequestration pathways and technologies). At recent COPs, IIPFCC (2023a, 2023b, 2024) underlines that while Indigenous peoples face increasing disasters, ravaged lands, and devastating physical and spiritual losses, states persistently fail to address root drivers (e.g., extractivism, hydrocarbons) or limit temperature rise. IIPFCC critiques climate governance as overtaken by problematic schemes and false solutions: carbon markets and offsets, geoengineering, carbon removal technologies, "net zero," so-called nature-based solutions, accelerated "transition" minerals extractivism, and so on (IIPFCC, 2023a, 2023b, 2024). IIPFCC (2023a, 2023b) underlines these do not cut emissions at the source and further "green" colonization, militarization, criminalization, and land loss. IIPFCC (2023a, 2024) warns about Article 6 (Paris Agreement) processes heading toward a global carbon market, threatening Indigenous peoples, and demands moratoria on aforementioned schemes and activities, which violate Indigenous rights.

Numerous Indigenous and allied actors share this stance (FOEI, 2024; HOME, 2024; PACT, 2023; Young, 2024). Evidence shows carbon market projects generate phantom offsets, land or "green" grabbing (upon environmental pretexts), and human and Indigenous rights violations (Dunne & Quiroz, 2023; Fairhead & Scoones, 2012; FOEI, 2022, 2023; Greenfield, 2023; McAfee, 2015; Rocheleau, 2015; Sovacool, 2021; Vigil, 2018; WRM, 2022, 2019). Offset project operators and brokers benefit from carbon trading, while actors causing climate change purchase credits, problematically claiming "progress" toward "net zero"; yet projects often fail to reduce emissions and perpetuate activities driving or exacerbating climate change and environmental degradation (Bragg, 2021; Stabinsky, 2021). Carbon Brief assessed 61 carbon offset projects: 43% of reports (EJAtlas.org and other sources) found projects overstating emissions reductions, and 72% involved harm—sometimes egregious—to Indigenous and local communities (Dunne & Quiroz, 2023). At the 2024 UN Permanent Forum on Indigenous Issues, the UN Special Rapporteur on the Rights of Indigenous Peoples, Francisco Calí Tzay, demanded a moratorium on carbon markets to stop abuses.

Others, like the Indigenous Environmental Network (2024), demand moratoria include carbon dioxide removal projects (e.g., carbon capture and storage); forest, soil, and ocean offsets; "nature-based solutions"; debt-for-nature swaps; biodiversity offsets; and geoengineering, which facilitate green colonialism, violate Indigenous rights, and greenwash destructive activities.

IIPFCC (2023b, 2023a, 2024) and IPGCC (2024) also warn that many "green/low-carbon (energy, technology) transitions" and some "just transitions" advance green grabbing and green colonialism, with negative socioenvironmental impacts (Brito-Millan et al., 2023; Figueroa-Helland, 2022; Scheidel et al., 2023; Sovacool, 2021). Transition minerals mining (lithium, copper, nickel, chromium, etc.), megadams, palm oil, biofuel and agrofuel, biomass and monocrop plantations, and industrial-scale wind and solar farms are often imposed at the expense of Indigenous territories and peoples, without FPIC, causing socio-ecological harms. Look at hydropower and dams (impacting and resisted by Sami, Adivasis, Cree, Inuit, Innu, and Lenca); agro/biofuels and bioenergy (Maya, Guarani, Dayak, and Mapuche), industrial-scale wind power (Ikoots, Sami, Adivasi, and pastoralists in Kenya), industrial-scale solar farms (Amazigh and Seneca), lithium (Yaqui, Likan Antay, Kolla, Paiute Shoshone, and Hualapai), rare earth, transition and technology minerals (Congo River basin peoples), industrial-scale geothermal (Maasai), and hydrogen (Pueblo, Diné/Navajo, Amazigh, and Saharawis) (Figueroa-Helland, 2022). IIPFCC (2023a, 2023b, 2024) demands address loss and damage, while prioritizing Indigenous-led transitions centering Indigenous rights, governance, sciences, agroecologies, and reciprocity practices with Earth.

IPGCC coconvened the 2024 Indigenous Summit on Just Transition (henceforth "Indigenous Summit"), concluding that "the just transition as ... currently presented is not true," as it implies "initiatives ... by ... corporations and States," involving

> imposition in our homelands of development projects promoted ... as "green/clean energy." ..., "green economy," and "climate change mitigation," including ... extraction of "transition" ... minerals, and ... resources ..., that displace Indigenous Peoples and contaminate our lands ..., due to ... land ... and water grabbing in the name of investment and conservation. (IPGCC, 2024)

Many such initiatives "consolidate neoliberalism," "deepen capitalist inequalities," and "fail to respect Mother Earth" when "carried out without FPIC," "violate Indigenous rights," and add "double harm" unto increasing hydrocarbon extraction (IPGCC, 2024). Thus, many "so-called green/clean projects ... do not constitute a transition, but ... a ... modern genocide against our worldviews, lifeways, and territorial governance" that perpetuates "ecocide" (IPGCC, 2024). Countering "models based on imposition, extraction, and colonialism," the Summit proposes "time-tested Indigenous knowledges ... and practices" for "sustainable food production, restoring soils, ecosystem protection, and true ... equitable just transition," which "means ... territorial governance according to our own ways." And demands "transitions" abide by *principles and protocols* like Indigenous self-determination, sovereignty, decolonization, reparations, land back, full restoration of lands, territories, waters, and biodiversity, and recognition and respect of Indigenous lifeways "as territorial caretakers/guardians, upon Indigenous laws" (IPGCC, 2024)

Thus, restituting sovereign Indigenous territorialities is the precondition for any real climate solution or sustainability transformation (Figueroa-Helland et al., 2021).

Regional Exemplars

Water Over Turtle Island

Turtle Island, an Indigenous identifier for North America, is territory of numerous and diverse Indigenous nations for whom interaction with environments, including water bodies, should be governed upon cultural-spiritual frameworks where homelands include all lifeforms, like the Dakota/Lakota principle, *mitakuye owasin*. Struggles of Indigenous "traditionalists" and leaders (activists) in Turtle Island embody matters central to all Indigenous peoples globally: land, water, and cultural sovereignty.

Observe how Chemehuevi traditional leader Matt Leivas joined forces with tribal nations and environmental organizations to resist ongoing (colonial) exploitation and destruction of Indigenous lifeways throughout the lower Colorado River region. The Chemehuevi established high desert reservation lands westward into California, and revitalized Salt Songs, that identify petroglyphs where water, sacred lands, and ancestral trails are found. This amazing cultural expression, recognizing sacred sites over lands in the vast Mojave Desert, supports an "activism" resisting water companies from tapping into a "10,000-year-old" aquifer. Their relatives, the Cahuilla, pressed western legal action through the Native American Land Conservancy sponsored by local tribes. These trail songs are "maps" of the land thousands of years ago, with ancient petroglyphs marked by even more ancient lakes. In modern times, artificially supported lakes led to an environmental disaster of dried, polluted lands and waters, with uranium exploitation, transportation, and storage in the Amargosa River basin, and rare mineral extraction, that Native coalitions resist with environmental groups.

Cahuilla and Luiseno people, Serrano and Tongva, gathered in 2018 to tell stories of understanding rivers, headwaters, and plant and animal life as wholes (Fenelon et al., 2020). These Native leaders provide stewardship of lands and peoples of the greater Santa Ana River region, like the Tongva (*Gabrieleno* or *Kizh*) (Sepulveda, 2018) maintained the Kuruvungna springs in west-central Los Angeles, connected to the L.A. River, also highly concretized (or "developed") like the Santa Ana (leaders resisted in the Robidoux project).

Dakota and Lakota prophecies also foretold great change in coming centuries like Black Snake visions of the Missouri River region (Estes, 2024) and in Dakota ceremony along the Mississippi that preceded a great cleansing in the 1993 flood,[3] overcoming local dams and bridges down to sea. Leaders called on the "forces of nature" *Tate Topa*, Four Winds, seen as a great *Oyate*, philosophical basis of land, water, and wind, heard in *Mni Wiconi* prayers and 2016 NoDAPL (No Dakota Access Pipeline) protest songs (Estes, 2024).

We also note *Lenape: hoking* peoples—federally "recognized," returning to "ethnic cleansed" homelands to reclaim rivers dammed for power companies (Fenelon et al., 2023), while others struggle for state recognition and over "land back" in New Jersey and New York. We see a small Lenape Indian Tribe of Delaware successfully involved in ecosphere restoration using traditional ways. The United States attempts to divide and delegitimize these tribes and defer and delay tribal sovereignty. The Wampanoag fight similar struggles for 400 years, over sea (Aquinnah), land (Mashpee), and water claims originating before the United States existed.

Indigenous resistance in Turtle Island is on the frontlines of defending and restoring water, territories, and Mother Earth against actors driving degradation and climate change. Observe Indigenous resistance against fossil fuel projects in Turtle Island that stopped or delayed greenhouse gas pollution equivalent to at least one quarter of annual US and Canadian emissions, as highlighted by the Indigenous Environmental Network (IEN, 2021). Resistance continues from fighting exploration and extraction in the Arctic National

Wildlife Refuge, Canadian Tar Sands, San Juan Basin, Chaco Canyon, and Permian Basin, to fighting the ever-growing pipeline network (e.g., Coastal Gaslink, Line 3, Keystone/XL, etc.).

Amazonia

Indigenous Amazonians protect and enrich Amazonia's biodiversity, carbon stocks, and forests. Their autonomous territorial governance embodies real climate and sustainability solutions, especially when unimpeded by colonial, state, and capitalist intrusions (Figueroa-Helland et al., 2025a). Indigenous Amazonians—Shipibo, Kukama, Kichwa, Wampis, Ashaninka, Bora, and so on—practice biodiversity and climate-enhancing territorial management, such as cyclical cultivation with long fallows and forest restoration, forest gardening, and nurturing forest patches at rainforest edges to expand rainforest extent—for example, Kayapó (Posey, 2002). We note Kichwa *integral chakras*: bi-culturally coevolving polycultural systems agroecologically combining crop mixes with forest gardening, with species too numerous and varied to estimate; over 40% of communal land remains forest—primary/undisturbed or secondary/human-regenerated. Indigenous-tended forest landscapes overlap and coconstitute Amazonian forests, undifferentiated (Coq Huelva et al., 2017). This governance promotes eco-biological-social health; food, material, cultural, and epistemic sovereignty; and autonomous communal social reproduction. Exemplified by Sarayaku Kichwa "life-plans" and *Kawsak Sacha*/Living Forest cosmovisions, Indigenous Amazonians advance autonomous bioculturally enriching territorial governance versus extractivism (Figueroa-Helland et al., 2024, 2025b).

In addition to "conventional" extractivism (hydrocarbons, pipelines, spills, logging, mining), Indigenous Amazonians confront "green colonialism" from policies supposedly mitigating climate change yet facilitating exclusionary conservation alongside carbon offsetting for polluters, while displacing or adversely incorporating Indigenous territories and peoples. In Peru, conservation areas and parks, and forest carbon projects, harm Indigenous nations, such as Shipibo (Imiria Conservation Area—now offered as Lot UC.XP 010 for hydrocarbon exploration and extraction), Kukama (Pacaya Samiria Reserve), Awajun (Alto Mayo Protected Forest), or Kichwa (Cordillera Escalera Conservation Area). The Kichwa communities confront the Cordillera Azul National Park (PNCAZ), its inclusion in the International Union for Conservation of Nature's "green list," and its REDD+ carbon-offsetting scheme. Kichwa organizations accuse PNCAZ of being created by erasing Indigenous peoples, facilitating exclusionary conservation and opaque carbon trading and offsetting, including massively selling carbon credits (cast as "nature-based solutions") to polluters (e.g., Shell, Total, BHP, Procter & Gamble, Delta Airlines, Bank of America, Scotiabank) (CEPKA, 2022, 2023; FPP, 2024a; Valderrama, 2022). Exclusionary conservation and forest carbon projects are often linked to ever-expanding extractivism. States, corporations, and nongovernmental organizations (NGOs) can include "green" schemes within territorial re-engineerings where conservation enclosures and carbon sequestration are claimed as offsets or compensations for ever-expanding extraction, destruction, and pollution elsewhere—claims Indigenous and climate/environmental justice approaches challenge (Figueroa-Helland, 2022; Figueroa-Helland et al., 2025a). Concerning PCNAZ, Kichwa organizations sued the government, park, and managing NGO in 2022, claiming dispossessed territory during its creation and forestry concessions, plus continuing extraction (logging, hydrocarbons) in and around it. After an initially favorable hearing, Kichwas warned that the government and park ignored joint agreements and stalled Indigenous territorial demarcation. Indigenous organizations declared SERNANP (the administering government agency) and CIMA (the managing NGO) *personae non gratae* in Kichwa territories, adding they will take all legal and physical measures to recover them (CEPKA, 2023)—for example, auto-demarcation alongside community patrols and guards.

Kichwa organizations contribute to powerful processes in Peru, such as Indigenous communal-defense *guardias* (Shipibo-Konibo, Kakataibo) and Indigenous autonomous territorial governments (IATGs) (Chapra, Awajún, Wampis, Achuar—the latter two halting the north-Peruvian pipeline expansion). Since 2015, the Wampis Autonomous Territorial Government has represented 83 communities sustainably governing 1.3 million ha, which upon reaffirmations of identity, knowledge-culture, and symbiotic human–nature interconnection toward *Tarimat Pujut* is reaching the final welfare stage where territorial integrity-communal autonomy coconstitute the *vida grata*/pleasant life. These real climate solutions defend territories versus colonialism and extractivism while regenerating autonomous biocultural territories.

Mesoamerica

Indigenous Mesoamericans, whose territories extend from north-central Mexico to northern Central America, have sustained and enhanced climate-positive bioculturally diverse territories while nurturing 15% of global edible plants. Peoples like Maya, Zapotec, Zoque, Nahua, Huastec, Totonac, Mazatec, and P'urhépecha, among others, practice integrated territorial governance sustaining biodiverse habitats, carbon stocks, food sovereignty, and social reproduction by employing diversity-enhancing management. Observe the polycultural milpas of 80–278 species complementing landscape tapestries, including home gardens, pasturelands, aquatic systems, managed forests, forest gardens, and primary forests, used in communal territories that support 1,300+ species (medicines, materials, foods, etc.) with habitat-inducing, carbon-enriching, and provisioning capabilities, ensuring biotic-community reproduction, climate stabilization, and communal sovereignty (Figueroa-Helland et al., 2025a; Toledo et al., 2003). Mesoamerican cosmovisions enable biocultural richness. Principles like Maya (Tstosil) *Ich'el ta muk' Balamil xchu'uk yol snich'on* (Honor Earth and all her children) prescribe *Lekil Kuxlejal*—"good living" from balance/complementarity/reciprocity, vitality, and diversity (Sanchez Álvarez & Figueroa-Helland, 2024). Cosmovisions inform agroecologies, such as Maya and Zoque forest garden habitats nourishing all species. Bioregionally adapted successional milpa-to-forest garden cycles, including soil-enriching controlled fires, enable high-performing food forests while stabilizing carbon cycling and regenerating nutrients for biodiversity (Ford & Nigh, 2015).

Colonial-extractivist empires, states, and corporations have assaulted Mesoamerica, including land grabs for settlement, infrastructure, minerals, hydrocarbons, agroindustry, logging, hydropower, biofuels, tourism, and megadevelopment. Upon climate, energy, conservation, and food pretexts, they are also guilty of green grabbing. Both reinforce and exploit colonial hierarchies, reengineering territories to benefit dominant races, classes and capital, and states. Megadevelopment plans impose patchwork geographies dividing territorial uses; "conventional" development and extraction (oil, gas, pipelines; mines; agroindustry; shipping and trade corridors; railroads and highways; export-processing zones; megatourism) *coexist* with "green" projects (e.g., conservation/biosphere reserves, forest carbon schemes, agro/biofuels, megahydro/wind, ecotourism). The "fog of greening" (Rocheleau, 2015) obscures exploitation, degradation, and dispossession and adverse incorporation of peoples and territories as commodified resources, labor, or "ecosystem services."

In southern Mexico, extractive and development megaprojects combine with conservation and biosphere reserves, forest carbon, and export-oriented industrial-scale low-carbon energy within land- and water-intensive megadevelopment plans, like the Transisthmian

Interoceanic Corridor and "Maya Train" (CNI, 2021; Garcia, 2024; EjAtlas.org, 2022, 2023; Sandoval, 2020). In the Tehuantepec Isthmus, Zoque, Ayuujk, Binnizá, Chontal, Mixe, and Ikoots peoples are affected. For centuries, US and Mexican governments and capital envisioned a reengineered resource-rich Isthmus. Now 117 mining concessions, oil expansion, and 10 industrial parks are implemented or planned, dispossessing communal lands. Over 2,100 wind turbines supplying energy to corporations (e.g., Walmart) displace Indigenous peoples from over 50,000 ha, to service the megacorridor. "El Itsmo es Nuestro"/"The Isthmus Is Ours" campaign marks Indigenous resistance.

Connected is the "Maya Train" megacorridor (Yucatan Peninsula) (CNI, 2021; EjAtlas.org, 2022; Sandoval, 2020). Actual Maya and other communities become territorially dispossessed. Mega pig farms and transgenic and chemical crops threaten sacred cenotes and lands. While state, corporate, and nongovernmental organization (NGO) actors tout the Calakmul Biosphere Reserve, the "Maya Train" corridor destroys Maya forests. For decades, conservation reserves—existing or proposed—divide, displace, or adversely incorporate communities, like Chiapas's Montes Azules (EJAtlas, 2023b), linked to government-corporate carbon schemes selling credits to polluters. Megaproject impacts are aggravated by organized crime/narco-violence colluded with private paramilitaries, military counterinsurgency, and acquiescent—or complicit—state agents (CDHFBDC, 2023).

Yet Indigenous Mesoamericans resist, advancing real solutions. Consider the autonomist Congreso Nacional Indígena (CNI) (Mexico), articulating territorial resistances and autonomies ("Resistance Mirrors") (CNI, 2021). Amplifying the Maya Zapatistas' movement with other Indigenous nations, CNI coalesces numerous Indigenous movements toward autonomy, territorial reclamation, and communal self-defense (CNI, 2021, 2024). Mesoamericans refigure ever-evolving traditions of sustainable autonomous governance: from Indigenous Oaxacan autonomous municipalities, to Lenca (Honduras) and Maya-Ixil (Guatemala) resurgences against US and Hispanic oligarchy-sponsored genocidal wars and dispossessions, to Cheráni P'urhépecha community guards evicting all political parties, corporations, resource pillagers, and narco-cartels to restore *anapu ireteri jurámukua* (autonomous communal government/structure), and more (CPG, 2011; Díaz & Parres, 2024). Such autonomies reaffirm Indigenous dignity with territorial defense toward sovereign decolonial futures protecting Mother Earth.

East Africa

Indigenous presence overlaps with and coconstitutes East Africa's least disturbed, most biodiverse and carbon-rich ecosystems (PINGO's Forum, 2020). Indigenous peoples inhabit Kenya, like Waata, forest-dwelling Ogiek, Sengwer, and Yiaku, and pastoralists like Maasai, Turkana, Rendille, Borana, Samburu, and Endeorios, among others (IWGIA, 2024). The Ogiek, meaning "caretakers of all" animals and plants (OPDP, 2015), inhabit the Mau Forest—Kenya's largest closed-canopy forest. Ogiek *Giiret* norms prohibit cutting certain tree species to preserve Indigenous forests—*timndo waonet*. Each clan protects a section; trees cannot be cut, only some bark and roots used, to prevent drying. Ogiek have community-protected forest areas, rehabilitation zones, and care/use practice calendars. Yiaku peoples, with similar norms and practices, reclaimed management of Mukogodo Forest, helping wildlife and tree cover recover from 52% to 70% (Kavilu, 2018). Pastoralists (e.g., Maasai, Samburu, Borana, and Turkana) sustain landscapes through migratory cycles, rotative grazing, and subsistence harvesting, thereby supporting dryland ecological richness, abundant species, and social reproduction; look at Borana *gada* and Samburu *mpaka* systems (Counsell, 2023).

Similar to the situation across Africa—from colonialism to present, in Kenya Indigenous peoples are harassed, dispossessed, and evicted from lands by extraction, development, conservation, forest enclosures, and carbon trading (e.g., Sengwer evicted by Kenya's Forest Service (KFS) from Embobut forest; Enderois evicted for tourist game reserves; Ogiek confronting eviction for timber and carbon credits) (ACHPR, 2017; AI, 2023, Barasa, 2024; Counsell 2021; FPP, 2024b; Langat, 2023). Consider Northern Rangeland Trust's (NRT) 39 conservation enclosures over 42,000 km^2, almost 8% of Kenya's land (Counsell, 2023; Mittal et al., 2021). Founded by Ian Craig, NRT's first conservancy came from his elite White-minority family-owned 62,000-acre ranch, obtained from British colonial rule. USAID, EU, Danish, and French development agencies and NGOs like Nature Conservancy fund NRT—epitomizing Western-led neocolonial conservation. Indigenous groups accuse NRT of employing armed security alongside Kenya Wildlife Service and KFS to enforce exclusionary conservation and carbon sequestration while violating human rights—including killings, abductions, disappearances, and torturing. Since 2013, NRT's 2-million-ha Northern Kenya Grassland Carbon Project (NKGCP) impacts 100,000 people, including Maasai, Samburu, Borana, and Rendille, with encroachment, adverse incorporation, or dispossession. Deemed a "model" by the European Commission's NaturAfrica, NKGCP sells carbon offsets with 6.7 million credits by 2023, 4.5 million sold to companies like Netflix and Meta. NRT claims it is the world's largest soil carbon removal project and first generating carbon credits from "modified" grazing. The project argues—problematically—that Indigenous pastoralist systems like the Borana's *gada* and Samburu's *mpaka* are unsustainable, needing replacement or modification toward centralized commercial ranching models supposedly increasing vegetation. But NRT's own maps of NKGCP areas show vegetation declining by almost 50% from 2012 to 2020, and it is unlikely soil carbon is retained (Counsell, 2023; Mittal et al., 2021).

Indigenous Maasai, Samburu, Turkana, Molo, and Rendille also confront green grabbing. Viewing the Lake Turkana Wind Power Project (LTWP) and industrial-scale geothermal power in the Great Rift Valley, integrating Kenya's Vision 2030, we see a complex landscape of land, resource, and energy-intensive infrastructure and industrial-scale power (oil, wind, and geothermal) connected to the Lamu Port South Sudan-Ethiopia (LAPSSET) megadevelopment and commerce corridor—a territorial reengineering megaproject massively impacting Indigenous peoples, territories, environments, and climate (Barasa, 2024; EJAtlas.org, 2022b, 2023c; IWGIA, 2019; Sena, 2021, 2013; Wiben, 2021). To build Olkaria Geothermal Power plants, Maasai were involuntarily relocated to places that cannot support their lifeway. Meanwhile, the LTWP Consortium includes British Aldwych International, Danish Vestas Wind Systems, and three European government development agencies, and it is financed by the European Union plus three European government investment banks. In 2009, LTWP leased 607 km^2 for 33 years—twice renewable; Vestas installed 365 turbines serving LAPSSET's massive energy demands. All are on community lands, without proper FPIC; most communities cannot access or afford this energy and face gendered violence from industrial "man camps" and some—like the Turkana of Sarima—are involuntarily relocated, yet resist via the Sarima Indigenous Peoples' Land Forum. In 2021, a Kenyan Court declared LTWP acquired the land illegally, but it did not require land devolution or project stoppage.

Yet Indigenous East Africans resist advancing climate-positive, community-based biocultural solutions. See the Hunter-Gatherers Forum–Kenya, Eastern and Southern Africa Pastoralist Network, Pastoralist Indigenous Peoples Non-Governmental Organization–Tanzania, Indigenous Peoples of Africa Coordinating Committee, and African Food Sovereignty Alliance. Indigenous peoples such as Waata, Sengwer, Yiaku, Samburu, and Enderois establish Biocultural Community Protocols to defend lands and lifeways and reaffirm or restitute climate- and biodiversity-positive, culturally based territorial governance (e.g., EWC, 2019; IIN, 2022). The Ogiek Peoples' Development Program successfully fought land- and green grabbing-induced evictions, while advancing biocultural community protocol-based territorial governance (OPDP, 2015) and community-led/centered microrenewable energy—a real climate/energy solution model for 63% of Kenyans who live off-grid and increasingly use microsolar (OPDP, 2023), versus megadevelopments, like LAPSSET serving unsustainable industrial demand.

Oceania

Māori of Aotearoa/New Zealand created similar cultures of environmental respect and (intergenerational) reciprocity, with oceanic features centering "foreshore and seabed" as defended in their Waitangi Treaty (1840), plus community revitalization efforts (Awarau, 2024). Colonial systems abused and distorted notions of sovereignty in translations skewed toward colonizing settler-states, without acknowledging "*tino rangatiratanga*" over all lifeforms in land, sea, and sky. "*Mana*" (Marsden, 2003) identifies both "authentic being" of self and spiritual concentric (outward) connections with *Hapū* (clan or community), *Iwi* (tribe or nation), and "*Waka-*" (canoe group or ethnonational descendancy)—similar to *Oceti Sakowin* of Dakota/Lakota people. Māori also recognize "animals" (land, sea, sky) and "plants" (land, sea) as relations or kin *whānau* (family); harming them for profiteering and exploitation is heretical. Māori "traditionalism" thus resists modern colonial takeovers of "environments"—resistances seen as contemporary "activism,"[4] yet carrying millennia of holistic relational worldviews centering spiritual environmental stewardship.

Māori identify European newcomers as *pakeha*, reflecting broader affirmations of autonomy and difference from the invaders' colonizing racism and English capitalism. Persisting Māori resistance plus a significant Māori population has compelled New Zealand to extend acknowledgments of Indigenous land, seas, and "resources," comparatively more than other settler-colonial states. Yet struggles persist against continuing settler colonialism, White supremacy and disproportionate incarceration of Māori, and for autonomy and sovereign decolonial futures (Kyne, 2024; Rata et al., 2023; Tan, 2021).

Kānaka Maoli (Native Hawaiians) are subject to late-stage colonization, quasi-conquest, and capitalist, cultural, military, and corporate domination by the United States, reducing Kanaka demographically to fractions of a percent, erasing histories and worldviews, allowing only superficial expressions.[5] Yet Kānaka Maoli resist and revitalize their relations, even restoring *ahupua'a* (holistic stewardship "from sky to sea"—from mountains and forests to oceans and reefs)[6]—a longtime resistance horizon seen as "environmental/climate activism."

Palau's Hatohobei also guard against ocean exploiters, patrolling waters for "poachers" and foreign interests who maximize extraction over sustainability.[7] Colonially miscast as "primitive," these Indigenous peoples confront "development" forces destroying environments and lifeways. The Iban Sea-Dayak and Kenyah peoples engage in resistances along Sarawak rivers, protesting dams and massive deforestation. One author of this chapter was far upriver in the 1980s when these peoples prophesied territorial destruction if large corporations invaded, which eventually happened. We talked to the Penan (Pun'an), deep forest people, who suggested invaders were huge predators who, unchecked, would devour worlds and communities (like the *Dakota Inyah* giant, or *Inktomi* uncivilized spiderman who can't control their appetites). So they sought Ipoh trees for poison, hoping to save "the world" and life in the forests from hydrocarbon ravages that would eventually become climate change.

Mapuche (Wallmapu)

The Mapuche cosmovision *Az-Mapu* shapes anticolonial resistance, climate solutions, and sustainable governance of ancestral *Wallmapu* territories in Abya Yala's Southern cone. "To be Mapuche is to be territory," states Beatriz Chocori of the Koz Mapuche Parliament. The territorially rooted *Az-Mapu* value system (Romero, 2021) has, for millennia, enhanced biocultural diversity and environmental conservation. *Az-Mapu*, for example, shapes sustainable forest tending, like regenerative nurturance and uses of biodiverse *Araucaria araucana* forests (Hermann, 2005). *Az-Mapu* charts paths to *Küme Mogen* (living well in balance) with *Itrofil Mogen* (sacred biodiversity), whose *Newen* (spiritual energy) connects all life. Transgressing *Itrofil Mogen* and *Küme Mogen* causes *kütran* (collective, environmental, and individual disease)—as has Spanish-Chilean-Argentinian colonialism (Anbleyth-Evants et al., 2021).

Mapuche political territorialities of *ixofillmongen* ("relationality of all lifeforms without exception") confront "colonialities of nature" (Mansilla-Quiñones et al., 2024). Dispossession of Mapuches replaces native governance of territories, forests, and waters, as in Chile, where annually 120,000 ha of native forest are degraded or replaced (Hermann, 2005). Given the biocultural territorial connection, onslaughts against Mapuches and native forests are the same. Continuing colonialism/settler colonialism drives territorial destruction, enabling capitalist accumulation secured by state repression and militarization (colonial-militarized-accumulation); its reversal requires land recovery, sovereign Indigenous control, and Mapuche autonomous governance (Nahuelpan et al., 2022). The Pinochet dictatorship's forestry model consolidated Chile's extractive economy; since 1974, biodiverse native landscapes have been replaced with industrial tree plantations—for example, eucalyptus, pine—now covering over 3 million ha (Gonzalez-Hidalgo & Reyes, 2023; Langman, 2023). Monoculture plantations are government-subsidized, and since 1986 they have been recast as "sustainable" afforestation. Plantation companies have received payments for ecosystem services, despite devastating native forests and ecology, and undermining carbon stocks (Heilmayr et al., 2020). Large-scale monoculture afforestation was remarketed in the 2000s/2010s—under the UNFCCC Kyoto Protocol's Clean Development Mechanism—as "green energy transition" (biomass to bioenergy with carbon capture) (Romero, 2021). Credits sold by forestry companies allow polluting companies and states to falsely claim "progress" toward "net zero." Tree plantations accelerate climate risks, destroy biodiversity, introduce invasive species, and heighten forest fire risks (Kimutai et al., 2024). The current government acknowledges tree plantations represent a greater fire threat than native forests, noting that half the area impacted by large fires during the 1985–2018 period was tree plantations, while only 20% were native forests (Langman, 2023).

A study of 133 environmental conflicts in Chile, Argentina, and Bolivia shows Mapuches—involved in 30% of cases—as the most active Indigenous group there resisting extractive drivers of climate change (Rodriguez-Vargas & Hidalgo, 2022). Mapuche organizations, particularly in "Chile," have for over two decades advanced land recoveries totaling thousands of hectares, establishing autonomous communes—lofs/lovs—with agricultural cooperatives and patrols to resist extractive companies (Gonzalez-Hidalgo & Reyes, 2023). Examples proliferate: Autonomous Lof Triftrifko (Galvarino), Lof Forrowe (Boroa), Lof Williche (Chiloé), Lof Temulemu (Traiguén), and so on. Consider Lof Paillakawe (Arauco Gulf) recovering lands once grabbed by companies like Arauco through Pinochet-era dictatorship concessions. Forest rangers burned Mapuche houses and persecuted communities. Still, Arauco commercialized products with Forest Stewardship Council "green" certification.

Mapuches continue confronting racist, colonial, state-capitalist dispossession, policing, surveillance and counterinsurgency strategies, and legislation enabling corporate profiteering through militarized accumulation, with territories subject to a "state of exception," Indigenous-recovered lands criminalized as "usurpation," and "terrorism" and leaders profiled, persecuted, criminalized, and imprisoned in a low-intensity war (Curaqueo & Almonacid, 2020; Nahuelpan et al., 2022). Yet Mapuches resist, with *Az-Mapu* as territorial life organization. Land recoveries persist with Mapuche organizing and mobilizing being dynamic, networked, diverse, decentralized, autonomous, and nonhierarchical across a diverse spectrum—from radical and militant (e.g., Coordinadora Arauco Malleco, Weichan Auca Mapu, Resistencia Lafkenche), to moderate (e.g., Identidad Territorial Lakkenche, Plataforma Política Mapuche), and in between (e.g., Consejo de Todas las Tierras, Alianza Territorial Mapuche).

Decolonizing Climate Change Mitigation—Indigenizing Climate Justice

Indigenous territories must be returned and recognized to restore ecosystems, mitigate climate change, and bring climate justice. From Turtle Island's reclaimed woodlands of the northwest and northeast, to Amazonian and Mesoamerican forest gardens and integral chacras and milpas, and Indian Warli and Gond Adivasi sacred groves, Indigenous peoples preserve and enrich biodiversity and carbon stocks (Figueroa-Helland et al., 2018, 2021; Hall & Fenelon, 2015). Indigenous territorial restoration revitalizes ancestral biocultural social ecologies while innovating for modern struggles and movements. As bison are reintroduced to Turtle Island's plains tribes, Standing Rock traditionalists renew buffalo honoring ceremonies for land stewardship and water protection. As Indigenous food systems are reclaimed, they join food sovereignty struggles toward antiextractivist, anti-imperialist, antipatriarchal political agroecologies (e.g., with La Via Campesina).

Indigenous movements embody long historical trajectories of identity reaffirmation and defense and recovery of territories, autonomy, and sovereignty, continuously refigured versus ever-renewing threats. Regardless of whether seen as "environmental/climate activism," Indigenous movements advance worldviews and reclamations confronting them with colonizing capitalist states and corporations occupying their territories, who are—not coincidentally—the main drivers of climate/ecological devastation. There are and have long been innumerable examples of Indigenous struggles requiring acknowledgment as long-standing—even prescient, forms of "climate/environmental activism," much preceding the all-too-late recognition by dominant society of the catastrophic Earth crises it has caused.

And still today Indigenous movements are treated as direct challenges to this modern, colonial dominant society, its states and corporations, since they carry living heritages of an *other* relationship to land and Earth—that is reciprocal, restorative, biocultural, and at odds with the violent objectification, appropriation, commodification, and accumulation of an elitist modern society where corporations maximize profiteering at the expense of people, planet, and life as a whole. Indigenous resistances are thus targeted as antithetical to the imperial logic of domineering capitalist states. Whether armed (e.g., Zapatista) or unarmed (e.g., NoDAPL Standing Rock Water protectors), Indigenous resistances embody "alternative" worlds—threats to hegemony that states similarly criminalize and repress.

For Indigenous movements, climate "justice" requires reverting colonial land theft and restoring Indigenous sovereignty so territories reflourish upon worldviews and governance honoring all lands and lifeforms, and Mother Earth's balance and diversity are restored. Climate justice means land back, where the dominant society recognizes Indigenous autonomy and sovereignty. Yet states and dominant actors often oppose Indigenous land reclamations and recognitions even when costing little or nothing to their interests (such as standoffs over sacred sites, like the Black Hills; or Maya ceremonial sites). One author of this chapter proposed that land attached to the college where he was a visiting professor be given to the Lenape, since the college could not manage or use it. Yet Trustees could not see wisdom in this; their dominant worldview, tied with financing from Vanguard and fossil fuel companies constrained them—even when climate changes fueled by these actors threaten their existence. And when Indigenous land recuperations are larger and more directly confront the strategic interests of dominant actors, like Maya-Zapatista or Mapuche recuperations, militarization, repression, and criminalization ensue.

Yet even when repressively constrained by colonial, state, corporate hegemony, Native nations, seeing all life of value, lead land defense to restore balance, biodiversity, and climate stability. Antisystemic mobilization strategies can tactically instrumentalize legal tools. For example, Indigenous nations recognize waters, lands, and sky as living persons; note the emblematic 2010 World People's Conference on Climate Change and the Rights of Mother Earth (WPCCC&RME, 2010). Affirming subjectivity for environmental bodies, Indigenous nations demand state and intergovernmental recognition of legal personhood for rivers, forests, mountains, and so on.[8] The Kukama (Peruvian Amazon), led by the women's organization Huaynakana Kamatahuara Kana, fought—successfully—for Marañón River as rights subject (against oil, shipping, agroindustry); Maori similarly with Whanganui River from Papatūānuku (Earth Mother) and Ranginui (Sky Father); Shasta Indian Nations with the Klamath River (northern California); Sarayaku Kichwa (Ecuador) demanding recognition for the Kawsak Sacha/Living Forest (against oil, logging); or Wixáritari (Huicholes, Mexico) obtaining legal recognition of sacred Wirikuta (against mining) (Perez-Aguilera, 2016). In innumerable examples, Indigenous "activism" centers care, defense, and restoration of environmental bodies, yet conflicts with the main climate change drivers: Eurocentric colonialism/coloniality, corporate capitalism, extractivist states, and neoliberalism.

Indigenous Conflict With Eurocentric/ Euro-American Climate Change Determinants

Socioenvironmental crises—including climate change—are profoundly determined by problems caused by the Eurocentric order and domination of Euro-American powers. We evaluate five such problems and how Indigenous (managing) models embody alternatives:

First problem: centralized authority and power in large states, interlocked with corporations prioritizing accumulation over community and ecology. In contrast, Indigenous systems often decentralize authority, democratize power, and practice communality and/or plurinationality. Consider communal assemblies and councils—inter alia, North American, Mesoamerica, Andean, Amazonian, and Adivasi. In Abya Yala/Americas, for instance, Indigenous assemblies involve prolonged collective decision making; consensus is pursued but cannot be imposed and requires extensive deliberations honoring the subjectivity and agency of all members and ensuring decisions safeguard the coupled reproduction of human and other-than-human territorial communities. Communality characterizes labor sharing and mutualism (e.g., Mesoamerican *tequio/ko'ol*, Andino-Amazonian *ayni*, *minga*, Mapuche *mingako*), seed/product sharing (Mapuche *trafkintun*, Andean *watunakuy*), ceremony (Mapuche *nguillatun*), and redistributive celebrations (Pacific Northwest *potlatch*; similarly, among Zapotec, Shipibo, Puebloan, etc.). Communal organization extends kinship relations across humans and other-than-humans, and respects members' agency—at personal, family, community, people/tribe, interpeople/intertribe, and interspecies levels. Look at Oceti Sakowin[9] nations (Seven Council Fires of four peoples—*Isantee, Tetonwan, Illahanktowan, Yanktowan*) including, traditionally, seven tribal nations in each of four directions for 28 nations. Therein, D/Lakota councils allow group choice to participate; "nations" participate in larger "council fires." Elsewhere, the plurinational Haudenosaunee Confederation (originally five and then six nations). Plurinationality, now adopted, for example, by Maori, has long been salient in Abya Yalan/Latin America Indigenous politics, yet contested as can be appropriated by states, which remaining extractivist also remain problematic, even when claiming to "include" Indigenous nations. Plurinationality need not be state-centric, and it is not exclusively Indigenous; upon it, diverse colonized people weave coalitions—like Indigenous-Afrodescendant-mestizo coalitions, for example, ASPROCIG, Colombia (Figueroa-Helland et al., 2025b).

Second problem: objectification, commodification, and capitalization of all life, relations, and entities—for example, land, water, forests, labor, trade, knowledge, communications, carbon, "environmental services," energy, health, Earth, and so on. This underpins economies of exploitation and monetary power over Earth and others (human and nonhuman) ontologically reduced to "resources" (extractive—consumptive, e.g., "natural/human resources"), particularly by corporate elites and consumer societies. Contrastingly, Indigenous communal economies sacralize land and prioritize relational reciprocity among all life, collective (re)distribution, labor mutualism, and rotating responsibilities. Observe Maya Tsotsil-Tseltal cosmovisions of *lekil kuxlejal* (existing/living well) informing *ko'ol komon koltombail* (equitable mutual/communal labor), such as other Mesoamerican cultures—for example, *tequio* (Nahuatl), *guetza* (Zapotec), and so on (Sanchez Álvarez & Figueroa-Helland, 2024). Indigenous peoples value living environments as relatives, recognizing personhood, subjectivity, and agency of land, plants, animal/physical life, and so on. Human-nonhuman interaction demands reciprocal nurturance and guardianship: People must care for, nurture, and guard territory and country as territory and country do for people (e.g., Aboriginal Australians). This gives distinctive meaning to Indigenous territorial community guards (e.g., Shipibo, Kakataibo, Wampis, Chapra, Kichwa, Mapuche, Maya, Cherani Purépecha), such as the Cofán (Ecuador), whose Forest Guardians patrol their over 1 million acres of forested homeland—from the Andes to Amazonia—against loggers, miners, corporations, and polluters, across the megabiodiverse Pantanal, the world's largest wetland: "If the world wants truly sustainable defense of ... endangered forests and waters ... then ... protective work of tribes like [ours,] the Cofán needs to be recognized" (Lucitante, 2024, p. 120).

Third problem: dominant laws and policies center and enforce property relations, while repressing non-property-centric—for example, Indigenous—societies. In state and private property regimes, ownership and (exclusionary) possession reign with nature becoming "resources" for accumulation and "development," serving elite power and wealth concentration, protected through punitive justice systems that safeguard structural inequities while sacrificing environmental and collective health and well-being. Contrastingly, Indigenous community-centric customary norms value (non-property-based) relations and carry legacies of communal tenure, equitable collective distribution, collective health, wealth, and well-being, consensual deliberation and adjudication, reciprocal repair, restorative justice, and communal healing, all honoring human-nonhuman personhood. Yet Indigenous systems are neither static nor exempt from colonial or precolonial formations requiring transformative internal critique. Indigenous processes where decolonization *demands* depatriarchalization of Indigenous governance and norms (e.g., Zapatistas; Congreso Nacional Indígena, Mexico) are strong examples of such systems.

Fourth problem: the dominant civilization's gargantuan, ever-expanding requirements for energy, resource, land, water, materials extraction, consumption, and throughput—for infrastructure, commodities, shipping, supply chains, transportation, and so on. This drives territorial dispossessions, ecological imperialism, ecosystem homogenization, carbon stock destruction, greenhouse gas emissions, agro/biodiversity loss, economic hyperspecialization, and expansive resource, energy, toxicity, and waste flows. Notorious in agroindustrial monocultures, hydrocarbons, and mining, this resurfaces in "green" technologies, transition minerals, and industrial-scale "low-carbon" energy (hydropower, agro/biofuels, solar/wind farms, etc.) which do not substitute but add to "conventional" sources. This violently undermines and supplants local and regional biocultural—including Indigenous—economies and bioregionally complementary distribution webs not reliant on exchange value and profit. Contrastingly, Indigenous economies embody time-tested locally organic biocultural models of polycultural agroecologies alongside bioregional webs of complementary distribution and reciprocal exchange, fostering highly diverse territories which ensure food sovereignty, materials provisioning, and communal sharing, while nurturing biodiversity, maintaining carbon stocks, respecting reproductive cycles of life forms, and ensuring ecosystem regeneration, which secures intergenerational provisioning.

Western paradigms undergird the entire system, reproduced in education and media, and must be countered by Indigenous paradigms of being good relatives in reflexivity with Earth and Sky.

Fifth problem: hegemony of Eurocentric and Abrahamic philosophy and worldviews, and techno-rationalist science, privileging anthropocentric, androcentric, and instrumental-rationalist thought, shaping social structures of human power stratification, dominion, and extraction. In dominant Eurocentric worldviews, Man reigns over "creation"/"nature"—to be mastered, known as object, exploited as resource, used as instrument, owned as property. Feminized "nature" is to be "taken" by patriarchal empires, states, armies, and corporations—actors enshrined with legal personhood and/or sovereign subjectivity. Contrastingly, Indigenous cultures carry relational nonanthropocentric worldviews centering community, valuing collective responsibility, and respecting all life—human and nonhuman—as relatives: reflexive subjects constituting intersubjective communities, each with personhood, not be owned, nor exploited. This shapes Indigenous spirituality, worldviews/cosmovisions, knowledges, and ethics-politics. Consider key Indigenous principles: sacralization of Earth as Mother (or Grandmother), honored and treated as origin of all life and ancestors; and recognition of other-than-human companions as ancestors and relations to be thanked and honored, not objectified or exploited for profit.

Such spiritualities, cosmovisions, and worldviews inform knowledge and social-ecological dispositions. We see acknowledgments of complementarity among the four Earth directions, or of the powerful life reproductive relationship between Sky (e.g., healthy climate, unpolluted air, healthy atmosphere, clean precipitation) and healthy Earth (e.g., unpolluted soils/subsoils, clean waters, toxicity-free organic matter and biochemistry) which engenders and reproduces healthy life and biosphere (i.e., thriving biotic communities and societies). Ceremonies and daily practices must (re)kindle this healthy Earth-Sky complementarity to reproduce strong communities where diverse life thrives with lush vitality, providing foundations for many generations to come.

Many Indigenous cultures exemplify this: Maya Tsotsil principles *Ich'el ta muk' Balamil xchu'uk yol snich'on* (honoring/dignifying Earth and all her descendants) where land is *not* owned, ordered, or managed by a reigning "Man"; instead, humans are coresponsible—alongside other lifeforms—for communal reciprocal care and nurturance of the face and contours of a living sacred Mother Earth (*yelov sat ch'ul Balamil*), as she engenders, cares, and nurtures us (Sanchez Álvarez & Figueroa-Helland, 2024).

Indigenous relationality weaves all our relations into Earth-rooted, spiritually nested kinship communities (e.g., family, territory, nation, Earth, cosmos). Inside *Lakota Oyate* (Fenelon, 1998; Fenelon & Alford, 2020), one identifies one's ancestry in relation to sacred ties with a sense of Earth as motherhood, observed in all relationships; *Hau Mitakuyepi*!—greetings to relatives; "*tiwaye*"—family (ti-pi, household); "*tiyospaye*"—extended relatives (community); *Unci Maka, Tunkashila*—grandmother, grandfathers; *Oyate* (*wanbli*, *pte*)—nation, belonging; (*unma oyate*—people outside, other); *Ikce Wicasa*—common (red) man (human being); and *Omidakuye owasin*—we are all related (global). Lakota practices center spirit-as-energy (*Wakan*) running through all life, and purification (*Oinikaga*) as releasing desires to control others, through making sacrifice for one's relatives (*takuyepi*); this relates community worship (*Wi-wanyang wacipi*, Sundance ceremony) as practiced spirituality, to collective economic practices honoring Sacred Lands (Deloria, 1999). In ceremonies started by Buffalo dancers greeting the morning sun, around a Sacred Tree (of life) with effigies (bison, elk, eagle, rabbit), dancers tie their ropes—umbilical cords to Mother Earth's roots and offshoots to ancestors called in offerings of sage, cedar, and sweetgrass, give flesh and blood from their bodies—the only real gift a person can make (unlike money or financed European gift-giving). When Dakota people restored Sun Dances along the Mississippi after 100 years since its banning (and their people sent to POW camps in Minnesota and the Dakotas), they prayed for a great cleansing, answered in rains pouring down historic lands of the *wo-Dakota oyate*, shutting down bridges and waterways from mid-America to the Gulf of Mexico.[10]

Many Indigenous ceremonies face repression, yet persist, often refigured in/as anticolonial resistance. Consider remote Cheyenne River Lakota Reservation communities holding sacred ceremonies even as US settler state militaries tried destroying them and then legally banning them in 1883. Banning ceremonies is a colonial usurpation of cultural sovereignty; practicing/reclaiming ceremony has a powerful anticolonial meaning. Ceremonies honoring Earth, waters, mountains, and lands resurge with struggles against destructive "development" projects harming territories, the atmosphere (e.g., climate, pollution), or waters—such as rivers endangered by dams or pipelines such as DAPL. Ceremonies communally honoring Earth and all life persist through histories of colonial, capitalist, and state onslaught. Look at Mapuche *ngillatun*, enacting respect for Earth, life, and community, in recovered territories and autonomous *lofs*/communes to reassert Wallmapu territorial and cultural sovereignty versus contemporary settler-state repression, militarization, and criminalization (Curaqueo & Almonacid, 2020; Nahuelpan et al., 2021, 2022). Ceremony ties Indigenous action to defense, recovery, care, and honoring of territories. Indigenous spiritualities thus live on, shaping survivance and resistance and anticolonial horizons.

Indigenous languages and philosophies embed these values as norms intimately tied to living territories, privileging relational ethics, embedded bioculturally with environments as kinship communities. Native American seven-generations principles, Anishinaabe and Cree *Mino-Bimaadiziwin*, Wampis *Tarimat Pujut*, Kichwa *Kawsak Sacha*, Maya *lekil kuxlejal*, Andean-Amazonian *suma qamaña/sumak kawsay*, Ogiek *timndo waonet*, and other philosophies, such as Maori *Kaitiakitanga*, are holistic worldviews of guardianship where genealogy (*whakapapa*) of sacred (*tapu*) relations (not "resources") roots cultural sovereignty (*tino rangatiratanga*) with land (Marsden, 2003). Also, Lakota languages communicate how D/Lakota see land (*makoche, ina*—whole like mother) and water (*mni-wiconi*—health, life, water is life), inclusive of all relations (*mitakuye oyasin*; we are all related)—plants, animals, stones, mountains. Countering Eurocentrism, Indigenous knowledges are key to sustainable futures in and beyond current crises:

> At the turn of the 20th century, settlers, pioneers, and the … government failed to recognize what Indigenous peoples have always known: … our understanding of and devotion to land remains constant. Indigenous knowledge is Indigenous innovation …, the ancestors were the first innovators. If we are to save ourselves, we must empower this knowledge with everything we have. (Haaland, 2024, p. 38)[11]

Thus, Indigenous worldviews, such as Standing Rock water protectors' *mni Wiconi* and *mitakuye owasin*, are indispensable.

Toward (Re)Asserting Sovereignty? Contrasting Frameworks of Indigenous Engagement—Consultation Versus Consent Versus Decolonization

Indigenous pathways constitute epistemologically and spiritually autonomous trajectories—distinct from Eurocentric development. Carrying their historicities, horizons, and futurities, Indigenous struggles mobilize visions to defend, restitute, and revitalize lifeways, territories, and biocultural relations on their own terms—with their own worldviews and principles. Indigenous "activism" includes climate and environmental action, weaving these internationally together (e.g., WPCCC&RME, 2010), as when the 2024 Indigenous Summit on Just Transition (mentioned earlier in the chapter) concluded:

> We … Indigenous Peoples … were born to our ancestral … territories … in a sacred … spiritual way. We belong to … Mother Earth, and she belongs to us as a mother and child belong … We are Earth, Earth is us. It was the Creator's intent that we live on these blessed homelands in perpetuity. We remain. We have lived our lives in … ways prescribed … from time immemorial, until our … worlds were shattered by intruder colonial and imperialist powers who claimed our … territories, and resources, subjecting us to unimaginable devastation.… We are still here. We rightfully maintain our belonging to our homelands. (IPGCC, 2024, p. 1)

Here we bring forward the survivance and revitalizing resistance of Indigenous people's reassertions of territorial authority that invoke three contrasting engagement frameworks: (1) *consultation*, (2) *consent*, and (3) *decolonial Indigenous sovereignty.*

Consultation occurs where corporations, states, or others hold meetings and information sessions—or other activities—with Indigenous actors, mainly to fulfill legal requirements or political expectations, but without subjecting projects to Indigenous decision making. Consultations can sidestep, marginalize, or subordinate Indigenous "authority to determine all that should occur to, with, on, and in [Indigenous] lands and territories" (IPGCC, 2024, p. 1). Whether well-intended or simulacra, consultations serve hegemonic legitimation.

The Summit adds: "[al]though concepts of 'co-management,' 'co-jurisdiction,' 'joint management,' or 'cross jurisdiction' … sometimes work … in many places they are used to undermine our inherent rights to our lands … and … resources" (IPGCC, 2024, p. 4). Consultations—sometimes even consent mechanisms—can be seen as constituting claims of innocence or good intentions (Valandra, 2006) by hegemonic actors and institutions who may "redwash" (Houle, 2022), continuing colonial/settler state, corporate domination by extending "inclusion," "participation," or "co-management" to Indigenous people *without* fully devolving decision-making power. Inclusion in hegemonic governance does not automatically bring autonomy and may turn Indigenous representation and knowledges into "supplements" to Western science and hegemonic governance (Whyte, 2015), while neutralizing autonomist anticolonial resistance and opposition. As "inclusion" can bring such adverse incorporation, it risks further eroding Indigenous sovereignty and deepening (colonial) assimilation.

Consent, comparatively, is often understood upon Indigenous claims to inherent self-determination rights (e.g., UN Declaration on the Rights of Indigenous Peoples [UNDRIP]; UN, 2007). Here, Indigenous nations do not just "have a seat at the table," but seek to reclaim some decision-making power, as found with the Indigenous Summit reaffirming:

> The right of Self-Determination is inherent and … prerequisite to … all other rights. This includes … right to FPIC prior to … approval of any project … , right to participate in decisions affecting our rights …, … rights to … territories, and resources traditionally-used, possessed, or acquired, [and] … right to determine our … priorities for … governance, management, protection, and use of our lands, waters, air … , and intangible and tangible resources. By … Self-Determination, Indigenous Peoples freely determine their … political, social, economic development, future, and rights. (IPGCC, 2024, p. 4)

Tensions between consultation and consent—notorious through colonial histories (e.g., treaty making)—pervade colonial presents of ecological distribution conflicts (EJAtlas.org) concerning Indigenous and climate/environmental matters (e.g., mining, hydrocarbons, biomass, agroindustry, nuclear, conservation, forest/soil/water/ocean carbon, dams [see Valandra, 2006], transition technologies/minerals, carbon sequestration, etc.). Evident in "conventional" development, like DAPL/Standing Rock or Coastal Gas Link/Unist'ot'en; it similarly marks "sustainable development," "green economy" transition, and climate change "mitigation" projects (IPGCC, 2024).

Yet even "consent" is constrained within the state-centric intergovernmental rights-based framework—the creature and instrument of states, and international legal apparatuses carrying legacies of imperial, colonial, hegemonic power. Rights-based frameworks remain subject to laws and jurisdictions of states and state-centric international regimes, which never allow questioning of state sovereignty and territoriality. Even UNDRIP—a milestone of Indigenous politics, still one operating *within* constraints of the hegemonic state-centric world order—had to concede in Article 46: "Nothing in this Declaration may be construed as authorizing … any action which would … impair, totally or in part, the territorial integrity or political unity of sovereign … States" (UN, 2007, pp. 28–29).

Without an anti/decolonizing assertion of "Reparations, Land Back, Full Restoration of Lands, Territories, Waters, and Biodiversity"—as asserts the Indigenous Summit (IPGCC,

2024), rights-based frameworks leave state sovereignty and territorial control unquestioned, keeping Indigenous decision-making subject to state jurisdictions and intergovernmental power. The Summit adds:

> A just transition must ensure return, recognition, and respect of Indigenous . . . territories, . . . ; unrestricted access, restoration, recognition, and respect of our rights to . . . ancestral . . . territories, . . . waters, and . . . resources . . . taken without our consent during . . . colonization . . . ; respect for our inherent sovereignty . . . ; full, unqualified implementation of our rights, including . . . Indigenous land tenure . . . , economies, jurisdiction, languages . . . , food systems, health, cultures, spirituality, natural world responsibilities, biodiversity . . . , knowing, and . . . lifeways. (IPGCC, 2024, p. 5)

Indigenous sovereignty is asserted, *unrestricted*, as necessarily decolonizing (e.g., *unrestricted* territorial restoration, *unqualified* jurisdiction). Such decolonial Indigenous sovereignty pathways remain indispensable in the 21st century vis-à-vis a colonial-state-capitalist order whose claims to territorial dominion and sovereign jurisdiction have enabled—to this day—centuries of exploitation of Earth as "resource" while undermining bioculturally sustainable Indigenous societies, causing climate/Anthropocene crises.

In current Indigenous politics and activism, constrained by state-centric hegemony (still a hegemony facing epochal, geological crisis and potential decline), there may *presently* be a useful tactical mix of rights-based with antidecolonizing movement frameworks. Yet compatibility and incompatibility between these remain to be deliberated among Indigenous actors if the strategic horizon aiming toward (unrestricted) futures of decolonization, Indigenous sovereignty, and full territorial restoration is to be realized.

Conclusion

Indigenous peoples and nations embody bioculturally diverse, territorially embedded social ecologies acting as keystone cultures in regenerating healthy ecosystems and climate, and thus constitute sustainable counterhegemonic alternatives (Fenelon, 2015; Figueroa-Helland et al., 2021). Wherever Indigenous territorial governance and knowledge systems are eroded, so is ecological health; wherever they are defended and restituted, so are ecosystem protection and reparation, including biodiversity-enabling and climate-stabilizing functions.

Indigenous activism makes Indigenous sovereignty, territorial restitution and defense, and decolonization beyond colonial, capitalist, state-centric power, preconditions for stabilizing the climate and restoring Mother Earth's health. It centers worldviews of sharing and land nurturance, holistic interactive relationships with all life, and movement organizing to "defend" and re-create healthy land–society relations. Indigenous lifeways, especially when liberated from coloniality and practiced autonomously upon Indigenous knowledges, governance, territories, and communality, embody *real solutions* for healing territories and Earth's climate, while sustainably reproducing living communities.

Yet Indigenous peoples and nations conflict with (neo-)imperial, settler-colonial and other states, corporations, and even NGOs, and their extractivism where living Mother Earth becomes "resources" for profiteering, including in territories where Indigenous peoples have lived for centuries or millennia. "Sustainable" and "green" turn(s) in hegemonic—state and capitalist—"development" do *not* alter this, as Indigenous peoples confront both "conventional" *and* "green" colonialism upon environmental/climate pretexts.

Whatever the conflicts, Indigenous struggles act internationally—in and beyond mainstream fora, and across borders. Consider when Amazonian Wampis and Achuar leaders, with territories impacted by oil (PetroPeru), traveled to Pennsylvania, a territory ethnically cleansed during US settler-colonial expansion (Fenelon et al., 2023). They protested at Vanguard's headquarters over its financing of and profiteering from oil destroying Indigenous territories, communities, and sovereignty (AW, 2023). Wampis and Achuar leaders, now representing Autonomous Territorial Governments (Figueroa-Helland et al., 2025a), noted how rivers and forests are their "supermarket"—the real wealth and sustenance, and states and corporations raiding them decimates communities, life, livelihoods, and Mother Earth, provoking climate upheavals.

Like Wampis, Achuar, Lakota, Mapuche, Maya, P'urhépecha, Zoque, Chemehuevi, Kichwa, Kukama, Wet'suwet'en, Maasai, Maori, Penan, Adivasis, and *many* more, Indigenous nations and peoples lead movements confronting the drivers and actors causing climate change and environmental degradation, to defend and restore territories and healthy ecologies. Such Indigenous "climate/environmental" struggles carry long-standing trajectories of resistance and liberation toward decolonization and resurgence that have always (re)asserted territorial autonomies upon biocultural community-centric notions of sovereignty. Thus, Indigenous "activism" already carries living historical legacies of sustainable decolonial futures, against and beyond the exploitations threatening all life on Mother Earth, recognizing the wisdom of our ancestors.

Notes

1. Mary Louise Defender Wilson (elder Dakotah, Hidatsa), representing Standing Rock's Landowner Association, arranged research meetings for *The Taken Land* pamphlet submitted to the US Senate Subcommittee for Indian Affairs (1987).
2. The Oahe dam/reservoir (1948–1962) was built and filled in the 1950s when Native people originally experienced its major impacts, continuing for over two generations.
3. Prayer circles started in 1992 (quincentennial of Columbus landing) to prepare to return the Sun Dance, asked in June 1993 for great cleansing (by rain, *mni wicozani*), which occurred exactly over *Dakota Oyate* between the Mississippi and Missouri Rivers, starting on the summer solstice ending of the ceremony.
4. Throughout the history of dominant (Eurocentric) social science, researchers (e.g., anthropologists) identified "traditional" societies as those with surviving cultural practices from precolonial times, such as Indigenous societies, but their actions (e.g., anticolonial, antistate, anticorporate struggles) were rarely seen (perhaps until recently) as "environmental activism"; while non-Indigenous actors mobilizing in environmental/climate politics were cast as ("pioneering"?) "activists."
5. These are contradictory yet coexisting phenomena—the dominant invasive society reproduces stereotypical cultural practices furthering dehumanization, even as so-called traditionalists maintain deep sociocultural practices (spiritual), often hiding them from colonizers, sometimes manifesting them as survivance when oppression is lifted, for example, D/Lakota Sun Dance (Fenelon, 1998, 2006).
6. "Ancient Practice, Future Promise" (2024) art by Kai'li Kaulukukui, (author of words), R. Wardley (author of the recognition of community of Ha'ena), Lei Wann, and Kwaika B. Winter (knowledge-keeper of *kupuna* [elders, ancestors]) (*National Geographic*, *246*(1), 92–95).
7. Raygorodetsky, G. (2024, January 28). Palau's waters are some of the most biodiverse in the world—Thanks to its defenders. *National Geographic*.
8. As we moved the chapter to publication, Native nations in North Carolina and California won these rights, dams were removed, and amazingly salmon have returned to these rivers, also experiencing reforestation.

9. When Dakota/Lakota water protector (protest) camps grew north of Standing Rock, a larger encampment on the DOI/ACE side of the Cannonball River was renamed "Oceti Sakowin," invoking ancient participatory councils in all decision making regarding territories and environments.
10. The 1993 Great Flood (1993 Great Mississippi and Missouri Rivers Flood) impacted the Midwestern United States from summer until October 1993, as among the costliest in US history.
11. We recognize Deb Haaland, specifically as Secretary of the Interior, has come under considerable criticism, including by Indigenous critics, for making compromises or negotiated settlements not always corresponding with Indigenous philosophies and with many arguments we state here (see, e.g., our section on "Consultation Versus Consent Versus Decolonization"). While as authors we may acknowledge and agree with some of these criticisms, the specific content from Haaland's statement communicates an important message. Separately, it should be noted Haaland assisted the NoDAPL movement at Standing Rock, and so we trust the Indigenous intentionality in the quotes.

References

African Commission on Human and Peoples' Rights. (ACHPR). (2017). *Extractive industries, land rights and Indigenous populations'/communities' rights*. IWGIA.

Amazon Watch (AW). (2023). Achuar, Wampis continue fight against oil exploration. https://amazonwatch.org/news/2023/0609-the-achuar-and-wampis-peoples-of-the-peruvian-amazon-continue-their-fight-against-oil-exploration-in-block-64

Amnesty International (AI). (2023). Forced evictions of Indigenous Peoples in Kenya. https://assets.survivalinternational.org/documents/2503/231129_Joint_statement_on_forced_evictions_of_Indigenous_Peoples_in_Kenya.pdf

Anbleyth-Evans, J., Ñanculef-Huaiquinao, J., Coúecar-Llancapani, Y., Araos Leiva, F., Riquelme Maulen, W., & Raymond, C. (2021). Mapuche *Az-Mapu* and nature's contribution to people. *Environmental Values, 32*(3), 291–314.

Awarau, A. (2024, January 28). The Māori saved their language from extinction. *National Geographic*.

Barasa, W. (2024). Skewed conservation policy and state validation of land dispossession of Indigenous peoples, Kenya. *Arizona Journal of Environmental Law & Policy, 14*, 45–55.

Bonneuil, C., & Fressoz, J.-B. (2016). *The shock of the Anthropocene*. Verso.

Bragg, J. (2021). *The big con: Big polluters advance "net zero" climate agenda to delay, deceive, deny*. GFC.

Brito-Millan, M., Figueroa Helland, L., Guzman Morales, J., Harrison, E., Ng, J., Quintanilla, L., Salomón, A., & Werner, B. T. (2023). Beyond green colonialism in global environmental & climate politics: Towards decolonial solidarity with Mother Earth's revolt. *Perspectives on Global Development and Technology, 21*(5–6), 366–386

Cardona-Calle, D. (2017). *Community forest management and agroecology*. FOEI.

Centro de Derechos Humanos Fray Bartolomé de Las Casas (CDHFBDC). (2023). *Chiapas—un desastre*. CDHFBDC.

Cherán P'urhépecha Government (CPG). (2011). *Manual: estructura y organización del nuevo gobierno*. CPG.

Congreso Nacional Indígena (CNI). (2021). *Espejos de la Resistencia*. CNI.

Congreso Nacional Indígena (CNI). (2024). Declaración: 5A Asamblea Nacional por el agua, vida y territorio. http://www.congresonacionalindigena.org/2024/08/18/declaracion-de-la-5a-asamblea-nacional-por-el-agua-la-vida-y-el-territorio/

Consejo Étnico de Pueblos Kichwa, Amazonia (CEPKA). (2022). Pueblo Kichwa Rechaza Conservación Excluyente y Negociado de Carbono del Parque Nacional Cordillera Azul. https://www.forestpeoples.org/sites/default/files/documents/Pronunciamiento%20Kichwa%20no%20al%20negociado%20del%20carbono%20del%20PNCAZ.%2031-05-2022.pdf

Consejo Étnico de Pueblos Kichwa, Amazonía (CEPKA). (2023). SERNANP again denies Kichwa territorial rights in Cordillera Azul National Park. https://www.forestpeoples.org/en/node/51056

Coq-Huelva, D., Higuchi, A., Alfalla-Luque, R., Burgos-Morán, R., & Arias-Gutiérrez, R. (2017). Co-evolution and bio-social construction: Kichwa agroforestry systems (chakras), Ecuadorian Amazonia. *Sustainability, 9*(10), 1920.

Counsell, S. (2021, April 16). Anatomy of a "nature-based solution." *REDDMonitor.*

Counsell, S. (2023). *Blood carbon: Carbon offset scheme makes millions from Indigenous land, Northern Kenya.* Survival International.

Curaqueo, J., & Almonacid, M. (2020). Resistencias contra extractivismo, racismo y Estado en Wallmapu. *Debates Indigenas.* https://www.servindi.org/actualidad-opinion/10/09/2020/el-pueblo-mapuche-contra-el-extractivismo-el-racismo-y-un-estado

Davis, H., & Todd, Z. (2017). On the importance of a date—Or decolonizing the Anthropocene. *ACME, 16*(4), 761–780.

Deloria, V. (1999). *Sprit & reason.* Fulcrum.

Dhillon, J. (2021). Indigenous resistance, planetary dystopia, and politics of environmental justice. *Globalizations, 18*(6), 898–911.

Díaz, C., & Parres, F. (2024, April 25). Cherán K'eri: pueblo p'urhépecha venció miedo por defensa de su bosque. *Avispa Midia.*

Dooley, K. (2018). Missing pathways to 1.5°C. CLARA. https://static1.squarespace.com/static/610ffde0dd5c39015edc6873/t/65524d91dc73dc2fbf4a68ae/1699892631263/Missing+Pathways+FULL+Report.pdf

Dunne, D., & Quiroz, Y. (2023, September 25). Mapped: Impacts of carbon-offset projects around the world. Carbon Brief. https://interactive.carbonbrief.org/carbon-offsets-2023/mapped.html

Enderois Welfare Council (EWC). (2019). *Endorois peoples' biocultural protocol.* EWC.

Environmental Justice Atlas (EJAtlas.org). (2022). Tren Maya, Yucatán, Quintana Roo, Campeche, Tabasco y Chiapas, Mexico. https://ejatlas.org/conflict/tren-maya-mexico

Environmental Justice Atlas (EJAtlas.org). (2022b). Lake Turkana Project in Indigenous Territories, Kenya. https://ejatlas.org/conflict/lake-turkana-project-in-indigenous-territories

Environmental Justice Atlas (EJAtlas.org). (2023). Corredor Interoceánico, Istmo de Tehuantepec, México. https://ejatlas.org/conflict/corredor-interoceanico-del-istmo-de-tehuantepec-mexico

Environmental Justice Atlas (EJAtlas.org). (2023b). REDD+ pilot, Lacandon Jungle, Chiapas, Mexico. https://ejatlas.org/conflict/redd-pilot-lacandon-jungle-chiapas-mexico

Environmental Justice Atlas (EJAtlas.org). (2023c). Lamu Port and related infrastructure, Kenya. https://ejatlas.org/conflict/lamu-port-and-related-infrastructure-kenya

Estes, N. (2024). *Our history is the future.* Haymarket Books.

European Union Copernicus Climate Change Service (EUC3S). (2024, July 10). June 2024 marks 12th month of global temperatures at 1.5°C above pre-industrial levels. *Monthly Climate Bulletin.*

Fairhead, J., Leach, M., & Scoones, I. (2012). Green grabbing: A new appropriation of nature? *Journal of Peasant Studies, 39*(2), 237–261.

Fenelon, J. V. (1998). *Culturicide, Resistance, and Survival of the Lakota.* Routledge.

Fenelon, J. V. (2006). Indigenous peoples and the modern state. *Contemporary Sociology, 35*(3), 249.

Fenelon, J. V. (2015). Indigenous alternatives to global crises of the modern world-system. In I. Wallerstein et al. (Eds.), *Overcoming global inequalities.* Routledge.

Fenelon, J. V. (2018). The struggle of Indigenous Americans: A socio-historical view. In P. Batur (Ed.), *Handbook of the sociology of racial and ethnic relations.* Springer.

Fenelon, J. V., & Alford, J. (2020). Envisioning Indigenous models for social and ecological change in the Anthropocene. *JWSR, 26*(2), 372–399.

Fenelon, J. V., Bogany, J., & Madrigal, L. (2020). Water talks of Native peoples (and nations) in northern L.A. and Ventura counties. Center for Indigenous Peoples Studies. USCR-Tribal-Needs-Assessment-Report-2023-FINAL.pdf

Fenelon, J. V., Everbach, C., Jeter, S., & Wilson-Becerril, M. (2023). Decolonizing ethnic cleansing: Lenape:hoking to Condorcanqui. Conference paper. American Sociological Association meeting, Philadelphia, August 2023.

Fenelon, J. V., & Hall, T. D. (2008). Revitalization and Indigenous resistance to globalization and neoliberalism. *American Behavioral Scientist, 51*(12), 1867–1901

Fenelon, J. V., & Norgaard, K. M. (2021). Towards an Indigenous environmental sociology. In B. S. Caniglia, A. Jorgenson, S. A. Malin, L. Peek, D. N. Pellow, & X. Huang (Eds.), *Handbook of environmental sociology* (pp. 477–494). Springer.

Fernández-Llamazares, A., Lepofsky, D., Lertzman, K., Armstrong, C., Brondizio, E., Gavin, M., Lyver, P., Nicholas, G., Pascua, P., Reo, N., Reyes-García, V., Turner, N., Yletyinen, J., Anderson, E., Balée, W., Cariño, J., David-Chavez, D., Dunn, C., Garnett, S., . . . Vaughan, M. B. (2021). Scientists' warning to humanity on threats to Indigenous and local knowledge systems. *Journal of Ethnobiology, 41*(2), 144–169.

Figueroa-Helland, L. (2022). Indigenous pathways beyond the "Anthropocene." *NYU Environmental Law Journal, 30*(3), 347–411.

Figueroa-Helland, L., Giacomini, T., Langner, L., & Kligler, T. (forthcoming, 2025a). Indigenous biocultural territories vs. the modern/colonial food system. In R. Emigh & D. McCourt (Eds.), *Oxford handbook of comparative historical sociology.* Oxford University Press.

Figueroa-Helland, L., Martinez, A., Gualinga, J., Reyes, N., Castro, A., López, J., Correa, M., & Yvaire, C. (forthcoming, 2025b). Nourishing communal territories of life. In M. Frame, F. Mantz, & S. Grant (Eds.), *Grassroots resistances, alternatives and solutions to 21st century climate and global ecological crises.* Bloomsbury.

Figueroa-Helland, L., Perez-Aguilera, A., & Mantz, F. (2021). Decolonize, reIndigenize. In S. McCullagh & L. Pradanos (Eds.), *Contesting extinctions* (pp. 23–70). Rowman & Littlefield.

Figueroa-Helland, L., Thomas, C., & Perez-Aguilera, A. (2018). Decolonizing food systems. *PGDT, 17*, 173–201.

Folke, C., Polasky, S., Rockström, J., Galaz, V., Westley, F., Lamont, M., Scheffer, M., Österblom, H., Carpenter, S., Chapin, F. S., Seto, K., Weber, E., Crona, B., Daily, G., Dasgupta, P., Gaffney, O., Gordon, L., Hoff, H., Levin, S., . . . Walker, B. H. (2021). Our future in the Anthropocene biosphere. *Ambio, 50*, 834–869.

Ford, A., & Nigh, R. (2015). *The Maya forest garden.* Left Coast.

Forest Peoples Programme (FPP). (2020). *Local biodiversity outlooks 2.* Moreton-in-Marsh: FPP. https://www.cbd.int/gbo/gbo5/publication/lbo-2-summary-en.pdf

Forest Peoples Programme (FPP). (2024a). Shilcayo Declaration: Kichwa Autonomy Confronts Exclusionary Conservation. https://www.forestpeoples.org/en/the-shilcayo-declaration

Forest Peoples Programme (FPP). (2024b). Illegal and forceful evictions of Sengwer Indigenous people from ancestral lands, Cherangany Hills, Kenya. https://www.forestpeoples.org/en/featured-topic/illegal-and-forceful-evictions-sengwer-indigenous-people-their-ancestral-lands#:~:text=In%20January%202014%20the%20Kenyan,seeking%20to%20stop%20them%20returning

Friends of the Earth International (FOEI). (2022). COP27 briefing: What's at stake regarding false solutions? https://www.foei.org/publication/cop27-briefing-false-climate-solutions/

Friends of the Earth International (FOEI). (2023). *Toolkit for fighting climate false solutions.* FOEI. https://www.foei.org/publication/toolkit-climate-false-solutions/

Friends of the Earth International (FOEI). (2024). *Rights for Indigenous Peoples and local communities to ensure well-being of forests and biodiversity.* FOEI. https://www.foei.org/publication/rights-for-indigenous-peoples-and-local-comunities-to-ensure-the-well-being-of-forests-and-biodiversity/

Gabay, M., & Alam, M. (2016). Community forestry and its mitigation potential in the Anthropocene. *Forest Policy and Economics, 79*, 26–35.

Garcia, M. A. (2024). *Corredor Interoceanico en el Itsmo de Tehuantepec.* CECCAM.

Global Forest Coalition (GFC). (2018). Community conservation resilience initiative. https://globalforestcoalition.org/ccri-reports/

Goldtooth, T. (2017). Indigenous peoples cosmovision, conflicts of conquest and need for humanity to come back to Mother Earth. In T. Goldtooth et al. (Eds.), *Rights of nature and Mother Earth* (pp. 15–19). IEN.

Gonzalez, C. (2024). Racial capitalism, climate change, and ecocide. *Wisconsin International Law Journal, 41*(4), 479–519.

Gonzalez-Hidalgo, M., & Huenchumán, P. (2023). Chile: Resistance to the forestry model in Wallmapu, Mapuche territory. *WRM Bulletin*, *268*, 24–27.

Greenfield. P. (2023, January 18). Revealed: More than 90% of rainforest carbon offsets by biggest certifier are worthless. *The Guardian*.

Haaland, D. (2024, January 28). A new era for tribal nations and the U.S. government. *National Geographic*.

Hall, T. D., & Fenelon, J. V. (2015). *Indigenous peoples and globalization*. Routledge.

Health of Mother Earth Foundation (HOME). (2024). Disastrous start to COP29, countries bulldoze with carbon markets paving way for risky geoengineering technofixes. https://handsoffmotherearth.org/resources/press-release-disastrous-start-to-cop29-as-countries-bulldoze-ahead-with-carbon-markets-paving-way-for-risky-geoengineering-technofixes/

Heilmayr, R., Echeverría, C., & Lambin, E. (2020). Impacts of Chilean forest subsidies on forest cover, carbon and biodiversity. *Nature Sustainability*, *3*, 701–709.

Hermann, T. (2005). Indigenous knowledge and management of Araucaria Araucana Forest, Chilean Andes. *Biodiversity and Conservation*, *15*, 647–662.

Houle, R. (2022). *Redwashing extraction*. Yellowhead Institute.

Indigenous Environmental Network. (IEN). (2021). *Indigenous resistance against carbon*. https://www.ienearth.org/indigenous-resistance-against-carbon/

Indigenous Environmental Network. (IEN). (2024). UNPFII "greening economy and false solutions" intervention. https://estatements.unmeetings.org/estatements/30.0290/20240419100000000/WPW2hMyz4AnB/AlgVwRMVHcjl_en.pdf

Indigenous Information Network. (IIN). (2022). *Waata bio-cultural protocol*. IIN.

Indigenous Peoples Global Coordinating Committee (IPGCC). (2024). Indigenous peoples principles and protocols for just transition. https://static1.squarespace.com/static/66a8dc6d84e4290916cd2be1/t/6717d1f90d6bfc19648e12c4/1729614330300/Indigenous+Peoples+Principles+and+Protocols+for+Just+Transition+ENG+%281%29.pdf

International Indigenous Peoples Forum on Climate Change (IIPFCC). (2023a). Opening COP28 statement. https://www.iipfccpavilion.org/stories/openingcop28

International Indigenous Peoples Forum on Climate Change (IIPFCC). (2023b). *Final advocacy paper on "Just Transition"—COP28*. Statement of IIPFCC at COP28. https://iipfcc.squarespace.com/blog/2024/2/10/international-indigenous-peoples-forum-on-climate-change-final-advocacy-paper-on-just-transition-for-unfccc-cop-28

International Indigenous Peoples Forum on Climate Change (IIPFCC). (2024). COP29 opening statement. https://www.iipfcc.org/blog/2024/11/12/international-indigenous-peoples-forum-on-climate-change-cop-29-opening-statement

International Work Group for Indigenous Affairs (IWGIA). (2019). *Impact of renewable energy projects on Indigenous communities, Kenya*. IWGIA.

International Work Group for Indigenous Affairs (IWGIA). (2024). *The Indigenous world 2024*. IWGIA.

Kavilu, S. (2018, November 28). A forest of their own—The Yiaku as Kenya's model forest stewards. *Mongabay*. https://news.mongabay.com/2018/11/a-forest-of-their-own-the-yiaku-as-kenyas-model-forest-stewards/

Kimutai, J., Carrasco-Escaff, T., & Garreaud, R. (2024). Despite coastal cooling trend, risk of deadly wildfires in central Chile increasing. Grantham Institute. https://spiral.imperial.ac.uk/entities/publication/b375bb86-9262-4097-b8fc-6d8771dfcbb1

Kyne, A. (2024, November 21). Renegotiating sovereignty? Māori resistance in the face of colonial erasure. Yellowhead Institute.

Langat, A. (2023, December 20). Violent evictions latest ordeal for Kenya's Ogiek seeking land rights." *Mongabay*. https://news.mongabay.com/2023/12/violent-evictions-are-latest-ordeal-for-kenyas-ogiek-seeking-land-rights/

Langman, J. (2023). Growing fire risk fuels tree-plantation doubts. *EcoAmericas*. https://www.ecoamericas.com/issues/article/2023/11/D2E78410-AB9C-4DD0-BBF0-2AF5A0ED90FB

Lucitante, H. (2024, January 28). "This is Cofán land"—The fight to save Amazonia from intruders. *National Geographic*.
Maffi, L., & Woodley, E. (2010). *Biocultural diversity conservation*. Earthscan.
Mansilla-Quiñones, P., Pehuén, M., & Curamil A. (2024). Confronting the coloniality of nature: Strategies to recover water and life in Mapuche territory. *Geoforum, 148*. 103922.
Marsden, M. (2003). *The woven universe: Selected Writings of Rev. Māori Marsden*. Edited by T. A. C. Royal. Estate of Rev. Māori Marsden.
McAfee, K. (2015). Green economy and carbon markets for conservation and development—A critical view. *International Environmental Agreements, 16*, 333–353.
McGregor, D., Whitaker, S., & Sritharan, M. (2020). Indigenous environmental justice and sustainability. *Current Opinion in Environmental Sustainability, 43*, 35–40.
Mittal, A, Moloo, Z., & Mousseau, F. (2021). *Stealth game: "Community" conservancies devastate land & lives in northern Kenya*. Oakland Institute.
Nahuelpán, H., Hofflinger, Á., Martínez, E., & Millalen, P. (2022). Colonial recognition and militarized accumulation in Mapuche Territory. *NACLA Report, 54*(4), 438–444.
Nahuelpán, H., Martínez, E. Hofflinger A., & Millalén, P. (2021). In Wallmapu, colonialism and capitalism realign. *NACLA Report, 53*(3), 296–303.
Nelson, M. K., & Shilling, D. (Eds.) (2018). *Traditional ecological knowledge: Learning from Indigenous practices for environmental sustainability*. Cambridge University Press.
Ogiek Peoples' Development Program (OPDP). (2015). *Ogiek bio-cultural community protocol (BCP)*. OPDP.
Ogiek Peoples' Development Program (OPDP). (2023). *Renewable energy tool-kit for rural development*. OPDP.
Pastoralists Indigenous Non-Governmental Organization's Forum—PINGO's Forum. (2020). *Climate change migrants and Indigenous peoples adaptation strategies*. PINGO's Forum.
Pathways Alliance for Change and Transformation (PACT). (2023). Indigenous peoples, local communities, and trade in terrestrial carbon markets. PACT.
Perez-Aguilera, D. A. (2016). Mining and Indigenous cosmopolitics: Wirikuta case. In M. Anderson & Z. Bora (Eds.), *Ecological crisis and cultural representation in Latin America* (pp. 179–198). Lexington Books.
Posey, D. (2002). *Kayapó ethnoecology and culture*. Routledge.
Rata, A., Brayne, G., & Barber, S. (2023). Māori sovereignty or death. *Decolonization of Criminology and Justice, 5*(1), 5–30.
Rocheleau, D. (2015). Networked, rooted and territorial: Green grabbing and resistance in Chiapas. *Journal of Peasant Studies, 42*(3–4), 695–723
Rodriguez-Vargas, M. F., & Checa-Hidalgo, D. (2022). Mapeo de Procesos de Resistencia Civil Indígenas ante Proyectos Extractivistas en Zona Sur de los Andes. *Historia Ambiental Latinoamerican y Caribeña, 12*(1), 209–245.
Romero, C. (2021). Carbon offsets, pulp mills and Indigenous resistance in Chile (Wallmapu). In A. Ernsting (Ed.), *Gender-justice and climate action* (pp. 22–25). Global Forest Coalition.
Sánchez Álvarez, M., & Figueroa-Helland, L. (2024). P'ijil k'opetik ta sk'elel, skoltael, yich'el ta muk' ch'ul osil-balamil/Cosmovisión Maya Tsotsil para Honrar y Defender a la Madre Tierra. *Bajo el Volcán, 5*(10), 433–466
Sandoval, D. (2020). *Tren Maya, Sembrando vida y Corredor Transistmico—Impactos en comunidades indígenas y campesinas*. CECCAM.
Scheidel, A., Fernández-Llamazares, A., Bara, A., Del Bene, D., & Whyte, K. P. (2023). Global impacts of extractive and industrial development projects on Indigenous peoples' lifeways, lands, and rights. *Science Advances, 9*(23), eade9557.
Sena, K. (2013). *Lamu Port-South Sudan-Ethiopia transport corridor and Indigenous peoples, Kenya*. IWGIA.
Sena, K. (2021). *Renewable energy projects and rights of marginalized/Indigenous communities, Kenya*. IWGIA.

Sepulveda, C. (2018). Our sacred waters: Theorizing Kuuyam as a decolonial possibility. *Decolonization, 7*(1), 40–58.

Sovacool, B. (2021). Who are the victims of low-carbon transitions? Political ecology of climate change mitigation. *Energy Research & Social Science, 7*, 101916.

Stabinsky, D. (2021). *Chasing carbon unicorns*. FOEI.

Tan, V. (2021). Land, race and state: The carceral state and mass imprisonment of Māori in Aotearoa within the settler-colonial landscape. In M. J. Coyle & M. Nagel (Eds.), *Contesting carceral logic*. Routledge.

Toledo, V., Ortiz-Espejel, B., Cortés, L., Moguel, P., & Ordoñez, M. (2003). Multiple uses of tropical forests by Indigenous peoples in Mexico. *Conservation Ecology, 7*(3), 9.

Toledo, V. M. (2009). ¿Por qué los pueblos indígenas son la memoria de la especie? *Papeles, 107*, 27–38.

United Nations (UN). (2007). *United Nations Declaration on the Rights of Indigenous Peoples*. https://www.un.org/development/desa/indigenouspeoples/wp-content/uploads/sites/19/2018/11/UNDRIP_E_web.pdf

Valandra, E. C. (2006). *Not without our consent: Lakota resistance to termination, 1950–59*. University of Illinois Press.

Valderrama, M. (2022). *Conservation without Indigenous peoples*. Forest Peoples Programme.

Vigil, S. (2018). Green grabbing-induced displacement. In R. McLeman & F. Gemenne (Eds.), *Handbook on environmental displacement and migration*. Routledge.

Whyte, K. (2015). What do Indigenous knowledges do for Indigenous peoples? In M. Nelson & D. Shilling (Eds.), *Traditional ecological knowledge* (pp. 57–82). Cambridge University Press.

Whyte, K. (2017). Indigenous climate change studies. *English Language Notes, 55*(1), 153–162.

Whyte, K. (2018). Indigenous science (fiction) for the Anthropocene. *Environment and Planning EI, 1*(1–2), 224.

Whyte, K. (2020). Too late for Indigenous climate justice. *WIREs, 11*(1), e603.

Wiben, M. (2021). Impacts of Lake Turkana wind power project on Kenya's Indigenous peoples. *Debates Indigenas*. https://rightenergypartnership.org/the-impact-of-the-lake-turkana-wind-power-project-on-kenyas-indigenous-peoples/

World Meteorological Organization (WMO). (2024). *Global annual to decadal climate update 2024–2028*. WMO.

World People's Conference on Climate Change and the Rights of Mother Earth (WPCCC&RME). (2010). People's agreement. https://pwccc.wordpress.com/2010/04/24/peoples-agreement/

World Rainforest Movement (WRM). (2019). *Offsets in the forests*. WRM.

World Rainforest Movement (WRM). (2022). *15 years of REDD—A mechanism rotten at the core*. WRM.

Young, D. (2024). *Beyond offsets*. Rainforest Foundation.

CHAPTER 19

Winning the Fossil Fuel Battle in Small-Town California

CAROLINE FARRELL AND INGRID BROSTROM

Introduction

One of the most impactful recent climate justice campaigns in California originated in an unlikely place: a small, predominantly Latino/a farmworker community in one of the most conservative areas in California, the City of Arvin in Kern County. It is the story of how residents in one of the most overburdened communities in the state identified a pressing health concern, defined a solution, and built power to fundamentally shift local and statewide policy. It also exemplifies that progressive policies do not need to originate in large urban areas. The campaign to ban oil and gas drilling near homes and schools in Arvin shows how community organizing, cross-sectoral coalition building across communities, legal strategy, and policy advocacy can work synergistically to create meaningful change.

On September 16, 2022, California Governor Gavin Newsom signed Senate Bill (SB) 1137 into law. SB 1137 prevents new permits for oil and gas drilling within 3,200 feet of homes, schools, hospitals, and other sensitive land uses. The bill also phases out existing drilling operations over time by prohibiting operators from redrilling or deepening their wells within the same 3,200-foot zone. These types of land use restrictions are commonly referred to as setbacks—a defined distance, measured from a structure or property line, in which certain land uses are prohibited. This new law catapulted California from having some of the nation's weakest setback requirements to having the largest statewide oil and gas setback in the nation. The legislative victory came 3 years after a broad coalition of environmental justice, public health, labor, youth, and environmental organizations first launched an effort to pass a statewide oil and gas setbacks policy. But the story did not start in the California legislature; it started about 300 miles south in a small, rural farmworker community in the heart of deep red Kern County. This legislation is not the end of the campaign for health protective oil and gas setbacks. But we can learn from the struggle to establish setbacks and assess what must still be done to achieve the goals of the community leaders who fought so hard for them.

Arvin

The City of Arvin is a small municipality in southern Kern County at the base of the Tehachapi Mountains with a population of 19,661. It is a Latino/a community (94.6%) with a median household income of $49,986 with 31.9% of residents living in poverty (U.S. Census, 2024). It is also one of the most environmentally impacted communities in the state, ranking in the top 15% of census tracts for environmental burdens and social vulnerabilities (Office of Environmental and Health Hazard Assessment, 2021). The State of California recognizes that Arvin is burdened by impacts from agriculture, warehouse distribution centers, and oil and gas operations, which contribute to the area's poor air quality, consistently ranked among the worst in the nation (California Air Resources Board, n.d.).

Arvin's History With Oil and Gas

The state has permitted more than 240,000 oil and gas wells throughout California, with the vast majority located in Kern County, and smaller clusters located in Los Angeles and Long Beach (Cart, 2024). Kern and Los Angeles counties account for more than 80% of the overall oil production in California, and thus bear the brunt of its health and safety hazards (Rogers & Villalvazo, 2022). Arvin sits atop the Mountain View Oil Field, a large, mature oil field that has been in production since the 1940s (Bruer & Porter, 1965).

Most of the oil producers are independent with wells scattered throughout the region—some adjacent to homes and schools, and some tucked away amidst agricultural fields (Shale XP, n.d.). Prior to 2022, California had no statewide rule establishing minimum distances between oil and gas drilling activities and sensitive land uses such as schools, homes, and hospitals—a regulatory gap that is rare among the nation's top oil-producing states (Ferrar, 2022).

In 2018, 13 wells were producing oil within Arvin's city limits, 11 of which were within 300 feet of sensitive receptors (Fractracker Alliance, 2018). One well is located directly across the street (150 feet) from a local elementary school, and another sits adjacent to a playground (Ferrar, 2021). In 2022, the Methane Task Force, a joint effort of the California Department of Conservation, the California Geologic Energy Management Division (CalGEM), the California Air Resources Board, and the San Joaquin Valley Air Pollution Control District (Air District), inspected 68 oil wells in the Arvin area, including the neighboring unincorporated community of Lamont. Of the 68 inspected wells, inspectors found 27 were leaking and ordered them repaired (San Joaquin Valley Air Pollution Control District, n.d.).

The Beginning of Oil and Gas Organizing in Arvin: Pipeline Leak

On March 18, 2014, eight families living on Nelson Court in Arvin were evacuated from their homes because of a gas leak. Residents near Arvin's high school began feeling sick. Gases leaking from an underground pipeline had saturated the soil and entered homes. Prior to the gas leak discovery, residents had complained of nausea, headaches, lightheadedness, and nosebleeds. One resident was pregnant at the time of the leak. Air sampling from inside homes on the street revealed levels of toxic gas 13 times higher than what was deemed safe by the Environmental Protection Agency (Brown, 2014). The cause: a leaky pipeline operated by Petro Capital Resources (PCR) that was used to carry raw natural gas. Ultimately, the Department of Conservation Division of Oil, Gas and Geothermal Resources fined the responsible oil company, PCR, $75,000 for failing to operate the pipelines according to good oil field practices (Gaspar, 2016).

This incident galvanized Arvin residents' concern about the oil and gas industry in and around the community. Residents on Nelson Court and throughout the community were concerned because Kern County, which oversaw the pipeline, left the evacuation response and remediation to the company responsible for the leaks. PCR housed residents in an apartment complex in nearby Bakersfield. The families evacuated the area for over 7 months. The families advocated for the county to verify when the levels were safe enough to return home. However, the county relied on PCR to determine when conditions were safe. The county did not require any ongoing health monitoring or testing, but instead left that responsibility to PCR. The company unsurprisingly decided against any ongoing monitoring (Aguilera, 2014). At that point, residents and the city's mayor, Jose Gurrola, wanted to ensure the City of Arvin had a larger role in regulating oil and gas within city limits.

History of Environmental Justice Organizing in Arvin Leading to Ordinance Updates

The Center on Race, Poverty & the Environment (CRPE) is one of the oldest environmental justice organizations in the country and applies a distinct model of "from the ground up" community organizing that invests in building the capacity of residents to lead and sustain their own community-based organizations and initiatives (Cole & Foster, 2001). CRPE helped Arvin residents form the Committee for a Better Arvin (CBA) in 2007 to address environmental issues in the community, specifically at that time, the Brown and Bryant Superfund site in Arvin. The United States Environmental Protection Agency's (EPA) clean-up plan for the abandoned pesticide manufacturing facility had languished for decades, and contamination had formed a plume that threatened a nearby drinking water well. Residents partnered with the CRPE to organize the community to improve the EPA clean-up plan, to mitigate the threat to drinking water, and to establish a group that would engage on environmental justice issues throughout the community.

Throughout its almost 20-year history, CBA has organized residents to reduce air pollution from nearby industrial waste operations, reduce pesticide exposure, increase investment in open space and complete streets, and reduce impacts from oil and gas. Once members of CBA learned of the gas leak and evacuation of residents in Nelson Court, the group began providing support to Nelson Court residents and engaged state and federal officials to provide oversight of the clean-up and response. In 2014, CBA began collecting signatures on a petition to the City of Arvin to adopt an ordinance that required city permitting for oil and gas operations and created a buffer zone or setback between oil and gas wells and sensitive receptors. The group ultimately collected over 2,500 signatures on the road to secure the ordinance (California Environmental Justice Alliance, n.d.).

Arvin Oil and Gas Ordinance

In 2016, Arvin unveiled proposed updates to its oil and gas ordinance, which had been written in the 1960s. The changes the city sought were modest. Among other protective measures, the proposed ordinance would prohibit new oil and gas drilling in residential zones and create a 300-foot setback between new drilling operations and homes, schools, and hospitals. California Attorney General Xavier Becerra supported the new ordinance and, in a letter to the City Council, affirmed his opinion that the changes were well-within the city's local authority and jurisdiction (Becerra, 2018). On July 17, 2018, before a packed city hall and dozens of residents who spoke passionately about the measure, the Arvin City Council passed the ordinance, the first of its kind to establish oil and gas setbacks in California. The purpose of the new ordinance was to protect the health, safety, and welfare of the community. It covered drilling, pipelines, storage, oil spills, and recovery equipment (City of Arvin Municipal Code, 2024). The ordinance prohibited the siting of new wells in specific zones (City of Arvin Municipal Code, 2024). In areas where wells were allowed, new oil development in the city required a Conditional Use Permit (City of Arvin Municipal Code, 2024). This was an important change. Rather than allowing oil well drilling by right, operators need to apply for a discretionary permit and undertake an environmental review process under the California Environmental Quality Act (CEQA) (California Environmental Quality Act Guidelines, n.d.). CEQA requires the city to undertake a public process before approving a project which would require public notice, the opportunity for comment, and the opportunity to oppose or improve proposed oil and gas development in the city (CEQA, n.d.). In addition, new oil wells are required to be located more than 300 feet from the property boundaries of schools, residences, long-term care facilities, and other sensitive receptors (City of Arvin Municipal Code, 2024).

Even though the municipal code provided modest protections, given the enormity of impact from oil and gas drilling operations, it was seen as an important step in oil and gas regulation. The oil industry and its allies strongly opposed the ordinance (Thompson, 2019). However, other communities saw hope in the opportunity to pass a local oil and gas ordinance in deep red Kern County. If a community in Kern County could impose regulations, there was a chance that other cities, and the state itself, could regulate the industry to improve public health (Bedoian, 2018).

Kern County Oil and Gas Ordinance

Kern County is the largest oil-producing county in California and one of the top oil-producing counties in the country (Brostrom et al., 2024). The oil and gas industry wields considerable power in Kern County and across the state, and it has consistently attempted to thwart new regulations, including public health protections such as setbacks. The Western States Petroleum Association (WSPA) and Chevron are currently the two top lobbying groups in the state (Vanteicher, 2025). WSPA commissioned the Kern Economic Development Foundation to study the importance of the oil and gas industry to Kern County. The study found that oil has been important to Kern County since it was discovered in the 1890s. Kern County was the seventh highest oil-producing county in the country in 2021. It accounted for 71% of the oil California produced and 3% of the country's output (Kern Economic Development Foundation, 2021). The study found that the oil and gas industry was a significant employer in the county and contributed $15 billion to the county's tax base (Kern Economic Development Foundation, 2021). The oil and gas industry is also culturally important. Every 5 years, the City of Taft in Kern County hosts the Oildorado Days, a tribute to the oil and gas heritage in the region, where the Maids of Petroleum compete to be named the Oildorado Queen (Taft Oildorado, n.d.).

Kern County opposed Arvin's Oil and Gas Ordinance, arguing that the county passed its own oil and gas ordinance in 2015. The county argued that Arvin should rely on the county's ordinance and environmental review (Thompson, 2019). However, the county ordinance was funded by and developed for the oil and gas industry—specifically the WSPA, the California Independent Petroleum Association, and the Independent Oil Producers' Agency—in response to mounting legal pressure on the state to require more robust environmental review processes prior to issuing oil and gas drilling permits (Kane, 2020). The Kern County Ordinance and accompanying Environmental Impact Report (EIR) aimed to create a permitting system that would fast-track future oil well permits and eliminate future environmental review and public participation for individual permits (Hanson, 2024). The Kern Oil and Gas Ordinance did not include setbacks.

CBA opposed the Kern County Ordinance because it was not focused on health. It was one of the reasons that residents advocated for the City of Arvin to adopt its own ordinance. CBA joined with other community groups throughout the county—Committee for a Better Shafter (CBS), Greenfield Walking Group, and the Comité Progresso de Lamont—to oppose the ordinance that would streamline permitting. CRPE worked with all these groups to address environmental justice issues in their respective areas. But several issues overlapped in these communities and created an opportunity to leverage their collective power. These communities joined together because enhanced oil drilling techniques, such as fracking, allowed oil wells to proliferate in and near their communities. Each of these groups supported CBA in its advocacy for the Arvin Oil and Gas Ordinance. They showed up to City Council hearings, spread the word, and provided testimony about their own experiences living near oil and gas extraction activities.

The Kern County Oil and Gas Ordinance created an opportunity for all the community groups to work together to advocate for a strong health protective ordinance. When the county first approved its oil and gas ordinance in 2015, CBA, CBS, Greenfield Walking Group, and Comité Progreso de Lamont joined with environmental groups such as the Center for Biological Diversity, the Natural Resources Defense Fund, and the Sierra Club to file a lawsuit under CEQA. In 2020, the Court of Appeals issued a decision agreeing with the groups that the county had not complied with CEQA. Specifically, the Court of Appeals issued a partially published opinion that found the county (1) inappropriately deferred, and failed to sufficiently address, the implementation of mitigation measures designed to address impacts to water supplies; (2) failed to sufficiently address or mitigate impacts from the project's conversion of agricultural land; and (3) failed to sufficiently address the project's noise impacts. In the unpublished portion of the opinion, the Court of Appeals held that the county: (1) inadequately addressed air quality impacts in relation to fine particulate matter (PM2.5); and (2) failed to include a cumulative health risk assessment in the draft EIR that the county conducted, preventing meaningful public review and comment on the assessment. The Court ordered the county to set aside the ordinance and EIR until it complied with CEQA (Marsh et al., 2020).

In 2021, Kern County purported to resolve the CEQA violations and reapproved the ordinance. However, the community and environmental groups challenged the EIR again for not complying with CEQA and the Court of Appeals' Order. For the second time, the Court of Appeal found that the county had not complied with CEQA. Specifically, the Court found that Kern County did not adequately analyze the cancer risks for people living near wells. The county's ordinance would allow drilling 200 feet from residences and 300 feet from schools. The Court also found that the county did not adequately address impacts to the water supply, given the industry's use of water, which was expected to lower water levels throughout the county, including in disadvantaged communities. The Court again set aside the ordinance and EIR until the county complied with CEQA (Hanson, 2024). As of June 2025, the county was considering the ordinance for a third time (Kern County Planning and Natural Resources Department, n.d.).

State Setback

While pushing Kern County to adopt a more health protective ordinance, the Kern community groups (CBA, CBS, Comité Progresso de Lamont, and Greenfield Walking Group) understood that they could not rely on the county to protect them. Instead, they worked with CRPE to turn to the state. To protect residents in areas where the local government would not act, the community groups wanted to establish a statewide setback between oil and gas wells and sensitive receptors. The common sentiment at the time was that such an endeavor was not feasible in California. That the "chippy" environmental justice organizations in California were no match for the powerful and well-funded oil and gas lobby and the powerful interests with which they aligned. The previous year saw well-funded industry opposition handily defeat Colorado's Proposition 112, which would have required new oil and gas development to be a minimum of 2,500 feet away from occupied buildings and other vulnerable areas (Coloradoan, 2018).

The effort faced other barriers as well. Some environmental organizations did not support setbacks because they thought the effort would delay phasing out oil and gas extraction in the state. However, groups living near oil and gas drilling faced immediate impacts that could not wait to be addressed. Environmental justice organizations in Los Angeles had been fighting for a local oil and gas setback for years and had been assured that Los Angeles County was poised to act. They were concerned that any state legislative effort might undermine the progress occurring in Los Angeles because environmental justice goals are often pitted against each other in a divide-and-conquer strategy. Kern advocates held off fighting for a state solution and worked on building local momentum in more progressive pockets, like Arvin, instead and building relationships with groups throughout the state concerned about oil and gas setbacks.

After years of promises with no sign of Los Angeles decision makers following through in sight, CRPE made a calculation based on three beliefs: (1) that despite the small yet meaningful victories in Arvin, Kern County, as a whole, would not take action to protect its residents from the harmful impacts of oil drilling; (2) that any momentum at the state level would help rather than impede progress in Los Angeles; and (3) that the persistent environmental justice movement in California could pull off a major policy win against the oil and gas industry by coalescing a big tent of supporters.

Based on this belief, in 2019, CRPE began its search for a state legislator who would be willing to introduce and author a bill to establish an oil and gas setback in California. At times, partnerships between environmental justice organizations and sympathetic actors inside the state can lead to progressive policy change (Harrison, 2019; Stearns & Almeida, 2004). Knowing the influence of the oil and gas industry in the Central Valley, CRPE approached legislators from other areas of the state ready to take on Big Oil. CRPE found its first champion in Assemblymember Al Muratsuchi, representing the refinery-heavy area of Torrance, California. Assemblymember Muratsuchi authored Assembly Bill 345, which would have mandated that California create a setback based on health and science data and environmental justice concerns by July 2022, and that it consider a 2,500-foot setback for schools, playgrounds, and other public places where children are present (LegisScan, n.d.).

Despite the pessimistic outlook about a statewide strategy from other advocacy organizations, CRPE and the Kern County community groups defied expectations when AB 345 passed the California Assembly in January 2020. This unexpected victory lit a fire, and organizations began to join the effort. By the time AB 345 reached the Senate Natural Resources Committee, it had amassed 270 organizational supporters and garnered the attention of media outlets and political pundits. The attention and widespread support were not enough to overcome the initial influence of the bill's industry opposition, as three moderate members of the committee, who had each received fossil fuel campaign contributions, sided with Republicans to block passage of the bill (Levin, 2020).

By this time, the slogan "No drilling where we're living" had become a rallying cry across the state. CRPE had launched a new statewide coalition of environmental justice organizations—Voices in Solidarity Against Oil in Neighborhoods (VISION)—to lead the legislative fight for a statewide setback. VISION joined forces with the Last Chance Alliance, a coalition of climate and environmental organizations focused on building a fossil fuel-free future for California. The California Environmental Justice Alliance, a policy powerhouse in California, elevated the issue as one of its top legislative priorities. A biweekly AB 345 supporters call to coordinate local efforts to support the policy steadily grew in attendance. Jane Fonda lent her support and celebrity status to the issue in her Fire Drill Fridays, a series of peaceful actions to call on elected leaders to act on the impending climate crisis.

Meanwhile, scientific evidence of the harmful effects of drilling in close proximity to homes mounted. CalGEM convened a Public Health Science Advisory Panel to assess the scientific evidence regarding the public health implications of oil and gas development. The resulting report concluded that epidemiological evidence indicates that close residential proximity to oil and gas development is associated with adverse perinatal and respiratory outcomes in populations living within and beyond 1 kilometer or 3,281 feet of oil and gas well sites, and those outcomes get worse where the density of wells is greater (Shonkoff et al., 2021). Around this same time, multiple studies concluded that pregnant individuals living near oil and gas drilling faced an increased risk of negative birth outcomes, such as preterm birth, low birth weights, and birth defects (Ferrar, 2022).

Based on the mounting support and scientific evidence, subsequent iterations of the bill were even more protective than the initial language in AB 345. In 2022, with support from the governor, Senator Lena Gonzalez and Senator Monique Limon authored SB 1137 to establish an even larger 3,200-foot buffer zone in neighborhoods and outright banned both new oil and gas wells and "reworks" of existing wells near homes, schools, and other "sensitive" areas such as hospitals (Senator Lena Gonzalez and Senator Monique Limon, 2022). In addition, the bill required operators of existing wells to take steps to improve their safety, such as installing leak detection equipment, testing water, and controlling dust and other pollutants. After years of advocacy in the state legislature, building coalitions, communicating with state agencies, developing strong narrative strategies, planning direct actions, and working with scientists, community leaders in Kern County and advocates across the state were rewarded with the passage and signing of SB 1137.

CRPE's initial analysis had proven correct—the legislative effort helped coalesce a big tent of cross-sectoral supporters that was able to effectively take on powerful industry interests. CRPE's two other predictions were also borne out: (1) Kern County doubled down on its efforts to fast track permitting and was only deepening its support of the oil and gas industry in the face of regulatory pressure from the state; and (2) the statewide campaign helped spur local efforts in Los Angeles. In 2021 and 2022, both the County and the City of Los Angeles voted to ban all new oil wells and phase out existing ones over the next two decades.

The Oil and Gas Industry Fights Back

Newsom signed SB 1137 on September 16, 2022. However, the fossil fuel industry did not give up without a fight. Just days after the bill's signing, the secretary of state approved an industry-backed referendum for circulation seeking to overturn the law on the 2024 ballot (Begert, 2022). The oil industry spent more than $20 million for a paid signature-gathering campaign and on February 3, 2023, officially qualified the referendum to appear on the November 2024 ballot (Gross, 2024). The referendum temporarily paused implementation of the law (Gross, 2024).

While the move was disappointing to advocates of the bill, the referendum provided yet more opportunities for the well-organized supporters of the bill to continue to uplift its importance and share concerns about the public health impacts of oil and gas drilling. Several political action committees emerged to raise funds to defeat the referendum. Public sentiment and momentum against the once-powerful oil and gas industry had already been shifting. The industry's attempt to sell a disgruntled public on the need to protect its right to extract oil and gas near schools, homes, and hospitals was risky. On June 27, 2024, just hours before the deadline, the industry-backed Californians for Energy Independence withdrew the referendum and removed it from the ballot. With the withdrawal, SB 1137 immediately went into effect. In total, the fossil fuel industry spent almost $61 million on the failed referendum effort (California Environmental Justice Alliance, 2024).

Industry representatives instead indicated plans to file a lawsuit to block the implementation of SB 1137. In April 2025, two California mineral rights owners sued the California Division of Geologic Energy Management, challenging the constitutionality of SB 1137, arguing that the bill's restrictions on the redrilling of existing wells within the setback zone amounts to an improper seizure of property rights (Cox, 2025). Despite this latest setback, SB 1137 remains in force as the litigation winds its way through the legal system.

A Final Note: Just Transition

While the fossil fuel industry has inflated concerns about the impact of health and safety setback zones on jobs and the economy in Kern County, the larger shift away from fossil fuel production in the state will have a significant impact on Kern County—both in terms of displaced workers and tax revenue losses. Kern's reliance on a few industry drivers, such as fossil fuel and agricultural production, makes it highly susceptible to market fluctuations. Kern County has the most volatile economy in the state (Flores, 2022). Much of this volatility is driven by the boom-and-bust cycles of oil production. In recent years, oil production employment has faced steep declines while the sector provides some of the highest wages in the county.

Community leaders in Kern have expressed a general desire to transition away from fossil fuel extraction as an economic driver in Kern based on the volatility of the industry, the unsustainable climate impacts of fossil fuel, and the impacts of fossil fuel extraction on public health and the environment. However, residents have also expressed the desire for the transition to be responsibly managed rather than left to market and regulatory forces that would leave the county with fewer jobs and less of a tax base to provide critical public services.

Recognizing the need to diversify the local economy and provide pathways for sustainable jobs, CRPE began developing a community-driven vision for environmentally and economically sustainable economic development with its Power to the People campaign back in 2007. The Power to the People campaign engaged residents in defining green jobs, where communities fit into the promise of the new green economy, and how their communities could develop in healthy and sustainable ways (Center on Race, Poverty & the Environment, 2011). The campaign identified several strategies for advancing the community's vision for a new green economic model:

- Invest in the human capital of low-income disadvantaged communities in the Valley;
- promote socially responsible green businesses and practices;
- address environmental impacts of economic development; and
- support meaningful public participation.

This approach closely mirrors the state's more recent approach to workforce and economic development. California defines "high road economic development" as a set of economic and workforce development strategies to achieve economic growth and equity, shared prosperity, and a clean environment (Unemployment Insurance Code § 14005(r); Zabin, 2020). The strategies include interventions that improve job quality and job access, including for people from underserved and underrepresented populations; meet the skill and profitability needs of employers; and meet the economic, social, and environmental needs of the community. These are principles consistent with Just Transition perspectives (see chapter by Kerrissey and Coon in this volume). CRPE worked with the UC Merced Community and Labor Center to develop the Kern Strategic Workforce Development Report and prepared its own People's Plan: A High Road Vision for Kern County. The reports highlight the need to focus on raising industry-wide working standards since few economic sectors in Kern currently provide jobs that pay a living wage, wage replacement and training programs for displaced workers, and prioritizing investments in sectors that advance sustainability, such as oil well remediation, electrification, building decarbonization, and public transportation (Brostrom et al., 2024).

Lessons Learned

As we reflect on the decade-long fight to reduce local impacts associated with oil and gas extraction activities in California communities, there are important lessons learned that can be useful to other climate and environmental justice campaigns.

The first is the importance of challenging assumptions. The common wisdom in rural areas, which tend to skew more conservative and business-friendly across the United States, is that progressive policies are unpassable. It is believed that these rural areas will be the last to see health and environmental protections that are often prioritized in larger, urban areas. However, as we saw in Arvin, small rural cities can provide fruitful avenues for innovative and progressive policies. Small, rural cities are eminently organizable. Engaged local community leaders hold considerable sway with their neighbors and the broader community. A few dozen votes in a local election can make a big difference in who holds decision-making authority on the city council. Industry influence can be moderated by the relatively low number of industry actors within a small city's boundaries.

Second is the importance of building a large multisectoral tent of support in passing difficult policies. The environmental justice movement has long fought for space at the table in important policy discussions. This has led to the development of strategies that center environmental justice organizations as key decision-makers on bills and policies that are priorities for the environmental justice movement. This can cause tension between the need for environmental justice organizations to maintain leadership and autonomy and the need to bring in powerful allies who can help move difficult policy. The oil and gas setback is a prime example of how environmental justice organizations maintained their leadership while welcoming and encouraging a broad base of supporters to activate their membership, leverage their relationships, and spread the message on the importance of health setbacks.

Finally, the history of achieving health protective oil and gas setbacks is a reminder that the fight for environmental and climate justice is long and not linear—there is no single fix. Rather, advocates must work with impacted communities on multiple levels (local, state, and beyond) and employ a variety of strategies, from organizing to legal representation to policy advocacy. It requires adaptability, persistence, coalition building, and a willingness to challenge powerful industries. While the story is not yet finished, the future is looking a little brighter for the many Californians who live in the shadows of oil and gas extraction.

References

Aguilera, D. (2014, October 28). Seven months after gas leak Arvin residents still can't return home. *KVPR*. https://www.kvpr.org/environment/2014-10-28/seven-months-after-gas-leak-arvin-residents-still-cant-return-home

Becerra, Xavier. (June 8, 2018). Letter from Xavier Becerra to Mayor, City Council Members and Planning Commission Members of the City of Arvin, Proposed Ordinance No. 18-XX to

Adopt Text Amendment No. 2017-04 and Ordinance for Regulation of Petroleum Facilities and Operations by Repealing Chapter 17.46, Title 17, and Adding Chapter 17.46 to Title 17 of the Arvin Municipal Code. https://www.oag.ca.gov/system/files/attachments/press-docs/comments-arvin-oil-and-gas-ordinance.pdf

Bedoian, V. (2018, September 19). Arvin's oil and gas ordinance points the way toward environmental justice in the valley. *Community Alliance.* https://fresnoalliance.com/arvins-oil-ordinance-points-the-way-toward-environmental-justice-in-thevalley/#:~:text=The%20oil%20industry%20opposed%20the%20ordinance%2C%20portraying%20the,major%20industry.%20In%20reality%2C%20the%20reforms%20are%20modest

Begert, B. (2022, December 20). California passed milestone law to stop neighborhood drilling. Now big oil has launched its counterattack. *Grist.* https://grist.org/equity/california-passed-a-milestone-law-to-stop-neighborhood-drilling-now-big-oil-has-launched-its-counterattack/

Brostrom, I., Flores, E., Alatriste-Diaz, R., Luna, K., Fonsah, E., Juarez, K., Guerrero, J., Arias Murillo, A., Arana E., & Almeida, P. (2024, February). Kern Strategic Workforce development report. UC Merced Community and Labor Center. https://clc.ucmerced.edu/sites/g/files/ufvvjh626/f/page/documents/kern_hrtp_report.pdf

Brown, R. (2014, April 26). Arvin gas leak reveals lack of oversight. *Bakersfield Californian.* https://www.bakersfield.com/news/arvin-gas-leak-reveals-lack-of-oversight/article_9c839848-1db0-516d-af8b-ec615157561b.html

Bruer, W. C., & Porter, F. C. (1965). Geology of Southeastern San Joaquin Valley, California, Kern River to Grapevine Canyon. https://archives.datapages.com/data/pacific/data/011/011001/16_ps0110016.htm

California Air Resources Board. (n.d). Arvin/Lamont. https://ww2.arb.ca.gov/capp/com/cip/arvin-lamont

California Environmental Justice Alliance. (n.d.). Committee for a Better Arvin—Oil and gas ordinance. https://ceja.org/what-we-do/green-zones/committee-for-a-better-arvin-oil-and-gas-ordinance/

California Environmental Justice Alliance. (2024, June 27). Grassroots environmental justice communities score a massive victory over big oil. https://ceja.org/2024/06/sb-1137-referendum-defeated-california-oil-gas-setbacks/

California Environmental Quality Act Guidelines, Discretionary Project. (n.d.). 14 CCR § 15257, https://www.law.cornell.edu/regulations/california/14-CCR-15357

California Environmental Quality Act, Guidelines, Public Participation. (n.d.). 14 CCR § 15201. https://www.law.cornell.edu/regulations/california/14-CCR-15201

Cart, J. (2024, July 7; updated 2024, September 24). Controversial measure overturning oil well restrictions won't be on California ballot. *CalMatters.* https://calmatters.org/environment/2024/06/oil-ballot-california

Center on Race, Poverty & the Environment. (2011). *The green paper: A community vision for environmentally and economically sustainable development.* https://crpe-ej.org/crpe-publications/

City of Arvin Municipal Code. (2024, October 9). Title 17, Chapter 46 Municipal Code, et seq. https://library.municode.com/ca/arvin/codes/code_of_ordinances?nodeId=TIT17ZO_CH17.46OIGACOy

Cole, L. W., & Foster, S. R. (2001). *From the ground up: Environmental racism and the rise of the environmental justice movement.* NYU Press.

Coloradoan, C. (2018, November 7). After Colorado's proposition 112 fails, what's next for oil and gas setbacks. *Journal of Petroleum Technology.* https://jpt.spe.org/after-colorados-proposition-112-failed-whats-next-oil-and-gas-setbacks

Cox, J. (2025, April 16). Lawsuit challenges state's landmark oil-field setbacks law. *Yahoo News.* https://www.yahoo.com/news/lawsuit-challenges-states-landmark-oil-035900267.html?guccounter=1&guce_referrer=aHR0cHM6Ly93d3cuYmluZy5jb20v&guce_referrer_sig=AQAAAN6nY25Q7np4SGSltMUv5qWR_2kSTMXaM7-7j3YX9bOQSky2UE43motip269OjKDpC6MVosINnTls84rblQPZuImvDm3-pmY9hI3iaAv9hrePJ8otMS-oossXQarmIeeaoNZIcyrYUfKQp5bYX5umDXWHiSwtJTNnOHHhxcDqEzf

Ferrar, K. (2021, February 9). Toxic emissions filmed at oil and gas wells in Arvin, CA. https://storymaps.arcgis.com/stories/5e5f228c6f034814b51003185ob32fe2

Ferrar, K. (2022, April 6). Implications for a 3,200-foot setback in California. *FrackTracker Alliance.* https://www.fractracker.org/2022/04/implications-of-a-3200-foot-setback-in-california/e

Flores, E. (2022, May). The future of the state: Kern County's young, growing, diverse population and dynamic economy. UC Merced Community and Labor Center. https://clc.ucmerced.edu/sites/g/files/ufvvjh626/f/page/documents/future_of_the_state.pdf

Fractracker Alliance. (2018, January 18). Arvin CA—a city in the most drilled county in the country—files of a setback ordinance. https://www.fractracker.org/2018/01/arvin-ca-setback-ordinance/

Gaspar, J. (2016, February 19). State fines Bakersfield company $75,000 for gas leak that displaced Arvin families. *Bakersfield Now Eyewitness News KBAK.* https://bakersfieldnow.com/news/local/state-fines-bakersfield-company-75k-for-gas-leak-that-displaced-arvin-families

Gross, L. (2024, July 1). California communities celebrate "massive" victory as oil industry drops unpopular referendum. *Inside Climate News.* https://insideclimatenews.org/news/01072024/california-celebrates-victory-as-oil-industry-drops-unpopular-referendum/ies

Hanson, N. (2024, March 7). State appeals court halts Kern County plan to fast-track oil drilling near homes, schools. *Courthouse News Service.* https://www.courthousenews.com/state-appeals-court-halts-kern-county-plan-to-fast-track-oil-drilling-near-homes-schools/

Harrison, J. L. (2019). *From the inside out: The fight for environmental justice within government agencies.* MIT Press.

Kane, J. (2020, August 7). Tire of wells threatening residents' health, a small California town takes on big oil. *InsideClimate News.* https://www.kqed.org/science/1968101/tired-of-wells-threatening-residents-health-a-small-california-town-takes-on-big-oil

Kern County Planning and Natural Resources Depart. (n.d.). Kern County oil and gas permitting. https://kernplanning.com/oil-gas-landing-main

Kern Economic Development Foundation. (2021, January n.d.). The economic contribution of the oil and gas industry in Kern County. https://kernedc.com/wp-content/uploads/2021/04/KEDF-Economic-Contribution-of-the-Oil-and-Gas-Industry-in-Kern-County_-2021.pdfunty_-2021.pdf

LegiScan (n.d.). Bill text: CA AB 345 (2019/2020) Regular session/amended. https://legiscan.com/CA/text/AB345/id/2109533:CAAB345|2019-2020|RegularSession|Amended|LegiScan

Levin, N. (2020, August 11). State fails to protect frontline communities from oil drilling. *KnockLA.* https://knock-la.com/ab345-wake-up-call-hertzberg-caballero-hueso-d131f038be1/

Marsh, C. L., Veasy, P. F., & Goldberg, S. H. (2020, March 12). Fifth Appellate District invalidates Kern County oil and gas ordinance. *CEQA Chronicles.* https://www.ceqachronicles.com/2020/03/fifth-appellate-district-invalidates-kern-county-oil-and-gas-ordinance/

Office of Environmental and Health Hazard Assessment. (2021, October 13). CalEnviroScreen 4.0. https://oehha.ca.gov/calenviroscreen/report/calenviroscreen-40

Rogers, V., & Villalvazo, V. (2022, September 6). California lawmakers approve 3,200 ft setbacks for new and existing wells. *South Kern Sol.* https://southkernsol.org/2022/09/06/california-lawmakers-approve-3200ft-setbacks-for-new-and-existing-wells/

San Joaquin Valley Air Pollution Control District. (n.d.). Methane taskforce inspection report for the Arvin/Lamont community. https://community.valleyair.org/media/idcjcyg2/final_methane-task-force-arvin-lamont-inspection-report.pdf

Senator Lena Gonzalez and Senator Monique Limon. (2022, August 26). SB 1137 (Gonzalez and Limón) health protection zones for frontline communities. https://sd33.senate.ca.gov/sites/sd33.senate.ca.gov/files/sb_1137_gonzalez_final_set_back_fact_sheet.pdf

Shale XP. (n.d.) Satellite map. https://www.shalexp.com/map?a1=on&f1=api_no&c1=contains&v1=&ay=35.19006199917641&ax=-118.92697857385791&zy=35.266063253380516&zx=-118.74639080530322&zoom=13

Shonkoff, S. B. C., Morello-Frosch, R., Casey, J., Deziel, N., DiGiulio, D. C., Foster, S., Harrison, R., Johnston, J., Kloc, K., McKenzie, L., McKone T., Miller M., & Polidori, A. (2021, October 1).

Re: Response to CalGem questions for the California Oil and Gas Public Health Rulemaking Scientific Advisory Panel. https://www.gov.ca.gov/wp-content/uploads/2021/10/Public-Health-Panel-Memo.pdf

Stearns, L. B., & Almeida, P. D. (2004). The formation of state actor-social movement coalitions and favorable policy outcomes. *Social Problems*, *51*(4), 478–504.

Taft Oildorado. (n.d.). Taft Oildorado. https://www.taftoildorado.com/

Thompson, G. (2019, July 12). Meet the millennial mayor who took on big oil-and won. *The Nation*. https://www.thenation.com/article/archive/arvin-california-oil-drilling-resistance/

Vanteicher, W. (2025, May 3). Oil and gas industry topped first-quarter lobbying spending in California. *E&E News Politico*. https://www.eenews.net/articles/oil-and-gas-industry-topped-first-quarter-lobbying-spending-in-california/O

U.S. Census. (2024, July 1). Quick facts Arvin city, CA. U.S. Census Bureau QuickFacts: Arvin city. https://www.census.gov/quickfacts/fact/table/arvincitycalifornia/PST045223California

Zabin, C. (2020, June 3). Putting California on the high road: A jobs and climate action plan for 2030. https://laborcenter.berkeley.edu/wp-content/uploads/2020/09/Putting-California-on-the-High-Road.pdf

CHAPTER 20

Social Movements, Industries, and Renewable Energy

ION BOGDAN VASI

Introduction

In recent years, an increasing number of individuals and organizations have recognized the fact that the world is facing a rapidly accelerating climate change emergency. Over 2,000 researchers from around the world have recently signed a letter urging fellow scientists to speak out about the climate crisis, and a growing number of scientists have also engaged in climate change activism (Fisher, 2024; Thompson, 2021; Valero, 2023). Additionally, a growing number of personalities have recognized this crisis. For example, in 2019, Pope Francis declared a global "climate emergency," and in 2020 President Joe Biden argued that climate change is "the number one issue facing humanity" (Harvey & Ambrose, 2019; Newburger, 2020). The Oxford Languages chose "climate emergency" as the word of the year in 2019 in recognition of "the rapid rise of climate emergency from relative obscurity to becoming one of the most prominent—and prominently debated—terms."[1] By early 2024, over 2,350 local governments from 40 countries have adopted a climate emergency declaration, which acknowledges that humanity is in a climate crisis.[2]

Preventing global temperatures from increasing too much is imperative, given the devastating impact this can have on modern civilization (The World Bank, 2012). Yet, despite the growing awareness about the climate change emergency, the world is not on track to meet the goals of the Paris Agreement, which aims to avoid the rise in global temperatures to more than 1.5 degrees Celsius. A recent report from the United Nations shows that, despite some progress toward global greenhouse gas emission reductions, current efforts are insufficient to limit global temperature rise to dangerous levels by the end of the century (UN Climate Change, 2023). In particular, it is crucial to accelerate not only the adoption of renewable energy technologies (solar photovoltaics, wind turbines, electric cars, etc.) but also the abandonment of fossil fuel technologies (coal and gas power plants, internal combustion engines, etc.). As the UN Secretary General argued, the world must prioritize the transformation of energy systems and speed up the shift to renewable energy "because without renewables, there can be no future."[3]

This chapter examines efforts to accelerate the transition to renewable energy by social movement actors and the resistance they encounter from fossil fuel industries. While most energy studies focus on the energy "hardware" (the technical elements of electricity generation, transmission, and distribution) rather than the human "software" behind energy-related behaviors (Sovacool, 2014), understanding the challenges of the global transition to renewable energy requires developing a sociological perspective and examining the interactions among collective actors such as environmental organizations, corporations, and governments. Building on the social movements, organizations, and the sociology of climate change literatures, this chapter argues that, to understand the current and future challenges of the global renewable energy transition, it is important to investigate the dynamic between environmental organizations and fossil fuel industries at micro, meso, and macro levels.

The chapter has three sections. The first section focuses on the environmental movement organizations' efforts to shape individuals' climate change and renewable energy attitudes and behaviors. This section also examines the factors that have contributed to the fossil fuel industries' success in preventing widespread changes in attitudes and behaviors related to energy and environment. The second section describes several environmental movement campaigns that aim to influence the energy-related decisions of various organizations, as well as countermobilizations by fossil fuel companies. The third section examines how environmental activism has influenced the adoption and implementation of policies related to low-carbon technologies, such as wind turbines, solar panels, and electric cars, and the abandonment of high-carbon technologies, such as coal power plants and internal combustion engines as a major form of climate action (Almeida et al., 2024). This section also explores several legislative setbacks due to polluting industries' political action strategies. The chapter provides examples from different geographical regions and economic sectors (electricity, transportation, commercial and residential, agriculture) and concludes by discussing implications for future research and action on climate change.

Changing Individuals' Energy Attitudes and Behaviors

The environmental movement is considered one of the most important social movements in many countries (Castells, 2009; Mertig & Dunlap, 2001). The movement has recognized the importance of the climate change issue since the late 1980s, and many environmental groups have attempted to change individual attitudes and behaviors toward climate change. Since the mid-1990s, environmental organizations have repeated and amplified messages from climate scientists about the impending climate change crisis. For example, Greenpeace has frequently published reports such as the *Climate Time Bomb: Signs of Climate Change From the Greenpeace Database*, which attempted to engage the public through images and text (Doyle, 2007).

The environmental movement worked to shape individual attitudes and behaviors toward energy by developing particular diagnostic, prognostic, and motivational frames (Benford & Snow, 2000; Snow & Benford, 1988). For example, since the mid-1990s, Greenpeace developed a diagnostic frame that attributed rising global temperatures to human activities, such as fossil fuel production and consumption, rather than to natural variation in Earth's climate. Greenpeace also created prognostic frames that articulated proposed solutions to the climate change problem: stopping new oil and gas exploration. In 1997, Greenpeace International published a report titled *Fossil Fuels and Climate Protection: The Carbon Logic*, which made the case that surpassing the Earth's atmospheric "carbon budget" can lead to irreversible climate change (Doyle, 2007).

The motivational frames used by environmental organizations represented a "call to arms" (Benford & Snow, 2000) and employed vocabularies that provided individuals with compelling accounts for engaging in climate change activism. Some reports and campaigns aimed to increase public support for renewable energy. For example, the report *New Power for Britain: A Strategy for a Renewable Energy Industry* published by Greenpeace in 1998 called for environmental movement activists and sympathizers to support the development of renewable energy industries, such as onshore and offshore wind farms. Other campaigns aimed to increase opposition to fossil fuels. Thus, the *StopEsso* campaign initiated in 2001 aimed to bring global awareness to the climate change issue and to highlight the role of oil and gas companies for the climate change crisis (Doyle, 2007). More recently, Greenpeace and other environmental groups have focused on communicating renewable energy solutions available at the local or regional level, and on promoting decentralized energy production.

The Sierra Club is another environmental organization that attempted to influence energy-related attitudes and behaviors. The Sierra Club has worked to increase public opposition against fossil fuels and support for renewable energy throughout the 1990s and early 2000s, but in 2005 it made climate change a top priority for the organization. Since then, the Sierra Club has launched several initiatives that aim to move the United States beyond fossil fuels and nuclear power toward a future made of 100% renewable energy sources. To reach this goal, the organization has employed prognostic frames that attempt to "shift the national narrative about clean energy from 'experimental, expensive, and unstable' to 'reliable, clean, and affordable'" (Sierra Club, 2015, p. 7). The organization has also launched a campaign to convince the public that fossil fuels are causing irreparable damage to the Earth's climate and, therefore, they should be kept in the ground.

Environmental nonprofit organizations have also tried to change consumer behaviors toward specific technologies, such as electric vehicles and solar panels. The Sierra Club has organized the National Drive Electric Week since 2011, a campaign that aims to increase consumer awareness of plug-in vehicles and highlight the benefits of all-electric and plug-in hybrid electric cars. The organization has also supported the adoption of solar photovoltaic technology by disseminating information about the environmental benefits of this technology and by partnering with local utilities to offer mail-in rebates for organizational members who install solar panels.[4] Similarly, Plug In America is a nonprofit organization with a mission to accelerate the transition to electric vehicles that works to both educate consumers and to advocate for electric cars. This organization has publicized research showing that electric cars powered by the grid "produce 65% less emissions than a conventional gasoline vehicle, even when accounting for emissions created by electricity production. Even better, installing solar panels on your roof drops your car's carbon pollution close to zero."[5]

As the climate change crisis has intensified, activists have founded environmental organizations dedicated primarily to organizing climate protests. The organization 350.org was formed in 2008 with the goals of mobilizing individuals against fossil fuel projects around the world, halting finance for fossil fuel extraction, and boosting investments in renewable energy projects (see chapter by Boeve, Bates, and Henn in this volume). This organization attempts to increase public opposition to fossil fuels and build public support for renewable energy through large global demonstrations such as the International Day of Climate Action, People's Climate March, and Global Climate Strike. The organization Fridays for Future was founded in 2018 by Greta Thunberg and other climate activists in order to mobilize young people to demand that governments accelerate the transition to renewable energy. And Extinction Rebellion was also founded in 2018 and aims to organize nonviolent civil disobedience actions to shift the public discourse on the climate crisis and to push governments to end the use of fossil fuels.

The growing climate change crisis has also contributed to a shift in the tactics used by environmental activists. While many early environmental movement campaigns were characterized by cooperation among environmental organizations that constituted relatively dense global and national networks focused on national and international actions (Fisher, 2004; Rohrschneider & Dalton, 2002), in recent years environmental groups have increasingly used confrontational tactics to provoke a strong emotional reaction among the general public and the political and business elites. For example, Climate Defiance has used tactics that aim to make "powerful politicians, government officials, and corporate executives uncomfortable" by labeling them "climate criminals" and giving them sarcastic awards such as "Big Oil Bestie."[6] Similarly, climate activists from the Last Generation have organized disruptive protests in which they glued their hands to the runways of local airports to draw attention to the impact of the airline industry on climate change, while activists from the organization Just Stop Oil have thrown soup at famous paintings in museums, spray-painted buildings, and disrupted traffic in major cities.[7]

Environmental groups have also attempted to influence attitudes toward the location of renewable energy projects. Unlike the adoption of decentralized, small-scale solar and wind projects, the adoption of centralized, large-scale solar and wind farms by utilities has generated some tensions with the environmental movement. Environmental groups have historically opposed large energy infrastructure projects, particularly in sensitive natural areas. Indeed, activists have organized site fights and battles around the placement of hydro-electric dams (Olson & Gareau, 2018), the siting of nuclear power plants (Aldrich, 2010), and natural gas facilities (McAdam & Boudet, 2012). They have also organized both not-in-my-backyard and not-in-anybody's-backyard campaigns against oil and gas drilling (Auyero et al., 2019; Buday, 2017; Dokshin, 2016; Jerolmack, 2021; Jerolmack & Walker, 2018; Ladd, 2018; Vasi et al., 2015) and pipelines (McAdam et al., 2010). Recently, the issue of prioritizing procedural justice in the development of industrial-scale renewable energy projects has created frictions between environmental groups that support the rapid development of renewable energy and those that argue for a slower process in order to avoid damaging Native American tribal nations' sacred cultural resources, wildlife, and communities' rights (Martinez, 2020; Mulvaney, 2013, 2017; Rignall, 2016).

While it is difficult to assess the exact impact of the environmental movement on energy and climate change attitudes, existing research suggests that the movement has increased public opposition against fossil fuels and support for clean energy. Scholars have argued that climate change and energy-related attitudes and planned behaviors can be explained by the values-beliefs-norms theory, which maintains that values such as altruism shape beliefs about climate change and that, in turn, these beliefs shape norms and decisions related to energy and climate change (Dietz, 2015; Poortinga et al., 2019; Steg, 2016). Empirical research has demonstrated that perceptions of the salience of the climate change problem are influenced by factors such as environmental values and beliefs (Corner et al., 2014; Crompton, 2011; Howell, 2013; Lee et al., 2015; Persson et al., 2015; Price et al., 2014). Moreover, studies show that there is a positive association between individuals' environmental values and their support for renewable energy policies and technologies (Fobissie, 2019; Perlaviciute & Steg, 2015; Thomas et al., 2022).

The environmental movement has contributed to a shift in individuals' attitudes toward climate change and renewable energy, but its impact has been uneven. The environmental movement's influence has been relatively weak for conservatives, given that the environmental movement's framing of energy and climate issues was counteracted by the conservative movement's use of counterframes (Benford & Snow, 2000). Rather than framing climate change as a major environmental problem and renewable energy as a solution to this problem, conservative think tanks have constructed the "non-problematicity" of climate change (McCright & Dunlap, 2003) and have played a key role for the spread of books denying anthropogenic climate change (Dunlap & Jacques, 2013; see also chapter by Barros Leal Farias in this volume). Moreover, in recent years conservative organizations such as the Manhattan Institute, the Heartland Institute, and the CATO Institute have increased

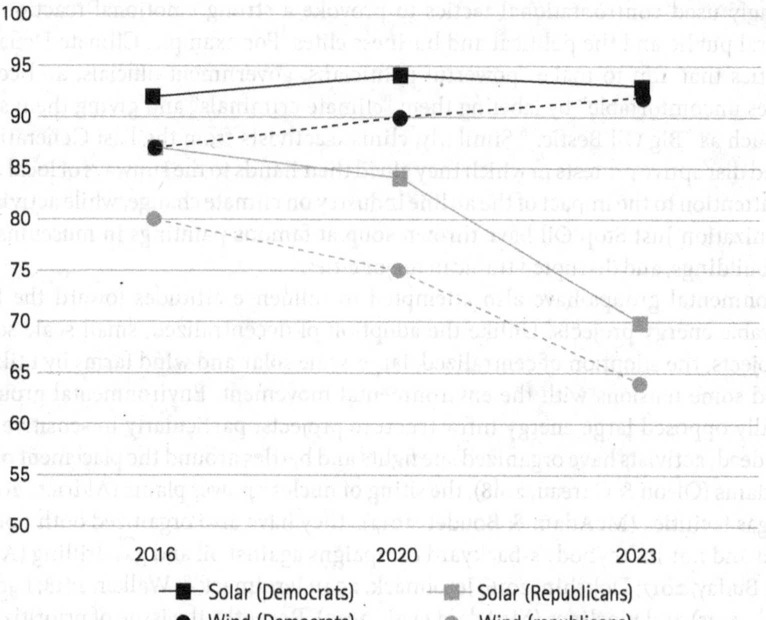

FIGURE 20.1. Growing gap between Democrats and Republicans over support for renewable energy (percent of US adults who favor more solar and wind in 2016, 2020, and 2023).

(*Source*: Pew Research Center 2023)

their attacks on renewable energy technologies and called the belief in a transition away from fossil fuels "magical thinking."[8]

Research shows that coordinated campaigns funded by fossil fuel industries have shaped the climate change attitudes of political elites and the general public (Dunlap & McCright, 2015; Farrell, 2016). Studies also demonstrate that political polarization has a negative effect on conservatives' concern about climate change and beliefs that climate change is happening (Smith et al., 2024). Political polarization also influences the adoption of energy efficiency technologies in the United States because conservatives place lower psychological value on reducing carbon emissions (Gromet et al., 2013). And opinion polls show that, as political polarization has increased in recent years, support for renewable energy sources such as solar and wind has increased slightly between 2016 and 2023 for those who identify as Democrats but has decreased for those who identify as Republicans (see Figure 20.1). Identifying communication strategies that can reduce the negative effects of political polarization on climate change and renewable energy attitudes and behaviors is an important area for future research.

Influencing Organizations' Energy Decisions

The environmental movement has also affected the transition toward renewable energy systems by pressuring organizations to engage in climate change mitigation. Organization scholars have argued that climate change is a "wicked problem" (Wijen, 2014) or a "grand challenge" (Ferraro et al., 2015), that is, a highly significant problem that affects vast numbers of individuals in profound ways. Studies have shown that organizations can successfully address grand challenges by pursuing bold, field-spanning ideas and by adopting less conventional solutions (Eisenhardt et al., 2016). Yet, while studies have argued that climate change is perhaps the most important grand challenge facing companies (George et al., 2016), research has shown that organizational attempts to address this challenge often end up in "business as usual" rather than in the development of new strategies (Wright & Nyberg, 2017). Because no central authority exists to deal with this global problem, corporations have few incentives to drastically reduce their greenhouse gas emissions, and internal climate change champions often find themselves isolated in arguing against the maximization of short-term profit that drives firms' strategy (Wright & Nyberg, 2017).

Organizations employ different strategies for dealing with the "uncomfortable knowledge" created by wicked problems or grand challenges such as climate change (Rayner, 2012). First, organizations may engage in *denial*, which is "the refusal or inability of organizations at any level to acknowledge information, even when external bodies or even individuals within the organization seek actively to bring it to the collective attention" (Rayner, 2012, p. 114). Other strategies include *dismissal*, which involves the simultaneous recognition and rejection of the wicked problem because it is seen as irrelevant or too distant; *diversion*, when organizations establish a decoy activity that distracts attention from the wicked problem; and *displacement*, when organizations substitute the actions required to address a wicked problem with more manageable surrogates.

To overcome these strategies and pressure organizations to reduce their emissions of greenhouse gases, some environmental nongovernmental organizations (NGOs) have used confrontational tactics. For example, the Sierra Club's "Beyond Coal" campaign has pressured utilities to retire their coal generators through what has been described as "the most extensive, expensive and effective campaign in the Club's 123-year history, and maybe the history of the environmental movement" (Grunwald, 2015). Recent research shows that this campaign had a significant impact on US utilities' decision to divest coal power plants, as well as on their decision to make exploratory investments in renewable energy (Schifeling et al., 2024). Similarly, research shows that environmental organizations can create a polluting technology stigma, which pressures electric utilities to develop green power programs (Vasi & King, 2019).

Other environmental nonprofits have used a collaborative approach and chosen to work together with companies toward the renewable energy transition. The World Resources Institute (WRI), an organization that works to transform the ways the world produces and consumes energy and food, has launched the Safe Climate, Sound Business campaign in 1998 and collaborated with companies such as General Motors, British Petroleum, and Monsanto to identify actions that can reduce emission of greenhouse gases such as investments in energy efficiency technologies and purchases of renewable energy. In 2000, WRI has created the Green Power Market Development Group, an initiative that was designed "to help corporations understand, explore, and evaluate renewable energy opportunities." This initiative was crucial in creating a market for commercial renewable energy; for example, between 2000 and 2005, the WRI's initiative resulted in 360 MW of renewable energy purchased by corporations.[9] The WRI played a crucial role for the growth of nonresidential market for renewable energy because it acted as a broker that connected companies with renewable energy developers and utilities, and as a negotiator that worked to obtain the best deals for companies that purchased renewable energy credits (Vasi, 2011).

The environmental movement has influenced companies' energy decisions in two ways. First, outsider activists who are members of environmental organizations have targeted companies with "naming and shaming" campaigns. Starting in 2018, a wave of climate change protests has targeted not only governments but also businesses with the goal of accelerating the transition to renewable energy systems. For example, the Fridays for Future protests initiated by Greta Thunberg have grown into a global protest movement that mobilized over 7 million people from over 180 countries in one week in 2019 (de Moor et al., 2021; Svensson & Wahlström, 2023; Wahlström et al., 2021).

Second, insider activists who are employees of various companies but identify or sympathize with the environmental movement have worked to change their companies from inside (Briscoe & Gupta, 2016). The rising tide of climate protests by outsider activists had a spillover effect on insider activism by employees of large companies such as Google, Facebook, Microsoft, and Amazon. For example, in 2019 more than 1,000 Google employees have formed a group called Google Workers for Action on Climate and signed a public letter calling on the company to commit to a company-wide climate plan that includes canceling contracts with the fossil fuel industry and halting its donations to climate change deniers.[10] Similarly, Amazon employees have formed a group called Amazon Employees for Climate Justice, which called on the company to reduce its carbon footprint, stop selling its cloud services to the oil and gas industry, and stop donating to politicians who deny anthropogenic climate change.[11]

Among the most significant impacts of climate change activism are the emergence of companies that are taking a leadership role in climate change mitigation and of networks of organizations that collaborate on decarbonization. Amazon, for example, has been criticized for generating pollution from its large warehousing operations (Emmons & Reese 2023), but it has also been recognized for becoming a leader on climate change mitigation in recent years. In 2019, the company has announced a partnership with the company Rivian to bring 100,000 electric delivery vehicles on the road by 2030 and announced a commitment to achieve net-zero carbon by 2040. Amazon also became a cofounder of The Climate Pledge, a coalition of businesses that work to "lead the charge on climate change" and work together to reach net-zero carbon emissions 10 years ahead of the Paris Accord's goals.[12] Companies that join the Climate Pledge commit to measure and report greenhouse gas emissions on a regular basis, to implement decarbonization strategies such as adopting renewable energy and increasing energy efficiency, and to neutralize remaining emissions with "quantifiable, real, permanent, and socially beneficial offsets."[13] Given current concerns about "carbonwashing" (i.e., carbon-related greenwashing) due to the absence of mandatory reporting frameworks (In & Schumacher, 2021; Trouwloon et al., 2023), it remains to be seen if the offsets used by these companies will have an effect on climate change mitigation.

The precise effect of the environmental movement's confrontational and collaborative tactics on organizations' energy-related decisions is difficult to assess. Nevertheless, several studies show a robust association between the presence of environmental organizations and the adoption of renewable energy technologies by organizations. Research shows that electric utilities are more likely to adopt solar photovoltaic technology if they are located in areas with a high density of environmental organizations (Vasi & Grant, 2023). Similarly, studies show that the installation of charging stations for electric cars in the United States is associated with the presence of local environmental organizations, but not of car manufacturing companies (Dutta et al., 2018; Vasi & Sauder, 2024). Studies also suggest that environmental activism has also contributed to many universities' decision to divest from fossil fuels and to invest in renewable energy (Barron et al., 2023).

In addition to influencing universities and electric utilities' decision to invest in renewable energy, the environmental movement has succeeded in increasing the prominence of sustainability and the climate change issue for many corporations. By targeting large companies such as Apple with sustained campaigns (e.g., "Green My Apple"), Greenpeace and other environmental groups forced corporations to increase their commitment to sustainability and pushed climate change mitigation to the top of the agenda for many companies. A 2024 report on sustainable practices within various industries around the globe has found that 65% of top managers consider that sustainability is a top-three priority, and 25% consider that it is number-one priority. Moreover, over 80% of the surveyed companies have a chief sustainability officer, who provides expert guidance and clear strategic direction on sustainability initiatives, such as purchasing renewable energy.[14]

The environmental movement also influenced companies' decision to invest in renewable energy by collaborating with the "environmental state," which is made of governmental organizations and institutions that work to provide "biophysical and social conditions required to support life and livelihoods into the imagined future" (Rea & Frickel, 2023, p. 256). The governmental organizations that comprise the environmental state are responsible not only for national but also subnational environmental policies (SNEPs) that have major consequences for managing climate change, addressing environmental injustices, and much more (Vasi & Walker, 2024). Environmental organizations such as the World Resources Institute have collaborated with governmental agencies such as the Environmental Protection Agency (EPA) to create the Green Power Leadership Awards, which have been awarded annually since 2001 by EPA's Green Power Partnership program.

The increasing pressure from insider and outsider activism has contributed to many corporations' decision to reduce greenhouse gas emissions and to participate in the Green Power Partnership program. This program has provided recognition for organizations that already work to accelerate the transition to a pollution-free electricity sector and has encouraged other organizations to start their decarbonization journey. The program also had a major impact on the solar, wind, and other renewable energy industries. By early 2024, the combined annual green power use of EPA's Top Fortune 500 Partners amounts to the annual electricity use of nearly 6.5 million average American homes.[15] As Figure 20.2 shows, 17 corporations purchased over 1 million MWh of renewable energy in 2023. Moreover, some companies have recently announced their intention to accelerate their decarbonization efforts. For example, Microsoft announced in 2024 that it will support the development of more than 10 gigawatts of new renewable energy capacity around the world by 2030.[16]

The growing markets for carbon removal create both a new challenge and an opportunity for the environmental movement. In early 2024, more than 3,100 companies approved a "net zero" target to reduce or completely offset their carbon emissions by 2050.[17] Because it is difficult for many companies to monitor their progress, environmental organizations such as the World Resources Institute (WRI) and World Wildlife Fund (WWF) have joined other business-focused climate nonprofit organizations (e.g., the Carbon Disclosure Project, We Mean Business Coalition) to form a new organization named Science Based Targets. This organization develops standards and tools that help companies set greenhouse gas emissions reductions targets in line with the goals of the Paris Accord. Since net-zero goals require not only reducing scope 1 and 2 emissions (i.e., emissions from sources controlled or owned by the companies, such as buildings and vehicles, and from the electricity they purchase) but also scope 3 emissions (i.e., emissions that are produced by companies' suppliers or products), many companies are using carbon offsetting. Because carbon offsetting, or the principle of paying other companies to reduce greenhouse gas emissions generated by a company, poses the risk of carbonwashing, environmental groups play an important role in monitoring and certifying the use of "high-quality" carbon offsets.

Although some energy companies responded to environmental activists' demands by adopting net-zero goals and purchasing renewable energy, others have adopted a more belligerent strategy. Many oil and gas companies involved in drilling in the Marcellus Shale area in the United States have supported industry associations that used diversionary reframing to attack the credibility of antifracking protesters (Walker & Vasi, 2024). Some companies

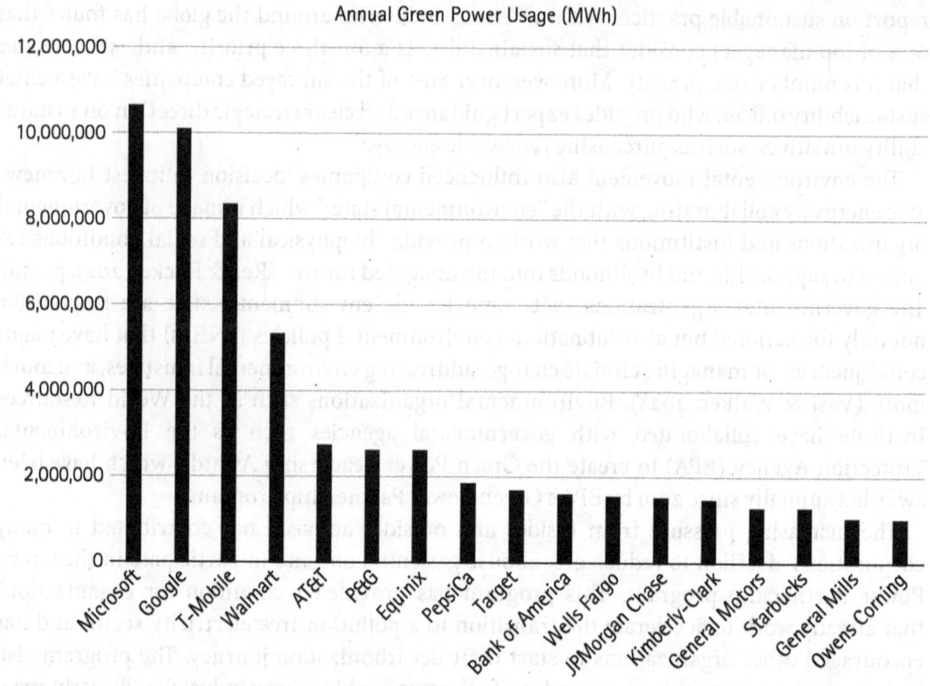

FIGURE 20.2. US companies that were purchasing over 1 million MWh of renewable energy in 2023.

(*Source*: EPA Green Power Partnership)

are also suing climate change activists in an attempt to intimidate them. For example, in 2023 Shell has launched a massive lawsuit against Greenpeace because it occupied an offshore oil drilling platform, prompting more than 100,000 members of the public to sign a letter that accused Shell of being "callous and vindictive" and calling on the company to "respect people's right to protest against climate change."[18]

Arguably the most contentious issue for environmental activists and corporations in recent years is that of future fossil fuel production. Echoing scientific reports about the need to make fossil fuels unextractable to prevent global warming from increasing over 1.5 degrees Celsius (Welsby et al., 2021), many environmental groups (e.g., 350.org, Friends of the Earth, Greenpeace) have called on energy companies to "keep fossil fuels in the ground."[19] In the United States, several environmental groups (e.g., Union of Concerned Scientists, Sierra Club) have organized campaigns for a phaseout of fossil fuels or to move "beyond coal," "beyond oil," and "beyond natural gas."[20] Similarly, in the European Union over 60 environmental nonprofits have created a coalition under the name Beyond Fossil Fuels, which aims to create a electricity sector that is predominantly powered by wind and solar energy and is free of fossil fuels by 2035.[21] But, given that a halt in fossil fuel exploration and expansion is perceived as an existential threat by fossil fuel companies, many of them are pushing back against environmental organizations' claims that the world needs to transition away from fossil fuels.

The way in which fossil fuel companies have responded to the environmental movement's pressure to divest from fossil fuels depends on several factors. First, it depends on the internal corporate opportunity structure, which includes changes in the executive team, recent mergers and acquisitions, and characteristics of the CEOs such as their ideology (Briscoe et al., 2014). For example, oil and gas firms may be more likely to invest in renewable energy projects if their CEOs and top managers have a more liberal political ideology, which correlates with the general political ideology in which companies are located. Hence, oil and gas companies from the European Union (e.g., BP, Shell, Total, Equinor) are more likely to invest in offshore wind farms, solar farms, and electric car charging stations because their top managers are more liberal. In contrast, fossil fuel companies located in more conservative regions are more likely to have conservative CEOs (e.g., Exxon, Saudi Aramco) and less likely to invest in renewable energy. Moreover, these companies often criticize environmentalists and proponents of energy transition, accusing them of willful ignorance. Indeed, Saudi Aramco's CEO stated in 2024 that the world should "abandon the fantasy of phasing out oil and gas" and that the energy transition is "visibly failing."[22]

Second, their response depends on their familiarity with and use of specific corporate political action strategies (Hillman & Hitt, 1999; Hillman et al., 1999). A common corporate political action strategy used by fossil fuel companies is to attack environmental activists indirectly, by supporting think tanks and public relations operations that oppose federal climate action (Brulle, 2014; Farrell, 2016). These efforts often take forms in which firms and industry groups engage in countermessaging (Walker & Vasi, 2022, 2024) and even mobilize constituencies directly (Mix & Waldo, 2015), following the model of "grassroots for hire" (Walker, 2014). Many companies support conservative think tanks such as the Competitive Enterprise Institute that have denigrated and harassed climate change scientists such as Michael Mann.[23] Some fossil fuel trade associations have also attacked plans to develop an infrastructure for electric vehicles and other alternative technologies. For example, the American Fuel and Petrochemical Manufacturers, a trade group for gasoline makers, has been waging a multimillion-dollar battle to suppress plans to build charging stations for electric cars across the country by creating a "Freedom to Drive" coalition (Bade, 2019).

Many fossil fuel companies are also involved in industry associations such as the American Petroleum Institute (API), an organization that funded organizations and individuals who downplayed the climate change crisis and was described by lobbyists as the industry's "whipping boy" who helps direct public and political criticism away from individual companies.[24] The American Petroleum Institute played a major role in defeating the 2009 cap-and-trade bill that attempted to limit greenhouse gas emissions and give polluting energy companies "allowances" to buy and sell the right to emit carbon dioxide by labeling the climate bill "cap-and-tax" and by mobilizing the conservative politicians against it.[25] Yet some oil and gas companies have recently stopped participating in industry associations because of their changing position on climate change and renewable energy. For example, in 2021, Total announced that it withdrew from API because this industry association does not support significant climate change mitigation initiatives, and that it changed its name to TotalEnergies because this name symbolizes its strategic transformation into a broad energy company.[26]

Another political actions strategy that is increasingly used by fossil fuel companies is the attack on institutional investors who want to shift investments from fossil fuels to renewable energy. Environmental activists have pressured firms for over two decades to disclose environmental risks through primary stakeholder (or shareholder) activism (Flammer, 2015; Flammer & Bansal, 2017; Vasi & King, 2012). Climate-related shareholder proposals, in particular, have increased in recent years, and research shows that these proposals increase the voluntary disclosure of climate change risks, especially when institutional investors introduced them (Flammer et al., 2021). In response, many oil and gas companies have lobbied conservative politicians who have "relentlessly pounded big banks and investment managers for being 'too woke' or 'hostile' to fossil fuel."[27] As a result, BlackRock, JP Morgan, and other asset manager companies have simultaneously reduced their investments in renewable energy and increased their investments in fossil fuel projects.[28] Future research is necessary to understand how environmental organizations and their allies can counteract fossil fuel companies' political action strategies, as well as their use of diversion and displacement tactics related to climate change (e.g., promoting carbon capture and storage technology instead of investing in solar and wind technologies).

Shaping Governmental Agendas and Energy Policies

National and subnational governments have been frequent targets of environmental activism. Climate change campaigns have attempted to move the climate change issue to the top of the agenda for governments around the world by using climate shaming tactics (Koliev et al., 2022). In some cases, climate change protesters were able to make governments recognize the urgency of the climate change problem. For example, Extinction Rebellion (XR), considered the largest grassroots environmental movement in the United Kingdom, has organized frequent nonviolent civil disobedience actions since 2018, which successfully pressured the government to declare climate change an emergency in 2021 (Berglund & Schmidt, 2020).

Environmental movement organizations have frequently targeted governments during climate change negotiations. Throughout the 1990s and early 2000s, many environmental organizations pressured governments to take stronger actions during the annual meeting of the UN Conference of the Parties (COP). Over 180 environmental NGOs participated in the Kyoto Protocol negotiations, an international treaty that aimed to commit countries to reduce greenhouse gas emissions and was adopted in 1997 and entered into force in 2005 (Betsill & Correll, 2008). The Climate Action Network (CAN), the world's largest network of environmental organizations with over 1,800 member organizations in more than 130 countries, has attempted to counteract the influence of the fossil fuel lobby at international climate change negotiations and to ensure that governments are held accountable to the Paris Agreement. CAN has also worked with municipal and regional governments to help them lower their greenhouse gas emissions by enabling peer support, disseminating information about effective practices, providing funding assistance and technical consultation, and creating partnerships (Betsill & Bulkeley, 2004; Davidson et al., 2019; Karhinen et al., 2021; Zapata Arango et al., 2024).

In the United States, CAN has also influenced the adoption of the Inflation Reduction Act (IRA), the most important national energy policy adopted in the United States. The adoption of IRA was supported by all environmental organizations that made fighting climate change and accelerating the transition to renewable energy systems a priority. After decades of failures to introduce climate change legislation, environmental groups and their allies have successfully introduced the IRA because they "worked on various provisions for months and months, so when the window of opportunity opened, the legislative language was ready to go and the hard negotiations had already happened."[29] Yet CAN organizations recognize that this historic victory is only the beginning, since an even greater challenge is making sure that the IRA is implemented "in a manner that is equally equitable and aggressive."[30] Unfortunately, President Donald J. Trump has issued an executive order to "freeze" spending on unspent IRA funds at the beginning of 2025, an action that was described by the Sierra Club as a slam of "the door on the lifeline that has created hundreds of thousands of jobs, is upgrading the country's infrastructure, and brought about the rebirth of American manufacturing."[31]

While CAN had succeeded in influencing the adoption of the IRA in the United States, it had modest success in shaping several energy policies. Environmental organizations that are part of CAN have focused on two types of energy policies at the national and subnational levels: fossil fuel subsidies and renewable energy incentives. Environmental groups have fought for decades to eliminate the national subsidies received by the fossil fuel industry, which can range from government revenues through tax exemptions and preferences, to transfer of financial liability to the public, to below-market provision of government goods or services. These subsidies had a major impact on the oil and gas industry's growth and profitability. According to one study that examined the potential effect of 16 subsidies and regulatory exemptions on the economics of US oil and gas between 2020 and 2030, these subsidies could increase the average rates of return of yet-to-be-developed oil and gas fields by at least 55% for oil and 68% for gas over unsubsidized levels, and over 96% of subsidy value is likely to flow to excess profits (Achakulwisut et al., 2021). Yet, given the extreme resources of the oil and gas lobby, environmental groups' efforts have been in vain: as one article has put it, "In Washington, it seems, oil and gas subsidies are the zombies of the tax code: impossible to kill" (Friedman, 2024).

Environmental organizations have also promoted renewable energy incentives at both the national and subnational levels (Vasi & Walker, 2024). When the Production Tax Credit (PTC), a policy that was developed by the American Wind Energy Association as a way to attract capital for wind farms and other renewable energy projects, was introduced in the early 1990s, virtually every major environmental group in the country supported this policy (Vasi, 2011). And when this policy was set to expire at the end of 1990s, a coalition of environmental organizations and renewable energy associations achieved an extension of the credit. Yet this policy expired several times since then because environmental groups could not overcome the opposition from the fossil fuel industries. These industries opposed the PTC policy because the tax credits were supposed to be paid by reducing some of the subsidies that the oil and gas industry were receiving.

Many environmental groups also supported the adoption of renewable portfolio standards (RPS) that required utilities to produce a certain amount of electricity from renewable energy. The adoption of RPS policies has started during the 1990s but accelerated in the early 2000s (see Figure 20.3). By 2023 a total of 36 states have adopted RPS policies, but the amount of renewable energy required by each policy varies greatly. Environmental organizations have not only advocated for the adoption of RPS policies but also defended their implementation. For example, when the Beacon Hill Institute (a think tank that has published many reports that aimed to discredit renewable energy standards) sought to repeal or weaken RPS policies in several states, the National Resources Defense Council (NRDC) and other organizations published research showing that the enactment of RPS is beneficial not only for the global environment but also for the local economies.[32] Moreover, due to the advocacy work of environmental organizations and their allies, since 2015 a total of 25 states have adopted net-zero policies that require 100% clean energy goals by 2050 or earlier (see Figure 20.3).

While the pro-renewable energy coalition composed of environmental organizations, renewable energy industry groups, progressive civic society organization, and others succeeded in protecting renewable energy policies from the attacks of the fossil fuel industries in some states, they failed in others. Several US states have experienced renewable energy policy retrenchment and declining investments in renewables in recent years (Stokes, 2020). For example, RPS policies experienced retrenchment in Texas and Kansas because fossil fuel companies used their influence over politicians and regulators not only to block reforms from passing in the legislature but also to prevent their implementation by organizing astroturf campaigns. Similarly, the Net-Metering policy was weakened in Arizona because of

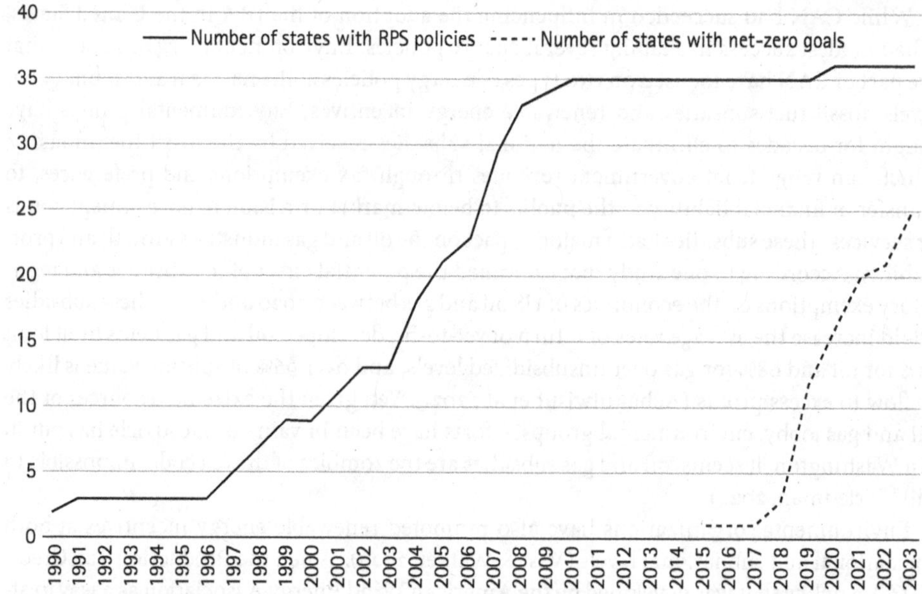

FIGURE 20.3. Number of US states with renewable portfolio standards (RPS) and net-zero policies between 1990 and 2023.

(*Sources:* NCLS; CESA)

regulatory capture, a process in which the regulator (Arizona Corporation Commission) was controlled by the entity it was supposed to regulate (Arizona Public Service). Fossil fuel interest groups used their vast resources to contribute to the election of Republican politicians with ties to the American Legislative Exchange Council (ALEC), an organization that promoted an anti-renewables agenda, for the Arizona Corporation Commission (Stokes, 2020). Moreover, the pro-renewable energy coalition's efforts to introduce a national RPS in 2007 were thwarted due to opposition from legislators with links to the fossil fuel industry.

In other countries, however, the environmental movement had more success in influencing climate and energy policies. In Denmark, for example, environmental activists have successfully mobilized in support of renewable energy since the 1970s. When the Danish government responded to the 1973 oil crisis by planning to build nuclear power plants, environmental activists founded the Organization for Information about Nuclear Power (OOA) and the Organization for Renewable Energy (OVE), two grassroots groups that organized not only protests against nuclear power but also campaigns that promoted renewable energy (Vasi, 2011; van Est, 2022). These organizations published an innovative Alternative Energy Plan in 1976 that promoted energy conservation and envisioned building wind farms across the country to take advantage of abundant wind resources. Together with other organizations (e.g., the Association of Danish Wind Turbine Owners, the Association of Danish Wind Turbine Manufacturers), OOA and OVE were instrumental for the adoption of a feed-in tariff policy that set the fair price for wind power at 85% of the retail electricity rate and provided guaranteed interconnection and power purchase of wind-based electricity (IRENA-GWEC, 2013).

Similarly, in Germany environmental groups such as Friends of the Earth Germany (BUND) and Greenpeace have successfully worked with renewable energy associations such as Eurosolar and the Wind Power Association to shape German environmental and energy policies. During the 1980s and 1990s, membership in the German environmental movement increased and the movement became more influential in shaping not only public debates about acid rain and climate change but also feed-in policies such as the Act on the Supply of Electricity from Renewable Energy Sources into the Grid from 1990, and the Act on Granting Priority to Renewable Energy from 2000 (Hake et al., 2015). Indeed, due to the influence of the environmental movement, the 2000 feed-in policy explicitly mentioned climate change and climate protection as an important reason for governmental support for renewable energy. This policy, in turn, is responsible for making Germany the largest market for solar and wind power in Europe at the beginning of the 21st century. Systematic comparisons of the social, political, and economic conditions that influence the environmental movement's impact on energy policymaking in different countries are essential to understand the heterogeneity in the rate of energy transition in different countries and regions.

Conclusion

This chapter shows that the environmental movement has created significant changes at micro, meso, and macro levels in recent decades. I briefly summarize the main arguments and highlight several promising directions for future research. First, by organizing large protests around the world, the movement has influenced environmental attitudes and raised awareness about the growing climate change crisis and the need to accelerate the transition to renewable energy systems (Fobissie, 2019; Perlaviciute & Steg, 2015; Thomas et al., 2022). The movement has also influenced behaviors related to renewable energy technologies, such as solar panels and electric cars (Vasi & Grant, 2023; Vasi & Sauder, 2024). However, the movement had less success in influencing climate change attitudes and in increasing support for solar, wind, and other types of renewable energy technologies among conservatives, particularly in the United States. This is because fossil fuel industries have sponsored misinformation campaigns against climate change science for decades, and they ramped up attacks on the energy transition in recent years (Dunlap & Jacques, 2013; Smith et al., 2024). Given the need to accelerate the transition to renewable energy systems, it is important for future research to examine communication strategies that can be used by environmental organizations and their allies to overcome skepticism about climate change science and low-carbon technologies, such as solar photovoltaics, electric cars, and heat pumps.

Second, by protesting against polluting corporations and by collaborating with organizations that were open to change, environmental groups were able to persuade many organizations that they need to address the climate change "grand challenge" (Wright & Nyberg, 2017). Environmental activism outside and inside organizations has contributed to many corporations' decision to invest in renewable energy technologies, such as solar and wind, and to sign the Climate Pledge. At the same time, while environmental activism has contributed to several organizations' decision to divest from fossil fuels (Barron et al., 2023), it had little success in decreasing major asset management firms' investment in fossil fuel projects. Moreover, the impact of campaigns such as Beyond Fossil Fuels varies greatly across geographic regions. For example, while some oil and gas companies from the European Union have invested over 10% of their total upstream investments in low-emissions sources, oil and gas companies from the United States and other countries have invested less than 5% in low-emissions sources, and these investments include controversial technologies, such as carbon capture and underground storage (International Energy Agency, 2023). Future research is necessary to investigate how environmental organizations can draw attention to the use of diversion and displacement tactics by some corporations (e.g., carbon capture and storage). Further work is also important to identify the conditions when environmental organizations can successfully collaborate with corporations and governments to accelerate the development of renewable energy markets.

Finally, environmental organizations have created a global network that constantly pressures governments to move the climate change issue to the top of their agenda. National governments have responded to the "naming and shaming" tactics used by environmental NGOs by committing to reduce greenhouse gas emissions through international negotiations (Betsill & Correll, 2008; Koliev et al., 2022). Regional and local governments have also frequently worked with environmental organizations to find ways to lower their greenhouse gas emissions (Betsill & Bulkeley, 2004; Davidson et al., 2019; Karhinen et al., 2021; Zapata Arango et al., 2024). Yet it is increasingly clear for environmental organizations that climate change mitigation requires not just the adoption of multiple energy policies but also the effective implementation of those policies. As a growing number of renewable energy projects are being delayed due to local opposition, excessive bureaucracy, and interest group politics, future research projects are necessary to identify the conditions when the environmental movement can contribute to the implementation of policies designed to accelerate the energy transition.

Notes

1. See https://languages.oup.com/word-of-the-year/2019
2. See https://www.cedamia.org/global/
3. See https://www.un.org/en/climatechange/raising-ambition/renewable-energy-transition
4. See https://www.sierraclub.org/ohio/solar-home-program. See also https://www.sierraclub.org/toiyabe/go-solar
5. See https://pluginamerica.org/learn/why-go-plugin/
6. See https://www.rollingstone.com/politics/politics-features/climate-defiance-activism-uncomfortable-1235004241/
7. See https://www.euronews.com/green/2023/10/12/german-climate-protesters-asked-to-pay-120000-in-damages-to-eurowings-after-airport-blocka. See also https://www.npr.org/2022/11/01/1133041550/the-activist-who-threw-soup-on-a-van-gogh-explains-why-they-did-it
8. See, for example, https://www.city-journal.org/article/the-magical-thinking-behind-the-energy-transition
9. See https://www.wri.org/research/green-power-market-development-group
10. See https://www.theguardian.com/technology/2019/nov/04/google-workers-climate-plan-letter
11. See https://www.vox.com/recode/2019/9/20/20874497/amazon-climate-change-walkout-google-microsoft-strike-tech-activism
12. See https://www.theclimatepledge.com/us/en
13. See https://www.theclimatepledge.com/us/en
14. See https://www.forbes.com/sites/forbes-research/2024/04/22/sustainability-climbs-the-corporate-agenda-forbes-releases-the-2024-state-of-sustainability-report
15. See https://www.epa.gov/greenpower/green-power-partnership-fortune-500r-partners-list
16. See https://www.cnbc.com/2024/05/01/microsoft-brookfield-to-develop-more-than-10point5-gigawatts-of-renewable-energy.html
17. See https://sciencebasedtargets.org/companies-taking-action
18. See https://www.theguardian.com/business/2024/apr/01/public-figures-urge-shell-to-drop-lawsuit-against-greenpeace
19. See https://keepitintheground.org/#read-the-letter
20. See https://www.ucsusa.org/ucs-fossil-fuel-phaseout. See also https://coal.sierraclub.org/
21. See https://beyondfossilfuels.org/about-us/
22. See https://www.nytimes.com/2024/03/19/climate/ceraweek-saudi-fantasy-energy-transition.html

23. For a description of the harassment campaign, see https://www.ucsusa.org/resources/how-fossil-fuel-industry-harassed-climate-scientist-michael-mann. See also https://www.nature.com/articles/d41586-024-00396-y
24. See https://www.theguardian.com/environment/2021/jul/19/big-oil-climate-crisis-lobby-group-api
25. See https://www.npr.org/2021/10/27/1047583610/once-again-the-u-s-has-failed-to-take-sweeping-climate-action-heres-why
26. See https://totalenergies.com/media/news/press-releases/total-withdraws-from-the-american-petroleum-institute
27. See https://www.ft.com/content/a76c7feb-7fa5-43d6-8e20-b4e496799167
28. See https://www.ft.com/content/9306c8f2-530d-45ca-a830-4d26e5a90509
29. See https://vitalsigns.edf.org/story/inflation-reduction-act-victory-climate-heres-what-comes-next
30. See https://chesapeakeclimate.org/the-inflation-reduction-act-today-benefits-for-justice-and-climate-action/
31. See: https://www.sierraclub.org/press-releases/2025/01/sierra-club-statement-trump-executive-order-freeze-unspent-ira-and-iija
32. See https://www.nrdc.org/resources/state-renewable-portfolio-standards-create-jobs-and-promote-clean-energy

References

Achakulwisut, P., Erickson, P., & Koplow, D. (2021). Effect of subsidies and regulatory exemptions on 2020–2030 oil and gas production and profits in the United States. *Environmental Research Letters, 16*, 084023. https://doi.org/10.1088/1748-9326/ac0a10

Aldrich, D. P. (2010). *Site fights: Divisive facilities and civil society in Japan and the West.* Cornell University Press.

Almeida, P., Gonzalez Marquez, L. R., & Fonsah, E. (2024). The forms of climate action. *Sociology Compass, 18*(2), e13177.

Auyero, J., Hernandez, M., & Stitt, M. E. (2019). Grassroots activism in the belly of the beast: A relational account of the campaign against urban fracking in Texas. *Social Problems, 66*, 28–50.

Bade, Gavin. 2019. The oil industry vs. the electric car. *Politico.* https://www.politico.com/story/2019/09/16/oil-industry-electric-car-1729429

Barron, A. R., Venator, R. C., Carlson, E. V. H., Andrews, J. K., Ding, J., & DeSwert, D. (2023). Fossil fuel divestment in U.S. higher education: Endowment dependence and temporal dynamics. *Elementa: Science of the Anthropocene, 11*(1), 00059. https://doi.org/10.1525/elementa.2023.00059

Benford, R., & Snow, D. (2000). Framing processes and social movements: An overview and assessment. *Annual Review of Sociology, 26*, 611–639.

Berglund O., & Schmidt D. 2020. *Extinction Rebellion and Climate Change Activism Breaking the Law to Change the World.* Palgrave Macmillan Cham. https://doi.org/10.1007/978-3-030-48359-3

Betsill, M., & Bulkeley, H. (2004). Transnational networks and global environmental governance: The cities for climate protection program. *International Studies Quarterly, 48*, 471–493.

Betsill, M., & Correll, E. (2008). *NGO diplomacy: The influence of nongovernmental organizations in international environmental negotiations.* The MIT Press.

Briscoe, F., &, Gupta, A. (2016). Social activism in and around organizations. *Academy of Management Annals, 10.* https://doi.org/10.5465/19416520.2016.1153261

Briscoe, F., Chin, M. K., & Hambrick, D. (2014). CEO ideology as an element of the corporate opportunity structure for social activists. *Academy of Management Journal, 57*, 1786–1809.

Brulle, R. J. (2014). Institutionalizing delay: Foundation funding and the creation of U.S. climate change counter-movement organizations. *Climatic Change, 122*(4), 681–694. https://doi.org/10.1007/s10584-013-1018-7

Buday, A. (2017). The home rule advantage: Motives and outcomes of local anti-fracking mobilization. *Social Currents, 4*, 575–593.

Castells, M. (2009). *The power of identity* (2nd ed.). Wiley-Blackwell.

Corner, A., Markowitz, E., & Pidgeon, N. (2014). Public engagement with climate change: The role of human values. *Wiley Interdisciplinary Reviews: Climate Change*, *5*(3), 411–422. https://doi.org/10.1002/wcc.269

Crompton, T. (2011). Values matter. *Nature Climate Change*, *1*(6), 276–277. http://dx.doi.org/10.1038/nclimate1196

Davidson, K., Coenen, L., Acuto, M., & Gleeson, B. (2019). Reconfiguring urban governance in an age of rising city networks: A research agenda. *Urban Studies*, *56*(16), 3540–3555. https://doi.org/10.1177/0042098018816010

de Moor J., De Vydt M., Uba K., & Wahlström M. (2021). New kids on the block: Taking stock of the recent cycle of climate activism. *Social Movement Studies*, *20*(5), 619–625. https://doi.org/10.1080/14742837.2020.1836617

Dietz, T. (2015). Environmental values. In Tobias Brosch T., and D. Sander (Eds.), *Handbook of Value: Perspectives from Economics, Neuroscience, Philosophy, Psychology and Sociology* (pp. 329–350). Oxford University Press.

Dokshin, F. A. (2016). Whose backyard and what's at issue? Spatial and ideological dynamics of local opposition to fracking in New York state, 2010 to 2013. *American Sociological Review*, *81*, 921–948.

Doyle, J. (2007). Picturing the clima(c)tic: Greenpeace and the representational politics of climate change communication. *Science as Culture*, *16*, 129–150.

Dunlap, R., & McCright, A. (2015). Challenging climate change: The denial countermovement. In Dunlap R., & R. Brulle (Eds.), *Climate change and society: Sociological perspectives* (pp. 300–332). New York: Oxford University Press.

Dunlap, R. E., & Jacques, P. J. (2013). Climate change denial books and conservative think tanks: Exploring the connection. *American Behavioral Scientist*, *57*(6), 699–731.

Dutta, S., Rao, H., & Vasi, I. B. (2018). Plug power. Social movements and electric vehicle charging stations in California, 1995–2012. *Research in the Sociology of Organizations*, *56*, 21–52.

Eisenhardt, K. M., Graebner, M. E., & Sonenshein, S. (2016). Grand challenges and inductive methods: Rigor without rigor mortis. *Academy of Management Journal*, *59*, 1113–1123.

Emmons, A. J., & Reese, E. (2023). *Unsustainable: Amazon, warehousing, and the politics of exploitation*. University of California Press.

Farrell, J. (2016). Corporate funding and ideological polarization about climate change. *Proceedings of the National Academy of Sciences*, *113*, 92–97. https://doi.org/10.1073/pnas.1509433112

Ferraro, F., Etzion, D., & Gehman, J. (2015). Tackling grand challenges pragmatically: Robust action revisited. *Organization Studies*, *36*, 363–390.

Fisher, D. (2004). *National governance and the global climate change regime*. Rowman & Littlefield.

Fisher, D. (2024). *Saving ourselves: From climate shocks to climate action*. Columbia University Press.

Flammer, C. (2015). Does corporate social responsibility lead to superior financial performance? A regression discontinuity approach. *Management Science*, *61*, 2549–2568.

Flammer, C., & Bansal, P. (2017). Does long-term orientation create value? Evidence from a regression discontinuity. *Strategic Management Journal*, *38*, 1827–1847.

Flammer, C., Toffel, M. W., & Viswanathan, K. (2021). Shareholder activism and firms' voluntary disclosure of climate change risk. *Strategic Management Journal*, *42*:1850–1879. DOI: 10.1002/smj.33131850.

Fobissie, E. (2019). The role of environmental values and political ideology on public support for renewable energy policy in Ottawa, Canada. *Energy Policy*, *134*, 110918. https://doi.org/10.1016/j.enpol.2019.110918

Friedman, L. 2024. The Zombies of the U.S. Tax Code: Why Fossil Fuels Subsidies Seem Impossible to Kill. *New York Times*. https://www.nytimes.com/2024/03/15/climate/tax-breaks-oil-gas-us.html

George, G., Howard-Grenville, J., Joshi, A., & Tihanyi, L. (2016). Understanding and tackling societal grand challenges through management research. *Academy of Management Journal*, *59*, 1880–1895.

Gromet, D. M., Kunreuther, H., & Larrick, R. P. (2013). Political ideology affects energy-efficiency attitudes and choices. *PNAS*, *110*(23), 9314–9319. https://doi.org/10.1073/pnas.1218453110

Grunwald, M. (2015). Inside the war on coal: How Mike Bloomberg, red-state businesses, a lot of Midwestern lawyers are changing American energy faster than you think. *Politico*. https://www.politico.com/agenda/story/2015/05/inside-war-on-coal-000002/
Hake, J-F., Fischer, W., Venghaus, S., & Weckenbrock, C. (2015). The German Energiewende: History and status quo. *Energy, 92*, 532–546.
Harvey, F., & Ambrose, J. (2019). Pope Francis declares "climate emergency" and urges action. *The Guardian*, June 14, 2019. https://www.theguardian.com/environment/2019/jun/14/pope-francis-declares-climate-emergency-and-urges-action
Hillman, A., & Hitt, M. (1999). Corporate political strategy formulation: A model of approach, participation, and strategy decisions. *Academy of Management Review, 24*, 825–842.
Hillman, A. J., Zardkoohi, A., & Bierman, L. (1999). Corporate political strategies and firm performance: Indications of firm-specific benefits from personal service in the U.S. government *Strategic Management Journal, 20*, 67–81.
Howell, R. (2013). It's not (just) "the environment, stupid!" Values, motivations, and routes to engagement of people adopting lower-carbon lifestyles. *Global Environmental Change, 23*, 281–290.
In, S. Y., & Schumacher, K. (2021). Carbonwashing: A new type of carbon data-related ESG greenwashing. Working paper. Stanford Sustainable Finance Institute. Precourt Institute for Energy. https://papers.ssrn.com/sol3/papers.cfm?abstract_id=3901278
International Energy Agency. 2023. World Energy Investment 2023. Overview and key findings. https://www.iea.org/reports/world-energy-investment-2023/overview-and-key-findings
IRENA-GWEC. (2013). 30 years of policies for wind energy. Lessons from 12 wind energy markets. https://www.irena.org/publications/2013/Jan/30-Years-of-Policies-for-Wind-Energy-Lessons-from-12-Wind-Energy-Markets
Jerolmack, C. (2021). *Up to heaven and down to hell: Fracking, freedom, and community in an American town*. Princeton University Press.
Jerolmack, C., & Walker, E. T. (2018). Please in my backyard: Quiet mobilization in support of fracking in an Appalachian community. *American Journal of Sociology, 124*, 479–516.
Karhinen, S., Peltomaa, J., Riekkinen, V., & Saikku, L. (2021). Impact of a climate network: The role of intermediaries in local level climate action. *Global Environmental Change, 67*, 102225. https://doi.org/10.1016/j.gloenvcha.2021.102225
Koliev, F., Park, B., & Duit, A. (2022). Climate shaming: Explaining environmental NGOs targeting practices. *Climate Policy, 23*(7), 845–858. https://doi.org/10.1080/14693062.2022.2143315
Ladd, A. E. (2018). *Fractured communities: Risk, impacts, and protest against hydraulic fracking in US shale regions*. Rutgers University Press.
Lee, T. M., Markowitz, E., Howe, P., Ko, C., & Leiserowitz, A. (2015). Predictors of public climate change awareness and risk perception around the world. *Nature Climate Change*, 5, 1014–1020. http://dx.doi.org/10.1038/nclimate2728
Martinez, N. (2020). Resisting renewables: The energy epistemics of social opposition in Mexico. *Energy Research and Social Science, 70*, 101632. DOI: 10.1016/j.erss.2020.101632
Mertig, A. G., & Dunlap, R. E. (2001). Environmentalism: Preservation and conservation. In N. J. Smelser & P. B. Baltes (Eds.), *International encyclopedia of the social & behavioral sciences* (pp. 4687–4693). Pergamon.
Mix, T. L., & Waldo, K. G. (2015). Know(ing) your power: Risk society, astroturf campaigns, and the battle over the red rock coal-fired plant. *The Sociological Quarterly, 56*, 125–151. DOI: 10.1111/tsq.12065
McAdam, D., & Boudet, H. (2012). *Putting social movements in their place: Explaining opposition to energy projects in the United States, 2000–2005*. Cambridge University Press.
McAdam, D., Boudet, H., Davis, J., Orr, R. J., Scott W. R., & Levitt, R. E. (2010). "Site fights": Explaining opposition to pipeline projects in the developing world." *Sociological Forum*, 25, 401–427.
McCright, A., & Dunlap, R. (2003). Defeating Kyoto: The conservative movement's impact on U.S. climate change policy. *Social Problems, 50*, 348–373.

Mulvaney, D. (2013). Opening the black box of solar energy technologies: Exploring tensions between innovation and environmental justice. *Science as Culture*, *22*(2), 230–237.
Mulvaney, D. (2017). Identifying the roots of green civil war over utility-scale solar energy projects on public lands across the American Southwest. *Journal of Land Use Science*, *12*, 493–515.
Newburger, E. (2020). Joe Biden: Climate change is "number one issue facing humanity." *CNBC*, October 24, 2020. https://www.cnbc.com/2020/10/24/joe-biden-climate-change-is-number-one-issue-facing-humanity.html
Olson, K. A., & Gareau, B. J. (2018). Hydro/power? Politics, discourse and neoliberalization in Laos's hydroelectric development. *Sociology of Development*, *4*, 94–118.
Perlaviciute, G., & Steg, L. (2015). The influence of values on evaluations of energy alternatives, *Renewable Energy*, *77*, 259–267.
Persson, J., Sahlin, N., & Wallin, A. (2015). Climate change, values, and the cultural cognition thesis. *Environmental Science & Policy*, *52*, 1–5.
Poortinga, W., Whitmarsh, L., Steg, L., Böhm, G., & Fisher, S. (2019). Climate change perceptions and their individual level determinants: A cross-European analysis. *Global Environmental Change*, *55*, 25–35.
Price, J., Walker, I., & Boschetti, F. (2014). Measuring cultural values and beliefs about environment to identify their role in climate change responses. *Journal of Environmental Psychology*, *37*, 8–20.
Rayner, S. (2012). Uncomfortable knowledge: The social construction of ignorance in science and environmental policy discourses. *Economy and Society*, *41*, 107–125.
Rea, C. M., & Frickel, S. (2023). The environmental state: Nature and the politics of environmental protection. *Sociological Theory*, *41*, 255–281.
Rignall, K. E. (2016). Solar power, state power, and the politics of energy transition in pre-Saharan Morocco. *Environment and Planning A*, *48*, 540–557.
Rohrschneider, R., & Dalton, R. (2002). A global network? Transnational cooperation among environmental groups. *Journal of Politics*, *64*, 510–533.
Schifeling, T., Lyon, T., & Vasi, I. B. (2024). The enfranchised stakeholder dilemma: Stakeholder management in the context on industry disruption. *Working paper.*
Sierra Club. (2015). Sierra Club strategic plan. Summary version. https://www.sierraclub.org/sites/www.sierraclub.org/files/BOD-Strategic%20Plan-Summary.pdf
Smith, E. K., Bognar, M. J., & Mayer, A. P. (2024). Polarisation of climate and environmental attitudes in the United States, 1973–2022. *NPJ Climate Action*, *3*, 2. https://doi.org/10.1038/s44168-023-00074-1
Snow, D. A., & Benford, R. D. (1988). Ideology, frame resonance, and participant mobilization. *International Social Movement Research*, *1*, 197–218.
Sovacool, B. K. (2014). Energy studies need social science. *Nature*, *511*, 529–530.
Steg, L. (2016). Values, norms, and intrinsic motivation to act proenvironmentally. *Annual Review of Environment and Resources*, *41*, 277–292.
Stokes, L. (2020). *Short circuiting policy. Interest Groups and the battle over clean energy and climate policy in the American states.* Oxford University Press.
Svensson, A., & Wahlström, M. (2023). Climate change or what? Prognostic framing by Fridays for Future protesters. *Social Movement Studies*, *22*, 1–22.
Thomas, M., DeCillia, B., Santos, J. B., & Thorlakson, L. (2022). Great expectations: Public opinion about energy transition. *Energy Policy*, *162*, 112777. https://doi.org/10.1016/j.enpol.2022.112777.
Thompson, T. (2021). Scientist rebellion: Researchers join protesters at COP26. *Nature News*. https://www.nature.com/articles/d41586-021-03430-5
Trouwloon, D., Streck, C., Chagas, T., & Martinus, G. (2023). Understanding the use of carbon credits by companies: A review of the defining elements of corporate climate claims. *Global Challenges*, *7*, 2200158. https://doi.org/10.1002/gch2.202200158
UN Climate Change. (2023). New analysis of national climate plans: Insufficient progress made, COP28 must set stage for immediate action. *UN Climate Press Release*. https://unfccc.int/news/new-analysis-of-national-climate-plans-insufficient-progress-made-cop28-must-set-stage-for-immediate

Valero, M. V. (2023). Outcry as scientists sanctioned for climate protest. *Nature News*. https://www.nature.com/articles/d41586-023-00440-3

van Est, R. (2022). The success of Danish wind energy innovation policy. In Caroline de la Porte et al. (Eds.), *Successful public policy in the Nordic countries* (pp. 45–65). Oxford University Press.

Vasi, I. B. (2011). *Winds of change. The environmental movement and the global development of the wind energy industry*. Oxford University Press.

Vasi, I. B., & Grant, D. (2023). Blocking the sun? Social movements, industries, and the adoption of solar energy in the United States. *Energy Research and Social Science*. doi.org/10.1016/j.erss.2023.103272

Vasi, I. B., & King, B. (2012). Social movements, risk perceptions, and economic outcomes: The effect of primary and secondary stakeholder activism on firms' perceived environmental risk and financial performance. *American Sociological Review*, *77*, 573–596.

Vasi, I. B., & King, B. (2019). How technological stigma affects the adoption and growth of green power programs. *Socio-Economic Review*, *17*, 37–61.

Vasi, I. B., & Sauder, M. (2024). Beyond petroleum? The diffusion of alternative fuels in the United States, 1994–2015. *Sociological Quarterly*, *65*, 427–447.

Vasi, I. B., & Walker, E. T. (2024). Subnational environmental policy: Trends and issues. *Annual Review of Sociology*, *50*, 319–339.

Vasi, I. B., Walker, E. T., Johnson, J. S., & Tan, H. F. (2015). "No fracking way!" Documentary film, discursive opportunity, and local opposition against hydraulic fracturing in the United States, 2010 to 2013. *American Sociological Review*, *80*, 934–959.

Wahlström M., de Moor J., Uba K., Wennerhag M., De Vydt M., Almeida P., Baukloh A., Bertuzzi N., Chironi D., Churchill B., Cisar O., Collin P., Nobel T., Daniel A., Davies S., della Porta D., Deutschman A., Ellefsen R., ... et al. (2021). Surveys of participants in Fridays For Future climate protests on 20–28 September, 2019, in 19 cities around the world. Open Society Foundations, August 18. DOI: 10.17605/OSF.IO/ASRUW

Walker, E. T. (2014). *Grassroots for hire: Public affairs consultants in American democracy*. Cambridge University Press.

Walker, E. T., & Vasi, I. B. (2022). Industry repertoires: How transgressive and conventional industry associations seek to counter contention. *American Behavioral Scientist*, *66*, 1238–1262. DOI: 10.1177/00027642211056277

Walker, E. T., & Vasi, I. B. (2024). Boosting your enemies to garner the sympathy of friends: Evidence from communication strategies by a contested industry. *Social Problems*, XX, 1–21. https://doi.org/10.1093/socpro/spae069

Welsby, D., Price, J., Pye, S., & Ekins, P. (2021). Unextractable fossil fuels in a 1.5°C world. *Nature*, *597*, 230–234.

Wijen, F. (2014). Means vs. ends in opaque institutional fields: Trading off compliance and achievement in sustainability standard adoption. *Academy of Management Review*, *39*, 302–323.

The World Bank. (2012). New report examines risks of 4 degree hotter world by end of century. www.worldbank.org/en/news/press-release/2012/11/18/new-report-examines-risks-of-degree-hotter-world-by-end-of-century

Wright, C., & Nyberg, D. (2017). An inconvenient truth: How organizations translate climate change into business as usual. *Academy of Management Journal*, *60*, 1633–1661.

Zapata, A. M., Hoppe, T., Itten, A., & Blok, K. (2024). The role of city climate networks in promoting citizen participation in municipalities: A critical multi-case analysis. *Energy, Sustainability, and Society*, *14*(5). https://doi.org/10.1186/s13705-023-00438-9

CHAPTER 21

Refugees, Displaced Persons, or Economic Migrants? Climate-Related Migration and Implications for Migration Classificatory Systems

NICK TINOCO AND CECILIA MENJÍVAR

Introduction

ENVIRONMENTAL migration scholarship increasingly recognizes the complexity of the relationship between climate change and human mobility. Despite frequent invocations of impending mass flows of environmental migrants within media and policy circles in the Global North, scholars routinely push back on alarmist "climate refugee" narratives, noting that the relationship between climate change and human migration is multidimensional and highly context-specific (Hunter et al., 2015; McLeman & Gemenne, 2018). Recent scholarship stresses that climate-related movement most often occurs within countries, rather than across borders, and is not always intended to be permanent (Kaczan & Orgill-Meyer, 2020). As a result, such trends most often lead those who move to be classified as internally displaced persons (IDPs). Internal displacement is more likely to be driven by sudden-onset climatic shocks such as flooding or wildfire, which might damage housing or livelihoods in the short term but are often met with recovery policies oriented toward reconstructing damaged homes and infrastructure. In contrast, cross-border climate-related moves are more closely associated with slow-onset environmental changes, such as drought or soil erosion, which gradually erode agricultural livelihoods and increase pressure for permanent out-migration. This pattern of movement has been found to be stronger in regions with a history of colonization (Omobowale et al., 2019) and well-established international migration histories and support networks (Riosmena et al., 2018).

While movements enacted in response to climate change can take many forms, frontline communities might also lack the means or desire to move, and instead remain immobile in vulnerable places with varying degrees of volition (Black et al., 2011; Zickgraf, 2018). Climate shocks can also transform long-term migration aspirations—for example, by suppressing return migration, prolonging or rendering permanent what would otherwise have been temporary labor migrations (Entwisle et al., 2020). Furthermore, the interplay between climate and conflict has been found to intensify out-migration, blurring the boundaries between economically motivated labor migration enacted in response to climate change and the forced migration of refugees seeking protection abroad (Abel et al., 2019). The migration process itself can shape the stages, duration, and direction of movement (including immobility) such that migration can take a U-shape especially in the context of long-term migration (Yoo & Agadjanian, 2024) or a stepwise trajectory (Paul, 2011), instead of a unidirectional flow that connects identifiable climatic shocks with an endpoint at the destination.

Recent global governance initiatives have acknowledged the importance of climate displacement and advocated for its centrality to the agenda of international climate policy. However, more ambitious efforts to modify the current refugee regime to secure avenues for legal protection to those displaced by climate change have largely fallen short of their aims (Van Der Vliet & Biermann, 2022). Initiatives like the 2018 UN Global Compact on Refugees emphasize the need for migrant-sending countries to tackle the root causes of migration—an objective that contradicts previous recognitions of global climate action needing to account for the "differentiated responsibility" between states based on their historic emissions, as outlined in the UN Framework Convention on Climate Change (Dickinson & Burton, 2021, p. 41). This inaction stems from the misalignment between the reality of climate-related migrations and the established classificatory systems that regulate entry. Thus, both environmental migrants and their compatriots fighting to remain in places with eroding ecological stability will be increasingly forced to draw upon collective resources and actions in the absence of more robust support and protection from state and global governance institutions.

In recent work, scholars of environmental migration are adopting a broader climate mobilities framework oriented toward understanding "the multiplicity of climate change-related human mobility (involving immobility, relocation, circular mobility, etc.), its embedding in ongoing patterns and histories of movement, and the material and political conditions under which it takes place" (Boas et al., 2022, p. 3366). Different forms of mobility are relational—taken together, (im)mobilities refers to the "intertwined coproduction of mobility for some and immobility for others, as well as the ways in which disempowered groups also experience forced displacements, evictions, deportations, and expulsions" (Sheller, 2020, p. 18). The "constellations of mobility" that arise in response to climate change are highly political; past and present power relations and sociopolitical structures shape the forms, motivations, experiences, and representations of human movement (Cresswell, 2010). Beyond solely examining how climate change influences patterns of migration, a climate mobilities research agenda interrogates a wider array of social, political, economic, and environmental factors that interact to form the landscape of mobility aspirations, processes, and outcomes as climate change heightens social vulnerability (Cundill et al., 2021).

Taken together, climate-related movements take many forms. Yet the categories by which we make sense of these migrations—those that we use to understand motivations for migration and those that are used to classify people at entry—touch and overlap in response to a changing climate (Menjívar, 2023). Research has shown that, rather than understanding drivers of migration through the voluntary/involuntary binary, a continuum seems more fitting (Abdelaaty & Hamlin, 2022; Arar & FitzGerald, 2022; Bartram, 2015; Bivand Erdal & Oeppen, 2018). Cases of environmental migration underscore even more forcefully the difficulty in determining when migration is voluntary and when it is forced; rather, at multiple stages in the migration process, the distinction between forced and voluntary movement becomes blurred and difficult to dichotomize into established categories. This ambiguity is also reflected in how migrants themselves self-identify, choosing to align with certain labels at different stages in their journeys depending on how those categories might help them navigate particular social, political, and legal contexts (Hack-Polay et al., 2021).

Following other scholars, we recognize that the binary classifications of voluntary versus involuntary often simplify the complexity of individuals' motivations and reduce drivers of migration to "political" or "economic" (Menjívar, 1993). Despite these recognized conceptual misalignments, taken-for-granted binaries are deployed in research to understand climate-related migration even when these migratory flows expose in sharp relief the shortcomings of binary concepts (see Menjívar, 2023). With this caveat in mind, given how the extant empirical work tends to follow convention, we organize this chapter accordingly and proceed as follows. The first and second sections generally organize the scholarship on climate migration following a voluntary versus involuntary structure, while the third and fourth sections depart from this classification by highlighting responses to pressures for mobility; the third section looks at what people do to resist climate-related displacement, while the last section covers how various actors in receiving societies mobilize in response. The four sections together underscore the complex links among climate change, mobility, immobility, mobilizations, and state responses through a dynamic lens. In the conclusion, we discuss potential alternatives to binary classifications that align more closely with complexities on the ground and the impetus behind mobilizations.

Voluntarily Enacted Climate Mobilities

Like most migrations, climate-related moves are often collectively enacted—both in the decision-making stage as perceptions of climate risks are shaped through social interaction (Norgaard, 2011) and in the transit stage when migrant social networks facilitate movement (Massey & España, 1987). When a climatic shock is a principal motivation for migration, scholars have often framed this decision as a means for a household to maintain a sustainable livelihood under increasing social and environmental stress (Van Praag & Timmerman, 2019). Instances of "migration for adaptation" have been identified in a variety of cases—migrant remittances were found to be critical assets for disaster recovery in Bangladesh and cultivating preparedness in Ghana and Burkina Faso (Afifi et al., 2016; Mohapatra et al., 2012). In their examination of Mexico–US migration, Riosmena, Nawrotzki, and Hunter (2018) identified a positive association between climate and migration only within the most vulnerable municipalities which were best connected with US labor markets. Their study found that rainfall variability alone was unlikely to drive mass migration from Mexico unless the particular locality in question already has an established migration culture with network ties to US cities and labor markets that can help facilitate movement and support migrants upon arrival (Riosmena et al., 2018). Moreover, new mobility patterns can evolve that enable residents to better manage their environmental vulnerabilities. For instance, Carte and colleagues (2019) found Nicaraguan farmers increasingly engaging in short-term, cyclical migration to El Salvador. Frequent short trips enabled farmers to earn supplemental income while still being able to maintain their long-standing small-scale agricultural practices back home amid both drought and economic and political instability.

Critics of migration for/as adaptation argue that such frameworks risk overlooking the broader political and economic factors that shape migration flows. For example, local adaptation strategies employed in the coffee-growing regions along the Guatemala-Mexico border tend to be overly focused on reviving and managing damaged crops, which does little to reduce out-migration pressures so long as the price of coffee in the international market remains insufficient to raise workers out of poverty (Ruiz-de-Oña et al., 2019). Likewise Wu and Bryan (2021) note that remittances sent home from Filipino migrants in Canada were critical assets to some households recovering from Typhoon Haiyan. Nevertheless, the gains in adaptive capacity remain precarious so long as they entail the continued participation of family members in unpredictable and exploitative international labor markets. Some scholars express apprehension toward characterizing environmental migration as simply labor migration or migration-for-adaptation, particularly as these understandings overlook the political and economic processes that expose some to mobility-producing risks more than others and risk individualizing environmental protection through household preparedness initiatives, rather than conceiving it as a universal right (Bettini et al., 2017).

Beyond decision making, migration itself can be collectively enacted. In the case of Central American migrant "caravans," collective mobility forces those outside to "bear witness to" the myriad challenges that encourage out-migration (e.g., environmental degradation and the erosion of rural livelihoods, in addition to violence and poverty) and foster solidarity geared toward cultivating resilience in the face of collectively experienced traumas (Wurtz, 2020). From this standpoint, these humanitarian crises can be conceptualized as political acts (Menjívar, 2022). In such instances of collective escape—from both climatic/nonclimatic shocks and a political-economic context ill-equipped to deal with those shocks—the organization of the migration process in and through "transnational social movements on the move" provides security for displaced populations encountering increasingly militarized mobility barriers (Rizzo Lara, 2021). Here we see the overlap between "voluntary" climate-related and "involuntarily" security-related push factors. If climate migration is likely to produce internal rural-urban moves, but the cities to which migrants arrive have few economic opportunities and high levels of violence—as is the case for many such migrants in the Central American Dry Corridor—then subsequent cross-border moves would prove far more complex and multicausal than the category of labor migration suggests (Huber et al., 2023).

While disaster-induced displacement can often be short in both distance and duration, when survivors are able to remain in place and rebuild, repeated exposure to disasters can erode or transform the attachments that bind people to vulnerable places and foster aspirations to relocate. For example, following Hurricane Sandy, Koslov (2016) found several neighborhoods in Staten Island to be hotbeds of organizing activity where residents pressured local officials to buyout and "undevelop" flooded properties, with the intention of restoring the natural wetland environment to better insulate surrounding neighborhoods from future floods. In these communities, a long history of flooding, and local community organizing for flood protection, facilitated the formation of several neighborhood buyout groups. According to Koslov, this case offers a vision of "another kind of movement in the context of climate change, one that aligns less with top-down interventions that displace people and abandon places in the name of progress than with a whole range of grassroots efforts to democratize and transform space and place" (Koslov, 2016, p. 381). However, evidence also suggests that people who participate in state-funded buyout programs tend to relocate locally and often remain exposed to similar hazard risks. Loughran and Elliott (2019) examined a dataset of over 3,000 buyouts in Houston, Texas, over the last two decades and argue that "what appears to be driving environmental mobility is less the environmental push of imminent risk than the social pull of 'upgrading' the socioeconomic status of one's neighborhood and by extension, its racial composition" (Loughran & Elliott, 2019, p. 67). Homeowners were found to mostly practice environmental mobility when such moves enabled them to maintain or increase the socioeconomic status of their neighborhood, even if those neighborhoods were situated on land with publicly recognized flood risk.

It is important to note that currently, buyout programs are relatively rare and often prove highly controversial. Numerous cases show residents resisting what are often interpreted as top-down plans, often developed with little community participation, which would break up long-standing neighborhoods. Buyout organizing committees like those in Staten Island were limited to several beachfront communities impacted by Hurricane Sandy. Surveying a wider array of Sandy-affected residents, Bukvic and Owen (2017) find that the majority of their respondents (63%) had an unfavorable view of relocation and prefer to explore other adaptation options before considering retreat (Bukvic & Owen, 2017). Moreover, the ability of resident-led organizing efforts to succeed in shaping resettlement policies are shaped by preexisting inequalities. Before the Isle de Jean Charles in coastal Louisiana became the first community to receive federal funding to finance resettlement due to sea-level rise in the United States; residents of the island's largely Indigenous population had already been considering relocation as a strategy to preserve their community for more than a decade. While residents' advocacy helped to secure some amendments to resettlement policy after government funds arrived, such as allowing residents to retain ownership of their properties after moving in order to maintain connections to the land (Simms et al., 2021), much of the transformative demands of community leaders advanced were sidelined. Jessee (2022) argues that what resulted was a "decontextualizing" of resettlement, reducing the framing of the project from one of Indigenous survival and self-determination to solely a project of environmental risk management (Jessee, 2022).

Involuntarily Enacted Climate Mobilities

The preponderance of "climate refugee" narratives in popular framings of climate-related movement reflects a general assumption that these moves are forced. While moves enacted in response to slow-onset environmental change, such as drought or soil erosion, might permit more agency for the migrant, forced climate-related displacement is more often associated with extreme weather events or disasters. These considerations make it difficult to categorize these flows as voluntary or involuntary. Furthermore, disasters do not operate in a social vacuum—scholars of disaster emphasize the importance of preexisting social inequalities in shaping the distribution and severity of disaster impacts (Tierney, 2019). Many note that disasters, rather than just a catalyst for new movement, disrupt the everyday mobilities of a given place which have evolved in accordance with local mobility regimes—particularly in a rural, agrarian context. For this reason, Kelley and colleagues (2022) propose a more appropriate framework to understand this pattern of movement which they term "cumulative socionatural displacements," to better understand patterns of climate-related movements in the Southeast Asian agrarian communities they study. Specifically, the authors note the importance of an array of seemingly nonclimatic factors such as "histories of dispossession via state territorialization, long-term deterioration of smallholder agro-ecologies, recent intensifications in export-oriented commodity production, [and] recent increases in labor migration," acting in concert with increasing climate vulnerability to shape patterns of mobility (Kelley et al., 2022, p. 669). Whether in an urban or rural context, a more complete understanding of the local social, political, and economic structures through which a disaster is refracted is essential to crafting effective recovery policies and providing aid and support to survivors.

One of the most studied cases of disaster-induced displacement is the aftermath of Hurricane Katrina. Decades of housing and urban development policies, in addition to endemic racism and economic inequality, laid the fault lines of social vulnerability that resulted in the displacement of hundreds of thousands from New Orleans alone (Horowitz, 2020). Elizabeth Fussell (2012) finds that the ability of social and community networks to withstand the disruptions wrought by Katrina were critical in securing and exchanging resources to facilitate evacuation and recovery. However, when displacement severed connections that might otherwise have provided social support, residents were compelled to seek out non-network support that was often insufficient or unavailable to meet their immediate needs, protracting experiences of displacement (Fussell, 2012). While strong network connections can facilitate return after displacement, these linkages have been found to be particularly critical in designing and implementing adaptation measures that enable community residents to remain in place. Such connections facilitate engagement with local and state bureaucracies to access funding, resources, and expertise which can be mobilized to strengthen local adaptive capacity to try and construct stopgaps to prevent future displacement (Adger, 2003; Aldrich, 2019).

Importantly, not all disasters generate displacement as approaches to emergency management can sometimes deliberately or inadvertently restrict the mobility of survivors. Although not a climate event, Sheller described Haiti in the aftermath of the 2010 earthquake as a "carceral space" where "victims were unable to move away from the disaster zone but were instead trapped in IDP camps, as these places of physical and spiritual incarceration are euphemistically called ... managed by foreign logistical teams of a wide range of extrastate agencies, from nongovernmental organizations and missionaries to Hollywood stars and foreign military forces" (Sheller, 2020, p. 19). Under the auspices of humanitarian assistance, foreign states and international aid organizations effectively instituted a restrictive mobility regime that limited survivors' mobility within Haiti and ability to leave the country. This case represents broader shifts in the landscape of humanitarian policy in the late 20th century that increasingly limit mobility—borders have become both more porous for aid and more restrictive to human mobility, with greater effort directed at aiding those in and around their home communities rather than encouraging resettlement elsewhere (Hyndman, 2000). In practice, the proliferation of refugee camps holds displaced people in a liminal position, serving the goals of receiving states of maintaining distance while at the same time creating the perception of accordance with international pressure to maintain humanitarian norms of charitability (FitzGerald, 2019). As a result, camps function as a mechanism barring entry and contribute to holding displaced populations in a state of prolonged temporariness which inhibits their social and economic participation in a host society and prevents any form of integration to take place (Arar & FitzGerald, 2022). In this context, individuals' abilities to shape their futures are constrained as they are restrained to a state of "agency in waiting," where the capacity to act in the present is critically tied to feelings of hope and resilience (Brun, 2015, p. 24).

In addition to disasters, both slow- and sudden-onset environmental shocks can intensify preexisting refugee flows when they occur in the context of ongoing conflict. Scholars have argued that forced migration during the Syrian civil war between 2011 and 2015 resulted from a confluence of political and environmental drivers—the immediate threat of violent conflict, combined with the added stress on livelihoods brought about by severe drought, produced forced migration at a greater magnitude. In the Syrian context, adaptation initiatives can also be conceptualized as a means for reducing the potential for or severity of conflict, and by extension, out-migration (Abel et al., 2019). Some studies point toward a relationship between temperature variation and asylum applications in the European Union (Missirian & Schlenker, 2017). Others note the role that political violence plays as a mediating factor in this relationship—it is not that climatic shocks alone heighten asylum applications, but rather the timing when those shocks strike regions embroiled in conflict with a limited state capacity, refugee flows are expected to intensify (Schutte et al., 2021). Upon arrival, host country policies of reception can leave refugees and asylum seekers exposed to new risks of environmental harm. For instance, after the Taliban returned to power in 2021, an estimated 1 million Afghans sought refuge in Iran. At the same time, Iran was reckoning with severe droughts which already were contributing to heightened internal migration among Iranian citizens. The mobility of Afghan refugees was limited by the state, and most were restricted to areas with high climate vulnerability and experiencing issues of water scarcity (Rostami & Paski, 2024).

Often, what is understood to be forced climate migration is not always as unpredictable or intractable as dominant images of disaster- or conflict-induced displacements suggest. In some instances, states are beginning to take proactive actions to resettle populations away from areas deemed too high-risk or no longer viable for human settlement through managed retreat policies. In practice, these policies can often resemble land grabs that call to mind development-induced displacement. While often framed as voluntary programs by states and planners, the degree of volition of those who relocate is difficult to determine in practice, particularly in Global South cases (Arnall, 2019). Ajibade (2019) examines case studies of managed retreat in Lagos and Manila and notes that these policies disproportionately affect the urban poor. Moreover, she characterized these policy efforts as more akin to urban renewal as some of these locales underwent redevelopment after former residents relocated (Ajibade, 2019). Thus, state efforts to manage population flows from vulnerable areas must recognize the social and historical processes that led often marginalized groups to inhabit such spaces (Marter-Kenyon, 2020). Without this care and attention, managed retreat policies can easily prove maladaptive when relocated communities lack the adequate support needed to prevent exacerbating preexisting social and economic inequalities (Ajibade & Siders, 2021).

Experiences and feelings of displacement can also negatively impact the well-being of those who remain in place, as the meanings, histories, and memories wrapped up in the construction of "home" can be irreversibly altered by cascading environmental harms, even in the absence of movement (Koslov et al., 2021; Tinoco, 2023). Recognizing the limitations of mobility-centric framings of climate change–induced displacement, some scholars proposed an alternative definition of displacement as "a phenomenon that sits beyond physical movement and mobility that instead is a condition that may occur without movement through space and across internal domains and international borders" (Askland et al., 2022, p. 268). The authors argue that by focusing predominantly on the material, dominant conceptualizations risk overlooking the symbolic and experiential dimensions of displacement that emerge through the rupture of one's subjective experience of feeling at home (Ramsay & Askland, 2022). Scholars adopting this perspective argue that displacement fundamentally entails a "loss of a space where individuals deploy their everyday practices and where they expect to experience the sense of being socially, economically, politically, and emotionally and existentially embedded" (Perez Murcia, 2021, p. 1366).

Staying in Place as Collective Action

Just as an array of economic, political, social, or environmental forces have been found to increase the likelihood of migration, so, too, can numerous cultural, social, or personal factors suppress movement; either mobility is constrained, or populations are exposed to escalating environmental hazards and desire to remain in place. Recent research has incorporated immobility into frameworks for understanding the environmental dimensions of migration. Nevertheless, this work has primarily analyzed involuntary immobility in an environmental context through the study of "trapped" populations—those who aspire to leave for their protection but are unable to do so because they lack the means to leave (DeWaard et al., 2022). This understanding of trapped figures has received criticism for being overly simplistic in its adoption of the language of victimhood, which takes for granted that immobility is either forced or voluntary, whereas in practice, these boundaries are often blurred as individuals still exercise agency in making mobility decisions, albeit in contexts where they face escalating constraints. Moreover, as scholars note, victimhood framings that characterize trapped populations as an ambiguous "humanitarian other" create the potential for white saviorism by serving as a justification for humanitarian intervention to protect and relocate trapped individuals (who in the literature are almost universally from the Global South) (Ayeb-Karlsson et al., 2018).

Climate migration discourses often take for granted that, when faced with displacement-inducing environmental disruptions that threaten human security and livelihoods, people will want to and try to relocate, ignoring that people might be just as likely to lack either the desire or the ability to move (Zickgraf, 2018). Rather, the gradual effect of climatic factors on migration decision-making processes over time leads prospective migrants to continually renegotiate the positive and negative dimensions of their place of residence alongside their capabilities to move. Moreover, the evolutionary nature of climate pressures may redirect migrants to other destinations, stay where they are, or return. A recent review of studies around environmental nonmigration in Africa finds place attachments, along with the social ties and obligations to people embedded in meaningful places, to be the most significant driver of aspirations to remain in place, effectively counteracting the economic and political pressures that might be expected by neoclassical migration theories to induce movement (Balgah & Kimengsi, 2022). In a Latin American context, Adams's study of voluntary immobility in high migrant-sending communities of the Peruvian highlands similarly concludes that strong senses of place attachment and feelings of obligation to family motivate decisions to stay in communities where livelihoods are increasingly threatened by environmental change (Adams, 2016).

Yet those who remain in places experiencing environmental decline are nevertheless forced to collectively reckon with indelible changes to their communities, which transform the nature of their relationships with the natural and social worlds they inhabit and might elicit feelings of disconnect and disorientation. Scholars increasingly stress the evolving nature of staying—it is an active and collective process that is continually being negotiated as affected residents adjust their way of life to manage and adapt to environmental changes to strengthen local capabilities to remain in place (Pemberton et al., 2021). The stress of uncertainty as populations remain in place can negatively impact residents' mental health and well-being (Koslov et al., 2021). Those with the means to do so might invest significant sums into strengthening their capabilities and confidence to stay alongside escalating risks in order to return to a predisaster sense of security—such as middle- and upper-class California communities that invest in developing volunteer fire brigades to protect homes from more frequent and severe wildfires (Tinoco, 2023).

Even in the face of existential, slow-onset changes like sea-level rise, populations assumed to become permanently displaced and forced to relocate engage in antidisplacement politics, resisting planned relocation efforts and adopting strategies of "re-emplacement." Re-emplacement projects aim to remake culturally significant sites that are often deemed past the point of saving by scientific and policy authorities. Such efforts often entail political contestations between local communities and state authorities encouraging relocation. Farbotko (2022) finds these antidisplacement mobilities increasingly common in the small island nation of Tuvalu, where there has been an increase in urban-rural movement as residents look to reconnect with and restore ancestral lands deemed unsalvageable by local authorities. Similar antidisplacement movements can be found along the rural Bangladeshi coast, where peasant-led social movements have organized to protect traditional rural livelihoods threatened by the development of shrimp aquaculture as a proposed climate-resilient industry, and push back against dominant political imaginaries of climate futures that take for granted the inevitability of rural-urban migration (Paprocki, 2020). Yet one challenge to movements resisting forced relocation is the need to contend with international aid organizations and adaptation funders that adopt a migration-as-adaptation framework. For example, officials and residents in the Republic of the Marshall Islands contest visions of the islands' future with adaptation funders who routinely "discount the cultural, political, and human rights implications of failing to adapt the islands to future climate conditions, assuming instead that migration is inevitable, economically rational, and even desirable," which ultimately undermines Marshallese sovereignty and the long-term survival of Marshallese identity (Bordner et al., 2020, p. 9).

How Receiving Societies Respond to Climate-Related Mobilities

Though climate change is a global problem, strategies aimed at addressing its impacts are likely to be articulated in national rather than global terms, as the geographic variation in social vulnerability leads climate change to manifest in different places with vastly different forms and degrees of severity. Indeed, as Boas and colleagues (2023) argue, "the concept of climate migration in itself is a process of bordering, emerging as a new category of migration that feeds the debates as to who are deserving migrants or which populations and landscapes worth saving in a changing climate" (Boas et al., 2023, p. 10).

Alarmist rhetoric of destabilizing mass climate-induced migrations—such as those routinely challenged by scholars—populate many security documents from Global North defense ministries, suggesting that the erosion of livelihoods in the Global South only matters insofar as it drives migration to wealthy countries (Vogler, 2023). Such securitized framings of climate change can prove an effective discourse for laundering social nativism. Romantic narratives stressing the aesthetic beauty of the natural world and the meaning it affords to the history and identities of its settlers can be invoked to distinguish between the citizen as the steward of the land from outsiders, usually immigrants, who not only lack the ability to appreciate nature but threaten to spoil it by the very nature of their presence alone (Hultgren, 2014). Much of this rhetoric is "often characterized by the use of aggregate numbers of future migrants and by hydrological metaphors of submersion such as torrents, streams, tides, and waves of migrants," and it tends to adopt a neo-Malthusian perspective that warns of the dangers of overpopulation, famine, resource scarcity, and natural disasters in straining the capacity of the state to operate (Piguet, 2013, p. 154). These narratives can be found in the Sierra Club's long-standing and previous stance on immigration, which prompted years of internal conflict within the organization as the social characteristics of those involved in environmental organizations do not often align politically with conservative, immigration restriction groups (King, 2008).

Most commonly, narratives espousing a green-tinged nationalism are invoked to restrict immigration and expand border security. Scholars have noted the increasing frequency of adoption of such narratives by far-right parties around the world, particularly in Europe, marking a departure from the previous right-wing environmental narratives grounded in outright climate denialism (Malm & Zetkin Collective, 2021; Vihma et al., 2021). Turner and Bailey (2022) use the term "eco-bordering" to refer to these discourses that respond to rising environmental concerns "by engendering fears about the supposedly active threat of immigration to previously 'pure' and 'sustainable' spaces of national nature, and thus presents borders as forms of environmental protection" (Turner & Bailey, 2022, p. 126). The authors identify two core strains of eco-bordering—one defining migration as a form of environmental plunder and the second defining migrants as environmental vandals—informing rhetoric used by the 15 out of 22 European far-right parties in their comparative analysis. The "nouvelle ecologie" or new ecology platform advanced by the National Front Party in France is emblematic of this trend. "Environmentalism [is] the natural child of patriotism, because it's the natural child of rootedness," says Nation Front leader Marine Le Pen "If you're a nomad, you're not an environmentalist.... Those who are nomadic ... do not care about the environment; they have no homeland" (Turner & Bailey, 2022, p. 120).

However, not all nationalist discourses incorporating environmental elements are as explicitly nativist as the eco-bordering narrative. Arguments for border restrictions can be embedded in critiques of neoliberal economic policies and mass consumerism driving environmental degradation, developing a framing of the climate crisis more likely to resonate with mainstream liberals and green activists (Hultgren, 2015). Those Hultgren calls "eco-communitarians" argue that immigration must be restricted not to stop ethnically coded "environmental vandals" from sullying the land, but rather to limit further development encroaching into wild places to accommodate a growing population. The agent of environmental destruction here is not immigrants themselves, but rather the transnational economic structures driving migratory flows. Yet whether explicitly or subtly nativist, these discourses ultimately appeal to the sovereign nation-state as the entity responsible for defending the nation's wilderness and environmental health, to establish boundaries of exclusion. Moreover, as immigration policy becomes increasingly politicized, limiting entry is argued by some to be a strategic move to prevent undermining the broad social solidarity required for effective climate action (Lieven, 2020).

Implicit within these green nationalist discourses is a concern for lifestyle preservation, expressed through appeals to both nature's aesthetic value and role as a site of recreation to enrich the lives of citizens. McLaren and Corry (2023) argue that these discourses represent calls for societal climate securitization—rather than referring to the survival of a state as the principal object of security, these narratives are centered on threats to the identities and lifestyles that climate migration or mitigation policies impose (McLaren & Corry, 2023). Should such framings become more prominent in the climate politics of the Global North, the approach to preserving lifestyles through restriction could resemble what some scholars term climate apartheid; a "co-produced system of privilege and precarity—a system that, while possessing some unique features, is built upon historical legacies of colonization, racial capitalism, and hetero-patriarchy," which, in practice, serves to maintain and intensify "the dispossession and expendability of historically oppressed communities, while simultaneously promoting exclusive landscapes, infrastructures, and policies for privileged populations" (Rice et al., 2022, p. 626).

Importantly, internally displaced persons also can encounter similar difficulties in obtaining aid and protection. Lori Peek (2012) examined the experiences of displaced survivors of Hurricane Katrina who sought refuge in Denver, Colorado. While initially welcomed with generosity and compassion, Peek illustrates how "Katrina fatigue" set in over time as local residents' frustrations grew and the needs of the displaced population proved difficult to meet. Peek characterizes this transformation as a "return to the familiar," as the disaster had only served to temporarily suspend the typical inattention to inequalities endemic in American life. She found that Denver residents' attitudes returned to the predisaster norm after several months, rendering displaced Katrina survivors vulnerable to discrimination after the initial warm welcome receded (Peek, 2012). Moreover, in contexts where there is a heightened racial distance between displaced people and receiving communities—such as among New Orleans residents displaced to Houston after Hurricane Katrina—critical assumptions similar to those informed by spatial or temporal distance from the displacing event can also harden attitudes toward new arrivals (Raker & Elliott, 2018). Thus, the subjective encounters survivors experience with the institutional, social, and labor market conditions in receiving communities largely inform decisions of whether to return or permanently relocate after being displaced (Asad, 2015).

Conclusion

Both internal and cross-border environmental migrations will constitute a preeminent form of climate action in the coming decades as the effects of climate change intensify. Current understandings of climate migration tend to emphasize that sudden-onset shocks are the climatic stimuli most likely to generate movement, though more often in the form of internal, rather than cross-border, displacement. Yet with many such shocks—including floods, wildfires, and heat waves—anticipated to strike with increasing frequency and intensity in the coming decades, the capacity of states to manage and resolve internal displacement will be tested. At the same time slower, iterative processes of environmental change will continue apace, threatening to further undermine livelihoods across the globe and continue to loosen the ties that link people to places.

The movements generated by these changing trends do not neatly align with the existing systems of categorization that states employ to classify migrants and manage admission policies, which are based on antiquated conceptualizations of forced displacement and an overly simplistic forced-voluntary dichotomy. Climate-induced migration calls for a rethinking of extant classification systems that account for current realities globally. Including immobility as part of a broader continuum of climate-related mobilities is conceptually capacious and more aligned with on-the-ground experiences of such movements, especially because people confronting escalating climate hazards and degradation might reject relocation and mobilize to remain in place (Arar & FitzGerald, 2022). Recognizing responsibility on the part of wealthier nations that have disproportionally impacted climate change trends to protect those most harmed would contribute a necessary reframing of climate-induced (in)mobility. In the absence of broader global recognition and support for the varied dimensions of enacted climate mobilities, those on the move will be compelled to mobilize networks and act politically to navigate and perhaps change an international landscape where climate destabilizations lead to increase border securitization and mobility restrictions. At the same time, social actors in receiving societies, confronted with apocalyptic narratives that paint climate change as creating uncontrolled migration flows, will also mobilize to demand their government to address the perceived problem.

References

Abdelaaty, L., & Hamlin, R. (2022). Introduction: The politics of the migrant/refugee binary. *Journal of Immigrant & Refugee Studies, 20*(2), 233–239. https://doi.org/10.1080/15562948.2022.2056669

Abel, G. J., Brottrager, M., Crespo Cuaresma, J., & Muttarak, R. (2019). Climate, conflict and forced migration. *Global Environmental Change, 54*, 239–249. https://doi.org/10.1016/j.gloenvcha.2018.12.003

Adams, H. (2016). Why populations persist: Mobility, place attachment and climate change. *Population and Environment, 37*(4), 429–448. https://doi.org/10.1007/s11111-015-0246-3

Adger, W. N. (2003). Social capital, collective action, and adaptation to climate change. *Economic Geography, 79*(4), 387–404.

Afifi, T., Milan, A., Etzold, B., Schraven, B., Rademacher-Schulz, C., Sakdapolrak, P., Reif, A., van der Geest, K., & Warner, K. (2016). Human mobility in response to rainfall variability: Opportunities for migration as a successful adaptation strategy in eight case studies. *Migration and Development, 5* (2), 254–274. https://doi.org/10.1080/21632324.2015.1022974

Ajibade, I. (2019). Planned retreat in Global South megacities: Disentangling policy, practice, and environmental justice. *Climatic Change, 157*(2), 299–317. https://doi.org/10.1007/s10584-019-02535-1

Ajibade, I. J., & Siders, A. R. (2021). *Global views on climate relocation and social justice* (1st ed.). Routledge. https://doi.org/10.4324/9781003141457

Aldrich, D. P. (2019). *Black wave: How networks and governance shaped Japan's 3/11 disasters.* University of Chicago Press. https://doi.org/10.7208/chicago/9780226638577.001.0001

Arar, R., & FitzGerald, D. S. (2022). *The refugee system: A sociological approach*. Polity Press.

Arnall, A. (2019). Resettlement as climate change adaptation: What can be learned from state-led relocation in rural Africa and Asia? *Climate and Development*, *11*(3), 253–263. https://doi.org/10.1080/17565529.2018.1442799

Asad, A. L. (2015). Contexts of reception, post-disaster migration and socioeconomic mobility. *Population and Environment*, *36*(3), 279–310.

Askland, H. H., Shannon, B., Chiong, R., Lockart, N., Maguire, A., Rich, J., & Groizard, J. (2022). Beyond migration: A critical review of climate change induced displacement. *Environmental Sociology*, *8*(3), 267–278. https://doi.org/10.1080/23251042.2022.2042888

Ayeb-Karlsson, S., Smith, C. D., & Kniveton, D. (2018). A discursive review of the textual use of 'trapped' in environmental migration studies: The conceptual birth and troubled teenage years of trapped populations. *Ambio*, *47*(5), 557–573. https://doi.org/10.1007/s13280-017-1007-6

Balgah, R. A., & Kimengsi, J. N. (2022). A review of drivers of environmental non-migration decisions in Africa. *Regional Environmental Change*, 22(4), 125. https://doi.org/10.1007/s10113-022-01970-8

Bartram, D. (2015). Forced migration and "rejected alternatives": A conceptual refinement. *Journal of Immigrant & Refugee Studies*, *13*(4), 439–456. https://doi.org/10.1080/15562948.2015.1030489

Bettini, G., Nash, S. L., & Gioli, G. (2017). One step forward, two steps back? The fading contours of (in)justice in competing discourses on climate migration. *The Geographical Journal*, *183*(4), 348–358. https://doi.org/10.1111/geoj.12192

Bivand Erdal, M., & Oeppen, C. (2018). Forced to leave? The discursive and analytical significance of describing migration as forced and voluntary. *Journal of Ethnic and Migration Studies*, *44*(6), 981–998. https://doi.org/10.1080/1369183X.2017.1384149

Black, R., Adger, W. N., Arnell, N. W., Dercon, S., Geddes, A., & Thomas, D. (2011). The effect of environmental change on human migration. *Global Environmental Change*, *21*, S3–S11. https://doi.org/10.1016/j.gloenvcha.2011.10.001

Boas, I., Farbotko, C., & Bukari, K. N. (2023). The bordering and rebordering of climate mobilities: Towards a plurality of relations. *Mobilities*, *19*(3), 1–16. https://doi.org/10.1080/17450101.2023.2279095

Boas, I., Wiegel, H., Farbotko, C., Warner, J., & Sheller, M. (2022). Climate mobilities: Migration, im/mobilities and mobility regimes in a changing climate. *Journal of Ethnic and Migration Studies*, *48*(14), 3365–3379. https://doi.org/10.1080/1369183X.2022.2066264

Bordner, A. S., Ferguson, C. E., & Ortolano, L. (2020). Colonial dynamics limit climate adaptation in Oceania: Perspectives from the Marshall Islands. *Global Environmental Change, 61* , 102054. https://doi.org/10.1016/j.gloenvcha.2020.102054

Brun, C. (2015). Active waiting and changing hopes: Toward a time perspective on protracted displacement. *Social Analysis*, *59*(1), 19–37. https://doi.org/10.3167/sa.2015.590102

Bukvic, A., & Owen, G. (2017). Attitudes towards relocation following Hurricane Sandy: Should we stay or should we go? *Disasters*, *41*(1), 101–123. https://doi.org/10.1111/disa.12186

Carte, L., Radel, C., & Schmook, B. (2019). Subsistence migration: Smallholder food security and the maintenance of agriculture through mobility in Nicaragua. *The Geographical Journal*, *185*(2), 180–193. https://doi.org/10.1111/geoj.12287

Cresswell, T. (2010). Towards a politics of mobility. *Environment and Planning D: Society and Space*, *28*(1), 17–31. https://doi.org/10.1068/d11407

Cundill, G., Singh, C., Adger, W. N., Safra De Campos, R., Vincent, K., Tebboth, M., & Maharjan, A. (2021). Toward a climate mobilities research agenda: Intersectionality, immobility, and policy responses. *Global Environmental Change*, *69*, 102315. https://doi.org/10.1016/j.gloenvcha.2021.102315

DeWaard, J., Hunter, L. M., Mathews, M. C., Quiñones, E. J., Riosmena, F., & Simon, D. H. (2022). Operationalizing and empirically identifying populations trapped in place by climate and

environmental stressors in Mexico. *Regional Environmental Change*, *22*(1), 29. https://doi.org/10.1007/s10113-022-01882-7
Dickinson, T., & Burton, I. (2021). The role of international governance to reduce maladaptive climate relocation. In I. J. Ajibade & A. R. Siders, *Global views on climate relocation and social justice* (1st ed., pp. 34–47). Routledge. https://doi.org/10.4324/9781003141457-3
Entwisle, B., Verdery, A., & Williams, N. (2020). Climate change and migration: New insights from a dynamic model of out-migration and return migration. *American Journal of Sociology*, *125*(6), 1469–1512. https://doi.org/10.1086/709463
Farbotko, C. (2022). Anti-displacement mobilities and re-emplacements: Alternative climate mobilities in Funafala. *Journal of Ethnic and Migration Studies*, *48*(14), 3380–3396. https://doi.org/10.1080/1369183X.2022.2066259
Fitzgerald, D. (2019). *Refuge beyond reach: How rich democracies repel asylum seekers*. Oxford University Press.
Fussell, E. (2012). Help from family, friends, and strangers during Hurricane Katrina: Finding the limits of social networks. In *Displaced: Life in the Katrina diaspora*, edited by Lynn Weber and Lori Peek; with *Social Science Research Council Research Network on Persons Displaced by Hurricane Katrina* (1st ed., pp. 150–166). University of Texas Press.
Hack-Polay, D., Mahmoud, A. B., Kordowicz, M., Madziva, R., & Kivunja, C. (2021). "Let us define ourselves": Forced migrants' use of multiple identities as a tactic for social navigation. *BMC Psychology*, *9*(1), 125. https://doi.org/10.1186/s40359-021-00630-6
Horowitz, A. (2020). *Katrina: A history, 1915–2015*. Harvard University Press.
Huber, J., Madurga-Lopez, I., Murray, U., McKeown, P. C., Pacillo, G., Laderach, P., & Spillane, C. (2023). Climate-related migration and the climate-security-migration nexus in the Central American Dry Corridor. *Climatic Change*, *176*(6), 79. https://doi.org/10.1007/s10584-023-03549-6
Hultgren, J. (2014). The "nature" of American immigration restrictionism. *New Political Science*, *36*(1), 52–75. https://doi.org/10.1080/07393148.2013.864899
Hultgren, J. (2015). *Border walls gone green: Nature and anti-immigrant politics in America*. University of Minnesota Press.
Hunter, L. M., Luna, J. K., & Norton, R. M. (2015). Environmental dimensions of migration. *Annual Review of Sociology*, *41*(1), 377–397. https://doi.org/10.1146/annurev-soc-073014-112223
Hyndman, J. (2000). *Managing displacement: Refugees and the politics of humanitarianism*. University of Minnesota Press.
Jessee, N. (2022). Reshaping Louisiana's coastal frontier: Managed retreat as colonial decontextualization. *Journal of Political Ecology*, *29*(1). https://doi.org/10.2458/jpe.2835
Kaczan, D. J., & Orgill-Meyer, J. (2020). The impact of climate change on migration: A synthesis of recent empirical insights. *Climatic Change*, *158*(3–4), 281–300. https://doi.org/10.1007/s10584-019-02560-0
Kelley, L. C., Shattuck, A., & Thomas, K. A. (2022). Cumulative socionatural displacements: Reconceptualizing climate displacements in a world already on the move. *Annals of the American Association of Geographers*, *112*(3), 664–673. https://doi.org/10.1080/24694452.2021.1960144
King, L. (2008). Ideology, strategy and conflict in a social movement organization: The Sierra Club immigration wars. *Mobilization: An International Quarterly*, *13*(1), 45–61.
Koslov, L. (2016). The case for retreat. *Public Culture*, *28*(2), 359–387. https://doi.org/10.1215/08992363-3427487
Koslov, L., Merdjanoff, A., Sulakshana, E., & Klinenberg, E. (2021). When rebuilding no longer means recovery: The stress of staying put after Hurricane Sandy. *Climatic Change*, *165*(3–4), 59. https://doi.org/10.1007/s10584-021-03069-1
Lieven, A. (2020). *Climate change and the nation state: The case for nationalism in a warming world*. Oxford University Press.
Loughran, K., & Elliott, J. R. (2019). Residential buyouts as environmental mobility: Examining where homeowners move to illuminate social inequities in climate adaptation. *Population and Environment*, *41*(1), 52–70. https://doi.org/10.1007/s11111-019-00324-7
Malm, A., & Zetkin Collective. (2021). *White skin, black fuel: On the dangers of fossil fascism*. Verso.

Marter-Kenyon, J. (2020). Origins and functions of climate-related relocation: An analytical review. *The Anthropocene Review, 7*(2), 159–188. https://doi.org/10.1177/2053019620915633

Massey, D. S., & España, F. G. (1987). The social process of international migration. *Science, 237*(4816), 733–738.

McLaren, D., & Corry, O. (2023). "Our way of life is not up for negotiation!": Climate interventions in the shadow of "societal security." *Global Studies Quarterly*, 3(3), 1–14. https://doi.org/10.1093/isagsq/ksad037

McLeman, R. A., & Gemenne, F. (2018). Environmental migration research: Evolution and current state of the science. In *Routledge handbook of environmental displacement and migration*, edited by Robert McLeman and François Gemenne (1st ed., pp. 3–16). Routledge.

Menjívar, C. (1993). History, economy and politics: Macro and micro-level factors in recent Salvadorean migration to the US. *Journal of Refugee Studies, 6*(4). https://doi.org/10.1093/jrs/6.4.350

Menjívar, C. (2022). Central American asylum seekers' caravans as a political act. In D. Della Porta, B. Klandermans, D. McAdam, & D. A. Snow (Eds.), *The Wiley-Blackwell encyclopedia of social and political movements* (1st ed., pp. 1–4). Wiley. https://doi.org/10.1002/9780470674871.wbespm593

Menjívar, C. (2023). State categories, bureaucracies of displacement, and possibilities from the margins. *American Sociological Review, 88*(1), 1–23. https://doi.org/10.1177/00031224221145727

Missirian, A., & Schlenker, W. (2017). Asylum applications respond to temperature fluctuations. *Science, 358*(6370), 1610–1614. https://doi.org/10.1126/science.aao0432

Mohapatra, S., Joseph, G., & Ratha, D. (2012). Remittances and natural disasters: Ex-post response and contribution to ex-ante preparedness. *Environment, Development and Sustainability, 14*(3), 365–387. https://doi.org/10.1007/s10668-011-9330-8

Norgaard, K. M. (2011). *Living in denial: Climate change, emotions, and everyday life*. MIT Press.

Omobowale, A. O., Akanle, O., Falase, O. S., & Omobowale, M. O. (2019). Migration and environmental crises in Africa. In C. Menjívar, M. Ruiz, & I. Ness (Eds.), *The Oxford handbook of migration crises* (pp. 314–330). Oxford University Press. https://doi.org/10.1093/oxfordhb/9780190856908.013.32

Paprocki, K. (2020). The climate change of your desires: Climate migration and imaginaries of urban and rural climate futures. *Environment and Planning D: Society and Space, 38*(2), 248–266. https://doi.org/10.1177/0263775819892600

Paul, A. M. (2011). Stepwise international migration: A multistage migration pattern for the aspiring migrant. *American Journal of Sociology, 116*(6), 1842–1886. https://doi.org/10.1086/659641

Peek, L. (2012). They call it "Katrina fatigue:" Displaced families and discrimination in Colorado. In *Displaced: Life in the Katrina diaspora*, edited by Lynn Weber and Lori Peek ; with *Social Science Research Council Research Network on Persons Displaced by Hurricane Katrina* (1st ed., pp. 31–46). University of Texas Press.

Pemberton, S., Tripathy Furlong, B., Scanlan, O., Koubi, V., Guhathakurta, M., Hossain, K., Warner, J., & Roth, D. (2021). "Staying" as climate change adaptation strategy: A proposed research agenda. *Geoforum, 121*, 192–196. https://doi.org/10.1016/j.geoforum.2021.02.004

Perez Murcia, L. E. (2021). "Physically sheltered but existentially homeless": Losing home in the aftermath of conflict and displacement. *Migration Studies, 9*(3), 1361–1379. https://doi.org/10.1093/migration/mnaa020

Piguet, E. (2013). From "primitive migration" to "climate refugees": The curious fate of the natural environment in migration studies. *Annals of the Association of American Geographers, 103*(1), 148–162. https://doi.org/10.1080/00045608.2012.696233

Raker, E. J., & Elliott, J. R. (2018). Attitudes toward mass arrivals: Variations by racial, spatial, and temporal distances to incoming disaster evacuees. *Social Science Quarterly, 99*(3), 1200–1213. https://doi.org/10.1111/ssqu.12498

Ramsay, G., & Askland, H. H. (2022). Displacement as condition: A refugee, a farmer and the teleology of life. *Ethnos, 87*(3), 600–621. https://doi.org/10.1080/00141844.2020.1804971

Rice, J. L., Long, J., & Levenda, A. (2022). Against climate apartheid: Confronting the persistent legacies of expendability for climate justice. *Environment and Planning E: Nature and Space, 5*(2), 625–645. https://doi.org/10.1177/2514848621999286

Riosmena, F., Nawrotzki, R., & Hunter, L. (2018). Climate migration at the height and end of the Great Mexican Emigration Era. *Population and Development Review, 44*(3), 455–488.
Rizzo Lara, R. D. L. L. (2021). La Caminata del Migrante: A social movement. *Journal of Ethnic and Migration Studies, 47*(17), 3891–3910. https://doi.org/10.1080/1369183X.2021.1940111
Rostami, J., & Paski, A. A. (2024). *Iran faces dwindling water and escalating climate pressures, aggravating displacement threats* (Migration Information Source). Migration Policy Institute. https://www.migrationpolicy.org/print/17906
Ruiz-de-Oña, C., Rivera-Castañeda, P., & Merlín-Uribe, Y. (2019). Coffee, migration and climatic changes: Challenging adaptation dichotomic narratives in a transborder region. *Social Sciences, 8*(12), 323. https://doi.org/10.3390/socsci8120323
Schutte, S., Vestby, J., Carling, J., & Buhaug, H. (2021). Climatic conditions are weak predictors of asylum migration. *Nature Communications, 12*(1), 2067. https://doi.org/10.1038/s41467-021-22255-4
Sheller, M. (2020). *Island futures: Caribbean survival in the Anthropocene*. Duke University Press.
Simms, J. R. Z., Waller, H. L., Brunet, C., & Jenkins, P. (2021). The long goodbye on a disappearing, ancestral island: A just retreat from Isle de Jean Charles. *Journal of Environmental Studies and Sciences, 11*(3), 316–328. https://doi.org/10.1007/s13412-021-00682-5
Tierney, K. (2019). *Disasters: A sociological approach*. Polity Press.
Tinoco, N. (2023). Post-disaster (im)mobility aspiration and capability formation: Case study of Southern California wildfire. *Population and Environment*, 45(2), 4. https://doi.org/10.1007/s11111-023-00416-5
Turner, J., & Bailey, D. (2022). "Eco-bordering": Casting immigration control as environmental protection. *Environmental Politics, 31*(1), 110–131. https://doi.org/10.1080/09644016.2021.1916197
Van Der Vliet, J., & Biermann, F. (2022). Global governance of climate migrants: A critical evaluation of the global compacts. In S. Behrman & A. Kent (Eds.), *Climate refugees* (1st ed., pp. 60–82). Cambridge University Press. https://doi.org/10.1017/9781108902991.004
Van Praag, L., & Timmerman, C. (2019). Environmental migration and displacement: A new theoretical framework for the study of migration aspirations in response to environmental changes. *Environmental Sociology, 5*(4), 352–361. https://doi.org/10.1080/23251042.2019.1613030
Vihma, A., Reischl, G., & Nonbo Andersen, A. (2021). A climate backlash: Comparing populist parties' climate policies in Denmark, Finland, and Sweden. *Journal of Environment & Development, 30*(3), 219–239. https://doi.org/10.1177/10704965211027748
Vogler, A. (2023). Barking up the tree wrongly? How national security strategies frame climate and other environmental change as security issues. *Political Geography, 105*, 1–11. https://doi.org/10.1016/j.polgeo.2023.102893
Wu, H., & Bryan, C. (2021). Mobile livelihoods and adaptive social protection: Can migrant workers foster resilience to climate change? In I. J. Ajibade & A. R. Siders, *Global views on climate relocation and social justice* (1st ed., pp. 180–193). Routledge. https://doi.org/10.4324/9781003141457-14
Wurtz, H. M. (2020). A movement in motion: Collective mobility and embodied practice in the central American migrant caravan. *Mobilities, 15*(6), 930–944. https://doi.org/10.1080/17450101.2020.1806511
Yoo, S. H., & Agadjanian, V. (2024). Drought and migration: A case of rural Mozambique. *Population and Environment, 46*(3), 1–22. https://doi.org/10.1007/s11111-023-00444-1
Zickgraf, C. (2018). Immobility. In *Routledge handbook of environmental displacement and migration*, edited by Robert McLeman and François Gemenne (1st ed., pp. 71–84). Routledge.

PART III

Institutionalized Climate Action I

Climate Action Planning

CHAPTER 22

Climate Action Planning in International Perspective

BENJAMIN LEFFEL

Introduction

Cities have become as instrumental in fighting climate change as nations, requiring a radical rethinking of bottom-up leadership as a formal fixture in global climate change governance. With a "Special Report on Climate Change and Cities" to be included in the Seventh Assessment Report (AR7) of the Intergovernmental Panel on Climate Change, research on climate action planning in an international perspective is becoming mainstream across environmental social sciences.

The necessity to study city contributions to climate mitigation is twofold. First, climate action planning by cities on an international scale has implications for the success or failure of nations in achieving decarbonization goals. National-level greenhouse gas (GHG) emissions reduction commitments fall short of what is needed to prevent catastrophic climate change conditions, leaving an emissions gap (UNEP, 2022). However, subnational authorities are making GHG reduction commitments independently of their national governments, as are businesses, and it is shown that if subnational actors fulfill these commitments, it will be sufficient to fill the national GHG gap (Kuramochi et al., 2019).

Second, many studies show significant patterns of national-level economic and policy factors driving national-level GHG emissions change (Givens et al., 2019; Huang, 2024; Jorgenson, 2016; Jorgenson & Clark, 2012; Jorgenson et al., 2019; Schofer & Hironaka, 2005; Shorette, 2012; Shorette et al., 2017). However, it has been shown that novel local-level policy action can have the same or greater impact on GHG reductions than national-level policy (Grant et al., 2023; Leffel, 2021), which begs the question whether the same may apply for economic and trade-based drivers. In other words, when local forces explain GHG change better than national-level forces, this translates to knowledge that is instrumental for more effective climate action. Greater variation in emissions change occurs at the subnational than national level, requiring that social science parses out if and to what extent forces driving change at the aggregate (national) level are better explained by local, bottom-up processes.

This represents the boundary problem in spatial analysis, where accurately estimating a spatial process depends on the spatial dimensions of the models (Barber, 1988; Cressie, 1992), meaning that analyzing environmental processes at an aggregate scale may do so at the expense of micro-scale detail (Weidmann & Minx, 2007). Cities account for 70% of global GHG emissions (Gurney et al., 2022; IEA, 2024), meaning that national emissions are, to an extent, a function of city emissions. Hence, gaining the most accurate understanding of decarbonization may require focusing on drivers of change at more granular geographic boundaries where emissions vary more meaningfully.

This requires that scholars respond to the call for greater use of institutional theory (see Patel and Shorette in this volume), which recognizes the role of institutional forces at multiple levels that drive environmental change (Hoffman & Jennings, 2015; Scott, 2001). Since cities in globalization have begun reasserting themselves on the world stage (Knight, 1989), there have been growing, broader calls for "scaling down" analysis to the city level, as continued decentralization of social processes globally suggests that a more accurate understanding of transformations within and across countries may be obtained by analyzing local phenomena (Sybblis & Centeno, 2017). This chapter focuses on two prominent aspects of subnational-level, international-scale climate action planning: international networks within the public sector and private sector forces in climate planning.

International Networks in the Public Sector

City governments have been adopting municipal climate action plans since the 1990s, some independently from national action in purely bottom-up form, and others doing so in compliance with national guidelines. Portland, Oregon, was an early adopter whose climate action plan has been updated periodically since 1993, in terms of the GHG emissions inventory and progress on actions being taken. Municipal climate action planning draws on multiple sources of knowledge, often involving external stakeholders. Portland established a public-private partnership called Clean Energy Works Portland, which included the municipal government and a diverse range of actors beyond contractors and environmental networks: the nonprofit community bank Shorebank Enterprise Cascadia, the national environmental nonprofit Green for All, Energy Trust of Oregon, Pacific Power, local labor unions, and community groups (City of Portland, 2015). This is fairly representative of the inherently local process of climate action plan formation and adoption (Almeida et al., 2024); however, the international perspective points toward new global organizational structures that both support these local processes and generate new translocal processes.

The United Nations Framework Convention on Climate Change (henceforth UN climate regime) has long been the chief organizing body of governments coordinating to govern climate change, which is designed for national governments to have negotiating power in determining climate governance efforts across borders. However, as subnational governments have taken to climate action plan adoption and similarly sought to extend their individual efforts beyond their own borders, the UN climate regime provided no support, given its nation-state-centric purview. Subnational authorities thus came to create their own plane of global environmental governance. This came to be known as polycentric systems of governance, a decentralized landscape where multiple centers of power, including organizations of cities and companies, work toward climate change governance goals (Ostrom, 2010). Organizations of subnational governments and other sub-/non-state actors cooperate independently of national governments in a horizontal, networked fashion as they orchestrate global bottom-up efforts to decarbonize (Carlisle & Gruby, 2017; Hsu et al., 2017). This serves to fill the governance gap in the multilateral system defined by the lack of representation for subnational government interests (Borja & Castells, 1997; Hale, 2016). The polycentric climate governance landscape represents a new complex institutional structure beyond the mono-institutional proxy of climate change governance, namely the UN climate regime (Zelli et al., 2023).

One of the key structures within polycentric governance is the transnational municipal network (TMN), or formalized membership-based organizations of city governments providing one another access to policy knowledge and related resources (Acuto & Leffel, 2020). Many TMNs are environmental or climate governance-focused and provide member cities with resources to reduce GHG emissions (Abbott, 2018; Betsill & Bulkeley, 2004; Schroder & Janda, 2009). The collective knowledge pool generated by networked cities facilitates the adoption of new best practices, reducing information asymmetries among local governments globally (Coe & Bunnell, 2003). Environmental TMNs were created by subnational authorities as a response to the failure of the traditional, nation-state-centric, multilateral system to provide cities access to climate policy resources (Tosun, 2018).

Environmental TMNs vary in focus, from the generalist urban sustainability and climate action foci of ICLEI—Local Governments for Sustainability (ICLEI) to the megacities focus of C40 Cities Climate Leadership Group (C40), but all provide policy knowledge comprising the core components of climate change governance (Bellinson & Chu, 2019; Wolfram et al., 2019). City governments join these TMNs to both learn how to increase climate action ambition and to leverage reputations as policy leaders, enabling investment attraction in the climate space (Lee, 2015).

Several key studies demonstrate how city membership in environmental TMNs translates to actual GHG reductions. Leffel (2021) links city memberships in environmental TMNs to collective urban-scale GHG reductions in cities globally, showing city climate action via TMN memberships are as important for urban decarbonization as national climate policy. Important mechanisms facilitating these outcomes are the provision of urban GHG inventorying software (such as ICLEI's ClearPath tool), step-by-step climate policy frameworks, model climate ordinances, access to climate finance, resources for feasibility studies, and group discounts on professionally provided retrofit services often used in urban decarbonization strategies. Various in-person and virtual mediums serve as delivery mechanisms of these resources, including through conferences, congresses, workshops, training sessions, reports, and the opportunity for individualized meetings with other member cities.

Grant et al. (2023) show that for the world's power plants, city climate action matters more for decarbonization than national energy policy, or specifically that power plant GHG reduction is more likely to occur if the plant is located within a city that is a member of environmental TMNs, rather than whether its nation is implementing climate mitigation policy. They also found that this impact was particularly acute in the developing world. Acuto et al. (2024) argue that as the impact of environmental TMNs continues to grow, it must integrate with broader city diplomacy efforts on climate justice, inclusive ecological transitions, housing affordability, migration, and other issues that urban leaders are using networked means to confront globally. Cities presently leading the charge are characterized politically by location within democratic nations, where they enjoy more autonomy in global engagement, but economically as the cities most heavily connected to the world economy (Leffel et al., 2023). The full landscape of TMNs is displayed in Figure 22.1 and is available publicly at Acuto and Leffel (2022).

These studies (Grant et al., 2023; Leffel, 2021) draw on the sociological tradition of world society theory to explain why this occurs (see chapter by Patel and Shorette in this volume). World society theory argues that environmental governance norms and policy models diffuse globally and reach national governments via such institutions as the United Nations (Frank et al., 2000a). National governments then implement these policy models in the form of environmental treaty ratification or any other policy measures (Longhofer & Schofer, 2010; Shorette, 2012), which ultimately translates to the subnational-level improvements and thus nationally improved environmental conditions (Shorette et al., 2017), including GHG emissions reductions (Schofer & Hironaka, 2005). While world society theorists have suggested this norm diffusion and environmental improvement process as a nationally led, top-down process (Frank et al., 2000a, 2000b; Schofer et al., 2012), Leffel (2021) and Grant et

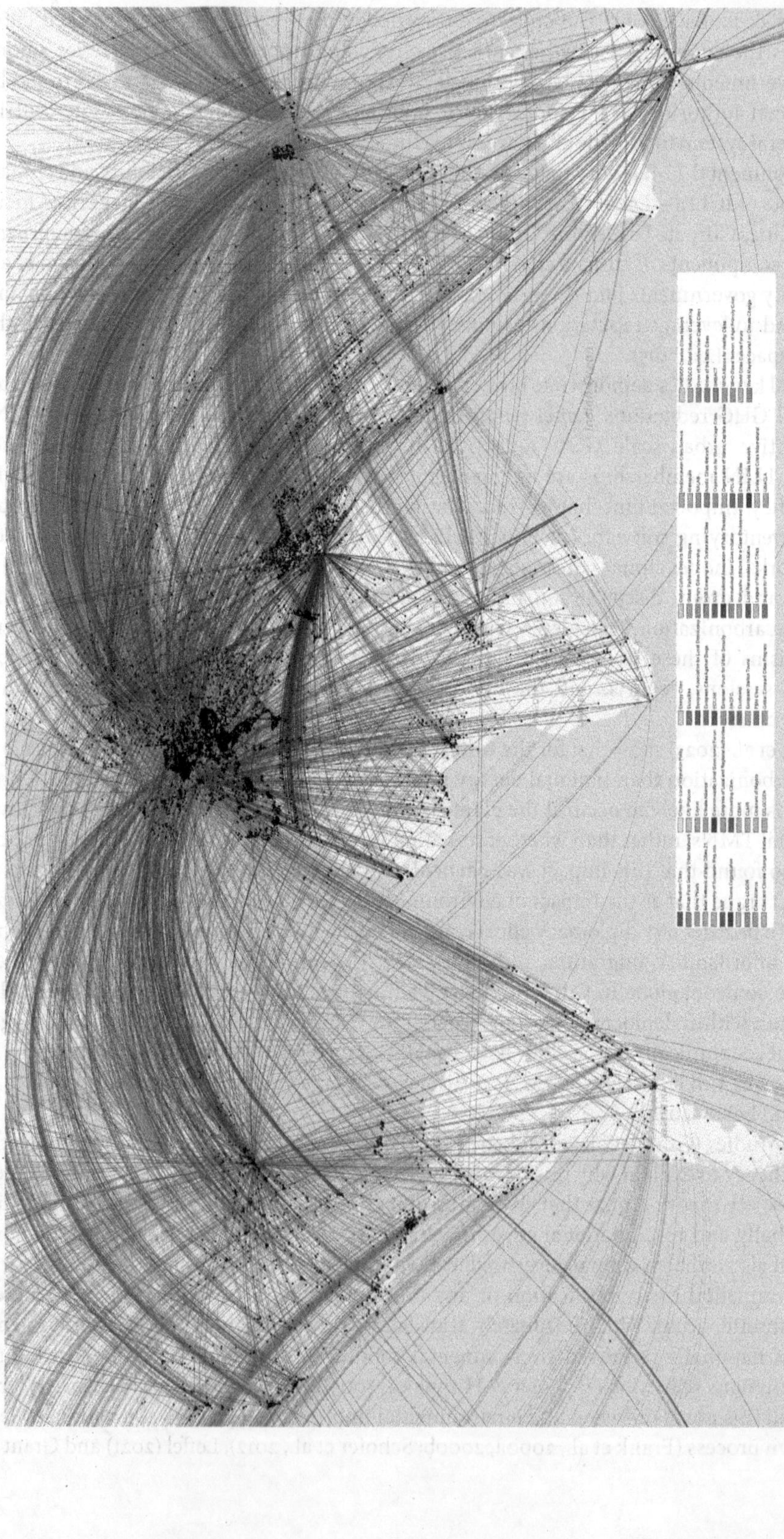

FIGURE 22.1. Geographic network of city memberships in transnational municipal networks.

al. (2023) showed that this process can also occur from the bottom up, with policy practices being diffused by environmental TMN horizontally among cities, independently of national governments. This theoretical frame should be used by future researchers in investigating the institutional processes of emergent subnational climate governance. City climate action via environmental TMNs is also shown to be meaningful beyond GHG reductions. Kim and Yun (2024) show that for Korean cities, ICLEI membership is associated with significant decreases in energy consumption per capita, especially when counties and states are members. This beckons researchers to explore further environmental impacts and implications which these networks have for international climate planning.

An initial picture of the structure of the polycentric landscape—measured specifically by TMNs—and its political-economic underpinnings has been produced in recent scholarship. The collective landscape of over 10,000 cities (as of 2017) that are members of TMNs often have multiple concurrent memberships, including in climate networks, which produce power inequalities in global urban governance. That is, cities with more concurrent TMN memberships have a greater ability to set TMN agendas, project policy ideas, and obtain policy best practices for use, including in climate governance.

The landscape of city TMN memberships mirrors the stratified structure of cities in the world economy. Leffel et al. (2022) showed that a powerful minority (4%) of cities structurally tie most of the world economy together, while the majority are only loosely connected to this structure. Leffel et al. (2023) showed that this same 4% constitutes the cities with the most concurrent TMN memberships, while the majority of cities have only one membership.

Hence, a political economy of cities in global governance emerges that mirrors a new core-periphery structure of cities in the world economy. An important theoretical implication is that this structure itself has diverged from national economies. That is, Leffel et al. (2022) show that national economic position of high, medium, and low centrality in world trade (or core, semi-periphery, and periphery in sociological terms) no longer neatly explains the global economic power of cities. Rather, city position in the world economy has diverged from that of their national economies, including world-system position of core, semi-periphery, and periphery. This means city participation in global climate governance is highly stratified and within a structure not defined by nations.

As detailed in the next section, economic forces intersect meaningfully with politics in driving environmental change. Indeed, the extent to which policy diffusion (world society) processes drive environmental change has been shown to depend upon national position in world trade—that is, if the nation is of high, medium, or low centrality in world trade (Shorette, 2012; Shorette et al., 2017). If these national political-economic structures are thought to impact environmental outcomes, but cities are forming political-economic geographies separate from nations (Leffel et al., 2022, 2023), then research must determine if and to what extent subnational political-economic forces impact environmental change in ways that national forces do not. This provides important future directions of inquiry for bottom-up forces in international climate action planning. Both studies (Leffel et al., 2022, 2023) provide open data for the world's cities so that future researchers may inquire further and test their own hypotheses.

It is important to note that while environmental TMNs and other mechanisms in the polycentric governance landscape are largely independent from the UN climate regime, they have consistently sought to build subnational capacity for climate governance contributions into the regime. Nation-states are the chief negotiating entities within the UN climate regime's key convening, the Conference of the Parties (COP); however, at the first COP in 1995, the Local Governments and Municipal Authorities (LGMA) constituency was launched and led by the largest environmental TMN, ICLEI (Borja & Castells, 1997). Over time, ICLEI and other networks have proposed via LGMA to have local governments be included in the design and implementation of climate action at all levels, including for the purpose of raising the national GHG reduction commitments or the nationally determined contributions (NDCs). The Non-State Actor Zone for Climate Action (NAZCA) was later created to incorporate subnational authorities' commitments to help nations increase NDC ambition, and multiple agreements—such as the Bonn-Fiji Commitment—have similarly advanced that goal. The Under2 Coalition has also emerged to build capacity for subnational governments to commit to GHG reductions and further push efforts to integrate subnational into national—and supranational, that is, UN climate regime—climate governance efforts (Bäckstrand & Kuyper, 2017; Chan et al., 2016). This remains an uphill battle, however, as there is no evidence yet that these efforts have meaningfully influenced national climate commitments.

To advance research on subnational climate planning—be it focused on TMNs or other bottom-up action—its relevance to local carbon performance itself and its otherwise unseen relevance to outcomes at the national level may require an institutional theory lens. Whether researchers conceptualize bottom-up environmental action as occurring due to world society processes within a polycentric governance landscape, or any other theoretical synthesis, the foundational reality is that environmental change is occurring within a multilevel context. Institutional theory recognizes organizations such as businesses, subnational governments, and national governments as embedded in multilevel normative and regulatory governance structures that are meaningful for one another, including for what forces stimulate GHG emissions reductions (Hoffman & Jennings, 2015; Scott, 2001). In this way, an institutional theory lens offers a useful conceptual pathway forward for investigating how subunits of nations play underappreciated roles in collective environmental change.

Private Sector Forces in Climate Planning

Economic dynamics, be they tied to urban industrial trends, the actions of individual businesses, or collaborative inroads between the public and private sector, are meaningful for climate action planning at the local level. Eminently relevant to climate action planning is regulating the activities of large corporations, which are the largest organizational sources of GHG emissions in the world. Many of these corporations are adopting meaningful climate action similar to that of governments. This includes voluntarily disclosing corporate GHG emissions, decarbonization efforts, and other climate-related metrics to CPD (formerly, Climate Disclosure Project) and other repositories for the strategic purposes of visibility to investors and other stakeholders, as well as performing beyond what regulatory compliance requires (Flammer et al., 2021; Hoffman, 2001).

For some of the largest corporations in the United States, subnational climate action has a stronger decarbonizing impact on corporate facilities than the internal initiatives of the companies themselves (Leffel et al., 2024). These findings were framed within an institutional theory lens to emphasize the multilevel nature of corporate carbon performance, meaning that performance improvements otherwise attributed only to firm-level efforts were in fact more attributable to another level. Specifically, mobilization of subnational government financial incentives, such as commercial energy efficiency rebates, was shown by Leffel and colleagues to have the strongest decarbonizing impact, due to their ability to reduce firms' often high costs of decarbonization. This process was confirmed following the publication of this study, as journalists interviewed businesses on their use of incentives and found that the availability of such incentives locally enabled them to expand the scope of their decarbonization efforts beyond what otherwise would have been possible (Gearino,

2024). This provides crucial guidance for local governments in the Inflation Reduction Act (IRA) era, as the IRA significantly expands the availability of funding which can be used to create financial incentive programs of the sort Leffel et al. (2024) show drives corporate decarbonization. These interventions will continue to be necessary, as corporate voluntary action remains insufficient to ensure that the private sector can align climate progress with the Paris Accord (Robiou du Pont et al., 2024). Future research should similarly use an institutional theory lens investigate how multilevel actions have relevance for environmental outcomes that are otherwise attributed to one level, including subnational versus national drivers of GHG emissions change.

The population of climate-active firms is also relevant for an emerging trend in local public–private partnerships, whereby urban and corporate sustainability are meaningfully converging. As firms push the envelope on higher environmental performance and innovative sustainability practices, so, too, are city governments as they form cross-sector partnerships. In particular, cities are coming to recognize that they cannot control all of their local emissions through regulation alone, as the local private sector is also a significant source of these emissions. Accordingly, new models are emerging whereby city governments that are implementing climate action are engaging with local corporate sustainability efforts, aiming to solve this problem (Shaver et al., 2021).

For instance, the City-Business Climate Alliance is a joint initiative of CDP, C40 Cities, and the World Business Council for Sustainable Development, which is assisting city governments that are implementing climate action plans to directly collaborate with local corporations who are doing the same. This allows cities and corporations to align and collaboratively increase their respective carbon targets, provide co-financing for one another's local climate mitigation projects, and build broader capacity to solve climate action problems (C40 et al., 2024). Collaborative areas, highlighted in CBCA case studies, can include but are not limited to sustainable procurement, decarbonization efforts, and electric vehicle fleet mobilization. The ultimate goal of these collaborations is to achieve greater local GHG reductions than otherwise would be possible, by engaging private emissions sources that city hall cannot abate through traditional climate planning alone.

In addition to the impact of environmental TMN memberships discussed in the previous section, Leffel (2021) showed that urban-scale GHG reductions were associated with the local presence of environmental consultancies, or private environmental consultancies, such as ERM, AECOM, and Ameresco. Specialized expertise has long spatially manifested itself as corporate offices of consultancies concentrating in cities (Knight, 1982), articulating the role of modern cities as global services centers (Kennedy, 2009). Such firms are commonly hired by local governments to assist in the design and implementation of climate mitigation projects and associated policies (Biagini & Miller, 2013; Keele, 2019). This industry is described as a global professional technocracy selling solutions for urban environmental governance problems (Feser & Runst, 2016; Mostafavi & Doherty, 2010) that diffuses policy models through private channels globally, which Leffel (2021) argues is explicable by world society theory.

Leffel (2022) showed that one specific mechanism through which consultancies affected urban decarbonization was the energy performance contract (EPC), in which the contracted consultancy develops, installs, and finances facility-level energy efficiency improvements. Rather than paying up front, the city only pays in accordance with the demonstrated energy savings achieved over time, enabling governments to achieve emissions reductions without making large budgetary investments (APEC, 2017; Capelo, 2011; Vine, 2005). EPCs are one of many services offered by local climate consultants and represent technical services often not available through environmental TMNs. Cities tend to contract to consultancies locally in order to invest in the growth of the local environmental services industry, thus pairing decarbonization efforts with local green economic growth goals (Leffel, 2022).

On the international scale, there is a Global North–South divide regarding access to private climate consultants. Less developed countries tend to have less access to consultancies because these professional environmental industries are more heavily concentrated in the Global North. For this reason, cities in less developed nations will join environmental TMNs as a more accessible option to obtain climate governance expertise. It is for this reason that Grant et al. (2023) attribute their finding that environmental TMN membership has a greater decarbonizing impact in developing world cities.

Air pollution abatement outcomes beyond GHGs in cities globally are also meaningfully impacted by urban economic growth trends across certain industries. Specifically, political-economic-environmental patterns previously observed only at the national level have also recently been observed at the city level. For example, the environmental Kuznet's curve (EKC) argues that increases in income via economic growth initially drive increasing pollution, but over time allow for pollution abatement (Orubu & Omotor, 2011), which occurs as a nation's industry transitions from a high-polluting, manufacturing-based economy to a more services-based economy (Khan et al., 2018).

Leffel (2021) shows that the same process is true for the world's largest metropolises, including for emissions of course particulate matter (PM_{10}), fine particulate matter ($PM_{2.5}$), nitrogen dioxide (NO_2), and sulfur dioxide (SO_2) over time. Specifically, Leffel and colleagues find that increases in local income, government expenditures on environmental and health services, and the value of the local real estate industry are associated with abatement across all four pollutants, while pollution increases are associated with growth in manufacturing and mining industries. This suggests that the EKC occurs in cities and extends to the city-level previous observations showing that real estate value increases are associated with pollution abatement (Anderson & Crocker, 1971; Nourse, 1967). This may suggest that real estate increases result in upgrades to residential and commercial buildings, and thus reduced pollution (Chen & Lee, 2020), or that polluting activities are simply pushed away from areas in which real estate is of high value.

Financing is an area of international climate action planning for which the subnational is increasingly crucial for national and global climate governance outcomes. Climate finance has become more available to cities over time, but it is also becoming more influenced by private sector entities (Bracking & Leffel, 2021), offering both promise and peril for financing the often high expenses associated with urban climate mitigation, adaptation, and resilience. The private influence is seen in the dominance of market rate debt as the primary form of climate finance available to cities, rather than grants and subsidies (Negreiros et al., 2021). Two prominent examples are direct urban climate loan allocations by the European Investment Bank (EIB) and municipal green bonds. Among international banks, the EIB has allocated some of the largest volumes of climate mitigation loans directly to cities, or relatively low-interest loans designated for climate governance purposes (CCFLA, 2015). Green bonds are similarly a debt-based instrument in which the issuer earmarks the use of proceeds for environmental protection purposes and designates the source of funds for repayment (CBI, 2020). A pervasive problem in climate finance is the investment trap, in which developing nations are considered higher risk and thus lenders impose higher lending costs upon them, restricting their access to climate finance (Ameli et al., 2021). At the city level, this problem is seen in the form of lower credit ratings in developing cities translating to higher costs of climate finance access, thus perpetuating the climate investment gap (Leffel, 2025). Future researchers should investigate if and to what extend greater climate finance access drives

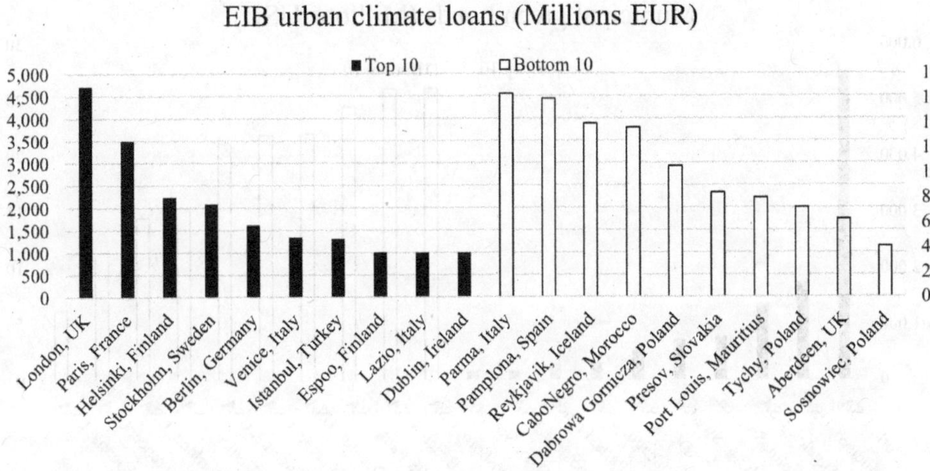

FIGURE 22.2. European Investment Bank (EIB) distribution of direct urban climate loans.

local decarbonization outcomes, and conversely, if restricted access prevents successful decarbonization.

As noted by Portes and Armony (2023), an important critique of global city literature is that empirically based studies often focus only on the top-ranked cities while ignoring those at the bottom, to the detriment of poor and third-world cities. To that end, I include EIB and municipal green bond statistics for both the top-10 and bottom-10 world cities here. Smaller cities and cities in developing countries (Morocco, Mauritius) are generally more likely to be among the lower end of EIB loan recipients, as shown in Figure 22.2, which may be more reflective of a lack of bankable projects and low credit scores than of the level of need. Figure 22.3 shows that the green bond market is considerably more restricted to developed countries, for which the underpinnings of the distribution are less obvious, which future researchers should investigate.

Last, there remains the important task of determining how economic forces in global urban climate governance map onto existing national-level economic theories of environmental change. A broadly industrial dimension of global environmental change is ecologically unequal exchange (see chapter by Jorgenson et al. in this volume), which shows that emissions reductions are more likely to take place in developed countries at the expense of developing nations, as the former extracts natural resources from and controls the export activity of the latter (Jorgenson, 2016). Researchers have not yet assessed whether this phenomenon has a uniquely local geography or if it is principally a national-level phenomenon, but they could do so by using PIERS data to obtain the bill of lading for all exported and imported shipments globally, which includes local information on source and destination.

Conclusion

National and global climate action goals cannot be achieved without successful city climate action planning, and drivers of national-level GHG emissions change cannot be fully understood without analyzing the subnational parts of which it is the whole. In so doing, future researchers should consider using an institutional theory lens to frame environmental change as driven by forces at multiple levels embedded within and meaningful for one

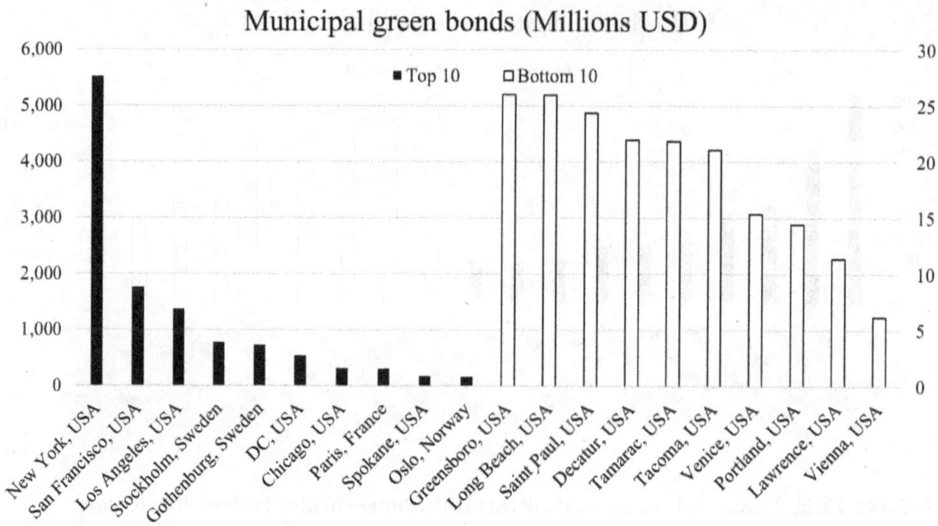

FIGURE 22.3. Municipal green bond distribution.

another. This may be further enriched by investigating the role of normative or policy diffusion processes such as world society theory, the political-economic forces of climate finance access, international trade patterns (ecologically unequal exchange), or other frameworks, as data relevant to these theories become accessible at the local level. Novel environmental attribute data at the subnational level are available from such sources as CDP and Climate TRACE, and open data on world cities' position in the world economy (Leffel et al., 2022) and role in global governance (Acuto & Leffel, 2022) also offer unique attribute data helpful for pursuing such research. Increasing data availability and the urbanization of the world population beckon researchers to perform more granular analysis of the sorts described here to overcome the limitations of existing nationally focused environmental studies.

Too often, nationally focused scholars will consider subnationally focused studies to be relevant only to the local level, and subnationally focused scholars will conceptualize their work only within a purely local purview. Such thinking stymies progress toward a fuller, more systemic understanding of climate action planning and progress, namely a multilevel integration of national and subnational environmental studies. The exponential increase of cities as central actors in global climate governance shows no signs of slowing, and scholars must understand its potential just as surely as world leaders must integrate them into existing supranational climate governance efforts.

To that end, I offer several critical questions that future researchers should consider answering: Does progress toward national-scale GHG reductions happen first within climate-active cities, or is it more evenly geographically distributed? In what other ways do bottom-up policy diffusion (world society) processes drive GHG reductions? When does top-down climate policy matter more than bottom-up? What role does climate justice and equity planning play in the internationalization of urban climate governance? Are economic and trade processes affecting environmental outcomes more so under the control of nations or cities, and do the governments or market actors control these outcomes? For that matter, is the relationship between international trade impact and environmental outcomes (ecologically unequal exchange) principally a national-level phenomenon, or are there meaningful local forces at play? How does climate finance access enable or constrain GHG reductions within cities, and how does this add up to collective national outcomes? How can cities ultimately gain a voice within the UN climate regime that enables them to influence NDC ambition? Answering these and similar questions will significantly advance the science and practices of international climate planning, which takes an increasingly multilevel form.

References

Abbott, K. W. (2018). Orchestration. In A. Jordan, D. Huitema, H. van Asselt, & J. Forster (Eds.), *Governing climate change: Polycentricity in action?* (pp. 188–209). Cambridge University Press. https://doi.org/10.1017/9781108284646.012

Acuto, M., Gordon, D., Mokhles, S., Martinez, R., Leffel, B., & Oke, C. (2024). What three decades of city networks tell us about city potential for climate action. *Nature Cities*, *1*(7), 451–456. https://doi.org/10.1038/s44284-024-00081-1

Acuto, M., & Leffel, B. (2020). Understanding the global ecosystem of city networks. *Urban Studies*, *58*(9), 1758–1774.

Acuto, M., & Leffel, B. (2022). *City networks membership dataset* [dataset]. https://doi.org/10.26188/21330996.v1

Almeida, P., González Márquez, L. R., & Fonsah, E. (2024). The forms of climate action. *Sociology Compass*, *18*(2), e13177. https://doi.org/10.1111/soc4.13177

Ameli, N., Dessens, O., Winning, M., Cronin, J., Chenet, H., Drummond, P., Calzadilla, A., Anandarajah, G., & Grubb, M. (2021). Higher cost of finance exacerbates a climate investment trap in developing economies. *Nature Communications*, *12*(1), 4046. https://doi.org/10.1038/s41467-021-24305-3

Anderson, R., & Crocker, T. (1971). Air pollution and residential property values. *Urban Studies*, *8*(3), 171–180.

Asia-Pacific Economic Cooperation Committee on Trade and Investments (APEC). (2017). *Sector study on environmental services: Energy efficiency businesses*. APEC Policy Support Unit.

Bäckstrand, K., & Kuyper, J. W. (2017). The democratic legitimacy of orchestration: The UNFCCC, non-state actors, and transnational climate governance. *Environmental Politics*, *26*(4), 764–788. https://doi.org/10.1080/09644016.2017.1323579

Barber, G. (1988). *Elementary statistics for geographers*. Guilford Press.

Bellinson, R., & Chu, E. (2019). Learning pathways and the governance of innovations in urban climate change resilience and adaptation. *Journal of Environmental Policy & Planning*, *21*(1), 76–89.

Betsill, M., & Bulkeley, H. (2004). Transnational networks and global environmental governance: The Cities for Climate Protection program. *International Studies Quarterly*, *48*(2), 471–493.

Biagini, B., & Miller, A. (2013). Engaging the private sector in adaptation to climate change in developing countries: Importance, status, and challenges. *Climate and Development*, *5*(3), 242–252.

Borja, J., & Castells, M. (1997). *Local and global: The management of cities in the information age*. Earthscan.

Bracking, S., & Leffel, B. (2021). Climate finance governance: Fit for purpose? *WIREs Climate Change*, e709. https://doi.org/10.1002/wcc.709

C40 Cities, CDP, and World Business Council for Sustainable Development (C40, CDP, and WBCSD). (2014). *City-business climate alliance*. https://www.city-businessclimatealliance.org/.

Capelo, C. (2011). Modeling the diffusion of energy performance contracting. *Proceedings of the 29th International Conference of the System Dynamics Society*, 1–20.

Carlisle, K., & Gruby, R. L. (2017). Polycentric systems of governance: A theoretical model for the commons. *Policy Studies Journal*, *47*(4), 927–952.

Chan, S., Brandi, C., & Bauer, S. (2016). Aligning transnational climate action with international climate governance: The road from Paris. *Review of European, Comparative & International Environmental Law*, *25*(2), 238–247. https://doi.org/10.1111/reel.12168

Chen, Y., & Lee, C.-C. (2020). The impact of real estate investment on air quality: Evidence from China. *Environmental Science and Pollution Research*, *27*(18), 22989–23001. https://doi.org/10.1007/s11356-020-08874-2

City of Portland. (2015). Climate action plan: Local strategies to address climate change. *2050 Vision for Portland and Multnomah County.* https://www.portland.gov/sites/default/files/2019-07/cap-2015_june30-2015_web_0.pdf

Climate Bonds Initiative (CBI). (2020). *Green bond market summary Q3*. https://www.climatebonds.net/resources/reports/green-bonds-market-summary-q3-2020

Coe, N. M., & Bunnell, T. G. (2003). "Spatializing" knowledge communities: Towards a conceptualization of transnational innovation networks. *Global Networks, 3*(4), 437–456. https://doi.org/10.1111/1471-0374.00071

Cressie, N. (1992). *Statistics for spatial data*. John Wiley and Sons.

Feser, D., & Runst, P. (2016). Energy efficiency consultants as change agents? Examining the reasons for EECs' limited success. *Energy Policy, 98*, 309–317.

Flammer, C., Toffel, M. W., & Viswanathan, K. (2021). Shareholder activism and firms' voluntary disclosure of climate change risks. *Strategic Management Journal, 42*(10), 1850–1879. https://doi.org/10.1002/smj.3313

Frank, D. J., Hironaka, A., & Schofer, E. (2000a). The nation-state and the natural environment over the twentieth century. *American Sociological Review, 65*(1), 96–116. https://doi.org/10.2307/2657291

Frank, D., Hironaka, A., & Schofer, E. (2000b). Environmental protection as a global institution. *American Sociological Review, 65*, 122–127.

Gearino, D. (2024). "A Boutique Hotel Helps Explain the Benefits of Businesses and Government Teaming Up to Conserve Energy," *Inside Climate News*. February 1.

Givens, J. E., Huang, X., & Jorgenson, A. K. (2019). Ecologically unequal exchange: A theory of global environmental injustice. *Sociology Compass, 13*(5), e12693.

Grant, D., Leffel, B., & Johnson, E. (2023). Can transnational municipal networks mitigate the carbon pollution of the world's power plants? An empirical analysis. *NPJ Climate Action, 2*(1), 38. https://doi.org/10.1038/s44168-023-00070-5

Gurney, K. R., Kılkış, Ş., Seto, K. C., Lwasa, S., Moran, D., Riahi, K., Keller, M., Rayner, P., & Luqman, M. (2022). Greenhouse gas emissions from global cities under SSP/RCP scenarios, 1990 to 2100. *Global Environmental Change, 73*, 102478. https://doi.org/10.1016/j.gloenvcha.2022.102478

Hale, T. (2016). "All hands on deck": The Paris Agreement and nonstate climate action. *Global Environmental Politics, 16*(3), 12–22. https://doi.org/10.1162/GLEP_a_00362

Hoffman, A. (2001). *From heresy to dogma: An institutional history of corporate environmentalism*. Stanford University Press.

Hoffman, A. J., & Jennings, P. D. (2015). Institutional theory and the natural environment. *Organization & Environment, 28*(1), 8–31.

Hsu, A., Weinfurter, A. J., & Xu, K. (2017). Aligning subnational climate actions for the new post-Paris climate regime. *Climatic Change, 142*(3), 419–432. https://doi.org/10.1007/s10584-017-1957-5

Huang, X. (2024). Harness the co-benefit and avoid the trade-off: The complex relationship between income inequality and carbon dioxide emissions. *Social Forces, 103*(2), 403–428. https://doi.org/10.1093/sf/soae074

International Energy Agency (IEA). (2024). *Empowering urban energy transitions*. https://www.iea.org/reports/empowering-urban-energy-transitions

Jorgenson, A. (2016). Environment, development, and ecologically unequal exchange. *Sustainability, 8*(227), 1–15.

Jorgenson, A. K., & Clark, B. (2012). Are the economy and the environment decoupling? A comparative international study, 1960–2005. *American Journal of Sociology, 118*(1), 1–44.

Jorgenson, A. K., Fiske, S., Hubacek, K., Li, J., McGovern, T., Rick, T., Schor, J. B., Solecki, W., York, R., & Zycherman, A. (2019). Social science perspectives on drivers of and responses to global climate change. *Wiley Interdisciplinary Reviews: Climate Change, 10*(1), e554. https://doi.org/10.1002/wcc.554

Keele, S. (2019). Consultants and the business of climate services: Implications of shifting from public to private science. *Climatic Change, 157*(1), 9–26.

Khan, A. Q., Saleem, N., & Fatima, S. T. (2018). Financial development, income inequality, and CO_2 emissions in Asian countries using STIRPAT model. *Environmental Science and Pollution Research, 25*(7), 6308–6319. https://doi.org/10.1007/s11356-017-0719-2

Kim, M., & Yun, S.-J. (2024). The impact of transnational municipal networks on local energy consumption. *Urban Climate*, *55*, 101861. https://doi.org/10.1016/j.uclim.2024.101861

Knight, R. V. (1982). City development in advanced industrial societies. In G. Gappert & R. Knight V. (Eds.), *Cities in the 21st century* (pp. 47–68). SAGE.

Knight, R. V. (1989). The emergent global society. In R. V. Knight & G. Gappert (Eds.), *Cities in a global society* (pp. 24–43). SAGE.

Kuramochi, T., & et al. (2019). *Global climate action from cities, regions and businesses: Impact of individual actors and cooperative initiatives on global and national emissions*. New Climate Institute.

Lee, T. (2015). *Global cities and climate change: The translocal relations of environmental governance*. Routledge.

Leffel, B. (2021). Toward global urban climate mitigation: Linking national and polycentric systems of environmental change. *Sociology of Development*, *8*(1), 1–27. https://doi.org/10.1525/sod.2021.0018

Leffel, B. (2022). Climate consultants and complementarity: Local procurement, green industry and decarbonization in Australia, Singapore, and the United States. *Energy Research & Social Science*, *88*, 102635. https://doi.org/10.1016/j.erss.2022.102635

Leffel, B. (2025). Climate creditors: How credit ratings widen the investment gap in climate finance access. In W. Solecki, M. Pathak, M. Barata, A. Barau, M. Dombrov, & C. Rosenzweig (Eds.), *Climate change and cities: Third assessment report of the Urban Climate Change Research Network* (pp. 1–3). Cambridge University Press.

Leffel, B., Derudder, B., Acuto, M., & van der Heijden, J. (2023). Not so polycentric: The stratified structure & national drivers of transnational municipal networks. *Cities*, *143*, 104597. https://doi.org/10.1016/j.cities.2023.104597

Leffel, B., Lyon, T. P., & Newell, J. P. (2024). Filling the climate governance gap: Do corporate decarbonization initiatives matter as much as state and local government policy? *Energy Research & Social Science*, *109*, 103376. https://doi.org/10.1016/j.erss.2023.103376

Leffel, B., Marahrens, H., & Alderson, A. S. (2023). Divergence of the world city system from national economies. *Global Networks*, *23*(2), 459–477. https://doi.org/10.1111/glob.12405

Leffel, B., Tavasoli, N., Liddle, B., Henderson, K., & Kiernan, S. (2022). Metropolitan air pollution abatement and industrial growth: Global urban panel analysis of PM_{10}, $PM_{2.5}$, NO_2 and SO_2. *Environmental Sociology*, *8*(1), 94–107. https://doi.org/10.1080/23251042.2021.1975349

Longhofer, W., & Schofer, E. (2010). National and global origins of environmental association. *American Sociological Review*, *75*(4), 505–533. https://doi.org/10.1177/0003122410374084

Mostafavi, M., & Doherty, G. (2010). *Ecological urbanism*. Lars Müller.

Negreiros, P., Furio, V., Falconer, A., Richmond, M., Yang, K., Jungman, L., Tonkonogy, B., Novikova, A., Pearson, M., & Skinner, I. (2021). The state of cities climate finance, Part 1: The landscape of urban climate finance. *Climate Policy Initiative*. https://www.citiesclimatefinance.org/2021/06/2021-state-of-cities-climate-finance/

Nourse, H. O. (1967). The effect of air pollution on house values. *Land Economics*, *43*(2), 181–189. https://doi.org/10.2307/3145241

Orubu, C. O., & Omotor, D. G. (2011). Environmental quality and economic growth: Searching for environmental Kuznets curves for air and water pollutants in Africa. *Special Section: Renewable Energy Policy and Development*, *39*(7), 4178–4188. https://doi.org/10.1016/j.enpol.2011.04.025

Ostrom, E. (2010). Polycentric systems for coping with collective action and global environmental change. *Global Environmental Change*, *20*(4), 550–557.

Portes, A., & Armony, A. (2023). *Emerging global cities: Origin, structure and significance*. Columbia University Press.

Robiou du Pont, Y., Rogelj, J., Hsu, A., van Vuuren, D., & Hoepner, A. G. F. (2024). Corporate emissions targets and the neglect of future innovators. *Science*, *384*(6694), 388–390. https://doi.org/10.1126/science.adl5081

Schofer, E., Frank, D., & Longhofer, W. (2012). Sociological institutionalism and world society. In E. Amenta, K. Nash, & A. Scott (Eds.), *The Wiley-Blackwell companion to political sociology* (pp. 57–68). Blackwell.

Schofer, E., & Hironaka, A. (2005). The effects of world society on environmental protection outcomes. *Social Forces, 84*(1), 25–47.

Schroder, H., & Janda, K. (2009). *Cities and climate change: The role of institutions, governance and urban planning.* Report prepared for the World Bank Urban Symposium on Climate Change.

Scott, W. R. (2001). *Institutions and organizations.* SAGE.

Shaver, L., Elwood, H., & Bonugli, C. (2021). *The power of collaboration: How U.S. cities and corporations can work together to advance renewable energy.* World Resources Institute. doi.org/10.46830/wriwp.20.00012

Shorette, K. (2012). Outcomes of global environmentalism: Longitudinal and cross-national trends in chemical fertilizer and pesticide use. *Social Forces, 91*(1), 299–325.

Shorette, K., Henderson, K., Sommer, J. M., & Longhofer, W. (2017). World society and the natural environment. *Sociology Compass, 11*(10), e12511.

Sybblis, M., & Centeno, M. (2017). Sub-nationalism. *American Behavioral Scientist, 61*(8), 799–807.

Tosun, J. (2018). Diffusion: An outcome of and an opportunity for polycentric activity? In A. Jordan, D. Huitema, H. van Asselt, & J. Forster (Eds.), *Governing climate change: Polycentricity in action?* (pp. 152–168). Cambridge University Press.

United Nations Environment Programme (UNEP). (2022). *Emissions Gap Report 2022: The closing window—Climate crisis calls for rapid transformation of societies.* https://www.unep.org/emissions-gap-report-2022

Vine, E. (2005). An international survey of the energy service company (ESCO) industry. *Energy Policy, 33*(5), 691–704.

Weidmann, T., & Minx, J. (2007). A definition of "carbon footprint." In *Ecological economics: Research trends* (pp. 1–11). Nova Science.

Wolfram, M., van der Heijden, J., Juhola, S., & Patterson, J. (2019). Learning in urban climate governance: Concepts, key issues and challenges. *Journal of Environmental Policy & Planning, 21*(1), 1–15.

Zelli, F., Gerrits, L., Moller, I., & Widerberg, O. (2023). Institutional complexity and political agency in polycentric governance. In F. Gadinger & J. Aart Scholte (Eds.), *Polycentrism: How governing works today* (pp. 73–97). Oxford University Press.

CHAPTER 23

Green Energy Transitions as Climate Action?

KATHRYN HOCHSTETLER

An emerging consensus among scientists and policymakers identifies green energy transition as a central strategy of climate mitigation, since it is seen as a comparatively efficient and inexpensive way of quickly reducing greenhouse gas (GHG) emissions (Intergovernmental Panel on Climate Change [IPCC], 2014, p. 20). The installation of new wind turbines or solar panels signifies climate action for almost everyone. This chapter begins by explaining what a green energy transition is and how it is a form of climate action, presenting evidence about the progress of the transition to date. A subsequent section explores the facilitating conditions for green energy transitions. The growth of green energy is often seen as offering substantial "co-benefits," which are positive outcomes for countries, firms, and households besides its decarbonizing function (Dubash, 2013; Karlsson et al., 2020). Yet as energy changes have unfolded, the transitions' limitations in both climate and social terms have also become clearer. The following section examines these. The Conclusion argues that while these limitations are concerning, green energy transition remains useful climate action and a clear improvement on the traditional, fossil fuel–based energy model.

Basic Dynamics of Green Energy Transition as Climate Action

Green energy transition is both simple and complex to define. With reference to climate change, a simple definition refers to a decarbonizing shift from the use of sources of energy that are high in the GHGs that cause climate change to the development and use of low-carbon and often renewable sources of energy.[1] Understood in this way, it is primarily a mitigation strategy, which is an effort to reduce the emission of the 10 primary GHGs into the atmosphere. Doing so prevents future climate change. Important in its own right, green energy transition is also a crucial step toward decarbonizing other activities and products that depend on energy, from transport to industrial production to buildings (IPCC, 2014, p. 20). Green energy transition can also be a strategy for adapting to climate change, especially when energy diversification allows a more flexible mix of sources of energy or when more climate-resilient systems are constructed (Hess, 2013). Energy supply shocks often motivate the first steps to transition (Aklin & Urpelainen, 2018, p. 12). This chapter focuses primarily on renewable energy and energy transition as climate mitigation action, as this is its most prominent and distinctive role to date.

Determining which energy sources are part of a green energy transition requires attention to the whole life cycle of the energy source. Fossil fuels like petroleum and coal are clearly on the high end of contributors to climate change. As a group, the fossil fuels produce far more

GHG emissions per unit of energy, with the high end of the estimates seven times the high end of the estimates of the carbon emissions of non-fossil fuel renewable sources of energy (Turconi et al., 2013, p. 559). The latter include energy sources such as hydropower, biofuels of various kinds, wind power, and solar power, with nuclear power based on a low-carbon, but not renewable, fuel.

Where emissions appear in the life cycle is very different for these two groups. With fossil fuels, 83–99% of their emissions come from their direct use in functions like electricity generation and transportation (Turconi et al., 2013, p. 562; see also Nugent & Sovacool, 2014). The latter renewable group have their emissions in indirect parts of their life cycle instead, with 97–99% of emissions in the mining, refining, and disposal of their infrastructural materials, for example, in the production and later disposal of solar cells and wind turbines. Their actual production of electricity is itself carbon-free (International Energy Agency, 2021; Turconi et al., 2013, p. 562). China's domination of solar component manufacturing and its reliance on coal-powered electricity to do so, for example, have significantly increased the GHG emissions associated with the life cycle of global solar power (International Energy Agency, 2022, p. 8), but those have been dropping as China transitions from coal power. Other energy sources like natural gas and biofuels are lower in carbon dioxide emissions than coal and petroleum when they are burned for fuel, but their production and transport emit other GHGs (methane and nitrous oxide, respectively). Biofuels depend on agricultural production, which often relies in turn on fossil fuels for cultivation and fertilizer and/or motivates deforestation and land-use change (Turconi et al., 2013, pp. 562–563).

"Green" energy may also refer to a broader set of environmental concerns beyond climate change. These can present "green vs. green" dilemmas, where energy sources that are low in GHG emissions may still have significant environmental impacts in other ways (Hochstetler, 2021; Yonk et al., 2013). Hydropower, for example, is notorious for the environmental and social impacts of its large reservoirs, with similar cumulative effects from aggregations of small dams (Bakken et al., 2012; Khagram, 2004). Some hydropower installations also emit methane, a GHG, especially in tropical countries (Barros et al., 2011). Nuclear power production results in radioactive waste that must be stored safely to protect ecosystem and human health (Aldrich, 2011). Wind and photovoltaic solar power and the batteries needed to balance their intermittent supply and power electric vehicles require considerably more minerals than do fossil fuel equivalents (International Energy Agency, 2021, p. 5); as they are expanded to meet climate goals, impacts on mining communities and their ecosystems will also expand (International Energy Agency, 2021, p. 12). These concerns are discussed in more detail in the section on "Limitations and Opposition."

There are many technical questions to be asked about the feasibility of a fully green energy transition, not least whether the inherent intermittency of many renewable sources of energy can meet all kinds and levels of demand and whether nuclear power should be included as green because of its non-climate environmental effects and nonrenewable fuels required (e.g., Clack et al., 2017; Jacobson et al., 2015). Integrated assessment models that quantitatively model the interaction of energy, environmental, and economic variables are a tool meant to inform policymakers about technical possibilities and requirements. They have been particularly prominent in the work of the Intergovernmental Panel on Climate Change (IPCC) (e.g., Intergovernmental Panel on Climate Change, 2014; Newell, 2021; Parson & Fisher-Vanden, 1997). The technical feasibility of clean energy has clearly increased over time, reflected in the IPCC's own regular reports to governmental negotiators on climate change.

As a result of the technical advances, the questions about the political and social feasibility of green energy transition increasingly appear to be its biggest potential challenge.

Fossil fuel–based energy systems have been fundamental to global economic and political history, laid down in physical infrastructure as well as a whole series of economic and social practices and power relations, including democracy itself (Katz-Rosene & Paterson, 2018; Mitchell, 2011; Oatley, 2021). To give just one example, "a rapid transition to electric vehicles will be shaped by questions of labor relations, inter-capitalist competition, intellectual property rights politics, and interstate competition" (Paterson, 2021, pp. 401–402). Electric vehicles also leave many unsustainable consumption patterns of wealthy individuals and countries intact (Paterson, 2021). Much of the rest of this chapter consequently explores the political, economic, and social dimensions of such a potential large and rapid change.

Whatever the drawbacks and contradictions of green energy transition noted here, the rise of renewable energy in electricity embodies one of the most significant arenas of recent global climate action. Most countries—164 of them, across all divides of comparative political economy—had some kind of renewable energy target by 2015 (International Renewable Energy Agency [IRENA], 2016, p. 8). Other aspects of energy transition and climate action are not advancing as quickly, but this one alone has already been transformative in its speed and scale. Wind and solar power led the change, growing from 0.2% of global electricity in 2020 to 13.4% in 2023 (Ember, 2024, p. 6). This rise brought renewables to 30% of the installed capacity of electricity worldwide, with hydropower providing much of the balance; nuclear power raised the total for low-carbon electricity to 39.4% (Ember, 2024, p. 11). The gains are widely shared, with 102 countries each showing renewable shares at 30% or more in 2023 (Ember, 2024, p. 11).

Growth has been much faster in the Global North than the Global South, excluding China, following uneven patterns of investments in energy transition. About 120 developing and emerging markets received 15% of renewable energy investments, well below their 50% share of the global population, and most of those investments went to Brazil, Chile, and India. The 45 least developed countries received an annual average of only 0.84% of total renewables investments from 2013 to 2020 (IRENA & CPI, 2023, p. 17). North American (excluding Mexico) and European investments per capita grew from 23 times the investments of sub-Saharan Africa in 2015 to 57 and 41 times the African levels in 2021. These developments largely reflected a halving of investments in sub-Saharan Africa over those years, so the gap is widening rather than narrowing (IRENA & CPI, 2023, p. 17). Part of this uneven dynamic reflects the heavy reliance on private investments, about 75% of the total from 2013 to 2020. These went mostly to mature technologies and economies (IRENA, 2023, p. 168). UN-based climate funds have not met either the promises of developed countries or the infrastructure needs of developing countries, hampering energy transition in the latter. The lack of agreed-upon accounting rules can make it difficult to know what exactly has been provided, but the gaps are clear (Roberts et al., 2021).

Even where there has been ample investment, the aggregate growth in low-carbon sources of electricity has functioned primarily to meet growing energy demand rather than displacing fossil fuels in the existing stock. Total annual global energy emissions have increased every year since 1990 except for years of severe dislocations, with the latter including the 2008–2009 global financial crisis and the years of the COVID-19 pandemic. However, the rate of increase has slowed considerably due to the expansion of renewables and decreases in GHG intensity (GHG emissions per unit of GDP), especially in the Global North and China: Global energy emissions increased 0.82 gigatons of CO_2E per year from 2000 to 2007 and only 0.37 gigatons annually from 2010 to 2019 (Climate Watch, 2024). Emissions from electricity itself have dropped even more quickly and seen annual declines in some recent years (Climate Watch, 2024).

Facilitating Conditions

Why has green energy transition, especially in electricity, advanced so much more quickly than other kinds of climate action? This section of the chapter examines the facilitating conditions for energy transition. Many of these derive from the fact that the expansion of green energy is seen as offering many "co-benefits." As described earlier, these are additional benefits that arise with energy transition alongside its climate benefits. Some initiatives promoting energy transition may be motivated primarily by these other outcomes but will still have their effects on GHG emissions. These extra benefits can be instrumental for building broader political coalitions to support climate action (Ansolabehere & Konisky, 2014; Meckling et al., 2015; Zysman & Huberty, 2014). The presence of the co-benefits contrasts to a common view that emphasizes the costs and limits imposed by climate action, reflected in the use of phrases like "burden-sharing" in international climate negotiations (Bhatti et al., 2010; Hochstetler & Milkoreit, 2015). As a result, green energy transition has received targeted support from both state and private actors, unlike many areas of potential climate action. Actors may "believe" in climate change to motivate their support for green energy transition—and many do (e.g., Verschoor et al., 2020)—but they do not have to. There are additional reasons to transform the energy system.

Notably, green energy transition requires building up new industrial sectors from renewable energy components to carbon capture and storage industry, which has proven to be a tantalizing prospect for public and private actors alike across the world (Aggarwal & Reddie, 2021; Chen & Lees, 2016; Gallagher, 2013; Rodrik, 2014). In addition to promising broader economic benefits like "economic growth, security of energy supply, and increasing employment," green energy has been seen as a particularly strong driver of innovation and growth acceleration (Hochstetler, 2021, p. 79; Nahm, 2021). Some of these benefits come in developing and refining the new sources of low-carbon electricity themselves, while other innovation impulses come from a subsequent turn to electrifying other sectors like vehicles now that renewable energy sources are increasingly available (Allan & Nahm, 2024; Zhao & Yuan, 2021). There are also technological fronts in the ongoing but more climate-friendly use of fossil fuels, such as carbon capture, usage, and storage (CCUS) strategies (Wettestad et al., 2024; see chapter by Yang in this volume).

Some of the most prominent approaches to understanding energy transition consequently see it through the lens of innovation studies (e.g., Lester & Hart, 2012; Nemet, 2019). The socio-technical transition approach associated with the work of Frank Geels provided an especially influential framework for understanding how what began as a small-scale and expensive experiment in developing renewable electricity in Europe could become a global phenomenon diffused in mainstream markets (e.g., Geels, 2002, 2014). While other frameworks exist, Geels's multilevel perspective has influenced many of them in the way that it traces interactions that might result in systemic change "from a *niche* technology and its supporters seeking to break into a market controlled by incumbent interests, thought to be part of a *regime* (often assumed to be operating at the national level), up to a series of *landscape* pressures that are often assumed to emanate from the global level" (Newell, 2021, pp. 28–29).

While earlier work on socio-technical transition often overlooked the role of the state, the "irreducibly political" character of transition increasingly came to the forefront as the framework was applied to climate change and energy transition (Meadowcroft, 2009, p. 323; see also Healy & Barry, 2017; Breetz et al., 2018). Among other political dimensions, the distributive nature of climate policy means conflict is likely among different social groupings that mobilize to promote their interests in the face of the changes associated with climate action (Aklin & Mildenberger, 2020). These cross-cut more traditional alliances, with some business and labor actors joining together to share a position on transition that depends on their industry's relationship to climate change—are they industries that accelerate climate change, are vulnerable to climate change, or benefit from action to address climate change?—while other business and labor actors will find other shared interests (Colgan et al., 2021; Hochstetler, 2021; Mildenberger, 2020; Nahm, 2021). Many of these critiques arise from an essentially liberal view of political contestation, while neo-Gramscian analyses of the political economy of energy transition ask further whether "politics and policy reinforce a market liberal approach to transitions *within* capitalism as opposed to more sweeping transformations of it" (Newell, 2021, p. 29).

For all understandings of energy transition as politicized climate action, the powerful resistance of actors who are dependent on assets that accelerate climate change like fossil fuel plants is a serious challenge. Their power is typically "locked-in" through numerous physical, economic, and political linkages that allow them to externalize the environmental and social costs of fossil fuel production and use (Aklin & Urpelainen, 2018; Hess, 2014; Moe, 2015; Stokes, 2020; Unruh, 2000). They deploy numerous political and economic strategies to try to undermine even seemingly socially beneficial outcomes from energy transition or to steer changes to have the least impact on their own power and position. Peter Newell calls this *transformismo* rather than transformation (Newell, 2021).

As a result of this likely distributional conflict with powerful incumbent actors in the energy sector, states have typically been required to intervene to advance green energy innovation and transition, a further political dimension. This requires significant state capacity, both positive and negative: states must have the ability to see "long-term strategic needs to address climate change ... even when they mean imposing costs on powerful actors with vested interests in the status quo" (Hochstetler, 2021, p. 31; Meckling & Nahm, 2017, p. 741; Pearson & Foxon, 2012, p. 121; see chapter by Madariaga et al. in this volume).

For green energy transition, many authors have argued that imposing such costs requires states to develop green industrial policies that will provide the necessary state coordination and resources to help nascent alternative low-carbon energies to develop and take hold (Allan et al., 2021; Allan & Nahm, 2024; Bradlow & Kentikelenis, 2024; Breetz et al., 2018; Colgan et al., 2021; Finnegan, 2022; Hochstetler, 2021; Juhász et al., 2024; Lewis, 2013; Meckling, 2021; Pegels, 2014; Rodrik, 2014). In doing so, they may be supporting new-product innovation, but they may also be simply promoting the localization of production and/or installation of the energy upstarts. As renewable energy has moved to the Global South, beyond China, the latter aims have most often moved to the forefront (Hochstetler, 2021; Lewis, 2013; Pegels & Altenburg, 2020).

Industrial policy for green energy transition carries potential direct economic benefits in innovation, industry development, and the critical metric of new kinds and sources of employment. The "good green jobs" of energy transition have been a very powerful motivator for states to initiate the transition in both the Global North and the Global South (Hess, 2012; Hochstetler, 2021). These economic co-benefits are thought to provide an additional positive stimulus for energy transition by building broad coalitions of actors who have an active interest in seeing the new forms of electricity take hold (Meckling et al., 2015). They are behind the logic of the various green new deals that seek to link climate to labor, housing, and other social issues (e.g., Aronoff et al., 2019). Early generations of renewable energy transition, mostly in the Global North, benefitted from at least partial "green spirals," where initial state support for development of the industries that manufactured components and constructed energy installations generated a larger pro-transition coalition where workers and corporate leaders in the sector pushed for yet more support and expansion from the state (Hochstetler, 2021; Huberty, 2014; Kelsey & Zysman, 2014).

Another area of hoped-for industrialization related to green energy transition has been in upgrading to more valuable levels of the production chain of transition-related minerals, for example, refining and processing minerals rather than simply extracting them (Gutiérrez et al., 2022; Irarrazaval & Carrasco, 2023; Torjesen, 2023). Communities that were able to own and sell renewable electricity to national grids, mostly in Germany and Denmark, were another important constituency for change (Morris & Jungjohann, 2016; Swilling, 2020).

While this is a promising dynamic for green energy transition, when the promised benefits do not appear—as they largely did not in South Africa—the lack of clear benefits can also move potential supporters into opposition to transition (Hochstetler, 2021, pp. 231–232). The risk of this failure to meet expectations is rising as the new sources of energy are no longer so new. The global value chains of wind and solar power, for example, have significantly matured, with Chinese manufacturing having even frozen out rich advanced countries that were the first-mover developers and innovators (Allan & Nahm, 2024, p. 3). Advanced industrial economies are reacting with statecraft and industrial policy initiatives to regain market position, putting extensive resources into initiatives like the United States' Inflation Reduction Act and the European Green Deal (Aggarwal & Reddie, 2021; McNamara, 2023). For the Global South, when even regional leaders like Brazil and South Africa struggle to find their footing, the prospects for other developing countries do not match the optimism with which many of them set out renewable energy ambitions hoping for such good green jobs (Hochstetler, 2021, pp. 231–232; Radley & Lehmann-Grube, 2022).

In addition, a full green energy transition eventually means displacing growth in traditional high-carbon sources or even shutting them down, bringing countervailing economic losses for the world economy as well as individual countries, communities, and firms (Ansari & Holz, 2020; Gaulin & Billon, 2020; Harrison & Bang, 2022; Hasan et al., 2024; Newell & Carter, 2024; Princen et al., 2015; van Asselt et al., 2024). Those losers have often mobilized against the coming changes. While more often studied in the Global South, comparative study has found very similar justice claims in both the Global South and the Global North (e.g., Busch et al., 2023; Krawchenko & Gordon, 2021). The social and economic losses from phasing out fossil fuels require additional attention to the distributional politics of energy transition, which are discussed in the next section as the "just transition" debate. The term "co-losses" has not had the same cultural trajectory as that of "co-benefits," but losses also feature alongside the gains of green energy transition.

To complete the picture, not all potential co-benefits are associated with the production of low-carbon energy as an industrial and economic advance. Others offer potential benefits to consumers. The ability to build solar power at very small scales, even at the household level ("distributed solar power"), means that it is a promising means of energy access in remote and off-grid locations, particularly in developing countries where many people still lack access to electricity (Brass et al., 2012; Roche et al., 2020; Sovacool & Drupady, 2012). Even in urban areas, the possibility of individual or community electricity self-provision is thought to democratize electricity access and control (Abromovay, 2014; Brass et al., 2012). A large literature debates whether significant additional benefits for work and education follow, which might in turn promote regional and national development (e.g., Aklin et al., 2017; Radley & Lehmann-Grube, 2022; Slough et al., 2015).

The quality of the resource may be quite low, however, with the offer of publicly provided distributed solar power in South Africa being so minimal that it was dismissed as itself an environmental injustice (Monyei et al., 2018). When wealthy customers move off the grid to provision their own solar power, common in South Africa and elsewhere, the public service faces a potential "utility death spiral" as it may no longer have the resources to sustain the physical network (Castaneda et al., 2017; Graffy & Kihm, 2014; Hochstetler, 2021, pp. 147–148; Konzen et al., 2024). As such, there may well be "co-losses" as well as co-benefits for consumers of the energy transition too.

Limitations and Opposition

As green energy transition has advanced since 2000, its limitations are emerging along with new forms of opposition. In this section I focus on two prominent clusters of concerns. One is around the uneven and slow reductions in fossil fuel production and use even as new greener sources of energy are developed. This raises questions about the completeness of the current green energy transition and its usefulness as climate action. The second cluster of concerns examines difficult questions about how the green energy transition can be carried out in an equitable and inclusive way. A comprehensive view needs to examine the justice dilemmas involved in moving away from fossil fuels as well as those emerging with the addition of the new greener sources of energy, especially as they grow in scale. Both dynamics bring new impacts and costs for local communities and environments, as well as for larger scales like economic sectors and nation-states. These questions are most pressing for already-vulnerable countries and groups within them. While a fuller view of climate justice definitions and debates can be found elsewhere (see Alvarez & Norton-Smith in this volume; Heffron & McCauley, 2018), my focus here is on the dimensions of justice specific to green energy transition that are already empirically unfolding. The dimensions are both compensatory/distributive and procedural (Bolet et al., 2024, p. 4).

To start with the first concern, the growth of decarbonized and renewable forms of energy has not yet been matched by significant decreases in the production and use of fossil fuels worldwide. Only one major oil producer, Venezuela, has experienced a drastic decline in oil production since 2015, and it has been an unplanned and painful decline that shows how important state capacity is for managing such a phaseout in a sustainable and just way (Rosales & Clark, 2023). In addition, one of the most successful energy transitions of recent years has been the shale oil revolution, where new hydraulic fracking technologies have allowed a huge expansion of natural gas extraction. This is one of the negative trends to which sociotechnical transitions analysis is rarely applied (Antal et al., 2020). As coal production and use have in fact been reduced, natural gas has largely taken coal's place as a result, changing the composition of GHG emitted but keeping total energy emissions high (IRENA, 2023). Some actors, including the European Union, have favored the turn to gas as a transitional fuel, as it produces less emissions than coal and petroleum at the point of burning and does not emit the very long-lasting carbon dioxide. However, there are extensive emissions of methane in supply chain processes and methane has very intense warming effects, if for a shorter time period (Alvarez et al., 2018; Holz et al., 2015).

As many others have pointed out, the absence of fossil fuel phaseout—an absence of green energy transition—is a failure of climate action. It is itself unjust to the countries and individuals most affected by climate change, in the present and even more in the future (Gupta et al., 2024; Newell, 2021). As the injustices of energy transition draw a great deal of current analytical attention, this basic fact should not be forgotten. The ongoing expansion of fossil fuel extraction and use is not an accident, but the product of concerted action. Fossil fuel producers, both countries and firms, have fought hard to keep mentions of fossil fuel phaseout out of international climate agreements and were wholly successful until the 26th UN Climate Change Conference of Parties (COP 26) in Glasgow in 2021. The next year in Egypt, one study counted 636 fossil fuel lobbyists present, larger than any single country delegation except the UAE (Barnett, 2008; Gupta et al., 2024, pp. 235–236). The UAE, an oil producer, reportedly even used its own host role the following year to pursue industry expansion (Carrington, 2023).

Global changes in fossil fuel production and use are beginning to arrive, however, comparatively quickly for coal and more slowly for petroleum. Between 1988 and 2017, 106 countries undertook 1,302 initiatives that would cut the supply of fossil fuels (Gaulin & Le Billon, 2020, p. 888; Harrison & Bang, 2022), such as Ecuador's well-known offer to leave oil reserves in the ground if sufficient international funds were granted (Alarcón, 2024; Martin, 2011; Riofrancos, 2020). Voters recently supported a referendum in Ecuador against new drilling, even after the governmental initiatives were withdrawn. A review of possible approaches to leaving fossil fuels underground found 28 distinct proposals, although only 12 were deemed environmentally effective and none of those were easily feasible, with half facing the likely opposition of powerful stakeholders (Rempel & Gupta, 2022).

Petrostates, dependent on production, consumption, and export of fossil fuels, would feel the economic pain most keenly (see chapter by Al-Sarihi in this volume). Even wealthy ones like Canada and Norway are reluctant to give up their fossil fuel–based prosperity (Harrison & Bang, 2022; Hasan et al., 2024). While there has been more transition out of coal mining and energy, the coal states remaining tend to have very strong mobilizations against phaseout that stress the losses for state-owned enterprises, public budgets, and export earnings, with ruling parties and heads of state often leading the determination to avoid transition (e.g., Ohlendorf et al., 2020; Ordonez et al., 2021; Tyler & Hochstetler, 2021). Given how much national decision makers in coal countries continue to demand coal expansion (e.g., Gallagher et al., 2021), it is likely that only the steady withdrawal of sources of financing for coal and eventually oil and gas will achieve true fossil fuel phaseout (but see Rempel & Gupta, 2022). Currently, they are even still heavily subsidized (Van de Graaf & Van Asselt, 2017).

If and when fossil fuels extraction and use are eventually reduced, that will bring costs and dislocations. Global scenario analysis of fossil fuel rollback shows significant GHG emissions reductions, higher prices for remaining fossil fuels produced, further growth for renewable energy, and negative global GDP effects, with all results magnified by more ambitious reductions, especially for petroleum (van Asselt et al., 2024). Some of these losses are captured by the concept of stranded assets—fossil fuel resources that will go untapped and the accompanying infrastructure, labor, and other assets that will lose value and position—which raise difficult questions for firms and countries about when and how to stop investing in fossil fuel extraction and refining (Heras & Gupta, 2024; van de Graaf and Bradshaw, 2018). Other dimensions are captured by the just transition dilemmas that I will focus on from here.

Justice issues arise at a global scale, especially for developing countries that have not yet been able to turn their fossil fuel production into sustainable economic growth because they are the ones that hold many of the remaining stores: "As of 2020, non-OECD countries (including Colombia and Mexico) had almost 85% of proved oil reserves, 89% of natural gas and 53% of coal" (Heras & Gupta, 2024, p. 2). This includes both countries that are actually low and middle income and also wealthier countries that often join them in the climate negotiations, at least in part on the grounds that their fossil fuels have not led to real development (Barnett, 2008). Academic scholars partially agree, with fossil fuels comprising the original resources of the "resource curse," where a series of political and economic knock-on effects prevent oil wealth from resulting in the economic growth and stability that it should (Karl, 1997; Ross, 2012). In climate negotiations, developing countries often argue that wealthy countries produced most of the historic emissions that are now generating climate change, growing wealthy in the process; they are now trying to prevent developing countries from doing the same (Barnett, 2008; Saha, 2024; Saha & Carter, 2022). Very few of the approaches for phaseout consider the fairness of these tools, particularly in a North-South context. Many do not even consider the Global South experiences, even for mainstream options like a carbon tax (Rempel & Gupta, 2022, p. 25).

Moving down the geographic scale, specific subgroups would also suffer heavy impacts. The fossil fuel labor force is a prominent example. It was the international labor movement that introduced the term "just transition" into global discussion in the 1980s, and it continues to be an important voice for those who are potentially left out by environmentally driven changes like green energy transition (Morena et al., 2020; Stevis & Felli, 2015; see chapter by Kerrissey and Coon in this volume). Coal workers in many countries, including South Africa and Spain, have mobilized to make sure that green energy transition takes account of those working in existing energy industries (Bolet et al., 2024; Hochstetler, 2021). While some individuals may be able to transition into the rising energy industries with retraining, more will need support through retirement and other packages that address the long-term health effects of life in coal mines or petroleum platforms (Morena et al., 2020). It is worth noting that while workers value these jobs, they also carried substantial risks to health and even life, with 1,549 coal miners dying in just one mine explosion in China in 1942 and India experiencing seven mine explosions with more than 50 fatalities (Sahu & Mishra, 2023, p. 1), to say nothing of longer-term health risks of participation in extraction.

Host communities have also often been both economically dependent on fossil fuel extraction *and* subject to severe health and environmental impacts from it (Bebbington & Bury, 2013; Riofrancos, 2020; Sawyer, 2004; Shah, 2023; Ukhurebor et al., 2023). As a result, they often share interests and concerns with the fossil fuel labor force, including the ambiguities of their positions. Like the workforce, the communities suffer the loss of many economic benefits with green energy transition while still continuing to face the legacies of the historic social, health, and environmental impacts (Sanz-Hernández et al., 2020).

The increasingly common solution to the energy-related dilemmas of both labor movements and host communities has been a negotiated just transition agreement of some kind, typically state-led but with participation from other actors. Among the most successful has been the 2018 Just Transition Agreement in Spain, brought in by the Spanish Socialist Party PSOE in response to an EU mandate (Bolet et al., 2024, p. 5). The Agreement was notable for both its generous compensation schemes for the 1,677 estimated workers who would lose their jobs and for affected municipalities. It was created via an inclusionary and trust-building consultation with business and labor participants (Bolet et al., 2024, p. 5). The PSOE's electoral support in 2019 in affected coalmining municipalities proved even higher than in similar non-coalmining municipalities. Interviews suggested that the clear support of unions who explained the process and result to their members contributed to this result (Bolet et al., 2024, p. 11). A broader review of the Just Transition Mechanism of the European Union finds uneven implementation in 25 countries and 74 regions, with inadequate funding and a lack of coordination between infrastructure and labor force planning at the root of many of the problems (Krawchenko & Gordon, 2021).

If the Spanish resolution shows the possibilities of such a negotiated approach, the South African experience offers cautions. South Africa has also seen extended efforts over more than a decade to negotiate a just transition arrangement. While offering consultation in numerous meetings, the South African processes have not engendered the same levels of trust and confidence. There have been deep divisions that derive from South Africa's racialized apartheid past, as well as more specific doubts about the proposed role of private (White) capital in its green energy transition (Lenferna, 2023; Tyler & Hochstetler, 2021). The meager jobs results of the initial rounds of renewable energy expansion also did not inspire confidence in the much large numbers still employed in the coal labor force, some 92,000 vulnerable jobs by union calculations (Hochstetler, 2021, p. 117). South Africa was the first country to negotiate an international Just Energy Transition Partnership, in 2021. While the Partnership was meant to provide international resources for an expensive transition, only 4% of the $8.5 billion was pledged as grants (Earsome, 2024) and only the same percentage of it was meant for just transition spending rather than support for the general electricity sector, mostly infrastructure (Lenferna, 2023).

Whatever the limitations of transition packages for fossil fuel workers and communities, there are currently no broad initiatives to address the other side of the just transition dilemma, the possible additional costs and impacts of the green energy transition itself. Scholars have drawn attention to two major areas of concern. The first is in the social and environmental impacts around the minerals and metals that will see their demand and extraction leap with the expansion of renewable energy, batteries, and an array of related features of the green energy transition. The second is with the communities that host the new infrastructures of renewable energy.

Expanding green energy sources will require significantly more minerals and more of scarce minerals than continuing with fossil fuel production (International Energy Association, 2021). This means an expansion of the "environmental degradation, adverse impacts to public health, marginalized communities and workers, and biodiversity damage" that has been essentially inherent in the mining of these (Sovacool et al., 2020, p. 30; see also Mulvaney, 2013; Sovacool, 2019). In rapidly growing academic, activist, and policy literatures, researchers note "place-based complaints and demands associated with distributive, procedural, and recognition justice" around lithium mining in Latin America (Escosteguy et al., 2024, p. 1; see also Barandiaran, 2019). Environmental impacts around critical minerals in Africa are similar (Boafo et al., 2024), while sociocultural losses are common across the Global South (Ash, 2023; Mantanzima & Loginova, 2024). Where minerals are mined in the Global North, the patterns are also very similar (e.g., Chaves et al., 2021; Kingsbury & Wilkinson, 2023; Phadke, 2024). Public and private regulations are being developed for these impacts, but they do not yet seem to be able to protect host ecosystems and communities (Kramarz et al., 2024).

The familiar coexistence of economic benefits with social and environmental costs makes it unsurprising that close studies of host communities find correspondingly complex politics of extraction. Kieran Gilfoy, for example, describes extractive communities in the Andes as living in "toxic endurance," seeing the poison and degradation of living near extraction but also, especially in early days, the promise of economic development (Gilfoy, 2022). The diversity and specificity of community mobilizations around mining are important to observe. For example, many mining conflicts in Peru are less stop-the-extraction conflicts than they are about improving compensation and settlements, although these may be "forms of (bitter) accommodation" with commodification of people and livelihoods (Orihuela et al., 2022, p. 1). In Chile, Indigenous groups also have been using contestation around lithium mining to make claims about compensation, but they have joined those claims on business actors with efforts to draw state attention to their territorial claims (O'Faircheallaigh & Babbidge, 2023). Activists may be acting through state participatory opportunities or around them (Jaskoski, 2022).

Are the justice considerations of the transition *to* low-carbon energy sources simple parallels to those of the transitions *from* fossil fuels? Clearly one differentiating factor is that the transitions to green energy generally involve prospective new impacts on societies and environments in both space and time, while the fossil fuel impacts tend to be long-standing and often partially accommodated (McAdam & Boudet, 2012). As such, they are more visible to both activists and academics. Beyond that, the impacts of the new and higher demand for minerals extraction largely in the Global South derive in many cases from actors in the Global North seeking solutions to climate change, a problem that they disproportionately caused. A recent study concludes, for example, that the "EU's energy goals might expose 15-89,000 African miners to modern slavery risks" and sums up the additional water that will be consumed in water-scarce regions (Berthet et al., 2024, p. 1). The displacement of harms to less responsible parties thus adds additional injustice to the equation. Relatedly, in Argentina and Chile, researchers have concluded that conflicts around lithium mining were particularly intense compared with other related conflicts from 2010 to 2019, arguing that economic opportunities and green discourses associated with energy transition apparently led state and private actors to comparatively neglect impacts on Indigenous communities (Ciftci & Lemaire, 2023).

One final area where justice considerations arise in a green energy transition is with the need to place the infrastructure of green energy installations in particular locations, where they might impact the human communities and natural environments that predate them. For this topic, too, there are both expectations of possible economic gains for the host communities and possible socioenvironmental losses (Hochstetler, 2021, pp. 175–177). Much of the non-economics social science literature has focused on possible local losses, most often on environmental or equity grounds (Avila, 2018). This is not least because host communities of wind and solar farms and the transmission lines that link them to larger grids have often mobilized in resistance to them (e.g., Mulvaney, 2017; Ogilvie & Rootes, 2015; Stokes, 2016; Yonk et al., 2013). Often dismissed as narrow NIMBY (not in my back yard) mobilizations (Dear, 1992), it is again important to remember that community mobilizations against renewable energy can be quite complex—when they happen.

In my own recent review of contestation around the full set of grid-scale wind and solar installations in Brazil and South Africa, I found considerably less community resistance than a review of the qualitative literature would suggest. In Brazil, for example, many studies had focused on the very conflictual municipality of Aracati and on the coastal northeastern region where it sits. But only 19 of 77 host cities of wind power reported any conflicts, and they were all in the northeast. No conflicts around solar power in either country were visible in the review, and residents of only one municipal area in South Africa contested wind power (Hochstetler, 2021, pp. 202–206). This is not to say that the local mobilizations that happened should be ignored. Many were in environmentally sensitive areas and/or brought significant disruptions to the "economic and cultural practices of local subsistence communities of African slave descent" (Hochstetler, 2021, p. 204). Similar patterns have appeared in Mexican wind power, concentrated in Indigenous communities with minimal consultation (Martinez, 2020). These are exactly the sorts of historically disadvantaged communities often thought to be targeted for disproportionate and disruptive infrastructural developments (Avila, 2018). Even so, the resistances of communities vary substantially, and the evaluations of the impact of renewable energy installations should, too.

Conclusions: Finding a Just Transition From an Unjust Status Quo

This chapter has aimed to take account of the complexity of green energy transition as climate action, paying due attention to both the positive possibilities and the negative dilemmas of this kind of action. In this conclusion, I will argue that the balance still clearly holds for viewing green energy transition as useful and even potentially just climate action.

The IPCC's regular assessments of the academic literature on climate change and climate action strategies have repeatedly underlined that green energy transition is a useful tool for decarbonization and mitigating the effects of future climate change. The transition to low-carbon and renewable sources of electricity is particularly valued for its ease and cost-effectiveness as climate action. As this chapter has shown, the growth alone of green energy has not yet proved sufficient for decarbonizing the whole energy sector, as fossil fuels have continued to be developed and used. This is an additional step needed for the completeness of green energy transition and for the completeness of such a transition as a form of climate action. The academic literature on fossil fuels has been particularly convincing on why this is such a difficult challenge, with the way forward requiring sustained struggle with very powerful incumbent energy actors.

The lack of full-scale green energy transition and corresponding climate action is an injustice to the many people of the present and future whose cultures, livelihoods, and even lives depend on avoiding the climate change that can still be avoided. This is an important framing thought for acknowledging that such a full-scale energy transition will itself bring other injustices. This chapter has identified multiple injustices, such as the economic losses of the countries of the Global South that will not be able to develop their fossil fuel resources as countries of the Global North did. However, fossil fuel development also brought many distortions to development in its resource curse and those, too, can potentially be avoided.

Below the national level, the chapter found some parallel injustices for both the communities and workers associated with the status quo of fossil fuel production and use and those associated with the new settings of renewable energy production and use. For all, there are economic promises (not always met) joined with social, environmental, health, and cultural impacts. Arguably, there is a common story about the injustice that comes from living and working with extraction and economic infrastructure. As such, green energy transition does not make a just situation unjust, but introduces new and different kinds of injustice into what is already a highly unjust status quo. This does not excuse the injustices associated with green energy transition, but it provides a historical caution against imagining a world of energy without exploitation. It also demands a full-scale and multifaceted attention to those various forms of exploitation. Just transition initiatives so far often have failed in both the compensation and the participatory processes they offer.

Just transition in a green economy transition is worth pursuing on its own grounds. But there is also the political statement that for green energy transition to be sustained and sustainable as climate action, it must be a just transition. Ignoring the social impacts of green energy transition can generate community or labor opposition with negative implications for low-carbon energy transition altogether. A blocking coalition with the actors already profiting from fossil fuel extraction and use is a very real possibility, while just transition strategies may build support for green energy transition and increasingly meaningful climate action.

Note

1. Following common practice, this chapter often uses carbon and decarbonization to refer to the full set of greenhouse gases. For example, CO_2E (carbon dioxide equivalent) is a measure of GHG that expresses the quantity of warming potential of all greenhouse gases in terms of their equivalence to the warming potential of carbon dioxide.

References

Abromovay, R. (2014). Innovations to democratize energy access without boosting emissions. *Ambiente e Sociedade, 17*(3), 1–18.

Aggarwal, V. K., & Reddie, A. W. (2021). Economic statecraft in the 21st century: Implications for the future of the global trade regime. *World Trade Review, 20*, 137–151.

Aklin, M., Bayer, P., Harish, S. P., & Urpelainen, J. (2017). Does basic energy access generate socio-economic benefits? A field experiment with off-grid solar power in India. *Science Advances, 3*(5), Article e1602153.

Aklin, M., & Mildenberger, M. (2020). Prisoners of the wrong dilemma: Why distributive conflict, not collective action, characterizes the politics of climate change. *Global Environmental Politics, 20*(4), 4–27.

Aklin, M., & Urpelainen, J. (2018). *Renewables: The politics of a global energy transition*. The MIT Press.

Alarcón, P. (2024). What's next for supply-side policy in the South: Emerging lessons from Ecuador's Yasuní Initiative. *International Environmental Agreements, 24*, 75–90.

Aldrich, D. P. (2011). *Site fights: Divisive facilities and civil society in Japan and the West*. Cornell University Press.

Allan, B. B., & Nahm, J. (2024). Strategies of green industrial policy: How states position firms in global supply chains. *American Political Science Review*, 1–15. https://doi.org/10.1017/S0003055424000364

Allan, B., Lewis, J. I., & Oatley, T. (2021). Green industrial policy and the global transformation of climate politics. *Global Environmental Politics, 21*(4), 1–19.

Alvarez, R. A., Zavala-Araiza, D., Lyon, D. R., Allen, D. T., Barkley, Z. R., Brandt, A. R., Davis, K. J., Herndon, S. C., Jacob, D. J., Karion, A., Kort, E. A., Lamb, B. K., Lauvaux, T., Maasakkers, J. D., Marchese, A. J., Omara, M., Pacala, S. W., Peischl, J., Robinson, A. L., … Hamburg, S. P. (2018). Assessment of methane emissions from the U.S. oil and gas supply chain. *Science, 361*(6398), 186–188.

Ansari, D., & Holz, F. (2020). Between stranded assets and green transformation: Fossil-fuel-producing countries towards 2055. *World Development, 130*, Article 104947.

Ansolabehere, S., & Konisky, D. M. (2014). *Cheap and clean: How Americans think about energy in the age of global warming*. The MIT Press.

Antal, M., Mattioli, G., & Rattle, R. (2020). Let's focus more on negative trends: A comment on the transitions research agenda. *Environmental Innovations and Societal Transitions, 34*, 359–362.

Aronoff, K., Battistoni, A., Aldana Cohen, D., & Riofrancos, T. (2019). *A planet to win: Why we need a green new deal*. Verso.

Ash, J. (2023). Social impacts of critical mineral exploration on indigenous peoples' lands: A case study from Solomon Islands. *The Extractive Industries and Society, 17*, Article 101439.

Avila, S. (2018). Environmental justice and the expanding geography of wind power conflicts. *Sustainability Science, 13*, 599–616.

Bakken, T. H., Sundt, H., Ruud, A., & Harby, A. (2012). Development of small versus large hydropower in Norway: Comparison of environmental impacts. *Energy Procedia, 20*(1), 185–99.

Barandiaran, J. (2019). Lithium and development imaginaries in Chile, Argentina, and Bolivia. *World Development, 113*, 381–391.

Barnett, J. (2008). The worst of friends: OPEC and G-77 in the climate regime. *Global Environmental Politics, 8*(4), 1–8.

Barros, N., Cole, J. J., Tranvik, L. J., Prairie, Y. T., Bastviken, D., Huszar, V. L. M., Del Giorgio, P., & Roland, F. (2011). Carbon emission from hydroelectric reservoirs linked to reservoir age and latitude. *Nature Geoscience, 4*(9), 593–596.

Bebbington, A. & Bury, J. (Eds.). (2013). *Subterranean struggles: New dynamics of mining, oil, and gas in Latin America*. University of Texas Press.

Berthet, E., Lavalley, J., Anquetil-Deck, C., Ballesteros, F., Stadler, K., Soytas, U., Hauschild, M., & Laurent, A. (2024). Assessing the social and environmental impacts of critical mineral supply chains for the energy transition in Europe. *Global Environmental Change*, *86*, Article 102841.

Bhatti, Y., Lindskow, K., & Holm Pedersen, L. (2010). Burden-sharing and global climate negotiations: The case of the Kyoto protocol. *Climate Policy*, *10*(2), 131–147.

Boafo, J., Obodai, J., Stemn, E., & Nkrumah, P. N. (2024). The race for critical minerals in Africa: A blessing or another resource curse? *Resources Policy*, *93*, Article 105046.

Bolet, D., Green, F., & González-Eguino, M. (2024). How to get coal country to vote for climate policy: The effect of a "just transition agreement" on Spanish election results. *American Political Science Review*, *118*(3), 1344–1359.

Bradlow, B. H., & Kentikelenis, A. (2024). Globalizing green industrial policy through technology transfers. *Nature Sustainability*, *7*, 685–687.

Brass, J. N., Carley, S., MacLean, L. M., & Baldwin, E. (2012). Power for development: A review of distributed generation projects in the developing world. *Annual Review of Environment and Resources*, *67*, 62–71.

Breetz, H., Mildenberger, M., & Stokes, L. (2018). The political logics of clean energy transitions. *Business and Politics*, *20*(Special Issue 4), 492–522.

Busch, H., Ramasar, V., Avila, S., Roy, B., van Ryneveld, T., Mandinic, A., & Brandstedt, E. (2023). Mining coal while digging for justice: Investigating justice claims against a coal-phase out in five countries. *The Extractive Industries and Society*, *15*, Article 101275.

Carrington, D. (2023, November 27). COP28 host UAE planning to promote oil deals during climate talks. *The Guardian*. https://www.theguardian.com/environment/2023/nov/27/cop28-host-uae-planned-promote-oil-deals-climate-talks

Castaneda, M., Franco, C. J., & Dyner, I. (2017). Evaluating the effect of technology transformation in the electricity utility industry. *Renewable and Sustainable Energy Reviews*, *80*, 341–351.

Chaves, C., Pereira, E., Ferreira, P., & Guerner Dias, A. (2021). Concerns about lithium extraction: A review and application for Portugal. *The Extractive Industries and Society*, *8*, Article 100928.

Chen, G., & Lees, C. (2016). Growing China's renewables sector: A developmental state approach. *New Political Economy*, *21*(6), 574–584.

Ciftci, M. M., & Lemaire, X. (2023). Deciphering the impacts of "green" energy transition on socio-environmental lithium conflicts: Evidence from Argentina and Chile. *The Extractive Industries and Society*, *16*, Article 101373.

Clack, C. T. M., Qvist, S. A., Apt, J., Bazilion, M., Brandt, A. R., Caldeira, K., David, S. J., Diakov, V., Handschy, M. A., Hines, P. D. H., Jaramillo, P., Kammen, D. M., Long, C. S., Granger Morgan, M., Reed, A., Sivarum, V., Sweeney, J., Tynan, G. R., Victor, D. G., ... Whitacre, J. F. (2017). Evaluation of a proposal for reliable low-cost grid power with 100% wind, water, and solar. *PNAS*, *114*(26), 6722–6767.

Climate Watch. (2024). Global historical emissions [Data set]. https://www.climatewatchdata.org/ghgemissions?breakBy=sector&calculation=ABSOLUTE_VALUE&end_year=2020&sectors=electricity-heat&start_year=1990

Colgan, J. D., Green, J. F., & Hale, T. N. (2021). Asset revaluation and the existential politics of climate change. *International Organization*, *75*(2), 586–610.

Dear, M. (1992). Understanding and overcoming the NIMBY syndrome. *Journal of the American Planning Association*, *58*(3), 288–300.

Dubash, N. (2013). The politics of climate change in India: Narratives of equity and cobenefits. *WIREs Climate Change*, *4*, 191–201.

Earsome, J. (2024). Fit for purpose? Just energy transition partnerships and accountability in international climate governance. *Global Policy*, *15*, 135–141.

Ember. 2024. *Global electricity review 2024*. https://ember-climate.org/insights/research/global-electricity-review-2024/

Escosteguy, M., Insaurralde, C. O., Días Paz, W. F., Brannstrom, B., Iribarnegaray, M. A., Hufty, M., & Seghezzo, L. (2024). The political ecology of justice outcomes of lithium mining: The case of Salar del Hombre Muerto, Argentina. *The Extractive Industries and Society*, *18*, Article 101477.

Finnegan, J. (2022). Institutions, climate change, and the foundations of long-term policymaking. *Comparative Political Studies, 55*(7), 1198–1235.

Gallagher, K. S. (2013). Why and how governments support renewable energy. *Daedalus, the Journal of the American Academy of Arts and Sciences, 142*(1), 59–77.

Gallagher, K. S., Bhandary, R., Narassimhan, E., & Nguyen, Q. T. (2021). Banking on coal? Drivers of demand for Chinese overseas investments in coal in Bangladesh, India, Indonesia, and Vietnam. *Energy Research & Social Science, 71*, Article 101827.

Gaulin, N., & Le Billon, P. (2020). Climate change and fossil fuel production cuts: Assessing global supply-side constraints and policy implications. *Climate Policy, 20*(8), 888–901.

Geels, F. W. (2002). Technological transitions as evolutionary reconfiguration processes: A multi-level process and a case-study. *Research Policy, 31*(8–9), 1257–1274.

Geels, F. W. (2014). Regime resistance against low-carbon transitions: Introducing politics and power into the multi-level perspective. *Theory, Culture & Society, 31*(5), 21–40.

Gilfoy, K. (2022). Toxic endurance and social becoming: Environmentalism in the shadows of Andean extraction. *The Extractive Industries and Society, 11*, Article 100930.

Graffy, E., & Kihm, S. (2014). Does disruptive competition mean a death spiral for electric utilities? *Energy Law Journal, 35*, 1–44.

Gupta, J., Chen, Y., Mckay, D. I. A., Fezzigna, P., Gentile, G., Karg, A., van Vliet, L., Lade, S. J., & Jacobson, L. (2024). Applying earth system justice to phase out fossil fuels: Learning from the injustice of adopting 1.5°C over 1°C. *International Environmental Agreements, 24*, 233–255.

Gutiérrez, D.-S., Paz, M. J., & Vite, A.-M. (2022). Industrialization of natural resources as a strategy to avoid the natural resource curse: Case of Chilean copper. *The Extractive Industries and Society, 11*, Article 101133.

Harrison, K., & Bang, G. (2022). Supply-side climate policies in major oil-producing countries: Norway's and Canada's struggles to align climate leadership with fossil fuel extraction. *Global Environmental Politics, 22*(4), 129–150.

Hasan, Q., Heffron, R. J., Mohtadi, S., Blankenship, B. D., Overland, I., & Urpelainen, J. (2024). Stepping into the just transition journey: The energy transition in petrostates. *Energy Research & Social Science, 113*, Article 103553.

Healy, N., & Barry, J. (2017). Politicising energy justice and energy system transitions: Fossil fuel divestment and a "just transition." *Energy Policy, 108*, 451–459.

Heras, A., & Gupta, J. (2024). Fossil fuels, stranded assets, and the energy transition in the Global South: A systematic literature review. *WIREs Climate Change, 15*, Article E866.

Heffron, R. J., & McCauley, D. (2018). What is the "just transition"? *Geoforum, 88*, 74–77.

Hess, D. J. (2012). *Good green jobs in a global economy: Making and keeping new industries in the United States*. MIT Press.

Hess, D. J. (2013). Transitions in energy systems: The mitigation-adaptation relationship. *Science as Culture, 22*(2), 197–203.

Hess, D. J. (2014). Sustainability transitions: A political coalition perspective. *Research Policy, 43*, 278–283.

Hochstetler, K. (2021). *Political economies of energy transition: Wind and solar power in Brazil and South Africa*. Cambridge University Press.

Hochstetler, K., & Milkoreit, M. (2015). Responsibilities in transition: Emerging powers in the climate change negotiations. *Global Governance, 21*(2), 205–226.

Holz, F., Richter, P. M., & Egging, R. (2015). A global perspective on the future of natural gas: Resources, trade, and climate constraints. *Review of Environmental Economics and Policy, 9*(1), 85–106.

Huberty, M. (2014). Motivating green growth: The political economy of energy systems transformation. In J. Zysman & M. Huberty (Eds.), *Can green sustain growth: From the religion to the reality of sustainable prosperity*, 26–41. Stanford Business Books.

Intergovernmental Panel on Climate Change. (2014). Summary for policymakers. In O. Edenhofer et al. (Eds.), *Climate change 2014: Mitigation of climate change: Contribution of Working Group III to the Fifth Assessment Report of the Intergovernmental Panel on Climate Change*. Cambridge

University Press. https://www.ipcc.ch/site/assets/uploads/2018/02/ipcc_wg3_ar5_summary-for-policymakers.pdf

International Energy Agency. (2021). *The role of critical minerals in energy transition.* International Energy Agency. https://www.iea.org/reports/the-role-of-critical-minerals-in-clean-energy-transitions

Irarrazaval, F., & Carrasco, S. (2023). One step forward, two steps back? Shifting accumulation strategies in the lithium production network in Chile. *The Extractive Industries and Society, 15,* 101327.

IRENA (International Renewable Energy Association). (2016). *Renewable energy and jobs 2018.* International Renewable Energy Agency. www.irena.org/publications/2018/May/Renewable-Energy-and-Jobs-Annual-Review-2018

IRENA (International Renewable Energy Agency). (2023). *World energy transitions outlook 2023, 1.5 C pathway.* International Renewable Energy Agency. https://mc-cd8320d4-36a1-40ac-83cc-3389-cdn-endpoint.azureedge.net/-/media/Files/IRENA/Agency/Publication/2023/Jun/IRENA_World_energy_transitions_outlook_2023.pdf?rev=db3ca01ecb4a4ef8accb31d017934e97

IRENA (International Renewable Energy Agency) & CPI (Climate Policy Initiative). (2023). *Global landscape of renewable energy finance 2023.* International Renewable Energy Agency. https://mc-cd8320d4-36a1-40ac-83cc-3389-cdn-endpoint.azureedge.net/-/media/Files/IRENA/Agency/Publication/2023/Feb/IRENA_CPI_Global_RE_finance_2023.pdf?rev=8668440314f34e588647d3994d94a785

Jacobson, M. Z., Delucchi, M. A., Cameron, M. A., & Frew, B. A. (2015). Low-cost solution to the grid reliability problem with 100% penetration of intermittent wind, water, and solar for all purposes. *PNAS, 112*(49), 15060–15065.

Jaskoski, M. (2022). *The politics of extraction: Territorial rights, participatory institutions, and conflict in Latin America.* Oxford University Press.

Juhász, R., Lane, N., & Rodrik, D. (2024). The new economics of industrial policy. *Annual Review of Economics, 16,* 213–242.

Karl, T. L. (1997). *The paradox of plenty: Oil booms and petro-states.* University of California Press.

Karlsson, M., Alfredsson, E., & Westling, N. (2020). Climate policy co-benefits: A review. *Climate Policy, 20*(3), 292–316.

Katz-Rosene, R., & Paterson, M. (2018). *Thinking ecologically about the global political economy.* Routledge Taylor and Francis Group.

Kelsey, N., & Zysman, J. (2014). The green spiral. In J. Zysman & M. Huberty (Eds.), *Can green sustain growth? From the religion to the reality of sustainable prosperity,* 79–88. Stanford Business Books.

Khagram, S. (2004). *Dams and development: Transnational struggles for water and power.* Cornell University Press.

Kingsbury, D., & Wilkinson, A. (2023). "We are a mining region": Lithium frontiers and extractivism in Abitibi-Témiscamingue, Canada. *The Extractive Industries and Society, 15,* Article 101330.

Konzen, G., Best, R., & de Castro, N. J. (2024). The energy injustice of household solar energy: A systematic review of distributional disparities in residential rooftop solar adoption. *Energy Research and Social Science, 111,* Article 103473.

Kramarz, T., Arias Mahiques, M. V., Allan, T., Escosteguy, M., Kingsbury, D., & Seghezzo, L. (2024). Redundancies, layers, and dilemmas: Comparing private standards and public regulations in lithium mining. *The Extractive Industries and Society, 18,* Article 101479.

Krawchenko, T. A., & Gordon, M. (2021). How do we manage a just transition? A comparative review of national and regional just transition initiatives. *Sustainability, 13*(11), Article 6070.

Lenferna, A. (2023). South Africa's unjust climate reparations: A critique of the just energy transition partnerships. *Review of African Political Economy, 50*(177–178), 491–501.

Lester, R. K., & Hart, D. M. (2012). *Unlocking energy innovation: How America can build a low-cost, low-carbon energy system.* The MIT Press.

Lewis, J. I. (2013). *Green innovation in China: China's wind power industry and the global transition to a low-carbon economy.* Columbia University Press.

Mantanzima, J., & Loginova, J. (2024). Sociocultural risks of resource extraction for the low-carbon energy transition: Evidence from the Global South. *The Extractive Industries and Society, 18*, Article 101478.

Martin, P. L. (2011). Global governance from the Amazon: Leaving oil underground in Yasuní National Park, Ecuador. *Global Environmental Governance, 11*(4), 22–42.

Martinez, N. (2020). Resisting renewables: The energy epistemics of social opposition in Mexico. *Energy Research & Social Science, 70*, Article 101632.

McAdam, D., & Boudet, H. S. (2012). *Putting social movements in their place: Explaining opposition to energy projects in the United States, 2000–2005*. Cambridge University Press.

McNamara, K. R. (2023). Transforming Europe? The EU's industrial policy and geopolitical turn. *Journal of European Public Policy, 31*(9), 2371–2396.

Meadowcroft, J. (2009). What about the politics? Sustainable development, transition management, and long term energy transition. *Policy Science, 42*, 323–340.

Meckling, J. (2021). Making industrial policy work for decarbonization. *Global Environmental Politics, 21*(4), 134–147.

Meckling, J., Kelsey, N., Biber, E., & Zysman, J. (2015). Winning coalitions for climate policy: Green industrial policy builds support for carbon regulation (perspectives). *Science, 349*(6253), 1170–1171.

Meckling, J., & Nahm, J. (2017). The power of process: State capacity and climate policy. *Governance, 31*, 741–757.

Mildenberger, M. (2020). *Carbon captured: How business and labor control climate politics*. The MIT Press.

Mitchell, T. (2011). *Carbon democracy: Political power in the age of oil*. Verso Press.

Moe, E. (2015). *Renewable energy transformation or fossil fuels backlash: Vested interests in the political economy*. Palgrave McMillan.

Monyei, C. G., Adewumi, A. O., & Jenkins, K. E. H. (2018). Energy (in)justice in off-grid rural electrification policy: South Africa in focus. *Energy Research & Social Science, 44*, 152–171.

Morena, E., Krause, D., & Stevis, D. (Eds.). (2020). *Just transitions: Social justice in the shift towards a low-carbon world*. Pluto Press.

Morris, C., & Jungjohann, A. (2016). *Energy democracy: Germany's ENERGIEWENDE to renewables*. Palgrave Macmillan and Springer Nature.

Mulvaney, D. (2013). Opening the black box of solar energy technologies: Exploring tensions between innovation and environmental justice. *Science as Culture, 22*(2), 230–237.

Mulvaney, D. (2017). Identifying the roots of green civil war over utility-scale solar energy projects on public lands across the American Southwest. *Journal of Land Use Science, 12*(6), 493–515.

Nahm, J. (2021). *Collaborative advantage: Forging green industries in the new global economy*. Oxford University Press.

Nemet, G. (2019). *How solar energy became cheap: A model for low-carbon innovation*. Routledge.

Newell, P. (2021). *Power shift: The global political economy of Energy Transitions*. Cambridge University Press.

Newell, P., & Carter, A. (2024) Introduction: What next for supply-side policy? *International Environmental Agreements, 24*, 1–6.

Nugent, D., & Sovacool, B. (2014). Assessing the lifecycle greenhouse gas emissions from solar PV and wind energy: A critical meta-survey. *Energy Policy, 65*, 229–244.

Oatley, T. (2021). Energy and the complexity of international order. *Global Environmental Politics, 21*(4), 20–41.

O'Faircheallaigh, C., & Babidge, S. (2023). Negotiated agreements, indigenous peoples and extractive industry in the Salar de Atacama, Chile: When is an agreement more than a contract? *Development and Change, 54*(3), 641–670.

Ogilvie, M., & Rootes, C. (2015). The impact of local campaigns against wind energy developments. *Environmental Politics, 24*(6), 875–893.

Ohlendorf, N., Jakob, M., & Steckel, J. C. (2020). The political economy of coal phase-out: Exploring the actors, objectives, and contextual factors shaping policies in eight major coal countries. *Energy Research & Social Science, 90*, Article 102590.

Ordonez, J. A., Jakob, M., Steckel, J. C., & Funfgeld, A. (2021). Coal, power and coal-powered politics in Indonesia. *Environmental Science & Policy, 123*, 44–57.
Orihuela, J. C., Pérez Cavero, C., & Contreras, C. (2022). Extractivism of the poor: Natural resource commodification and its discontents. *The Extractive Industries and Society, 9*, Article 100986.
Parson, E. A., & Fisher-Vanden, K. (1997). Integrated assessment models of global climate change. *Annual Review of Energy and the Environment, 22*, 589–628.
Paterson, M. (2021). Climate change and international political economy: Between collapse and transformation. *Review of International Political Economy, 28*(2), 394–405.
Pearson, P. J. G., & Foxon, T. J. (2012). A low carbon industrial revolution? Institutions and challenges from past technology and economic transformations. *Energy Policy, 50*, 117–127.
Pegels, A. (Ed.). (2014). *Green industrial policy in emerging countries*. Routledge.
Pegels, A., & Altenburg, T. (2020). Latecomer development in a "greening" world: Introduction to the special issue. *World Development, 135*, Article 105084.
Phadke, R. (2024). Green nickel, electric vehicles and mining governance challenges in the U.S. *The Extractive Industries and Society, 18*, Article 101469.
Princen, T., Manno, J. P., & Martin, P. L. (Eds.). (2015). *Ending the fossil fuel era*. The MIT Press.
Radley, B., & Lehmann-Grube, P. (2022). Off-grid solar expansion and economic development in the Global South: A critical review and research agenda. *Energy Research and Social Science, 89*, Article 102673.
Rempel, A., & Gupta, J. (2022). Equitable, effective, and feasible approaches for a prospective fossil fuel transition. *WIREs Climate Change, 13*(2), Article e756.
Riofrancos, T. (2020). *Resource radicals: From petro-nationalism to post-extractivism in Ecuador*. Duke University Press.
Roberts, J. T., Weikmans, R., Robinson, S-A., Ciplet, D., Khan, M., & Falzon, D. (2021). Rebooting a failed promise of climate finance. *Nature Climate Change, 11*, 180–182.
Roche, M. Y., Verolme, H., Agbaegbu, C., Binnington, T., Fischedick, M., & Oladipo, E. O. (2020). Achieving sustainable development goals in Nigeria's power sector: Assessment of transition pathways. *Climate Policy, 20*(7), Article 8460865.
Rodrik, D. (2014). Green industrial policy. *Oxford Review of Economic Policy, 30*(3), 469–491.
Rosales, A., & Clark, P. (2023). State capacity and the unplanned decline of Venezuela's petro-state: Reflections for sustainable transitions and the Green New Deal. *Global Political Economy, 2*(2), 225–245.
Ross, M. L. (2012). *The oil curse: How petroleum wealth shapes the development of nations*. Princeton University Press.
Saha, C. K. (2024). Least developed countries versus fossil fuel incumbents: Strategies, divisions, and barriers at the United Nations climate negotiations. *International Environmental Agreements, 24*, 91–120.
Saha, C. K., & Carter, A. V. (2022). Phase-out or lock-in fossil fuels? Least developed countries' burning dilemma. *The Extractive Industries and Society 11*, Article 101140.
Sahu, A., & Mishra, D. P. (2023). Coal mine explosions in India: Management failure, safety lapses and mitigative measures. *The Extractive Industries and Society, 14*, Article 101233.
Sanz-Hernandez, A., Ferrer, C., López-Rodríguez, M. E., & Marco-Fondevila, M. (2020). Visions, innovations, and justice? Transition contracts in Spain as policy mix instruments. *Energy Research & Social Science, 70*, Article 101762.
Sawyer, S. (2004). *Crude chronicles: Indigenous politics, multinational oil, and neoliberalism in Ecuador*. Duke University Press.
Shah, A. (2023). Rethinking "just transitions" from coal: The dynamics of land and labour in anti-coal struggles. *Journal of Peasant Studies, 50*(6), 2145–2164.
Slough, T., Urpelainen, J., & Yang, J. (2015). Light for all? Evaluating Brazil's rural electrification progress, 2000–2010. *Energy Policy, 86*, 315–327.
Sovacool, B. (2019). The precarious political economy of cobalt: Balancing prosperity, poverty, and brutality in artisanal and industrial mining in the Democratic Republic of the Congo. *The Extractive Industries and Society, 6*, 915–939.

Sovacool, B., & Drupady, I. M. (2012). *Energy access, poverty, and development: The governance of small-scale renewable energy in developing Asia*. Ashgate.

Sovacool, B. K., Ali, S. H., Bazilian, M., Radley, B., Nemery, B., Okatz, J., & Mulvaney, D. (2020). Sustainable mineral and metals for a low-carbon future. *Science, 367*(6473), 30–33.

Stevis, D., & Felli, R. (2015). Global labour unions and just transition to a green economy. *International Environmental Agreements, 15*(1), 29–43.

Stokes, L. (2016). Electoral backlash against climate policy: A natural experiment on retrospective voting and local resistance to public policy. *American Journal of Political Science, 60*(4), 958–974.

Stokes, L. C. (2020). *Shortcircuiting policy: Interest groups and the battle over clean energy and climate policy in the American states*. Oxford University Press.

Swilling, M. (2020). *The age of sustainability: Just transitions in a complex world*. Routledge.

Torjesen, S. (2023). A battery bonanza for the Global South? Prospects for economic upgrading in lithium-ion battery value chains in the context of strategic capitalism. *The Extractive Industries and Society, 17*, Article 101375.

Turconi, R., Boldrin, A., & Astrup, T. (2013). Life cycle assessment (LCA) of electricity generation technologies: Overview, comparability and limitations. *Renewable and Sustainable Energy Reviews, 28*, 555–565.

Tyler, E., & Hochstetler, K. (2021). Institutionalising decarbonisation in South Africa: Navigating climate mitigation and socio-economic transformation. *Environmental Politics, 30*(Suppl. 1), 184–205.

Ukhurebor, K. E., Ngonso, B. F., Egielewa, P. E., Cirella, G. T., Akinsehinde, B. O., & Balogune, V. A. (2023). Petroleum spills and the communicative response from petroleum agencies and companies: Impact assessment from the Niger Delta region of Nigeria. *The Extractive Industries and Society, 15*, Article 101331.

Unruh, G. C. (2000). Escaping carbon lock-in. *Energy Policy, 30*(4), 317–325.

Van Asselt, H., Fragkos, P., Peterson, L., & Fragkiadakis, K. (2024). The environmental and economic effects of international cooperation on restricting fossil fuel supply. *International Environmental Agreements, 24*, 141–166.

Van de Graaf, T., & Bradshaw, M. (2018). Stranded wealth: Rethinking the politics of oil in an age of abundance. *International Affairs, 94*(6), 1309–1328.

Van de Graaf, T., & van Asselt, H. (2017). Introduction to the special issue: Energy subsidies at the intersection of climate, energy, and trade governance. *International Environmental Agreements: Politics, Law and Economics, 17*, 313–326.

Vershoor, M., Albers, C., Poortinga, W., Böhm, G., & Steg, L. (2020). Exploring relationships between climate change beliefs and energy preferences: A network analysis of the European Social Survey. *Journal of Environmental Psychology, 70*, Article 101435.

Wettestad, J., Inderberg, T. H. J., & Gulbrandsen, L. H. (2024). Exploring paths and innovation in Norwegian carbon capture and storage policy. *Environmental Policy and Governance, 34*(2), 125–136.

Yonk, R. M., Simmons, R. T., & Stead, B. C. (2013). *Green vs. green: The political, legal and administrative pitfalls facing green energy production*. Routledge.

Zhao, Q., & Yuan, C-H. (2021). Can China's industrial policies enhance the green competitiveness of the manufacturing industry? *PLoS ONE, 16*(6), Article e0253774.

Zysman, J., & Huberty, M. (Eds.). (2014). *Can green sustain growth? From the religion to the reality of sustainable prosperity*. Stanford University Press.

CHAPTER 24

Toward Gender and Climate Justice

Women, Climate Action Projects for Accelerated Transition, and Transformative Change in the Global South

JOYEE SHAIREE CHATTERJEE, SIRAYUTH THONGPRASERT, PARIMITA MOHANTY, AND JOYASHREE ROY

Introduction

THE United Nations Economic and Social Council (ECOSOC) (2022) reports that climate and environmental crises, combined with gender inequality, are a major hindrance to sustainable development. The global goal calls for faster, deeper, and scaled-up climate actions. Unsustainable land use and energy use have contributed to net emissions and biodiversity change—climate change is adversely impacting agricultural productivity and growth. However, land- and energy-sector-related responses can also lead to climate adaptation, climate mitigation, food security, and sustainability. Both adaptation gap reduction and adoption of low-emission technologies need to happen at an accelerated rate in the Global South with adequate financial flows (IPCC, 2019, 2023).

The Intergovernmental Panel on Climate Change (IPCC) Sixth Assessment Report (AR6)—Working Group II report (IPCC, 2022), the IPCC synthesis report (IPCC, 2023), and Roy et al.(2022) detail numerous, intersecting, and compounding gender and climate vulnerabilities across sectors and social groups. Among women, geographical location, cultural norms, and social stratification by age, class, caste, race, and/or religion intersect and deepen vulnerabilities. Further, the needs and vulnerabilities of those of diverse genders and sexual orientations (e.g., LGBTQ+ communities) in the context of climate action continues to be a research and data gap (Kilpatrick et al., 2023; Roy et al., 2022; Sultana, 2010). In climate change literature, the latter challenge is compounded when approaches continue to use the terms "gender" and "women" interchangeably, collapse terminology of sex (biological difference) and gender (socially constructed roles and identities), and/or perpetuate gendered data gaps by limiting data and interventions to a binary/heteronormative understanding of gender and society (Chatterjee et al., in press). It is thus time for climate change experts and gender experts to come together to strengthen the literature and effective intervention assessment by addressing climate justice and gender equity in an integrated manner.

We conceptualize climate justice and gender justice as the realization of human rights and equity for all—ensuring that we address power relations and imbalances not only between men and women but going beyond the binary, heteronormative/cisgender paradigm to recognize the diversity and plurality of sexual orientation, gender identity, gender expression, and sex characteristics (SOGIESC) in social contexts with due cognizance to historical injustice. This also aligns with moving from a procedural justice paradigm to distributive justice and retributive justice paradigms in a climate justice context (Hayward & Roy, 2019). Further, transformative social change requires attention to how diverse gendered identities and marginalizations intersect with other structural marginalizations (such as race, class, caste, and ableism) to create multifaceted and unique systemic challenges for disparate groups (see also chapter by Alvarez and Norton-Smith in this volume). In this chapter we specifically focus on climate action projects that have attempted to address women's inclusion and empowerment within the broader domain of gender justice and transformative change in the context of the escalating climate crisis. We limit the scope both for a deeper understanding of this specific domain but also due to practical challenges and the continued paucity of data and interventions focused on diverse SOGIESC groups.

The regions of Southeast Asia, East Asia, and South Asia are experiencing frequent and intensive climate impacts and are identified as highly vulnerable regions due to climate-induced disasters (Future Earth et al., 2024). The adverse impact of these disasters has been extensively documented, especially for marginalized or vulnerable groups, including pregnant women and children (Future Earth et al., 2024). The Asian continent faces temperature increases at a rate that is double the global average, resulting in intensifying natural disasters (UN Women & UNEP, 2023), including typhoons, super cyclones, and tsunamis. In 2022, floods accounted for 74.4% of the disaster events in the Asia-Pacific region, and the region accounted for 88.4% of disaster-related total global deaths (UN Women & UNEP, 2023, p. 2). In the ranking of total fatalities related to extreme weather caused by climate change across the globe, India is first, Bangladesh is seventh, Myanmar is twenty-fourth, Vietnam is twenty-fifth, and Thailand is thirty-sixth (Eckstein et al., 2021, pp. 37, 39, 40–42). In 2024, there were approximately 600 typhoon-related deaths across Thailand, Laos, the Philippines, and Vietnam (Center for Disaster Philanthropy, 2024, para. 4).

Concurrently in climate change literature, researchers continue to underline the persistence of discriminatory gender norms and slow progress toward the realization of gender equity in the region and globally (ASEAN Secretariat, 2021; Future Earth et al., 2024; Goodrich et al., 2022; Han et al., 2022; Prakash et al., 2024; UN Women & IUCN, 2022; UN Women & UNEP, 2023). Patriarchal norms and structural barriers for women's workforce participation, restrictions on mobility, lack of access to land ownership, capital (assets and finance), education, and entrenched gender hierarchies continue to be challenges in the region which have ongoing implications for climate interventions and climate justice (Goodrich et al., 2022; Han et al., 2022; Khan, 2023; UN Women & UNEP, 2023). These gendered disparities and forms of discrimination experienced by women are evidenced in two of the most important climate-change-inducing and impacted sectors: the energy and land sectors. In the energy sector in India, women historically have had less access to clean energy sources, appliances, and technical knowledge than men (Khosla, 2021). Similarly in Vietnam, women, the economically poor, and ethnic groups experience energy inequality and energy poverty (Nguyen et al., 2019; Son & Yoon, 2020). In the agriculture sector in Bangladesh, women are found to be more reliant on homestead agriculture, and they lack land ownership and access to financial and infrastructural resources, as well as access to training, technology, and decision-making roles. All of these exclusions place women at a disadvantage in dealing with the impacts of climate change (UN Women & IUCN, 2022).

Gender inequality is rife in the region. The Organisation for Economic Co-operation and Development (OECD) and Association of Southeast Asian Nations (ASEAN) Secretariat (2021, p. 9) report that, excluding Singapore, an estimated 50% of working-age women in ASEAN countries are living in rural areas with experiences of multiple and intersecting social and economic issues. The Food and Agriculture Organization (FAO) (2023b) reports that half of the countries in Sustainable Development Goals (SDG) Indicator 5.a.2—"Proportion of countries where the legal framework (including customary law) guarantees women's equal rights to land ownership and/or control women's land rights" are not legally well protected. Even in more developed Taiwan, where women have comparatively higher political and social status, inequality between men and women is evidenced in disaster risk reduction (DRR), such as in the existence of guidelines for gender awareness but which in application were found to be gender neutral and determined by individual leaderships (of mostly male leaders) rather than an institutionalized process (Lee et al., 2022). Other contemporary examples from the region include the Bay of Bengal Initiative for Multi-Sectoral Technical and Economic Cooperation's (BIMSTEC) Sectoral Objectives on Environment and Climate Change in which gender and social development goals are absent (Goodrich et al., 2022).

A review of climate change and gender-based violence and sexual reproductive health commitments in the nationally determined contributions (NDCs) by the United Nations Population Fund (UNFPA) and Queen Mary University (2023) found that multiple countries in the Asia and Pacific regions recognize or mention elements of gender-based violence and sexual reproductive health. However, while some had detailed solutions, they lacked meaningful implementation in their programs. Almost all the countries of South, Southeast, and East Asia have NDCs that address gender and vulnerable groups. The exceptions are Mongolia with no NDCs for gender and Japan with no NDCs for vulnerable groups. Further, only Cambodia, Laos, Nepal, Sri Lanka, and Vietnam have NDCs that address gender-based violence (UNFPA & Queen Mary University, 2023).

Examples of intersectional vulnerabilities experienced by women are also evidenced in other geographical areas, such as in sub-Saharan Africa. Awiti (2022) found that women across the continent disproportionately experience negative impacts in relation to food security, agriculture, disaster, migration, health, and conflict. Further, climate justice and gender-based responses of the region face difficulties from the shortage of appropriate gender disaggregated data and limited research on the topic overall. Likewise, in the Global North, women experience greater climate impacts, such as in Europe, where women account for 56% of heat-related deaths (Tandon, 2023, p. 1857).

The United Nations (2022) reports that despite some progress and multiple interventions, with the current trajectory, Sustainable Development Goal 5 on gender equality will not be met by the target date of 2030. An examination of the implemented projects specifically on gender and climate intervention reveals mixed results with shortcomings, especially in sectors of coastal and mountain ecosystems, poverty, livelihood, and sustainable development (Roy et al., 2022). Women of the regions are both affected by vulnerabilities but are also simultaneously potential drivers of change. For example, across ASEAN, it is found that women are responsible for household energy management, thus possessing the knowledge and skills that could make them agents of change for green energy development (ASEAN Centre for Energy, 2022).

The need for intervention projects is greater than ever as the threat of climate change escalates and social inequities widen, demanding immediate and effective responses from all levels (Awiti, 2022; IPCC, 2023). Therefore, this warrants discussion on the factors behind shortcomings of gender and climate interventions and the need for course correction (Roy et al., 2022) by including aspects of gender inequity in NDCs is being suggested in newly published reports (Queen Mary University of London, 2024).

Climate action planning is multilayered and driven by multiple actors—that is, they happen at national, subnational, and project levels with diverse priorities and frameworks. Further, it is well accepted in the literature that climate action is not detached from wider social, economic, political, and non-climate environmental goals (Honegger et al., 2021; IPCC,

2023; Prakash et al., 2022; Roy et al., 2018). Here we focus on the energy- and land-based agriculture sectors, as it is anticipated that these two sectors are among those that experience the most rapid transitions. Additionally, we also cover equity issues in the context of DRR programs as disasters are increasingly exacerbated due to climate change, and research demonstrates that developmental outcomes are not gender neutral in the case of planning and response to catastrophes (Queen Mary University of London, 2024).

To critically analyze the interventions and identify the emergent barriers and enablers on transformative change from these selected climate projects, we use the framework of a Continuum of Approaches on Gender and Climate Action (Chatterjee et al., in press) to classify actions across a continuum ranging from exploitative to accommodating to transformative (see Figure 24.1). The framework recognizes that some climate-specific interventions (and policies) may unintentionally or blindly exacerbate gender-related climate vulnerabilities and inequities and further widen gaps among genders, communities, locations, and nations (exploitative). Interventions can maintain the structural causes of gender-related climate vulnerabilities and inequities, even though they might procedurally acknowledge inequalities and to varying degrees work to be gender inclusive (accommodating). Actions and interventions can be designed to lead to systemic and sustainable change, addressing the root causes of gender-related climate vulnerabilities and inequities (transformative). A transformative approach we envision addresses gender and climate challenges in an integrative and systemic way so that the outcome is *both* greater gender equity and greater climate resilience (Chatterjee et al., in press). This approach is also in line with the 28th United Nations Climate Change Conference (COP 28)—United Arab Emirates (UAE) or the COP28-UAE Framework for Global Climate Resilience.

Applying this conceptual framework, this chapter analyzes how gender equity has been dealt with in climate intervention projects that have attempted to address women's inclusion and empowerment in South, Southeast, and East Asia since 2021. We identify and draw exemplars from both academic and other secondary sources (e.g., published reports from nongovernmental organizations [NGOs] and quasi-government bodies) focusing on studies/projects across regional, national, and local scales. We also concentrate on specific sectors—namely energy, agriculture, and DRR—where there has been a strong focus on climate

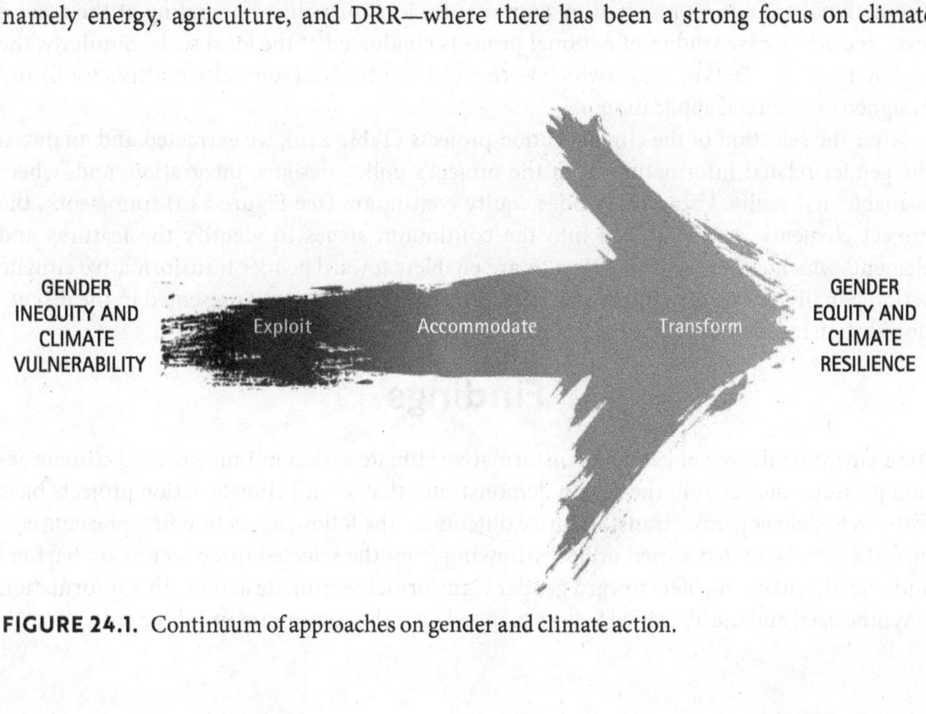

FIGURE 24.1. Continuum of approaches on gender and climate action.

adaptive and climate mitigative action and sustainable transition. Our goal is to understand gendered climate justice implications and enablers and barriers toward gender transformative climate goals, especially in these sectors where transitions are occurring at an accelerated pace. This chapter also covers equity issues in DRR programs because catastrophes are happening with greater frequency due to climate change, and developmental outcomes in such situations are not gender neutral (ESCAP, 2023; Goodrich et al., 2022; Han et al., 2022; Khan, 2023; UN Women & IUCN, 2022; UN Women & UNEP, 2023).

Methodology

For this review we conducted a purposive search to identify examples of contemporary climate action projects (since 2021) having a specifically designated gender component across the three chosen sectors of agriculture, energy, and DRR. These identified interventions were geographically in South and Southeast Asia and could have been at the local, national, or regional scale. Both Google Scholar (for journal articles) and Google search engine (for project reports) were used. A supplementary targeted search was conducted across organizational websites of the UN Women and UNEP's EmPower project, the International Centre for Integrated Mountain Development (ICIMOD), FAO, Care Evaluations, and Carbon Brief to ensure that each sector was represented. Examples of keywords used for search include "climate change," "energy," "disaster," "agriculture," "climate change and gender," "women," and "project evaluation." Supporting information was also gathered from academic literature databases like Science Direct through searches with a more expanded time frame to provide national and regional context.

All 39 projects and policies identified matched our sectoral, geographic, and thematic criteria for use in our analysis (see Table 24.1 for a list of project/policy names). Table 24.2 provides an overview of the range of projects across sector and scale. For the local scale, we focused on selecting examples which were locally situated and led and implemented by an international development organization or international development programs—that is, UN bodies and projects such as EmPower or regional organizations like ASEAN or ICIMOD. The national scale is represented by domestic projects or policies operating at the federal level, including case studies of national projects conducted at the local scale. Similarly, the regional-scale projects are those which were multicountry (and some also multisectoral) and designed for regional application.

After the selection of the climate action projects (Table 24.1), we extracted and analyzed the gender-related information from the project's policy designs, integration, and, where available, its results. Using the gender equity continuum (see Figure 24.1) framework, the project elements were classified into the continuum stages to identify the features and elements which either create barriers or are enablers toward gender transformative climate action. Finally, the results of this analysis were synthesized and are presented in the discussion section in Table 24.3.

Findings

Assessment on degree of gender transformative climate action in implemented climate action projects since 2021 in the region demonstrates that not all climate action projects have features to deliver gender transformative outcomes. The following section first presents empirical exemplars from varied projects drawing from the selected three sectors on barriers and then discusses enablers toward gender transformative climate action. This information is synthesized and the identified factors mapped onto the continuum in Table 24.3.

TABLE 24.1 List of Projects/Polices Reviewed

No.	Project/Policy Name	Sectors	Country	References
1	Impacts being and lacking ASEAN agricultural policies overview	Agriculture	ASEAN in general	Han et al., 2022
2	Bangladeshi Agriculture Extension Policy 2021	Agriculture	Bangladesh	UN Women & IUCN, 2022
3	Cambodian energy policies overview	Energy	Cambodia	Han et al., 2022
4	Clean and Improved Cooking program	Energy	Cambodia	Han et al., 2022; SNV, n.d.
5	Closing the STEM gender gap: Training women in Earth observation and geospatial information technology	DRR	Pakistan, Nepal, Bhutan, Bangladesh	Tripathi et al., 2022
6	Filippino energy policies overview	Energy	Philippines	Han et al., 2022
7	Gender Resource Group	Multisectoral	China, India, and Pakistan	Khan, 2023
8	Indonesian energy policies overview	Energy	Indonesia	Han et al., 2022
9	InfoAct	Agriculture	Vietnam	CARE, 2021b
10	Koshi Gender Portal	Multisectoral	China, Nepal, and India	Khan, 2023
11	National Agricultural Extension Policy, 2020	Agriculture	Bangladesh	UN Women & IUCN, 2022
12	EmPower: Women for Climate Resilient Societies solar-powered goat farming projects	Energy	Bangladesh	Islam, 2023; Zinc & Mills, 2022
13	EmPower: Women for Climate Resilient Societies solar energy project	Energy	Bangladesh	UNEP, 2018, 2020
14	Thailand's Climate Change Action Plan on Agriculture (CCAPA 2023–2027)	Agriculture	Thailand	FAO, 2023a
15	The ASEAN Agreement on Disaster Management and Emergency Response (AADMER) Work Programme (2021–2025)	DRR	ASEAN in general	Han et al., 2022

(continued)

TABLE 24.1 Continued

No.	Project/Policy Name	Sectors	Country	References
16	The Cambodia Horticulture Advancing Income and Nutrition (CHAIN)	Agriculture	Cambodia	Han et al., 2022; SNV, n.d.
17	The Ide dan Analitika Indonesia (IDEA) and YAPPIKA-ActionAid's DRR projects in general	DRR	Indonesia	Han et al., 2022; YAPPIKA-ActionAid, 2022
18	The Sindh Emergency Service Rescue 1122	DRR	Pakistan	Khan, 2023
19	The Woman Champion in Disaster Risk Reduction and Climate Change	DRR	Cambodia	ActionAid Cambodia, 2021; Han et al., 2022
20	Vietnamese energy policies overview	Energy	Vietnam	Han et al., 2022; Nguyen et al., 2019
21	WeCaN's Gender-Responsive Pasture Management Practices	Agriculture	Mongolia	FAO, 2024
22	YAPPIKA-ActionAid Tsunami Emergency Response and Recovery Program in Central Sulawesi Indonesia	DRR	Indonesia	Lacasna & Wardani, 2020
23	Islamabad's urban DRR readiness and policies	DRR	Pakistan	Saad, 2021
24	Global Platform for Disaster Risk Reduction 2019	DRR	India, Pakistan, Bangladesh, Philippines, Sri Lanka	Duryog Nivaran & Asian Disaster Preparedness Centre, 2021
25	Overview of DRR policies from India	DRR	India	Chaiya et al., 2022
26	Overview of DRR policies from China	DRR	China	Chaiya et al., 2022
27	Overview of DRR policies from the Philippines	DRR	The Philippines	Chaiya et al., 2022
28	Overview of DRR policies from Thailand	DRR	Thailand	Chaiya et al., 2022
29	SEWA CSR Gujarat Solar Park CSR policies	Energy	India	Stock, 2021
30	Bhadla Solar Park impacts	Energy	India	Stock & Sovacool, 2024
31	Micro-Hydro Power projects of West Java and East Java	Energy	Indonesia	Hermawati et al., 2023

TABLE 24.1 Continued

No.	Project/Policy Name	Sectors	Country	References
32	Thailand Climate Change (CC), Gender, and Social Inclusion (GSI) into Planning and Budgeting in Thailand	Agriculture, Energy, DRR	Thailand	Suriyasarn & Talerngsri, 2024
33	Climate Change Benefit Analysis in Maha Sarakham province	Agriculture, Energy, DRR	Thailand	Suriyasarn & Talerngsri, 2024
34	The SAARC food bank	Agriculture	SAARC in general	Goodrich et al., 2022
35	The SAARC seed bank	Agriculture	SAARC in general	Goodrich et al., 2022
36	The SAARC Energy Centre (SEC)	Energy	BIMSTEC in general	Goodrich et al., 2022
37	ASEAN RE-Gender Roadmap	Energy	ASEAN in general	Han et al., 2022
38	Market Access Through Digital Innovation in India (MANDI)	Agriculture	India	Johnson, 2021
39	Women's Economic Empowerment and Equality, Land Rights, and Agricultural Engagement in the PepsiCo Potato Supply Chain in West Bengal, India	Agriculture	India	Johnson, 2021

DRR, disaster risk reduction.

TABLE 24.2 Overview of Sectoral and Implementation Scale of Selected Projects

Sector Scale	Agriculture	Energy	DRR
Local	5	3	4
National	5*+	10*+	11*+
Regional	5#$	4#$	5#$

* Thailand climate change (CC), gender, and social inclusion (GSI) into planning and budgeting in Thailand.

+ Climate change benefit analysis in Maha Sarakham province, Thailand.

Gender resource group (multisectoral—energy, agriculture, DRR; multi-country—China, India, and Pakistan).

$ Koshi Gender Portal (multisectoral—energy, agriculture, DRR; multi-country—China, Nepal, and India).

Barriers Toward Gender Transformative Change in Climate Action

Despite the recognition in multiple frameworks and policies of the need for gender and social inclusion in climate action, which is a positive step toward transformative change, in general ASEAN member countries face the challenge of gender being siloed to its own sector separate from climate initiatives or having weak coordination and mainstreaming in sectoral climate initiatives (Han et al., 2022). Lack of gender disaggregated data as well as monitoring and evaluation mechanisms on the integration of gender perspective emerge as overarching challenges across sectors and countries.

Even in policies or guidelines where gender is explicitly centered, there continue to be gaps in implementation and realization. The ASEAN Renewable Energy Gender Roadmap to promote gender equity and climate change in the renewable energy sector has four phases—awareness and creating a database for gender in renewable energy; gender integration into policies; frameworks for monitoring and evaluation; and implementation and oversight of policies (ASEAN Centre for Energy, 2022). However, gender awareness was found to be poor and participation of women was low in the sector, despite such guidelines and roadmaps being in place (ASEAN Centre for Energy, 2022). Possible causes identified for these include lack of training, capacity building, and opportunities for women. More importantly, there is a lack of guidance for gender integration in the policies and an absence of budget allocation for women to be addressed in the sectoral initiatives, highlighting some of the barriers toward structural changes even in the presence of policy or goal toward inclusive change (ASEAN Centre for Energy, 2022).

Traditionally in the energy sector, a male-dominated sector, there is a popular assumption of gender neutrality, and almost all focus is on the economics of productivity. On a national scale, the main issue is the exclusion of marginalized groups or communities. For instance, in Vietnam, the government's renewable energy initiatives are large-scale projects targeted at businesses and traditionally male-dominated sectors, such as construction, industry, and commercial agriculture but lacking support for decentralized renewable energy grids, that is, solar mini-grids (Han et al., 2022). This is problematic because economically poor Vietnamese, including lower class women and ethnic groups, rely on coal and biomass which have negative health and environmental impacts. These groups would benefit from access to micro-grids (Nguyen et al., 2019). Similarly, in Indonesia and the Philippines large-scale renewable energy projects are popular while smaller scale projects, which are of greater benefit to local/marginalized groups, are fewer and their planning has no references to gender, resulting in the inequality in the distribution of resources and budget toward materials and infrastructure but not toward social development and potential empowerment of women and other minority groups. Likewise, Cambodia has multiple ethnic groups, but it is reported that they were absent in the renewable energy decision-making process, leaving their demands and contributions unheard (Han et al., 2022).

Trends of gender and social exclusion were also found in the DRR sector with a lack of attention to the role of women, including reporting on the gendered impact of disasters. For example, in Pakistan, Islamabad's urban DRR readiness and policies were found to miss out on the slum population, and first responders did not have adequate training in disaster response with limited knowledge on early warning systems (Saad, 2021). More significantly, there was a lack of female representation or participation in disaster planning and women lacked attention in situation reports prepared during disaster situations (Saad, 2021).

In the *agriculture sector*, looking at examples of national-level policies, the Bangladeshi national policies such as the Agriculture Extension Policy have raised awareness and recognition of the importance of women's participation in agriculture and climate interventions, with the promotion of women-based agrarian groups and small and medium-sized enterprises (UN Women & IUCN, 2022). However, they also evidenced lack of comprehensive gender analysis and related challenges in understanding gendered impacts of climate change.

Exemplar of Interventions With Barriers Leading to Exploitative Outcomes

Examples of challenges and inequitable outcomes across marginalized groups were found in the Gujarat Solar Park (GSP) in India (Levien, 2018; Stock et al., 2019). Farming lands were taken away from local communities who were dependent on them (Stock & Birkenholtz, 2021), including sites for women to gather firewood (Stock & Birkenholtz, 2020). In their place, low-paying labor and technical jobs were made available, but mostly to men (Stock, 2021). Stock (2021, p. 4) surveyed four villages and only found six women who earned income via the initiative. The corporate social responsibility scheme put in place to redress such identified imbalances, however, reproduced pitfalls from development approaches that focus on women's productivity and ignore existing gender power relations. Only women of middle and upper classes and castes were able to benefit from their scheme, while people of lower social status and those without land missed out on the benefits despite being the numerical majority in the area (Stock, 2021). This phenomenon is found in other Indian solar park projects such as the Bhadla Solar Park, where it was found that the acquisition of land increased women's burden because they needed to travel further to gather firewood (Stock & Sovacool, 2024). In effect, such outcomes therefore reproduce, reinforce, or even intensify existing social inequalities and do not result in *just* green energy transitions.

Enablers Toward Gender Transformative Change in Climate Action

In contrast to the barriers and challenges discussed earlier, promising trends, positive outcomes, and multiple examples of enablers toward transformative change across sectors are also identified. Success emerged where platforms and opportunities were created for multisectoral initiatives, and multiple stakeholders were engaged to share gender and climate insights, including sex disaggregated data, information on vulnerabilities, and success stories used to inform policies and development efforts (Khan, 2023). In general, across sectors interventions by nongovernment and civil society organizations show more successful integration of gender perspectives. Intervention especially at the local level, including support by local governments toward green microfinance, could empower women and support eco-friendly enterprise (Atahau et al., 2021). Examples of gender-sensitive and gender-inclusive intervention designs which showcase transformative changes typically also include strong components of continuous monitoring and evaluation throughout the project cycle, capacity building, and leadership training for women with features such as building of soft skills, networking, knowledge development, providing mentorship from multiple disciplines, and post-training/intervention engagement (Tripathi et al., 2022). Further, initiatives with multisectoral engagement and integration and alignment among regional, national, and local priorities and frameworks helped enhance momentum toward transformative change.

For instance, CARE International's InfoAct project in Vietnam is a multisectoral project aimed to build climate resiliency for economically poor ethnic women through creating a climate information system alongside a village loan program to enhance livelihood opportunities, and this also acts as a safe space and networking platform for the women (CARE, 2021a). The project takes inputs from men and women while coordinating with local communities and authorities. All projects under CARE include gender-sensitive indicators such as tracking the percentage of ethnic women and men who received the disseminated information and are able to apply it, local (tailored) language use in intervention materials, use of appropriate communication channels for the intended audience, reduction in their household expenditures, and so on—these types of indicators extend to monitoring in budgeting processes, collaborations with other NGOs and intergovernmental organization, and in externally conducted evaluation (CARE, 2021b).

Sectoral examples of intervention from the *energy sector* include Indonesia, where under the Micro-Hydro Power of West Java and East Java, multiple local NGOs and cooperatives were able to conduct development projects, enable access to power from the plants, and provide knowledge and skills, which helped women earn higher incomes and as a result receive greater recognition from their community peers signaling transformative change (Hermawati et al., 2023). Additionally, the powering of streetlights and public infrastructure such as health centers increased perceptions of safety at night and allowed communities to engage in communal activities beyond the hours of daylight. However, the complexity of outcomes when considered from the twin perspectives of gender equity and climate resilience is highlighted with the project also evidencing how such social and infrastructural changes concurrently create shifts in demand toward more consumerist and materialistic lifestyles (Hermawati et al., 2023).

Similarly, in the agricultural sector, examples of projects with enablers toward gender transformative climate outcomes are the projects under the Cambodia Horticulture Advancing Income and Nutrition (CHAIN) initiative. Here, women were among the main beneficiaries and received recognition for their contributions to the project and their households, with enhanced skills and knowledge in climate smart farming technology and techniques, including the use of solar pumps and drip irrigations and increased climate coping capacity (Agrifood Consulting International, 2022). Their successes derived from a consistent monitoring system that (despite critiques of its indicator being too complicated and incomplete at times) covers the entire work process from start to finish, including the engagements with stakeholders. This resulted in early identification of challenges and course corrections as needed. Active collaboration with NGO partners on the ground led to increased learning and networking capacities among the beneficiaries with local agricultural and market networks. A notable strength was the successful alignment with the Cambodian government's goals, particularly in line with the Strategic Plan for Climate Change Adaptation 2019–2023 and Gender Mainstreaming Policy and Strategic Framework in Agriculture (Agrifood Consulting International, 2022). However, the project also noted that not all categories of farmers participated/benefitted equally with semicommercial and commercial farmers being more involved compared to homestead farmers. Gender-inclusive planning and implementation are also evidenced in Mongolia, where the pastoral herders have felt the effect of climate change with extreme weather, and less rain and plants, reducing their productivity. The Environment and Development Association JASIL empowers women pastoralists for leadership by taking account of their perspectives and needs and developing guidelines that aligned with national environment laws for the women to get legal recognition and co-management of their natural resources (FAO, 2024).

Likewise, an example from the *disaster risk reduction* sector which demonstrates enablers toward gender transformative climate action is ActionAid Cambodia's The Woman Champion in Disaster Risk Reduction and Climate Change project on providing resiliency for women by creating women "champions" for DRR leadership. The project started by identifying potential candidates from varied backgrounds, including local women, women community councilors, and committee members through the help of NGOs and local organizations. The women underwent skill trainings in DRR and advocacy resulting in them becoming leaders in DRR for their local communities (ActionAid Cambodia, 2021). The project also provided insight on the challenges in women's participation due to their multiple productive and reproductive burdens, noting that it may take a long time and the necessity of engaging and involving male household members toward enacting normative and transformative societal change (Han et al., 2022). Another transformative example is from Pakistan in the Sindh Emergency Service Rescue intervention, which was a multistakeholder effort with international and local entities involved in enhancing local disaster response efforts and capacity. Here women were trained as first responders and emergency call center operators (Khan, 2023). Their presence, participation, and effort challenged biased gender norms and negative perceptions of women's strength and dedication (Khan, 2023). Participants reported gaining recognition in the family and community and a sense of pride for their service and in representing the organization (Khan, 2023).

Exemplar of Interventions With Enablers Leading to Transformative Outcomes

An example on a national scale of multisectoral, gender, and climate change interventions includes the Thai national effort to integrate gender and social inclusion into climate projects (called the CC-GSI), which comes with extensive guidelines that include definitions of gender and climate change, checklists for identification of issues and resolution, data collection and analysis, stakeholders and related climate impacts, and mechanisms of budgeting, monitoring, and evaluation (Suriyasarn & Talerngsri, 2024). The plan recognizes intersectional gender roles and inequalities in land, forestry, and natural resources usage driven by intersectional marginalizations (Suriyasarn & Talerngsri, 2024). Sectoral departments of the Thai government, such as the Thai Ministry of Agriculture and Cooperatives (MoAC), have also made efforts to integrate gender mainstreaming and social inclusion into their national- and local-level planning (e.g., ensuring the presence of women in climate-smart agriculture projects) (FAO, 2023a). Further, part of the initiative is a climate change benefit analysis in the Maha Sarakham province where a multistakeholder team identified issues at multiple levels from youth to local leaders and state agents; provided training workshops on gender and climate change; and gave postintervention reporting and evaluation to government agencies and community (Suriyasarn & Talerngsri, 2024).

Discussion

Worldwide there are efforts to push gender equity into national policies with a declaration of commitment toward gender, but what follows nationally and subnationally varies or has mixed results (Han et al., 2022; UN Women & IUCN, 2022). Applying the analytical framework of *the Continuum of Approaches on Gender and Climate Action* (Chatterjee et al., in press) across initiatives in our three chosen climate action sectors, we observe that many are limited to the exploitative end of the continuum (being either gender blind or insensitive—including projects with no gender lens which were excluded from this review), some are toward the middle and in the accommodating stage, and some positively moving toward the transformative end of the continuum. Additionally, across the Asian region (and globally) foundational gender disparities remain unresolved with constricting and patriarchal gender norms limiting women's property and land ownership, consumption of renewable energy, participation, and skills related to climate and political activities, among other issues (Duerto-Valero et al., 2021; Han et al., 2022; Tan & Uprasen, 2021; UNDP Bangladesh, 2023).

Table 24.3 summarizes the enablers and barriers toward gender transformative change in climate action identified through our review before explicating on the synthesis of our findings. Some common challenges we identify which move climate action project interventions toward the gender and climate inequity end of the continuum (either exploitative or moving closer toward accommodating) relate to governance and conscious implementation of policies. Governance is traditionally technocratic and top-down oriented. A consequence of this seen from the energy sector, for example, is that local and marginalized groups may have less voice while powerful stakeholders may take all the benefits (Han et al., 2022). As found in Bangladesh, local government agencies are not able to impact the implementation of climate policies that are designed at higher levels (Mahmud & Roy, 2021, 2023; Stock et al., 2021). This type of lack of vertically integrated governance system and decision making is more likely to support unjust transitions (Johnson et al., 2020). Evidence from Indonesia and Fiji reveals that climate finance goes to large-scale projects with the fewest risks that are highly bankable (Anantharajah & Setyowati, 2022). In some Thai climate and social development plans that were criticized for not delivering their policy goals, problems included having an ambiguous definition for "vulnerable groups," gender disaggregated data were only available on household heads, and no data were available on resiliency and impact; there was also a lack of participation of national social development agencies and no budget or plans for gender and social equity (Suriyasarn & Talerngsri, 2024).

Securing long-term solid financial foundation is important for integrated gender activities to be carried out in a climate intervention. For government agencies, the IPCC (2022) recommends institutionalizing the project into the budget and policy planning cycle. This ensures that there will be a baseline funding dedicated to the projects as seen with The Women Budget from the Philippines (Philippines Commission on Women, n.d.). Additionally, there must be mechanisms to ensure that the budget and resources are spent correctly and not overly spent in one area (Han et al., 2022). Thus, there should be gender and climate sensitive monitoring and evaluation mechanisms that would run from the inception of the project to its conclusion (Goodrich et al., 2022; UN Women & IUCN, 2022).

Multiple reviewed projects have suffered inconsistent staffing in quality and retention, so it is critical to have dedicated staff that are capable of carrying out multidisciplinary projects across their areas of expertise—that is, professionals from gender and climate disciplines who can execute gender-climate projects (Allwood, 2014). In one example from Bangladesh, the government agricultural extension unit was found to be male dominated,

TABLE 24.3 A Summary of Enablers and Barriers Toward Gender Transformative Change in Climate Action: Summary Results of Key Features Identified Through Review of the Projects Reports

Barriers Leading to Exploitative Outcomes	Enablers Leading to Accommodating (Incremental) Outcomes	Enablers Leading to Transformative Outcome
Work in silos: siloed approach to gender and climate change	Addresses gender data gap: Initiation of actions for sex/gender disaggregated data to be collected, information on vulnerabilities and success stories to inform policies and development efforts	Promotion of gender-equitable norms: Engagement of both men and women to enable acceptance of gender equitable norms, new technologies, and better environments
Gender-neutral/invisible approach: Gender neutrality assumed (especially common in male-dominated sectors) or lack of any gender analysis/assessment	Inclusion of women: Designing projects to enable participation of women and building knowledge and skills	Women's leadership: Connecting women with businesses, community leadership, and public centers to enable the acceptance of women leadership and women supportive technologies. Building capacity and creating women role model to be leaders with expertise in their relevant areas
Lack of attention to gender roles: Project designs fail to account for women's multiple burdens—productive and reproductive—which limit their participation capacity and/or further burden them	Consistent gender-sensitive implementation monitoring throughout all phases and coordination with all stakeholders	Community mobilization: Cooperate with local communities, practitioners, nongovernmental organizations, and religious leadership, to promote gender equality and to help finding and legitimizing capable women to be leaders
Gender data gap: Lack of gender disaggregated data as well as data on women's activities and practical and strategic needs such as land and asset ownership	Initiation of processes to identify the minimum categories to be included in database	Multisectoral engagement integrating gender and climate action
Lack of intersectional approach: Women treated as a homogenous group and needs and disproportional impact on women in minority/marginalized communities overlooked compounding social health and environmental impact		

which limited social interactions due to cultural gender barriers and missed the demands of women agriculturalists (UN Women & IUCN, 2022).

Likewise, national projects in the ASEAN region have similar patterns: The application of gender and climate change in national projects is sporadic and not systemic (ASEAN Secretariat, 2021; Han et al., 2022). Across sectors many policies are not gender responsive, women lack access to knowledge and skill training in the sectors considered in relation to climate change knowledge, and there is poor staff retention and insufficient budget allocation, among others (ASEAN Secretariat, 2021; Duerto-Valero et al., 2021; Han et al., 2022). Additionally, policies are nonbinding by nature, and as a result, the application of gender and climate change depends on the individual state and can lead to mixed results (Han et al., 2022).

The most common issue encountered by projects and policies is the siloing of gender and climate change or weak linkages. Therefore, there is a need for institutional arrangements enabling and incentivizing interdisciplinary cooperation and intersectoral linkages. For example, the Philippines' Gender and Development budget (GAD), a result of the Republic Act 9710: The Magna Carta of Women, mandates that all government agencies and departments dedicate 5% of their total annual budget toward gender actions and plans (Philippines Commission on Women, n.d., p.1). At the national or regional scale, it is recommended that to avoid siloing of gender, state entities and their stakeholders should work through a nexus approach where a platform like a regional organization helps facilitate multidisciplinary cooperation (Goodrich et al., 2022; Khan, 2023).

Conclusion

In all sectors within the multilevel climate governance architecture (Roy et al., 2019), any local action is embedded in a hierarchical policy structure where local and national policies are intricately related and flow top-down from national policies to local policies. In such scenarios, sometimes national climate policies and interventions are insensitive to local voices, demands, and needs due to various intranational political power structures, social stereotypes, and social stratification that serve vested interest groups. Therefore, it is important to identify a lack of alignment between vertically linked policy frameworks and social and political actors at various levels to identify sensitive entry points for accelerating transformative change in relationships of power and social structures through synergies between gender equity outcomes and climate action goals. To be transformative, project designs need to incorporate the voices of marginalized groups, ensuring equal access, use, and distribution—all of which requires local voices and input (Han et al., 2022; Roy et al., 2019).

At a local scale, international NGOs appear to have multiple successes that engage women and aim to make transformative changes. The driving factors for successes of these local-scale projects include inclusive, bottom-up approaches, and identifying and working with community and women's groups and authorities to facilitate the project—the latter (working with local stakeholders) is critical as women's participation is affected by local gender-specific cultural and social norms and the power structure in the community.

Finally, maladaptation must be avoided such as relying on and reinforcing gender stereotypes that women should be included in development work because of their pure and altruistic nature (rather than focusing on inclusion, social justice, and rights). In such cases, the consequences include increasing the burden of work and responsibilities on women and/or creating unrealistic multitasking expectation from them (Chant & Sweetman, 2012; Davids et al., 2014; Goetz, 2007). Additionally, in the context of increasing burdens of women, there needs to be more mitigation-based intervention instead of merely adaptation-based because the latter relies on women to be the main labor for changes; this is not inherently beneficial because women typically deal with multiple productive and reproductive burdens to begin with. Lack of an intersectional approach is also an aspect of maladaptation, for instance, where low-carbon energy intervention may negatively impact gender inequities not only between men and women but also among different communities of women (Han et al., 2022; Johnson et al., 2020).

The current chapter strengthens the need for action and policies for upscaling and accelerating gender transformative action programs. We must go beyond SDG5 goals (Roy, Prakash, et al., 2022) and move from a procedural justice paradigm to distributive justice and retributive justice paradigms. These can be the guiding principles for developing indicators for the UAE framework for climate resilience and the UAE—Belém work program on indicators. However, the foremost important step in this direction is the necessity of joint research efforts with equal roles of communities and those with specialization on climate change and gender mainstreaming. COP 28 has provided a strong spotlight on the need to address these concurrently, and this cannot be another missed opportunity. The research outcomes on indicators need also to be taken to the communities and practitioners for piloting and refining so that substantive progress can be made on building relevant indicators on gender and climate action over the next 2 years with a long-term vision for just and equitable (i.e., transformative) change.

Acknowledgments

Authors from AIT acknowledge funding under the EDITS-AIT project. The EDITS-AIT project at the Asian Institute of Technology, Thailand, received funding from the Energy Demand changes Induced by Technological and Social innovations (EDITS) project, which is part of the initiative coordinated by the Research Institute of Innovative Technology for the Earth (RITE) and the International Institute for Applied Systems Analysis (IIASA) (and funded by Ministry of Economy, Trade, and Industry [METI], Japan)

References

ActionAid Cambodia. (2021). *Woman champion in disaster risk reduction and climate change.* ActionAid Cambodia. https://cambodia.actionaid.org/sites/cambodia/files/publications/Project%20directory_Woman%20Champion_2021.pdf

Agrifood Consulting International. (2022). *Final evaluation of SDC's Cambodia horticulture advancing income and nutrition (CHAIN) project*. Agrifood Consulting International. https://www.aramis.admin.ch/Default?DocumentID=69880&Load=true

Allwood, G. (2014). Gender mainstreaming and EU climate change policy. *European Integration Online Papers, 2014–006*, 1–26. https://doi.org/10.1695/2014006

Anantharajah, K., & Setyowati, A. B. (2022). Beyond promises: Realities of climate finance justice and energy transitions in Asia and the Pacific. *Energy Research & Social Science, 89*, 102550. https://doi.org/10.1016/j.erss.2022.102550

ASEAN Centre for Energy. (2022). *Roadmap on accelerating ASEAN's renewable energy deployment through gender-responsive energy policy*. ASEAN Centre for Energy. https://asiapacific.unwomen.org/sites/default/files/2023-06/asean-re-gender-roadmape6b5.pdf

ASEAN Secretariat. (2021). *ASEAN state of climate change report*. ASEAN Secretariat. https://asean.org/wp-content/uploads/2021/10/ASCCR-e-publication-Correction_8-June.pdf

Atahau, A. D. R., Sakti, I. M., Huruta, A. D., & Kim, M.-S. (2021). Gender and renewable energy integration: The mediating role of green-microfinance. *Journal of Cleaner Production, 318*, 128536. https://doi.org/10.1016/j.jclepro.2021.128536

Awiti, A. O. (2022). Climate change and gender in Africa: A review of impact and gender-responsive solutions. *Frontiers in Climate, 4*, 895950. https://doi.org/10.3389/fclim.2022.895950

CARE. (2021a). *Information for adaptation in Vietnam (InfoAct)*. CARE. https://www.care.org.vn/wp-content/uploads/2018/08/InfoAct-project-brief-English-1.pdf

CARE. (2021b). *Information for adaptation in Vietnam (InfoAct) project's final evaluation*. CARE International. https://reliefweb.int/report/viet-nam/enhanced-livelihoods-and-increased-resilience-poor-ethnic-minority-women-and-men

Center for Disaster Philanthropy. (2024, November 26). *2024 Super Typhoon Yagi*. https://disasterphilanthropy.org/disasters/2024-super-typhoon-yagi/

Chaiya, C., Pal, I., & Pinthong, P. (2022). Paradigm of disaster risk reduction: A comparative study of five Asian countries. *Social Science Asia*, *8*(3). 1–33. https://doi.nrct.go.th/ListDoi/listDetail?Resolve_DOI=10.14456/ssa.2022.17

Chant, S., & Sweetman, C. (2012). Fixing women or fixing the world? "Smart economics," efficiency approaches, and gender equality in development. *Gender & Development*, *20*(3), 517–529. https://doi.org/10.1080/13552074.2012.731812

Chatterjee, J. S., Roy, J., Thongprasert, S., & Datta, S. (in press). Gender inclusive climate action: How far is the international community from transformative action? In R. Matthew & A. LeHolland (Eds.), *Climate change and international relations* (pp. 1–33). Edward Elgar.

Davids, T., Van Driel, F., & Parren, F. (2014). Feminist change revisited: Gender mainstreaming as slow revolution. *Journal of International Development*, *26*(3), 396–408. https://doi.org/10.1002/jid.2945

Duerto-Valero, S., Kaul, S., & Chanchai, R. (2021). *ASEAN gender outlook*. ASEAN, ASEAN Committee on Women, UN Women. https://data.unwomen.org/sites/default/files/documents/Publications/ASEAN/ASEAN%20Gender%20Outlook_final.pdf

Duryog Nivaran, & Asian Disaster Preparedness Centre. (2021). *Gender responsive disaster risk management status review and recommendations for implementing the Sendai framework for DRR in the Asia Pacific*. Duryog Nivaran, Asian Disaster Preparedness Centre, UNDRR.

Eckstein, D., Künzel, V., & Schäfer, L. (2021). *Global climate risk index 2021: Who suffers most from extreme weather events? Weather-related loss events in 2019 and 2000–2019*. Germanwatch. https://www.germanwatch.org/sites/default/files/Global%20Climate%20Risk%20Index%202021_2.pdf

ECOSOC. (2022). *Achieving gender equality and the empowerment of all women and girls in the context of climate change, environmental and disaster risk reduction policies and programmes* [Report of the Secretary-General]. ECOSOC. https://digitallibrary.un.org/record/3956348/files/E_CN.6_2022_3-EN.pdf?ln=en

ESCAP. (2023). *Impacts of climate change on population and development in Asia and the Pacific*. ESCAP. https://www.unescap.org/sites/default/d8files/event-documents/ESCAP_APPC7.3_E.pdf

FAO. (2023a). *Celebrating women's leadership in climate action*. FAO. https://www.fao.org/in-action/scala/news/news/news-detail/celebrating-women-s-leadership-in-climate-action/en

FAO. (2023b). *The status of women in agrifood systems*. FAO. https://doi.org/10.4060/cc5060en

FAO. (2024). *WeCaN's gender-responsive pasture management practices*. FAO. https://www.fao.org/in-action/dryland-sustainable-landscapes/news-and-events/news-detail/wecan-s-gender-responsive-pasture-management-practices/en

Future Earth, the Earth League, & WCRP. (2024). *10 new insights in climate science 2024/2025*. Future Earth, the Earth League, the World Climate Research Programme. https://doi.org/10.5281/ZENODO.13950098

Goetz, A. M. (2007). Political cleaners: Women as the new anti-corruption force? *Development and Change*, *38*(1), 87–105. https://doi.org/10.1111/j.1467-7660.2007.00404.x

Goodrich, C. G., Hussain, A., Pasakhala, B., Bano, K., Bhuchar, S., Chitale, V. S., Bisht, S., Bastola, A., & Silpakar, S. (2022). *State of gender equality and climate change in South Asia and the Hindu Kush Himalaya*. ICIMOD, UNEP, UN Women. https://asiapacific.unwomen.org/sites/default/files/2023-05/state-of-ge-and-cc-in-sa-and-hkh-2d933.pdf

Han, Y.-C. J., Pross, C., Agarwal, R., Torre, R. A., & Stockholm Environment Institute. (2022). *State of gender equality and climate change in ASEAN*. ASEAN, UN Women. https://asiapacific.unwomen.org/sites/default/files/2022-09/State-of-Gender-Equality-and-Climate-Change-in-ASEAN-2.pdf

Hayward, B., & Roy, J. (2019). Sustainable living: Bridging the North-South divide in lifestyles and consumption debates. *Annual Review of Environment and Resources*, *44*(1), 157–175. https://doi.org/10.1146/annurev-environ-101718-033119

Hermawati, W., Ririh, K. R., Ariyani, L., Helmi, R. L., & Rosaira, I. (2023). Sustainable and green energy development to support women's empowerment in rural areas of Indonesia: Case of micro-hydro power implementation. *Energy for Sustainable Development, 73*, 218–231. https://doi.org/10.1016/j.esd.2023.02.001

Honegger, M., Michaelowa, A., & Roy, J. (2021). Potential implications of carbon dioxide removal for the sustainable development goals. *Climate Policy, 21*(5), 678–698. https://doi.org/10.1080/14693062.2020.1843388

IPCC. (2019). Summary for policymakers. In P. R. Shukla et al. (Eds.), *Climate change and land: An IPCC special report on climate change, desertification, land degradation, sustainable land management, food security, and greenhouse gas fluxes in terrestrial ecosystems.* Cambridge University Press. https://doi.org/10.1017/9781009157988.001

IPCC. (2022). *Climate change 2022: Impacts, adaptation, and vulnerability. Contribution of Working Group II to the Sixth Assessment Report of the Intergovernmental Panel on Climate Change* (H. O. Pörtner, D. C. Roberts, M. Tignor, E. S. Poloczanska, K. Mintenbeck, A. Alegría, M. Craig, S. Langsdorf, S. Löschke, V. Möller, A. Okem, & B. Rama, Eds.). Cambridge University Press.

IPCC. (2023). *Climate change 2023: Synthesis report. Contribution of working groups I, II and III to the Sixth Assessment Report of the Intergovernmental Panel on Climate Change* (Core Writing Team, H. Lee, & J. Romero, Eds.). Intergovernmental Panel on Climate Change (IPCC). https://doi.org/10.59327/IPCC/AR6-9789291691647

Islam, S. (2023). *Women show how they are building climate-resilient agricultural communities in Bangladesh.* United Nations Bangladesh. https://asiapacific.unwomen.org/en/stories/feature-story/2023/07/women-show-how-they-are-building-climate-resilient-agricultural-communities-in-bangladesh

Johnson, N. (2021). *Evaluation of the gender, agriculture, and assets project, Phase 2 gender and empowerment frameworks and tools.* International Food Policy Research Institute. https://doi.org/10.2499/p15738coll2.134453

Johnson, O. W., Han, J. Y.-C., Knight, A.-L., Mortensen, S., Aung, M. T., Boyland, M., & Resurrección, B. P. (2020). Intersectionality and energy transitions: A review of gender, social equity and low-carbon energy. *Energy Research & Social Science, 70*, 101774. https://doi.org/10.1016/j.erss.2020.101774

Khan, S. R. (2023, February 21). *Three projects in South Asia putting gender at the centre of responses to climate change.* ICIMOD. https://www.icimod.org/article/three-projects-in-south-asia-putting-gender-at-the-centre-of-responses-to-climate-change/

Khosla, R. (2021, March 26). *Guest post: How energy demand for cooling in India's cities is changing.* Carbon Brief. https://www.carbonbrief.org/guest-post-how-energy-demand-for-cooling-in-indias-cities-is-changing/

Kilpatrick, C., Higgins, K., Atkin, S., & Dahl, S. (2023). A rapid review of the impacts of climate change on the queer community. *Environmental Justice, 8*(5), 306–315. https://doi.org/10.1089/env.2023.0010

Lacasna, Y., & Wardani, N. (2020). *Final evaluation of YAPPIKA-ActionAid tsunami emergency response and recovery programme in Central Sulawesi Indonesia—Executive summary.* Yappika ActionAid. https://reliefweb.int/attachments/8c1fb416-34b5-3798-ae27-40bcfofadfe4/ActionAid_DEC_SulawesiTsunami_FinalEvaluation_ExecutiveSummary.pdf

Lee, C.-C., Huang, K.-C., Kuo, S.-Y., Lin, Y.-J., Ke, K.-Y., Pan, T.-Y., Tai, L.-L., Cheng, C.-K., Shih, Y.-L., Lai, H.-T., & Ke, B.-H. (2022). Gender matters: The role of women in community-based disaster risk management in Taiwan. *International Journal of Disaster Risk Reduction, 77*, 103046. https://doi.org/10.1016/j.ijdrr.2022.103046

Levien, M. (2018). *Dispossession without development* (Vol. 1). Oxford University Press. https://doi.org/10.1093/oso/9780190859152.001.0001

Mahmud, H., & Roy, J. (2021). Barriers to overcome in accelerating renewable energy penetration in Bangladesh. *Sustainability, 13*(14), 7694. https://doi.org/10.3390/su13147694

Mahmud, H., & Roy, J. (2023). What to focus on in order to accelerate access to modern energy services and energy use efficiency in Bangladesh. *Asia-Pacific Sustainable Development Journal, 30*(1), 89–120.

Nguyen, T. T., Nguyen, T.-T., Hoang, V.-N., Wilson, C., & Managi, S. (2019). Energy transition, poverty and inequality in Vietnam. *Energy Policy*, *132*, 536–548. https://doi.org/10.1016/j.enpol.2019.06.001

Philippines Commission on Women. (n.d.). GAD budget. Philippine Commission on Women. https://pcw.gov.ph/gad-budget-report/

Prakash, A., Ley, D., & Thamari, M. (2024). How gender-sensitive are environmental institutions, climate adaptation, and mitigation actions? A narrative from the Global South. *Annual Review of Environment and Resources*, *49*(1), 449–474. https://doi.org/10.1146/annurev-environ-121322-073202

Prakash, A., McGlade, K., Roxy, M. K., Roy, J., Some, S., & Rao, N. (2022). Climate adaptation interventions in coastal areas: A rapid review of social and gender dimensions. *Frontiers in Climate*, *4*, 785212. https://doi.org/10.3389/fclim.2022.785212

Queen Mary University of London. (2024, June 11). *UNFPA and Queen Mary publish regional reports to help countries across the world meet COP commitments to tackle gender injustice from climate change*. Queen Mary University of London. https://www.qmul.ac.uk/media/news/2024/fmd/unfpa-and-queen-mary-publish-regional-reports-to-help-countries-across-the-world-meet-cop-commitments-to-tackle-gender-injustice-from-climate-change-.html?utm_medium=third_party_advert&utm_source=research_gate&utm_campaign=qmrb_2024_resgate_pilot&utm_content=sustainability

Roy, J., Dasgupta, S., Ghosh, D., Das, N., Chakravarty, D., Chakraborty, D., & De, S. (2019). Governing national actions for global climate stabilisation: Examples from India. In A. Barua, V. Narain, & S. Vij (Eds.), *Climate change governance and adaptation: Case studies from South Asia* (pp. 131–152). CRC Press, Taylor & Francis Group.

Roy, J., Prakash, A., Some, S., Singh, C., Bezner Kerr, R., Caretta, M. A., Conde, C., Ferre, M. R., Schuster-Wallace, C., Tirado-von Der Pahlen, M. C., Totin, E., Vij, S., Baker, E., Dean, G., Hillenbrand, E., Irvine, A., Islam, F., McGlade, K., Nyantakyi-Frimpong, H., . . . Tandon, I. (2022). Synergies and trade-offs between climate change adaptation options and gender equality: A review of the global literature. *Humanities and Social Sciences Communications*, *9*(1), 251. https://doi.org/10.1057/s41599-022-01266-6

Roy, J., Tschakert, P., Waisman, H., Halim, S. A., Antwi-Agyei, P., Dasgupta, P., Hayward, B., Kanninen, M., Liverman, D., Okereke, C., Pinho, P. F., Riahi, K., & Rodriguez, A. G. S. (2018). Sustainable development, poverty eradication and reducing inequalities. In V. Masson-Delmotte et al. (Eds.), *Global warming of 1.5°C: IPCC Special Report on impacts of global warming of 1.5°C above pre-industrial levels in context of strengthening response to climate change, sustainable development, and efforts to eradicate poverty* (1st ed., pp. 445–538). Cambridge University Press. https://doi.org/10.1017/9781009157940

Saad, R. (2021). Women and DRR in urban slums—Building resilience through development. *International Journal of Disaster Risk Reduction*, *60*, 102264. https://doi.org/10.1016/j.ijdrr.2021.102264

SNV. (n.d.). *Cambodia Horticulture Advancing Income and Nutrition (CHAIN)*. https://snv.org/project/cambodia-horticulture-advancing-income-and-nutrition-chain

Son, H., & Yoon, S. (2020). Reducing energy poverty: Characteristics of household electricity use in Vietnam. *Energy for Sustainable Development*, *59*, 62–70. https://doi.org/10.1016/j.esd.2020.08.007

Stock, R. (2021). Bright as night: Illuminating the antinomies of "gender positive" solar development. *World Development*, *138*, 105196. https://doi.org/10.1016/j.worlddev.2020.105196

Stock, R., & Birkenholtz, T. (2020). Photons vs. firewood: Female (dis)empowerment by solar power in India. *Gender, Place & Culture*, *27*(11), 1628–1651. https://doi.org/10.1080/0966369X.2020.1811208

Stock, R., & Birkenholtz, T. (2021). The sun and the scythe: Energy dispossessions and the agrarian question of labor in solar parks. *Journal of Peasant Studies*, *48*(5), 984–1007. https://doi.org/10.1080/03066150.2019.1683002

Stock, R., Birkenholtz, T., & Garg, A. (2019). Let the people speak: Improving regional adaptation policy by combining adaptive capacity assessments with vulnerability perceptions of farmers in Gujarat, India. *Climate and Development*, 11(2), 138–152. https://doi.org/10.1080/17565529.2017.1410089

Stock, R., & Sovacool, B. K. (2024). Blinded by sunspots: Revealing the multidimensional and intersectional inequities of solar energy in India. *Global Environmental Change*, 84, 102796. https://doi.org/10.1016/j.gloenvcha.2023.102796

Stock, R., Vij, S., & Ishtiaque, A. (2021). Powering and puzzling: Climate change adaptation policies in Bangladesh and India. *Environment, Development and Sustainability*, 23(2), 2314–2336. https://doi.org/10.1007/s10668-020-00676-3

Sultana, F. (2010). Living in hazardous waterscapes: Gendered vulnerabilities and experiences of floods and disasters. *Environmental Hazards*, 9(1), 43–53. https://doi.org/10.3763/ehaz.2010.SI02

Suriyasarn, B., & Talerngsri, P. (2024). *Integrating climate change (CC), gender, and social inclusion (GSI) into planning and budgeting in Thailand.* UNDP, Office of Natural Resources and Environmental Policy and Planning. https://www.undp.org/sites/g/files/zskgke326/files/2023-03/UNDP_CC-GSI%20Handbook_Eng_Final.pdf

Tan, Y., & Uprasen, U. (2021). Carbon neutrality potential of the ASEAN-5 countries: Implications from asymmetric effects of income inequality on renewable energy consumption. *Journal of Environmental Management*, 299, 113635. https://doi.org/10.1016/j.jenvman.2021.113635

Tandon, A. (2023, July 11). *Heat-related deaths "56% higher among women" during record-breaking 2022 European summer.* Carbon Brief. https://www.carbonbrief.org/heat-related-deaths-56-higher-among-women-during-record-breaking-2022-european-summer/

Tripathi, P., Thapa, R. B., Bajracharya, B., & Maden, U. (2022). *Closing the STEM gender gap: Training women in Earth observation and geospatial information technology.* International Centre for Integrated Mountain Development (ICIMOD). https://doi.org/10.53055/ICIMOD.1005

UN. (2022). *The sustainable development goals report 2022.* UN. https://unstats.un.org/sdgs/report/2022/The-Sustainable-Development-Goals-Report-2022.pdf

UN Women & IUCN. (2022). *State of gender equality and climate change in Bangladesh.* UN Women, IUCN. https://asiapacific.unwomen.org/sites/default/files/2022-09/State-of-GE-and-CC-in-Bangladesh.pdf

UN Women & UNEP. (2023). *EmPower in Bangladesh.* UN Women, UNEP. https://asiapacific.unwomen.org/sites/default/files/2023-08/empower-programme-in-bangladesh_final_21-aug.pdf

UNDP Bangladesh. (2023). *UNDP Bangladesh 2023–2026 gender equality strategy.* UNDP Bangladesh. https://www.undp.org/sites/g/files/zskgke326/files/2023-03/GES%202023-2026%20Final%20Draft.pdf

UNEP. (2018, December 20). *Solar power a one-stop solution in rural Bangladesh.* UNEP. https://www.unep.org/news-and-stories/story/solar-power-one-stop-solution-rural-bangladesh

UNEP. (2020, March 9). *Can renewable energy give climate-displaced women in Bangladesh a new beginning?* https://www.unep.org/news-and-stories/story/can-renewable-energy-give-climate-displaced-women-bangladesh-new-beginning

UNFPA & Queen Mary University. (2023). *Taking stock: Sexual and reproductive health and rights in climate commitments—An Asia and the Pacific review.* UNFPA. https://www.unfpa.org/sites/default/files/pub-pdf/2553-UNFPA-NDC-Asia%20Pacific%20Report.pdf

YAPPIKA-ActionAid. (2022). *YAPPIKA-ActionAid Annual Report 2022.* YAPPIKA-ActionAid. https://yappika-actionaid.or.id/uploads/downloads/Annual_Report_of_YAPPIKA-ActionAid_2022.pdf

Zinc, I., & Mills, Z. (2022, June 29). *Aid activists and a goat farmer tell of how a Sweden-funded project has changed lives in disaster-prone Bangladesh.* UN Women Asia-Pacific. https://asiapacific.unwomen.org/en/stories/feature-story/2022/06/aid-activists-and-a-goat-farmer-tell-of-how-a-sweden-funded-project-has-changed-lives

CHAPTER 25

Climate Diplomacy

Why It Remains So Intractable

MIZAN R. KHAN AND FARHANA R. KHAN

Introduction

HUMAN-INDUCED climate change is a reality today. The world, both rich and poor, is experiencing increasing climate impacts. The year 2024 is reported to be the hottest one on record, with August witnessing a temperature of 1.51°C above preindustrial levels (Climate Action, June 9, 2024). The Sixth Assessment Report (AR6) of the Intergovernmental Panel on Climate Change (IPCC) cautions us of an impending runaway climate change, unless immediate actions are taken (IPCC, 2021, 2022). Many international agencies are coming up with different estimates of developing country needs and gaps in climate finance, and these are likely to increase in the future due to lack of mitigation and an adaptation deficit (CPI, 2024a; SCF, 2024; UNEP, 2023, 2024).

After the end of the Cold War in the 1990s, perhaps no other issue has commanded so much attention from the global community as climate change. But very little progress has been achieved during the last three decades (Fisher, 2024). It continues to remain as intractable as ever. Greenhouse gas (GHG) emissions have almost doubled now compared to 1992 when the Convention was adopted. Though emissions in the Global North are stabilizing, they continue to rise in the Global South, for justifiable reasons. Still, the North remains historically the largest emitter, with its global share of around 60%, down from 75% since 1992 (WRI, 2024). As the second largest emitter (13% of global emissions in 2023), the United States' share stands at 25% in cumulative terms since the Industrial Revolution. Against this, China is the largest emitter today (30% of the global total), with a cumulative share of around 12% (WRI, 2023). As climate change is a stock, not an annual flow problem, the disproportionate share of antecedent depositions by the Global North is an issue of debate in climate diplomacy, because accumulated emissions are having a ratchet effect, destabilizing the climate system and adding increasing loss and damage, particularly to low-income countries. As the UN Secretary General cogently argued: "the rich cause the problem, the poor pay the price" (UN Secretary General, 2024a). This was the rationale of including Article 3.1 of the United Nations Framework Convention on Climate Change (UNFCCC)—the principle of equity and common but differential responsibilities, based on respective capabilities (CBDR + RC).

Accordingly, the climate convention adopted in 1992 envisaged mitigation as the primary solution. But over the last two decades, adaptation came to the fore because of lack of mitigation and increased frequency and magnitude of climate disasters (Khan & Roberts, 2013). Again, a continued deficit in adaptation was resulting in increasing loss and damage

(L&D) to national economies, particularly in the Global South. The result was the inclusion of articles on adaptation and on L&D (Articles 7 and 8, respectively) in the Paris Agreement. Vanderheiden (2011) cogently argues that as mitigation is not taking place, adaptation must include prevention of harm and ex post compensation for unavoidable L&D. But Decision 1/CP.21 (para. 51) forecloses the claims of liability and compensation under L&D. However, it does not stop civil society organizations or citizens from suing the polluters, as is happening worldwide, particularly in the European Union and the United States.

The Loss and Damage Response Fund (LDRF) established at Conference of the Parties (COP) 27 was moving toward its full operationalization at COP28 and COP29. Still, it commands resources less than US$800 million. Another development taking place outside the UNFCCC process seeks advisory opinions (AOs) from several international courts. Can these AOs impel stronger climate action in all three areas—mitigation, adaptation, and L&D? As COP decisions proceed at a snail's pace against the repeated warnings of science, what is the potential of L&D to crack the intractability of climate diplomacy?

In view of the this backdrop, we argue that climate change as a systemic problem is sustained by the prevailing political order, which does not allow for mobilizing the required investments immediately needed. Climate change as a multi-decade, long-term problem does not yet command sufficient attention of countries and voters. Our mainstream economic model—the neoliberal market system upon which climate regime is founded (Almeida & Perez, 2022)—does not encourage investments for mitigation, the benefits of which are viewed to be distant. But impacts continue to escalate. Then, why do adaptation and L&D command utterly inadequate resources? Given this context, the chapter includes the following sections: the temporal and spatial dimensions of climate change, which render our political system dysfunctional to solving climate change, followed by a section reviewing the potential of the L&D agenda advancing in future negotiations. Finally, the chapter concludes with potential pathways forward.

Nature of Climate Change—Temporal Dimension

Climate change is the result of impersonal activities of all citizens and economic agents across the globe. Unlike the ozone regime, which involved a limited number of chemicals and the impacts were likely to affect industrial countries more (Khan, 2014), the climate regime touches the very foundations of modern life. The complexity is compounded by the fact that there are *temporal* and *spatial* dimensions. It is a "stock," rather than a "flow" problem. Annual emissions do not have immediate impacts, but over a 20- to 30-year period, the flows turn into stocks that push global temperatures higher. The already deposited GHG emissions will cause some warming. These antecedent depositions from industrial countries are reinforced by today's rapidly growing emissions from the Global South. Citing the latest data while speaking at the Earth Day on June 5, 2024, in New York, the UN Secretary General argued that "the remaining carbon budget to limit long-term warming to 1.5 degrees will be breached even before we reach 2030" (UN Secretary General, 2024b). Even a full compliance of the nationally determined contributions (NDCs) submitted so far will reduce only 6% of GHG emissions (WRI, 2023), but a reduction of 42% is needed by 2030 and 57% by 2035 to stay on track to sustain 1.5°C. Without ambitious mitigation in the next round of NDCs to be submitted early next year, the world will witness a temperature rise of 2.6°C–3.1°C over the course of this century (UNEP, 2024), causing debilitating impacts to people, planet, and economies.

The temporal dimension of climate change will manifest more severely in the future, though we already live in a climate-changed world. Future generations are predicted to suffer more (IPCC, 2018, 2021). But the prevailing economic and political dynamics respond only to immediate concerns, dictated by the short-term political cycles and short-term gains. Our economic system does not encourage investment in longer term actions, because of discounting the future benefits. But with climate change, costs are upfront, while benefits from mitigation are a little distant. So, resource mobilization for addressing climate change must be based on social discount rates (Daly & Foley, 2011). This is not the case yet.

Second, climate change is universally recognized as a market failure, where the true cost of emission externalities is not reflected in pricing of goods and services, produced by inputs that emit GHGs. The most fundamental lesson from mainstream economics is to internalize the externalities of our production and consumption processes, through the polluter-pays-principle (PPP). The principle of CBDR RC (Article 3.1 of the UNFCCC) implicitly recognizes the PPP (Khan, 2015; Rajamani, 2010). While many countries, including several developing countries, apply the PPP as a tax or levy in their domestic context (Carbon Tax Center, 2024), there is yet no international application. Here is the moral corruption in global climate policymaking (Gardiner, 2011; Giddens, 2011). While solid waste dumping is increasingly costly in the industrial countries, atmospheric dumping is treated as free of cost (Nordhaus, 2005; Young, 2010), because GHG emissions straddle across borders.

Third, while fossil fuels were responsible for 91% of CO_2 emissions in 2022, globally this sector is extremely powerful, with their financial resources and political power, enjoying more than $1.7 trillion in direct public support, including a subsidy of $1.3/1.4 trillion, which is double that of the previous year (CPI, 2024b; Laan et al., 2023). The fossil fuel industries are not supportive of the diffusion of clean technology internationally, because decarbonization warrants a phase-out of fossil fuels. But fossil powers led by Saudi Arabia obstructed any action at COP29 in this direction. The result is a wide gap in investment needs for mitigation. One silver lining is that investment flows in clean energy are approaching US$2 trillion each year, almost double the combined amount spent on new oil, gas, and coal supply—and costs for most clean technologies, particularly of solar and wind, are showing a downward trend (IEA, 2024).

Climate stability or its sink capacity may better be conceptualized as a common pool resource (CPR), which has become a congestible public good, for which economic instruments, such as PPP, are the ideal solution. But because of fractured global politics, no consensus exists on the global application of PPP. We live in a post-Paris world, which re-emphasizes the role of the private sector, assuming the market will provide the common good, at the cost of de-emphasizing the role of the public sector (Khan et al., 2019). But the private sector does not lead in providing public goods. The contradiction is that the basic neoliberal market solution—the internalization of externalities—is avoided, while country-parties are supposed to promote this system under the climate regime. This is a double standard, causing dysfunctionality of the prevailing politico-economic order. Another barrier observed in the negotiations process is that the contribution to the problem is unequal. The G20 major economies emit more than 80% of global GHGs. Even countries of the industrial world have different responses to the needs of mitigation, with the EU countries playing a progressive role, though their net emissions are going down. Emissions in developing countries, led by major emitters of the G20 countries, are going up for justifiable reasons, and they do not have enough resources to provide the common good of mitigation.

Fourth, climate change faces the typical dilemma of the tragedy of the commons (Hardin, 1968). Perception of free riding is a problem, for which the United States did not join the Kyoto Protocol back in the late 1990s, though the then US Vice President Al Gore was a leading architect of the Protocol.[1] As the problem is inherently global, both in its cause-effect dimensions, the problem must be solved within an agreed-upon global framework.

Finally, the economic order that was established 80 years ago under the Bretton Woods system appears unsuited for the purpose of addressing climate change. The multilateral financial institutions (MFIs), led by the World Bank, continue to promote investments in market terms. But funding for climate adaptation does not in most cases bring immediate benefits. The private sector as the main vehicle of development, which the World Bank system promotes, is not interested in investments in adaptation. The Climate Policy Initiative found annual average financial flows of $4.7 billion from the private sector to adaptation-relevant activities (CPI, 2024b). Climate justice as an issue still is not an agenda under the World Bank, though it has been made as the trustee of the Loss and Damage Response Fund (LDRF). This will be very challenging because it does not have any tradition of devolution of funding to local communities. But the historic COP28 decision establishing the LDRF stipulates devising local, community-level funding mechanisms (paragraph 125). COP27 for the first time urged reforming the MFIs. The Bridgetown initiatives also continue hammering on such reforms, particularly in terms of addressing debt distress in developing countries, which is fueled by loans to address climate change (Bridgetown Initiative, 2024). Referring to MFIs, the UN Secretary General Guterres cogently argued in September 2024: "We can't build a future for our grandchildren with a system built for our grandparents" (UN Secretary General, 2024b).

Spatial Dimension of Climate Change Stands in the Way of Support to Developing Countries

The *spatial* dimension of climate change manifests in a disconnect between the cause and effect, happening in physically distant places. Developed countries, historically the main generator of climate problems, are also impacted by increasing climate disasters, but they have better adaptive capacity. Least developed countries (LDCs) and small island developing nations (SIDS) as nano-emitters suffer disproportionately much more, with the least adaptive capacity. The LDCs have contributed about 1% of global emissions but have witnessed a fivefold increase in climate-related disasters since the 1970s (UNCTAD, 2022). Over 70% of deaths caused by climate disasters globally have taken place in the LDCs during the last 50 years (WMO, 2020). About 8.5 million people in the LDCs were displaced due to climate disasters in 2020 (IDMC, 2021). Around one third of the crop and livestock loss in the LDCs can be traced to recurrent droughts, costing the sectors a loss of about $37 billion during 2008–2018 (FAO, 2021).

These multifaceted effects of climate change, including sea level rise (SLR) and salinity intrusion, will render increasing shares of land located in tropical zones going under water and make territories uninhabitable (UNDRR, 2023). This already is engendering climate-induced displacement, particularly in sub-Saharan Africa and South Asia (IDMC, 2021; Khan et al., 2021; Singh et al., 2020). Many SIDs are likely to face "watery death" due to SLR. Because of this spatial disconnect between the cause of the problem and main effects occurring far away in distant shores, the concept of "vulnerability interdependence" (Volger, 1995) is not working yet. This disconnect is reflected in the poor response to support adaptation in the developing countries (GCA, 2019). The LDCs and SIDs receive less than $10 billion a year in adaptation support against the disruption of their economies (OXFAM, 2023). This is in stark contrast to rapid mobilization of initial pledges of over $10 billion to address the flood impacts in Central Europe in early September 2024 (Reuters, September 19, 2024). This is natural in prioritizing support for nations proximate to funding sources, but developed countries must not forget their legal responsibility assumed under Articles 3 and 4 of the UNFCCC and Article 9 of the Paris Agreement. Article 7.4 of the Paris Agreement establishes a cause-effect relationship between the need for adaptation and the level of mitigation. Some authors differentiate between "luxury emissions" in developed countries against "survival emissions" in the low-income countries (Caney, 2010). IPCC AR6 (2022) argues that the global financial system is awash with money, but investments do not flow where needed most. $2.4 trillion is spent a year for addressing the "perceived" military threats, manufactured by powerful governments (SIPRI, 2024). Runaway climate change is a "real" existential threat for the whole of humanity, but it is not afforded the level of attention and action relative to the urgency.

Again, there is a conceptual lacuna in framing adaptation under the climate governance regime: It is often viewed from a narrow economic perspective, as if it provides only local, national, or at best some regional benefits (Barrett, 2008; Zedillio, 2008). Article 3.3 of the Climate Convention stipulates that any climate action has to contribute to global benefits, though in practice, funding from the Global Environment Facility has been allowed gradually with an expanded understanding of adaptation (Khan, 2014). This is a contradiction: Continued undersupply of mitigation enhances the need for adaptation, though antecedent depositions of GHGs will continue having a rachet effect, requiring financing for adaptation for quite some time (Article 7.4 of the Paris Agreement). But, if climate change and its impacts are regarded as a global public bad (GPB), then adaptation to address those GPBs should be regarded as a GPG. If more than 100 vulnerable countries—including low-income populous countries like Bangladesh, India, and Pakistan—go down the development spiral, the global economic, financial, trade, and security systems will be seriously impaired. Therefore, adaptation in the developing countries should be considered as a GPG, to be funded by levies imposed on the GPBs (Khan & Munira, 2021; Netherlands Department of Environment, 2011). Adam Smith argued that market forces would operate as the "invisible hand" only if the economic agents behave ethically (Smith, 1759/1982; 1776/2008). Only in that case private interests will maximize social and collective good. Had this been the case, climate change would have had no dearth of investments; it could have been avoided before it reached this threshold level of danger. Even with increasing transboundary and second-order climate impacts, the framing of adaptation as a GPG is slowly moving in that direction (Banda et al., 2018), but it is not gaining enough traction yet at the political level (Hedlund et al., 2018).

A number of other reasons can be ascribed to the poor state of adaptation support (Khan, 2014, 2020), including (a) adaptation is blurred with development, and developed countries consider that developing countries for their own interests have to climate-proof their development; (b) middle and higher-income developing countries focus more on mitigation and related technologies than adaptation; and (c) adaptation does not stop climate change, and continuing such investments in developed countries inhibits their ambitious mitigation targets, making developing countries worse off (Michaelowa, 2001).

Sengupta and Bearak in a recent analysis (2024) argue that three issues together stand in the way of global climate cooperation: Though China is now the biggest emitter, burning the highest amount of coal, its leadership and dominance in the renewable energy supply chain fuel protectionist backlash in the Western countries, for fear of falling farther behind. They have imposed substantial tariffs on China's electric vehicles, including elimination of Chinese-processed metals from their own factories. The second issue is the failure of complying with their obligations to support the developing countries with a minimum level of climate finance, discussed earlier. The third issue is the lack of US leadership in coordinating global climate cooperation. The process of disengagement will be reinforced by another potential withdrawal from the Paris Agreement under the second Trump presidency.

The geopolitical divide fostered by wars and conflicts in the Middle East and Eastern Europe diverts global attention away from climate diplomacy. Climate change should be considered as a different ball game compared to other global issues, in terms of the level of genuine interdependence among nations. The tragedy is that our political system has sparked a protectionist backlash even in clean technologies, even though the United States and China were leading supporters of the Paris Agreement in 2015.

This reversal is a testimony that our divisive political order remains anchored to the "Westphalian" model of sovereignty and national interests, which viewed global cooperation as a "zero-sum," not as a "positive sum/win-win" game. There are no serious cooperation initiatives among major emitters, particularly between China and the United States. Promoting cooperation as a win-win option based on "enlightened self-interest/smart sovereignty" by nations to generate "responsible internationalism" has been proposed as an alternative pathway (Kaul, 2017; Nye, 2016; Rosenthal, 2022).

Adam Smith, the father of classical economics, taught us that interests are indivisible, either at individual or national levels, and that one's interests are bound with those of others (Smith, 1759/1982). Nothing can be more true or urgent than the case of climate change, where the stakes of the whole of humanity are bound in one thread—our survival from the existential threats that climate change increasingly poses. We may also recall that achieving global social justice in an interdependent world was a central goal of Franklin Delano Roosevelt's New Deal against economic disruptions caused by the Great Depression in the 1930s. Roosevelt stated at the opening of the Bretton Woods Conference in 1944 that "Economic diseases are highly communicable, [i]t follows, therefore, that the economic health of every country is a proper matter of concern to all its neighbors, near and distant" (Roosevelt, 1944). The geopolitical reality is different today as UN Secretary General Guterres laments that we continue living in a dysfunctional world order (Crawford, 2022).

Loss and Damage Agenda: Does It Have the Potential of Turning the Tide?

Looking back to 1991, Vanuatu on behalf of SIDS submitted a proposal at the first Intergovernmental Negotiating Committee (INC), charged for drafting the UNFCCC, for an insurance mechanism to compensate for the impacts of sea level rise (INC, 1991). Instead of endorsing the proposal, the Convention included "insurance" as an adaptation instrument in Article 4.8. Then the Bali Action Plan at COP13 mandated Parties to explore "means to address loss and damage associated with climate change impacts in developing countries that are particularly vulnerable to the adverse effects of climate change" (para. 1(c)(iii), Decision1/CP.13).

With some developments since then, COP19 agreed for some institutional infrastructure to address loss and damage (L&D), including the Warsaw International Mechanism, with its Executive Committee to further study the issues of L&D. Finally, at COP21, the Paris Agreement included Article 8 on L&D, with nine areas of action. The first author of this chapter argued, even before adoption of the Agreement, that several factors were pushing both the Global North and Global South for adoption of this article (Khan, 2014): (a) increased frequency and magnitude of climate disasters in the last decade that exceeded even the IPCC AR4 predictions; (b) sheer lack of progress in emissions reduction; (c) the widening gap in delivery of pledged finance for adaptation; (d) avoiding the liability/compensation that the issue of loss and damage has the potential to germinate; and finally (e) understanding by the developing countries of the difficulty of proving specific causation (as of now) to specific climate change impacts. Therefore, a negotiated settlement appeared to be a better option for both the industrial and developing countries for promoting adaptation as a global responsibility.

There remains conceptual ambiguity in understanding L&D. Article 8.3 of the Paris Agreement discusses climate action and support for "averting, minimizing and addressing" loss and damage. Averting and minimizing L&D can be achieved largely through mitigation and adaptation. Well-planned adaptation as anticipatory actions can minimize L&D, but not the residual impacts, which cannot be adapted to, or which entail prohibitive costs. Khan (2023) argues there are mainly two conceptualizations, as evident from the available literature: the first one is that L&D entails liability and compensation, and the other is about risk management and risk transfer through insurance. However, developing countries have lost the fight under the Paris Agreement for realizing the first conception. We may recall that under Article 4.8 of the UNFCCC, insurance was adopted as a response to climate change impacts, but insurance for the poor and uninsurable is still not available. Insurance can cover only rapid-onset disasters, such as cyclones, floods, storm surges, with unknown probabilities, but not the slow-onset events with known probabilities like land degradation, sea level rise, loss of biodiversity, thawing of permafrost, melting of glaciers, and so on. Developing countries are relatively unified in defining L&D as "beyond adaptation," that is, residual damages which cannot be adapted to. But as a broader concept, it also includes noneconomic L&D, that is, loss of lives, loss of habitats, loss of culture, and mental health suffering. These noneconomic L&D issues cannot be valued in economic terms, though economists keep trying through concepts like disability-adjusted life years (Varbanova et al., 2023).

Finally, at COP27 in Sharm el Sheikh, a decision was adopted to establish a Loss and Damage Response Fund (LDRF) and its funding arrangements. A 26-member Board was also established, with 14 members representing the Global South, and 12 from the North. At COP28, it was agreed that the World Bank will serve as the interim trustee. The Board has already met four times to operationalize the LDRF. Donor countries, including the United Arab Emirates, at the COP28 pledged about $700 million as initial capitalization. COP29 did not see any substantive progress other than Australia, New Zealand, and Sweden pledging around $55 million. In Baku there was the launching of the Annual High-Level Dialogue on complementarity and coherence co-convened by the LDRF and the UN Secretary General. At its opening, the UN Secretary General Guterres stated: "[t]he creation of the Loss and Damage Fund is a victory for developing countries, for multilateralism, and for justice," but he underscored that "[I]ts initial capitalization of US$700 million doesn't come close to righting the wrong inflicted on the vulnerable. US$700 million is roughly the annual earnings of the world's ten best-paid footballers. . . . We must get serious about the level of finance required (UN Secretary General, 2024c). Fearing that public climate finance will continue to be insufficient, the Secretary General urged nations to explore innovative sources like solidarity levies on shipping, aviation, and fossil fuel extraction.

Another challenge in financing L&D involves reaching out to local communities. Decision paragraph 125 on Stocktake, which served as the cover decision of COP28, stipulates that the World Bank as the interim trustee must devise mechanisms of decentralizing finance support to local vulnerable communities. But the Bank does not have a culture of working directly with local communities at the subnational level. This will likely pose a serious test for the Bank as an intergovernmental institution, because many of the developing country governments are very centralized in their decision-making processes. The success of LDRF lies in how much the Bank can reform and decentralize itself to ensure climate justice in its operation through reaching out to climate-vulnerable communities at the local level. In this context, there may be a better argumentative and discursive power in conceptualizing the L&D issues in moral/ethical terms for wider public legitimacy. Arguments for grants may be stronger in addressing L&D issues on moral/ethical grounds. Debt distress in developing and low-income countries is a significant agenda item globally now, since over two thirds of adaptation finance comes as loans even for the LDCs, and this is creating a new "climate debt trap."

The challenge is this: How do we inject the ethical considerations into individual and social consciousness across countries? This is where there is a need for alternative framings of L&D in the UNFCCC process. Economic rationality to address climate change is not taking us far enough, as experience of the last two decades and the earlier discussion show. The "polluter pays" principle, the most cardinal solution, though applied nationally in many countries in different ways, is not applied at the global level. Adam Smith's lesson of "indivisibility of interests" is not considered yet by the major emitters.

Therefore, an alternative framing of L&D grounded on moral reasoning about the harms being inflicted by major emitters on the vulnerable countries and communities in the Global South may resonate better (Khan, 2023). Increasing L&D from extreme events as the new normal is violating in many ways basic development and human rights of low-income nations. Climate impacts harm the vital interests of the poor, against the "trivial interests" of the rich (Caney, 2010). This reasoning is likely to have better and deeper resonance and strike the altruistic, humane chords of empathy, individually and socially, and engender greater global solidarity and action. A body of literature demonstrates that ethical or moral reasoning, rather than a rational actor model, has more influence on generating pro-environmental behavior, because when issues are viewed as moral, impetus for action is often higher (Turaga et al., 2010).

The ongoing hearings at the international courts about their AOs can contribute to this line of reasoning. With over 62 written submissions, more than 90 comments on those submissions and the early December 2024 public hearings by more than 100 states and international organizations (UN News, 2024), and the AO from the International Court of Justice (ICJ) may add fuel to such a discourse about state responsibility on climate change. This is perhaps the most attended hearing ever heard in the ICJ's history. The historic consensus Resolution adopted by the UN General Assembly on March 29, 2023, requested the ICJ to deliver its AO on climate change. Vanuatu as a Pacific SIDS requested the UNGA for the AO.

We may cite the first AO given by the International Tribunal of the Law of the Sea (ITLOS) under the UN Convention on the Law of the Sea (UNCLOS). On May 2024, the first ground-breaking AO on the specific obligations of states was issued by the ITLOS, which provided interpretations of the UNCLOS's role on ambitious climate action (Heri, 2024; Paine, 2024). ITLOS was requested by the Commission of Small Island States (COSIS) for clear interpretations of the specific obligations of parties to UNCLOS arising from Article 192 to protect and preserve the marine environment and their obligations under Article 194 to prevent, reduce, and control "marine pollution." ITLOS AO detailed the links between UNFCCC and UNCLOS, arguing that under UNCLOS, countries must reduce GHG emissions to protect the marine environment and its resources. Based on a review of select international regimes, the first author of this chapter argued elsewhere that UNCLOS has stronger provisions on state responsibility for extra-territorial pollution, albeit in a marine environment (Khan, 2014). This ruling is the first of the trilogy—from an international tribunal about state obligations regarding climate change mitigation. It also emphasized that

Article 194.1 of UNCLOS imposes a duty of care and due diligence on states in combating climate change. The AO is also in order, to be issued hopefully by mid-2025 after due process, from the Inter-American Court of Human Rights (IACHR), initiated by Chile and Colombia in March 2023.

The mission of all these AOs is to seek explication of the state obligations under related legal regimes toward climate change. While the legal nature of AOs is nonbinding, it provides authoritative interpretations of obligations under global regimes. Together with ITLOS AO, positive AOs from the ICJ and IACHRs will surely galvanize global public opinion to impel speedier actions, and the global youth is likely to capitalize on the outcomes as a political opportunity for pressurizing their politicians to act and not just talk in the COP jamborees. Reports suggest that such AOs are being sought by countries and civil society organizations in other regional courts, including the African Court of Human Rights (Yusra & Marie, 2024). The more such AOs that are issued from global and regional courts, the better for the countries and communities facing existential threats from planetary warming.

In recent years, the youth, both in the Global North and South, are increasingly becoming active in climate negotiations. They are engaging themselves under different platforms, such as the World Youth for Climate Justice (leaders in seeking AOs), Fridays for Future, led by Greta Thunberg of Sweden, and the associated Youth Climate Strikes (see chapter by Uba in this volume). The Children and Youth Pavilion in the COPs now are given space for engaging with policymakers and are recognized as change agents. Their lobbying contributed to establishing the Loss and Damage Fund. As a result of this mobilized advocacy, there are more programs and funding for promotion of youth leadership in climate policymaking.

Conclusion—Some Ways Forward

This chapter was devoted to unpacking the root causes of the intractability of climate diplomacy for the last two decades. We have discussed that two disconnects between the cause and effect of climate change stand in the way of progress. A temporal disconnect is manifest in the time lag between emissions and its effects in the form of climate impacts on national economies and societies. Here our mainstream economic model is no help to bridge this disconnect, as it is based on the concept of net present value, which values the present over the future. But with global warming, cost is upfront, and benefits are a little distant. Added to this are issues of unequal contributions to the problem and the potential of free riding by parties. Collective action to arrest the collapse of climate stability is not taken at a speed needed to bridge the gap between science and policy. Global public support, including the application of social discounting, has not materialized yet.

The spatial disconnect between the main sufferers and the main causers of climate change is manifest in poor adaptation support to the vulnerable developing countries, particularly to the LDCs and SIDS. Several conditions impede adequate climate support for the Global South nations, including a conceptual lacuna in framing adaptation narrowly under neoclassical economics, the overlapping relationship between adaptation and development, and the inefficacy of private-sector participation in adaptation. Increasingly expansive patterns of climate disasters across all regions, both rich and poor, and their transboundary and secondary impacts are yet to bring vulnerability interdependence into shaping global and national policies. The suggested framing of adaptation as a GPG, to be funded by solidarity levies on high-emitting sectors, is not gaining the required momentum necessary. A reformed global financial system led by the World Bank can lead this process, to ensure climate justice, but it all depends on what kind of reforms it may undergo. Many negotiators from developing countries in the UNFCCC still have reservations about the Bank's ability to ensure climate justice.

Against this continued intractability, the third plank of climate policy, loss and damage have been brought in to see whether they may make a difference. As traditional economic-instrumental rationalities are not likely to work, as experience shows, the nonlitigious process of advisory opinions of international courts regarding state responsibility to address climate change and their positive recommendations as high-level moral/ethical pressure may impel stronger climate action. But such AOs on realizing reparations caused by increasing loss and damage will not work, because such outcomes have been foreclosed under the Paris Agreement. The impacted communities across the globe, led by the increasingly active youth leadership, must push the social movement to a critical mass, to force politicians particularly of the G20 countries for actions warranted by climate science. We may conclude with a statement of W. Churchill, a veteran British politician in the House of Commons on November 12, 1936, which sounds as if it had been written for today's intractable climate negotiations:

> So they [the Government] go on in strange paradox, decided only to be undecided, resolved to be irresolute, adamant for drift, solid for fluidity, all-powerful to be impotent.... Owing to past neglect, in the face of the plainest warnings, we have entered upon a period of danger.... The era of procrastination, of half measures, of soothing and baffling expedience of delays, is coming to its close. In its place we are entering a period of consequences.... We cannot avoid this period, we are in it now. (http://www.churchill-society-London.org.uk/Locusts.html.

Note

1. The Clinton administration initially signed the protocol, but the US Congress failed to ratify it.

References

Almeida, P., & Martín, A. P. (2022). *Collective resistance to neoliberalism*. Cambridge University Press.

Banda, M. L. (2018). Climate adaptation law: Governing multi-level public goods across borders. *Vanderbilt Journal of Transnational Law*, *51*, 1027–1074.

Barrett, S. (2008). Climate treaties and the imperative of enforcement. *Oxford Review of Economic Policy*, *24*(2), 239–258.

Bridgetown Initiative. (2024). Bridgetown initiative on the reform of the *International Development and Climate Finance Architecture*. Government of Barbados. https://www.bridgetown-initiative.org/bridgetown-initiative-3-0/

Caney, S. (2010). Climate change, human rights, and moral thresholds. In S. Humphrey (Ed.), *Human rights and climate change* (pp. 69–90). Cambridge University Press.

Carbon Tax Center. (2024). Where carbon is taxed. https://www.carbontax.org/where-carbon-is-taxed-some-individual-countries/

Climate Action. (2024). 2024 on track to be warmest and first year above 1.5°C, 10 December. https://www.climateaction.org/news/2024-on-track-to-be-warmest-and-first-year-above-1.5c.

CPI. (2024a). *Global landscape of climate finance: Insights for COP 29*. climatepolicyinitiative.org/publication/global-landscape-of-climate-finance-2024

CPI. (2024b). *Tracking and mobilizing private sector climate adaptation finance*. chrom=extension://efaidnbmnnnibpcajpcglclefindmkaj/https://www.climatepolicyinitiative.org/wp-content/uploads/2024/09/Tracking-and-Mobilizing-Private-Sector-Climate-Adaptation-Finance-2024.pdf

Crawford, S. K. (2022, September 1). UN chief warns of "colossal global dysfunction" but urges world to unite on sweeping solutions. ABC News. https://abcnews.go.com/Politics/chief-warns-colossal-global-dysfunction-urgesworld-unite/story?id=90215597.

Daly, H., & Farley, J. (2011). *Ecological economics: Principles and applications* (2nd ed.). Island Press.

FAO. (2021). *The impact of disasters and crises on agriculture and food security*. FAO.

Fisher, D. R. (2024). *Saving ourselves: From climate shocks to climate action*. Columbia University Press.

Gardiner, S. M. (2011). *A perfect moral storm: The ethical tragedy of climate change*. Oxford University Press.

GCA. (2019) *Adapt Now: A global call for leadership in climate resilience*. Amsterdam.

Giddens, A. (2011). *The politics of climate change*. 2nd ed. Polity Press.

Hardin, G. (1968). Tragedy of the commons. *Science*, *162*, 1234–1348.

Hedlund, J., Fick, S., Carlsen, H., & Benzie, M. (2018). Quantifying transnational climate impact exposure: New perspectives on the global distribution of climate risk. *Global Environmental Change*, *52*(Sept.), 75–85.

Heri, C. (2024, December 21). An interim report on ICJ's climate advisory opinion. https://www.justsecurity.org/105917/interim-report-icj-climate-opinion/

IDMC. (2021). *Global report on internal displacement*. IDMC, Norwegian Refugee Council.

Intergovernmental Negotiating Committee (INC). (1991). Report of the INC, First session, 4–14 February 1991, UN Doc.A/AC237/6,6f. INC (1991b). A/AC.237/ Misc.1/Add.3.UNFCCC, Second session, 19–28 June. INC.

International Energy Agency. (2024). *World energy outlook 2024*. IEA.

IPCC. (2018) *Summary for policymakers: Global warming of 1.5°C*. World Meteorological Organization.

IPCC. (2022). *AR6: WGII report: Impacts, vulnerability and adaptation*. World Meteorological Organization.

IPCC WGI. (2021). *Climate change 2021: The physical science basic summary for policy-makers*. World Meteorological Organization.

Kaul, I. (2017). Putting climate finance into context: A global public goods perspective. In A. Markandya, I. Galarraga, & D. Rübbelke (Eds.), *Climate finance* (pp. 129–156). World Scientific.

Khan, M. (2023, October 6). Loss and damage must be defined on moral grounds. *The Daily Star*. https://www.thedailystar.net/author/mizan-r-khan-0

Khan, M. R. (2020). *Debt for adaptation swap—Investment in adaptation and resilience*. Mimeo.

Khan, M. R. (2014). *Toward a binding climate change adaptation regime*. Routledge.

Khan, M. R. (2015). Polluter-pays-principle: The cardinal instrument for addressing climate change. *Laws*, *4*(3), 638–653.

Khan, M. R., Huq, S., Risha, A. N., & Alam, S. S. (2021). High density population and displacement in Bangladesh. *Science*, *372*(6538).

Khan, M. R., & Munira, S. (2021). Climate change adaptation as a global public good: Implications for financing. *Climatic Change*, 167, 50.

Khan, M. R., & Roberts, J. T. (2013). Towards a binding adaptation regime: Few levers and instruments. In S. Moser S. & B. Max, B (Eds.), *Toward successful adaptation* (pp. 132–148). Routledge.

Khan, M., Robinson, S.-A., Weikmans, R., Ciplet, D., & Roberts, J. T. (2019). Twenty-five years of adaptation finance through a climate justice lens, Special issue. https://doi.org/10.1007/s10584-019-02563-x

Laan, T., Geddes, A., Do, N., Cameron, L., Goel, S., & Jones, N. (2023). Burning billions: Record public money for fossil fuels impeding climate action. IISD. https://www.energypolicytracker.org/burning-billions-record-fossil-fuels-support-2022/

Michaelowa, A. (2001). Mitigation versus adaptation: The political economy of competition between climate policy strategies and the consequences for developing countries. HWWA Discussion Paper 153. Hamburg Institute of International Economics.

Netherlands Environmental Assessment Agency. (2011). *A global public goods perspective on environment and poverty reduction: Implications for Dutch foreign policy*. The Hague.

Nordhaus, W. D. (2005). *Paul Samuelson and global public goods: A commemorative essay for Paul Samuelson*. Yale University Press.

Nye, J. S., Jr. (2016). Politicians say American leadership is in decline: They are wrong,' op. ed. *The Washington Post*, January 29. Available at: https://www.washingtonpost.com/news/in-theory/wp/2016/01/29/politicians-say-american-leadership-is-in-decline-theyre-wrong/

Oxfam. (2023). *Oxfam climate finance shadow report 2023*.

Paine, J. (2024, June 3). The ITLOS advisory opinion on climate change: Selected issues of treaty interpretation. https://www.ejiltalk.org/the-itlos-advisory-opinion-on-climate-change-selected-issues-of-treaty-interpretation/

Rajamani, L. (2010). The increasing currency and relevance of rights-based perspective in the international negotiations on climate change. *Journal of Environmental Law*, *22*(3), 391–429.

Reuters. (2024, September 19). EU chief pledges billions in aid for flood relief. https://www.yahoo.com/news/eu-chief-pledges-billions-aid-223256461.html?fr=yhssrp_catchall.

Roosevelt, F. D. (1944). *United Nations Monetary and Financial Conference: Bretton Woods, final act and related documents, New Hampshire, July 1 to July 22, 1944* (Department of State, Ed.). United States Government Printing Office.

Rosenthal, J. H. (2022). Autocrats, oligarchs, and us: A moment of crisis for responsible internationalism. Carnegie Council, September 23. https://www.carnegiecouncil.org/media/series/presidents-desk/autocrats-oligarchs-crisis-responsible-internationalism

Sengupta, S., & Bearak, M. (2024, September 23). The world is a mess: That makes the climate crisis harder to solve. *The New York Times*. https://www.nytimes.com/2024/09/23/climate/climate-diplomacy-fracture.html

Singh, H., Faleiro, J., Anderson, T., & Vashist, S. (2020). Cost of climate inaction: Displacement and disaster migration. *Actiona*. XANSA.id.

SIPRI. (2024). Global military spending surges amid war, rising tensions and insecurity. https://doi.org/10.55163/BQGA2180

Smith, A. (1759/1982). *The theory of moral sentiments* (D. D. Raphael & A. L. Macfie, Eds.). Liberty Fund.

Smith, A. (1776/2008). *Wealth on nations* (K. Sutherland, Ed., 6th ed.). Oxford University Press.

Standing Committee on Finance. (2024). *Report of the Standing Committee on Finance*. UNFCCC.

Turaga, R. M., Howarth, R. B., & Borsuk, M. E. (2010). *Pro-environmental behavior: Rational choice meets moral motivation*. Annals of the New York Academy of Sciences.

UNCTAD. (2022). *The least developed countries report*. https://www.undrr.org/implementing-sendai-framework/sendai-framework-action/disaster-risk-reduction-least-developed-countries

UNDRR. (2023). Disaster risk reduction in the least developed countries. https://www.undrr.org/implementing-sendai-framework/sendai-framework-action/disaster-risk-reduction-least-developed-countries.

UNEP. (2023). *Adaptation gap report, Nairobi*. UNEP.

UNEP. (2024). *Emissions gap report, Nairobi*. UNEP.

UN News. (2024, December 23). UN World Court concludes landmark hearings on states' responsibility for climate change. https://news.un.org/en/story/2024/12/1158476

UN Secretary General. (2024a, January 17). At Davos Forum, Secretary-General warns of global norms collapsing, highlights need to rebuild trust, reform governance. https://press.un.org/en/2024/sgsm22109.doc.htm

UN Secretary General. (2024b). Statement of the UN Secretary General at Earth Day 2024 in New York, 5 June. https://press.un.org/en/2024/sgsm22255.doc.htm

UN Secretary General. (2024c). Statement of the UN Secretary General at the high-level dialogue on LDRF, COP29, Baku, 12 November 2024. Available at: https://press.un.org/en/2024/sgsm22448.doc.htm

Vanderheiden, S. (2011). Globalizing responsibility for climate change. *Ethics and International Affairs*, *25*(1), 66–84.

Varbanova, V., Hens, N., & Beutels, P. (2023). Determinants of life-expectancy and disability-adjusted life years (DALYs) in European and Organisation for Economic Co-operation and

Development (OECD) countries: A longitudinal analysis (1990–2019), Vol. 24. SSM: Population Health.

Volger, J. (1995). *The global commons: A regime analysis*. John Wiley & Sons.

WMO. (2020). *Atlas of mortality and economic losses from weather, climate and water-related extremes, 1970–2019*. WMO.

WRI. (2023). *The state of climate action 2023*. WRI Digest.

WRI. (2024). *The history of carbon dioxide emissions*. WRI Digest. https://www.wri.org/insights/history-carbon-dioxide-emissions.

Young, O. (2010). *Institutional dynamics: Emerging patterns in international environmental governance*. MIT Press.

Yusra, S., & Marie, F. (2024). Climate change litigation before the African human rights system: Prospects and pitfalls. *Journal of Human Rights Practice, 16*(1), 146–159.

Zedillo, E. (Ed.). (2008). *Global warming: Looking beyond Kyoto*. Center for the Study of the Globe, Yale University; Brookings Institution Press.

CHAPTER 26

Climate Change Actions in the Sahel

A Look at Past Actions and Future Nature-Based Solutions

TERENCE EPULE EPULE AND VINCENT POIRIER

Introduction

The African continent is rich in natural resources and provides numerous benefits to its inhabitants as well as all human beings living on Earth. In addition, the African continent stores about 209 Pg of carbon in above- and below-ground biomass and soils (Ernst et al., 2024), representing about 10% of the carbon that can be found in Earth's terrestrial biosphere (i.e., vegetation and soils; Canadell et al., 2021). Still, despite the huge importance played by African ecosystems in regulating global climate, the number of published studies on climate action in Africa remains relatively low. Basic research using the keywords "climate" AND "action" AND "Africa" on the Scopus database revealed fewer than 2,000 publications over more than 60 years, with 98% of them being published in the last 25 years. Although the topic has become more popular recently (i.e., >1,000 articles published in the last 5 years), this observation reveals a disconnect between the importance of Africa in regulating climate change and the research efforts carried out on the continent.

The evidence to support the reality of climate change is enormous, as evidenced by global temperature and precipitation changes and increased emissions of greenhouse gases (GHGs) (Intergovernmental Panel for Climate Change [IPCC], 2022). Globally, the effects of climate change are unequally distributed. Across Africa, the situation is daunting as the continent is more affected by and vulnerable to the effects of climate change despite emitting the lowest GHGs and its limited contribution to the global crisis (Adepoju, 2021; Epule et al., 2021; Mosberg & Eriksen, 2015). One of such sectors that is significantly affected due to its high level of vulnerability is agriculture (Mbuli et al., 2021; Recha et al., 2016; Reenberg, 2010). To ensure food security by 2050 for a global population of over 9 billion, global food production needs to be increased by 70–100% (Dubois, 2011). If concerted actions are not taken, global food production will decline by about 1% by 2050 (Bruinsma, 2009). These global changes will continue to have more devastating effects in Africa because about 70% of all agricultural production is in the hands of small-scale farmers who are poorly adapted to climate change (Donatti et al., 2018; Epule et al., 2021).

In Africa, the hotspots of vulnerability to climate change are also unevenly distributed. Among the most vulnerable regions of Africa is the Sahel (Epule et al., 2014). This semiarid strip of land covers about 3,053,200 km^2 located between the tropical rainforest in the south and the arid north of Africa and is home to about 60 million inhabitants (Olsson & Mryka, 2008). The mean annual precipitation of the Sahel varies from less than 200 mm of rainfall at the 17° isohyet (northern boundary) to 450–500 mm of rainfall at the 15° isohyet (southern boundary) (Wang et al., 2005; Zeng, 2003). During April to July, the temperatures are about 32°C–35°C throughout the region (Wang et al., 2005). Prominent droughts have affected this region, particularly between the end of the 1960s and the mid-1980s (Opiyo et al., 2015; Anyamba & Tucker, 2005).

Consequently, by virtue of the high vulnerability of Africa and the Sahel in particular, a lot of global attention is given to climate change adaptation actions in the region. The latter are placed as priorities of political and environmental agendas of governments, non-governmental organizations, international organizations, and civil society groups (Ford et al., 2015; Noble & Huq, 2014). Still, substantial divergence exists among stakeholders as to what climate change adaptation actions really mean (Berrang-Ford et al., 2014; Dupuis & Biesbroek, 2013).

In this chapter, we begin by reviewing past works that identified climate change adaptation actions taken across the Sahel during the 1975–2015 period. This period was chosen since it marks the era of recurrent droughts that ravaged most of the Sahel. The action-based approached used (Ford et al., 2013, 2014) enabled us to compare regions of the Sahel and identify general patterns in climate change adaptation actions over the 40 years covered (Ford et al., 2013, 2014; Mannke 2011). However, many adaptation actions are under-represented in peer-reviewed literature. The patterns reported should thus be considered solely as a baseline. In addition, several adaptation actions have taken place outside the range (1975–2015) used to identify spatial variations and properties of climate change adaptation actions across the Sahel. Several actions involving nature-based solutions have emerged over the last decade since the emergence of the "climate-smart agriculture" concept (Lipper et al., 2014; Mannke, 2011). Therefore, in the second part of this chapter, we present potential actions involving nature-based solutions that could be implemented in the Sahel to increase carbon sequestration and help meet the challenges of climate change.

Climate Change in the Sahel: A Snapshot of Actions Over the 1975–2015 Period

Geographical Distribution of Climate Change Adaptation Actions in the Sahel

The Sahel is witnessing climate change (Epule et al., 2013a, 2017; Lesnikowski et al., 2013, 2016; Mertz et al., 2009). The past 40 years across the Sahel have shown increased temperatures between 0.2°C and 2.0°C (IPCC, 2007). In addition, recent reports show that the region has witnessed high-intensity precipitation of short duration that does not tally with the growing season of several crops (IPCC, 2022). Epule et al. (2017) have identified >400 adaptation actions reported in 70 peer review journal articles. The latter study reports that across the Sahel, Kenya, recorded the largest total number of climate change adaptation actions, followed closely by Senegal (Table 26.1). Kenya was previously reported as one of the top reporting highest climate change adaptation actions among other countries in Africa and Asia (Ford et al., 2015). From a regional point of view, West African Sahel recorded about two times more climate change adaptation actions than in the East African Sahel, even though Kenya in East African Sahel records the highest number of climate change adaptation actions on an individual country basis (Table 26.1). This observation could be linked to the fact that West African Sahel is closer to the Sahelian desertification belt and efforts such as the African green wall started in this area.

Broad Categories of Climate Change Adaptation Actions Taken

Climate change adaptation actions were grouped into four key categories by Epule et al. (2017). The first category gathers technically related climate adaptation actions that are anchored in technology and science. Though they originate from the scientific community and external stakeholders, these adaptation actions are often used by the communities. Climate monitoring and forecasting, soil conservation, water harnessing, use of high-yielding crops and varieties that tolerate droughts and changing planting schedules are examples of actions included in this category. The second one brings together Indigenous problem-solving climate change adaptation actions defined as ones anchored on Indigenous knowledge and originate from local communities. For example, actions that fall in this category could be related to preservation and storage of products, migration, fallowing, destocking, and dietary change. The third category combines socially related climate change adaptation actions that have social foundations or safety nets. Help from next of kin, spirituality or prayers, investing in kids' education, and giving up kids are examples of climate adaptation actions that would be grouped in this third category. The fourth one concerns economically related adaptation actions. The latter are anchored on economic and financial safety. These actions are mostly used to adapt to climate change, such as money transfers, access to credits, savings, and sale of products (e.g., cattle, grain).

Based on the data from Epule et al. (2017), we can see that technically based climate adaptation actions represented half of the actions documented in the Sahel (Table 26.1). The second most important ones related to Indigenous problem-solving adaptation and represented slightly less than 42% of all actions documented. Socially related and economically related adaptation actions thus represented minor fractions (i.e., about 5% and 3%, respectively) of total actions identified by Epule et al. (2017) (Table 26.1). Although technically related climate adaptation actions have the highest rates of usage, in the two countries having the highest number of total adaptation actions (i.e., Kenya and Senegal), those related to Indigenous problem-solving adaptation are the most frequent. This suggests that in these countries as in those where adaptations actions mostly relate to Indigenous problem-solving, the introduction of new technology to help adapt to climate change must be approved and accepted by local communities to be adopted by them. The West African Sahel records approximately 66% of all technically based adaptation actions, followed by East Africa (Table 26.1). This is consistent with the fact that technical efforts such as the African green wall started in West Africa due to the proximity of the region with the desertification belt of the Sahel.

Specific Adaptation Actions Used to Adapt to Climate Change in the Sahel

A total of 60 specific actions taken by Sahelian countries were identified in the literature (Epule et al., 2017). Of these, the 10 most reported ones are presented in Table 26.2. Altogether, they represent >75% of all climate change adaptation actions reported by Epule et al. (2017). Income diversification dominates the types of specific actions taken. Examples of such actions include (a) off-farm employment (Elmqvist & Olsson, 2006; Fleuret, 1986; Mertz et al., 2009; Morand et al., 2012), (b) working for development projects (Moretimore, 2010; Nielsen & Reenberg, 2010), (c) small-scale commerce (Nielsen & Reenberg, 2010), (d) non-farm income (Burnham & Ma, 2016; Mertz et al., 2011; Reardon et al., 1988), or (e) selling of personal belongings (Brockhaus et al., 2013; Opondo, 2013; Schaer, 2015). Income diversification recorded the highest usage (i.e., 21 publications) during the period 2008–2015 (Table 26.2). However, it was also recorded in 18 publications between 1985 and 1996. The fact that income diversification is the most used adaptation action is consistent with previous works that argue that many African communities resort to this strategy to help absorb the shocks of climate change and environmental protection (Epule et al., 2013b; Le et al., 2012).

Water harnessing actions are the second most important type of specific actions identified by Epule et al. (2017) (Table 26.2). They include (a) rainwater harvesting (Barbier et al., 2009; Bryan et al., 2009; Zampaligre et al., 2014), (b) irrigation (Bryan et al., 2009; Deressa et al., 2009; Fleuret, 1986), (c) water management (Burnham & Ma, 2016; Douxchamps et al., 2016; Rockstrom, 2003), (d) boreholes (Mbow et al., 2008), and (e) construction of dams and drainage systems (Magistro & Lo, 2001). Water harnessing strategies were very important between 1985 and 1996 during which 27 peer-reviewed studies were recorded, while 15 peer-reviewed publications were reported during the 2008–2015 period. The finding that water harnessing related actions are the second most reported ones is also important. The Sahel faces recurrent droughts and water stress throughout the year; its population thus tends to develop adaptation strategies to absorb the shocks caused by water deficit through water management and harnessing.

According to Epule et al. (2017), using soil conservation practices is in third position in terms of the importance of specific actions identified in peer-reviewed literature to adapt to climate change in the Sahel (Table 26.2). Such actions include (a) mulching to protect soil surface (Burnham & Ma, 2016), (b) change in tillage practices and crop rotation (Bryan et al., 2009), (c) use of half-moon and stone dykes techniques to preserve soil moisture (Zampaligre et al., 2014), (d) planting shade trees to reduce the temperature at soil surface (Bryan et al., 2009; Burnham & Ma, 2016; Deressa et al., 2009), (e) fallowing to allow the soil to rest (Gebrehiwot & Van der Veen, 2013; Nyong et al., 2007), and (f) methods to control soil erosion and stabilize dunes (Brockhaus et al., 2013; Burham & Ma, 2016; Okoye, 1998). Soil conservation–related options recorded their highest usage during the decade 1985–1996 (Table 26.2). Although specific soil conservation actions arrive third, the number of published studies on the topic is about 2.0 and 1.8 times less important than that of specific actions related to income diversification and water harnessing (Table 26.2). Still, actions related to water harnessing are closely related to soil conservation ones; to produce food in the face of climate change, it is essential to protect the soils to preserve their moisture so that they can provide crops with the water resources they need.

The number of peer-reviewed publications identifying water harnessing and specific soil conservation actions was the highest during the 1986–1996 decade (see Table 26.2). It could be argued that this is linked to the fact that this decade followed a period that witnessed some of the most ravaging droughts in the history of the Sahel (Anyamba & Tucker, 2005; Maranz, 2009). Consequently, governments and other stakeholders have made efforts to enhance adaptations (Epule et al., 2013a, 2014, 2015; Zeng, 2003). The adaptation efforts yielded fruit, as there were more reports of increased greening in the Sahel (Anyamba & Eastman, 1996; Eklundh & Olsson, 2003). The observed greening of the Sahel results from a complex combination of social and environmental factors and suggests a potential for improved agricultural productivity and human welfare (Olsson et al., 2005).

Among the specific adaptation actions identified, those related to population migrations, adding farm inputs and using high-yielding crops and animals, arrive in fourth, fifth, and sixth positions (Table 26.2) (Epule et al., 2017). Examples of specific farm input actions include the addition of (a) manure (Marenya & Barrett, 2007; Mertz et al., 2009; Wood et al., 2014) and (b) fertilizers (Croppenstedt et al., 2003; Mertz et al., 2009, 2011), either inorganic (Marenya & Barrett, 2007; Wubeneh & Sander, 2006) or organic forms (Epule & Bryant,

TABLE 26.1 Total and Categorized Number of Adaptation Actions in Countries and Regions of the Sahel Reported in Peer-Reviewed Literature Published Between 1975 and 2015

Countries	Total Number of Adaptation Actions	Technically Related Adaptation Actions	Indigenous Problem-Solving Adaptation Actions	Socially Related Problem-Solving Adaptation Actions	Economics Related Adaptation Actions
Kenya	75	25	39	7	4
Somalia	1	0	1	0	0
Ethiopia	46	38	6	1	1
Northern Eritrea	0	0	0	0	0
Central and South Sudan	5	1	4	0	0
Central African Republic	0	0	0	0	0
North Cameroon	19	6	12	1	0
Central Chad	0	0	0	0	0
Northern Nigeria	58	33	22	2	1
Southern Niger	36	14	18	2	2
Northern Burkina Faso	58	34	20	3	1
Central Mali	39	25	11	2	1
Southern Mauritania	5	1	2	1	1
Senegal	65	26	34	2	3
Studies on the entire Sahel	7	3	4	0	0
Total	414	206	173	21	14
Regions					
Eastern African Sahel	127	64	50	8	5
Central African Sahel	19	6	12	1	0
West African Sahel	261	133	107	12	9
Studies on the entire Sahel	7	3	4	0	0
Total	414	206	173	21	14

Source: Data taken from Epule et al., 2017.

TABLE 26.2 Most Popular Specific Adaptation Actions Taken in Sahelian Countries Over Four Decades

Specific Actions	1975–1985	1986–1996	1997–2007	2008–2016	Total	Rank
	Number of Peer-Reviewed Studies					
Income diversification	6	18	8	21	53	1
Water harnessing/ irrigation	3	27	3	15	48	2
Soil conservation	1	19	1	6	27	3
Migration	6	7	5	8	26	4
Farm input	2	5	6	11	24	5
High-yielding crops/animals	1	8	1	13	23	6
Reforestation/ afforestation	1	7	1	8	17	7
Change planting dates	0	8	1	8	17	7
Mixed cropping	1	8	3	3	15	8
Increase/reduce livestock	0	4	1	10	15	8
Pasture/forage management	0	3	1	5	9	9
Embankments/ dams	2	2	2	3	9	9
Help from next of kin	5	0	0	4	9	9
Buy food/get food aid	0	6	0	2	8	10
Spirituality and prayers	1	6	1	0	8	10
Drought tolerant crops/animals	4	3	0	1	8	10

Source: Data from Epule et al., 2017.

2016), such as compost (Somda et al., 2002). With regard to specific actions related to high-yielding crops and animals, examples identified include (a) use of high-yielding crop varieties from a general standpoint (Bryan et al., 2009; Deressa et al., 2009; Fleuret, 1986) and more specifically for producing potatoes (Abebe et al., 2013; Burnham & Ma, 2016; Thuo et al., 2011), sorghum (Adesina & Baidu-Forson, 1995), and maize (De Groote et al., 2013); and (b) improved animal husbandry (Zander et al., 2013). These types of specific actions recorded their highest frequency in the last decades of the period covered, that is, during 2008–2016.

Global Perspective on Adaptation Actions Identified in the Sahel

The studies reviewed earlier indicate that adaptations are mostly driven by climate change. Verifying the shocks to which people are adapting provides further information regarding whether climate adaptation actions are driven by climate or nonclimatic variables. The latter are important because, in many cases, adaptation actions are taken also in response to other issues (e.g., poverty, health, literacy, financial, and sociocultural disparities), which can either enhance or reduce adaptive capacity (Dupuis & Biesbroek, 2013; Ford et al., 2015). Berrang-Ford et al. (2011) found that considerable research has been done with a greater focus on intentions than groundwork, and that motivations for adapting to climate change are increasingly linked to nonclimatic factors. Across Africa and the Sahel, evidence shows that most of the problems faced by populations (e.g., food insecurity, pandemic and epidemics, and poverty) result from nonclimatic causes such as deforestation, migrations, wars, and land degradation, with climate change only playing a reinforcing role (Araos et al., 2016; Giannini et al., 2003, 2005, 2008; Lu & Delworth, 2005; Reynolds et al., 2007).

The findings presented earlier are consistent and vary compared with other studies. As such, Ford et al. (2015) also verified climate change adaptation status in Africa. However, their study also included Asia and was based on both peer review and gray literature published from 2006, while results from the Epule et al. (2017) study reviewed earlier were based on peer review literature published on the Sahel only and covered the period from 1975 to 2015. Berrang-Ford et al. (2011), who used the systematic review approach and focused on peer review literature only, reported results that were consistent with ours. However, their latter study departed from ours as it reviewed papers submitted between 2008 and 2012, and it was more of a global perspective to adaptation to climate change. Lwasa (2015) looked at gray literature and peer review papers published from 2006 to investigate climate change adaptations policy and practice across Africa and Asia, but they focused only on the context of deltas. The overall status of climate change adaptation actions in the Sahel can hardly be assessed solely based on the adaptation actions reported in the peer review literature. According to Epule et al. (2017), the Sahel remains highly vulnerable in its capacity to adapt to climate change, and much more scientific research needs to be carried out to better understand and increase the adaptive capacities of Sahelian populations.

Climate Actions in the Sahel: Looking to the Future With Nature-Based Solutions

Using Plants to Restore Degraded Soils

A recent review by Massoukou Pamba et al. (2023) revealed that plants could help restore degraded tropical soils through five key mechanisms, the first one being nitrogen fixation. Legume plants (i.e., Fabaceae) can restore degraded soils because they can fix atmospheric nitrogen into soil nitrogen by symbiotically associating with Rhizobium bacteria (Duponnois et al., 2001; Graham & Vance, 2003). Actinorhizal plants are also capable of fixing atmospheric nitrogen into soil nitrogen by forming symbiotic associations with actinomycete bacteria of the genus Frankia (Hocher et al., 2009). By increasing the amount of nitrogen in the soil, such plants thus have the capacity to help regenerate soil fertility and improve crop yields. The second mechanism is related to organic carbon sequestration (Massoukou Pamba et al., 2023). According to Lal (2015), the minimal critical threshold of soil organic carbon

(SOC) concentration is estimated to be 10 to 15 g C kg^{-1}. Any soil conservation practices that would increase SOC concentration above this critical threshold would help reverse degradation trends and restore degraded soils (Lal, 2002, 2015). The third mechanism implies organic matter addition on or in the soil, such as farm inputs like manure or compost, mulching, or litter from herbaceous plants (Massoukou Pamba et al., 2023). However, in the absence of exogenous organic matter such as compost or manure, the addition of branches, leaves, roots, and crop residues to the soil surface will nourish soil microbial communities, preserve soil moisture, and help maintain or increase soil fertility (Bonneau, 2005; Janssen et al., 1990; Njoukam et al., 1999; Yan et al., 2018). The stabilization of soil structure is the fourth mechanism identified by Massoukou Pamba et al. (2023) by which plants can help restore degraded soils. This is especially important in the context of sandy soils where the cohesion among coarse soil particles is limited and often insufficient to create aggregates. In this context, plants are very helpful. Their roots influence the biotic mechanisms that lead to the formation and stabilization of soil aggregates (Bardgett et al., 2014; Poirier et al., 2018a, 2018b; Six et al., 2004), while their aerial parts attenuate the impact of raindrops on the soil, which helps reduce the destruction of surface soil structure. The reduction of soil erosion by wind and water is the fifth mechanism identified by Massoukou Pamba et al. (2023) by which plants contribute to restore degraded soils. By slowing down rainfall and water movement on the soil surface, plant biomass reduces soil losses through water erosion and promotes water infiltration (Rey et al., 2004). Planting windbreaks and adding mulch on the soil surface also reduce soil losses trough wind erosion (Lamers et al., 1995).

Improving Fallows and Agroforestry Systems

Agroforestry systems in the Sahel have the potential to increase carbon sequestration in the plant-soil system and consequently improve soil fertility and increase the capacity of local population to adapt to climate change. Indeed, by selling carbon credits to industrialized countries, small producers from Sahel would improve their income and benefit from the carbon sequestered by agroforestry systems (Marone et al., 2017; Takimoto et al., 2008). In the semiarid zone of Niayes, Senegal, Marone et al. (2017) showed that the amount of carbon stored in the soil and plant system was greater in agroforestry systems using the fallow. Moreover, these authors found the tree species *Faidherbia albida* and *Acacia raddiana* stored more carbon in their biomass than the other species studied. This could be because these species can fix atmospheric nitrogen in the soil and thus potentially increase their fertility. Using nitrogen-fixing species such as those from the genus Acacia can be helpful in restoring degraded soils and improving their fertility. As such, Kachaka et al. (2023) observed that the establishment of *Acacia auriculiformis* agroforestry fallow improved the soil's physicochemical properties and increased its carbon stocks compared to traditional savanna fallow. These authors suggested the scaling up of the improved fallow with *A. auriculiformis* in agroforestry systems as a soil regenerative technology. Jean-Pierre (2023) showed that fast-growing trees such as *Acacia mangium* and *Leucena Leucocephala* could grow on poor and degraded soils and improve their carbon stocks and fertility 18 months after their plantation. While the amount of carbon stored in plant biomass was more important for *Acacia mangium*, the amount stored in the soil was greater under *Leucena Leucocephala* (Jean-Pierre, 2023). Another species that can rapidly improve the fertility of degraded tropical soils is *Jatropha curcas* L. Indeed, Massoukou Pamba et al. (2024) showed that this plant was capable of increasing soil carbon and nitrogen concentrations by 1.4 and 1.8 times those of the degraded soils prior to *J. curcas* plantation, 3 years earlier. Although the study of Massoukou

Pamba et al. (2024) was conducted in a tropical humid climate, *J. curcas* is recognized for its capacity to grow under Sahelian conditions (Minengu et al., 2014).

Producing Biochar for Soil Improvement and Waste Management

According to the International Biochar Initiative (2015), biochar can be defined as "a solid material obtained from the thermochemical conversion of biomass in an oxygen-limited environment." Application of biochar to the soil could increase soil carbon stocks and potentially results in a net CO_2 removal from the atmosphere; this practice is seen as a strategy to implement climate-smart soils in agriculture (Paustian et al., 2016). Globally this approach would potentially deliver emission reduction of 3.4 to 6.3 Pg CO_2 (Lehmann et al., 2021) at rates in the order of 0.17 to 0.32 Gt C yr^{-1} (Sirén, 2024). In the context of adapting to climate change in the Sahel, this technology could be helpful by valorizing organic wastes to produce energy and biochar capable of improving soil fertility and sequestering carbon. Zubairu et al. (2023) reviewed studies on the impact of biochar addition to the soil in Sahelian conditions and found that it could enhance soil fertility and increase crop yields. According to Adepoju (2021), this technology could be achieved at low cost and has the potential to be a highly effective nature-based solution for climate change mitigation in the context of the developing world. Dassi Djoukouo et al. (2024) designed a small-scale pyrolizer model in Cameroon that was adapted to recycling plastic wastes into fuel. Although more work must be done to produce fuel that will fully meet European and Cameroonian diesel fuel specifications, their approach showed high potential. Lontsi et al. (2024), also in Cameroon, used a locally made pyrolyser to produce biochar with different agriculture wastes such as corncobs (*Zea maize*), bean straws (*Phaseolus vulgaris*), cocoa pod husks (*Theobroma cacao*), coconut shells (*Cocos nucifera*), palm kernel shells (*Elaeis guineensis*), palm mesocarp fibers (*Elaeis guineensis*), rice husks (*Oryza sativa*), and cassava peels (*Manihot esculenta*). Their study revealed that the most promising biochars as amendments were those from bean straw and cocoa pod husks, while the biochar that had the highest CO_2 reduction potential was the one produced from coconut shells.

Conclusion

Moving forward, there are areas of research that need to be given attention. The first is a review study on the entire African continent using both peer review and gray literature covering the 1970s to present. This will help provide a regional picture of the state of climatic change adaptations across Africa from both country by country and a regional perspective. An understanding of where most of the adaptation options are being reported should also be perfected (whether it is in the peer review or gray literature). Secondly, a lot of attention is being focused on climate change adaptation across Africa using the systematic review; the climate shocks such as winds, droughts, and floods that are driving adaptations need to be given attention and verified. It is still unclear how these shocks are spatially and temporally distributed in the Sahel and across Africa, and this is a major knowledge gap to be filled. Furthermore, more studies that cover both actions and groundwork in the context of adaptations reported in the peer review and gray literature need to be carried out. There is a need to also dig deeper into nature-based solutions and verify how the actions have varied over time and space. This will help shed more light on the disparities between adaptation plans in the context of groundwork and real or concrete actions. Current civic activism to enhance the acceptance of nature-based solutions across Africa involves participatory design or cocreation of nature-based solution product—an approach that elicits the involvement of the population in the conceptualization, design, execution, and evaluation of such projects (Diep et al., 2022). In addition, media outlets ranging from audiovisual and print media, including social media, are increasingly being used by stakeholders to promote nature-based solutions (Willemen et al., 2015). Finally, there is strong activism geared toward integrating nature-based solutions education in public and private education curriculum across Africa (Chavula et al., 2024).

References

Abebe, G. K., Bijman, J., Pascucci, S., & Omta, O. (2013). Adoption of improved potato varieties in Ethiopia: The role of agricultural knowledge and innovation system and smallholder farmers' quality assessment. *Agricultural Systems, 122*, 22–32. https://doi.org/10.1016/j.agsy.2013.07.008

Adepoju, P. (2021, November 7). Africa worst hit by climate change impacts, COP26 told. *Nature Africa*. https://doi.org/10.1038/d44148-021-00107-z

Adesina, A., & Baidu-Forson, J. (1995). Farmers' perceptions and adoption of new agricultural technology: Evidence from analysis in Burkina Faso and Guinea, West Africa. *Agricultural Economics, 13*, 1–9.

Anyamba, A., & Eastman, J. (1996). Inter-annual variability of NDVI over Africa and its relation to El Niño/southern oscillation. *International Journal of Remote Sensing, 17*, 2533–2548.

Anyamba, A., & Tucker, C. J. (2005). Analysis of Sahelian vegetation dynamics using NOAA-AVHRR NDVI data from 1981–2003. *Journal of Arid Environments, 63*, 595–614.

Araos, M., Berrang-Ford, L., Ford, J. D., Austin, S. E., Biesbroek, R., & Lesnikowski, A. 2016. Climate change adaptation in large cities: A systematic global assessment. *Environmental Science and Policy, 66*, 375–382.

Barbier, B., Yacouba, H., Karambiri, H., Zorome, M., & Some, B. (2009). Human vulnerability to climate variability in the Sahel: Farmers' adaptation strategies in northern Burkina Faso. *Environmental Management, 43*(5), 790–803. https://doi.org/10.1007/s00267-008-9237-9

Bardgett, R. D., Mommer, L., & De Vries, F. T. (2014). Going underground: Root traits as drivers of ecosystem processes. *Trends in Ecology and Evolution, 29*, 692–699.

Berrang-Ford, L., Ford, J. D., Lesnikowski, A., Poutianinen, C., Barrera, M., & Heymann, S. J. (2014). What drives national adaptation? A global assessment. *Climate Change, 124*, 441–450. https://doi.org/10.1007/s10584-014-1078-3

Berrang-Ford, L., Ford, J. D., & Paterson, J. (2011). Are we adapting to climate change? *Global Environmental Change, 21*, 25–33. https://doi.org/10.1016/j.gloenvch.2010.09.012

Bonneau, M. (2005). Evolution of the mineral fertility of an acidic soil during a period of ten years in the Vosges mountains (France). Impact of humus mineralisation. *Annals of Forest Science, 62*, 253–260.

Brockhaus, M., Djoudi, H., & Locatelli, B. (2013). Envisioning the future and learning from the past: Adapting to a changing environment in northern Mali. *Environmental Science & Policy, 25*, 94–106. https://doi.org/10.1016/j.envsci.2012.08.008

Bruinsma, J. (2009). The resource outlook to 2050. In *How to Feed the World in 2050*. Rome, Italy. June 24–26, 2009. FAO. http://www.fao.org/3/a-ak971e.pdf

Bryan, E., Deressa, T., Gbetibouo, G., & Ringer, C. (2009). Adaptation to climate change in Ethiopia and South Africa: options and constraints. *Environmental Science and Policy, 12*, 413–426.

Burnham, M., & Ma, Z. (2016). Linking smallholder farmer climate change adaptation decisions to development. *Climate and Development, 8*(4), 289–311. https://doi.org/10.1080/17565529.2015.1067180

Canadell, J. G., Monteiro, P. M. S., Costa, M. H., Cotrim da Cunha, L., Cox, P. M., Eliseev, A.V., Henson, S., Ishii, M., Jaccard, S., Koven, C., Lohila, A., Patra, P. K., Piao, S., Rogelj, J., Syampungani, S., Zaehle, S., & Zickfeld, K. (2021). Global carbon and other biogeochemical cycles and feedbacks. In V. Masson-Delmotte et al. (Eds.), *Climate change 2021: The physical science basis*. Contribution of Working Group I to the Sixth Assessment Report of the Intergovernmental Panel on Climate Change (pp. 673–816). Cambridge University Press. https://doi.org/10.1017/9781009157896.007

Chavula, P., Umer, Y., Abdi, E., Uwimbabazi, A., Habowa, C., Mensah, G. B., Ntezimana, G. M., Amanzi, L., Lungu, G., & Kayusi, F. (2024). Bridging environmental education and sustainable development: An integrated approach for a greener future. *Asian Basic and Applied Research Journal*, *6*(1), 161–171. https://jofresearch.com/index.php/ABAARJ/article/view/150

Croppenstedt, A. Demeke, M., & Meschi, M. M. (2003). Technology adoption in the presence of constraints: The case of fertilizer demand in Ethiopia. *Review of Development Economics*, *7*(1), 58–70.

Dassi Djoukouo, N. H., Djousse, B. M. K., Doukeng, H. G., Egbe, D. A. M., Tangka, J. K., & Tchoffo, M. (2024). Design of a pyrolyser model for the conversion of thermoplastics into fuels. *Heliyon*, *10*, e26702.

De Groote, H., Dema, G., Sonda, G. B., & Gitonga, Z. M. (2013). Maize for food and feed in East Africa: The farmers' perspective. *Field Crops Research*, *153*, 22–36. https://doi.org/10.1016/j.fcr.2013.04.005

Deressa, T. T., Hassan, R. M., Ringler, C., Alemu, T., & Yesuf, M. (2009). Determinants of farmers' choice of adaptation methods to climate change in the Nile Basin of Ethiopia. *Global Environmental Change*, *19*(2), 248–255. https://doi.org/10.1016/j.gloenvcha.2009.01.002

Diep, L., Mulligan, J., Oloo, M. A., Guthmann, L., Raido, M., & Ndezi, T. (2022). Co-building trust in urban nature: Learning from participatory design and construction of nature-based solutions in informal settlements in East Africa. *Frontiers in Sustainable Cities*, *4*, 927723.

Donatti, C. I., Harvey, C. A., Martinez-Rodrigues, M. R., Vignola, R., & Rodriguez, C. M. (2018). Vulnerability of smallholder farmers to climate change in Central America and Mexico: Current knowledge and research gaps. *Climate & Development*, *11*(3), 264–286.

Douxchamps, S., Van Wijk, M. T., Silvestri, S., Moussa, A. S., Quiros, C., Ndour, N. Y., Buah, S., Somé, L., Herrero, M., Kristjanson, P., Ouedraogo, M., & Rufino, M. C. (2016). Linking agricultural adaptation strategies, food security and vulnerability: Evidence from West Africa. *Regional Environmental Change*, *16*(5), 1305–1317. https://doi.org/10.1007/s10113-015-0838-6

Dubois, O. (2011). The state of the world's land and water resources for food and agriculture: Managing systems at risk. FAO. http://www.fao.org/3/i1688e/i1688e00.pdf

Duponnois, R., Plenchette, C., & Bâ, A. M. (2001). Growth stimulation of seventeen fallow leguminous plants inoculated with *Glomus aggregatum* in Senegal. *European Journal of Soil Biology*, *37*, 181–186.

Dupuis, J., Biesbroek, R. (2013). Comparing apples and oranges: The dependent variable problem in comparing and evaluating climate change adaptation policies *Global Environmental Change*, *23*, 1476–1487.

Eklundh, L., & Olsson, L. (2003). Vegetation index trends for the African Sahel 1982–1999. *Geophysical Research Letters*, *30*(8), 1430. https://doi.org/10.1029/2002GL016772

Elmqvist, B., & Olsson, L. (2006). Livelihood diversification: Continuity and change in the Sahel. *GeoJournal*, *67*(3), 167–180. https://doi.org/10.1007/s10708-007-9043-6

Epule, T. E, & Bryant, C. (2016). Small scale farmers' Indigenous agricultural adaptation options in the face of declining or stagnant crop yields in the Fako and Meme divisions of Cameroon. *Agriculture*, *6*(2), 22. https://doi.org/10.3390/agriculture6020022

Epule, T. E., Chehbouni, A., Dhiba, D., & Moto, M. W. (2021). The readiness index for climate change adaptation in Africa: The role of climate and adaptive capacity proxies. *Applied Sciences*, *11*(20), 9413.

Epule, T. E., Ford, J. D., Lwasa, S., & Lepage, L. (2017). Climate change adaptation in the Sahel. *Environmental Science & Policy*, *75*, 121–137.

Epule, T. E., Peng, C., & Lepage, L. (2015). Environmental refugees in sub-Saharan Africa: A review of perspectives on trends, causes, challenges and way forward. *Geojournal*, *80*, 79–92. https://doi.org/10.1007/s10708-014-9528-z

Epule, T. E., Peng, C., Lepage, L., & Chen, Z. (2013a). The causes, effects and challenges of the Sahelian droughts: A critical review. *Regional Environmental Change*, *14*, 145–156. https://doi.org/10.1007/s10113-013-0473-z

Epule, T. E., Peng, C., Lepage, L., & Chen, Z. (2013b). Enabling conditions for successful greening of public spaces: The case of Touroua, Cameroon based on perceptions. *Small Scale Forestry, 13*, 143–161. https://doi.org/10.1007/s11842-013-9246-2

Epule, T. E., Peng, C., Lepage, L., & Chen, Z. (2014). Policy options towards deforestation reduction in Cameroon: An analysis based on a systematic approach. *Land Use Policy, 36*, 405–415. https://doi.org/10.1016/j.landusepol.2013.09.004

Ernst, Y., Archibald, S., Balzter, H., Chevallier, F., Ciais, P., Gonzalez Fischer, C., Gaubert, B., Higginbottom, T., Higgins, S., Lawal, S., Lacroix, F., Lauerwald, R., Lourenco, M., Martens, C., Mengistu, A. G., Merbold, L., Mitchard, E., Moyo, M., Nguyen, H., ... Scholes, R. J. (2024). The African regional greenhouse gases budget (2010–2019). *Global Biogeochemical Cycles, 38*(4), 1–32. https://doi.org/10.1029/2023GB008016

Fleuret, A. (1986). Indigenous responses to drought in sub-Saharan Africa. *Disaster, 10*(3), 224–229.

Ford, J. D., Berrang-Ford, L., Bunce, A., McKay, C., Irwin, M., & Pearce, T. (2015). The status of climate change adaptation in Africa and Asia. *Regional Environmental Change, 15*, 801–814.

Ford, J. D., Berrang-Ford, L., Lesnikowski, A., Barrera, M., & Heymann, S. J. (2013). How to track adaptation to climate change: A typology of approaches for national-level application. *Ecology and Society, 18*, 40.

Ford, J. D., McDowell, G., & Jones, J. (2014). The state of climate change adaptation in the Arctic. *Environmental Research Letters, 9*, 104005. https://doi.org/10.1088/1748-9326/9/10/104005

Gebrehiwot, T., & van der Veen, A. (2013). Farm level adaptation to climate change: The case of farmers in the Ethiopian highlands. *Environ Manage, 52*(1), 29–44. https://doi.org/10.1007/s00267-013-0039-3

Giannini, A., Biasutti, M., Held, I. M., & Sobel, A. H. (2008). A global perspective on African climate. *Climate Change, 90*, 359–383. https://doi.org/10.1007/s10584-008-9396-y

Giannini, A., Saravanan, R., & Chang, P. (2003). Oceanic forcing of Sahel rainfall on inter-annual to inter-decadal timescales. *Science, 302*, 1027–1030.

Giannini, A., Saravanan, R., & Chang, P. (2005). Dynamics of the boreal summer African monsoon in the NSIPP1 atmospheric model. *Climate Dynamics, 25*, 517–535. https://doi.org/10.1007/s00382-005-0056-x

Graham, P. H., & Vance, C. P. (2003) Legumes: Importance and constraints to greater use. *Plant Physiology, 131*, 872–877.

Hocher, V., Auguy, F., Bogusz, D., Doumas, P., Franche, C., Gherbi, H., Laplaze, L., Obertello, M., & Svistoonoff, S. (2009). Les symbioses actinorhiziennes fixatrices d'azote: Un exemple d'adaptation aux contraintes abiotiques du sol. *Cahiers Agricultures, 18*, 498–505.

International Biochar Initiative (IBI). (2015). Standardized product definition and product testing guidelines for biochar that is used in soil—version 2.1. Document IBI-STD-2.1. http://www.biochar-international.org/characterizationstandard

Intergovernmental Panel for Climate Change (IPCC). (2007). *Climate change 2007: impacts, adaptations and vulnerability. Contributions of the working group 2 to the fourth assessment report of the IPCC*. Cambridge University Press.

Intergovernmental Panel for Climate Change (IPCC). (2022). Impacts, adaptation, and vulnerability. IPCC Sixth Assessment Report: Summary for Policy Makers. https://www.ipcc.ch/report/ar6/wg2/

Janssen, B. H., Guiking, F., van der Eijk, D., Smaling, E. M., Wolf, J., & van Reuler, H. (1990). A system for quantitative evaluation of the fertility of tropical soils (QUEFTS). *Geoderma, 46*, 299–318.

Jean-Pierre, J. (2023). Évaluation du carbone et de l'azote stockés par trois légumineuses *(Acacia mangium, Leucaena leucocephala* et *Prosopis juliflora)* dans les sols dégradés d'Haïti 18 mois après leur plantation. [Master's thesis, Université du Québec en Abitibi-Témiscamingue, Rouyn-Noranda, Canada].

Kachaka, E. Y., Poirier, V., Munson, A. D., & Khasa, D. P. (2023). *Acacia auriculiformis* agroforestry fallows of different ages improve soil physico-chemical properties and carbon stocks on the Batéké plateau, Democratic Republic of Congo. *Geoderma Regional, 34*, e00691.

Lal, R. (2002). Soil carbon dynamics in cropland and rangeland. *Environmental Pollution, 116*, 353–362.

Lal, R. (2015). Restoring soil quality to mitigate soil degradation. *Sustainability, 7*, 5875–5895.

Lamers, J., & Feil, P. (1995). Farmers' knowledge and management of spatial soil and crop growth variability in Niger, West Africa. *Netherland Journal of Agricultural Science, 43*, 375–389.

Le, H. D., Smith, C., Herbohn, J., & Harrison, S. (2012). More than just trees: Assessing reforestation success in tropical developing countries. *Journal of Rural Studies, 28*(1), 5–19.

Lehmann, J., Cowie, A., Masiello, C. A., Kammann, C., Woolf, D., Amonette, J. E., Cayuela, MéL., Camps-Arbestain, M., & Whitman, T. (2021). Biochar in climate change mitigation. *Nature Geoscience, 14*, 883–894.

Lesnikowski, A., Ford, J., Biesbroek, R., Berrang-Ford, L., & Heymann, S. J. (2016). National-level progress on 816 adaptations. *Nature Climate Change, 6*, 261–264. https://doi.org/10.1038/nclimate2863

Lesnikowski, A. C., Ford, J. D., Berrang-Ford, L., Barrera, M., Berry, P., Henderson, J., & Heymann, S. J. (2013). National-level factors affecting planned, public adaptation to health impacts of climate change. *Global Environmental Change, 23*, 1153–1163. https://doi.org/10.1016/j.gloenvcha.2013.04.008

Lipper, L., Thornton, P., Campbell, B. M., Baedeker, T., Braimoh, A., Martin Bwalya, M., Caron, P., Cattaneo, A., Garrity, D., Henry, K., Hottle, R., Jackson, L., Jarvis, A., Kossam, F., Mann, W., McCarthy, N., Meybeck, A., Neufeldt, H., Remington, T., ... Torquebiau, E. F. (2014). Climate-smart agriculture for food security. *Nature Climate Change, 4*, 1068–1072.

Lontsi, G. R. M., Djousse, B. M. K., Pernot, C., & Munson, A. D. (2024) Production and characterization of eight biochars originating from agricultural waste in Cameroon. *Journal of Material Cycles and Waste Management, 26*, 3280–3290.

Lu, J., & Delworth, T. L. (2005). Oceanic forcing of the late 20th century Sahel drought. *Geophysical Research Letters, 32*, 1–5. L22706. https://doi.org/10.1029/2005GL023316

Lwasa, S. (2015). A systematic review of research on climate change adaptation policy and practice in Africa and South Asia deltas. *Regional Environmental Change, 15*, 815–825. https://doi.org/10.1007/s10113-014-0715-8

Magistro, J., & Lo, M. (2001). Historical and human dimensions of climate variability and water resource constraint in the Senegal River valley. *Climate Research, 19*(2), 133–147.

Mannke, F. (2011). Key themes of local adaptation to climate change: lessons from mapping community-based initiatives in Africa. In Leal W. Filho (Ed.), *Experiences of climate change adaptation in Africa* (pp. 17–32). Springer.

Maranz, S. (2009). Tree mortality in the African Sahel indicates an anthropogenic ecosystem displaced by climate change. *Journal of Biogeography, 36*, 1181–1193.

Marenya, P. P., & Barrett, C. B. (2007). Household-level determinants of adoption of improved natural resources management practices among smallholder farmers in western Kenya. *Food Policy, 32*(4), 515–536. https://doi.org/10.1016/j.foodpol.2006.10.002

Marone, D., Poirier, V., Coyea, M., Olivier, A., & Munson, A. D. (2017) Carbon storage in agroforestry systems in the semi-arid zones of Niyaes, Senegal. *Agroforestry Systems, 91*, 941–954.

Massoukou Pamba, R., Poirier, V., Nguema Ndoutoumou, P., & Epule, T. E. (2023). How can plants help restore degraded tropical soils? *Land, 12*, 2147. https://doi.org/10.3390/land12122147

Massoukou Pamba, R., Poirier, V., Nguema Ndoutoumou, P., and Epule, T. E. (2024). Growing *Jatropha curcas* L. improves the chemical characteristics of degraded tropical soils. *Forest*, 15, 1709.

Mbow, C., Mertz, O., Diouf, A., Rasmussen, K., & Reenberg, A. (2008). The history of environmental change and adaptation in eastern Saloum–Senegal: Driving forces and perceptions. *Global and Planetary Change, 64*(3–4), 210–221. https://doi.org/10.1016/j.gloplacha.2008.09.008

Mbuli, C., Lotsmart, S., Fonjong, N., & Fletcher, A. J. 2021. Climate change and small farmers' vulnerability to food insecurity in Cameroon. *Sustainability, 13*(3), 1523.

Mertz, O., Mbow, C., Reenberg, A., & Diouf, A. (2009). Farmers' perceptions of climate change and agricultural adaptation strategies in rural Sahel. *Environ Manage, 43*(5), 804–816. https://doi.org/10.1007/s00267-008-9197-0

Mertz, O., Mbow, C., Reenberg, A., Genesio, L., Lambin, E. F., D'haen, S., Zorom, M., Rasmussen, K., Diallo, D., Barbier, B., Moussa, I. B., & Sandholt, I. (2011). Adaptation strategies and climate vulnerability in the Sudano-Sahelian region of West Africa. *Atmospheric Science Letters, 12*(1), 104–108. https://doi.org/10.1002/asl.314

Minengu, J. D. D., Mobambo, P., & Mergeai, G. (2014). Influence de l'environnement et des pratiques culturales sur la productivité de *Jatropha curcas* L. en Afrique subsaharienne (synthèse bibliographique). *Biotechnologie, Agronomie, Société et Environnement, 18*, 290–300.

Morand, P., Kodio, A., Andrew, N., Sinaba, F., Lemoalle, J., & Béné, C. (2012). Vulnerability and adaptation of African rural populations to hydro-climate change: experience from fishing communities in the Inner Niger Delta (Mali). *Climatic Change, 115*(3–4), 463–483. https://doi.org/10.1007/s10584-012-0492-7

Moretimore, M. (2010). Adapting to drought in the Sahel: Lessons for climate change. *WIREs Climate Change, 1*, 134–143.

Mosberg, M., & Eriksen, S. H. (2015). Responding to climate variability and change in dryland Kenya: The role of illicit coping strategies in the politics of adaptation. *Global Environmental Change, 35*, 545–557. https://doi.org/10.1016/j.gloenvcha.2015.09.006

Noble, I., & Huq, S. (2014). *Adaptation needs and options.* Cambridge University Press.

Nielsen, J. Ø., & Reenberg, A. (2010). Cultural barriers to climate change adaptation: A case study from Northern Burkina Faso. *Global Environmental Change, 20*(1), 142–152. https://doi.org/10.1016/j.gloenvcha.2009.10.002

Njoukam, R., Oliver, R., & Peltier, R. (1999). Restitution minérale au sol par la litière dans les plantations de *Pinus kesiya* Royle ex-Gordon dans l'ouest du Cameroun. *Annals of Forest Science, 56*, 431–439.

Nyong, A., Adesina, F., & Osman Elasha, B. (2007). The value of indigenous knowledge in climate change mitigation and adaptation strategies in the African Sahel. *Mitigation and Adaptation Strategies for Global Change, 12*(5), 787–797. https://doi.org/10.1007/s11027-007-9099-0

Okoye, C. U. (1998). Comparative analysis of factors in the adoption of traditional and recommended soil erosion control practices in Nigeria. *Soil and Tillage Research, 45*, 251–263.

Olsson, L., Eklundh, L., & Ardö, J. (2005) A recent greening of the Sahel—Trends, patterns and potential causes. *Journal of Arid Environments, 63*, 556–566.

Olsson, L., & Mryka, H. (2008). Greening of the Sahel. *The Encyclopedia of Earth*. http://www.eoearth.org/article/Greening_of_the_Sahel

Opiyo, F., Wasonga, O., Nyangito, M., Schilling, J., & Munang, R. (2015). Drought adaptation and coping strategies among the Turkana pastoralists of northern Kenya. *International Journal of Disaster Risk Science, 6*(3), 295–309. https://doi.org/10.1007/s13753-015-0063-4

Opondo, D. O. (2013). Erosive coping after the 2011 floods in Kenya. *International journal of Global Warming, 5*(4), 452–466.

Paustian, K., Lehmann, J., Ogle, S., Reay, S., Robertson, G. P., & Smith, P. (2016). Climate-smart soils. *Nature, 532*, 49–57.

Poirier, V., Roumet, C., Angers, D. A., & Munson, A. D (2018a). Species and root traits impact macroaggregation in the rhizospheric soil of a Mediterranean common garden experiment. *Plant & Soil, 424*, 289–302.

Poirier, V., Roumet, C., & Munson, A. D. (2018b). The root of the matter: Linking root traits and soil organic matter stabilization processes. *Soil Biology & Biochemistry, 120*, 246–259.

Reardon, T., Malton, P., & Delgado, C. (1988). Coping with household-level food insecurity in drought-affected areas of Burkina Faso. *World Development, 16*(9), 1065–1074.

Recha, J. W., Mati, B. M., Nyasimi, M., Kimeli, P. K., Kinyangi, J. M., & Radeny, M. (2016). Changing rainfall patterns and farmers' adaptation through soil water management practices in semi-arid eastern Kenya. *Arid Land Research and Management, 30*(3), 229–238. https://doi.org/10.1080/15324982.2015.1091398

Reenberg, A. (2010). Land-use dynamics in the Sahelian zone in eastern Niger-monitoring change in cultivation strategies in drought prone areas. *Journal of Arid Environments, 27*, 179–192.

Rey, F., Ballais, J.-L., Marre, A., & Rovéra, G. (2004). Rôle de la végétation dans la protection contre l'érosion hydrique de surface. *Comptes Rendus Geoscience*, *336*, 991–998.

Reynolds, J. F., Stafford, S. D. M., Lambin, E. F., Turner, B. L., Mortimore, M., Batterbury, S. P., Downing, T. E., Dowlatabadi, H., Fernandez, R. J., Herrick, J. E., Huber-Sannwald, E., Jiang, H., Leemans, R., Lynam, T., Maestre, F. T., Ayarza, M., & Walker, B. (2007). Global desertification: Building a science for dryland development. *Science*, *316*, 847–851.

Rockström, J. (2003). Resilience building and water demand management for drought mitigation. *Physics and Chemistry of the Earth, Parts A/B/C*, *28*(20–27), 869–877. https://doi.org/10.1016/j.pce.2003.08.009

Schaer, C. (2015). Condemned to live with one's feet in water? *International Journal of Climate Change Strategies and Management*, *7*(4), 534–551. https://doi.org/10.1108/ijccsm-03-2014-0038

Sirén, A. H. (2024). The global potential for carbon removal through biochar in shifting cultivation systems. *Mitigation and Adaptation Strategies for Global Change*, *29*, 75.

Six, J., Bossuyt, H., Degryze, S., & Denef, K. (2004). A history of research on the link between (micro) aggregates, soil biota, and soil organic matter dynamics. *Soil & Tillage Research*, *79*, 7–31.

Somda, J., Nianogo, A. J., Nassa, S., & Sanou, S. (2002). Soil fertility management and socio-economic factors in crop-livestock systems in Burkina Faso: A case study of composting technology. *Ecological Economics*, *43*, 175–183.

Takimoto, A., Nair, V. D., & Nair, P. K. R. (2008) Carbon stock and sequestration potential of traditional and improved agroforestry systems in the West African Sahel. *Agriculture, Ecosystems and Environment*, *125*, 159–166.

Thuo, M., Bravo-Ureta, B., Hathie, I., & Obeng-Asiedu, P. (2011). Adoption of chemical fertilizer by smallholder farmers in the peanut basin of Senegal. *African Journal of Agricultural Resource Economics*, *6*(1), 1–6. http://purl.umn.edu/156960

Wang, B., Ding, Q., Fu, X., Jin, K., Shukla, J., & Doblas-Reyes, F (2005). Fundamental challenge in simulation and prediction of summer monsoon rainfall. *Geophysical Research Letters*, *32*, 15711. https://doi.org/10.1029/ 2005GL022734

Willemen, L., Cottam, A. J., Drakou, E. G., & Burgess, N. D. (2015). Using social media to measure the contribution of red list species to the nature-based tourism potential of African protected areas. *PloS One*, *10*(6), e0129785.

Wood, S. A., Jina, A. S., Jain, M., Kristjanson, P., & DeFries, R. S. (2014). Smallholder farmer cropping decisions related to climate variability across multiple regions. *Global Environmental Change*, *25*, 163–172. https://doi.org/10.1016/j.gloenvcha.2013.12.011

Wubeneh, N. G., & Sanders, J. H. (2006). Farm-level adoption of sorghum technologies in Tigray, Ethiopia. *Agricultural Systems*, *91*(1–2), 122–134. https://doi.org/10.1016/j.agsy.2006.02.002

Yan, J., Wang, L., Hu, Y., Tsang, Y. F., Zhang, Y., Wu, J., Fu, X., & Sun, Y. (2018). Plant litter composition selects different soil microbial structures and in turn drives different litter decomposition pattern and soil carbon sequestration capability. *Geoderma*, *319*, 194–203.

Zampaligré, N., Dossa, L. H., & Schlecht, E. (2014). Climate change and variability: perception and adaptation strategies of pastoralists and agro-pastoralists across different zones of Burkina Faso. *Regional Environmental Change*, *14*(2), 769–783. https://doi.org/10.1007/s10113-013-0532-5

Zander, K. K., Mwacharo, J. M., Drucker, A. G., & Garnett, S. T. (2013). Constraints to effective adoption of innovative livestock production technologies in the Rift Valley (Kenya). *Journal of Arid Environments*, *96*, 9–18. https://doi.org/10.1016/j.jaridenv.2013.03.017

Zeng, N. (2003). Drought in the Sahel. *Science*, *302*, 999–1000.

Zubairu, A. M., Michéli, E., Ocansey, C. M., Boros, N., Rétháti, G., Lehoczky, É., & Gulyás, M. (2023). Biochar improves soil fertility and crop performance: A case study of Nigeria. *Soil Sytems*, *7*, 105.

CHAPTER 27

New Possibilities for Climate Action in East Asia? The Role of Non-State Actors in China, South Korea, and Japan

YOUNGMIN OH AND DAVID A. SONNENFELD

Introduction

It was a blessing in disguise. The failure of the 2009 Copenhagen Global Climate Conference to yield a successor to the Kyoto Protocol was a critical juncture to pave the way for a shift in climate change strategies to formally include non-state actors' engagement, culminating in the 2015 Paris Agreement. The initial primary focus of climate change science, policymaking, and actions was on a global scale (Stone et al., 2012). Recent analyses have highlighted the critical need to address policy options and climate actions at the municipal or metropolitan government level, as well at the level of businesses and civil society.

In this chapter, and in international climate governance more generally, "non-state actors" are defined as any actors other than national governments who are the sole formal parties to the United Nations Framework Convention on Climate Change (UNFCCC). The term "non-state actor" thus encompasses a variety of sectors and organizations, including from civil society, the private sector, financial institutions, cities and other subnational authorities, local communities, and groups of Indigenous people (UNFCCC, 2015).[1]

Non-state actors' participation in climate action increased strikingly in the first two decades of the 21st century (Bäckstrand et al., 2017; Lövbrand et al., 2017). This did not happen overnight. Over the long course of climate negotiations, non-state actors explored approaches outside the UNFCCC, for example, through smaller intergovernmental clubs and transnational coalitions among business and municipal governments (Betsill & Bulkeley, 2006; Hale, 2016).[2] Together with their extraordinary efforts, the practical objectives of non-state actors with respect to climate change were worked out in tandem with those of state actors and sometimes in a coalition with them (Stearns & Almeida, 2004). These actors outside of national governments strive to harness climate change mitigation potentials and narrow the gap between parties' commitments to reduce global emissions and further actions necessary to limit global climate change to +2°C (Hale, 2018).[3] Since the Paris Agreement at the 21st Conference of Parties (COP 21) in 2015, the international climate change regime has shifted from a regulatory, "top-down" approach to a nonregulatory, "bottom-up" approach with a pledge and review system (Chan et al., 2018). This change, in turn, has spurred more non-state actors' active engagement in climate action (Hale, 2016).

Since COP 21, the UNFCCC has acknowledged the formal affiliation of non-state actors for climate action.[4] An initiative called the Marrakesh Partnership for Global Climate Action (GCA) was launched at COP 22 in Morrocco in 2016 to mobilize non-state actors' climate activities (Bäckstrand et al., 2017). The post-Paris international climate regime established by the UNFCCC underscores the increasing significance of non-state actors and established institutional mechanisms to involve and coordinate their contributions. The UNFCCC and the UN Secretary-General's Lima-Paris Action Agenda (LPAA) supports official recognition and tracking of non-state actors' climate actions through the UNFCCC's online portal and the CDP-ICLEI unified reporting system.[5]

Compared to global trends, non-state climate actions in China, South Korea, and Japan have been less vibrant, reflecting less participatory policymaking environments and shorter histories of environmental movements. This chapter explores the role of non-state actors in climate action in China, the Republic of Korea (South Korea), and Japan. The three neighboring states are critical players in addressing global climate change, considering their contributions to world greenhouse gas (GHG) emissions, together accounting for 23% of the world economy (World Bank, 2021) and approximately 30% of GHG emissions (Climate Watch, 2019; World Bank, 2020). Surprisingly, the role of non-state climate actions in the three countries has been understudied. Prior to the UNFCCC's COP 21 in Paris, non-state climate actions in the three countries were limited to assisting state actors in developing or obtaining technologies and knowledge, and/or in implementing national regulations. Since receiving official recognition at the Paris COP, however, non-state actors in the three countries cannot be underestimated.

This chapter reviews the historical background of non-state climate action in the three countries, analyzes non-state actors' officially registered climate actions in the three countries via the UNFCCC's online data portal, and discusses possibilities for future climate action by non-state actors in the region and beyond. This allows for a comparison of diverse dynamics between developing and developed nations (such as Japan categorized as an Annex I developed country and China and South Korea as Non-Annex I countries under the Kyoto Protocol) and a discussion of how non-state actors in authoritarian or state-centric countries engage in non-state climate actions.[6] A better understanding of the character of non-state climate actions in the varying political and legislative contexts in China, South Korea, and Japan will contribute to a deeper comprehension of possible future climate action in the region and globally.

Historical Aspects of Non-State Climate Action in East Asia

The significance of the three countries' climate actions stems not only from their relative importance in the global economy and climate actions but also from stark differences of their historical responsibilities in climate change. Prior to the Paris Climate Agreement, while these three nations all rank among the top-10 global emitters of GHGs, Japan was the sole Annex I country in Asia obligated to mandatory GHG reduction under the Kyoto regime. In contrast, as developing nations, China and South Korea were exempt from such mandatory commitments. The post–Paris Agreement marked a pivotal moment in the international climate regime by shifting from a top-down, mandatory approach for developed countries to a bottom-up, voluntary contribution scheme applicable to all UNFCCC parties, irrespective of their development stage. Thus, countries including China and South Korea also now bear the same voluntary responsibility of GHG reductions. This is particularly critical, considering the divergence between their approaches before and after the Paris Climate Summit. Thus, the new climate regime has been characterized as one of multipolarities (Bäckstrand et al., 2017).

Despite different political and legislative contexts, there are both commonalities and divergences among non-state climate actors in the three nations. China has a single-party-led, communist political system where subnational governments and civil society are strictly under the control of central government. South Korea and Japan also experience government-centric climate action, their democratic political systems notwithstanding. This section elucidates the historically expected roles of non-state actors in the three countries, and it explores why their initial participation in climate action was largely confined to specific aspects, such as educational and informational activities, rather than engaging in decision-making processes (Loher, 2012).

China

As of 2018, China accounted for 25% of global CO_2 emissions (World Bank, 2020), showing a substantial increase in its absolute and relative emissions from 11% in 1990. Despite their innovative approaches to green economic transition outlined in their 11th and 12th Five-Year Plans (FYP),[7] the dominance of national government-driven strategies persists. China is known for its state-centric governance and top-down lines of control (Carter & Mol, 2006). Some scholars characterize it as "environmental authoritarianism," emphasizing a centralized regulatory authority in government agencies with limited non-state actor participation (Engels, 2018; Gilley, 2012; Zinda, 2023). While the central government continues to be the predominant actor in China, nongovernmental and civil society actors have played a peripheral role in the formulation and execution of climate change policies (Wang et al., 2018). As Guttman et al. (2018) argue, the term "non-state actors" in China is not analogous to its meaning in Western countries with historically vibrant grassroot movements. In China, a great majority of these actors are not entirely autonomous and often align their strategies with government preferences. This complicates the delineation among the public sector, private sector, and civil society.

Local Governments

Research on municipal governments in politically stable and economically advanced countries affiliated with the market-oriented Organization for Economic Co-operation and Development (OECD) is extensive. While subject to growing attention, the climate actions of Chinese provinces and cities have been less well studied, with such research focusing primarily on energy governance, policies, and emission reduction policies (Mai & Francesch-Huidobro, 2015, pp. 33–34; Rock, 2002; Westman & Broto, 2018; Zhang & Li, 2018). In China, local government involvement in climate action primarily revolves around the roles assigned by the central government for executing specific policies and objectives. For instance, provincial governments, metropolitan areas, cities, and industrial sectors are tasked with formulating region-specific policies (e.g., regional 5-year plans) that aligned with the national Five-Year Plans produced by China's National Development and Reform Commission (NDRC) (Hale, 2018; Wang et al., 2018). Local governments have engaged in climate actions since 2007 when the central Chinese government promulgated national strategy for responding to climate change as a major national policy guide. In accordance with the NDRC's guidelines, local plans were developed by the Provincial Development and Reform Commissions (Li & Song, 2016).

Despite such administrative limitations, since 2010, Guangdong, Shenzhen, Guangzhou, and other municipalities have developed explicit climate change programs, transitioning from promotional support to the central government, to setting specific low-carbon development pathways, with key areas identified for pilot programs in energy and land use (Mai & Francesch-Huidobro, 2015, pp. 80–83). Unlike Japan's emphasis on local renewable energy cooperatives, Chinese municipalities have focused on energy monitoring, efficiency improvements, and consumption savings to align with regional targets assigned by the NDRC (Guttman et al., 2018; Kuramochi et al., 2019; Liu et al., 2017; Mai & Francesch-Huidobro, 2015; Wang et al., 2018; Westman & Broto, 2018).

Initially, financial benefits, technology transfer, and law enforcement were key motivations for Chinese municipalities to promote climate actions. Now, municipalities play larger roles in climate action, supported by partnerships among government, the market, and civil society, which enhance access to information, technology, funding, and resources (Westman & Broto, 2018). Still, they tend to reluctantly participate in international networks or disclose inventory emissions. As of July 2024, only six Chinese city governments are members of ICLEI, and 13 are members of C40 (C40, 2024; ICLEI, 2024). Rather, they prefer climate actions in developing sustainable infrastructure or green development (Westman & Broto, 2018) and participating in global initiatives that do not require direct emission reductions but other indirect mitigation efforts such as energy saving campaigns, knowledge dissemination, and technical implementations (Nilsson et al., 2021).

Business Actors

Beginning in the early 2000s, responsibility for climate change mitigation gradually expanded from ministries and local governments to industrial associations and state-owned enterprises, accompanied by privatization under local governments. Chinese business sectors are participating more actively in global decarbonization initiatives, as evidenced by their gradual expansion of membership in initiatives like Renewable Energy 100 (RE100) and the Science Based Target initiative (SBTi) (Nilsson et al., 2021). Chinese business actors have participated in standard development process. For instance, the China Business Council for Sustainable Development (CBCSD) played a key role in formulating GHG emissions guidelines, later acknowledged as the national standard.[8] The Innovation Center for Energy and Transportation (*i*CET) collaborated on developing national fuel standards for the transport sector (Liu et al., 2017).

Since enactment of the Renewable Energy Law in 2005, China's renewable energy industry has experienced substantial growth, signaling a shift toward a pro-climate energy transition from a coal-oriented energy system. Notably, during the 12th Five-Year Plan (2011–2015), business actors began playing a more active role beyond merely supporting the central government (Engels, 2018). During the 11th Five-Year Plan (2006–2010), China first launched the "One Thousand Firms Energy Saving" program for nine industries, targeting firms with annual energy consumption exceeding 180,000 tons of coal (or equivalent). In the subsequent 12th Five-Year Plan (2011–2015), this expanded into the "Ten Thousand Firms Energy Saving and Low-Carbon Development" program, involving over 16,000 firms across various sectors. The goal was to reduce energy consumption and CO_2 emissions per unit of GDP by 16% and 17%, respectively, between 2011 and 2015 (Wang et al., 2018). The China Business Climate Action Initiative, launched in 2018, urges 800,000 entities, including industry associations such as the China Chain-Store & Franchise Association and China Textile Industry Federation, to integrate climate change into their corporate strategies (Kuramochi et al., 2019). Major Chinese companies, including the state-owned China National Petroleum Corporation (CNCP), China Petroleum & Chemical

Corporation (Sinopec), and the China National Offshore Oil Corporation (CNOOC), have pledged carbon neutrality, aiming to achieve it before China's national target of 2060 (CSIS, 2021). Despite recent active participation, industries responsible for a large portion of CO_2 emissions in China, most of which are state-owned, remain dependent on the will of Chinese political authorities (Nilsson et al., 2021). China also operates an emission trading system (ETS), which was nationally adopted in 2021 after testing in eight pilot regions. Yet it only applies to around 2,000 companies in the power sector with more than 26,000 tons of CO_2 per year (ICAP, 2024).[9]

Civil Society Actors

Since the late 1970s, economic reforms in China have spurred the growth of private enterprises and nongovernmental organizations (NGOs), often referred to as "social organizations." While not part of the formal government structure, these entities, known as government-organized NGOs (GONGOs), maintain close ties with the government, both formally and informally (Ho, 2001). Other than GONGOs, there are different forms of social organizations such as student groups, civic NGOs, and international NGOs (Liu et al., 2017; Tang & Zhan, 2008; Wang et al., 2018). Such organizations tend to be weak, disorganized, involve minimal public participation, and be closely linked with the state, prioritizing avoidance of misunderstanding over challenging government policies (Ho & Edmonds, 2008). There is a growing trend in the number of climate-related civic NGOs in China since 2007, marked by the establishment of groups such as "Society's Response to Climate Change," the China Civil Climate Action Network (CCCAN), and the China Youth Climate Action Network (Liu et al., 2017) (see Box 27.1). Still, only 40 Chinese NGOs have joined the international Climate Action Network (CAN), which boasts over 1,800 NGO members from more than 130 countries as of April 2024 (CAN, 2024).

BOX 27.1 Examples of Formally Recognized, Non-State Climate Actors in China

- China Association for NGO Cooperation (CANGO)
  - Domestic members of organizations and individuals
  - Project cooperation with foreign NGOS, bilateral, and multilateral organizations
  - Charity projects across China, including on climate and environment[10]
- Greenovation Hub (GHub)
  - Independent think tank focused on "research and policy discussion and innovation for a net-zero, climate-resilient and nature-positive future"
  - "Seeks innovative solutions to quality growth, climate resilience and carbon neutrality in China and beyond"[11]
- Vanke Foundation
  - "A national non-public fundraising foundation supervised by the Ministry of Civil Affairs"
  - "committed to promoting sustainable development from the perspectives of policy, legislation, market, and international platforms"[12]

South Korea

South Korea's climate actions can be characterized as having three key elements: seeing climate policies as a new growth engine; strong corporatist ties between the government and business; and limited engagement from civil society (Kalinowski, 2021; E. Kim, 2016; S. Y. Kim & Thurbon, 2015). Some scholars argue that these characteristics are related to the developmentalist state legacy from the Park Jung Hee administration (E. Kim, 2016; H. R. Kim, 2000; S. Y. Kim & Thurbon, 2015).

Since 2008, South Korea's climate policies were embodied in the Green Growth Plan under the Lee Myung-Bak administration, under the conservative Grand National Party. The government's promotion of pro-climate policies during this period stood out. During Lee's presidency (2008–2013), South Korea implemented an energy target management scheme and established an emission trading scheme through the 2010 Framework Act on Low Carbon Green Growth. This legislative framework continues to shape South Korea's climate policy, including key initiatives such as the National Strategy for Green Growth (2009–2050) and the First Five-Year Plan for Green Growth (2009–2013). The Presidential Committee on Green Growth (PCGG) was formed to oversee and guide these initiatives. In 2022, the Framework Act was replaced by the "Framework Act on Carbon Neutrality and Green Growth for Coping with Climate Crisis" (hereinafter, the Carbon Neutrality Act). The latter act promulgated many new responsibilities for local governments (Articles 11 and 12), based on South Korea's national climate plan.

Local Governments

From the early stage of climate policies in South Korea, local governments have been more engaged in climate adaptation than mitigation actions. This is partially because the Low Carbon Green Growth Act stimulated local governments' actions for adaptation planning and educational efforts.[13] Adaptation actions provide local governments opportunities to secure central government budget support and to reinvest in infrastructure for local benefits. Since 2011, South Korea's central government has had a series of National Adaption Plans requiring provinces and municipalities to develop their own, local adaptation plans. Lee and Kim studied first local climate change adaptation plans of six metropolitan areas (Seoul, Busan, Daegu, Daejeon, Incheon, and Ulsan), most of which were related to health, infrastructure, and forest management (J.-S. Lee & Kim, 2018).

With the enactment of South Korea's Carbon Neutrality Act in 2022, local governments were also required to submit local implementation plans every 5–10 years depending on their size and hierarchy.[14] The Act includes regional GHG emissions reduction targets, impact assessment, cooperation among local governments, promotion of green technologies, and so on (KLIC, 2022).[15] These plans commonly hold municipal policies for systematic climate education and public awareness, establishment and operation of sector-specific adaptation councils, strengthening of linkage and cooperation between the national and local governments, and support for the operation of climate change adaptation urban councils as well (Koh, 2017). Governance within sectors related to adaptation, such as disaster, health, and water management, is typically well-established, with each sector exhibiting distinct characteristics. For instance, local governments' approach to addressing heat waves primarily revolves around the welfare delivery system, although these services are temporary and focus on vulnerable groups, while storm and flood damage involve a comprehensive disaster response system, albeit its limitation regarding intensity and uncertainty of the event (Koh, 2017).

However, mitigation actions have been mandated only recently, although local governments are also required to report their GHG emissions to the central government as part of "the public sector GHG target control program" promulgated in the Green Growth Act and Carbon Neutrality Act (KLIC, 2022).[16] To name a few, ChungNam-Do Province, home to nearly 50% of South Korea's 58 coal-fired power plants, announced in 2021 its plan to phase out coal plants by 2036 and promote a transition to renewable energy (ChungNam-Do Provincial Council, 2023). JeonNam-Do Province proclaimed a Renewable Energy Roadmap including the world's largest offshore wind power generation of 25 GW and solar power generation of 5 GW (PCCNGG, 2023). These recent initiatives are partially influenced by the Carbon Neutrality Act, which places emphasis on local governments' proactive engagements.

Business Actors

According to a 2023 poll of 408 export companies conducted by the Korea International Trade Association (KITA), 66.7% recognized the importance of climate actions in their business, with 85% of them acknowledging that climate change has an influence in their operations (KITA, 2023). Given that the export of goods and services accounts for 42% and 48% of South Korea's GDP as of 2021 and 2022 (OECD, 2024; World Bank, 2024), the climate actions undertaken by business actors are crucial not only for their own sustainability but also for the broader South Korean economy.

Since the Lee administration (2008–2013), business actors have become key negotiation partners in establishing early-stage climate policy frameworks. During the Moon administration (2017–2022) and the Yoon administration (2022–present), companies and financial investors actively participated in setting national CO_2 reduction targets and policymaking processes. This increased involvement was partly driven by international pressure from various voluntary or quasi-regulatory initiatives, reflecting a growing voluntary commitment to responsible climate actions.

These initiatives include the adoption of higher environment, social, and governance (ESG) standards,[17] endorsement of the Task Force on Climate-Related Financial Disclosures (TCFD) guidelines,[18] and voluntary participation in movements like RE100 within the business sector. These climate initiatives were partially influenced by the bottom-up approach of the post-Paris regime and international market barriers, where noncompliance with these initiatives could weaken business competitiveness both domestically and internationally.

Despite initial opposition to adopting a carbon pricing system, most large Korean companies, including Samsung Electronics, Hyundai Motors, and POSCO, have been subject to the Korean Emission Trading System (K-ETS) since its adoption in 2015. This includes regulated industries such as power and energy, steel, and semiconductors (Howie et al., 2020; J. Lee & Yu, 2017). According to South Korea's Ministry of Environment, there are around 680 companies and entities subject to allocation of national GHG emissions allowance under the 3rd K-ETS plan[19] (2021–2022) (GIR, 2021; MOE Korea, 2020).[20] Many South Korean large companies announced their carbon neutrality pledges by 2050, including the Korea Electric Power Corporation (KEPCO), POSCO, Hyundai Heavy Industries, and Samsung Electronics (Climate Action 100+, 2022; Climate Group, 2023; Samsung, 2022). Lim and Kim pointed out that Korean public companies' climate actions center on organizational agility to extreme weather effects (Lim & Kim, 2015) from the perspective of cost-efficient business management.

Civil Society Actors

Non-state actors' engagement in climate actions in South Korea has been relatively low, even though the country's civil society is not weak on environmental issues despite its late start in the 1980s and revitalization in 2000s.[21] Environmentalism began as an antipollution movement, sparked by the rapid economic development in South Korea (Kern, 2010). According to a survey conducted by S. J. Yun and colleagues, central government organizations[22] were identified as the most influential climate action actors in South Korea, while civil society organizations were the least (Yun et al., 2014). Also, the survey indicated that governmental organizations (GOs) primarily sought advice from other GOs, followed by collaboration with business organizations (BOs), while collaborations between BOs and civic organizations were infrequent.

During the Lee administration in the realm of non-state actors, entities like the Korea Chamber of Commerce and Industry (KCCI), *chaebols*, investment banks, and commercial banks actively participated in the climate policymaking process. In contrast, Kim argues that environmental NGOs' engagement was relatively limited in these discussions (E. Kim, 2016). Some argue that this dynamic is akin to the case of Japan (Kalinowski, 2021), in that vibrant civil activism and strong government-business ties coexist with a government-centric approach to climate action. In effect, throughout the climate legislative preparation process in Korea, there were ongoing interactions between heavily affected industries, such as steel, electronics, shipbuilding, and energy companies, and the government (E. Kim, 2016). In the early climate policymaking period in South Korea, environmental NGOs were actively engaged in behavioral change campaigns and policy promotional activities but played a less prominent role in the legislative process (E. Kim, 2016; S. Y. Kim & Thurbon, 2015; Moon et al., 2021; Yun et al., 2014, 2022).

The Park Geun-hye administration (2013-2017), which succeeded the Lee administration from the same conservative party, known as the New Frontier Party, arguably retreated from the pro-climate policy stance of its predecessor by reducing domestic GHG reduction targets (E. Kim, 2016). After the Park administration, the advent of the Moon administration (2017–2022) marked a significant expansion of non-state climate actors' roles and an emphasis on renewable energy and carbon neutrality policies (Yun et al., 2022). Pro-climate NGOs not only acted as vocal watchdogs but also directly participated in administrative positions within the government, including in the areas of renewable energy, the environment, and policy planning in the presidential office and other government branches during this period. After the promulgation of the Carbon Neutrality Act in 2022, pro-environment civic organizations formed coalitions around a variety of climate-related topics, including renewable energy, green economy, and green finance, to name a few, to promote the presence of their campaigns. Among those groups, the Korean Federation for Environmental Movements (KFEM), Youth 4 Climate Action, and Green Korea United launched nationwide campaigns (see Box 27.2).

Japan

Non-state climate action has been unexpectedly limited in Japan, where the dominant discourse on climate change has been characterized by interministerial politics and the so-called iron triangle (Kameyama, 2021). The notion of an iron triangle implies a close relationship among industries, economic ministries within the Japanese government, and the Japanese Diet (legislature). The interministerial discussion was also centralized within Japan's three main ministries: MOFA (Ministry of Foreign Affairs), MOE (Ministry of the Environment), and METI (Ministry of Economy, Trade, and Industry).[26] Historically, actors outside the iron triangle enjoyed only limited access to Japan's government-centered climate policymaking (Reimann, 2002). This resulted in a focus largely on economic and energy security matters framing overall Japanese climate discourse (Kameyama, 2021; Kawashima,

BOX 27.2 Examples of Formally Recognized, Non-State Climate Actors in South Korea

- Korean Federation for Environmental Movements (KFEM)
  - One of the largest and oldest civil environmental organizations in South Korea
  - Key activities include phasing out coal power plants, promoting renewables, and forming networks with laborers and local communities to influence local and central policymaking[23]
- Youth 4 Climate Action
  - Started as a small gathering and grew into a full-scale movement, with school protests linked to Fridays For Future
  - Filed a climate constitutional petition, arguing that the government's inadequate climate response violates basic rights and demanding stronger GHG reduction goals[24]
- Green Korea United
  - Has led significant environmental efforts, including wildlife conservation in the DMZ, lifestyle change movements, and climate actions focused on renewable energy
  - One of Korea's largest and oldest environmental groups, with nine local branches and four suborganizations[25]

2001), akin to the case of South Korea. Until its recent commitment to achieving net-zero emissions by 2050, announced in October 2020, Japan had maintained a less progressive stance, prioritizing feasible policies for industries (Nakazawa et al., 2023). In this earlier political context, the involvement of non-state actors in climate action in Japan was notably limited.

Three key events served as milestones in the growing prominence of Japanese non-state climate actions: the issuance in 1998 of laws for nonprofit organizations (NPOs) regarding official registration guidelines as a legal entity; a surge of national demand in clean energy transition following the 2011 Fukushima nuclear plant disaster; and the advancement of renewable energy regulations in 2012 (Fraser et al., 2020; Fraser & Aldrich, 2020; Fraser & Temocin, 2021; Hager & Hamagami, 2020; Nakazawa et al., 2023).[27] Since the 1990s, the relative power of the iron triangle with respect to climate action in Japan has diminished, influenced by factors such as government restructuring, the strengthening of advocacy coalitions, and international norm changes. Today in Japan, corporatist ties among government, business actors, and civil society are dominant in local energy initiatives and village-level sustainability movements. The partial shift away from top-down, government-centered climate action is more pronounced in the post–Paris Agreement era, with Japanese firms and local governments setting ambitious emission reduction targets and prioritizing a larger share of clean energy, surpassing national goals (Fraser et al., 2020; Fraser & Temocin, 2021; Kameyama, 2021).

Local Governments

Japanese local governments are increasingly inclined to take independent climate actions, moving away from reliance on national government initiatives (Nakazawa et al., 2023). Burgeoning municipal participation in decentralized energy production and distribution contributed to these trends, stimulated by the 2012 Act on Special Measures for Renewable

Energy Electricity Procurement by Electricity Companies and the Fukushima disaster, and abetted by rising antinuclear sentiment in the public. Hager and Hamagami (2020) suggest that further factors guiding pro-climate policy include corporatist governance of powerful conventional utilities and less influence of the traditionally powerful government ministries governing for economy, industry, and energy matters (Haddad, 2010; Hager & Hamagami, 2020). The Eco-Model City program adopted in 2008 by the national government also promoted a wide participation of city governments in low-carbon living and energy independence together with community revitalization purposes (FutureCity Initiative, 2016). Many local governments such as prefectures and municipalities have developed climate action programs for their jurisdiction without direct links to either national targets or central government plans (Zhou & Mori, 2011).

There has been increased non-state engagement at the municipal level, driven by local infrastructure transitions, green city movements, and active citizen participation in renewable projects. Companies' climate actions, including utilities of all sizes, grew with the 2012 Act on Feed-in-Tariff scheme for solar and wind power. Japanese renewable energy projects often relied on cooperation with government or industry partners (Hager & Hamagami, 2020). As is seen in the case of Miyama City's local energy company model (Miyama Smart Energy), local utilities provide renewable energy in the municipality and socioeconomic benefits to participating citizens. This is a stark difference from Japan's traditional utility model of national government supported local-level subpolicies and programs (Hager & Hamagami, 2020).

The involvement of municipal governments in Japan in other climate actions has varied in terms of size and effectiveness. According to Japan's Act on Promotion of Global Warming Countermeasures (MOJ, 2006), all jurisdictions with a population exceeding 200,000 are required to draft a local GHG reduction plan (Nakazawa et al., 2023). Past research indicates that the environmental governance of municipal governments, especially those with stronger budget capacities, such as Tokyo and Kyoto, tends to result in more effective and active climate actions (Fraser et al., 2020). There are studies that identified factors affecting municipal governments' capacity to engage in climate actions: existence of a power coalition between constellations of decision-makers, including elected officials and heads of major local businesses; learning ability and readiness to adapt good practices of comparable municipalities; impact of the Fukushima nuclear plant disaster; socioeconomic disparities across different municipalities; citizens' carbon footprint; and economic dependence on carbon. Local governments share a corporatist style of governance with powerful traditional utilities and economics ministries that until recently showed little interest in promoting renewable energy (Hager & Hamagami, 2020). Against this backdrop, as of May 2024, there are more than 1,000 local governments who have declared net-zero ambition by 2050 in Japan (MOE Japan, 2024)

Business Actors

Kameyama (2021) advocates that industrial capital was highly representative of climate actions in Japan through an iron triangle, especially via METI with its industrial counterpart in the Energy industry (Kameyama, 2021). For example, Keidanren,[28] Japan's business federation, engaged in internal discussions on climate change from the early 1990s. Published in 1997, its Voluntary Action Plan emphasized emission reductions from the industrial sector to stabilize CO_2 emissions to 1990 levels by 2010. While appearing cooperative with the government, it also served as a trade-off to dissuade further demands for reductions. Another business group, Keizai Doyukai, also pledged individual and industry responsibility, internalization of environmental costs, and innovative technology development (Kameyama,

2016, pp. 82–83). These proactive climate actions were also influenced by COP 3 in Kyoto and pressure from environmental activists criticizing the Japanese government (Reimann, 2002).

As non-state climate actions gain more importance in Japan, large private companies such as Toyota began diverging from Keidanren's position. Toyota set a 2050 target to reduce CO_2 emissions from newly sold cars by 90% from the 2015 level, considering its global image and preferences of overseas consumers (Kameyama, 2016, p. 163). As of July 2024, around 600 business actors are members of the Japan Climate Initiative, one of the largest non-state climate networks in Japan (Japan Climate Initiative, 2024). The shift in international norms within the business and trade environment has also fostered Japan's business sector to embrace climate-conscious initiatives like RE100[29] and the TCFD guidelines, aligning their strategies with ESG criteria as well. This trend is akin to the case of South Korea with an economic emphasis of climate actions.

The engagement of non-state actors in climate action in Japan continues in a pattern of working collaboratively with the national government to promote more robust policies. The Japan Climate Initiative, for example, represents a coalition of domestic stakeholders from various sectors, dedicated to advancing and expediting endeavors aimed at establishing a decarbonized society (Hale, 2018). Corporate actors in Japan have been crucial partners in renewable energy projects, such as microgrid energy distribution with METI's support. The government prioritizes domestic technology and industrial competitiveness for the 2030 energy mix and 2050 decarbonization goals. It emphasizes collaborative projects, investment-friendly policies, and dialogue with the financial sector to strengthen the energy industry (METI Japan, 2018).

Civil Society Actors

Prior to the 2011 Fukushima disaster, Japan's social movements were relatively weak with respect to climate action resource mobilization, cultural framing, organizational foundations, and the involvement of experts (Hasegawa, 2022). In Japan, it was not until 1998 when the Japanese Diet passed legislation to officially incorporate civic organizations as a legal entity. Before this legislation, there were regulatory obstacles that hampered advocacy groups from burgeoning in Japan, which includes resource mobilization, and formation of legitimate organizations (Reimann, 2002). Scholars generally agree that Japan's environmental regime has evolved through stages since the 1960s. Even after massive protests over pollution in the 1960s and early 1970s, domestic and international pressure on the Japanese government gradually increased in the 1980s and 1990s as civic environmental awareness grew alongside looming issues such as ozone depletion, biodiversity, and global warming (Broadbent, 2002). Scholars also pointed out that the degree of openness of political structure affects non-state actors to actively engage in policymaking process and political discussion, which was unfavorable in Japan. In the realm of civil society, the Kiko Network and Japan Climate Initiative (JCI) are among the most known climate advocacy groups (see Box 27.3). JCI was inaugurated in July 2018, attracting almost 800 membership from companies, local governments, universities, and various organizations as of July 2024 (Japan Climate Initiative, 2024).

Formally Registered Non-State Climate Actions in East Asia

This section analyzes climate actions in China, South Korea, and Japan, as formally registered by non-state actors in the UNFCCC Non-State Actor Zone for Climate Action

BOX 27.3 Examples of Formally Recognized, Non-State Climate Actors in Japan

- Edogawa Citizens Network for Climate Change
  - Focus on citizen initiatives to reduce CO_2 emissions in Edogawa Ward
  - Promoting climate actions at the grassroots level, practical activities, and closely monitoring the slow progress of policies and international negotiation[30]
- Japan Climate Initiative (JCI)
  - A climate action network of companies, local governments, organizations, and NGOs
  - Steering committee includes representatives from ICLEI Japan, Network of Business Leaders and Entrepreneurs for a Sustainable Business and Energy Future, CDP Worldwide-Japan, Renewable Energy Institute, WWF Japan, Japan Climate Leaders' Partnership (Japan-CLP), and the Frontier Network[31]
- Kiko Network
  - An environmental NGO dedicated to addressing climate change through significant societal transformations
  - Missions include achieving a zero-carbon Japan, transitioning to non-fossil fuel energy, promoting regional decarbonization, and enhancing climate policymaking with public participation.[32]

(NAZCA) portal.[33] Climate actions involve commitments from various actors such as companies and cities, as well as collaborative efforts with multiple stakeholders in diverse formats, including carbon reporting, local regulations, and grassroots mobilization (Bäckstrand et al., 2017; Chan et al., 2018). These actions can be categorized based on institutionality, such as policy planning and civil movements, or by focus areas like energy or finance (Almeida et al., 2024). Despite the variety and complexity, non-state actors' climate actions can be officially acknowledged and tracked through the UNFCCC NAZCA portal. In this section, we follow the UNFCCC's categorizations of types of non-state climate actors and actions to enhance consistency with the data sources. To enhance traceability and comparability, NAZCA categorizes non-state actors as either companies, investors, organizations, regions, or cities (UNFCCC, 2015). In turn, these actors' types of engagement in climate action are grouped into four, nonexclusive categories: commitment, including emission reduction targets; climate action plans; actions undertaken; and initiative participation. There are three focus areas in their climate actions: mitigation, adaptation, and finance. And each type of climate action may apply to as many as seven themes: land use, oceans and costal zones, water, human settlement, transport, energy, and industry. In what follows, we categorize all the formally registered actions according to engagement types and actor types, regardless of their focus areas and themes.

As of February 2024, the NAZCA portal had documented over 34,636 climate actors globally, with 3,178 actors from China, South Korea, and Japan combined. Specifically, there are 1,281 registered actors in China, 277 in South Korea, and 1,620 in Japan. Differences in the engagement of non-state actors in climate actions exist among the three countries, varying in both quantity and quality. These disparities may stem from the unique domestic contexts of each country, including historical responsibility for cumulative emissions and distinct political and legislative environments.

There are several interesting points about climate actions in NAZCA. First, companies and investors in this portal can be either international or domestic entities whose headquarters or branches are located in the country where their climate actions are registered, regardless of their legal nationality.[34] For instance, a climate action registered in China may be undertaken by a European company with a headquarters in China. Second, the NAZCA portal counts actions by the number of actors, not the actions themselves. Consequently, one actor can register multiple actions, and they are nonexclusive. Therefore, the number of actions registered can exceed that of actors, which may cause some level of confusion and conflict to actor-centered state climate actions (see Figure 27.1).

Third, the data reveal that companies are the most active in registering climate actions, both globally and in the three countries. Yet cities in the three countries show lower levels of registration in contrast to their high volume of global registration.[35] Cities in the global data set account for one third of registrations, while in China, South Korea, and Japan, they make up less than one tenth. Specifically, China has 13 cities and 2 regions registered; South Korea has 26 cities and 1 region registered; and Japan has 169 cities and 5 regions registered.[36,37] Globally, out of 34,636 registered actors, 11,366 are cities and 283 are regions. Only 8 regions across the three countries have registered their climate actions. In Japan, these include the prefectures of Gifu and Kyoto. In South Korea, it is the province of Chungcheongnam-Do (commitment), and in China, it is the provinces of Sichuan and Jiangsu.

Fourth, actors in the three countries show different preferences over their climate actions. For example, most of Chinese companies and investors in the NAZCA platform have registered their participation in existing global initiatives (e.g., the Agricultural Bank of China registered two climate finance actions by participating in an existing global initiative on Green Bonds). Yet Japanese non-state actors, particularly companies, show active registration in the form of commitment, actions undertaken and initiatives, while Korea shows limited registrations across all categories. China has a high number of actors in the category of commitment, comparable to that of Japan. It is evident that companies are the most dominant actor type across the three countries (see Figure 27.2).

Limitations and Possibilities for Future Non-State Climate Action

Non-state climate actions can aid states in acquiring the technology, expertise, and public engagement needed to create and implement more ambitious contributions to addressing global climate change. Given the diverse and complex landscape of climate actions, however, there are several limitations regarding the transitional movement of non-state actors: First, *the multitude of non-state initiatives have diverse origins, scopes, and objectives.* The current UNFCCC categorization of non-state climate actors and actions does not account for this variation. Second, *climate action outcomes are not guaranteed*, even though commitments are selected for additional trackability, with regular progress reports. Due to the vast quantity and variety of actors and actions, it is extremely hard for the present tracking mechanism to encompass the overall impacts. Moreover, a simple registration is not accompanied by an official methodology to measure their cumulative efforts. And third, *the relation between non-state and state actors' climate actions is not structured in a way that avoids double-counting.*

Yet there are also potential silver linings. First, despite being historically less responsible, *major emitters such as China and South Korea voluntarily engaged in more substantial*

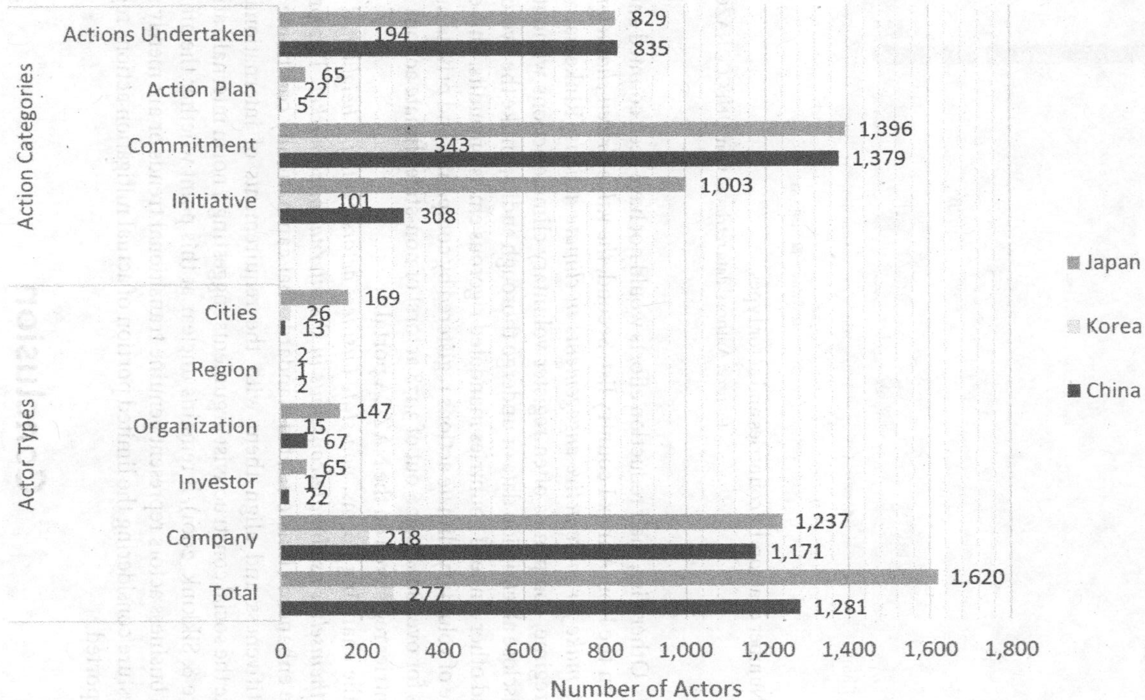

FIGURE 27.1. Number of actors by actor types and action categories.

(*Source*: Authors' data retrieved from UNFCCC NAZCA portal)

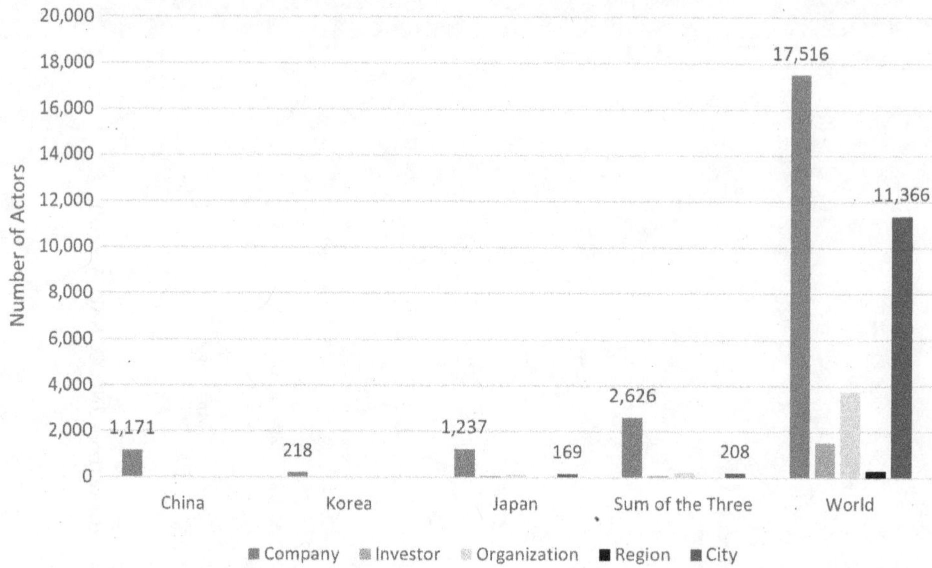

FIGURE 27.2. Number of actors by countries and actor types.

(*Source*: Authors' data retrieved from UNFCCC NAZCA portal)

reduction efforts. Otherwise, their reduction efforts would not have been counted and would have remained on the Non-Annex I country list. Second, *the value-driven characteristics of non-state actors make for more active engagements in climate actions* (Rinkema, 2014). In the post-Paris regime, companies often register voluntary climate actions without feeling compelled to disclose all emission data or undergo thorough audits, unlike the Kyoto regime where Japan and other Annex I countries mandated rigorous emission audits. This explains the high volume of voluntary climate actions registered by companies and private investors, which accounts for over 80% (2,626 out of 3,178 actors) of non-state climate actions in three East Asian countries registered on the NAZCA portal.

Third, with the changing norms, *non-state business actors exhibit a heightened disposition to present themselves as climate-conscious in the international market*. This proclivity is driven by the enhanced climate-related standards that can naturally contribute to their market competitiveness and align them with the requirements of international trade. However, unlike the social constructivist arguments suggesting a norm internalization process (Finnemore & Sikkink, 2001), it remains unclear at this point whether the affirmative responses from business actors represent genuine transitional trends or are merely a superficial climate gesture considering the limited portion of actual mitigation actions registered and progress reported.

Conclusion

There is a noticeable trend of close collaboration between non-state actors and national governments in all three East Asian nations discussed here. Civil society involvement is relatively limited, while local governments and business actors play less restricted roles in collaborating closely with the national government. This trend aligns with the prevailing approach to climate policy development, which emphasizes government involvement. This is, in part, because the administrative frameworks of these three East Asian nations exhibit a notably pronounced vertical structure in comparison to their Western counterparts. It is particularly evident in the case of China, a nation governed by the Communist Party, where subnational authorities experience considerable constraints in both budgetary allocation and policymaking autonomy.

Certain types of actions, such as climate mitigation, require optional traceability of progress, posing an additional burden for non-state actors who typically prefer a fully voluntary and widely flexible participation. For this reason, many non-state climate actions primarily involve participation in international initiatives that are already widely adopted in the global trade and business markets. Interestingly, the percentage of cities registered in the NAZCA portal is lower in China compared to Korea and Japan amid the low registrations shown commonly in the three countries compared to that of global statistics. Such difference may be partially influenced by the environmental authoritarianism in China, as opposed to Korea and Japan, albeit their common inactivism.

The participation of international and domestic non-state actors in climate actions has increasingly grown in China, South Korea, and Japan, as evidenced by our analysis of their historical background and current practices via the UNFCCC NAZCA portal. It is worth noting that the UNFCCC registration process is rigidly categorized in terms of actor types and actions, potentially hindering even more dynamic registration activities. With the increasing involvement of non-state actors and the importance of these three nations in the post-Paris international climate regime, improving the quantity and quality of diverse actions recognized and registered on the UN portal is crucial. Enhancing the sophistication of non-state climate actions and combining them with national efforts are essential steps toward advancing more ambitious climate goals.

Acknowledgments

The authors acknowledge and thank Koichi Hasegawa, Chao Jiang, Ryoichi Terada, and Lei Zhang for helpful comments and suggestions in the early stages of development of this chapter. All responsibility for interpretation and arguments presented herein is that of the authors alone.

Notes

1. This definition is articulated in the draft COP decisions, which were prepared by the co-chairs of the UNFCCC's Ad Hoc Working Group on the Durban Platform for Enhanced Action (ADP) as part of the Paris Package (UNFCCC, 2015). This book chapter adopts the UNFCCC's definition of non-state actors. The status of cities and regions can be categorized as subnational authorities, distinguishing them from other non-state actors.
2. Non-state actors have been part of the climate action since 1992 when the UN Convention on Climate Change was first signed in Rio Environment and Sustainable Development Summit. Yet, before Paris, their efforts were not officially considered in the UN system where the state actors are the only officially recognized but as observers or secondary contributors at first.
3. For instance, there is a study to show that in the case of Japan, individual commitments from non-state actors have the potential to reduce emissions by 8% to 12% below current policy projections by 2030. Moreover, this approach could lead to emissions levels up to 70 $MtCO_2e$/year lower than the targeted levels outlined in the nationally determined contributions (NDCs) by 2030.
4. The Peruvian presidency of COP 20, the French presidency of COP 21, and the Secretary-General of the United Nations launched the Lima-Paris Action Agenda. The purposes were to contribute to the less than 2 degrees Celsius goal, to further increase ambitions before 2020, and to engage private and non-state actions to these ends.

5. The Carbon Disclosure Project (CDP)—ICLEI Local Governments for Sustainability reporting system is not official reporting portal. It provides additional information in detail and is linked to the official UNFCCC reporting portal (NAZCA).
6. The UNFCCC categorizes countries into three main groups according to differing commitments: (1) *Annex I* for industrialized countries of OECD members in 1992 plus countries with economies in transition (the EIT Parties), including the Russian Federation, the Baltic States, and several Central and Eastern European states; (2) *Annex II* for the OECD members of Annex I (except the EIT Parties) who are required to provide financial resources to developing countries; and (3) *Non-Annex I* for developing countries.
7. During the 11th Five-Year Plan (2006–2011), China nearly achieved its target of a 20% reduction in national energy intensity. The subsequent 12th Five-Year Plan (2011–2015) dedicated climate policies as its central theme in national planning.
8. The calculation method of CO_2 emission in petrochemical production.
9. Allocation is based on the carbon intensity of each unit of electricity generated by a power plant instead of national or sectoral emission cap.
10. CANGO website, https://www.cango.vip/jgjj
11. GHub website, https://en.ghub.org.cn/about#sct_1
12. Vanke Foundation website, http://en.vankefoundation.com/about.aspx?type=6
13. Article 38 of the Enforcement Decree of the Framework Act on Low Carbon and Green Growth [저탄소녹색성장 기본법시행령 제38조], which was replaced by Carbon Neutrality Act in 2022 (effective) later.
14. Five years for cities and provinces (Do) and 10 years for towns or counties (Si/Gun/Gu).
15. Article 11 and 12 of South Korea's Carbon Neutrality Act.
16. Article 26 of South Korea's Carbon Neutrality Act.
17. The idea originated from similar initiatives endorsed by the United Nations, such as the Principles for Responsible Investment (PRI) and the United Nations Environment Programme Finance Initiative (UNEP FI).
18. Established in 2015, the TCFD under the Financial Stability Board of G20 developed climate-related financial risk disclosures for voluntary adoption by companies, aiming to enhance transparency and decision making on climate-related financial risks.
19. The 3rd Period National Greenhouse Gas Emission Allowance Allocation Plan (2021–2022).
20. In accordance with South Korea's Act on Allocation and Trading of Greenhouse Gas Emission Permits (hereinafter referred to as the "Emissions Permits Act"), businesses and entities subject to allocation are those that have emitted more than 125,000 tons of greenhouse gases per year on average over the past 3 years or those that one or more business sites with emissions of 25,000 tons of greenhouse gases per year.
21. According to Kim Hyuk Rae, the revitalization of civil society in Korea commenced as the authoritarian regime receded following the 1987 Democratic Movement. During the Roh and Moon Administrations, South Korea's civil society experienced substantial growth in both quality and quantity, attaining significant positions in governmental and business sectors.
22. Ministry of Environment, Ministry of Knowledge and Economy, and PCGG were the top-three most influential organizations in South Korean Central Government.
23. Korean Federation for Environmental Movements website, https://kfem.org/
24. Youth 4 Climate Action website, https://youth4climateaction.org.
25. Green Korea United website, https://www.greenkorea.org/english
26. This dynamic was evident in behind-the-scenes policymaking during internal negotiations in 1990 to formulate the Action Plan to Arrest Global Warming.
27. The Democratic Party of Japan (DPJ) became more outspoken against nuclear energy after the Fukushima disaster in 2011 and shifted policy support away from traditional utilities who were close to its rival, the Liberal Democratic Party (LDP) of Japan (Aldrich, 2012; Hughes, 2016).
28. The Keidanren, also known as the Japan Business Federation, stands as Japan's largest economic and business federation, boasting over 1,300 member companies, 109 nationwide industrial associations, and 47 regional economic organizations, as of June 2015. In contrast, the

Keizai Doyukai, or Japan Association of Corporate Executives, founded in 1946, consists of approximately 1,300 top executives from about 940 corporations, with a specific emphasis on less energy-intensive businesses. This distinction occasionally results in differing policy positions compared to the Keidanren.

29. The international initiative RE100 originated in the United Kingdom in 2014. It advocates for voluntary pledges from companies to utilize 100% renewable energy resources across their entire supply chain and manufacturing process.
30. Edogawa Citizens Network for Climate Change website, http://www.sokuon-net.org/
31. Japan Climate Initiative website, https://japanclimate.org/english/
32. Kiko Network website, https://kikonet.org/en/
33. All datasets utilized herein were reconstructed by the authors, based on the UNFCCC NAZCA dataset (available at https://climateaction.unfccc.int/). Commitment and action plans were linked to the CDP-ICLEI portal (https://www.cdp.net/en/cities-discloser), for detailed analysis, scoring, and achievement tracking.
34. In some instances, these entities are owned or controlled by their respective governments. This characteristic is particularly pronounced in the case of China, given its status as a communist country.
35. "Cities" throughout this report generally refer to administrative units, which include municipalities, towns, urban communities, districts, and counties defined by the actors themselves.
36. "Regions," including South Korean provinces, Japanese prefectures, and Chinese provinces, are larger administrative units that are generally broader in geographic scope and population than cities. They usually have separate governing bodies from national governments but encompass lower administrative levels of government; often, they are the first administrative level below the national government.
37. However, it is not easy to define a region because all the three countries use different terms for designating a region: province for China; prefecture for Japan; province for Korea. To note, we triangulated the definition of OECD reports, each country's national official website, and US CIA's country profile. Japan is geographically divided into 10 regions, further subdivided into 47 smaller prefectures. There are 10 regions in South Korea as in "Do," a provincial unit and one metropolitan area of Seoul. There are 23 regions as in "Sheung," a provincial unit.

References

Aldrich, D. P. (2012). Post-crisis Japanese nuclear policy: From top-down directives to bottom-up activism. *Asia Pacific Issues, 103*, 1–12.

Almeida, P., González Márquez, L. R., & Fonsah, E. (2024). The forms of climate action. *Sociology Compass, 18*(2), e13177. https://doi.org/10.1111/soc4.13177

Bäckstrand, K., Kuyper, J. W., Linnér, B.-O., & Lövbrand, E. (2017). Non-state actors in global climate governance: From Copenhagen to Paris and beyond. *Environmental Politics, 26*(4), 561–579. https://doi.org/10.1080/09644016.2017.1327485

Betsill, M. M., & Bulkeley, H. (2006). Cities and the multilevel governance of global climate change. *Global Governance: A Review of Multilateralism and International Organizations, 12*(2), 141–159.

Broadbent, J. (2002). Japan's environmental regime: The political dynamics of change. In D. Uday (Ed.), *Environmental politics and policy in industrialized countries* (pp. 295–355). MIT Press.

C40. (2024). Climate action and innovation in Chinese cities. https://www.c40.org/news/climate-action-innovation-chinese-cities/

CAN (Climate Action Network International). (2024, April 9). Climate Action Network International about CAN. https://climatenetwork.org/overview/members/

Carter, N., & Mol, A. (2006). China and the environment: Domestic and transnational dynamics of a future hegemon. *Environmental Politics, 15*(2), 330–344. https://doi.org/10.1080/0964401060 0562294

Chan, S., Falkner, R., Goldberg, M., & van Asselt, H. (2018). Effective and geographically balanced? An output-based assessment of non-state climate actions. *Climate Policy, 18*(1), 24–35. https://doi.org/10.1080/14693062.2016.1248343

ChungNam-Do Provincial Council. (2023). *Legislative debate to prepare alternative energy transition plan following closure of thermal power plant in South Chungcheong Province.* ChungNam-Do Council.

Climate Action 100+, N. (2022, January 31). KEPCO commits to carbon neutrality and coal phase out by 2050. https://www.climateaction100.org/news/kepco-commits-to-carbon-neutrality-and-coal-phase-out-by-2050/

Climate Group. (2023, October 12). Net zero in South Korea: Leaders gather in Chungnam to drive forward commitment to decarbonisation. *SteelZero.* https://www.theclimategroup.org/our-work/news/net-zero-south-korea-leaders-gather-chungnam-drive-forward-commitment-decarbonisation

Climate Watch. (2019). Global historical emissions. https://www.climatewatchdata.org/ghg-emissions?end_year=2019&start_year=1990

CSIS (Center for Strategic and International Studies). (2021, August 26). Chinese national oil companies face the energy transition. https://www.csis.org/analysis/chinese-national-oil-companies-face-energy-transition

Engels, A. (2018). Understanding how China is championing climate change mitigation. *Palgrave Communications, 4*(1), 101. https://doi.org/10.1057/s41599-018-0150-4

Finnemore, M., & Sikkink, K. (2001). Taking stock: The constructivist research program in international relations and comparative politics. *Annual Review of Political Science, 4*(1), 391–416.

Fraser, T., & Aldrich, D. P. (2020). The Fukushima effect at home: The changing role of domestic actors in Japanese energy policy. Wiley Interdisciplinary Reviews: Climate Change, 11(5), e655.

Fraser, T., Cunningham, L., Bancroft, M., Hunt, A., Lee, E., & Nasongo, A. (2020). Climate crisis at City Hall: How Japanese communities mobilize to eliminate emissions. *Environmental Innovation and Societal Transitions, 37,* 361–380. https://doi.org/10.1016/j.eist.2020.09.006

Fraser, T., & Temocin, P. (2021). Grassroots vs. greenhouse: The role of environmental organizations in reducing carbon emissions. *Climatic Change, 169*(3–4), 22. https://doi.org/10.1007/s10584-021-03260-4

FutureCity Initiative. (2016). *FutureCity Initiative: Urban development for the future, get it rolling, regional wisdom, power, and resources will strengthen cities.* Tokyo: Promotion Council for the FutureCity Initiative. https://future-city.go.jp/doc/pdf/pclcc/pamphlet2016_e_n001.pdf

Gilley, B. (2012). Authoritarian environmentalism and China's response to climate change. *Environmental Politics, 21*(2), 287–307. https://doi.org/10.1080/09644016.2012.651904

GIR (Greenhouse Gas Inventory and Research Center). (2021). Completion of emission allowance allocation for the 3rd planning period of the greenhouse gas emissions trading system. https://www.gir.go.kr/home/board/read.do;jsessionid=DSNRjyvhJpJymJr2UlPkCrokkmZkCv1NgsSfcyFRrlbeMFWIxB6aAq9qQUm36Y5q.og_was1_servlet_engine1?pagerOffset=40&maxPageItems=10&maxIndexPages=10&searchKey=&searchValue=&menuId=10&boardMasterId=4&boardId=152

Guttman, D., Young, O., Jing, Y., Bramble, B., Bu, M., Chen, C., Furst, K., Hu, T., Li, Y., Logan, K., Liu, L., Price, L., Spencer, M., Suh, S., Sun, X., Tan, B., Wang, H., Wang, X., Zhang, J., … Zeidan, R. (2018). Environmental governance in China: Interactions between the state and "nonstate actors." *Journal of Environmental Management, 220,* 126–135. https://doi.org/10.1016/j.jenvman.2018.04.104

Haddad, M. A. (2010). The state-in-society approach to democratization with examples from Japan. *Democratization, 17*(5), 997–1023.

Hager, C., & Hamagami, N. (2020). Local renewable energy initiatives in Germany and Japan in a changing national policy environment. *Review of Policy Research, 37*(3), 386–411. https://doi.org/10.1111/ropr.12372

Hale, T. (2016). "All hands on deck": The Paris Agreement and nonstate climate action. *Global Environmental Politics, 16*(3), 12–22. https://doi.org/10.1162/GLEP_a_00362

Hale, T. (2018). *The role of sub-state and non-state actors in international climate processes*. The Royal Institute of International Affairs Chatham House.

Hasegawa, K. (2022, October). Fridays for Future: A social movements perspective. *Global Dialogue, Magazine of the International Sociological Association*. https://globaldialogue.isa-sociology.org/articles/fridays-for-future-a-social-movements-perspective

Ho, P. (2001). Greening without conflict? Environmentalism, NGOs and civil society in China. *Development and Change, 32*(5), 893–921. https://doi.org/10.1111/1467-7660.00231

Ho, P., & Edmonds, R. (Eds.). (2008). *China's embedded activism: Opportunities and constraints of a social movement*. Routledge.

Howie, P., Gupta, S., Park, H., & Akmetov, D. (2020). Evaluating policy success of emissions trading schemes in emerging economies: Comparing the experiences of Korea and Kazakhstan. *Climate Policy, 20*(5), 577–592. https://doi.org/10.1080/14693062.2020.1751030

Hughes, L. (2016). Renegotiating Japan's energy contract. In C. Hager & C. H. Stefes (Eds.), *Germany's energy transition: A comparative perspective* (pp. 165–184). New York, NY: Palgrave.

ICAP (International Carbon Action Partnership). (2024). China national ETS. https://icapcarbonaction.com/en/ets/china-national-ets

ICLEI. (2024). Membership in East Asia. https://eastasia.iclei.org/members/

Japan Climate Initiative. (2024). Joining the front line of global push for decarbonization from Japan. https://japanclimate.org/english/

Kalinowski, T. (2021). The politics of climate change in a neo-developmental state: The case of South Korea. *International Political Science Review, 42*(1), 48–63.

Kameyama, Y. (2016). *Climate change policy in Japan*. Routledge. https://doi.org/10.4324/9781315733920

Kameyama, Y. (2021). Climate change policy: Can new actors affect Japan's policy-making in the Paris Agreement era? *Social Science Japan Journal, 24*, 67–84. https://doi.org/10.1093/ssjj/jyaa051

Kawashima, Y. (2001). Japan and climate change: Responses and explanations. *Energy & Environment, 12*(2–3), 167–179. https://doi.org/10.1260/0958305011500689

Kern, T. (2010). Translating global values into national contexts: The rise of environmentalism in South Korea. *International Sociology, 25*(6), 869–896. https://doi.org/10.1177/0268580910378139

Kim, E. (2016). The politics of climate change policy design in Korea. *Environmental Politics, 25*(3), 454–474. https://doi.org/10.1080/09644016.2015.1104804

Kim, H. R. (2000). The state and civil society in transition: The role of non-governmental organizations in South Korea. *The Pacific Review, 13*(4), 595–613. https://doi.org/10.1080/09512740045534

Kim, S. Y., & Thurbon, E. (2015). Developmental environmentalism: Explaining South Korea's ambitious pursuit of green growth. *Politics & Society, 43*(2), 213–240. https://doi.org/10.1177/0032329215571287

KITA (Korea International Trade Association). (2023). Trade brief: Current status and implications of climate change response by export companies. https://www.kita.net/researchTrade/report/tradeBrief/tradeBriefDetail.do?no=2500&ContentsID=KI_cmercReport_01_2500&query=%EA%B8%B0%ED%9B%84%EB%B3%80%ED%99%94%20%EC%88%98%EC%B6%9C%EA%B8%B0%EC%97%85

KLIC (Korean Law Information Center). (2022, March). Framework Act on carbon neutrality and green growth for coping with climate crisis. https://www.law.go.kr/LSW/eng/engLsSc.do?menuId=2&section=lawNm&query=carbon+neutrality&x=0&y=0#liBgcolor1

Koh, J. (2017). A study on change in climate change adaptation governance of Korean local governments: Focusing on the process of developing the climate change adaptation action plan and its implementation stage. *Journal of Climate Change Research, 8*(2), 99–108. http://dx.doi.org/10.15531/KSCCR.2017.8.2.99

Kuramochi, T., Swithin, L., Höhne, N., Smit, S., Casas, M., Hans, F., Nascimento, L., Tanguy, P., Hsu, A., Weinfurter, A., Yeo, Z.-Y. Kim, Y., Raghavan, M., Krummenacher, C. I., Xie, Y., Roelfsema, M., Chan, S., Hale, T., & New Climate Institute, Data-Driven Lab, PBL Netherlands Environmental Assessment Agency, German Development Institute(Deutsches Institut für Entwicklungspolitik (DIE), Blavatnik School of Government, University of Oxford. (2019). *Global climate action from cities, regions and businesses: Impact of individual actors and cooperative initiatives on global and national emissions.* Oxford, UK (for the Blavatnik School of Government, University of Oxford) and Chapel Hill, NC (for the Data-Driven Envirolab).

Lee, J., & Yu, J. (2017). Market analysis during the first year of Korea emission trading scheme. *Energies, 10*(12), 1–13. https://doi.org/10.3390/en10121974

Lee, J.-S., & Kim, J. (2018). Assessing strategies for urban climate change adaptation: The case of six metropolitan cities in South Korea. *Sustainability, 10*(6), 2065. https://doi.org/10.3390/su10062065

Li, C., & Song, Y. (2016). Government response to climate change in China: A study of provincial and municipal plans. *Journal of Environmental Planning and Management, 59*(9), 1679–1710. https://doi.org/10.1080/09640568.2015.1085840

Lim, S.-U., & Kim, J. (2015). Evaluation model of climate change in public enterprises. *Journal of Climate Change Research, 10*(2), 89–96. http://dx.doi.org/10.14383/cri.2015.10.2.89

Liu, L., Wang, P., & Wu, T. (2017). The role of nongovernmental organizations in China's climate change governance. *WIREs Climate Change, 8*(6), e483. https://doi.org/10.1002/wcc.483

Loher, J. F. (2012). Domestic climate change governance in East Asia: A comparative analysis of Japan, China and South Korea. *Vienna Journal of East Asian Studies, 3*(1), 73–107. https://doi.org/10.2478/vjeas-2012-0003

Lövbrand, E., Hjerpe, M., & Linnér, B.-O. (2017). Making climate governance global: How UN climate summitry comes to matter in a complex climate regime. *Environmental Politics, 26*(4), 580–599. https://doi.org/10.1080/09644016.2017.1319019

Mai, Q., & Francesch-Huidobro, M. (2015). *Climate change governance in Chinese cities.* Routledge.

METI Japan (Ministry of Economy, Trade, and Industry of Japan). (2018). The fifth strategic energy plan. https://www.enecho.meti.go.jp/en/category/others/basic_plan/5th/pdf/strategic_energy_plan.pdf

MOE Japan (Ministry of Environment of Japan). (2024, March 29). Ministry of Environment of Japan: Global environment, net-zero. https://www.env.go.jp/en/earth/cc/2050_zero_carbon_cities_in_japan.html

MOE Korea (Ministry of Environment of Korea). (2020). *The 3rd Period National Greenhouse Gas Emission Allowance Allocation Plan (2021–2022).* Seoul: Ministry of Environment of Korea.

MOJ (Ministry of Justice of Japan). (2006). *Act on Promotion of Global Warming Countermeasure (Act No. 117 of 1998).* Tokyo: Ministry of Justice of Japan. https://www.japaneselawtranslation.go.jp/en/laws/view/3796/en#je_ch1at4

Moon, T. H., Chae, Y., Lee, D.-S., Kim, D.-H., & Kim, H. (2021). Analyzing climate change impacts on health, energy, water resources, and biodiversity sectors for effective climate change policy in South Korea. *Scientific Reports, 11*(1), 18512. https://doi.org/10.1038/s41598-021-97108-7

Nakazawa, T., Satoh, K., Trencher, G., Tatsumi, T., & Hasegawa, K. (2023). Net-zero carbon declarations by Japanese local governments: What caused the domino-like diffusion? *Review of Policy Research, 40*(6), 853–893.

Nilsson, A., Smit, S., & Kuramochi, T. (2021). *Non-state and subnational climate action in China: An overview of the current landscape, emission reduction potential and implementation* [Working Paper]. New Climate Institute. newclimate.org/sites/default/files/2022/01/NewClimate_ChinaNSA_Jan22.pdf

OECD (Organization for Economic Co-operation and Development). (2024). OECD data: Trade in goods and services. https://data.oecd.org/trade/trade-in-goods-and-services.htm

PCCNGG (Presidential Commission on Carbon Neutrality and Green Growth). (2023). Discussion on ways to supply and expand renewable energy such as local government offshore wind power and solar power. https://www.2050cnc.go.kr/base/board/read?boardManagementNo=3&boardNo=2141&page=4&searchCategory=&searchType=&searchWord=&menuLevel=3&menuNo=9

Reimann, K. D. (2002). Building networks from the outside in: Japanese NGOs and the Kyoto Climate Change Conference. In *Political Science Faculty publications of Georgia State University* (Vol. 6, pp. 173–187). Georgia State University Department of Political Science. https://scholarworks.gsu.edu/political_science_facpub/6

Rinkema, R. A. (2014). *Environmental agreements, non-state actors, and the Kyoto Protocol: A third way for international climate action*. Penn Law Scholarship Repository.

Rock, M. T. (2002). Integrating environmental and economic policy making in China and Taiwan. *American Behavioral Scientist, 45*(9), 1435–1455.

Samsung, E. (2022, September 15). Samsung Electronics announces new environmental strategy. *Samsung Newsroom*. https://news.samsung.com/global/samsung-electronics-announces-new-environmental-strategy

Stearns, L. B., & Almeida, P. D. (2004). The formation of state actor-social movement coalitions and favorable policy outcomes. *Social Problems, 51*(4), 478–504. https://doi.org/10.1525/sp.2004.51.4.478

Stone, B., Vargo, J., & Habeeb, D. (2012). Managing climate change in cities: Will climate action plans work? *Landscape and Urban Planning, 107*(3), 263–271. https://doi.org/10.1016/j.landurbplan.2012.05.014

Tang, S.-Y., & Zhan, X. (2008). Civic environmental NGOs, civil society, and democratisation in China. *Journal of Development Studies, 44*(3), 425–448. https://doi.org/10.1080/00220380701848541

UNFCCC (United Nations Framework Convention for Climate Change). (2015). Non-paper. Note by the Co-Chairs. The Ad Hoc Working Group on the Durban Platform for Enhanced Action. http://unfccc.int/resource/docs/2015/adp2/eng/8infnot.pdf.

Wang, P., Liu, L., & Wu, T. (2018). A review of China's climate governance: State, market and civil society. *Climate Policy, 18*(5), 664–679. https://doi.org/10.1080/14693062.2017.1331903

Westman, L., & Broto, V. C. (2018). Climate governance through partnerships: A study of 150 urban initiatives in China. *Global Environmental Change, 50*, 212–221. https://doi.org/10.1016/j.gloenvcha.2018.04.008

World Bank. (2020). Total greenhouse gas emissions (kt of CO2 equivalent)—China, Korea, Rep., Japan. https://data.worldbank.org/indicator/EN.ATM.GHGT.KT.CE?locations=CN-KR-JP

World Bank. (2021). GDP (current US$)—China, Korea, Rep., Japan, World Bank national accounts data, and OECD National Accounts data files. https://data.worldbank.org/indicator/NY.GDP.MKTP.CD?end=2021&locations=CN-KR-JP-1W&most_recent_year_desc=true&start=1960&view=chart

World Bank. (2024). *Korea, Rep. Exports of goods and services, in % of GDP 2017–2021*. World Integrated Trade Solutions. https://wits.worldbank.org/CountryProfile/en/Country/KOR/StartYear/2017/EndYear/2021/Indicator/NE-EXP-GNFS-ZS

Yun, S.-J., Ahn, S., & Soh, R. Y. (2022). Energy democracy for energy transition in South Korea? Focusing on politicization of media. In J. Asuka & D. Jin (Eds.), *Energy transition and energy democracy in East Asia* (pp. 25–49). Singapore: Springer. https://doi.org/10.1007/978-981-19-0280-2_3

Yun, S.-J., Ku, D., & Han, J.-Y. (2014). Climate policy networks in South Korea: Alliances and conflicts. *Climate Policy, 14*(2), 283–301. https://doi.org/10.1080/14693062.2013.831240

Zhang, L., & Li, X. (2018). Changing institutions for environmental policy and politics in New Era China. *Chinese Journal of Population Resources and Environment, 16*(3), 242–251. https://doi.org/10.1080/10042857.2018.1496702

Zhou, X., & Mori, H. (2011). National institutional response to climate change and stakeholder participation: A comparative study for Asia. *International Environmental Agreements: Politics, Law and Economics, 11*(4), 297–319. https://doi.org/10.1007/s10784-010-9127-5

Zinda, J. A. (2023). Unpacking authoritarian environmental governance: Gauging authoritarian elaboration in China and beyond. *Sociology of Development, 9*(2), 195–216. https://doi.org/10.1525/sod.2023.9.2.195

CHAPTER 28

Climate Planning Across Latin America

ALDO MADARIAGA, ANTOINE MAILLET, AND DANIELA TAPIA

Introduction

THE Kyoto Protocol (1997) and the Paris Agreement (2016) are key milestones that proposed action to address the effects of climate change. Two main strategies were devised: mitigation, which is the adoption of low-emission trajectories to reduce greenhouse gas (GHG) emissions; and adaptation, which refers to measures to moderate the consequences of climate change (Füssel, 2007). The signatory countries of these agreements committed to initiate a planning process to decide what climate action measures they would adopt, called nationally determined contributions (NDCs) (European Commission, 2020). In this chapter, we base our exploration of climate planning in Latin America with a detailed analysis of NDCs.

NDCs contain mitigation and adaptation targets defined by a particular country according to its capacities and circumstances, including sectors considered priorities and requirements for external support (Samaniego et al., 2019). However, the content and form of climate planning can vary considerably among countries depending on the definitions in their NDCs and the varying degrees of progress in meeting them. The NDC Partnership provides assistance to countries in developing their climate action plans and reports that the most frequently requested area of support is related to finance (NDC Partnership, 2023). First, we review the conceptualization of climate planning and its development in Latin America. Then we explain the methodology for an empirical analysis and present a brief overview of the region's advances in implementation. Finally, to better understand the factors behind country advances in climate planning, we introduce four cases, two nations with an advanced level of planning (Chile and Colombia) and two with a moderate trajectory (Brazil and Uruguay).

Climate Planning

"Climate planning" is an expression that is widely used by public actors of different kinds, but it is still loosely defined. Alongside Rojas et al. (2021, p. 14), we define climate planning as "an action that prioritizes the most conducive strategies to design and implement policies, plans, programs, projects that strengthen multi-stakeholder articulation and climate governance." Almeida et al. (2024) point out that climate action planning frequently occurs at the municipal level, and that nongovernmental entities and professionals collaborate with state actors for its design.

The work of Almeida et al. (2024) also classifies the different forms of climate action that exist as institutional or noninstitutional. Institutional climate actions are activities "initiated by organizations and groups using routine and formalized procedures to achieve goals of reducing greenhouse gas (GHG) emissions and preparing for adaptation and resilience" (Almeida et al., 2024, p. 2). The authors classify green transition and renewable energy policies, cap and trade/offsets, carbon markets, and climate action planning as institutional climate actions. However, noninstitutional climate actions involve civil society and the use of disruptive strategies such as climate-based social movements and local struggles over fossil fuel extraction. In this chapter, we focus on the former.

Climate action planning during the 1990s and early 2000s emphasized the incorporation of climate change mitigation, implementing top-down international policy initiatives that reduce GHG emissions, such as the proposals of the United Nations Framework Convention on Climate Change (UNFCCC) (Almeida et al., 2024; Preston et al., 2010). However, the planning of these years was criticized for "carbon reductionism" (Méndez, 2020) and for ignoring other dimensions of the climate crisis (Almeida et al., 2024). Lately, adaptation measures have been given greater prominence as a risk management strategy, which can be attributed to the recognition that certain effects of climate change are unavoidable regardless of the trajectory of emissions (Preston et al., 2010, p. 408).

A challenge for planning relates to the roles of various actors at different levels. Mimura et al. (2014) mention horizontal coordination among different agencies and departments, as well as vertical coordination among national, regional, and local stakeholders, as a problem. In addition, there is the need for effective leadership during the planning process and the potential for conflicting objectives between planning and other policies (Thomas et al., 2019). Planning also entails dealing with instruments (strategies, plans, etc.) that can have a high level of heterogeneity, which makes their implementation difficult.

National governments often take on a coordinating role at the subnational and local levels. This coordination involves the provision of risk information, public policy frameworks that guide decision making on sectoral activity and resource use, and financial support (Mimura et al., 2014). However, formulating climate planning in developing countries is complex due to the lack of institutional capacity of governments and local authorities, deficiency of resources and infrastructure, and the irregular growth of urban areas and difficulty in regulating land use (Campos Pérez, 2019).

The first NDCs submitted by Latin American countries failed to incorporate key ministries in the development of climate action objectives and lacked public consultation and deliberation processes prior to their drafting (Cárdenas et al., 2021). Subsequently, some countries updated their NDCs, and as of 2020, a second round of NDCs began to be presented. On this occasion, countries reported carrying out a more participatory decision-making process, involving a variety of actors with different levels of commitment (Cárdenas et al., 2021). Due to increased political support, the new versions of the NDCs featured better prioritization, financing, and mainstreaming. However, several evaluations indicate that the integration of environmental and economic guidelines into climate policy is still required for NDCs to truly represent a comprehensive plan for national and sectoral climate action (Cárdenas et al., 2021). Indeed, several countries still require support to design robust climate finance strategies and to coherently incorporate NDC objectives and guidelines into national policy and planning (NDC Partnership, 2023).

Methodology

We compiled and systematized the climate planning processes of 19 Latin American countries.[1] We reviewed the planning of the countries of the region, mainly through the latest NDC submitted by each nation. From this information, we built a classification system in terms of the level of advancement of climate planning.

The review of country planning was based on information and methodology extracted from the Low-Emission Development Strategy for Latin America and the Caribbean (LEDS LAC), the regional platform of the Global Climate Action Partnership (GCAP), including LEDSenLAC 2019 (LEDS LAC, 2020) and LEDSenLAC 2022 (LEDS LAC, 2023), two reports published by LEDS LAC. These reports compile and evaluate the progress of the implementation of the NDCs and were prepared with the support of the Inter-American Development Bank (IDB), the United Nations Development Program (UNDP) and the EUROCLIMA+ Program of the European Union. In the 2019 publication, the objective was to identify the level of progress made by countries with respect to the development status of their current NDCs. To determine this level of progress, this report organized the information and data into four stages, each with its respective components (Table 28.1).

The 2022 report aimed to present progress regarding the implementation of the first-generation NDCs and the content of the second-generation NDCs (LEDS LAC, 2023). In this version, it analyzed the quality of progress and evaluated the integration of climate goals into national and sectoral planning. Unlike the previous report, instead of implementation stages,

TABLE 28.1 Climate Planning Stages and Components

Stage	Description	Components
1. Preparation	"Efforts made by countries to raise citizen awareness, establish institutional arrangements and develop technical capacities and information that can be used to plan the implementation of their NDCs" (LEDS LAC, 2020).	- National inventory system - Analysis of spending and climate investment
2. Planning Implementation and Mobilization of Resources	"Progress that the countries have made regarding the development of an implementation plan for their NDCs, as well as the development of a financial strategy to mobilize the resources required for the implementation of the aforementioned plan" (LEDS LAC, 2020).	- Implementation plan or roadmap - Financing strategy - National adaptation plan
3. NDC Implementation and Progress Monitoring	"Construction of enabling conditions and implementation actions carried out to meet the goals set out in their NDCs. The progress that countries have made is evaluated regarding the establishment of a monitoring, reporting and verification system that allows evidence of the progress of implementation of their NDCs" (LEDS LAC, 2020).	- Climate change law - Carbon tax - Monitoring, reporting, and verification system
4. Review and Update of the NDC and Future Cycles	"Review and adjustment of implementation plans according to results and whether they have a long-term development instrument that frames and institutionalizes the process of updating and increasing the ambition of future NDCs" (LEDS LAC, 2020).	- NDC updated - Long-term low emissions development strategy

Source: LEDS LAC (2020).

the following dimensions were used: (1) ownership—the NDC is developed with a participatory and inclusive approach; (2) solidity—commitments are framed and supported by robust and long-term strategies; (3) ambition—improvements in commitments are shown compared to the previous version of the NDC; and (4) feasibility—commitments are supported by the necessary conditions to be implemented (resources, legal support, etc.). Even though the evaluated components and their organization differ between each report, the type of content of the NDC does not change, so that the same components (national inventory system, roadmap, adaptation plan, etc.) are present in both.

In addition to this, we used the NDC LAC platform, which compiles information on commitments and the status of progress in the fulfillment of the NDCs.[2] This tool shows the level of progress in the implementation of NDCs in Latin America and the Caribbean with the stages presented in the 2020 report and updated information from the 2023 report. However, unlike these reports, the platform assigns a classification or status to the level of progress of each instrument or component: Completed, Advanced, Started, and Not Started. We used this classification to categorize the level of advancement in planning and updated the data with information available on respective government websites.

We assigned a score for the level of progress in each category, ranging from 0 for "Not Started" to 3 for "Completed" (Table 28.2).

TABLE 28.2 Classification of the Level of Progress in Climate Planning

Classification	Criteria	Score
Completed	Strategy, planning, and/or instrument are designed and officially published, with the information and mechanisms necessary to be implemented. It is possible that in the future it can be improved to meet new goals that are established, but the component in question has already been designed and/or is being implemented. Example: - The publication of a "National Adaptation Plan for period X." (Once published, it can be updated.) - The creation and operation of the National Inventory System. (It is required to be constantly running.)	3
In progress	The publication of the already formulated component (e.g., a law) has been announced; the component is in the process of implementation, or it has been implemented but there are no concrete mechanisms for it to function. Example: - The publication of a "National Adaptation Plan for period X" is announced or is in the discussion stage to be approved. - A Plan is published, but it does not have concrete actions or previous instruments that support it (e.g., cost analysis).	2
Initiated	The formulation stage has begun. It has been announced that the strategy, planning, and/or instrument will be formulated in a certain period.	1
Not started	There is no mention of the strategy, planning, and/or instrument in the NDC or other official documents. There is no information available from official sources.	0

Results

Tables 28.3 and 28.4 present the classification of climate planning for all Latin American countries, which allows us to have a cross-cutting view of climate planning in the continent. Table 28.3 shows in detail the scores for each component, while Table 28.4 shows the aggregate summary.

The tables reveal that there is variation not only in the progress of total climate planning among countries but also in the components. Chile and Colombia get a total score of 26, the closest to the maximum score. Chile's performance stands out in stage 2 of planning and resource mobilization, and stage 3 of NDC and progress monitoring. However, Colombia scores full in stage 1 of preparation for planning. Alongside the total score, there are countries that have more advanced stages than others. For example, Ecuador has high scores in the first and second stages, but has 0 in the third. This internal variation is likely to be related to the course of domestic politics. However, the tables show that there is no linear development between one stage and the other, but that countries can advance in the different stages without necessarily completing the previous one.[3] For example, Mexico has the highest score in stages 3 and 4 (9 and 6 points, respectively), but only half the score in stages 1 and 2 (3 and 5 points). Similarly, Uruguay scores the highest in stage 4, but it has mid-range scores in all other stages. Later we analyze in detail the cases of Brazil and Uruguay to inquire more about this.

To identify possible factors that explain these differences, we examine the relationship between climate planning scores according to three factors that could explain them: emissions, gross domestic product (GDP), and state capacity.[4] Figure 28.1 illustrates the relationship between per capita GHG emissions (in tons) and the level of climate planning indicated in Table 28.4 for each country. Countries with lower GHG emissions tend to have a better level of climate planning (medium-high and high), with Colombia, Chile, the Dominican Republic, Costa Rica, and Mexico standing out. On the contrary, countries such as Bolivia and Brazil generate more emissions and have a lower level of planning (medium-low or low). The case of Uruguay stands out due to its high emissions and a medium level of planning.

Figure 28.2 shows the relationship between GDP per capita and the level of climate planning shown (Table 28.4). The figure shows that the higher the GDP per capita, the higher the level of climate planning. Uruguay once again stands out as an outlier due to having a lower level of planning than expected by its GDP level.

Figure 28.3 illustrates the relationship between the level of climate planning and state capacity. To do this, we use the World Bank's Government Effectiveness Index as a reference. This index incorporates the degree of independence of government from political pressures, and perceptions around the quality of public services, the civil service, and policy formulation and implementation, among others.[5] As with Figure 28.2, the greater the state capacity, the higher the level of climate planning. A closer look, however, shows interesting results. First, Chile and Colombia are the countries with the highest score, but with very different levels of state capacity. This warrants further inspection of the case of Colombia and how it has managed to overcome state capacity problems to advance in climate planning. Second, the case of Uruguay stands out once again. With a level of state capacity similar to that of Chile, it should be closer to this country in terms of climate planning, yet it is well below the region's average advancement, and it is behind countries with serious state capacity problems like Ecuador and Honduras. Finally, we highlight the case of Brazil. Although this country shows the expected advancement by its level of state capacity, until recently Brazil was a pioneer in the institutionalization of environmental considerations in the state and in

TABLE 28.3 Climate Planning Progress (Score Detail)

Country	Stage 1: Preparation		Stage 2: Planning Implementation and Mobilization of Resources			Stage 3: NDC Implementation and Progress Monitoring			Stage 4: Review and Update of the NDC and Future Cycles	
	National Inventory System	Analysis of Spending and Climate Investment	Implementation Plan or Roadmap	Financing Strategy	National Adaptation Plan	Climate Change Law	Carbon Tax	Monitoring, Reporting, and Verification System (MRV)	NDC Update	Long-Term Low Emissions Development Strategy
Argentina	In progress	In progress	Initiated	Initiated	Initiated	Completed	Completed	Initiated	Completed	Initiated
Bolivia	Not started	Not started	Initiated	In progress	Not started	Completed	Not started	Not started	Completed	Not started
Brazil	Completed	Not started	Not started	Not started	Completed	Completed	Initiated	Not started	Completed	Not started
Chile	Completed	Not started	Completed	Completed	In progress	Completed	Completed	Completed	Completed	Completed
Colombia	Completed	Completed	In progress	Completed	Initiated	Completed	Completed	In progress	Completed	Completed
Costa Rica	Completed	In progress	In progress	In progress	Completed	Not started	Not started	In progress	Completed	Completed
Cuba	Initiated	Completed	Completed	In progress	Initiated	In progress	Not started	Initiated	Completed	Not started
Ecuador	Completed	Completed	In progress	Completed	Completed	Not started	Not started	Not started	In progress	Initiated
El Salvador	Not started	Initiated	Initiated	Completed	Initiated	Initiated	Not started	Completed	Completed	Initiated
Guatemala	Initiated	In progress	Completed	In progress	Completed	Completed	Not started	Initiated	Completed	Completed
Honduras	Not started	Not started	Completed	Initiated	Completed	Completed	Not started	Completed	Completed	Initiated
Mexico	In progress	Initiated	Initiated	In progress	In progress	Completed	Completed	Completed	Completed	Completed
Nicaragua	In progress	In progress	Not started	Completed	In progress	In progress	Not started	Not started	Completed	Not started
Panama	Completed	Initiated	Completed	Not started	In progress	In progress	Not started	In progress	Completed	Initiated
Paraguay	Initiated	Not started	In progress	Initiated	Completed	Completed	Not started	Initiated	Completed	Not started
Peru	Initiated	Not started	Completed	Initiated	Completed	Completed	Not started	Completed	Completed	In progress
Dominican Republic	Completed	Initiated	In progress	In progress	Completed	Completed	Completed	Completed	Completed	Not started
Uruguay	Initiated	Initiated	Not started	Not started	Completed	Not started	In progress	Completed	Completed	Completed
Venezuela	In progress	Initiated	Completed	Not started	Not started	Not started	Not started	Not started	Completed	Not started

Source: Authors' elaboration based on LEDS LAC.

TABLE 28.4 Level of Climate Planning (Total Scores)

Country	Stage 1: Preparation	Stage 2: Planning Implementation and Mobilization of Resources	Stage 3: NDC Implementation and Progress Monitoring	Stage 4: Review and Update of the NDC and Future Cycles	**TOTAL**
	6 p max.	9 p max.	9 p max.	6 p max.	30 p max.
Argentina	4	3	7	4	18
Bolivia	0	3	3	3	9
Brazil	3	3	4	3	13
Chile	3	8	9	6	26
Colombia	6	6	8	6	26
Costa Rica	5	7	2	6	20
Cuba	4	6	3	3	16
Ecuador	6	8	0	3	17
El Salvador	1	5	4	4	14
Guatemala	3	8	4	6	21
Honduras	0	7	6	4	17
Mexico	3	5	9	6	23
Nicaragua	4	5	2	3	14
Panama	4	5	4	4	17
Paraguay	1	6	4	3	14
Peru	1	7	6	5	19
Dominican Republic	4	7	9	3	23
Uruguay	2	3	5	6	16
Venezuela	3	3	0	3	9

Source: Authors' elaboration based on LEDS LAC.

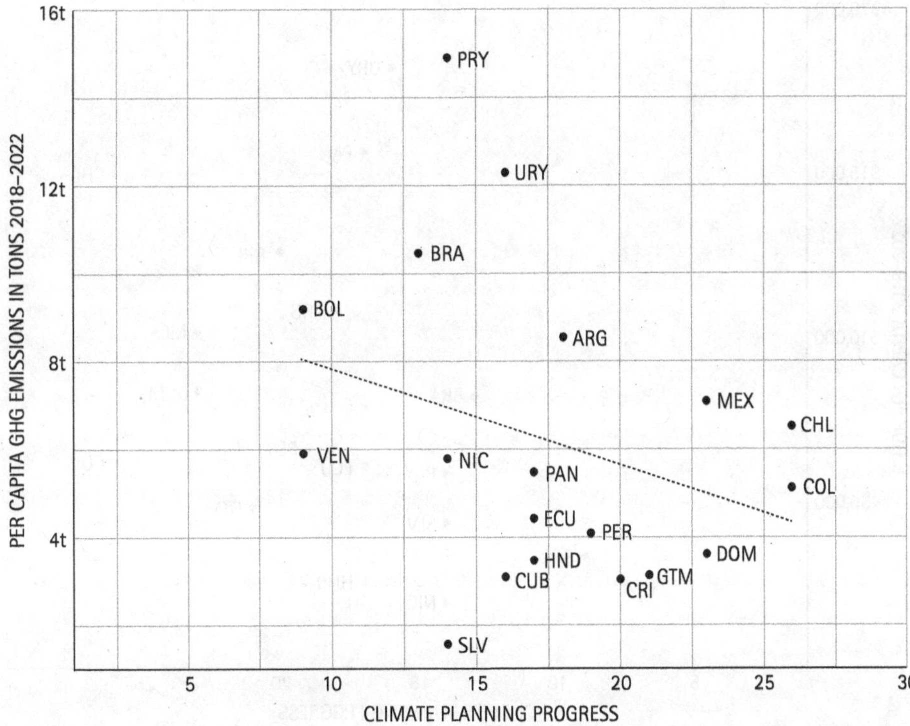

FIGURE 28.1. Greenhouse gas (GHG) emissions versus climate planning progress.
Note: GHG emissions correspond to domestic consumption.

(*Source*: Authors' elaboration based on Table 4 and Our World in Data; https://ourworldindata.org/).

the advancement of environmental policy (Orihuela, 2014). Its place among the laggards of climate planning deserves therefore more attention.

Cases

To explore the factors that explain the variation in the levels of climate planning in the region, and based on existing literature, we analyzed four cases: Chile and Colombia among high performers, and Brazil and Uruguay among low performers.

Chile

Following the overall evolution of environmental institutions in Chile, the construction of Chile's environmental institutions and climate action began to a large extent to comply with international standards and requirements (Aedo & Parker, 2020; Estenssoro, 2012; Madariaga, 2019). It also took advantage of the opportunities generated by the Clean Development Mechanism from the Kyoto Protocol (Barton, 2013a). The management of climate issues was initially led by the Ministry of Foreign Affairs (MINREL), which positioned climate change as a global problem rather than a local one, giving priority to the area of mitigation. Hence, priority was given to the mitigation dimension and tools such as the National Emissions Inventory.

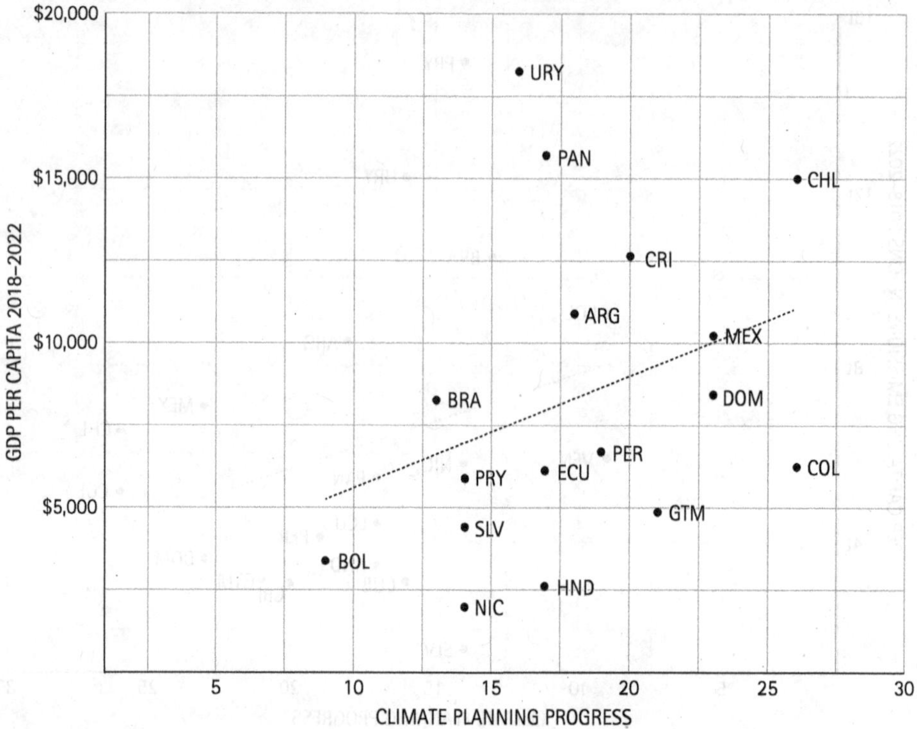

FIGURE 28.2. Gross domestic product (GDP) versus climate planning progress.

(*Source*: Authors' elaboration based on Table 4 and World Bank, 2024).

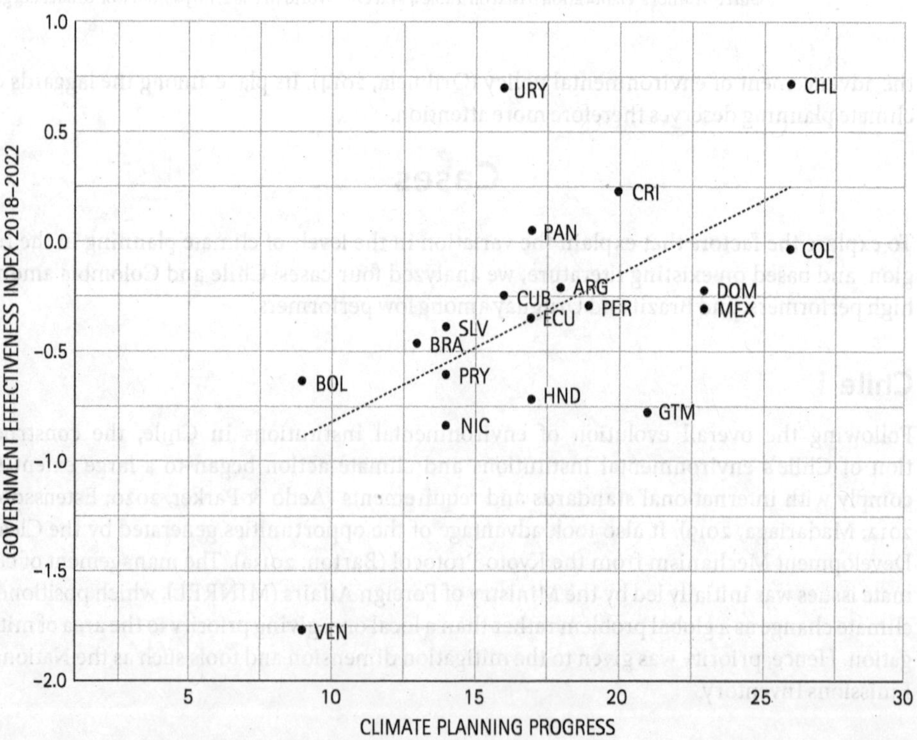

FIGURE 28.3. State capacity and climate planning progress.

(*Source*: Authors' elaboration based on Table 28.4 and data from Worldwide Governance Indicators [World Bank, 2023]).

The first step toward climate planning in Chile was the establishment of the National Committee for Climate Change in 1996. However, it was not until 2006, under Michelle Bachelet's first presidency, that a national climate action strategy began to be developed, a process that culminated in the presentation of the National Plan at COP14 (Barton, 2013a). The first National Action Plan on Climate Change 2008–2012 (PANCC I) was "an articulating instrument with action lines to be developed by various ministries focused on catching information on vulnerability and adaptive capacity by different resources and economic sectors" (Villamizar et al., 2016, p. 13). In addition to the Ministry of the Environment (MMA), the institutions assigned to vulnerability management and planning were sectoral (e.g., the Ministry of Agriculture and the Ministry of Public Works). The proposed initiatives were short-term and focused on "hard" responses such as coastal defense infrastructure or water supply for mining and agriculture, to the detriment of "soft" responses focused on social and local issues (Barton, 2013a).

Following this action plan, ministries with a closer link to spatial planning gradually became involved in planning discussions, as they perceived a necessity to incorporate climate change into existing urban planning instruments (Barton, 2013b). These institutions include the Ministry of Housing and Urban Planning (MINVIU), the Undersecretariat of Regional and Administrative Development (SUBDERE), the Ministry of Social Development and Family (MDSF) (MIDEPLAN at the time), and regional governments. However, an evaluation of the implementation of PANCC I showed that allocated funding covered only 25% of what would have been required for its execution (Villamizar et al., 2016). This inconsistency is attributed to a lack of coordination between the approval of the designed plan and the approval of the budget, a situation aggravated by the fact that the Ministry of Finance did not participate in the process.

In 2012, the MAPS-Chile project was launched. This was part of the Mitigation Action Plans and Scenarios (MAPS) program, a South-South cooperation initiative that promoted a process of knowledge co-production involving a variety of actors to investigate and evaluate mitigation scenarios and measures. This program provided support to four Latin American countries, including Chile. Through international funding, the MAPS-Chile project formed a public–private partnership that incorporated the private sector, academia, and the public sector, under the direction of the state. This project allowed Chile to gain experience and develop technical and methodological capacities that were later applied to other planning and public policy processes (Calfucoy, 2016).

The second National Action Plan on Climate Change 2017–2022 (PANCC II) also had a sectoral nature, and it defined mitigation, adaptation, and governance measures in accordance with international agreements on issues of transport, territory, infrastructure, energy, water, and clean technologies (Mendes, 2020). After an exhaustive evaluation of the first PANCC in 2014, officials determined that the design of the second version should strengthen the participatory dimension, increasing the diversity of actors and consultative bodies (Arriagada et al., 2018). Thus, the participation of civil society, nongovernmental organizations (NGOs), the private sector, academia, and local and national government agencies was strengthened. For example, meetings were held in different regions of the country. In this way, PANCC II managed to increase the diversity of actors involved in planning, both in terms of sectors and levels (Arriagada et al., 2018).

Later for the construction of the NDC, a participatory process was designed both for the information generation stage and for public consultation of the final proposal. Arriagada et al. (2018) recognize MAPS as crucial in this first stage of NDC development, specifying that, for example, both the technical and political teams of MAPS subsequently participated in the formulation of the NDC. In addition, they highlight a high level of involvement of political actors, such as then President Michelle Bachelet and the Minister of Foreign Affairs (MINREL). In the process, MINREL and MMA pushed for fulfilling international standards, sectoral ministries advocated for the measures to be feasible in the long term within their respective sectors, and NGOs lobbied for the implementation of the commitments. Despite this, authors have recognized that the mechanisms through which climate planning is built in Chile remain fragmented (Billi et al., 2021). From an institutional perspective, Lehmann et al. (2015) point out the need for a coordinating institution to provide constantly updated information, with mechanisms to generate and distribute less time-consuming climate change data. The authors explain that long-term planning is especially complex, because the rotation of public officials following government changes hinders the generation of institutional memory, interrupting the process of capacity building. In addition, a lack of resources leads to a reliance on external projects or international NGOs for information, while the financial constraints of national policies and centralization restrict an ambitious planning process.

Lucas Garín (2018) argues that a unitary regime such as Chile's is more conducive to the adoption of climate measures compared to states with a federal regime, since centralized power facilitates policy change. This is consistent with the description of the climate planning process in Chile, where top-down measures were established early on. Despite this, coordination remained mostly horizontal and the connection with subnational regions weak (Arriagada et al., 2018). In this context, the state used its capacity to achieve goals, while at the same time incorporating new actors into the process. In fact, climate planning in Chile has gradually incorporated participatory mechanisms and diversified the nature of actors and sectors involved in the decision-making process. Despite the fact that climate policy falls on different ministries, especially in the case of the NDC, the country established coordination in decision making through different mechanisms (Arriagada et al., 2018). Improved communication channels between stakeholders also increased trust in the planning process (Arriagada et al., 2018; Calfucoy, 2016). It is also worth mentioning that climate action has advanced gradually without experiencing significant setbacks, and that there is in general continuity in the underlying projects beyond a particular government. This notwithstanding, there is questioning about the ambition of climate measures and their dependence on the interests of productive sectors (Madariaga, 2019; Maillet & Carrasco, 2021).

Colombia

Climate change has been gradually incorporated into Colombian politics since the 2000s. In addition to international factors such as the Kyoto Treaty, commentators on Colombia point to internal factors linked to extreme weather phenomena as key triggers for climate action (Howland & Francois, 2022). At the end of 2010, the La Niña phenomenon triggered a series of natural disasters in the country referred to as the "Winter Wave of 2010–2011." The rains and subsequent floods induced by La Niña caused a significant number of deaths and material damage, which led to the incorporation of climate change as an issue on the national agenda (Valdivieso Cervera & Sandoval, 2021). It prompted the incorporation in the "National Development Plan 2010–2014: Prosperity for All" of a chapter on environmental sustainability and risk prevention, which proposes to develop an institutional framework to address climate change (Fisher, 2024; Howland & Francois, 2022).

Billi et al. (2020) classify Colombia's climate governance as hierarchical, as it is centrally coordinated by the Intersectoral Commission on Climate Change (CICC) and the National Climate Change System (SISCLIMA). As mentioned earlier, climate policy has been mostly built with a reactive logic, with the National Plan for Adaptation to Climate Change (PANCC) being the first of the instruments to be developed as a response to the Winter Wave of 2010 mentioned earlier. There are different views regarding the level of participation in this process. On the one hand, Valdivieso Cervera and Sandoval (2021) interviewed a series of officials involved in the PANCC formulation process, who confirmed this reactive logic and low participation as one of the characteristics of this process. On the other hand, the authors conclude that participation occurs unevenly in different areas, but where participation occurs, the influence of the involved actors tends to be high (Valdivieso Cervera & Sandoval, 2021).

Along these lines, Campos Pérez (2019) explores how, in the Colombian context, participatory mechanisms serve to address problems of low institutional capacity and limited resources. The institutional framework allows community-proposed initiatives to be incorporated, and the use of these mechanisms has increased. However, there are obstacles to the successful deployment of participatory mechanisms, as it is difficult for the population to know how to participate in these instances. A successful case of changes at the local level that are scaled up and adopted at the national level includes the Tayrona National Park. The establishment of the Park contributed to connecting climate information with decision making at the national level to manage tourism in protected areas, particularly to improve water monitoring according to climate variability (Múnera & van Kerkhof, 2019).

Like Chile, Colombia received support from MAPS, a process through which it developed its Colombian Low Carbon Development Strategy (CBDC) and established a dynamic of discussion between government and academia that was used in the formulation of the NDC (De Pinto et al., 2017). The Ministry of Environment and Sustainable Development (MADS) was appointed as the leader of the process, while MAPS connected government agencies with actors from different productive sectors, academia, and international organizations. The regional climate change nodes (NRCC) verify the coordination of regional initiatives with national plans, emphasizing the verticality of the system, while evaluation and monitoring are carried out centrally by the executive and legislative branches (Billi et al., 2020).

The overall assessment of the NDC formulation process is positive, highlighting the participation of academia, government, and private institutions. Samaniego et al. (2019) attribute this result to ensuring a certain level of transparency with open processes and interaction between all stakeholders, breaking down disciplinary and institutional barriers. However, there is still a lack of trust toward the activities carried out by the local government, with a perception that it is corrupt and inefficient (Campos Pérez, 2019). In this sense, the involvement of external institutions and NGOs has helped to reduce this reluctance to participate in planning. The participation of civil society in local processes such as at the neighborhood level allows for suggested and adopted planning measures that can be implemented at the local level, laying a foundation for capacity building.

De Pinto et al. (2017) identify four elements that were fundamental to conducting the research needed to formulate the NDC: sustained interaction and openness to all stakeholders; usable science and willingness to break down institutional and interdisciplinary barriers; one institution acted as the leader of the project; and flexibility and availability of resources. The author describes that during the process, multiple meetings were held with policymakers, interest groups, and activists to discuss the objectives, methods, and results of the process, and there was also a willingness to consider different approaches to the problem. In this way, a degree of trust was generated between the parties that improved collaboration between disciplines and institutions; this multidisciplinary approach built a sufficiently credible and legitimate scientific information base for the formulation of the process. The parties developed the ability to operate beyond institutional boundaries and, therefore, maintained access to sufficient resources (funding, staff, time) allowed for full collaboration.

Nevertheless, obtaining quality data and information for monitoring climate action goals remains a problem in Colombia (De Pinto et al., 2017).

As a result, Colombia's NDC is recognized as ambitious and comprehensive. It is not only ambitious in terms of the goals and commitments it sets but also in the implementation mechanisms it proposes at the sectoral and territorial levels (Baumgartner et al., 2022). Colombia effectively involved the ministries of planning and finance in the formulation of NDCs. The mobilization of resources from the private sector and preventive planning make it possible to establish a long-term climate strategy on which the country can build for future NDCs (Cárdenas et al., 2021). Despite the hierarchical nature of the system, there are also increasing levels of participation and mechanisms to promote it, reducing mistrust, particularly at the local level.

Brazil

Climate institutions in Brazil were initially developed in line with international negotiations, similar to the case of Chile, where international influences prevailed when it came to setting the process in motion. In the first stage, from 1990 to 2000, issues related to climate change were the responsibility of the Ministry of Science and Technology and Innovation (MCTI) and the Ministry of Foreign Affairs, which determined the nature of the climate problem (Hochstetler, 2021). Brazil was a participant in the negotiations that founded the Clean Development Mechanism (CDM), and it was the MCTI that was in charge of reporting and compiling the country's official climate data. This has influenced, for example, the long-held perception that deforestation, an environmental problem of great national concern, is separate from the climate change issue.

From 2003 to 2010, under Lula's first presidency, Brazilian climate policy went through a second stage, in which the Ministry of the Environment acquired greater relevance in climate issues, beginning to have better sources of information for planning and monitoring. In particular, the years 2009–2010 were the time when Brazil reached its maximum climate commitment with rapid and profound progress in the field (Franchini et al., 2020). The Climate Change Law and the National Climate Change Policy were approved, which established the federal bureaucratic structure to address climate change, by creating a carbon market, sectoral mitigation plans, among others. This was the first general climate law in Latin America and one of the first outside the OECD. Franchini et al. (2020) highlight two elements: first, several small-scale initiatives were deployed without attracting political attention; and second, various responsibilities for climate issues were assigned to institutions designed for other purposes. In other words, climate policy in Brazil lies on institutions not specifically designed for that purpose (Hochstetler, 2021).

In a third stage, which covers the period from 2011 to 2021, progress on climate issues suffered a setback linked to the conflict with the agriculture sector and the consequences it has in terms of deforestation, the reduction of the budget allocated to climate planning, and the weakening of support for climate institutions (Hochstetler, 2021). For example, President Dilma Rousseff (2011–2016) eliminated the gasoline tax on the same day that the Rio+20 negotiations concluded, while her successor, Michel Temer (2016–2018), was closely linked to the oil industry and agribusiness; likewise, Jair Bolsonaro (2019–2022) claimed on several occasions that climate change had no scientific basis and that it was a commercial game (Franchini et al., 2020).

It is important to understand the emergence of the conflict between climate planning and agribusiness interests. During the first government of president Lula da Silva (2003–2011), a series of initiatives were implemented to address deforestation, and these were strongly rejected by the so-called ruralist caucus in Congress,[6] linked to the powerful agricultural sector. Then, both presidents Rousseff and Temer cut the budget for climate action in response to such pressures. Likewise, President Jair Bolsonaro was elected, directly representing the "ruralist" sector. Thus, in contrast to the Chilean case in which the productive sectors chose to participate in decision making around climate policies to influence their content, in Brazil a more radical position was adopted that directly sought to undermine existing policy and stop further developments.

This institutional dynamic affected climate planning and its formulation, as can be seen in the design process of Brazil's National Adaptation Plan (NAP). It had an evidence-based perspective and a working group coordinated by the Ministry of Environment in 2013. While this group consisted mainly of ministry officials, other public institutions, the scientific community, NGOs, and the private sector also participated in meetings (Milhorance et al., 2021). Funding was also obtained from the federal government and international agencies. Originally, the NAP tried to implement sectoral risk management strategies in a coordinated manner, including agriculture; biodiversity and ecosystems; cities; natural disasters; industry and mining; infrastructure; vulnerable population; water resources; health; food safety; and coastal areas (Milhorance et al., 2021). Accordingly, a technical advisory body and another group were established to develop a monitoring system (in 2017). The aim was to incorporate adaptation targets into existing policies and instruments, rather than new ones.

However, a number of factors caused the NAP to lose political support and its priority position on the agenda (Milhorance et al., 2021). First, Brazil faced a political and economic crisis during the government of Dilma Rousseff that led to budget cuts, reduced requirements for environmental licenses, reduced protected areas, and a strengthened conservative political coalition. The Ministry of Environment failed to get all the sectors involved to incorporate adaptation as a planning priority, each having its own perception of the problem and needs. Therefore, after the publication of the NAP, each sector kept its own priorities intact.

As in the cases of Chile and Colombia, MAPS also provided support to Brazil to carry out an information-gathering process involving a variety of actors and sectors. Through this program, Brazil built models to identify the cost-benefit of actions to reduce GHG emissions. However, during the process it was difficult to reach a broad consensus among stakeholders on how to develop mitigation scenarios, and therefore, public officials decided to start another process to generate the information required to formulate the NDC (De Pinto et al., 2017).

Do Nascimento Nadruz et al. (2018) state that there is a gap between climate policy and regional and sectoral planning in Brazil. According to the authors, Law 12.187/2009 and Federal Decree 7390/2010, which contain the National Policy for Climate Change and deployed various measures to fulfil GHG reduction goals, have a very weak influence on planning formulation, resulting in low implementation of Environmental Assessment Service (SEA) guidelines. Although climate change-related issues should be integrated into plans in various sectors through the SEA, the Brazilian SEA considers these issues in a very limited way (Do Nascimento Nadruz et al., 2018). Nor is there any recognition of a dialogue between climate change and other environmental policies. Although the law explicitly recommends the participation of the SEA in the planning process, its use is voluntary (Do Nascimento Nadruz et al., 2018).

Another element identified as deficient for planning is the availability of information (Torres et al., 2021) on impacts, vulnerability and resilience analysis, long-term projections, historical climate information, and so on. In the same vein, there is a lack of greater centralization and adaptation of its language for civil servants and public administrators.

Finally, due to the decentralized nature of climate planning in Brazil, regulations tend to be nonbinding, nonsanctionable, and without a specific deadline for implementation (Billi et al., 2020). This strengthens the argument of Garín (2018), who warns that countries with a federal regime will face greater obstacles to make environmental and/or climate competencies compatible in each regional state.

Despite these issues, there is a chance that Brazil can correct course and return to the forefront of climate planning in the region. Indeed, since Lula's re-election as president in 2023, the government's intentions to "correct" or reverse some changes made by the Bolsonaro government were communicated. For example, the Inter-Ministerial Committee on Climate Change (CIM) approved several resolutions, including correcting the objectives of the NDC that the previous government had changed (Ministério do Meio Ambiente e Mudança do Clima, 2023).

Uruguay

Uruguay is a unitary and decentralized state with 19 departments, each with its own administration. There are also municipalities, which do not necessarily correspond to the capital of a department (Piperno et al., 2019). Through Law 16.517 of 1994 and Law 17.279 of 2000, Uruguay ratified and entered the United Nations Framework Convention on Climate Change (UNFCCC) and the Kyoto Protocol, respectively. The concept of sustainable development was incorporated into the General Law on Environmental Protection of 2000, as well as management principles and instruments that served as the basis for climate issues (Stuhldreher, 2021). Thus, departments and municipalities have climate measures for emergency situations, replacement of technologies, and infrastructure in their local plans (Goñi Mazzitelli et al., 2019).

In this country, climate change has caused an increase in sea level that has affected coastal areas, and it has also led to an increase in the duration of rains, droughts, and heat waves that leave the health of the population vulnerable. Stuhldreher (2021) points out that, despite these circumstances, Uruguay has a low level of risk according to the Climate Change Vulnerability Index (IVCC). This can be explained by the fact that Uruguay has a high adaptive capacity, which can be attributed to a well-developed institutional system for climate decision making. In addition, it is necessary to take into account that Uruguay's geography is characterized by pampas and plains that facilitate agricultural and cattle-raising activities. In this sense, with agriculture being the country's main productive sector, the Uruguayan government has developed systems to adapt and mitigate rural territories, thus protecting them from climatic events (Goñi Mazzitelli et al., 2019). In contrast to the Brazilian case, given that agriculture does not require the clearing of native forests for its expansion, the sector has not adopted an opposing position to climate action; rather, it has sought innovation to address its challenges.

The Ministry of Housing, Territorial Planning, and Environment (MVOTMA) was the national authority in charge of applying the Kyoto Protocol, which strengthened capacities to develop GHG inventories and to implement mitigation and adaptation pilot projects (Goñi Mazzitelli et al., 2019). Uruguay has worked to develop interinstitutional networks that allow different entities to collaborate to carry out particular tasks, exchange information, and improve the follow-up process. Piperno et al. (2019) point to the National System for Response to Climate Change and Variability (SNRCC) created in 2009, the National Emergency System (SINAE), the process of formulating National Adaptation Plans, and to some extent, the National Water Directorate (DINAGUA) as examples of networking. One of the objectives of these interinstitutional networks is to address the deficiency in the quality and quantity of information available to carry out planning and management activities (Piperno et al., 2019).

Notwithstanding, the rigidity of government structures and the dealignment between the functioning and deadlines and political times are an obstacle. Villamizar et al. (2016) identify the deficiency in information and efficient monitoring systems as a limit to adaptation in coastal countries such as Uruguay, adding the poverty of territories as an aggravating factor. The low position of climate change in the country's agenda is also problematic.

While it is true that Uruguay's climate governance and policies have been mostly evaluated positively since the creation of the SNRCC in 2009, showing a high capacity for decision making, the need to improve coordination between levels of government persists. There are several disputes between ministries of the same rank within the SNRCC and a lack of participation of the territories within it and independent of the IC (Stuhldreher, 2021). Likewise, as is often the case in unitary countries, there is a strong centralism in the public policy process, and therefore a deficiency in the articulation of departments that hinders collective action. This characterization shows that, although there is a high state capacity coming from a climate institutionality that delimits policy, at the subnational level this capacity is uneven. Such heterogeneity means that, despite the fact that decisions are made in line with international standards, civil participation and monitoring capacity are deficient.

Finally, a variation between regions is identified in the way in which different actors interact with institutions. In places with greater activism, there is greater commitment to institutional processes and participation in planning is more active. In sectors where there is a lack of knowledge about the functioning of these decision-making processes, even among incumbent actors, the fragmentation of institutions and planning is accentuated. In general, problems with the availability of information occur at all levels and dimensions, and there are no clear guidelines governing citizen participation (Stuhldreher, 2021). Private actors charge that they also cannot have information that allows them to analyze the "implications and opportunities" of climate change. Even among subnational governments, there is a deficiency in channels of communication and dissemination of information. Although the evaluation of Uruguayan climate policies is overall positive, the shortcomings of information and communication systems prevail and prevent the reduction of gaps between territories.

Conclusion

In this chapter, we have advanced the analysis of climate planning in Latin America. For this, we focused on NDCs. The NDCs contain the mitigation and adaptation measures that each country promises to comply with, and thus represent an overview of planning for climate action. The NDC makes it possible to distinguish the planning component from climate action in general and, since they are formulated based on international standards, they facilitate the process of evaluating the variegated advancement by countries.

We have analyzed the differences in the progress of climate planning in Latin America using data from the LEDS LAC and proposed a categorization of this advance. We have observed a negative relationship between emissions and planning progress, and a positive relationship between planning progress and GDP per capita and between planning progress and state capacity. Still, there are meaningful differences in the latter that we explore in detail. We selected four country cases that represent different combinations of state capacity and planning progress: Chile, Colombia, Brazil, and Uruguay.

In all four cases, different levels of deficiency of state capacity were identified with respect to participation and information generation. Chile and Uruguay have gradually incorporated participation mechanisms, but as unitary regimes their deployment remains limited.

In the case of Uruguay in particular, conflicts between the participating entities (e.g., different views between ministries) are mentioned. In the cases of Brazil and Colombia, it is observed that the uneven distribution of resources between territories and distrust in institutions due to corruption are obstacles to participation. With regard to information systems, Chile still needs to improve the process through which information is collected and disseminated to the entities that need it. Similarly, in Uruguay, deficiencies are pointed out in the channels of communication between institutions and between the private sector and the government, with fragmented institutions. For Brazil and Colombia, it is necessary to improve mechanisms for generating information in a generalized way, reducing the gap between territories. As mentioned, centralized states are more able to incorporate climate action into public policies with fewer obstacles, while federal systems have the task of balancing the environmental competencies of each province for the correct functioning of a policy. Regardless, change in public policy does not imply effective implementation, and the gap in the capacities of each province or municipality is yet the main challenge to be solved.

We also identified key factors to understand the greater progress in Chile and Colombia, compared to Brazil and Uruguay. First, external support has helped build state capacity to carry out planning processes, facilitating multistakeholder and multisector dialogue and producing reliable information. In turn, participation fosters the diversity of co-production, building capacity to solve local problems and give greater legitimacy to the process.

Particularly important from this point of view was the participation in the MAPS project, which helped address important gaps in state capacity in Chile and Colombia. The information generated in the program and the experience acquired by the participants were used for the formulation of the NDC in Chile and Colombia. This allowed them to make significant progress in their climate planning. As for Brazil, although it also collaborated with MAPS, the information-gathering process was weakened due to the lack of stakeholder consensus in the previous process. Uruguay, however, was not part of the project, which explains why the participatory processes they carried out lacked clear and concrete guidelines. Thus, establishing a coordinating institution has the potential to allow access to constantly updated data, facilitate collaboration among actors, and, most importantly, provide stability and projection to the capacity-building process needed for climate planning.

The interests of productive sectors and the political willingness to support climate action initiatives also played an important role. With respect to the government's initiative, the government's position and support for climate action allowed for continuity between governments in the Chilean case and led to setbacks in Brazil. This is also linked to the kind of climate problems that are important for the country and the predominant productive sectors. In Chile, the productive sectors participated in various processes of discussion and then in the formulation of the NDC, while in Brazil they represented opposition to these processes and then modified certain elements of the NDC. Even so, it stands out from the latter case that the responsibilities for climate action fall on entities that are not necessarily related to climate change. In the case of Chile, the participation of several productive sectors has allowed planning to have a strong component in cost analysis and financing, so its planning is usually evaluated as concrete and ready to be implemented. In this vein, we believe that more research should be devoted to understanding the role of the private sector in promoting, withholding, or blocking climate planning at different levels (e.g., Madariaga et al., 2021).

Climate planning still faces important challenges. Although it has significantly advanced through the efforts of Latin American states at different levels, as we have shown in this chapter, it still needs to prove that it is capable of producing concrete and relevant outcomes in terms of climate change mitigation and adaptation. Will climate planning generate actions that effectively curb CO_2 emissions and prevent major environmental crises in the future? The study of environmental policy and politics can help answer this question. This is why keeping track with planning actions, and especially the results of their implementation in terms of climate mitigation and adaptation, is an important agenda for the study of climate action in Latin America and beyond.

Acknowledgments

The authors acknowledge financial support from the following grants: Fondecyt 1220048, Fondecyt 1240812, and COES (ANID/FONDAP/1523A0005).

Notes

1. Countries include Argentina, Bolivia, Brazil, Chile, Colombia, Costa Rica, Cuba, the Dominican Republic, Ecuador, El Salvador, Guatemala, Honduras, Mexico, Nicaragua, Panama, Paraguay, Peru, Uruguay, and Venezuela.
2. https://ndc-lac.org/es/paises
3. Earlier stages are not necessary to comply with more advanced ones; therefore, despite having zero points in the preparation stage, a country can still have points in higher stages. For example, Honduras has zero points on stage 1 (elaboration of National Inventory System and Analysis of spending and climate investment). However, the climate change law, which does not require these components, is complete, and therefore has 7 points on stage 2.
4. A myriad of other factors could have been considered, but they are beyond the scope of our mostly descriptive analysis. Further exploring the drivers of climate planning in Latin America is an important agenda that entails looking at other factors such as political orientation of the government or level of democracy, the presence of green civil society organizations or social movements, and being part of international environmental treaties, among others.
5. Available at www.govindicators.org
6. More detailed information on the "ruralist" caucus in Hochstetler (2021), Sauer (2019), and Aamodt (2018).

References

Aamodt, S. (2018). The ability to influence: A comparative analysis of the rise of advocacy coalitions in Brazilian climate politics. *Review of Policy Research*, *35*(3), 372–397. https://doi.org/10.1111/ropr.12282

Aedo, M., & Parker, C. (2020). Funcionarios públicos y evaluación ambiental en Chile: tensiones en la construcción de una gobernanza ambiental democrática. *Revista mexicana de ciencias políticas y sociales*, *65*(239), 379–393. https://doi.org/10.22201/fcpys.2448492xe.2020.239.67896

Almeida, P., González Márquez, L. R., & Fonsah, E. (2024). The forms of climate action. *Sociology Compass*, *18*(2), e13177. https://doi.org/10.1111/soc4.13177

Arriagada, R., Aldunce, P., Blanco, G., Ibarra, C., Moraga, P., Nahuelhual, L., O'Ryan, R., Urquiza, A., & Gallardo, L. (2018). Climate change governance in the Anthropocene: Emergence of polycentrism in Chile. *Elementa: Science of the Anthropocene*, *6*, 68. https://doi.org/10.1525/elementa.329

Avendaño Castro, W. R., Luna Pereira, H. O., & Rueda Vera, G. (2021). Gobernanza climática y planificación del desarrollo: una mirada desde la academia. *Revista Boletín Redipe*, *10*(11), 529–543.

Barton, J. R. (2013a). Climate change adaptation and socio-ecological justice in Chile's metropolitan areas: The role of spatial planning instruments. In C. Boone & M. Fragkias (Eds.), *Urbanization and sustainability. Human-environment interactions* (Vol. 3, pp. 137–157). Springer. https://doi.org/10.1007/978-94-007-5666-3_9

Barton, J. R. (2013b). Climate change adaptive capacity in Santiago de Chile: Creating a governance regime for sustainability planning. *International Journal of Urban and Regional Research*, *37*(6), 1916–1933. https://doi.org/10.1111/1468-2427.12033

Baumgartner, L., Carman, R., & Jo, Y. (2022). *El Estado de la Ambición Climática*. Programa de las Naciones Unidas para el Desarrollo (PNUD).

Billi, M., Delgado, V., Jiménez, G., Morales, B., Neira, C. I., Silva, M. I., & Urquiza, A. (2020). Gobernanza policéntrica para la resiliencia al cambio climático: análisis legislativo comparado y Ley Marco de Cambio en Chile. *Estudios Públicos*, *160*, 7–53. https://doi.org/10.38178/07183089/1028191015.

Billi, M., Moraga, P., Aliste, E., Maillet, A., O'Ryan, R., Sapiains, R., Bórquez, R., Aldunce, P., Azócar, G., Blanco, G., Carrasco, N., Galleguillos, M., Hervé, D., Ibarra, C., Ibarra, C., Gallardo, L., Inostroza, V., Lambert, F., Manuschevich, D., & Mart, A. (2021). *Gobernanza Climática de los Elementos*, 67. https://www.cr2.cl/wp-content/uploads/2021/08/Informe_a_las_Naciones_Gobernanza_de_los_Elementos.pdf

Calfucoy, P. (2016). Building institutional capacities for climate change policies: The case of MAPS Chile. *Development*, *59*(3–4), 335–341. https://doi.org/10.1057/s41301-017-0103-z

Campos Pérez, J. E. (2019). Planning for climate change in Cartagena, Colombia: Institutionalizing alternative approaches. *Economía & Región*, *6*(2), 53–96. https://revistas.utb.edu.co/economiayregion/article/view/27

Cárdenas, M., Bonilla, J. P., & Brusa, F. (2021). *Políticas climáticas en América Latina y el Caribe: casos exitosos y desafíos en la lucha contra el cambio climático*. Banco Interamericano de Desarrollo (BID).

De Pinto, A., Loboguerrero, A., Londoño, M., Ovalle Sanabria, K., & Suarez Castaño, R. (2017). Informing climate policy through institutional collaboration: Reflections on the preparation of Colombia's nationally determined contribution. *Climate Policy 18*(5), 612–626. https://doi.org/10.1080/14693062.2017.1321521

Do Nascimento Nadruz, V., Lucia Casteli Figueiredo Gallardo, A., Montaño, M., Ramos, H. R., & Ruiz, M. S. (2018). Identifying the missing link between climate change policies and sectoral/regional planning supported by strategic environmental assessment in emergent economies: Lessons from Brazil. *Renewable and Sustainable Energy Reviews*, *88*, 46–53. https://doi.org/10.1016/j.rser.2018.02.006

Estenssoro, F. (2012). La dimensión ambiental de la política exterior de Chile. In C. Ross & M. Artaza (Eds.), *La Política Exterior de Chile de los últimos 20 años* (pp. 335–365). RIL Editores.

European Commission. (2020). *Progress on climate action in Latin America: Nationally determined contributions as of 2019*. EUROCLIMA+ Programme, DG for International Development and Cooperation—EuropeAid, European Commission, Brussels, Belgium.

Fisher, D. R. (2024). *Saving ourselves: From climate shocks to climate action*. Columbia University Press.

Franchini, M., Evangelista Mauad, A., & Viola, E. (2020). De Lula a Bolsonaro: una década de degradación de la gobernanza climática en Brasil. *Análisis Político*, *33*(99), 81–100.

Füssel, H. M. (2007). Adaptation planning for climate change: Concepts, assessment approaches, and key lessons. *Sustainability Science*, *2*(2), 265–275. https://doi.org/10.1007/s11625-007-0032-y

Goñi Mazzitelli, A., Roche, I., Sierra, P., & Gadino, I. (2019). Incorporating resilience and adaptive strategies to climate change in urban and territorial planning in Uruguay. In C. Henríquez & H. Romero (Eds.), *Urban climates in Latin America* (pp. 355–377). Springer. https://doi.org/10.1007/978-3-319-97013-4_14

Hochstetler, K. (2021). Climate institutions in Brazil: Three decades of building and dismantling climate capacity. *Environmental Politics 30*(suppl.1), 49–70. https://doi.org/10.1080/09644016.2021.1957614

Howland, F., & Francois, J. (2022). Disaster risk management, or adaptation to climate change? The elaboration of climate policies related to agriculture in Colombia. *Geoforum*, *131*, 163–172.

LEDS LAC. (2020). LEDSenLAC 2019: Hacia un desarrollo resiliente y bajo en emisiones en Latinoamérica y el Caribe: Progreso en la implementación de las Contribuciones Nacionalmente

Determinadas (NDC). https://ledslac.org/wp-content/uploads/2020/04/LEDSenLAC_2019_v1.3.pdf

LEDS LAC. (2023). LEDSenLAC 2022: Hacia un desarrollo bajo en emisiones y resiliente al clima en Latinoamérica y el Caribe: Avances en las estrategias nacionales. https://www.ledslac.org/download/26198/?tmstv=1679074430

Lehmann, P., Brenck, M., Gebhardt, O., Schaller, S., & Süßbauer, E. (2015). Barriers and opportunities for urban adaptation planning: Analytical framework and evidence from cities in Latin America and Germany. *Mitigation and Adaptation Strategies for Global Change*, *20*, 75–97. https://doi.org/10.1007/s11027-013-9480-0

Lucas Garín, A. (2018). Más de una década de cambio climático en Chile: análisis de las políticas domésticas. *Boletín mexicano de derecho comparado*, *51*(153), 587–625. https://doi.org/10.22201/iij.24484873e.2018.153.13651

Madariaga, A. (2019). From "green laggard" to regional leader: Explaining the recent development of environmental policy in Chile. *Bulletin of Latin American Research*, *38*(4), 453–470. https://doi.org/10.1111/blar.12841

Madariaga, A., Maillet, A., & Rozas, J. (2021). Multilevel business power in environmental politics: the avocado boom and water scarcity in Chile. *Environmental Politics*, *30*(7), 1174–1195. https://doi.org/10.1080/09644016.2021.1892981

Maillet, A., & Carrasco, S. (2021). Between environmental subsystem change and extractive regime resilience. In G. Damonte & B. Schorr (Eds.), *Andean states and the resource curse* (pp. 158–176). Routledge. https://doi.org/10.4324/9781003179559-11

Mendes, F., Romero, H., & Ferreira da Silva, D. (2020). Cambio Climático adverso provocado por la urbanización sin planificación ni evaluación ambiental en Santiago de Chile. *Revista de geografía Norte Grande*, *77*, 191–210. https://dx.doi.org/10.4067/S0718-34022020000300191

Méndez, M. (2020). *Climate change from the streets: How conflict and collaboration strengthen the environmental justice movement*. Yale University Press.

Milhorance, C., Sabourin, E., Chechi, L., & Mendes, P. (2021). The politics of climate change adaptation in Brazil: Framings and policy outcomes for the rural sector. *Environmental Politics*, 31(2), 183–204. https://doi.org/10.1080/09644016.2021.1907097

Mimura, N., Pulwarty, R. S., Duc, D. M., Elshinnawy, I., Redsteer, M. H., Huang, H. Q., Nkem, J. N., & Sanchez Rodriguez, R. A. (2014). Adaptation planning and implementation. In C. B. Field et al. (Eds.), *Climate change 2014: Impacts, adaptation, and vulnerability. Part A: Global and sectoral aspects* (pp. 869–898). Contribution of Working Group II to the Fifth Assessment Report of the Intergovernmental Panel on Climate Change. Cambridge University Press.

Ministério do Meio Ambiente e Mudança do Clima. (2023). Comitê Interministerial sobre Mudança do Clima aprova cinco resoluções. Publicado el 15 de septiembre de 2023. https://www.gov.br/mma/pt-br/noticias/comite-interministerial-sobre-mudanca-do-clima-aprova-cinco-resolucoes

Múnera, C., & van Kerkhof, L. (2019). Diversifying knowledge governance for climate adaptation in protected areas in Colombia. *Environmental Science & Policy*, *94*, 39–48. https://doi.org/10.1016/j.envsci.2019.01.004

NDC Partnership. (2023). Finance strategy for the implementation of the NDC Partnership Work Program 2021–2025. https://ndcpartnership.org/sites/default/files/2023-09/ndc-partnership-finance-strategy.pdf

Orihuela, J. C. (2014). Converging divergence: The diffusion of the green state in Latin America. *Studies in Comparative International Development* *49*, 242–265. https://doi.org/10.1007/s12116-013-9147-6

Piperno, A., Benenati, S., Emanueli, C., Cuadrado, A., Martínez, J. P., García, P., & Alonso, D. (2019). Ciudades resilientes en LAC: aportes desde la gestión de las aguas urbanas en Uruguay. *Aqua-LAC*, *11*(2), 1–16. https://doi.org/10.29104/phi-aqualac/2019-v11-2-01

Preston, B. L., Westaway, R. M., & Yuen, E. J. (2010). Climate adaptation planning in practice: An evaluation of adaptation plans from three developed nations. *Mitigation and Adaptation Strategies for Global Change*, *16*(4), 407–438. https://doi.org/10.1007/s11027-010-9270-x

Rojas, Y., Calero, D. & Carrión, A. (2021). *Planificación participativa y gobernanza climática: gestión local del cambio climático en ciudades de América Latina [guía pedagógica]*. FLACSO Ecuador; IDRC-Canadá.

Samaniego, J., Alatorre, J., Ferrer, J., Reyes, O., Arpaia, L., & Muñoz, L. (2019). *Panorama de las contribuciones determinadas a nivel nacional en América Latina y el Caribe, 2019: avances para el cumplimiento del Acuerdo de París (LC/TS.2019/89-P)*. Comisión Económica para América Latina y el Caribe (CEPAL).

Sauer, S. (2019). Rural Brazil during the Lula administration: Agreements with agribusiness and disputes in agrarian policies. *Latin American Perspectives*, *46*(4), 103–121. https://doi.org/10.1177/0094582X16685176

Stuhldreher, A. (2021). Cambio climático y gobernanza multinivel en Uruguay: percepciones tras diez años del Sistema Nacional de Respuesta al Cambio Climático. *Perfiles latinoamericanos*, *29*(58), 00004. https://doi.org/10.18504/pl2958-004-2021

Thomas, A., Shooya, O., Rokitzki, M. Bertrand, M., & Lissner, T. (2019). Climate change adaptation planning in practice: Insights from the Caribbean. *Regional Environmental Change*, *19*, 2013–2025. https://doi.org/10.1007/s10113-019-01540-5

Torres, P. H. C., Jacobi, P. R., Momm, S., & Leonel, A. L. (2021). Data and knowledge matters: Urban adaptation planning in São Paulo, Brazil. *Urban Climate*, *36*, 100808. https://doi.org/10.1016/j.uclim.2021.100808

Valdivieso Cervera, G., & Sandoval, A. (2021). La coproducción de la adaptación al cambio climático en Colombia (2012–2019). In M. I. Gomez Lee (Ed.), *Ahora o nunca: gobernanza, coproducción y bioeconomía contra el cambio climático* (pp. 240–279). Universidad Externado de Colombia.

Villamizar, A., Gutiérrez, M. E., Nagy, G. J., Caffera, R. M., & Leal Filho, W. (2016). Climate adaptation in South America with emphasis in coastal areas: The state-of-the-art and case studies from Venezuela and Uruguay. *Climate and Development*, *9*(4), 364–382. https://doi.org/10.1080/17565529.2016.1146120

CHAPTER 29

Climate Action in Central Asia

NYGMET IBADILDIN AND JAMILYA UKUDEEVA

Introduction

This chapter examines the climate crisis faced by the five Central Asian nations—Kazakhstan, Kyrgyzstan, Tajikistan, Turkmenistan, and Uzbekistan—and actions taken by the government and civil society in response to the climate crisis. These countries face severe climate crises, including acute water scarcity, desertification, glaciers melting, weather extremes, and heavy air pollution. Methane emissions, population growth, industrial waste, and intensive exploitation of subsoil resources are the primary causes of the region's climate crisis. Urgent climate mitigation actions are needed as projections indicate that by the end of the 21st century, the temperature in the region will rise by 7°C relative to the 1950–1981 period (Central Asian Analytical Network, 2016). Central Asia, a region in the heart of the Eurasian continent, shares borders with Russia to the North, China to the East, the Caspian Sea to the West, and Afghanistan and Iran to the South. The region features expansive deserts, vast steppes, and towering mountain ranges. The climate is consistently arid across the region, characterized by cold winters and hot summers. Forests grow primarily in the northern and eastern parts of Central Asia. The Amu Darya and the Syr Darya Rivers, originating in the Pamir and Tian Shan Mountains, are the two principal rivers that connect the region. Mountains and glaciers are the primary suppliers of drinking water. A significant proportion of the region's rivers originate beyond the political borders of the five Central Asian states, highlighting the transboundary nature of the region's water resources (Janusz-Pawletta & Gubaidullina, 2015). The largely impoverished population remains mostly unaware of the climate crisis threats. Zhusipbek and Nagayeva (2018) write that "the prevailing reality in Kazakhstan is the low level of environmental awareness and lack of pro-environmental thinking among the majority of the population. Environmental protection is perceived as something distant ("somewhere in the West"), incomprehensible (pursued by "freaks"), or irrelevant to daily life (a concern for those with "nothing better to do"). An analysis of studies published between 1991 and 2021 concluded that the subject of climate change has been mostly overlooked by scientists in the region (Vakulchuk et al., 2022). The lack of information on climate change in the local languages (e.g., Uzbek and Tajik languages) is particularly acute (Times of Central Asia, 2024a) and violates the principle of environmental justice.

The Freedom House report (Smeltzer & Karppi, 2024) classifies all five nations as consolidated authoritarian regimes. Restrictions on freedom of speech and assembly impede social mobilization even on benign issues such as climate crisis and environment (Critical Ecosystem Partnership Fund, 2017, pp. 96–97; Ibraeva, 2024). Peyrouse (2022) argues that Central Asian authoritarianism contradicts official climate-friendly policies by "minimizing, denying, or even censoring" information on climate change. With the region's average score of 27 out of 100 on the corruption scale, all five nations struggle with systemic corruption (Transparency International, 2023). For example, at the domestic level, local laws protecting biodiversity and air quality are undermined by bribes that allow illegal hunting, logging, and construction. Domestic actors with self-serving motives sabotage investments in green projects (Critical Ecosystem Partnership Fund, 2017). The authorities consistently prioritize political stability and other immediate concerns over climate change issues. This approach reflects the balancing act of addressing multiple national priorities while navigating the complexities of the climate crisis. Skalamera (2024) highlights the conflicting energy policies of the two Central Asian petrostates, Kazakhstan and Uzbekistan. Internationally, their governments make ambitious commitments to combat climate change, while domestically, they continue to invest in fossil fuels. Skalamera (2024) argues that the Central Asian governments' bold pledges to achieve carbon neutrality are disingenuous due to the influence of political and business elites with strong ties to the fossil fuel sector. Skalamera holds that the governments of these states address the climate crisis through energy diversification rather than decarbonization, prioritizing the maintenance of their fossil fuel-dependent economies over climate action.

Current Climate Issues

Central Asian region's urgent climate change issues include extreme weather events, water shortages, soil degradation, air pollution, glacial melting, and declining biodiversity. The ClimaMeter's investigation of weather anomalies in Central Asia identifies a trend marked by increasing frequency and intensity of warm spells during winter and heat waves in summer (ClimaMeter, 2024). The study ascribes the region's extreme weather events to human-caused climate change. Based on research from other parts of the world, ClimaMeter predicts disruptions in regional agriculture, animal life, and water supply. The Intergovernmental Panel on Climate Change (IPCC) also found that the temperatures in Central Asia rise faster than the global average, resulting in more desertification and glacial melting (Hoegh-Guldberg et al., 2018, p. 177). Extreme weather events cause disastrous floods and droughts, forcing people to relocate to safer areas. According to the World Bank's 2021 report (Clement et al., 2021), if governments fail to act promptly, the damages from droughts and floods in Central Asia are projected to reach 1.3% of gross domestic product (GDP) annually; crop yields could decrease by a staggering 30% by 2050, potentially resulting in about 5.1 million internal climate refugees.

Water Shortages

Water plays an important role in the region's economy as agriculture in Kyrgyzstan, Tajikistan, Turkmenistan, and Uzbekistan relies on irrigation systems, while hydropower stations generate about 90% of electricity consumed in Kyrgyzstan and Tajikistan (see Table 29.1). As a renewable energy source, hydropower offers a cleaner alternative to coal, significantly reducing greenhouse gas emissions and environmental pollution. Heavy reliance on hydropower as a source of electricity in Kyrgyzstan and Tajikistan makes water supply a sensitive political and economic issue. Considering the central role of water in the region's economy, water shortage is arguably the most significant threat facing the region. The water shortage stems from geographical, climatic, political, and managerial factors. Since the early 1990s, annual interstate disputes have occurred in winter when the upstream countries release water to generate hydropower for their own domestic heating needs, at times when it does not coincide with the irrigation needs of the downstream countries. The reverse situation happens in summer, when the upstream country collects water in the dams for its

TABLE 29.1 Sources of Consumed Electricity in Central Asia, 2022 (%)

	Fossil Fuels			Renewables	
Country	Oil	Natural Gas	Coal	Hydro	Wind and Solar
Kazakhstan	1	25	62	8	4
Kyrgyzstan	Negligible	1	13	86	Negligible
Tajikistan	Negligible	1	6	93	Negligible
Turkmenistan	Negligible	100	Negligible	Negligible	Negligible
Uzbekistan	1	82	8	9	Negligible

Source: From the International Energy Agency (n.d.a). Energy Systems of Eurasia. https://www.iea.org/regions/eurasia

future use in winter, while the downstream country demands the release of water for irrigation. Water disputes in the region's fertile Fergana Valley tend to escalate into ethnic and transboundary violence (Climate Diplomacy, n.d.; see chapter by Montoya in this volume). The frequency and intensity of water disputes point to the failure of the five Central Asian republics to develop a new energy and water-sharing network to replace the old resource-sharing system that collapsed along with the Soviet Union. Moreover, climate change and drought exacerbate water scarcity, further intensifying transboundary conflicts over limited water resources.

Fossil fuels, natural gas, and coal generate most of the region's electricity using outdated technology. The two most populous countries in the region, Kazakhstan and Uzbekistan, consume electricity generated primarily by fossil fuels—88% and 91%, respectively. In contrast, the less populous Kyrgyzstan and Tajikistan rely on hydropower and thus are less dependent on fossil fuels for electricity consumption. Only Kazakhstan has developed noticeable solar and wind production power, 4% of the total electricity consumption.

Air Pollution

The most common type of pollution in Central Asia is particulate matter (PM2.5). Exposure to PM2.5 has been associated with cardiovascular and respiratory diseases, particularly in children. Despite the decline in carbon emissions following the collapse of the Soviet Union in 1991 (Schierhorn et al., 2019), recent years have seen a resurgence in these emissions (Zhao et al., 2023). In 2021, four of the Central Asian capitals ranked within the top 44 most polluted cities globally (Tursumbayeva et al., 2023). While the region is a mix of urban and rural living (Tajikistan and Kyrgyzstan are less urbanized than the other three nations), the moderate rates of urbanization and population growth further drive up the air pollution rates. Reliance on government-subsidized fossil fuels to meet domestic energy needs drives the region's high pollution rates in the two largest countries of Central Asia—Kazakhstan and Uzbekistan. Much of the energy, especially heat, is still generated by outdated fossil fuel–burning plants, contributing to uncomfortable living conditions in urban areas, especially in winter. As shown in Table 29.2, fossil fuels are the region's primary energy source, with Tajikistan being the only country in the region where hydropower is the main energy source (42%). The watchdog group Global Energy Monitor reports that Central Asian countries, except Turkmenistan, are doubling down on the construction of coal-burning power plants

TABLE 29.2 Energy Supply by Source Type in Central Asia, 2022 (%)

Country	Oil	Natural Gas	Coal	Hydro
Kazakhstan	23	26	50	1
Kyrgyzstan	34	9	33	24
Tajikistan	30	4	24	42
Turkmenistan	13	87	Negligible	Negligible
Uzbekistan	10	83	6	1

Source: From International Energy Agency (n.d.a). Energy Systems of Eurasia. https://www.iea.org/regions/eurasia

and have not set a target year to phase out coal (Global Energy Monitor, 2024, pp. 5–6). While the transition to renewable energy sources has started, progress remains slow despite the region's vast solar and wind energy production potential.

Glacial Melting

Glacial melting represents a challenge specific to mountainous regions. Saidrasul Sanginov, Uzbekistan's expert on climate change, was cited as saying that more than 30% of the glacier mass has been lost in the last 40 years. He forecasts a decrease in the water flow of the Amu Darya and Syr Darya Rivers by 10% to 15% in the next 10 years (Olimova, 2023). The effects of melting glaciers are profound and far-reaching. Glacial melting disrupts ecosystems and biodiversity, impacts community livelihoods, and creates poverty and internal climate refugees. It also threatens food security and sustainable agriculture by affecting irrigation and water supply across the region. These cascading impacts underscore the urgent need for adaptive strategies to mitigate the climate crisis.

Biodiversity

Climate change and economic activities are increasingly threatening the biodiversity of Central Asia, which some experts suggest may already be irreversible (Olimova, 2023). The prolonged droughts in the region's southern parts have negatively affected key natural reserves, such as the Koytendag Reserve in Turkmenistan and the Amu Darya wetlands. Forests and steppes in Kazakhstan face increased wildfire risks due to rising temperatures (Novikov & Kato, 2024, pp. 4–6). Novikov and Kato are concerned that rising temperatures and evaporation will further lower water levels, harming the Caspian Sea's biodiversity. Despite these challenges, the region's biodiversity conservation efforts have had some notable successes. For example, species such as the snow leopard and saiga have shown signs of recovery, and deforestation projects have stabilized forests in some regions. The government remains the region's primary source of funding for conservation projects, covering salaries and operational costs. International organizations provide critical funding for new technologies and capacity-building efforts (Critical Ecosystem Partnership Fund, 2017). The report by Novikov and Kato (2024) highlights promising government efforts, such as developing national biodiversity strategies, increased participation in international agreements, climate action planning, and cooperation between countries.

Institutional Climate Actions

To evaluate the effectiveness of different climate crisis actions, scholars categorize climate actions into two types: institutional and noninstitutional (Almeida et al., 2023). Institutional actions refer to actions by governmental bodies, international organizations, and other formal institutions that involve policymaking, treaty signing, passing regulations to reduce greenhouse gas emissions, transitioning to renewable energy, and participating in cap-and-trade markets. These actions are typically supported by substantial financial resources and formal mandates. The most common form of institutional climate action is climate action planning, which is further categorized into mitigation, adaptation, and resiliency strategies. In contrast, noninstitutional climate actions are driven by grassroots actors outside the formal structures of government and large organizations, such as civil society activists, nongovernmental organizations (NGOs), private sector leaders, Indigenous groups, and faith-based groups. Often, noninstitutional forms of climate action are disruptive and leverage social media to mobilize support. Common forms of noninstitutional climate actions include transnational climate movements, local-level mobilizations against extraction, and fossil fuel divestment campaigns. Noninstitutional actions raise public awareness about the climate emergency and pressure governments to take institutional actions (Almeida & Stearns, 1998).

The Central Asian countries have taken a range of institutional actions, including adopting legal frameworks and signing international treaties to combat the climate crisis, such as the United Nations Framework Convention on Climate Change (UNFCCC), Kyoto Protocol, and Paris Agreement. All five countries participate in achieving the United Nations's Sustainable Development Goals. Kazakhstan, Turkmenistan, and Uzbekistan are among the signatories of the UN Economic Commission for Europe's (UNECE) Water Convention. All five nations have endorsed the Integrated Water Resources Management (IWRM) process to recognize the need for a coordinated approach to sustainable water governance. The region's record on interstate cooperation on climate crisis action is mixed. While there have been some positive collaborative efforts, such as regional initiatives for water resource management and environmental protection, the "Navruz" project to monitor water quality in Syr Darya and Amu Darya, and cooperation between countries' Academies of Science, these have often been hindered and in some cases entirely halted by political and economic tensions. Other positive signs of growing regional cooperation on climate change is the creation of the Central Asia Regional Environmental Center (CAREC) in 2001 and the signing of the "Green Agenda" program for Central Asia in 2022. However, achieving effective climate actions on the regional level remains a challenge due to conflicting national priorities and low and unclear levels of commitment to climate action.

Each country in Central Asia has enacted laws to address climate change, but still they vary on the levels of commitment to the climate agenda, the time frames, and indicators of climate actions. The main legal aspects in addressing the climate action in the five Central Asian republics include the following:

- Adaptation to climate change/crisis in general laws
- Water, floods, a deficit of water, and desertification
- Decarbonization of the economy and carbon markets
- Development of green financing systems
- Preservation of biodiversity

While the number of climate actions taken by the five governments at international and domestic levels is a positive trend, climate actions at the interstate level are lagging. The governments in the region face challenges in sharing data on glacial melting and aligning policies on water and energy management. The pursuit of self-reliance in energy and water resources hampers climate action and cooperation at the interstate level. Sebastien Peyrouse (2022) has documented the disappointing gap between government proclamations on climate action and the actual implementation of the projects.

Noninstitutional Climate Actions

Most Central Asian public views climate-related issues as distant and secondary to economic well-being and stability. Even though public awareness about the climate crisis is slowly growing, public participation in climate actions remains low, primarily due to the lack of economic resources and information on the climate crisis (Limon.kg, 2023). Restricted access to information on environmental violations and lack of freedom of speech and assembly halt local-level mobilization against harmful practices. Peyrouse (2022) points at the striking need for more dialogue on climate change between civil society and government. The region's most common type of noninstitutional climate action is conducting training workshops on sustainable farming practices. More disruptive climate actions, such as picketing or criticism of the government on social media, can be too costly for the participants, resulting in the dismissal from a job, being labeled as "an enemy of the people," or being subjected to excessive administrative inspections.

On the bright side, training workshops by local NGOs and international donors, along with the influence of social media, are slowly improving public awareness of the climate crisis. A new emerging type of noninstitutional climate action is climate crisis resilience campaigns, such as eco-activism, anti-littering campaigns, and participation in eco-fairs. Civil society and the business community lead the movement to promote the culture of reducing, recycling, and reusing household waste. Eco-activists, such as Maria Kolesnikova of MoveGreen in Kyrgyzstan and Pakizat Sailaubekova of Recycle Birge in Kazakhstan, use social media to promote sustainable lifestyles, such as bicycling to work and reducing plastic waste (Gizzat, 2023). Eco-entrepreneurs launch commercial projects with an environmental agenda, such as organic farming and open waste-free stores, such as HelloEco in Kazakhstan and Fluid in Kyrgyzstan (Tskhay, 2023). The issues of climate change and the environment resonate more with the younger demographic group. However, within the authoritarian political environment of the region—a region that experiences a pronounced "youth bulge" (Goldstone, 1993)—the ability of young people to effectively mobilize and advocate, not just on climate-related matters, but across a range of civic issues, appears to be considerably constrained.

Central Asia Country Case Studies

Kazakhstan

Kazakhstan's main regulatory act in environmental protection is the Environmental Code of the Republic of Kazakhstan, adopted in 2021. This code stipulates that Kazakhstan aims to reduce its carbon balance by 2030 by at least 15% from 1990 levels. To combat climate change, Kazakhstan has pledged to achieve the Sustainable Development Goals (SDGs). It has ratified the UNFCCC and the Paris Agreement, integrating these goals into its national policies. Specifically, Goal 13, which calls for urgent action to combat climate change and its impacts, is incorporated into government programs that monitor SDG progress (Bureau of National Statistics, n.d.). The Republic of Kazakhstan is a member of the Global Green Growth Institute (GGGI), an international organization dedicated to promoting sustainable development and green growth in developing countries. The GGGI achieves this mission through green economy projects, conducting research to foster resource-efficient investments and promoting innovation in production and consumption. Kazakhstan joined the GGGI as its 45th member in December 2022. At the World Climate Action Summit in December 2023, the president of Kazakhstan announced that the country would assume the chairmanship of the Fund for Saving the Aral Sea, underscoring Kazakhstan's commitment to combating climate change and natural disasters (Astana Times, 2024). President Tokayev further emphasized this commitment by supporting the United Nations' call for concrete action to preserve the environment for future generations. He highlighted Kazakhstan's leadership in the region, citing the country's status as the first in the region to ratify the Paris Agreement and pledge to achieve carbon neutrality by 2060.

Despite strengthening its long-term climate ambitions by committing to a net-zero target by 2060, Kazakhstan faces challenges meeting its climate goals. Projections by the Climate Action Tracker (CAT), a NGO that monitors 42 countries across four dimensions—policy and action, conditional NDC target, unconditional NDC target, and climate finance—show that under current policies, emissions are set to rise until at least 2035, leading CAT to assess Kazakhstan's climate action as "insufficient" or "weak." This rating is comparable with the rating assigned to the United States, Great Britain, Switzerland, and Australia (Climate Action Tracker, 2023). While insufficient, Kazakhstan's rating is considered more favorable than the "critically insufficient" assessments given to Iran, Turkey, and Russia and the "extremely insufficient" ratings assigned to South Korea and Egypt. Kazakhstan's low rating stems from the country's heavy reliance on fossil fuels for energy, industry, and exports, resulting in Kazakhstan ranking 16th globally in per capita emissions and fourth in emissions per GDP. To address this challenge, Kazakhstan has developed the "Concept of Transition to Green Economy," a long-term roadmap aimed at transforming all sectors of its economy toward sustainability (Garafutdinova & Tang, 2024). The comprehensive plan emphasizes optimizing resources, integrating new technologies and production methods, and adopting renewable energy sources. The plan identifies key areas for action, such as sustainable water management, efficient agriculture, energy conservation, efficiency improvements, waste management, air pollution reduction, and ecosystem conservation. Additionally, Kazakhstan established a carbon credit system, codified in the 2021 Environmental Code, which provides legal definitions and benchmarks on carbon units, quotas, trade, and thresholds (Government of the Republic of Kazakhstan, 2021).

Despite the presence of legal frameworks and international treaties, Kazakhstan continues to experience many human-caused climate-related disasters. In the summer of 2023, scientists detected a massive methane leak in western Kazakhstan (Silva et al., 2024). Satellite imagery captured the eruption of substantial methane plumes, a finding corroborated by experts from the Netherlands Institute for Space Research (SRON). This incident involved the release of a gas-water mixture, which escalated into an uncontrolled gas fountain. Other environmental disasters include the Semey nuclear testing site, the disappearance of the Aral Sea, and improper handling of radioactive and industrial tailings. These disasters not only have direct links to climate crises but also serve as reminders of past exploitive colonial practices dating back to the Soviet era.

The shrinking of the Aral Sea starkly illustrates the urgency of the water shortage plaguing Central Asia. The Aral Sea's disappearance is directly related to human activities and the broader climate change crisis (see Figure 29.1). Historically, the Aral Sea's disappearance

FIGURE 29.1. Water levels in the Almaty region are returning to those in previous years. This image, taken by the author Nygmet Ibadildin, was submitted to Kazakhstan's 2020 national contest, "Climate is changing. What about you?"

was driven by large-scale irrigation projects when the Soviet Union diverted the Amu Darya and Syr Darya Rivers for cotton and rice fields. Rising temperature, increased evaporation rates, intensified droughts, and reduced precipitation in the region driven by climate change made the Aral Sea shrink even further. As a result, while the Aral Sea's shrinking started due to poor water management, climate change has turned this disaster into a vicious cycle. The disappearance of such a large water reservoir has triggered a local climate crisis, resulting in salt storms, soil salinization, dust storms, loss of vegetation, and the expansion of desertified land (Columbia University, n.d.). Efforts to mitigate the situation include planting drought-tolerant trees in the dried-up seabed to combat desertification and building dams to retain the water level. However, the recent Kosh Tope water canal launch in Afghanistan, drawing water from the Pyanj River, could further exacerbate the water shortage issue. Replacing water-intensive cotton and rice cultivation with more water-efficient crops in Central Asia could create a more sustainable solution. Kazakhstan's assumption of the chairmanship of the International Fund for Saving the Aral Sea (IFAS) has allowed it to focus more intently on the problem of the Aral Sea. The Fund's activities are monitored and supported by the United Nations, the Shanghai Cooperation Organization, and environmental NGOs in Europe and Asia. The IFAS aims to address the Aral Sea crisis by supporting water management, environmental protection, and socioeconomic development projects.

Noninstitutional actions and adaptation practices often occur in cosmopolitan urban areas, driven by higher exposure to global trends, media, and social networks. The growing number of eco-fairs, book crossings, and clothing exchange events among young people exemplifies this trend. Authorities increasingly seek to align themselves with these grassroots movements, reflecting a growing influence of youth-led initiatives (Aladina, 2023). Tree-planting events under the World Cleanup Day campaign resonated with many residents of Astana, Kazakhstan's capital. With government support and media coverage, these grassroots events attracted volunteers, including high-ranking government officials (Government of Kazakhstan, 2020). Despite government involvement, these actions remain noninstitutional due to their sporadic grassroots nature, volunteer-based structure, and appeal to civic duty.

Kyrgyzstan

Kyrgyzstan's direct contribution to the global climate crisis is relatively small due to its reliance on hydropower and its small industrial sector. Kyrgyzstan has a comprehensive legislative framework to address climate change. The country ratified the UNFCCC in 2000, the Kyoto Protocol in 2003, and the Paris Agreement in 2019. It regularly participates in the UNFCCC's annual Conference of the Parties. International donors, such as the World Bank, Asian Development Bank, and United Nations Development Program, carry out multiple projects on climate change adaptation and mitigation. Kyrgyzstan's greenhouse gas (GHG) emissions are relatively minimal, comprising just 0.03% of the global GHG emissions. Kyrgyzstan's government has pledged to reduce its GHG emissions unconditionally (i.e., on its own without help from outside donors) by almost 16% and by 44% with international support. This pledge is part of Kyrgyzstan's broader strategy to achieve carbon neutrality by 2050 (UNDP, 2021). Kyrgyzstan's geography creates favorable conditions for renewable energy, but most of this potential remains untapped. The country has committed to expanding its renewable energy production capacity from the current 1.5% to 10% by 2030, which would help reduce reliance on polluting fossil fuels and improve air quality (United Nations, 2024).

Rising temperatures in the region have accelerated the melting of glaciers, a crucial water source for irrigation. The World Bank forecasts that by 2050, Kyrgyzstan's glaciers may shrink by 50% and disappear entirely by 2100 (Radio Free Europe/Radio Liberty, 2024). Since 70% of the country's farmland requires irrigation, the glacial retreat driven by climate change directly threatens Kyrgyzstan's water and food security (The World Bank, 2019). To mitigate the impact of glacial loss, the Kyrgyz government has adopted climate-focused policies, including the National Strategy on Climate Change (2013) and the National Water Resources Management Strategy (2017). Despite the limited state funding, scientists continue researching the effects of climate change using outdated research facilities and occasional grants from foreign counterparts (Kanbakova, 2024). Notably, in the early 2020s, business entrepreneurs and social activists used artificial intelligence (AI) technology and digital tools to launch the country's first climate monitoring initiatives, marking an advancement in climate adaptation (Internet Society Kyrgyz Chapter, n.d.; Kabar News, 2023). Around the same time, a diverse coalition of business activists created the Green Alliance association, unifying about 40 environmental NGOs to foster climate resilience through training, research, and social media campaigns (Green Alliance, n.d.).

Hydropower generates 86% of consumed electricity in Kyrgyzstan (see Table 29.1). The potential of Kyrgyzstan's rivers is estimated to be around 10 times higher than what is currently utilized (International Energy Agency, n.d.c). To achieve carbon neutrality, Kyrgyzstan aims to expand its hydropower capacity by seeking investments from international donors, foreign businesses, and regional neighbors (Shambetov, 2023). Kyrgyzstan already operates seven large-scale hydropower stations built between the 1970s and 1980s, all needing modernization. Lack of funds has delayed the construction of six new large-scale hydropower facilities for decades, while the operation of the existing hydropower dams and reservoirs is mired in regional disputes. As the primary manager of the reservoirs, Kyrgyzstan incurs substantial costs, while the benefits of the water supply extend beyond its borders. The Kyrgyz government's proposed cost-sharing mechanism has become a source of regional contention.

Fossil fuels account for three fourths of Kyrgyzstan's total energy supply, while hydropower generates the remaining one fourth (see Table 29.2). While planning to transition to cleaner energy, the government recognizes that drought and reduced winter snowfall make hydropower less reliable. To mitigate extreme weather fluctuations, Kyrgyzstan attracts foreign direct investments to diversify its energy sources by developing new coal mines (Kamchybekova, 2024), coal-burning thermal plants (Pavlenko, 2023), and the first nuclear power facilities (Cabar, 2024; Rickleton, 2024). Biomass remains an important fuel source in remote rural areas, although official data on its use are unavailable. Rural residents, particularly those practicing nomadic lifestyle, traditionally use biomass, such as dry wood and dry cow dung, as their principal fuel. In 2023, Kyrgyzstan took its first steps toward joining the carbon market by launching carbon-offset initiatives in collaboration with two international agroforestry institutions: the Asian Forest Cooperation Organization and Rabobank (Acorn, n.d.; Han, 2023). The two projects intend to help farmers to earn carbon credits through sustainable agroforestry practices. Also, in 2023, to facilitate access to the carbon emissions market, the Kyrgyz government partnered with a local tech company to digitize and manage carbon sequestration (Kabar News, 2023).

A network of young activists has emerged in Kyrgyzstan who engage in noninstitutional climate actions around the issues of air pollution, conservation, and biodiversity. The first incidents of grassroots mobilization go back to the 1998 cyanide spill accident at the Kumtor gold mine, which underscored failures in resource extraction practices, with direct implications for climate resilience (Pannier, 2005). The most successful environmental campaign in Kyrgyzstan was in the spring of 2019 to oppose uranium mining in Tash Bulak. News of the proposed mining project sparked public outcry due to fears of radioactive waste contaminating soil and water essential for adapting to changing climate. Young activists launched a social media campaign to oppose the uranium mine. Within 2 weeks, the petition opposing the uranium mine garnered 30,000 signatures. A special campaign profile frame, "I Am Against Uranium Mine," was launched among Facebook users, further amplifying the movement's reach. Government picketings by groups of 50–300 people included business leaders, academics, artists, opposition leaders, activists, and environmentalists (Pannier, 2019). After months of sustained protests, the internally split government elites yielded to public pressure and passed a bill banning uranium mining until 2024. However, following the election of a more autocratic president, the ban was overturned, opening the door for new uranium exploration and construction of nuclear power stations. The history of careless disposal of radioactive materials has galvanized the residents to oppose uranium mining and the construction of nuclear power stations despite the government's promise of clean energy and more local jobs. The authorities cited climatic pressure on hydroelectric power and the prolonged drought spells as the reasons for the construction of the nuclear plant (Cabar, 2024; Rickleton, 2024).

The capital Bishkek's reliance on a coal-burning heating plant contributes to local air pollution and global climate change. Bishkek has been ranked among the most polluted cities in the world (Today.kg, 2022). In December 2019, the MoveGreen youth group organized the first peaceful anti-pollution protest in front of the Parliament building, with about a hundred activists. Despite COVID-19 restrictions, two more peaceful protests were held the following year, attracting wide media coverage. MoveGreen gained popularity in 2017 after installing over 100 air quality monitors nationwide and developing a free phone app to provide real-time air quality data. Today, the databases on air quality, water resources, waste management, and land resources are publicly available on the group's website. MoveGreen has expanded its agenda to promote global trends such as recycling and bicycle infrastructure (MoveGreen, 2024). The country has committed to expanding its renewable energy production capacity from the present 1.5% to 10% by 2030, which would help reduce reliance on polluting fossil fuels and improve air quality (United Nations, 2024).

Another example of a noninstitutional climate adaptation and resilience action is the climate data-collection project by the Internet Society Kyrgyzstan Chapter, an Internet advocacy NGO led by Talant Sultanov. With rising temperatures and melting glaciers, mudflows and floods pose a growing threat to remote mountain communities. According to the Ministry of Emergency Situations of Kyrgyzstan, natural disasters occur in the Kyrgyz Mountains almost weekly, constituting one third of all emergencies in the country. For forecasting and monitoring purposes, the Internet Society Kyrgyzstan Chapter, in partnership with the International Centre for Theoretical Physics, installs sensors in mountainous regions to collect data on precipitation, temperature, water levels, and other meteorological variables. The NGO's next project is to employ AI technology to refine climate emergency responses through enhanced modeling and simulation capabilities (Internet Society Kyrgyzstan Chapter, 2024).

Public awareness of the climate crisis in Kyrgyzstan remained low until 2019 when extreme weather conditions placed climate crisis at the center of public discourse. The country's relatively open civil liberties at the time allowed Kyrgyzstan's civil society to organize around the climate change. However, the rise of authoritarianism since 2020 has kept tabs on these efforts. The government has increasingly employed laws against "incitement of unrest" and "spread of false information" to suppress activism, creating a climate of fear and self-censorship. Furthermore, the law passed in 2024, declaring NGOs receiving foreign funding as "foreign representatives," has raised concerns about the future of the local civil society working with foreign counterparts on the climate change agenda.

Tajikistan

Although Tajikistan's GHG emissions are among the lowest in the world, accounting for only 0.02% of the global total, the country is significantly impacted by climate change, especially in rural areas. As the country is still recovering from the Civil War in the 1990s, many international donors have stepped in to fund the country's climate crisis mitigation and adaptation efforts. Tajikistan ratified the UNFCCC in 1998, joined the Kyoto Protocol in 2008, and signed the Paris Agreement in 2016. Tajikistan has pledged an unconditional 60–70% reduction in GHG emissions compared to the 1990 levels by 2030. Tajikistan's conditional goal is a 50–60% reduction by 2030, contingent on receiving international support (The Republic of Tajikistan, 2022, p. 5). Agriculture and energy, Tajikistan's two major economic sectors, are the primary sources of emissions (The Republic of Tajikistan, 2022, p. 3). In the early 2000s, the rapid expansion in animal husbandry (cattle, goat, and sheep) increased the amount of enteric fermentation and manure, leading to an approximate 27% increase in emissions from agriculture. Emissions from the energy sector have been growing since 2010 at an annual rate of 18%. Due to limited access to natural gas and the unreliability of year-round electricity across seasons, biomass, mainly wood, supplies 35–85% of energy needs in rural areas (United States Agency for International Development, 2019). According to the Tajikistan government, the main risks caused by the climate crisis are flooding, landslides, mudflows, avalanches, and drought. To adapt to the climate crisis, the government outlined three main areas of action: developing renewable energy sources, minimizing the transport sector's impact on the environment, and developing "green employment" and capacity training for civil servants in climate change adaptation and governance.

Given the climate-sensitive nature of its two major industries, agriculture and energy, Tajikistan's entire economy is vulnerable to climate change. In agriculture water shortages are projected to cause crop failures, livestock losses, and deforestation, negatively affecting walnut, pistachio, and fruit production farmers (Pfefferle et al., 2020, p. 2). As Table 29.1 demonstrates, hydropower supplies 93% of Tajikistan's consumed electricity. Still, Tajikistan utilizes only about 5% of its total hydropower potential (Ministry of Foreign Affairs of the Republic of Tajikistan, 2019). These numbers demonstrate two points: First, Tajikistan has great untapped potential for renewable energy; second, Tajikistan's energy supply is highly vulnerable to climate change. Increased droughts and reduced snow caps can adversely affect Tajikistan's energy production. Tajikistan's adaptation strategy includes a long-term project to modernize its existing hydropower facilities and expand its wind and solar energy capacity (Masdar, 2023).

As a mountainous country, Tajikistan is prone to natural disasters, such as mud floods, avalanches, and landslides, that cause significant human and economic losses. Rising temperatures and precipitation are forecasted to weaken road infrastructure, bringing traffic disruptions. As part of its mitigation strategy, the government has restricted the movement of heavy vehicles during hot summer days. Tajikistan is collaborating with the Asian Development Bank and the World Bank to improve its roads to adapt to the climate crisis. (Asian Development Bank, 2021). The increase in severe floods also urged Tajikistan's government to start modernization of its dams, especially their spillway capacities. To address water supply issues, the government partners with the Islamic Development Bank, World Bank, and Asian Infrastructure Investment Bank to improve water management in agriculture by modernizing irrigation pumping stations and constructing distribution networks. Tajikistan also maintains bilateral cooperation on adaptation to climate change with the European Bank for Reconstruction and Development, the Global Environment Facility, the International Fund for Agricultural Development, and the German Corporation for International Cooperation.

Around the world, younger activists are at the forefront of the climate debate (see chapters by Uba and Karsgaard in this volume). Tajikistan is one of the world's youngest countries, with 60% of its population under 30. Young people in Tajikistan report ageism as a major issue, with authorities dismissing young people's grievances (Democracy Fund, 2023). Abibulloeva and Amanbaiuly (2024) point to the lack of awareness among Tajikistan's youth about the urgency of the climate crisis. They cite the lack of education, information, and resources as obstacles to youth climate action. The climate change subject is absent from the school's curriculum and is not covered sufficiently in the media. The few examples of Tajikistan's youth engagement in climate action are overly formal, as the government tightly controls the civil society. Most of Tajikistan's climate experts are part of (government-organized NGOs (GONGOs). Only a handful of independent environmental NGOs are registered in Tajikistan, which often operate ad hoc. The few existing environmental youth movements and NGOs have failed to identify a common goal for cooperation. An unfriendly business environment and red tape discourage the creation of eco-businesses and green companies. Government restrictions on civil society have a cooling effect on climate change activism.

Turkmenistan

Turkmenistan has one of the world's most repressive and totalitarian regimes, which has isolated the country from the rest of the world (Human Rights Watch, 2024). Cotton farming and fossil fuel mining dominate the country's economy. Turkmenistan has the region's flattest landscape and driest climate, with the Karakum Desert covering more than 70% of the territory (The Government of Turkmenistan, 2022, p. 12). The two main rivers, Amu Darya and Murghab, are heavily used for irrigation via an extensive network of outdated and mismanaged canals. Turkmenistan ratified the UNFCCC in 1995, the Kyoto Protocol in 1998, and the Paris Agreement in 2016. The government pledged to reduce GHG emissions by 20% by 2030 relative to 2010. Turkmenistan adopted the National Strategy on Climate Change in 2012, which includes climate change adaptation measures, such as afforestation, water management, public awareness, reduction of greenhouse emissions, renewable energy, and pasture management (UNDP, 2011). The country faces several climate challenges, including desertification, water shortages, extreme weather events (heat waves, droughts, and dust storms), and biodiversity loss. The country's GHG emissions consist of carbon dioxide (56%) and methane (41%) (The Government of Turkmenistan, 2022, p. 32). The combustion of fossil fuels accounts for 85% of Turkmenistan's GHG emissions, and agriculture for 12% (The Government of Turkmenistan, 2022, pp. 16–25). Turkmenistan's vast deposits of natural gas and oil allow the government to provide electricity and gas for household consumption at little to no cost. The heavily subsidized energy in Turkmenistan contributes to the country's political stability but also incentivizes wasteful energy consumption. Natural gas accounts for approximately 87% of the country's total energy consumption, while oil contributes 13%. The consumption of renewable energy sources is negligible (International Energy Agency, n.d.b).

In 2022, a leak in Turkmenistan's pipeline caused the world's worst methane-emitting event. In response to negative global publicity and international pressure, Turkmenistan joined the Global Methane Pledge and launched projects to cut methane leaks from the country's oil and gas industry (Carrington, 2023). The Darvaza gas crater, known as "Gateway to Hell," which has been burning uncontrollably since 1971, is another source of massive methane emissions. This disaster contributes to the country's air pollution, destruction of biodiversity, and desertification. The government's approach to the burning gas crater has been a mix of leveraging the crisis as a tourism site while exploring ways to mitigate its environmental impact. The tightly controlled local media avoids reporting negative stories, while attempts by foreign news outlets to obtain comments from Turkmen authorities have been unsuccessful (The Guardian, 2023; Radio Azatlyk, 2023).

Turkmenistan's agriculture is plagued by inefficient and outdated infrastructure. Only 5% of Turkmenistan's land is arable, and all of it requires irrigation. The country specializes in cultivating cotton and wheat, which use 95% of the nation's water (Stanchin & Lerman, 2007). The Karakum Canal diverts 45% of Amu Darya's flow to irrigate the two water-intensive crops, but 30% of its water is lost to evaporation and seepage. This inefficient operation of the leaky Karakum Canal has been one of the key causes of the Aral Sea's disappearance. Since 2020, the Global Environmental Facility, the Adaptation Fund, and USAID have funded water-saving technologies, such as drip irrigation, rainwater harvesting, and water consumption metering systems. Large public investments in existing and new water facilities, such as water reservoirs and irrigation systems, are taking place across Turkmenistan (The Government of Turkmenistan, 2022, p. 18). As part of the United Nations' "Trees in

Cities" initiative, the government planted 25–30 million seedlings nationwide in 2020 and 2021 (The Government of Turkmenistan, 2022, p. 43). However, the poor track record of trees planted under similar past projects has raised concerns about the initiative's efficiency (Eurasianet, 2013).

International donors operating in Turkmenistan adopt a positive outlook, praising the government for its actions and encouraging it to stay on track with its promises. Since Turkmenistan's climatic conditions are exceptionally favorable for the widespread use of renewable energy sources, international donors encourage Turkmenistan to research and develop alternative energy technology, such as solar and wind. In 2022, the UAE's Masdar company signed an agreement with the government of Turkmenistan to build the country's first solar plant with a projected capacity of 100 MW. The Turkmenistan State Energy Institute houses the "Renewable Energy Sources" research center to train local specialists in photovoltaic technology, biomass production, and electricity storage (Kasyanova, 2023). The World Bank and Asian Development Bank provide technical and financial assistance to support the country's move toward energy efficiency and alternative energy sources. The government has already built a new power plant in Mary City using the latest technology with high energy efficiency standards and environmental protection requirements (Turkmenportal, 2023).

Lack of civil liberties, government surveillance, and tight control over many aspects of life makes it difficult, if not impossible, for Turkmenistan's civil society to organize on climate crisis. Participants in unauthorized gatherings risk arrest and imprisonment. The government closely monitors the use of satellite dishes, Internet access is limited to a small portion of the population, and social media is banned. Public awareness about climate crises is low, as media or school curriculum do not cover climate change. As of 2024, out of 134 officially registered GONGOs, three focus on the environment. The largest youth organization, Bosfor, runs projects on biohumus and ecology. Eco-Durmush and the Nature Conservation Society offer training in environmental protection (Asian Development Bank, 2023). Unprecedented protests took place in 2020 in Turkmenistan in the aftermath of a hurricane that damaged infrastructure and killed 30 people (Muhamedov, n.d., p. 13). About a thousand protestors blocked the main road and distributed photos of the damages. Businesses and ordinary people donated food and clothing for the victims. The authorities responded by arresting the distributors of the photos and shutting down the collection of humanitarian aid (Denber, 2020).

Uzbekistan

Uzbekistan has many ecosystems, including deserts, steppes, and mountains. Climate change challenges include water shortage, soil degradation, loss of biodiversity, and desertification, particularly around the Aral Sea. Uzbekistan joined the UNFCCC in 1993, ratified the Kyoto Protocol in 1999, and ratified the Paris Agreement in 2018 (Republic of Uzbekistan, 2021). Uzbekistan's GHG emissions constitute 0.39% of total global emissions. By 2030, Uzbekistan set a target of reducing GHG emissions by 35% below 2010 levels. In 2021, the European Bank of Reconstruction and Development and the Government of Japan supported Uzbekistan's pledge to achieve carbon neutrality by 2050, helping it design the roadmap to carbon neutrality (Gazeta.uz, 2021). As one of the world's top cotton-producing countries, Uzbekistan's economy heavily depends on this water-intensive crop. Projected frequency of droughts, temperature increase of 1.6°C–1.9°C by 2041–2060, and glacial decline threaten the country's economic growth (USAID, 2024). The absence of integrated water management practices on the domestic and regional levels and the use of the outdated leaky irrigation system exacerbate the impact of climate change on biodiversity and desertification. The most striking example of this is the shrinking of the Aral Sea. To mitigate the Aral Sea crisis, the government planted drought-resistant trees (saxaul) at the bottom of the dry seabed. The project is expected to reduce dust storms, desalinate the nearby soils, and protect biodiversity. The authorities also use "The Green Nation Initiative" to mobilize its citizens to plant 1 billion saplings over 5 years. The project aims to grow the green areas in cities from 8% to 30% and to reverse desertification in rural areas (Dimovska, 2022). Under the Clean Development Mechanism of the Kyoto Protocol, Uzbekistan has completed 15 projects and introduced water-saving technologies to 27% of agricultural land. As part of its adaptation strategy, by 2030, the country aims to plant 26,200 hectares of forests and cultivate drought-resistant crops (Sabirova, 2024).

As Table 29.2 shows, fossil fuels generate 99% of Uzbekistan's energy. To diversify its energy sources, Uzbekistan aims to increase renewable electricity production to 40% by 2030 (Mirzamakhmudov, 2024), including the production of 7 GW of solar and 5 GW of wind capacity (Masdar, n.d.b). Since 2020, the UAE-owned Masdar company has been financing and building seven renewable energy stations—one wind power plant and six photovoltaic power (PV) plants, making it the largest wind and solar development project in Central Asia. In 2021, Uzbekistan put into operation its first PV plant in Navoi. One of the PV plants includes the battery energy storage system (BESS) helping Uzbekistan to address the intermittency of renewable energy. When completed, the six PV and one wind power plants are expected to displace over 1 million tons of carbon dioxide annually and power more than 1 million homes (Masdar, n.d.a). Another step toward energy diversification took place in 2024 when Uzbekistan signed an agreement with Rosatom, the Russian state-owned corporation, to build the region's first nuclear plant comprised of six reactors with a total capacity of 330 MW (Gazeta.uz, 2024b). In the same year, with the investment from USAID, Uzbekistan announced the building of the Green Hydrogen Hub in Tashkent, the capital of Uzbekistan. The Green Hydrogen Hub will assist with the storage and long-distance transportation of renewable energy. The Hub also includes an educational center to train local specialists in green technology (Gazeta.uz, 2024a). To unite the region in addressing climate change, Uzbekistan proposed two regional initiatives—the Regional Green Agenda Program for Central Asia and the Regional Strategy for Adaptation to Climate Change. Recognizing that water shortages, energy problems, and food security issues could potentially lead to conflict between Central Asian countries, the government of Uzbekistan declared that solving climate change–related problems is its top priority. In 2024, Uzbekistan's government established the University of Environmental and Climate Change Studies (Green University), the region's first university with a curriculum focused on environment and climate change (Times of Central Asia, 2024b).

To improve public awareness, the Uzbek government uses media, press conferences, and round tables to promote campaigns, such as "Car-Free Day," "European Mobility Week," essay and video competitions on climate change, mobile apps promoting water-saving irrigation techniques, training for bloggers and journalists covering climate-related issues, litter-picking events, "Go Wasteless" campaigns, small grants program, and environmental education for preschoolers (UNECE, 2024, p. 79). In 2023, to draw attention to climate change, the Uzbek authorities conducted a live broadcast of a marathon across the dry seabed of the disappearing Aral Sea (Gizzat, 2023). In 2019, the largest environmental GONGO, the Ecological Movement of Uzbekistan transformed into the pro-government Ecological Party of Uzbekistan (UNECE, 2024, p. 83). The extensive network of GONGOs in Uzbekistan constrains the development of a genuinely independent civil society focused on the climate crisis. Among the independent NGOs working on climate-related issues, Ecoforum is the most impactful and long-standing. Its expertise includes integrated water and land resource management, ecotourism, renewable energy, and developing an environmental journalists' network. The Khorezm Rural Advisory Support Service is another grassroots group of experts on rural livelihood and environmental sustainability. The Uzbekistan Society for the Protection of Birds organizes activists to conserve and rehabilitate the gene pools of birds of Uzbekistan (Asian Development Bank, p. 6). In 2018, the government eased some of its restrictions on civil society by allowing independent NGOs to receive funding from foreign institutions and simplifying the registration process later in 2020. Even though the operating environment for independent NGOs remains constrained, environmentalists are among the most active groups in civil society (Asian Development Bank, 2024, p. 4).

The five Central Asian countries take different climate action approaches, as shown in Tables 29.1 and 29.2. Table 29.1 illustrates the reliance on fossil fuels in electricity consumption across the region, with Kazakhstan and Uzbekistan depending heavily on coal and natural gas, while Kyrgyzstan and Tajikistan mostly use hydropower, demonstrating their renewable energy potential. Table 29.2 highlights the broader energy supply trends, reaffirming that fossil fuels remain the primary source for most countries, except for Tajikistan, as it significantly relies on hydropower. Kazakhstan uses wind and solar energy for 4% of its electricity, leading the region in renewables, though its use is still limited. These differences in energy sources highlight each country's challenges in tackling climate change, such as water shortages in Kyrgyzstan and Tajikistan and outdated energy systems in Uzbekistan and Turkmenistan. The region's small yet increasing efforts in expanding renewable energy sources and green initiatives are important in addressing shared climate issues.

Conclusion

Climate actions in Central Asia are expanding, driven by the growing awareness of the urgency of climate change. Legislatures are adopting laws on climate action, while NGOs and media are advocating for further actions. International treaties are getting ratified and, in some cases, enforced. Central Asian countries face different climate change challenges, including air pollution, water shortage, extreme weather, and glacier melting. While Central Asian governments and civil societies acknowledge the climate change paradigm and environmental problems, their actions often lag. The region's authoritarian governments can resist social pressures and, as such, have not yet faced significant public demands for a robust response to the climate crisis; this may not bode well for other parts of the globe where democratic backsliding is on the rise. Furthermore, widespread poverty limits society's capacity to prioritize climate concerns and exert pressure on governments. Patron–client relationships, characterized by societal reliance on the state, further limit climate actions. Despite these challenges, Central Asian governments are starting to adopt climate-related programs in water management, biodiversity conservation, carbon emissions reduction, green transition, and green finance. The encouragement from international agencies partly drives this shift. Market forces and technological advancements also fuel the private sector's interest in green transition. A comparison of the five Central Asian countries reveals a trend toward the growing adoption of legal norms, initiatives, and ecological projects, often in collaboration with international organizations such as the United Nations Environment Programme (UNEP) and the United Nations Framework Convention on Climate Change (UNFCCC).

Water issues are paramount for Central Asian states due to the region's arid climate, rising summer temperatures, divergent interests of the upstream and downstream countries, and substantial rural populations. Effective water management requires political decisions at the local, national, and regional levels. While water-saving technologies are gaining traction, more comprehensive action is needed. Water shortages have forced people to migrate, creating broader demographic and social instabilities. Furthermore, the pressure of population growth in some parts of the region exacerbates the existing water problems. Kazakhstan has made strides in promoting a green economy and conserving water resources. However, the country's high carbon dioxide emissions contribute to rising temperatures and melting glaciers. Kyrgyzstan has Central Asia's most sophisticated, detailed, and policy-oriented climate change legislation. This development can be attributed, in part, to the relatively strong role of civil society and the relative openness of the state to public pressure. This trend may be reversed as Kyrgyzstan exhibits increasing authoritarian tendencies in recent years. Despite its untapped wind and solar energy potential, Kyrgyzstan invests in coal mining to reduce its reliance on hydropower and its vulnerability to droughts. Tajikistan has recently developed its climate change strategy and adaptation initiative, encompassing various frameworks. This initiative received support from the local population, who are participating in protecting flora and fauna from climate stress. One notable project within this framework aims to mitigate the rise in global temperature and conserve freshwater resources. While Turkmenistan remains closed to changes compared to other Central Asian countries, it actively participates in United Nations campaigns for climate conservation and developing climate change strategies. Despite joining the Kyoto Protocol and the Paris Agreement relatively later than other Central Asian countries, Uzbekistan is actively attracting foreign investment into renewable energy sources.

While water problems, biodiversity, and carbon emissions are addressed in the region with some country-specific differences, green finance remains at an early stage of development and is limited to conference debates. Green finance is a relatively novel and unfamiliar concept in the region. Finally, the region has to overcome two main obstacles to address climate change. The first is widespread poverty, which often prevents individuals from engaging in long-term planning. The second is the authoritarian nature of the governments, which is more responsive to international pressure than to internal demands, potentially leading governments to overlook climate actions originating from vulnerable domestic groups.

APPENDIX

There are two tables for the five Central Asian countries. Table 29.3 includes domestic sources of law for each of the five issues: decarbonization, climate action, water and glaciers, green finance systems, and biodiversity. Table 29.4 includes international sources of law (conventions and treaties) and each countries' membership in international organizations for each issue.

TABLE 29.3 Legal Frameworks in Central Asian Countries to Main Issues of Climate Action

State / Issue	Kazakhstan	Kyrgyzstan	Uzbekistan	Turkmenistan	Tajikistan
Decarbonization of the economy and carbon markets	On approval of the strategy for achieving carbon neutrality of the Republic of Kazakhstan until 2060	Law of the Kyrgyz Republic, dated June 30, 2022, No. 49 About renewable energy sources (as amended by the Law of the Kyrgyz Republic, dated August 9, 2023, No. 174)	Resolution of the president of the Republic of Uzbekistan On measures to increase the effectiveness of reforms aimed at transitioning the Republic of Uzbekistan to a "green" economy until 2030	About renewable energy sources from 2021	Law of the Republic of Tajikistan on the Use of Renewable Sources of Energy 2015
Climate crisis and action	Decree of the Government of the Republic of Kazakhstan dated April 19, 2023, No. 313 On approval of the updated national contribution of the Republic of Kazakhstan to the global response to climate change	Resolution of the Cabinet of Ministers of the Kyrgyz Republic dated May 19, 2021, No. 11 On issues of the State Committee on Ecology and Climate of the Kyrgyz Republic	Law of the Republic of Uzbekistan About environmental control 2013	About nature conservation	National Strategy for Adaptation to Climate Change of the Republic of Tajikistan for the period until 2030, dated 2019
Water/glaciers	Water Code of the Republic of Kazakhstan (with changes and additions as of May 1, 2023)	Law of the Kyrgyz Republic, January 14, 1994, No. 1422-XII About water, April 5, 2019, No. 44	Law of the Republic of Uzbekistan "On amendments and additions to some legislative acts of the Republic of Uzbekistan in connection with increased liability for offenses in the field of environmental protection, rational use of natural resources, including drinking water" (2018)	Code of Turkmenistan "On Water"	Water Code of the Republic of Tajikistan of 2020

Green financial system	On approval of the classification (taxonomy) of "green" projects to be financed through "green" bonds and "green" loans from 2021	(to the resolution of the Government of the Kyrgyz Republic dated November 14, 2019, No. 605) Green Economy Development Program in the Kyrgyz Republic for 2019–2023	Resolution of the president of the Republic of Uzbekistan On measures to increase the effectiveness of reforms aimed at transitioning the Republic of Uzbekistan to a "green" economy until 2030	???	Resolution of the Government of the Republic of Tajikistan dated September 30, 2022, No. 482 Strategy for the development of the green economy in the Republic of Tajikistan for 2023–2037
Biodiversity	Government order on Approval of the Rules for Implementing Compensation for Biodiversity Loss	Decision of March 17, 2014, No. 131 On the Priorities for the conservation of biological diversity of the Kyrgyz Republic for the period until 2024 and the action plan for the implementation of the priorities for the conservation of biological diversity of the Kyrgyz Republic for 2014–2020	Resolution of the Cabinet of Ministers of the Republic of Uzbekistan on approval of the strategy for the conservation of biological diversity in the Republic of Uzbekistan for the period 2019–2028	Law of Turkmenistan on nature protection of 2017	Law of the Republic of Tajikistan "On Biological Farming and Production of 2019."

TABLE 29.4 Participation in Conventions

State / Issue	Kazakhstan	Kyrgyzstan	Uzbekistan	Turkmenistan	Tajikistan
Decarbonization of the economy	Global Decarbonization in Fossil Fuel Export-Dependent Economies	Mentioned, but not as participant	Global Decarbonization in Fossil Fuel Export-Dependent Economies	Global Decarbonization in Fossil Fuel Export-Dependent Economies	Mentioned, but not as participant
Climate crisis and action	United Nations Framework Convention on Climate Change	United Nations Framework Convention on Climate Change	United Nations Framework Convention on Climate Change	United Nations Framework Convention on Climate Change	United Nations Framework Convention on Climate Change
Water/glaciers	Convention on the Protection and Use of Transboundary Watercourses and International Lakes Member of the Bureau of the Water Convention National representatives for the National Policy Dialogues	National focal points of the Protocol on Water and Health National representatives for the National Policy Dialogues	Convention on the Protection and Use of Transboundary Watercourses and International Lakes Member of the Bureau of the Water Convention	Convention on the Protection and Use of Transboundary Watercourses and International Lakes Member of the Bureau of the Water Convention National representatives for the National Policy Dialogues	National Focal points of the Protocol on Water and Health National representatives for the National Policy Dialogues
Green financial system	UNDP Green Project	UNDP Green Project	UNDP Green Project	UNDP Green Project	UNDP Green Project
Biodiversity	Convention on Biological Diversity	Convention on Biological Diversity	Convention on Biological Diversity	Convention on Biological Diversity	Convention on Biological Diversity

References

Abibulloeva, A., & Amanbaiuly, M. (2024). Youth involvement in climate issues in Kazakhstan and Tajikistan: Illusion or reality? https://cabar.asia/en/youth-involvement-in-climate-issues-in-kazakhstan-and-tajikistan-illusion-or-reality

Acorn. (n.d.). Kyrgyzstan: Fertile ground for Acorn. https://acorn.rabobank.com/en/blog/kyrgyzstan-fertile-ground-for-acorn/

Aladina, T. (2023, May 1). Darmarka proshla v Almaty [A free market took place in Almaty]. https://www.inalmaty.kz/news/3588754/darmarka-prosla-v-almaty

Almeida, P., González Márquez, L. R., & Fonsah, E. (2023). The forms of climate action. *Sociology Compass*, *17*(11), e13177. https://doi.org/10.1111/soc4.13177

Almeida, P., & Stearns, L. B. (1998). Political opportunities and local grassroots environmental movements: The case of Minamata. *Social Problems*, *45*(1), 37–60. https://doi.org/10.2307/3097142

Asian Development Bank. (2021). Tajikistan transport sector assessment. https://www.adb.org/sites/default/files/institutional-document/755636/tajikistan-transport-sector-assessment.pdf

Asian Development Bank. (2023). Civil society brief: Turkmenistan. https://www.adb.org/sites/default/files/publication/928046/civil-society-brief-turkmenistan.pdf

Asian Development Bank. (2024). Civil society brief: Uzbekistan. https://www.adb.org/sites/default/files/publication/753866/civil-society-brief-uzbekistan.pdf

Astana Times. (2024, January 1). Kazakhstan takes over chairmanship at International Fund for Saving of Aral Sea. *Astana Times*. https://astanatimes.com/2024/01/kazakhstan-takes-over-chairmanship-at-international-fund-for-saving-of-aral-sea

Bureau of National Statistics (n.d.). Tzel' 13. Bor'ba s izmeneniem klimata. [Goal 13: Climate Action]. *Qazstat*. https://stat.gov.kz/ru/sustainable-development-goals/goal/13/

Central Asian Analytical Network. (2016). Changes in climate in Central Asia: What are the consequences? https://www.caa-network.org/archives/8007

Cabar. (2024, June 4). AES v Kyrgyzstane: bol'shie vygody i znachitel'nye riski [Nuclear power plants in Kyrgyzstan: Great benefits and significant risks]. *Cabar*. https://cabar.asia/ru/aes-v-kyrgyzstane-bolshie-vygody-i-znachitelnye-riski

Carrington, D. (2023, June 13). Turkmenistan moves towards plugging massive methane leaks. *The Guardian*.https://www.theguardian.com/environment/2023/jun/13/turkmenistan-moves-towards-plugging-massive-methane-leaks

Clement, V., Rigaud, K. K., de Sherbinin, A., Jones, B., Adamo, S., Schewe, J., Sadiq, N., & Shabahat, E. (2021). *Groundswell Part 2: Acting on internal climate migration*. World Bank. http://hdl.handle.net/10986/36248

ClimaMeter. (2024, January 11). 2023/11/24–30 Central Asia heatwave. https://www.climameter.org/20231124-30-central-asia-heatwave

Climate Action Tracker. (2023). Kazakhstan. https://climateactiontracker.org/countries/kazakhstan/

Climate Diplomacy. (n.d.). Dam conflict between Kyrgyzstan and Uzbekistan. https://climate-diplomacy.org/case-studies/dam-conflict-between-kyrgyzstan-and-uzbekistan

Columbia University. (n.d.). Environmental impacts. The Aral Sea crisis. https://www.columbia.edu/~tmt2120/environmental%20impacts.htm

Critical Ecosystem Partnership Fund. (2017). Ochag bioraznoobrazija v Central'noaziatskom gornom regione [Biodiversity hotspot in the Central Asian mountain region]. https://www.cepf.net/sites/default/files/mountains-central-asia-ecosystem-profile-rus.pdf

Democracy Fund. (2023). Strengthening young people's political skills in Tajikistan. https://www.un.org/democracyfund/news/strengthening-young-peoples-political-skills-tajikistan#:~:text=With%20more%20than%2060%20per,of%20the%20world's%20youngest%20countries

Denber, R. (2020, May 4). Turkmenistan government's deafening silence after hurricane: Storm survivors need help, not censorship. Human Rights Watch. https://www.hrw.org/news/2020/05/04/turkmenistan-governments-deafening-silence-after-hurricane

Dimovska, M. (2022, May 27). Ensuring a green future for Uzbekistan through 1 billion trees. Climate Promise. UNDP. https://climatepromise.undp.org/news-and-stories/ensuring-green-future-uzbekistan-through-1-billion-trees

Eurasianet. (2013, March 12). Turkmenistan attempts quixotic greening of the desert. Eurasianet. https://eurasianet.org/turkmenistan-attempts-quixotic-greening-of-the-desert

Garafutdinova, D & Tang, D. (2024). Green finance and industrial low-carbon transition: A case study on green economy policy in Kazakhstan. *Sustainability*, *16*(17), 7731. https://doi.org/10.3390/su16177731

Gazeta.uz. (2021, February 2). Uzbekistan planiruet pereyti k uglerodno-neytralnoy energetike k 2050 godu [Uzbekistan plans to transition to carbon-neutral energy by 2050]. Gazeta.uz. https://www.gazeta.uz/ru/2021/02/02/carbon-neutrality/

Gazeta.Uz. (2024a March 2). Tsentr zelenogo vodoroda sozdadut v Uzbekistane pri podderzhke SShA [A green hydrogen hub will be created in Uzbekistan with U.S. support]. Gazeta.uz. https://www.gazeta.uz/ru/2024/03/02/green-hydrogen-hub/

Gazeta.uz. (2024b, May 27). Stroit' maluyu AES nachnut letom v Dzhizakskoi oblasti [Constructions of a small nuclear plant will start this summer in Jizak region]. https://www.gazeta.uz/ru/2024/05/27/nuclear-station-uzbekistan/

Gizzat, M. (2023, October 19). Central Asian climate heroes: Creative climate advocacy in my region. Bow Seat. https://bowseat.org/news/central-asian-climate-heroes-creative-climate-advocacy-in-my-region/

Global Energy Monitor et al. (2024) Bum i krakh uglya [Boom and bust of coal]. https://globalenergymonitor.org/wp-content/uploads/2024/04/Boom-Bust-Coal-2024-Russian.pdf

Goldstone, J. A. (1993). *Revolution and rebellion in the early modern world*. University of California Press.

Government of Kazakhstan. (2020, September 19). Today, as part of the environmental campaign "World Cleanup Day," tree planting took place in Nur-Sultan. https://www.gov.kz/memleket/entities/ecogeo/press/news/details/100459?lang=ru

Government of the Republic of Kazakhstan. (2021). Environmental Code of the Republic of Kazakhstan (Code of the Republic of Kazakhstan No. 400-VI ZRK dated January 2, 2021) [in Russian]. https://adilet.zan.kz/rus/docs/K2100000400

Government of Turkmenistan. (2022). *Nationally determined contribution of Turkmenistan*. United Nations Framework Convention on Climate Change. https://unfccc.int/sites/default/files/NDC/2023-01/NDC_Turkmenistan_12-05-2022_approv.%20by%20Decree_Eng.pdf

Green Alliance. (n.d.). O nas [About us]. https://green-alliance.kg/o-nas/

The Guardian. (2023, May 9). Mind-boggling methane emissions from Turkmenistan revealed. *The Guardian*. https://www.theguardian.com/world/2023/may/09/mind-boggling-methane-emissions-from-turkmenistan-revealed

Han, Y. (2023, December 2). Advancing forest carbon projects spanning 600,000 hectares in Kyrgyzstan. *AFOCO*. https://afocosec.org/newsroom/news/advancing-forest-carbon-projects-spanning-600000-hectares-in-kyrgyzstan/

Hoegh-Guldberg, O., Jacob, D., Taylor, M., Bindi, M., Brown, S., Camilloni, I., Diedhiou, A., Djalante, R., Ebi, K. L., Engelbrecht, F., Guiot, J., Hijioka, Y., Mehrotra, S., Payne, A., Seneviratne, S. I., Thomas, A., Warren, R., & Zhou, G. (2018). Impacts of 1.5°C global warming on natural and human systems. In V. Masson-Delmotte et al. (Eds.), *Global warming of 1.5°C. An IPCC Special Report on the impacts of global warming of 1.5°C above pre-industrial levels and related global greenhouse gas emission pathways, in the context of strengthening the global response to the threat of climate change, sustainable development, and efforts to eradicate poverty* (pp. 175–311). Intergovernmental Panel on Climate Change. https://www.ipcc.ch/site/assets/uploads/sites/2/2019/02/SR15_Chapter3_Low_Res.pdf

Human Rights Watch. (2024). Turkmenistan. Human Rights Watch. https://www.hrw.org/europe/central-asia/turkmenistan

Ibraeva, O. (2024, September 30). Bolee desjati aktivistov dostavili v policiju v Almaty [More than ten activists were taken to the police in Almaty]. *Orda*. https://orda.kz/bolee-desjati-aktivistov-dostavili-v-policiju-v-almaty-392449/

International Energy Agency (n.d.a). Energy systems of Eurasia. https://www.iea.org/regions/eurasia

International Energy Agency. (n.d.b). Energy system of Turkmenistan. https://www.iea.org/countries/turkmenistan
International Energy Agency. (n.d.c). Kyrgyzstan energy profile: Energy security. https://www.iea.org/reports/kyrgyzstan-energy-profile/energy-security
Internet Society Kyrgyzstan Chapter. (n.d.). Creating an open and secure IoT infrastructure for monitoring and preventing emergencies in landlocked mountainous countries: The case of Kyrgyzstan. https://isoc.kg/iot-climatechange-monitoring-research/
Internet Society Kyrgyzstan Chapter. (2024, November 10). The role of IoT in forecasting natural disasters explained in the AI Summit for Climate Action in Istanbul. https://isoc.kg/news/the-role-of-iot-in-forecasting-natural-disasters-explained-in-the-ai-summit-for-climate-action-in-istanbul/
Janusz-Pawletta, B., & Gubaidullina, M. (2015). Transboundary water management in Central Asia. *Cahiers d'Asie centrale*, 25, 195–215. http://journals.openedition.org/asiecentrale/3180
Kabar News. (2023, November 14). KFB budet prodvigat na birzhe uglerodnye kredity [Kyrgyz Stock Exchange will launch carbon credit market]. *Kabar*. https://ru.archive.kabar.kg/news/kfb-budet-prodvigat-na-birzhe-uglerodnye-kredity/
Kamchybekova, I. (2024, August 22). V Kyrgyzstane k 31 avgusta sostoyitsya otkrytiye okolo 40 promyshlennykh predpriyatiy [In Kyrgyzstan, the opening of about 40 industrial enterprises is scheduled for August 31]. *Kabar*. https://kabar.kg/news/v-kyrgyzstane-k-31-avgusta-sostoitsia-otkrytie-okolo-40-promyshlennykh-predpriiatii/
Kanbakova, D. (2024, September 18). Ledniki v Central'noy Azii tayut s ugrozhayushchey skorost'yu [Glaciers in Central Asia are melting at an alarming rate]. Check-Point. https://www.check-point.kz/opinion?id=2160
Kasyanova, A. (2023, September 20). Predstaviteli OAE posetili energeticheskii vuz Turkmenistana [UAE's representatives visited Turkmenistan's energy university]. *Zolotoi Vek*. https://turkmenistan.gov.tm/ru/post/75746/predstaviteli-oae-posetili-energeticheskij-vuz-turkmenistana
Limon.kg. (2023, May 27). Kyrgyzstan i izmenenie klimata: kak molodye rebyata iz KR boryutsya s global'noy problemoy [Kyrgyzstan and climate change: How young people from Kyrgyzstan are tackling a global issue]. *Limon*. https://limon.kg/ru/news:77481
Masdar. (n.d.a). Zarafshan wind farm. https://masdar.ae/en/renewables/our-projects/zarafshan-wind-farm
Masdar. (n.d.b). Jizzakh solar project. Masdar. https://masdar.ae/en/renewables/our-projects/jizzakh-solar-project
Masdar. (2023, October 3). Masdar enters Tajikistan market. https://masdar.ae/en/news/newsroom/masdar-enters-tajikistan-market
Ministry of Foreign Affairs of the Republic of Tajikistan. (2019, June 7). Obzor energeticheskogo sektora Respubliki Tajikistan [Review of the energy sector of the Republic of Tajikistan]. https://mfa.tj/ru/main/view/185/obzor-energeticheskogo-sektora-respubliki-tadzhikistan
Mirzamakhmudov, J. (2024, September 30). Uzbekistan's electricity demand to grow steadily by 7.5% annually until 2030—Energy minister. *Gazeta.uz*. https://www.gazeta.uz/en/2024/09/30/jurabek-mirzamakhmudov/#:~:text=Uzbekistan%20has%20adopted%20a%20program,will%20increase%20it%20to%2015%25.
MoveGreen. (2024, February 7). Istoriya Bishkekskogo smoga: za 7 let—ot polnogo otritsaniya problemy do popytok zapreta prodazhi uglya [The history of Bishkek's smog: Over 7 years—From total denial of the problem to attempts to ban coal sales]. MoveGreen. https://movegreen.kg/2024/02/07/istoriya-bishkekskogo-smoga-za-7-let-ot-polnogo-otritsaniya-problemy-do-popytok-zapreta-prodazhi-uglya/
Muhamedov, R. (n.d.). *Online civic activism in politically restricted space: A case study on Turkmenistan*. Institut für Europäische Politik (IEP). https://iep-berlin.de/site/assets/files/2161/turkmenistan_policy_paper_final_en.pdf
Novikov, V., & Kato, T. (2024). Klyuchevye faktory utraty bioraznoobraziya i finansirovanie mer sokhraneniya zhivoi prirody v Central'noi Azii [Key drivers of biodiversity loss and funding for nature conservation measures in Central Asia]. Zoï Environment Network. https://zoinet.org/wp-content/uploads/2024/05/CA-biodiv-financing_RU.pdf

Olimova, L. (2023, December 15). Central Asian ecosystems are rapidly collapsing. Central Asian Bureau for Analytical Reporting. https://cabar.asia/en/central-asian-ecosystems-are-rapidly-collapsing

Pannier, B. (2005, August 3). Kyrgyzstan: A commercial tragedy revisited. *Radio Free Europe/Radio Liberty*. https://www.rferl.org/a/1060387.html

Pannier, B. (2019, May 6). Keeping Kyrgyzstan's uranium in the ground. *Radio Free Europe/Radio Liberty*. https://www.rferl.org/a/keeping-kyrgyzstan-s-uranium-in-the-ground-/29924122.html

Pavlenko, A. (2023, October 18). Rossiya podderzhit proyekt stroitel'stva ugol'noy TES v Kyrgyzstane [Russia will support the project for the construction of a coal-fired thermal power plant in Kyrgyzstan]. *E-CIS*. https://e-cis.info/news/568/112910/

Peyrouse, S. (2022). *The nexus of environmental issues, poverty, and political authoritarianism in Central Asia: What role for civil society organisations and the international community?* International Partnership for Human Rights (IPHR). https://www.iphronline.org/wp-content/uploads/2022/06/IPHRCSP-CA-report-by-Sebastien-Peyrouse-ENG.pdf

Pfefferle, N., Zolly, M., & Bierkandt, R. (2020). Climate change profile: Tajikistan. ReliefWeb. https://reliefweb.int/report/tajikistan/climate-change-profile-tajikistan

Radio Azatlyk. (2023, March 8). V Turkmenistane za chas v vozdukh prosochilsya metan ekvivalentnyy vybrosam 67 millionov avtomobiley [In Turkmenistan, methane equivalent to the emissions of 67 million cars leaked into the air within an hour]. *Radio Azatlyk*. https://rus.azathabar.com/a/32308219.html

Radio Free Europe/Radio Liberty. (2024, June 10). Kyrgyzstan battles climate change as glaciers and rivers disappear. *Radio Free Europe/Radio Liberty*. https://www.rferl.org/a/kyrgyzstan-climate-change-glaciers-rivers-/32734186.html

Republic of Tajikistan. (2022). *The updated NDC of the Republic of Tajikistan*. UNFCCC. https://unfccc.int/sites/default/files/NDC/2022-06/NDC_TAJIKISTAN_ENG.pdf

Republic of Uzbekistan. (2021). *Updated Nationally Determined Contribution (NDC) of Uzbekistan*. United Nations Framework Convention on Climate Change. https://unfccc.int/sites/default/files/NDC/2022-06/Uzbekistan_Updated%20NDC_2021_EN.pdf

Rickleton, C. (2024, March 2). Kyrgyzstan uranium mining project sparks concerns amid Rosatom's nuclear plans. *Radio Free Europe/Radio Liberty*. https://www.rferl.org/a/kyrgyzstan-uranium-mining-rosatom-russia-nuclear-plant/32845058.html

Sabirova, K. (2024, January 6). Uzbekistan's contribution to the global fight against climate change. United Nations Permanent Mission of Uzbekistan. https://www.un.int/uzbekistan/news/uzbekistans-contribution-global-fight-against-climate-change#:~:text=By%202030%2C%20the%20country%20aims,up%20of%20the%20Aral%20Sea

Schierhorn, F., Kastner, T., Kuemmerle, T., Meyfroidt, P., Kurganova, I., Prishchepov, A. V., Erb, K.-H., Houghton, R. A., & Müller, D. (2019). Large greenhouse gas savings due to changes in the post-Soviet food systems. *Environmental Research Letters*, *14*(6), 065009. https://doi.org/10.1088/1748-9326/ab1cf1

Shambetov, T. (2023, August 1). Kyrgyzstan turns to Chinese companies for major hydropower project. *Radio Free Europe/Radio Liberty*. https://www.rferl.org/a/kyrgyzstan-hydropower-megaproject-china/32529949.html

Silva, M., Palumbo, D., & Rivault, E. (2024, February 15). Kazakhstan: Methane mega-leak went on for months. *BBC*. https://www.bbc.com/news/world-asia-68166298

Skalamera, M. (2024). The varying levels of contrasting adaptation in Central Asia's climate change politics. *Central Asian Survey*, *44*, 1–20. https://doi.org/10.1080/02634937.2024.2373086

Smeltzer, M., & Karppi, A. (2024). *Nations in transit 2024*. Freedom House. https://nationsintransit.org

Stanchin, I., & Lerman, Z. (2007). *Water in Turkmenistan I*. Hebrew University of Jerusalem, Department of Agricultural Economics and Management, Discussion Papers. https://www.researchgate.net/publication/46471390_Water_In_Turkmenistan_I/citation/download

Times of Central Asia. (2024a, February 6). Harnessing social media for social change: An interview with Tashkent-based eco-blogger Mutabar Khushvaktova. *Times of Central Asia*. https://timesca.com/harnessing-social-media-for-social-change-an-interview-with-tashkent-based-eco-blog ger-mutabar-khushvaktova/#:~:text=More%20and%20more%20people%20in,inspire%20others%20to%20take%20action.

Times of Central Asia. (2024b, February 13). Central Asia's first green university opens in Uzbekistan. *The Times of Central Asia*. https://timesca.com/central-asias-first-green-university-opens-in-uzbekistan/

Today.kg. (2022, December 12). Bishkek snova vozglavil spisok gorodov s samym gryaznym vozdukhom [Bishkek once again tops the list of cities with the most polluted air]. *Today.kg*. https://today.kg/news/735978/?utm_source=cat&hl=ru

Transparency International. (2023). *Corruption Perceptions Index*. Transparency International. https://www.transparency.org/en/cpi/2023

Tskhay, A. (2023). The culture of recycling, re-use and reduction: Eco-activism and entrepreneurship in Central Asia. In R. Sabyrbekov, I. Overland, & R. Vakulchuk (Eds.), *Climate change in Central Asia* (pp. 161–170). Springer. https://doi.org/10.1007/978-3-031-29831-8_13

Turkmenportal. (2023, October 29). V Turkmenistane vedutsya podgotovitelnye raboty k vvody v stroy LEP Mary-Akhal [In Turkmenistan, preparatory work is conducted for the launch of the Mary-Akhal power line]. Turkmenportal. https://turkmenportal.com/blog/69130/v-turkmenistane-vedutsya-podgotovitelnye-raboty-k-vvodu-v-stroi-lep-maryahal

Tursumbayeva, M., Muratuly, A., Baimatova, N., Karaca, F., & Kerimray, A. (2023). Cities of Central Asia: New hotspots of air pollution in the world. *Atmospheric Environment*, *309*, 119959. https://doi.org/10.1016/j.atmosenv.2023.119959

UNDP. (2011). National strategy of Turkmenistan on climate change. UNDP. https://www.undp.org/sites/g/files/zskgke326/files/migration/tm/undp-tm-NCCS_ENG.pdf

UNDP Kyrgyzstan. (2021, October 26). Kyrgyzstan presented new national climate pledge under Paris Agreement to wide public. UNDP Kyrgyzstan. https://www.undp.org/kyrgyzstan/press-releases/kyrgyzstan-presented-new-national-climate-pledge-under-paris-agreement-wide-public

UNECE. (2024). Uzbekistan: State of the environment. UNECE. https://unece.org/sites/default/files/2024-02/uzbekistan-state-of-the-environment-en.pdf

United Nations. (2024, January 20). Kyrgyzstan: zagrjaznenie vozdukha—odna iz prichin prezhdevremennoy smerti i problem so zdorov'yem [Kyrgyzstan: Air pollution—One of the causes of premature death and health problems]. United Nations. https://news.un.org/ru/story/2024/01/1448732

United States Agency for International Development. (2019). Tajikistan climate change risk profile. https://www.climatelinks.org/sites/default/files/asset/document/2019_USAID_Tajikistan%20Factsheet.pdf

USAID. (2024). Climate change risk profile. Uzbekistan. USAID. https://www.climatelinks.org/sites/default/files/asset/document/2024-10/CASA_Uzbekistan_CRP_2024.pdf

Vakulchuk, R., Daloz, A. S., Overland, I., Sagbakken, H. F., & Standal, K. (2022). A void in Central Asia research: Climate change. *Central Asian Survey*, *42*(1), 1–20. https://doi.org/10.1080/02634937.2022.2059447

The World Bank. (2019). Climate-smart agriculture in the Kyrgyz Republic. Climate Knowledge Portal. The World Bank. https://climateknowledgeportal.worldbank.org/sites/default/files/2019-06/CSA%20_Profile_The%20Kyrgyz%20Republic.pdf

Zhao, C., Liu, B., Wang, J., Xue, R., Shan, Y., Cui, C., Dong, X., & Dong, K. (2023). Emission accounting and drivers in Central Asian countries. *Environmental Science and Pollution Research*, *30*, 102894–102909. https://doi.org/10.1007/s11356-023-29608-0

Zhusipbek, G., & Nagayeva, Z. (2018, November 18). Bez ekologicheskogo soznaniya ne reshit' ekologicheskie problemy v stranakh Central'noy Azii [Without environmental awareness, environmental problems in Central Asian countries cannot be solved]. Central Asian Analytical Network. https://www.caa-network.org/archives/14565

CHAPTER 30

Subnational Climate Policy Action in the United States and Canada

Past Progress and Future Insufficiency

JOSHUA A. BASSECHES

Introduction

THIS chapter deals with "institutional" forms of climate action (Almeida et al., 2023a); specifically, its subject is policies to *mitigate* (as opposed to adapt to) the impacts of climate change. The geographical focus is on the United States and Canada. More specifically, the focus is on *subnational* (state and provincial) climate action: the role it has played historically, the impact it has had, and the advantages and disadvantages of it relative to national-level action in the political and institutional contexts of these two North American countries. The chapter concludes with a provocation. I argue that, while subnational climate action has been indispensable in enabling progress despite gridlock at the national level (especially in the United States), going forward, it will yield diminishing returns. We have reached a moment in which the scale and magnitude of the climate crisis demand a level of centralized planning and coordination that national governments are best positioned to deliver. I illustrate this with the case of the US electricity sector, the cornerstone of President Biden's strategy to "electrify everything" (transport, buildings, industry, etc.) as a means of achieving economy-wide decarbonization (Harper et al., 2023).

In what follows, I first provide background about the United States and Canada that illustrates their importance individually and together when it comes to action to mitigate climate change in the face of a global climate crisis. Next, I review some of the major subnational policy achievements in each country, highlighting the significance of subnational action in the two political and institutional contexts. Finally, I provide evidence from the US electricity sector of the challenges moving forward that I argue will yield diminishing returns if action continues to be pursued mostly or entirely at the subnational level rather than at the national (and ideally, international) one.

The United States and Canada: Two Wealthy, High-Emitting Neighbors With Empowered Subnational Governments

The United States and Canada are the world's second and third largest greenhouse gas (GHG) emitters per capita, respectively (Government of Canada, 2024a).[1] Both countries are also extremely wealthy. In gross domestic product (GDP) per capita, the United States is twelfth and Canada is twenty-first (World Bank, 2024). In both countries, the transportation sector generates a high percentage of the country's total emissions, about 28% in the US case and about 22% in Canada. In the United States, the plurality of emissions comes from transportation, followed closely by electricity (25%), where coal-fired and gas-fired power plants continue to drive emissions in this sector despite recent progress in adding unprecedented quantities of renewables to the grid. In Canada, the oil and gas sector constitutes the plurality of emissions (31%) and electricity makes up a much smaller share (7%) due to the fact that the vast majority of Canada's electricity is generated from nonemitting hydropower and nuclear sources (Basseches & Ikenze, 2022; Government of Canada, 2024b; US Environmental Protection Agency, 2024c).

The two countries are also worth examining together because their climate policy destinies are intertwined in several ways. First, they are physically interconnected in terms of their electricity grids, with electrons (whether "clean" or "fossil fueled") routinely crossing the political border—which happens to be the longest national border in the world (Basseches & Ikenze, 2022). Second, although Canada's government is a parliamentary system while the United States' is a presidential system, both are characterized by two dominant political parties with often opposing views on climate policy action: Democrats and Republicans, in the case of the United States, and the Liberal Party and Conservative Party, in Canada (Basseches et al., 2022; Macdonald, 2020).[2] Third, there is a similar dynamic in terms of the political geography of their energy production and consumption, whereby in both countries energy production tends to take place in their rural areas while energy consumption tends to take place in their urban and suburban areas. Given the size of the two countries, as well as correlates between geography and the demographic strongholds of the two major, opposing political parties, this tends to create patterns of mutual resentment among their populaces, and correspondingly, among each party's political leadership (Macdonald, 2020; Oatley, 2023).

Over the years, there have been numerous examples of cross-national coordination on climate and energy policy, between the two national governments but perhaps even more so between *subnational* jurisdictions within each country (VanNijnatten & McWhinney, 2022). For instance, California and Quebec have a linked cap-and-trade program (California Air Resources Board, 2024). Since 1973, the six New England states[3] and the five eastern Canadian provinces[4] have coordinated through the New England Governors and Eastern Canadian Premiers Working Group, which includes a "Climate Change Steering Committee" (CONEG, 2024). Also, recently a "green grid" planning task force has been assembled with numerous stakeholders from these same jurisdictions (Sosland, 2023).

The United States and Canada also have parallel—though certainly not identical—dynamics in terms of their systems of federalism; that is, "a system of constitutionally derived and apportioned authority where state and national governments retain sovereignty and yet at the same time are interdependent" (Scheberle, 2013, p. 395). Of course, in the Canadian case we are referring to "provinces," not states, but conceptually, they are parallel forms of subnational government. Constitutionally, Canada's national government is stronger relative to its provinces than the United States' is relative to its states. However, Macdonald (2020) documents how this has shifted over time in favor of subnational provinces having more power vis-à-vis the national government. Furthermore, in the realm of energy specifically, even the Canadian Constitution is explicit that the primary governing role is reserved for the provinces (Gattinger, 2015); however, the Canadian federal government maintains jurisdiction over nuclear/atomic energy (Macdonald, 2020).

In the United States, the Tenth Amendment, among other constitutional and jurisprudential guidance, establishes a strong role for state governments in general (Rosenthal & Joseph,

2021). In the domain of energy and climate policy in particular, certain states have seized upon this constitutional autonomy (e.g., Basseches et al., 2022; Bromley-Trujillo et al., 2016; Rabe, 2011). Although many states acted on climate decades before the US Congress finally did so with the recent passage of the Inflation Reduction Act in 2022, and Rabe (2004) writes about features of state-level policymaking that may have facilitated this early leadership, it is important to note that state-level climate policymaking has been highly uneven across the states (Basseches et al., 2022). In general, it has tended to be wealthier, less energy-intensive, and Democratic Party-controlled states that have taken on the mantle of state-level climate policy action leaders, and the inverse has tended to be state-level climate policy laggards (e.g., Bromley-Trujillo et al., 2016; Coley & Hess, 2012; Trachtman & Meckling, 2022). In the next section, I provide some specific examples of subnational leadership—even in the absence of national government climate action.

Subnational Climate Policy Leadership in the United States and Canada

In both the United States and Canada, certain subnational governments have attempted a whole-of-government approach at tackling the climate crisis through the policy instrument of economy-wide greenhouse gas (GHG) emissions stabilization/reduction targets. While the sectoral scope of such policies is broad (they are "economy-wide"), their enforceability and the degree to which subnational governments also implement other complementary policies in pursuit of these high-level goals vary considerably from one jurisdiction to the next (Glasgow et al., 2021). In Canada, Quebec was the first province to adopt such a target in 1992, aiming to stabilize overall provincial GHG emissions by returning them to their 1990 levels by 2000. British Columbia then did the same in 1995. By 2012, all the Canadian provinces had one of these economy-wide GHG emissions targets in place except for the "carbon province" of Alberta, "which refused to give itself an absolute limit on emissions" (Macdonald, 2020, p. 77). Several of these provincial targets were amended over the years to become more ambitious (Macdonald, 2020). Similarly, in the case of the United States, 23 states have adopted economy-wide GHG targets. Of these, 13 were adopted legislatively and 10 were adopted through executive order (Center for Climate and Energy Solutions, 2024). The ones that were adopted legislatively are generally more "durable" (Rabe, 2016), though even their quality varies considerably from one adopting state to the next (Basseches, 2024b).

The most prevalent state-level climate policy instrument in the United States is renewable portfolio standards (RPS), which, having been adopted in 37 states,[5] are targeted squarely at electricity-sector emissions. The design of these policies varies considerably from one state to the next, but broadly speaking they mandate that certain percentages of utilities' electricity load originate from renewable or clean energy sources (Basseches, 2024a). In Canada, the preferred electricity sector provincial policy instruments have been Ontario's coal phase-out (Macdonald, 2020) and feed-in tariffs (Stokes, 2013). The RPS and feed-in tariff approach each have their advantages and disadvantages, but one advantage of the latter is that it tends to be more transmission-dependent (Rowlands, 2007), whereas RPS policies have often required additional, supportive policies to ensure adequate transmission; an example would be the competitive renewable energy zones (CREZ) policy in Texas (Hurlbut, 2008).

Cap-and-trade policies—although controversial (Cullenward & Victor, 2020), especially from a climate justice perspective (Basseches et al., 2021; Mendez, 2020)—have also been implemented by subnational jurisdictions in both countries. While Ontario failed to adopt a durable cap-and-trade program (Raymond, 2020), Quebec and California each have linked cap-and-trade programs. The Regional Greenhouse Gas Initiative (RGGI), which includes several states in the US Northeast, is regarded as being especially successful in its allowance allocation methods, which rely heavily on auctions (Raymond, 2016). By contrast, the California-Quebec program has more loopholes and exemptions, and the California regulations provide for a significant share of free or discounted allowances, which can undermine program effectiveness. On the other hand, the California program has greater sectoral coverage, whereas RGGI applies only to the electricity sector (Cullenward & Victor, 2020). Washington State recently adopted a statewide cap-and-trade program as well, though it currently faces the possibility of repeal through a ballot measure (Mulkern, 2024).

A variety of other subnational climate policy instruments have also been used. These include various forms of tax incentives, net metering and interconnection standards, and energy efficiency policies (Carley, 2011). Energy efficiency policies can take multiple forms, including public benefit funds (PBFs), which Prasad (2023) considers a form of a carbon tax, and energy efficiency resource standards (EERS), whose design more closely resembles an RPS than a PBF (Carley, 2011). The majority of these subnational-level policy instruments are focused on the electricity sector, GHG emissions from which Cullenward and Victor (2020, p. 64) argue governments have a "larger toolkit" to address.

One especially promising electricity sector policy, provided in California and a handful of other states, is known as community choice aggregation (CCA). Under this policy, municipalities can leverage their greater purchasing power of bulk electricity to assemble a greener portfolio of resources than the incumbent utility might be offering but retain access to the transmission and distribution system of the utility (US Environmental Protection Agency, 2024a). This policy innovation, which was actually born in Massachusetts as part of its 1997 electric sector restructuring legislation, is also notable for its "bottom-up" approach to energy policy reform, empowering local communities to lead on the energy transition (Hess, 2019; Hsu, 2022).

By comparison, given the significant share of total emissions in both countries that come from the transportation sector, there have been far fewer policy instruments to address these emissions. The exception would be low-carbon fuel standards (LCFS) in states such as California and Oregon (Center for Climate and Energy Solutions, 2024), as well as subsidies and tax credits for electric vehicle (EV) purchases (Huether et al., 2023). There is also the California waiver of the US Environmental Protection Agency (EPA)'s vehicle emissions standards under the Clean Air Act (US Environmental Protection Agency, 2024b)—something which has been the subject of ongoing litigation (Solis, 2022).

The next frontier of state-level climate policy instruments targeted at transportation sector emissions will undoubtedly have to do with promoting widespread adoption of EVs. This has already been incentivized through the Inflation Reduction Act (The White House, 2023), and there are some initial indications that these incentives have been effective (Heatmap News, 2024). However, many barriers to widespread adoption remain, and some states will undoubtedly work to lower them through state-level climate policy action. For example, in 2021, Massachusetts passed a state law requiring adjustments to utility rate making intended to lure private investment in EV charging infrastructure to the state (Kaiser, 2024). At the same time, mirroring the general pattern by which states have diverged into being leaders or laggards based largely on factors like partisanship, certain "red states" have responded to pro-EV policies with backlash. For example, the Louisiana legislature recently passed a bill that would prevent state agencies from "discriminating" against certain types of vehicles based on their fuel source, a move intended to "protect the 'freedom of choice' to buy gasoline vehicles" (Aton, 2024). There are also concerns about whether such policies will be designed with equity in mind, as preliminary data suggest that the tax benefits of the Inflation Reduction Act have flown primarily to wealthier Americans (Pontecorvo & Meyer, 2024).

It is also worth noting that subnational climate action in both the United States and Canada takes place at the local and/or municipal level, in addition to the state/provincial level. There are numerous examples of this in both countries (Hughes, 2019; Krause & Hawkins, 2021). An interesting line of research involves the *process* by which such local action comes about, and who participates in decision making. Almeida et al. (2023b) examine participation in local climate policymaking meetings in Fresno, California, and find that ties to existing forms of civic engagement, such as through unions and community-based organizations, were important in facilitating participation in such local climate policy meetings.

Given the multitude of political and economic conditions confounding the relationship between subnational climate policies and overall GHG emissions, as well as heterogeneity in policy design itself, measuring the effectiveness of any single policy instrument or combination of policy instruments is challenging. Still, Martin and Saikawa (2017) undertook an analysis of the impact of specific policy instruments on electricity sector carbon emissions and found that certain mandatory state-level policies such as mandatory GHG reporting and power plant–level emissions performance standards are significantly associated with a decline in electricity sector emissions while others are not. Bergquist and Warshaw (2023) created a quantitative index which aggregates 25 individual state-level policies to develop a stringency score for the overall "climate policy regime" of each of the 50 US states. They find that increased stringency of a state's overall climate policy regime was associated with declining per capita carbon emissions both in the electricity sector and economy-wide.

Clearly, the climate has benefitted from subnational mitigation policy action in the absence of national action and despite occasional rollbacks of subnational policies after their initial adoption (Stokes, 2020). Given the decades of gridlock, climate denial, and climate delay that have resulted from intentional obstruction on the part of a well-resourced network of fossil fuel interests (Brulle & Downie, 2022; Fisher, 2024), the climate has benefitted from leadership on the part of *subnational* governments in both the United States and Canada. Indeed, Rabe (2004, 2008) makes note of the unique benefits of states' ability to act as "laboratories of democracy" in the climate policy domain.

Although research finds social movement influence over the finer details of policy design to be limited (Basseches, 2019), there is no question that grassroots mobilization has been helpful in getting climate action on the policy agenda for those subnational governments that have led in this area (Basseches, 2024b). Moreover, social movement scholars have used empirical research to generate recommendations for further innovation in tactics and strategy which may be even more likely to yield policy results (Fisher, 2024).

However, climate and energy federalism are not without their pitfalls as well. Karapin (2020), who refers to energy federalism as a "double-edged sword," acknowledges the advantages of state-led climate policy in the United States but also highlights its drawbacks, concluding that overall its impact on renewable energy policy is "ambivalent" rather than positive (p. 29). Although recent growth in US renewable energy has been remarkable, it has lagged that of European countries like Spain, Germany, and the United Kingdom. Moreover, it has been highly uneven, led by a handful of states, including California.

This mirrors the uneven subnational policy dynamics by which certain states' policy action has been met with backlash from other states, and the national government has failed to play the strong, guiding role it has elsewhere. Karapin notes that state-led action "is limited by economic competitiveness concerns, because [state-level policies] create costs that are borne by business as well as residential consumers of electricity" (Karapin, 2020, p. 27). This leads to the (sometimes credible and other times not) threat of businesses fleeing to other states, with laxer climate and energy regulations. This concern is especially acute with firms that are energy-intensive and/or trade-exposed (Basseches, 2024b). Not only has the lack of national-level leadership created economic competitiveness concerns for those states that have elected to take action, thereby limiting the ambition of that action, but also it has created a situation in which state governments that have *not* acted on climate have become enclaves in which anti–climate action forces such as the fossil fuel industry have been able to take refuge and consolidate and grow their power. From an institutional perspective, the number of veto points afforded by the US system of federalism has hindered the potential that the best ideas from certain states can be adopted on a necessarily large enough scale (Karapin, 2020).

In Canada, Macdonald (2020) points to the pitfalls created by the system of strong federalism there. As in the United States, there is an unfortunate interaction between political economy and federalist institutions, making overall climate progress more difficult. Specifically, in Canada, Macdonald argues that political economy divides its provinces into two categories: "carbon provinces" such as Alberta and Saskatchewan, in which the operations of the oil and gas industries are concentrated and in which overall GHG emissions rose significantly between 1990 and 2017, and "hydro provinces" such as Ontario and Quebec, in which vast amounts of clean electricity from hydropower fuel more service-based and information-based economies (as opposed to natural resource extraction economies) and in which overall GHG emissions fell between 1990 and 2017. The climate ambition of the Canadian national government, not unlike in the US case, has oscillated back and forth depending on the political party in power in Ottawa, but, Macdonald argues, even the plans of climate-forward national governments like Trudeau's are held back by Canadian federalism. As Macdonald puts it, "A coordinated federal-provincial [policy] program is the only means of achieving a given [GHG emissions reduction] target" (Macdonald, 2020, p. 6).

Both Karapin and Macdonald take a bird's-eye view of the challenges that US and Canadian federalism have presented for meaningful climate action, looking at overall renewable energy growth (in Karapin's case) and at overall, economy-wide GHG emissions reduction targets (in Macdonald's case). However, the governance and policy coordination challenges that federalism poses for addressing the climate crisis can also be seen at a very technical and granular level when we look at electricity policy. In the next section, I will focus on the US case (the case that I know best from my own empirical research) to provide an important example of how, despite subnational climate policy action having historically been far preferable to no action at all, it is not sufficient for getting us where we need to go next, which is achieving 100% clean electricity so that as we begin to "electrify" the rest of our economy (transportation, buildings, industry, etc.) and consequently demand on the electricity grid surges, we can meet that demand without burning more fossil fuels.

State-Level Renewable Portfolio Standards Policies Meet an Electricity Grid Without Political Borders

Let us begin with a stubborn set of five facts. First, electrons are physically indistinguishable from one another once they enter onto the electricity grid; some originated from fossil-fueled sources such as coal and gas plants while others originated from clean or renewable sources such as solar, wind, and hydropower, but once they enter the grid, they become physically indistinguishable from one another. Policymakers that want to track their source because they are concerned about their origin for environmental reasons do so through an accounting system called renewable energy credits (RECs). One REC is produced per unit of electricity generated from a qualifying source. Depending on the state, RECs can be bought, sold, or traded independently from the electrons that created them to comply with state policy; these are known as "unbundled RECs" (Basseches, 2024a).

Second, the most prevalent state-level climate policy instrument is renewable portfolio standards (RPS) policies, which require that specified percentages of the electricity loads of utilities delivered *within* the state originate from renewable generation sources, tracked through RECs (Basseches, 2024a). A state legislature which would pass an RPS to be implemented through its public utility commission (PUC) has jurisdiction over the regulated electric utilities within its borders. However, third, the electricity system—one of the greatest engineering feats of the 20th century—is divided into three physical components: generation, transmission, and distribution; all three are necessary for a functioning electricity grid, and yet only one or two of them (depending on the state[6]) actually fall within a state's regulatory jurisdiction (DeLosa et al., 2024).

Fourth, most transmission, and especially most of the transmission that will be needed in the future is *interstate* transmission, which is regulated by the Federal Energy Regulatory Commission (FERC), not state governments. Through its Orders 888/889 and 2000, FERC authorized the formation of voluntary, quasi-governmental independent system operators (ISOs) and regional transmission organizations (RTOs), to which it delegated governance of electricity transmission in many parts of the country; in the rest of the country, the transmission system is not operated by state governments but rather by incumbent monopoly utilities (many but not all of which are privatized) that are ostensibly regulated by the state PUCs (DeLosa et al., 2024), though concerns about regulatory capture of these PUCs abound (Basseches, 2024b). Figure 30.1 shows which regions of the United States fall under the purview of these ISOs/RTOs; as the figure shows, these regions transcend state political borders.

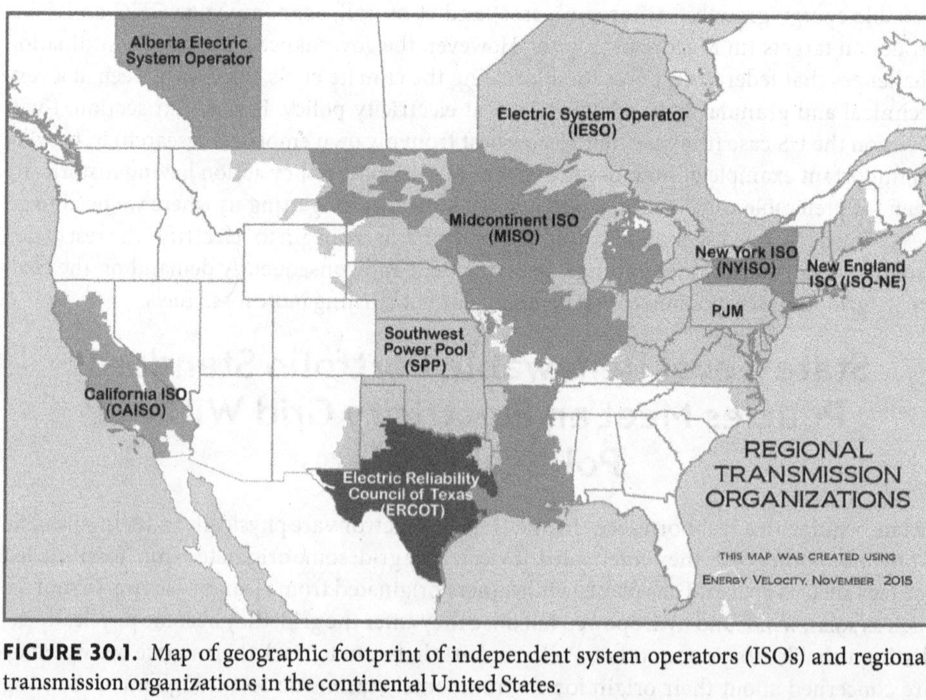

FIGURE 30.1. Map of geographic footprint of independent system operators (ISOs) and regional transmission organizations in the continental United States.

(*Source*: https://www.epa.gov/green-power-markets/power-market-structure)

Fifth, since the passage of the US Inflation Reduction Act, the single biggest hurdle the United States faces when it comes to greening the electricity grid is the lack of sufficient high-voltage transmission capacity which would be necessary to bring clean-sourced electricity from the (more often than not) rural areas in which it is generated to the (more often than not) urban and suburban areas in which it is consumed; this requires transmission lines that need to be constructed from scratch and that will inevitably cross state political borders (Harper et al., 2023). Taken together, these five facts help to illustrate why we are fast approaching a ceiling—at least in the electricity sector—in terms of subnational climate policy action's potential to take us from where we are right now to where we must go in the future to further combat the climate crisis, from a mitigation policy perspective.

To be sure, these state-level RPS policies in the approximately 30 states that have adopted durable ones have been critically important, and successful, in bringing us to where we are now. They have worked to lower the cost of renewable power and to transform the market for renewable electricity generation, since the RECs are in effect a subsidy available exclusively to nonemitting generators (Fischlein & Smith, 2013). As studies have shown, these state-level policies have contributed to real emissions reductions in the power sector, one of the largest sources of overall emissions in the economy (Bergquist & Warshaw, 2023).

And yet, because of the five facts just mentioned, and importantly for the purposes of this chapter, because RPS policies' jurisdiction is limited to the political borders of particular states, they are an instrument whose effect on the electricity system as a whole will always be limited. By themselves, they are unable to fully deliver upon the promise of an interconnected, effectively coordinated, and 100% clean-powered electricity grid because they cannot ensure adequate interstate transmission. In the broader context of challenges to effective electricity system governance that go far beyond the limitations of RPS policies (see Klass et al., 2024), I now turn to elaborate specifically upon how these five facts will limit the utility of strictly subnational clean electricity policy (in the form of RPSs) moving forward.

At its core, the target-setting aspect of state-level RPS policies is a political exercise, so that politicians can signal to (typically co-partisan) voters and organized constituencies such as environmental interest groups and social movement organizations (SMOs) that the state prioritizes renewable or "clean" electricity generation sources. However, this target-setting process, which typically occurs in state legislatures, is often divorced from the mechanics of operating the grid in real time as well as from the complex regulatory and market structures that govern things like transmission access, wholesale electricity pricing, cost allocation to ratepayers, and so on. In other words, this target setting can often occur in a silo that is separate from the nuts and bolts of how the actual grid functions and is governed. While RPS targets are typically put forward by state legislatures where members are directly accountable to voters, how the grid actually functions as a coherent whole, in which electricity supply and demand are always perfectly synchronized, is currently determined by a patchwork of regulatory and market structures, including state PUCs, ISOs/RTOs, and FERC (Angwin, 2020; DeLosa et al., 2024; Klass et al., 2024).

Governance of the entire physical grid is siloed, with different actors responding to different incentives (whether political, economic, or both) and focusing only on their one piece of the puzzle without seeing how the different parts of the overall system known as the electricity grid relate to one another and are dependent on one another for optimal functioning (Basseches & Ikenze, 2022; Klass et al., 2024). During the last few decades, in which grid planning that had historically operated with only least-cost-service and reliability principles in mind has begun to also incorporate environmental concerns, pushed to do so by individual state policies, state-level RPS policy has been helpful *in the absence* of any national action. However, now that the federal government has finally acted in the form of the Inflation

Reduction Act of 2022, the optimal implementation of that law from a climate perspective—which would involve centralizing grid planning with climate goals in mind rather than *just* least-cost-service and reliability—is being hamstrung by the very siloed nature of grid governance that allowed past climate action on the part of only *some* subnational governments to precede national action.

Given the blind spots of state-level RPS policies (e.g., building *interstate* transmission capacity sufficient to accommodate the demand for new sources of local generation made possible by such policies as RPS and tax incentives), we have now arrived at a moment in which the national government must take the reins and impose on states the optimal policies for clean electricity grid function. Because the Inflation Reduction Act was merely a budget reconciliation bill, adhering to the US Senate's Byrd Rule in order to allow it to be passed exclusively with Democratic votes rather than the filibuster-proof 60 votes required of other major legislation to pass the Senate (Davenport & Friedman, 2022), it was unable to impose the sorts of regulatory reforms that would be necessary to truly transform our electricity grid.

When national policy finally (and hopefully) does so, it would ideally also address the distributional consequences of the renewable energy transition (Mildenberger, 2020). That is, rather than only concerning itself with rapid decarbonization, it should *also* address the monopoly power that investor-owned utilities currently exercise, which enables them to shift financial costs and risks from themselves onto ratepayers (Basseches, 2024b). This will be the bare minimum necessary to ensure an *equitable* transition.

In the absence of congressional action, FERC recently took a major positive step with its adoption of Order 1920 governing interstate transmission planning. Still, this rule is necessary but insufficient (DiGangi, 2024). Furthermore, it has sparked a backlash by state regulators and their sympathizers, who see it as a power grab by the national government at the expense of state autonomy (Howland, 2024a, 2024b). This illustrates how the durability of institutions of US federalism, which have historically been a great asset to climate policy action, now may pose one of the greatest threats.

Conclusion

In focusing on policy outcomes rather than the social processes by which policy emerges, this chapter has illustrated only one narrow piece of the many forms of climate action. Moreover, this chapter relates to just one subset of climate actions having to do with what Almeida and coauthors call "institutional climate actions" (Almeida et al., 2023a). The sociology of climate change is underdeveloped (Scoville & McCumber, 2023), and this volume thus offers an important corrective. While sociologists have rightly focused on many other aspects of climate change, the discipline has some catching up to do when it comes to a focus on climate *policy* action, and specifically, *subnational* policy action (Vasi & Walker, 2024).

In this chapter, I have argued that indeed subnational policy has been an important form of climate action. I have shown how far it has gotten us, in both Canada and the United States, in the face of powerful forces that have obstructed national action (Brulle & Downie, 2022). There are even some promising state- and local-level policy tools, such as CCA, that have the potential to combat the power that monopoly investor-owned utilities exert over grid planning and resource ownership.

Nevertheless, I have also pointed out—using the example of the US electricity sector—the limits of relying exclusively on subnational policy action as the *only* policy action. If we are serious about combatting the climate crisis, we must continue to encourage state ambition while *also* emphasizing the importance of cohesive *national* and *international* climate policy that can bring with its larger jurisdiction the type of coordination and foresight necessary to take on the "super wicked problem" of climate change (Levin et al., 2012).

Notes

1. Overall, the United States is the second largest after China at 6001.2 $MtCO_2e$, and Canada is tenth at 736.9 $MtCO_2e$, but given Canada's much smaller population, the two countries are much closer together when measured per capita.
2. However, third parties in Canada such as the New Democratic Party (NDP) have achieved much greater gains than third parties in the United States (Johnston, 2017).
3. Connecticut, Maine, Massachusetts, New Hampshire, Rhode Island, and Vermont.
4. New Brunswick, Newfoundland and Labrador, Nova Scotia, Prince Edward Island, and Québec.
5. Some are purely voluntary (see Vasseur, 2014), and others have since been repealed.
6. Specifically, depending upon whether the state restructured its electric utility sector (Isser, 2015); in restructured states, generation is "deregulated" and state governments only have authority over distribution and certain *intra*-state transmission, while in traditionally regulated states, the state government has authority over generation as well.

References

Almeida, P., González Márquez, L. R., & Fonsah, E. (2023a). The forms of climate action. *Sociology Compass, 18*(2), 1–15.

Almeida, P., Rubén González, L., Orozco Flores, E., Curry, V., & Padilla, A. (2023b). The building blocks of community participation in local climate meetings. *NPJ Climate Action*, 2(1), 37.

Angwin, M. (2020). *Shorting the grid: The hidden fragility of our electric grid.* Carnot Communications.

Aton, A. (2024, May 24). Louisiana set to protect "freedom" to buy gasoline-powered cars. *E&E News.*

Basseches, J. A. (2019). "It happened behind closed doors": Legislative buffering as an informal mechanism of political mediation. *Mobilization: An International Quarterly, 24*(3), 365–388.

Basseches, J. A. (2024a). Renewable portfolio/clean energy standards. In D. J. Fiorino (Ed.), *Elgar encyclopedia of climate policy* (pp. 356–360). Edward Elgar.

Basseches, J. A. (2024b). Who pays for environmental policy? Business power and the design of state-level climate policies. *Politics and Society, 52*(3), 409–451.

Basseches, J. A., Bromley-Trujillo, R., Boykoff, M. T., Culhane, T., Hall, G., Healy, N., Hess, D. J., Hsu, D., Krause, R. M., Prechel, H., Roberts, J. T., & Stephens, J. C. (2022). Climate policy conflict in the U.S. states: A critical review and way forward. *Climatic Change, 170*(3), 32.

Basseches, J. A., & Ikenze, N. (2022). *The U.S.-Canada (clean) electricity relationship: challenges and opportunities in policy design and coordination.* North American Colloquium, University of Michigan.

Basseches, J. A., Rubinstein, K., & Kulaga, S. M. (2021). Coalitions that clash: California's climate leadership and the perpetuation of environmental inequality. In D. Pettinicchio (Ed.), *Research in political sociology* (pp. 23–44). Emerald.

Bergquist, P., & Warshaw, C. (2023). How climate policy commitments influence energy systems and the economies of U.S. states. *Nature Communications, 14*(1).

Bromley-Trujillo, R., Butler, J. S., Poe, J., & Davis, W. (2016). The spreading of innovation: State adoptions of energy and climate change policy. *Review of Policy Research, 33*(5), 544–565.

Brulle, R., & Downie, C. (2022). Following the money: Trade associations, political activity, and climate change. *Climatic Change, 175*(3), 1–19.

California Air Resources Board. (2024). Program linkage. https://ww2.arb.ca.gov/our-work/programs/cap-and-trade-program/program-linkage

Carley, S. (2011). The era of state energy policy innovation: A review of policy instruments. *Review of Policy Research, 28*(3), 265–294.

Center for Climate and Energy Solutions. (2024). State climate policy maps. https://www.c2es.org/content/state-climate-policy/

Coley, J. S., & Hess, D. J. (2012). Green energy laws and Republican legislators in the United States. *Energy Policy*, *48*, 576–583.

CONEG. (2024). NEG/ECP. https://www.coneg.org/neg-ecp/

Cullenward, D., & Victor, D. G. (2020). *Making climate policy work*. Polity Press.

Davenport, C., & Friedman, L. (2022). Five decades in the making: Why it took Congress so long to act on climate. *New York Times*, August 8.

DeLosa III, J., Pfeifenberger, J. P., & Joskow, P. L. (2024). Regulation of access, pricing, and planning of high voltage transmission in the U.S. MIT CEEPR Working Paper Series, CEEPR WP 2024-03.

DiGangi, D. (2024, May 14). Clean energy groups praise FERC transmission order but want more reform. *Utility Dive*.

Fischlein, M., & Smith, T. M. (2013). Revisiting renewable portfolio standard effectiveness: Policy design and outcome specification matter. *Policy Sciences*, *46*(3), 277–310.

Fisher, D. R. (2024). *Saving ourselves: From climate shocks to climate action*. Columbia University Press.

Gattinger, M. (2015). A national energy strategy for Canada: Golden Age or Golden Cage of energy federalism? In A. Juneau, C. H. Tuohy, & L. Berdhal (Eds.), *Canada: The state of the federation* (pp. 39–69). Queens University Press.

Glasgow, D., Zhao, S., & Rai, S. (2021). Rethinking climate change leadership: An analysis of the ambitiousness of state GHG targets. *Review of Policy Research*, *38*(4), 398–426.

Government of Canada. (2024a). Global greenhouse gas emissions. https://www.canada.ca/en/environment-climate-change/services/environmental-indicators/global-greenhouse-gas-emissions.html

Government of Canada. (2024b). Greenhouse gas emissions. https://www.canada.ca/en/environment-climate-change/services/environmental-indicators/greenhouse-gas-emissions.html

Harper, C., Krasnow, S., Stokes, L. C., Lynch, L., Ricketts, S., Levin, A., Schulman, D., Slyfield, J., & Walsh, C. (2023). *Powering toward 100 percent clean power by 2035: The path to carbon-free electricity after the Inflation Reduction Act*. Evergreen Action and NRDC.

Heatmap News. (2024, February 23). Transcript: Is Biden's climate law actually working? *Shift Key* [Podcast].

Hess, D. J. (2019). Coalitions, framing, and the politics of energy transitions: Local democracy and community choice in California. *Energy Research and Social Science*, *50*, 38–50.

Howland, E. (2024a). States shouldn't have to pay for transmission driven by other states' policies: FERC's Christie. *Utility Dive*, March 26.

Howland, E. (2024b). FERC's transmission rule likely boon to consumers, but Christie dissent is a blueprint for litigation, analysts say. *Utility Dive*, May 16.

Hsu, D. (2022). Straight out of Cape Cod: The origin of community choice aggregation and its spread to other states. *Energy Research and Social Science*, *86*.

Huether, P., Cohn, C., Jennings, B., Mah, J., Tolentino, C., & Vaidyanathan, S. (2023). *State transportation electrification scorecard*. American Council for an Energy Efficient Economy (ACEEE). https://www.acee.org/research-report/t2301

Hughes, S. 2019. *Repowering cities: Governing climate change mitigation in New York City, Los Angeles, and Toronto*. Cornell University Press.

Hurlbut, D. (2008). A look behind the Texas renewable portfolio standard: A case study. *Natural Resources Journal*, *48*(1), 129–161.

Isser, S. (2015). *Electricity restructuring in the United States: Markets and policy from the 1978 energy policy act to present*. Cambridge University Press.

Johnston, R. (2017). *The Canadian party system: An analytic history*. University of British Columbia Press.

Kaiser, C. (2024, January 5). To spur more EV chargers, regulators must get creative with demand charges. *Utility Dive*.

Karapin, R. (2020). Federalism as a double-edged sword: The slow energy transition in the United States. *Journal of Environment and Development*, *29*(1), 26–50.

Klass, A., Macey, J., Welton, S., & Wiseman, H. (2024, March). *The key to electric grid reliability: Modernizing governance*. Center for Applied Environmental Law and Policy.

Krause, R. M., & Hawkins, C. V. (2021). *Implementing city sustainability: Overcoming administrative silos to achieve functional collective action*. Temple University Press.

Levin, K., Cashore, B., Bernstein, S., & Auld, G. (2012). Overcoming the tragedy of super wicked problems: Constraining our future selves to ameliorate global climate change. *Policy Sciences*, *45*(2), 123–152.

Macdonald, D. (2020). *Carbon province, hydro province: The challenge of Canadian energy and climate federalism*. University of Toronto Press.

Martin, G., & Saikawa, E. (2017). Effectiveness of state climate and energy policies in reducing power-sector CO_2 emissions. *Nature Climate Change*, *7*(12), 912–919.

Mendez, M. (2020). *Climate change from the streets*. Yale University Press.

Mildenberger, M. (2020). *Carbon captured: How business and labor control climate politics*. MIT Press.

Mulkern, A. C. (2024, April 5). Bid to repeal Washington cap-and-trade hits a nerve in California. *E&E News*.

Oatley, T. (2023). The dual economy, climate change, and the polarization of American politics. *Socio-Economic Review*.

Pontecorvo, E., & Meyer, R. (2024, August 7). The first IRA tax credit data is in. *Heatmap News*.

Prasad, M. (2023). A carbon tax by any other name: Public benefit funds in the American states. *PLOS Climate*, *2*(2), 1–12.

Rabe, B. G. (2004). *Statehouse and greenhouse*. Brookings Institution Press.

Rabe, B. G. (2008). States on steroids: The intergovernmental odyssey of American climate policy. *Review of Policy Research*, *25*(2), 105–128.

Rabe, B. G. (2011). Contested federalism and American climate policy. *Publius: The Journal of Federalism*, *41*(3), 494–521.

Rabe, B. G. (2016). The durability of carbon cap-and-trade policy. *Governance*, *29*(1), 103–119.

Raymond, L. (2016). *Reclaiming the atmospheric commons: The Regional Greenhouse Gas Initiative and a new model of emissions trading*. The MIT Press.

Raymond, L. (2020). Carbon pricing and economic populism: The case of Ontario. *Climate Policy*, *20*(9), 1127–1140.

Rosenthal, L. H., & Joseph, G. P. (2021). Foundations of U.S. federalism. *Judicature*, *101*(1), 39–50.

Rowlands, I. H. (2007). The development of renewable electricity policy in the province of Ontario. *Review of Policy Research*, *24*(3), 185–207.

Scheberle, D. (2013). Environmental federalism and the role of state and local governments. In M. E. Kraft & S. Kamieniecki (Eds.), *The Oxford Handbook of U.S. Environmental Policy* (pp. 394–412). Oxford University Press.

Scoville, C., & McCumber, A. (2023). Climate silence in sociology? How elite American sociology, environmental sociology, and science and technology studies treat climate change. *Sociological Perspectives*, *66*, 888–913.

Solis, N. (2022, May 13). 17 states push EPA to revoke California's ability to set its own vehicle emission standards. *LA Times*.

Sosland, D. L. (2023). International clean energy grid coordination in the Northeast and Eastern Canada: Starting the Northeast grid planning forum dialogue. https://acadiacenter.org/international-clean-energy-grid-coordination-in-the-northeast-and-eastern-canada-starting-the-northeast-grid-planning-forum-dialogue/

Stokes, L. C. (2013). The politics of renewable energy policies: The case of feed-in tariffs in Ontario, Canada. *Energy Policy*, *56*, 490–500.

Stokes, L. C. (2020). *Short-circuiting policy: Interest groups and the battle over clean energy and climate policy in the American states*. Oxford University Press.

Trachtman, S., & Meckling, J. (2022). The climate advocacy gap. *Climatic Change, 172*(24), 1–7.
US Environmental Protection Agency. (2024a). Community choice aggregation. https://www.epa.gov/green-power-markets/community-choice-aggregation
US Environmental Protection Agency. (2024b). Vehicle emissions California waivers and authorizations. https://www.epa.gov/state-and-local-transportation/vehicle-emissions-california-waivers-and-authorizations
US Environmental Protection Agency. (2024c). Sources of greenhouse gas emissions. https://www.epa.gov/ghgemissions/sources-greenhouse-gas-emissions
VanNijnatten, D., & McWhinney, M. (2022). *Canada-U.S. green bilateralism: Targeting cooperation for climate mitigation*. North American Colloquium, University of Michigan.
Vasi, I. B., & Walker, E. T. (2024). Subnational environmental policy: Trends and issues. *Annual Review of Sociology, 50*, 319–339.
Vasseur, M. (2014). Convergence and divergence in renewable energy policy among U.S. states from 1998 to 2011. *Social Forces, 92*(4), 1637–1657.
The White House. (2023). *Building a clean energy economy: A guidebook to the Inflation Reduction Act's investments in clean energy and climate action*.
World Bank. (2024). GDP per capita. https://data.worldbank.org/indicator/NY.GDP.PCAP.CD?most_recent_value_desc=true

CHAPTER 31

Climate Action in the Framework of the European Green Deal

JALE TOSUN

Introduction

The European Union (EU) constitutes a distinctive political system characterized by a unique distribution of policymaking authority that defies conventional categorizations of either federal or intergovernmental governance. Within this sui generis institutional framework (Hix, 1998), member states have delegated certain policymaking competences entirely to the EU level (such policies relating to the customs union), retained exclusive control over others within their national jurisdictions (including areas like education and direct taxation), and established shared competences in numerous policy domains where both EU institutions and member states exercise concurrent authority (such as climate and environmental policy). This complex competence framework creates a dynamic system of multilevel governance where policy outcomes result from intricate interactions between supranational EU institutions and national governments, each operating within their respective spheres of authority while navigating areas of overlapping jurisdiction (Hooghe & Marks, 2001).

Climate and environmental policy fall within the shared competences of the EU and its member states. EU-level policies in these areas result from the ordinary legislative procedure, a process where the European Commission, the European Parliament, and the Council of the EU collaborate on legislation. The European Commission, which serves as the EU's executive branch, holds the exclusive right to propose EU legislation. The European Parliament, composed of directly elected members, and the Council of the EU, composed of member state sectoral ministers, function as co-legislators in the policymaking process (Dupont et al., 2024).

As the "motor of integration," the European Commission has played an important role in advancing the EU's policies addressing climate change and environmental degradation (Knill & Liefferink, 2007), which results not only through its exclusive right to initiate legislation but also other mechanisms such as by its ability to shape the EU's broader policy agenda and coalition building (Bürgin, 2023). Its most recent initiative in the domains of climate and environmental policy is the European Green Deal (EGD), which was launched by the first Commission led by Ursula von der Leyen in December 2019. The EGD is even more than that: It is the EU's flagship green growth strategy, represents a comprehensive policy framework that encompasses, and integrates goals related to both environmental protection and climate action. The EGD's holistic approach also aims to achieve a "just transition" through the "leave no one behind" principle (Sandmann et al., 2024). This chapter offers an overview of the EGD, exploring the different types of climate and environmental action it includes. It begins with a broad overview of EU climate and environmental policy and then provides a detailed examination of the EGD framework. The analysis evaluates different components of the EGD to demonstrate how they target distinct stakeholder groups and policy areas. The chapter concludes with reflections on the overall approach and its implications.

EU Climate and Environmental Policy

The development of European environmental policy began in 1972 with the adoption of the first Environmental Action Program (EAP). Covering the period from 1973 to 1976, this inaugural EAP emerged as a direct response to the landmark United Nations Conference on the Human Environment held in Stockholm in 1972, reflecting the growing international awareness of environmental degradation and the urgent need for coordinated policy responses, especially also in view of the transboundary economic activities in the common market. The first EAP catalyzed the creation of a formal organizational structure for environmental policy implementation in 1973 through the establishment of the Environmental Unit within the Directorate-General (DG) for Industrial Policies (Knill & Liefferink, 2007). This modest beginning represented the EU's first dedicated institutional capacity for environmental matters. Recognizing the increasing importance and complexity of environmental challenges, the Environmental Unit underwent significant transformation and was elevated to become a fully fledged DG for environmental policies in 1981, providing it with greater administrative weight and policymaking authority.

As successive EAPs became progressively more comprehensive and ambitious in scope, addressing broader environmental concerns, and setting more stringent environmental protection targets, the areas of competence and influence of DG Environment expanded correspondingly. This institutional growth reflected the EU's evolving understanding of environmental challenges as interconnected issues requiring integrated solutions. A particularly significant milestone occurred in 1999 with the formal adoption of the Environmental Policy Integration (EPI) principle, which fundamentally transformed environmental considerations from a sectoral concern into a mainstream element of EU policymaking across all policy domains (Jordan & Lenschow, 2010). This principle is enshrined in Article 11 of the Treaty on the Functioning of the European Union (TFEU), which states that "environmental protection requirements must be integrated into the definition and implementation of the Union's policies and activities."

The principle includes both horizontal and vertical policy integration (Lafferty & Hovden, 2003). Horizontal policy integration involves ensuring that different policy areas at the same governance level coordinate to address environmental concerns. For example, according to the horizontal dimension of EPI, the EU's common agricultural policy should consider environmental issues (Feindt, 2010). Vertical policy integration is about ensuring that environmental considerations flow through different levels of governance, that is, from the EU level down to the member states and within the member states to the subnational level (Schleyer et al., 2015). Beyond Article 11 of the TFEU, EPI is also supported by secondary legislation, such as the EU's Environmental Impact Assessment Directive, which stipulates that major building or development projects in the EU must first be assessed for their impact on the environment. The EGP represents one of the EU's most ambitious attempts to achieve EPI across all EU policies.

Beyond the original concept of EPI, the EU has expanded it to include the integration of climate policy into other sectoral policies (Dupont, 2015; Dupont et al., 2024). Climate change emerged as a priority area receiving enhanced attention from the late 1990s onward, as scientific evidence of anthropogenic climate impacts became increasingly compelling. Initially, the EU's climate policy was formulated and coordinated by DG Environment, treating climate action as a subset of broader environmental policy. This institutional arrangement resulted in ambitious climate policies such as the EU Emissions Trading System, which was launched in 2005. However, recognizing the distinctive nature and urgency of climate challenges, the EU undertook a significant institutional reorganization in 2010 by establishing a dedicated DG for Climate Action, to which climate policy responsibilities were transferred (Rayner & Jordan, 2016). This structural change demonstrated the EU's recognition of climate change as a policy area requiring specialized expertise and focused attention.

The current coexistence of both DG Environment and DG Climate Action illustrates the EU's comprehensive commitment to environmental protection and climate action, while acknowledging the distinct yet complementary nature of these policy domains as reflected by the EGD. This dual institutional framework enables the EU to address both traditional environmental concerns and the specific challenges posed by climate change through specialized yet coordinated approaches.

European Green Deal

The EGD articulates three fundamental and ambitious objectives: achieving climate neutrality through zero net emissions of greenhouse gases by 2050, fundamentally decoupling economic growth from resource consumption and environmental degradation, and ensuring a just transition that leaves no one behind in the transformation toward sustainability (European Commission, 2019).

These interconnected goals reflect the EU's recognition that environmental sustainability cannot be achieved without addressing climate change, and that both objectives must be pursued in a manner that promotes social equity and economic prosperity, which aligns with international policy agendas such as most importantly the United Nations' Sustainable Development Goals adopted in 2015 (Koundouri et al., 2024). The global awareness of sustainable development and the momentum for climate action generated by the 2015 Paris Agreement, and the rise of climate movements across the EU and beyond, help explain why the EGD is not unique—similar multidimensional policy agendas have emerged in other jurisdictions as well. Remarkably, while the US government under Joe Biden did not formally adopt a Green New Deal, it launched ambitious climate initiatives including the Inflation Reduction Act in 2022 that incorporated similar principles. The Green New Deal, a proposal primarily associated with progressive Democrats that was not enacted, shares "striking similarities" in its novel policy architecture (Bloomfield & Steward, 2020, p. 773).

However, even by European standard, the EGD is considered more disruptive than previous policy attempts which were marked by incremental steps and developments. Based on an in-depth review of the literature, Dupont et al. (2024) state that the EGD indeed breaks with the past patterns of EU climate policy and governance, while also maintaining some of its elements. The authors contend that the quality of integrated policy approaches exceeds past efforts as the EGD highlights that all policies and projects of the EU need to integrate the overarching goal of climate neutrality (European Commission, 2019). The authors continue by arguing that "the aim to achieve a just transition is perhaps the clearest example of the expansion of the integrated approach under the EGD—it connects EU climate (mitigation) policy to social policy, an area where the EU has limited policymaking competence and few available policy tools" (Dupont et al., 2024, p. 6).

Before engaging more with the EGD's approach to just transition, it is worth outlining its main policy dimensions, which comprise the following (European Commission, 2019):

- Increasing the EU's climate ambition for 2030 and 2050
- Supplying clean, affordable, and secure energy

- Mobilizing industry for a clean and circular economy
- Building and renovating in an energy- and resource-efficient way
- Accelerating the shift to sustainable and smart mobility
- From "Farm to Fork": designing a fair, healthy, and environmentally friendly food system
- Preserving and restoring ecosystems and biodiversity
- A zero pollution ambition for a toxic-free environment

Delivering on these policy dimensions requires legislative action initiated by the European Commission. Regarding the EU's climate ambitions for 2030 and 2050, the "Fit for 55" legislative package is particularly relevant. This package includes measures designed to achieve the 2030 target of reducing greenhouse gas emissions by at least 55% compared to 1990 levels (Oberthür & Kulovesi, 2025). It represents the largest collection of climate governance policy measures ever published by the Commission (Dupont et al., 2024). Key components of the package were adopted in April 2023.

The goal of supplying clean, affordable, and secure energy has proven more challenging to implement due to Russia's war of aggression against Ukraine and its impact on natural gas deliveries to the EU. Consequently, beginning in 2022, the European Commission focused on securing natural gas supplies by diversifying sources. To this end, it adopted the RePowerEU plan as a comprehensive strategy (Homeyer et al., 2022). However, particularly in 2022, the EU temporarily increased its share of electricity from coal-fired power plants to replace Russian gas supplies (Tosun, 2023).

The European Commission adopted the new circular economy action plan in March 2020 as one of several measures to deliver on the third policy dimensions. Comprising 35 actions, the action plan announces initiatives along the entire life cycle of products and targets how products are designed (European Commission, 2020). Furthermore, it promotes circular economy processes, encourages sustainable consumption, and seeks to ensure that waste is prevented and the resources used are kept in the EU economy for as long as possible.

Regarding the fourth policy dimension, the EU, inter alia, revised the Energy Performance of Buildings Directive in 2024. The goal of the revised directive is to ensure all buildings are more energy-efficient and climate-resilient. The measures target both new and old buildings. Regarding new buildings, the directive stipulates that they all reach zero emissions by 2030. For old building, the introduction of renovation passports aims to support building owners as they plan staged renovations.

Turning to the mobility sector, the EU institutions reached a landmark agreement in 2023 to phase out sales of new combustion engine cars by 2035, marking a decisive shift toward electric vehicle adoption. This policy was initially hailed as a crucial step in achieving the EU's climate neutrality goals. However, this decision has since become increasingly contested amid changing political and economic circumstances. The EU's economic stagnation has intensified debates about the timing and feasibility of such transformative policies (Politico, 2025). Simultaneously, the electoral success of right-wing and populist parties across numerous member states has created new political dynamics that challenge the consensus behind ambitious climate policies. These parties often frame such measures as economically burdensome and potentially harmful to traditional industries and workers, thereby mobilizing opposition to what was once considered settled policy within EU climate governance (Paterson et al., 2024; Tosun, 2023).

Even more controversy accompanied the delivery of the Farm to Fork strategy, which aims at transforming European agriculture toward sustainability. The strategy outlines ambitious goals to reduce the environmental and climate footprint of food production, such as halving pesticide use and risk, reducing nutrient losses by at least 50%, lowering antimicrobial sales for farmed animals, and increasing organic farming to 25% of EU agricultural land by 2030. Originally regarded as an initiative of a potentially game-changing nature (Schebesta & Candel, 2020), it has been met with opposition from farmers, industry groups, and political groups, most importantly the European People's Party Group, leading to delays and modifications in proposed policies (Finger et al., 2024; Tosun, 2023; Tosun et al., 2024).

The policy dimension on ecosystems and biodiversity is closely interlinked with the Nature Restoration Law, which stipulates restoration objectives for the recovery of nature in the land and sea areas of the EU. The law not only concerns biodiversity but also aims to tackle climate change and introduce ecosystem services to prevent natural disasters and increase food security. The passage of this legislation was anything but smooth. The Members of the European Parliament affiliated with the right-leaning European People's Party have been vocal critics of the proposal. Because of this opposition, the proposal failed in three parliamentary committees but survived the vote in the parliamentary plenary and was adopted in an amended version by a knife's edge. Subsequently, the proposal got stuck in the Council of the EU as it failed to gain the qualified majority votes necessary for adoption. Surprisingly, the Council then passed it in June 2024 and now needs to be implemented by the EU member states (Tosun et al., 2024).

Regarding the final policy dimension, in May 2021, the Commission adopted the EU Action Plan entitled "Towards Zero Pollution for Air, Water and Soil." This action plan is particularly noteworthy because it is implemented through a Zero Pollution Stakeholder Platform. The platform aims to connect stakeholders and experts from different policy areas, including health, agriculture, research and innovation, transport, digitalization, and the environment. It represents one of several elements of the EGD that offers participation opportunities for a wide range of organizations. The Zero Pollution Stakeholder Platform welcomes members from academia, business, civil society, cities and regions, representatives of EU member states, and observers from international organizations. Stakeholders can interact through various formats, such as the "Zero Pollution Talks," which facilitate knowledge exchange, raise awareness, and identify key issues that must be addressed to achieve zero pollution. We will return to the use of participatory governance in subsequent sections.[1]

The European Commission (2019) acknowledges that realizing the ambitious objectives of the EGD will require substantial and sustained investment. This need is particularly pronounced in the context of the EU's commitment to ensuring a just transition. The EGD explicitly recognizes that the green transition can only be successful if it is implemented in a fair, inclusive, and socially balanced manner, which the literature associates with worker- and equity-centered processes and outcomes (Kerrissey & Coon, this volume). This involves addressing the differentiated impacts that climate and environmental policies may have on individuals, communities, economic sectors, and regions across the EU.

To operationalize this commitment, the EU has established the Just Transition Mechanism, which includes the Just Transition Fund as one of its core components (Sandmann et al., 2024). This mechanism is specifically designed to mitigate the socioeconomic consequences of the transition, with the overarching objective of ensuring that no one is left behind. It aims to provide targeted support to the most affected regions and stakeholders, thereby reinforcing social cohesion and maintaining public support for the broader climate agenda. The Just Transition Fund offers targets support for different groups. One of them is workers who are transitioning out of carbon-intensive sectors. One way of supporting displaced workers is by creating new job opportunities in other (green) sectors as well as to re- or up-skill them (Busemeyer et al., 2025). Another target group is regions undergoing industrial transformation and the small- and medium-sized enterprises located in them. These regions may, for instance, require funding for clean energy projects.

The European Commission (2019) characterizes both the financing of the transition and the commitment to a just transition as key instruments for embedding sustainability across all EU policies. These mechanisms are not merely supportive measures; rather, they play a strategic, instrumental role in advancing the integration of climate and environmental considerations into the broader policy framework of the European Union. In essence, they are fundamental to achieving what is often referred to as climate and environmental policy integration.

In addition to these core instruments, another significant—though by no means the only—tool for mainstreaming sustainability involves the mobilization of research and innovation, particularly in the development and deployment of clean technologies. This approach underscores the EU's recognition that technological advancement and scientific research are vital drivers of the systemic transformation needed to meet long-term climate goals and ensure a sustainable future. Both financing the transition and the commitment to just transition are characterized by the EU Commission as instruments for achieving the mainstreaming of sustainability across all EU policies (European Commission, 2019). In other words, they serve an instrumental role in achieving climate and environmental policy integration. Another mainstreaming instrument worth noting, though not the only additional one, refers to the need to mobilize research and foster clean technologies.

In addition to outlining specific policy objectives and mainstreaming instruments, the European Commission also emphasizes the importance of both internal and external governance in achieving the goals of the EGD. The external dimension pertains to the EU's role in international cooperation, highlighting the Union's commitment to advancing a "green deal diplomacy" (European Commission, 2019). This approach underscores the EU's intention to promote global climate action and sustainable development through multilateral engagement and strategic partnerships. Historically, the EU has positioned itself as a proactive actor in global climate governance, often portraying itself as a normative power with aspirations for climate leadership even prior to the adoption of the EGD (Oberthür & Dupont, 2021). With the launch of the EGD, the Commission signals its intention to consolidate and amplify these efforts, seeking to reinforce the EU's leadership role on the global stage by aligning its foreign policy more closely with its environmental and climate objectives.

The internal governance dimension is addressed by the European Climate Pact, with which the Commission has launched a bottom-up format to facilitate climate action across the EU and contribute to ensuring a just transition (Tosun, 2022c; Tosun, Geese, & Lorenzoni, 2023; Tosun, Pollex, & Crumbie, 2023). At its heart lie the European Climate Pact Ambassadors (henceforth: Pact Ambassadors), who are volunteers who "inform, inspire and support climate policy and action in their communities and networks."[2]

To become a Pact Ambassador, interested individuals must apply through an open call. Eligible candidates include leaders of organizations or associations, community representatives, members of informal groups or movements, as well as opinion leaders and influencers who reside in an EU member state. The program also accepts public office holders, such as members of Parliament or mayors, as Pact Ambassadors. While a Pact Ambassador must be an individual, the person can represent an organization. As stated on the website, since the Commission regards the Climate Pact as an instrument to achieve just transition, it encourages applications from individuals with experience in activities related to social inclusion, poverty eradication, and support for vulnerable groups. When selected by the Commission, Pact Ambassadors hold this mandate for 1 year, with the opportunity to apply for renewal each subsequent year.

The Commission perceives the Climate Pact as a format for climate action. In this regard, it expects Pact Ambassadors to lead by example—that is, to demonstrate behavior that addresses climate and environmental issues, for instance, by using public transportation instead of driving cars or installing solar panels in their homes. Furthermore, Pact Ambassadors are expected to inspire others to take climate action, for example, by suggesting more sustainable materials or processes in their workplaces. The third type of climate action involves communication with others, either independently or by collaborating with other individuals or organizations, as well as partnerships with stakeholders (Tosun, Pollex, & Crumbie, 2023).

How the European Green Deal Shapes Climate Action

The EGD, as outlined in the previous section, is delivered through the adoption of EU-level policies, which must then be implemented by the individual member states. While the member states have some leeway in how they implement the EU policies in their national context, the goal of EU-level policies is to ensure that the approach adopted by the individual member states is coherent (Knill & Liefferink, 2007). The European Commission oversees the implementation of these policies, with support from the European Court of Justice. When provisions of the relevant legislation are not followed, the Commission can take legal action through the European Court of Justice and initiate sanctions (Bürgin, 2023; Knill & Liefferink, 2007).

However, EU-level policies are not the only form of policy-based climate action emerging from the EGD. Indeed, climate action more broadly encompasses many different forms and approaches (Almeida et al., 2024). In addition to binding EU policies, member states can also adopt national policy measures that are inspired by the EGD, even if they are not directly mandated by it. These national initiatives reflect a political or strategic alignment with the objectives of the EGD, rather than a legal obligation to comply. A prominent example of such an initiative is France's Green Planning Strategy launched in 2022.

Crucially, the adoption of policies is not an end in itself. The ultimate objective of policies is to drive changes in the behavior of targeted actors and sectors (Boasson et al., 2025). As discussed earlier, the EGD addresses a broad range of sectors, including energy, transport, agriculture, and industry, all with the overarching goal of achieving climate neutrality by 2050. To that end, the policies established under the EGD employ a variety of instruments, such as enforceable targets (e.g., greenhouse gas emissions reduction goals), regulatory measures (e.g., the phase-out of internal combustion engine vehicles), and financial tools (e.g., the Just Transition Fund).

A policy is considered effective when it results in the desired behavioral change among its target groups. For instance, a policy aimed at decarbonizing the transport sector would be deemed effective if it leads to a measurable shift from combustion engine vehicles to electric alternatives. When defining target groups within the context of public policy, it is important to recognize that these groups can be delineated according to various criteria. Target groups may correspond to broad sectors of public life—such as the economy, civil society, or educational institutions—but they can also refer to more specific subsectors within these domains. Additionally, target groups can be defined based on the social roles that organizations or individuals occupy. For instance, individuals may be addressed by public policy in different capacities: as consumers, prosumers, citizens, or stakeholders. These roles shape how policies affect them and how they are expected to respond to policy measures (Nielsen et al., 2021).

Importantly, such behavioral change by those targeted by a policy is not voluntary climate action; it is action induced by policy mandates and incentives. Therefore, in assessing the effectiveness of climate action delivered through public policy, it is essential to examine the extent to which the intentions of adopted policies are translated into concrete action—and to analyze the real-world consequences of those actions for mitigating climate change (Boasson et al., 2025; Fransen et al., 2023).

However, climate action is not confined to top-down policymaking, or what Almeida et al. (2024) term "institutional" forms of climate action. It also encompasses a diverse array of bottom-up initiatives driven by civil society, community groups, and individual actors, which complement and, at times, catalyze institutional processes (Tosun, 2022a). These forms of engagement, which Almeida et al. (2024) call "noninstitutional" climate action, are essential for multiple reasons: They foster engagement with climate governance, generate innovative practices that respond to local contexts and lived experiences, and enhance the implementation of public policies addressing climate change. The EGD does not only stimulate the adoption of policies, but it also aims to stimulated different types of climate action via bottom-up initiative. The best example for that is the EU Climate Pact, which—as we have illustrated—offers a framework within which citizens may serve as Pact Ambassadors, assuming a communicative and mobilizing role within their local communities.

Pact Ambassadors can interpret and perform their roles in a variety of ways (Tosun, 2022c; Tosun, Geese, & Lorenzoni, 2023; Tosun, Pollex, & Crumbie, 2023). For one, they may act as intermediaries between policymakers and grassroots communities, helping to translate abstract policy objectives into accessible and actionable practices at the local level. In this role, they serve as facilitators of public engagement, bridging the gap between high-level policy and everyday citizen involvement.

Alternatively, some Pact Ambassadors see themselves primarily as policy advocates, engaging directly with policymakers across different levels of the EU's political system to push for more ambitious climate policies or to highlight specific aspects of the EGD, such as just transition (Tosun, Pollex, & Crumbie, 2023). For example, the Zero Pollution Stakeholder Platform can other an opportunity for policy advocacy carried out by Pact Ambassadors. Others adopt the role of knowledge brokers (Tosun, Geese, & Lorenzoni, 2023)—individuals who actively produce and/or disseminate knowledge about climate change and climate action. As knowledge brokers, they may either generate their own information or help circulate research and data produced by others (Galford et al., 2016). In doing so, they support various stakeholders by providing the information and understanding necessary to enable informed climate action.

Pact Ambassadors as well as other individuals or organizations striving to take climate action can use governance tools that exist beyond the EGD. In fact, the EU has been eager to strengthen its tools for deliberative and participatory modes of governance, allowing citizens to directly or indirectly influence policy development (Parks et al., 2023). Among these, an important governance tool is the European Citizens' Initiative (ECI), which empowers EU citizens to propose new legislation on matters within the EU's competencies, including climate and environmental policy (Tosun, 2022b). The ECI process begins with forming a citizens' committee of at least seven EU citizens from seven different member states. The European Commission then checks whether the proposal falls within its powers and aligns with EU values. If approved, organizers have 12 months to collect at least one million signatures from seven member states, meeting a minimum threshold in each. National authorities verify the signatures, after which the Commission publishes the initiative and holds meetings with the organizers and a hearing in the European Parliament. Finally, the Commission decides whether to take legislative action, though it is not obligated to do so and may opt for alternative measures instead.

Past ECIs have addressed a range of climate-related issues, including fossil fuel divestment, protection of biodiversity, and regulation of environmental pollutants. Most notably, the ECI titled "Actions on Climate Emergency" was launched by the social movement Fridays for Future to urge the EU Commission to adopt more ambitious climate policies in line with the 1.5°C global warming limit set by the Paris Agreement. The initiative did not meet the formal requirement of collecting one million signatures within the official timeframe. Consequently, the European Commission did not proceed to the subsequent steps of the ECI process.[3]

Even when proposals do not lead to the adoption of EU policies, they often serve as catalysts for public debate, media attention, and greater institutional responsiveness. This was the case with the ECI "Right2Water," which asked the European Commission to ensure universal access to safe drinking water, which has resulted, inter alia, that the member states make an effort to provide water fountains in the public space (Tosun et al., 2025). Research has also shown that, even if the European Commission does not act on the demands of an ECI, national or subnational governments within member states sometimes respond to them (Tosun et al., 2022).

In addition to the ECI, several other participatory governance tools exist within the EU, some of which date back many years. For instance, in 1993, the EU granted citizens the right to submit petitions to the European Parliament. Since then, additional avenues for engagement have been introduced, including the right to take part in public consultations and citizens' dialogues, allowing individuals to contribute to policymaking processes and voice their concerns more directly (Parks et al., 2023). Beyond the governance tools mentioned in the previous sections, a growing range of citizen-led and community-based climate initiatives has emerged across Europe, spanning a wide spectrum of approaches. While not all are direct outcomes of the EGD, it has played a role in raising their visibility and facilitating access to funding.

In several EU member states, climate assemblies and citizens' panels have emerged as participatory forums that bring together randomly selected citizens to engage in informed deliberations and formulate nonbinding policy recommendations on climate-related issues. Although not directly connected to the EGD, the first initiative of this kind was Ireland's Citizens' Assembly, held between 2016 and 2018. Since then, similar forums have been established in Austria, Denmark, France, Germany, Luxembourg, Spain, Sweden, and the United Kingdom (Lorenzoni et al., 2025). These assemblies did not direct result from EGD, but the EGD still played a role for their diffusion across the EU member states since the EGD stresses the importance of broad public participation in climate and environmental action (European Commission, 2019). This aligns directly with the idea behind climate assemblies and citizens' panels, which can be important to give legitimacy to climate actions that might otherwise face public resistance.

In addition to national formats, the EU launched the EU Citizens' Panels as part of the Conference on the Future of Europe. These panels bring together randomly selected citizens from all EU member states, reflecting demographic and geographic diversity. Each panel focuses on specific topics, including climate action, which is directly influenced by the policy agenda put forth by the EGD. The panels serve as a platform for citizen input, and their recommendations are considered by EU institutions and can influence legislative proposals.[4] The use of digital technologies has significantly broadened the scope and reach of bottom-up engagement initiatives, such as citizen-driven "climate hackathons."

These hackathons serve as dynamic co-production formats where participants from diverse fields, competencies, and areas of expertise collaborate intensively—typically over a 24- to 48-hour period—to develop innovative solutions to predefined challenges (Kvamsås et al., 2021). By fostering interdisciplinary dialogue and rapid ideation, hackathons help translate complex climate issues into actionable strategies. A noteworthy example is the "EU Sparks for Climate" initiative, where the European Commission actively invites both citizens and researchers to jointly co-create practical, scalable solutions to pressing climate challenges, emphasizing inclusivity and participatory innovation in climate governance.[5]

Another governance tool used in several cities in the EU is participatory budgeting and planning for climate projects. With such formats, citizens are invited to propose, deliberate, and vote on projects—ranging from green infrastructure to sustainable mobility—ensuring that local climate action reflects community priorities and generates a sense of ownership over the urban climate agenda. While participatory budgeting and planning is not formally mandated by the EGD, it is a citizen-centered tool that can help fulfill its goals. The EGD creates a favorable environment for such initiatives to flourish, especially at municipal and regional levels. In this regard, it is worth noting that the Just Transition Mechanism creates funds, which can be used for local-level projects that align with the aims and scope of the EGD. An illustrative example of participatory budgeting for climate action is the Climate Air Energy Plan of the French city of Grenoble, which gives citizens the right to inform the authorities about the actions they are taking or they would be willing to take to fight climate change.[6]

Since 2016, the EU has been revising its energy policy framework to increasingly recognize energy communities as a vital component of the transition toward more sustainable energy systems (López et al., 2024). These energy communities are typically citizen-owned solar and wind projects that empower local populations to collectively produce, manage, and consume renewable energy. To facilitate the creation and growth of such initiatives, the European Commission established the Citizen Energy Advisory Hub, providing guidance and support to a range of citizen-led energy projects. In 2024, it further reinforced its commitment by launching the European Energy Communities Facility, a dedicated mechanism aimed at promoting the development and scaling of energy communities across the EU.[7] Given this strong political backing, it is unsurprising that energy communities have proliferated across the region. As reported by (Koltunov et al., 2023), nearly 4,000 energy communities comprising approximately 900,000 members were active in the EU at the time of writing. Germany, the Netherlands, and Denmark stand out as the member states with the highest concentration of these initiatives.

Conclusion

This chapter has provided a comprehensive overview of the EGD and its multifaceted approach to facilitating both policy-based and non-policy-based climate action across the EU. The analysis reveals that the EGD has established a new paradigm for EU climate action that extends beyond traditional regulatory approaches. By consciously integrating social justice concerns through mechanisms such as the Just Transition Fund, expanding participatory governance through innovative citizen engagement tools, and maintaining ambitious decarbonization and environmental protection targets despite significant political and economic challenges, the EGD has created a distinctive template for transformative climate and environmental governance that seeks to balance institutional innovation with democratic legitimacy and social cohesion. At the same time, the launch of the EGD was itself facilitated by enhanced societal demand for greater environmental and climate action, supported by corresponding bottom-up initiatives.

The EGD's significance lies not merely in its unprecedented comprehensive sectoral scope—spanning energy transition, sustainable transport, agricultural transformation, industrial decarbonization, and ecosystem restoration—and its commitment to fostering both horizontal policy integration across different sectors and vertical integration between governance levels. Equally important is its innovative approach to participatory and deliberative governance, which has systematically created various institutional venues and informal opportunities for diverse forms of climate action. These participatory mechanisms serve multiple functions: They operate not only as direct vehicles for grassroots climate action but also as important sources of democratic legitimacy for potentially contentious policy measures, helping to bridge the gap between EU-level ambitions and local implementation realities.

Climate action as it manifests across the contemporary EU comprises a remarkably diverse array of forms and formats. Citizens and organizations can engage through serving as Pact Ambassadors who advocate for climate concerns within their communities, participating in deliberative forums such as climate assemblies and citizens' panels that generate policy recommendations, launching or supporting ECIs that can influence the EU's legislative agenda, actively participating in energy communities that democratize renewable energy production and consumption, and utilizing innovative co-production formats such as climate hackathons that foster technological and social innovation. This multiplicity of engagement pathways reflects the EGD's recognition that achieving systemic transformation requires mobilizing diverse forms of expertise, commitment, and action across all levels of society.

However, as this chapter has demonstrated, the path from ambitious policy proposals to implemented reality has proven complex and contested. The European Commission has proposed numerous initiatives, but not all have become binding EU policies. Even among formally adopted measures, implementation remains uncertain. These policies may face significant modifications due to political pressure from two key sources: the growing electoral strength of right-wing populist parties in both the European Parliament and national parliaments, and mobilization of important stakeholders such as farmers against EU policies. A particularly illustrative case is the controversial goal to end sales of cars with internal combustion engines by 2035. Since the formal adoption of this target, numerous—mostly right-wing or right-wing populist—politicians from EU member states as well as Members of the European Parliament have publicly called for either postponing the measure or abandoning it altogether, citing concerns about economic competitiveness, technological readiness, and social acceptability. Similarly, the wave of farmers' protests that swept across the EU in late 2023 and early 2024 vividly demonstrated that agricultural stakeholders are not uniformly willing to accept those elements of the EGD—particularly the Farm to Fork strategy—which they perceive as instituting additional regulatory burdens that threaten their livelihoods and traditional practices (Nagel et al., 2025).

These implementation challenges and sources of resistance suggest that despite the considerable measures already undertaken to enhance democratic participation and stakeholder engagement, the EU may need to develop even more robust and inclusive opportunities for meaningful citizen engagement, more effective stakeholder consultation processes, and more sophisticated conflict resolution mechanisms that can address competing interests and values. The experience with contentious policies indicates that technical solutions alone are insufficient; successful climate governance requires ongoing negotiation, adaptation, and compromise among diverse societal groups.

The ultimate success of the EGD approach will depend not only on the EU institutions' technical and administrative capacity to deliver on ambitious policy commitments within prescribed timeframes, but perhaps more critically on their political and social capacity to sustain broad-based public support for the green transition across different social groups, economic sectors, and geographic regions. The diverse forms of climate action documented throughout this chapter—ranging from mandatory regulatory compliance by businesses and member states to voluntary citizen-led initiatives and community-based projects—suggest that the EGD has begun to cultivate the kind of multilevel, multistakeholder engagement that scholars and practitioners increasingly recognize as necessary for achieving systemic sociotechnical transformation.

However, it remains to be seen how the EU's approach will develop further, considering that in 2025 the European Commission launched the Clean Industrial Deal (CID) as the next phase of the EGD. While building directly on the main objectives of the EGD, the CID shifts focus toward industrial competitiveness. It takes a narrower approach than the EGD—focusing primarily on energy-intensive industries and decarbonization rather than pursuing a holistic view of sustainable transformation across various aspects of society (Gurreck, 2025). The role that societal actors can play in this new phase, and the strategies they need to employ to remain influential, are important questions for future research.

Notes

1. Information retrieved on June 4, 2025, from the following website: https://environment.ec.europa.eu/strategy/zero-pollution-stakeholder-platform/get-involved_en#become-a-member``
2. Information retrieved on June 4, 2025, from the following website: https://climate-pact.europa.eu/get-involved/become-pact-ambassador_en
3. For information on the Commission's reaction to the initiative, see the following website: https://citizens-initiative.europa.eu/initiatives/details/2019/000015_en
4. Two panels directly connected to the EGD are the "Energy Efficiency Panel" and the "European Citizens' Food Waste Panel." Information retrieved on June 4, 2025, from the following website: https://citizens.ec.europa.eu/european-citizens-panels_en
5. Information retrieved on June 4, 2025, from the following website: https://eusparks.eu/
6. Information retrieved on June 4, 2025, from the following website: https://oidp.net/en/practice.php?id=1260
7. Information retrieved on June 23, 2025, from the following website: https://energy.ec.europa.eu/topics/markets-and-consumers/energy-consumers-and-prosumers/energy-communities_en

References

Almeida, P., González Márquez, L. R., & Fonsah, E. (2024). The forms of climate action. *Sociology Compass*, *18*(2), Article e13177. https://doi.org/10.1111/soc4.13177

Bloomfield, J., & Steward, F. (2020). The politics of the Green New Deal. *The Political Quarterly*, *91*(4), 770–779. https://doi.org/10.1111/1467-923X.12917

Boasson, E. L., Peters, G. P., & Tosun, J. (2025). Policy-driven acceleration of climate action. *PLOS Climate*, *4*(5), e0000626. https://doi.org/10.1371/journal.pclm.0000626

Bürgin, A. (2023). The European Commission: A climate policy entrepreneur. In T. Rayner, K. Szulecki, A. J. Jordan, & S. Oberthür (Eds.), *Handbook on European Union climate change policy and politics* (pp. 23–37). Edward Elgar. https://doi.org/10.4337/9781789906981.00013

Busemeyer, M. R., Carstensen, M. B., Kemmerling, A., & Tosun, J. (2025). Foundations for a green economy: How institutions shape green skills. *Npj Climate Action*, *4*(27), 1–5. https://doi.org/10.1038/s44168-025-00234-5

Dupont, C. (2015). *Climate policy integration into EU energy policy.* Routledge. https://doi.org/10.4324/9781315751665

Dupont, C., Moore, B., Boasson, E. L., Gravey, V., Jordan, A., Kivimaa, P., Kulovesi, K., Kuzemko, C., Oberthür, S., Panchuk, D., Rosamond, J., Torney, D., Tosun, J., & Homeyer, I. von. (2024). Three decades of EU climate policy: Racing toward climate neutrality? *WIREs Climate Change, 15*(1), Article e863, 1–12. https://doi.org/10.1002/wcc.863

European Commission. (2019). *Communication from the Commission to the European Parliament, the Council, the European Economic and Social Committee and the Committee of the Regions: The European Green Deal.* European Commission.

European Commission. (2020). *Communication from the Commission to the European Parliament, the Council, the European Economic and Social Committee and the Committee of the Regions: A new Circular Economy Action Plan for a cleaner and more competitive Europe.* European Commission.

Feindt, P. H. (2010). Policy-learning and environmental policy integration in the Common Agricultural Policy, 1973–2003. *Public Administration, 88*(2), 296–314. https://doi.org/10.1111/j.1467-9299.2010.01833.x

Finger, R., Fabry, A., Kammer, M., Candel, J., Dalhaus, T., & Meemken, E. M. (2024). Farmer Protests in Europe 2023–2024. *EuroChoices, 80*(2), Article 1746-692X.12452, 314. https://doi.org/10.1111/1746-692X.12452

Fransen, T., Meckling, J., Stünzi, A., Schmidt, T. S., Egli, F., Schmid, N., & Beaton, C. (2023). Taking stock of the implementation gap in climate policy. *Nature Climate Change, 13*(8), 752–755. https://doi.org/10.1038/s41558-023-01755-9

Galford, G. L., Nash, J., Betts, A. K., Carlson, S., Ford, S., Hoogenboom, A., Markowitz, D., Nash, A., Palchak, E., Pears, S., & Underwood, K. L. (2016). Bridging the climate information gap: A framework for engaging knowledge brokers and decision makers in state climate assessments. *Climatic Change, 138*(3–4), 383–395. https://doi.org/10.1007/s10584-016-1756-4

Gurreck, M. (2025). *Simplification or deregulation? Evaluating the EU's Clean Industrial Deal and Action Plan for Affordable Energy.* https://www.greendealnet.eu/Simplification-or-deregulation-EU-Industrial-Clean-Deal

Hix, S. (1998). The study of the European Union II: The "new governance" agenda and its rival. *Journal of European Public Policy, 5*(1), 38–65. https://doi.org/10.1080/13501768880000031

Homeyer, I. von, Oberthür, S., & Dupont, C. (2022). Implementing the European Green Deal during the evolving energy crisis. *JCMS: Journal of Common Market Studies, 60*(S1), 125–136. https://doi.org/10.1111/jcms.13397

Hooghe, L., & Marks, G. (2001). *Multi-level governance and European integration.* Rowman & Littlefield.

Jordan, A., & Lenschow, A. (2010). Environmental policy integration: A state of the art review. *Environmental Policy and Governance, 20*(3), 147–158. https://doi.org/10.1002/eet.539

Knill, C., & Liefferink, D. (2007). *Environmental politics in the European Union: Policy-making, implementation and patterns of multi-level governance.* Manchester University Press.

Koltunov, M., Pezzutto, S., Bisello, A., Lettner, G., Hiesl, A., van Sark, W., Louwen, A., & Wilczynski, E. (2023). Mapping of energy communities in Europe: Status quo and review of existing classifications. *Sustainability, 15*(10), 8201. https://doi.org/10.3390/su15108201

Koundouri, P., Alamanos, A., Plataniotis, A., Stavridis, C., Perifanos, K., & Devves, S. (2024). Assessing the sustainability of the European Green Deal and its interlin kages with the SDGs. *Npj Climate Action, 3*(1), 291. https://doi.org/10.1038/s44168-024-00104-6

Kvamsås, H., Neby, S., Haarstad, H., Stiller-Reeve, M., & Schrage, J. (2021). Using collaborative hackathons to coproduce knowledge on local climate adaptation governance. *Current Research in Environmental Sustainability, 3*, 100023. https://doi.org/10.1016/j.crsust.2020.100023

Lafferty, W., & Hovden, E. (2003). Environmental policy integration: Towards an analytical framework. *Environmental Politics, 12*(3), 1–22. https://doi.org/10.1080/09644010412331308254

López, I., Goitia-Zabaleta, N., Milo, A., Gómez-Cornejo, J., Aranzabal, I., Gaztañaga, H., & Fernandez, E. (2024). European energy communities: Characteristics, trends, business models

and legal framework. *Renewable and Sustainable Energy Reviews*, *197*, 114403. https://doi.org/10.1016/j.rser.2024.114403
Lorenzoni, I., Jordan, A. J., Sullivan-Thomsett, C., & Geese, L. (2025). A review of National Citizens' Climate Assemblies: Learning from deliberative events. *Climate Policy* (early view), 1–17. https://doi.org/10.1080/14693062.2024.2449416#d1e205
Nagel, M., Gall, A., & Tosun, J. (2025). The "Hottest Ever January" in Germany: Farmers' Protests and the Discourse on Agriculture and Food Production. *Politics and Governance*, *13*(1), 1–23. https://doi.org/10.17645/pag.9830
Nielsen, K. S., Clayton, S., Stern, P. C., Dietz, T., Capstick, S., & Whitmarsh, L. (2021). How psychology can help limit climate change. *The American Psychologist*, *76*(1), 130–144. https://doi.org/10.1037/amp0000624
Oberthür, S., & Dupont, C. (2021). The European Union's international climate leadership: Towards a grand climate strategy? *Journal of European Public Policy*, *28*(7), 1095–1114. https://doi.org/10.1080/13501763.2021.1918218
Oberthür, S., & Kulovesi, K. (2025). Accelerating the EU's climate transformation: The European Green Deal's Fit for 55 Package unpacked. *Review of European, Comparative & International Environmental Law*, *34*(1), 7–22. https://doi.org/10.1111/reel.12596
Parks, L., Della Porta, D., & Portos, M. (2023). Environmental and climate activism and advocacy in the EU. In T. Rayner, K. Szulecki, A. J. Jordan, & S. Oberthür (Eds.), *Handbook on European Union climate change policy and politics* (pp. 98–112). Edward Elgar. https://doi.org/10.4337/9781789906981.00018
Paterson, M., Wilshire, S., & Tobin, P. (2024). The rise of anti-net zero populism in the UK: Comparing rhetorical strategies for climate policy dismantling. *Journal of Comparative Policy Analysis: Research and Practice*, *26*(3–4), 332–350. https://doi.org/10.1080/13876988.2023.2242799
Politico. (2025). *The EU ban on combustion car engines is in trouble*. https://www.politico.eu/article/why-eu-combustion-car-ban-is-in-trouble-greenhouse-gas-climate-change/
Rayner, T., & Jordan, A. (2016). Climate Change Policy in the European Union. In H. v. Storch (Ed.), *Oxford research encyclopedias. Oxford research encyclopedia of climate science*. Oxford University Press. https://doi.org/10.1093/acrefore/9780190228620.013.47
Sandmann, L., Bülbül, E., Castaño-Rosa, R., Hanke, F., Großmann, K., Guyet, R., Jiglau, G., Laakso, S., Nuorivaara, E., & Vornicu, A. (2024). The European Green Deal and its translation into action: Multilevel governance perspectives on just transition. *Energy Research & Social Science*, *115*, 103659. https://doi.org/10.1016/j.erss.2024.103659
Schebesta, H., & Candel, J. J. L. (2020). Game-changing potential of the EU's Farm to Fork Strategy. *Nature Food*, *1*(10), 586–588. https://doi.org/10.1038/s43016-020-00166-9
Schleyer, C., Görg, C., Hauck, J., & Winkler, K. J. (2015). Opportunities and challenges for mainstreaming the ecosystem services concept in the multi-level policy-making within the EU. *Ecosystem Services*, *16*, 174–181. https://doi.org/10.1016/j.ecoser.2015.10.014
Tosun, J. (2022a). Addressing climate change through climate action. *Climate Action*, *1*(1), 1–8. https://doi.org/10.1007/s44168-022-00003-8
Tosun, J. (2022b). European Citizens' Initiative. In P. R. Graziano & J. Tosun (Eds.), *Elgar encyclopedia of European Union public policy* (pp. 130–137). Edward Elgar. https://doi.org/10.4337/9781800881112.ch14
Tosun, J. (2022c). What role for climate pact ambassadors? A policy process perspective. *European View*, *21*(2), 171–177. https://doi.org/10.1177/17816858221128719
Tosun, J. (2023). The European Union's climate and environmental policy in times of geopolitical crisis. *JCMS: Journal of Common Market Studies*, *61*(S1), 147–156. https://doi.org/10.1111/jcms.13530
Tosun, J., Béland, D., & Papadopoulos, Y. (2022). The impact of direct democracy on policy change: insights from European citizens' initiatives. *Policy & Politics*, *50*(3), 323–340. https://doi.org/10.1332/030557321X16476244758073
Tosun, J., Debus, M., Schaub, S., Klein, S., Scherer, U., Goeschl, T., & Horn, H. (2025). Tap water consumption choices in out-of-home settings: Insights from a survey of German adults. *International*

Journal of Water Resources Development, *41*(3), 647–666. https://doi.org/10.1080/07900627.2024.2423742

Tosun, J., Geese, L., & Lorenzoni, I. (2023). For young and future generations? Insights from the web profiles of European Climate Pact Ambassadors. *European Journal of Risk Regulation*, 1–13. https://doi.org/10.1017/err.2023.53

Tosun, J., Pollex, J., & Crumbie, L. (2023). European climate pact citizen volunteers: Strategies for deepening engagement and impact. *Policy Design and Practice*, *6*(3), 344–356. https://doi.org/10.1080/25741292.2023.2199961

Tosun, J., Schaub, S., Marek, C., Kellermann, L., & Koch, M. (2024). Attributing responsibility to farmers for environmental protection and climate action: Insights from the European Union. *Journal of Environmental Studies and Sciences* (early view). https://doi.org/10.1007/s13412-024-00981-7

PART IV

Institutionalized Climate Action II

The Search for Alternatives

CHAPTER 32

Carbon Capture and Storage as Climate Action

XIAOLIANG YANG

Introduction

THE urgency of addressing climate change necessitates a comprehensive portfolio of solutions to curb greenhouse gas emissions and achieve global targets. Carbon capture and storage (CCS) has emerged as a critical technology in this portfolio since society began to concern itself with global warming.[1] CCS offers a direct approach by capturing carbon dioxide (CO_2) at its source, such as industrial facilities or power plants, thereby preventing its release into the atmosphere. Alongside CCS, carbon removal technologies, including direct air capture with carbon storage (DACCS) and bioenergy with CCS (BECCS) have been gaining traction in the recent years. These technologies aim to remove existing CO_2 from the atmosphere, offering an additional tool for mitigating climate change. The captured CO_2 from air or industrial sources then undergoes transportation to designated geological formations for secure, long-term storage. While CCS has been a recognized climate mitigation technology for decades, it garnered increased interest and support following the 2020 COP26 conference, and the history of CCS is not without its complexities. Economic, social, and political challenges have impeded its widespread adoption. This chapter delves into the multifaceted nature of CCS as climate action. It examines its scientific underpinnings, historical development, and the social and economic considerations surrounding its implementation. Additionally, it explores the intricate relationship between CCS and international climate politics. By analyzing these aspects, this chapter can form a more comprehensive picture of CCS and its potential contribution to serving climate goals.

Scientific Assessment of CCS's Climate Contribution

CCS involves capturing CO_2 emissions at their source, such as power plants or industrial facilities. The captured CO_2 is then transported, often via pipelines, ships, or trucks, to designated geological sites for storage. By preventing this anthropogenic CO_2 from entering the atmosphere, CCS offers a valuable approach to climate change mitigation (IPCC, 2005). Renowned scientific organizations like the Intergovernmental Panel on Climate Change (IPCC) and the International Energy Agency (IEA) have consistently incorporated CCS into their climate and energy models since the early 2000s to assess its potential contribution to achieving specific climate goals. These works have become the scientific underpinnings of the role of CCS in global climate governance.

Established in 1988 by the United Nations, the IPCC, the most influential international source on climate change, publishes comprehensive reports on the science, the impacts, and the potential solutions of climate change on a regular basis. IPCC's "2005 Special Report on Carbon Dioxide Capture and Storage" identified CCS as a technology with significant potential for mitigating climate change (IPCC, 2005). This 442-page report is the first comprehensive scientific publication that offers a holistic introduction to CCS with a global perspective. The content ranges from technical feasibility to global storage capacity assessment, and cost and economic challenges. It also emphasizes the importance of developing methods to estimate and report the quantities of emissions that are reduced from CCS. This report has had a lasting impact on the climate community, establishing a foundational understanding of the role of CCS. However, since its release, significant new analysis and improvements in CCS have emerged.

The IPCC's "special report on limiting global warming to 1.5°C" explored various mitigation strategies, almost of which all scenarios require use of CCS to some degree. The amount of fossil- and biomass CO_2 needed varies greatly, with potential cumulative CO_2 storage ranging from zero to 300 billion tons by 2050 in all pathways (IPCC, 2018). While technically possible to entirely avoid CCS, such a scenario would require a drastic and unrealistic reduction in global energy demand. The Sixth IPCC Assessment Report, which is the latest major report, reinforces the importance of CCS for achieving the 1.5°C goal. The report indicates that "the technical geological storage capacity is estimated to be on the order of 1000 Gt CO_2, which is more than the CO_2 storage requirement though 2100 to limit global warming to 1.5°C" (IPCC, 2023, p. 86).

Drawing on its World Energy Outlook and Energy Technology Perspective, the IEA has also delved into scenarios for limiting CO_2 emissions. These studies offer quantitative insights into the efficacy of CCS, positioning it as a potential climate solution. In the IEA's Sustainable Development Scenario, achieving net-zero emissions by 2070 heavily relies on CCS, which cumulatively captures 15% of the energy sector's emissions (IEA, 2020). Initially, the focus lies on retrofitting existing power plants and exploiting low-cost capture opportunities such as hydrogen production. The IEA's 2023 Net Zero Roadmap envisages CCS playing a significant role in decarbonizing the energy sector, contributing approximately 8% of the total CO_2 mitigation between 2022 and 2050 in its Net Zero Scenario (IEA, 2023b).

The Increasing Scientific Focus on Carbon Removal Technologies

The IPCC (2023) highlights that limiting global warming to below 2°C and achieving net-zero CO_2 emissions require removing anthropogenic CO_2 from the atmosphere, such as DACCS and BECCS. This is crucial for offsetting emissions from hard-to-decarbonize sectors like industrial manufacturing, transportation, and agriculture. The IPCC's mitigation scenarios project a significant increase in global deployment of carbon dioxide removal (CDR) technologies.[2] To limit warming to 2°C with a greater than 67% probability, cumulative CO_2 removal from 2020 to 2100 is estimated to be 328 Gt for BECCS, 29 Gt for DACCS, and 252 Gt for afforestation, reforestation, and improved forest management (IPCC, 2022). The IEA's Net Zero Emissions Scenario estimates that DACCS and BECCS are expected to rapidly scale up, reaching 0.6 Gt in 2035 and 1.7 Gt in 2050 (IEA, 2023b). By 2050, total energy sector CO_2 emissions are anticipated to reach net zero, with any remaining emissions balanced by removals through BECCS and DACCS. The United States' long-term climate strategy underscores the need to remove around 1 billion tons of CO_2 annually by midcentury to achieve net-zero emissions, even with rapid emission reduction efforts (US Government, 2021).

Recognized by leading scientific and research organizations like the IPCC and IEA, CCS and CDR offer a versatile and essential approach for reducing carbon emissions across various sectors and to achieve net-zero emissions. While the specific deployment levels may vary in different scenarios, studies consistently highlight its potential for achieving ambitious climate goals. However, all these scientific publications have clearly emphasized that CCS is just one piece of the climate puzzle. In summary, the understanding of CCS and CDR in the climate circle is more advanced than that of the two solutions from a technology standpoint. The future contribution of CCS and CDR hinges on the speed of global energy transition and the development of other mitigation approaches.

The History of CCS Development

Technological Roots

Although CCS is a relatively new concept from a decarbonization perspective, its components have had long-standing applications in various industries. The origins of carbon capture technologies can be traced back to the 1930s, when the oil and gas industry first began separating CO_2 during natural gas processing (Kearns et al., 2021). This early application laid the technological foundation for contemporary carbon capture approaches. The 1970s marked another significant advancement with the introduction of captured CO_2 for enhanced oil recovery (EOR) by the oil and gas sector. EOR is pivotal for large-scale underground CO_2 storage in modern CCS (IEA, 2016). Established in Texas in 1972, the Terrell Natural Gas Processing Plant stands as one of the world's oldest operating CCS projects, extracting CO_2 from methane to meet pipeline specifications (Loria & Bright, 2021). Subsequently, the captured CO_2 is transported for EOR, enhancing oil production while preventing CO_2 emissions into the atmosphere. Although EOR created a market for captured CO_2 from industrial sources, it is worth noting the vast majority of CO_2 for EOR comes from natural CO_2 domes.

Early Support

The 1980s marked a pivotal moment for CCS. As scientific evidence confirming human-caused climate change solidified, CCS emerged as a promising technology for reducing carbon emissions. Researchers began exploring the geological storage of CO_2 as a strategy for mitigating climate change. Early pioneers like Marchetti (1977) and Steinberg and Albanese (1980) were among the first to conceptualize and discuss the application of CCS for climate mitigation. Subsequently, Audus et al. (1997) elaborated on the application of capture and storage as a climate protection measure. Around this time, national governments initiated R&D programs to support CCS technology development in pursuit of cleaner energy transitions. Notably, in 1997, the United States established the world's first comprehensive R&D program for CCS at the Department of Energy, aimed at accelerating innovation in CCS technologies. This program has since expanded significantly, as documented by Herzog (2016).

The Sleipner project, located in the North Sea off the coast of Norway, is another landmark achievement in CCS's early history. Operational since 1996, it's one of the first large-scale ventures in CCS for decarbonization purposes. The project captures 1 million tons of CO_2

annually during natural gas processing. The captured CO_2 is injected deep underground into a suitable geological formation, demonstrating the viability of CCS for long-term CO_2 sequestration and offering a crucial strategy for mitigating climate change. Sleipner's ongoing success over the past two decades provides valuable data and experience for the safe and effective geological storage of captured emissions (Furre et al., 2017).

Early Enthusiasm and Setbacks

The turn of the millennium ushered in an era of significant advancements in CCS. This coincided with heightened public awareness of climate change, which fueled increased investment in CCS research and development. Governments also began exploring policy frameworks to incentivize CCS deployment. US President George Bush in 2003 announced FutureGen 1.0 Project, a high-profile CCS project integrated with a coal-fired electrical generating station in Illinois. A primary goal of FutureGen was to pave the way for decarbonized fossil fuel power plants which would enable the continued use of coal. This initiative was a partnership between the US Government and an alliance of coal-related corporations. Five years later, in 2008, the US Department of Energy decided to pull its funding for this project due to higher-than-expected costs (Tollefson, 2008).

In 2008, G8 leaders made a significant commitment at the Hokkaido Toyako Summit, stating their strong support for the launch of 20 large-scale CCS demonstration projects globally by 2010, with the goal of initiating broad deployment of CCS by 2020 (G8 Hokkaido Toyako Summit, 2020). This declaration generated political momentum and global support for CCS initiatives. Subsequently, in 2009, the Government of Canada announced the Clean Energy Research Fund, aimed at reducing Canada's total greenhouse gas emissions by 20% from 2006 levels by 2020. Of the $1 billion budget allocated to the CEF in 2009, $650 million was designated for large-scale CCS demonstration projects (Natural Resources Canada, 2014). In the same year, Australia launched the Carbon Capture and Storage Flagships Program, providing support for demonstration projects focused on CO_2 capture, transportation, and storage. On the international stage, the Global Carbon Capture and Storage Institute was launched by Prime Minister Kevin Rudd at an event jointly hosted by President Barack Obama and Italy's Prime Minister in L'Aquila, Italy, on October 7, 2009.

Despite international coordination efforts, challenges facing CCS persisted throughout the 2010s. Between 2010 and 2016, over 20 large-scale CCS projects were cancelled, while funding commitments were reduced or withdrawn across Europe, the United States, and Australia. In Australia, the AUD 1.9 billion CCS Flagships program, initiated in 2009, was gradually scaled back to around AUD 300 million by 2016 (IEA, 2016). Similarly, in the United Kingdom, the cancellation of the GBP 1 billion CCS Commercialisation Programme in 2015, shortly before COP21, dealt a significant blow to CCS efforts. This occurred during the final stages of project selection and led to the termination of two promising projects—White Rose and Peterhead. The Peterhead CCS project, notable for its application to gas-fired power generation, and White Rose, showcasing oxy-fuel capture technology, were affected. In the United States, the Department of Energy suspended funding for the FutureGen 2.0 project due to insufficient time to meet deadlines mandated by the 2009 federal stimulus package (Marshall, 2015). This marked another setback for the FutureGen initiative, following the cancellation of the 1.0 version in 2008. The FutureGen 2.0 project aimed to validate the technical feasibility and economic viability of near-zero emission energy from coal, while also verifying the effectiveness, safety, and permanence of CO_2 sequestration in a saline formation.

The Petra Nova Project, situated near Thompsons, Texas, stands as a beacon of hope for CCS technology, particularly within the coal power sector, as the sole survivor amid numerous setbacks in that period. Operational since 2014, it captures 1.4 million tons of CO_2 emissions annually from an existing coal-fired plant, utilizing it for enhanced oil recovery. This $1 billion project received significant public funding, including a $190 million grant from the US Department of Energy. Serving as a testament to advancements in CCS for coal plants, Petra Nova provides invaluable insights for the future widespread deployment of this decarbonization technology. This plant was placed in reserve shutdown in May 2020 due to the collapse of oil and power prices, and it was brought back online in September 2023. It remains the world's largest post-combustion carbon capture system in the power sector.

Renewed Interest With the Net-Zero Momentum

The ambitious climate goal set forth by the 2015 Paris Agreement reignited global interest in all potential mitigation strategies, including CCS. With the requirement for countries to submit nationally determined contributions (NDCs) for the United Nations Framework Convention on Climate Change (UNFCCC), some nations, such as Australia, Canada, China, Japan, the United Kingdom, and the United States, began incorporating CCS into their official national climate strategies. The momentum toward net-zero commitments was further accelerated at the UK COP26 in 2021, where a significant increase in countries announcing net-zero targets was observed, along with more ambitious timelines for achieving them. As of November 2023, approximately 145 countries had announced or were considering net-zero targets, covering nearly 90% of global emissions (Climate Action Tracker, 2023). This convergence of rising net-zero ambitions, supportive climate policy, and ongoing technological advancements has led to a significant increase in CCS project announcements and development efforts worldwide. The number of projects in the development pipeline has been steadily rising since 2017, reaching an all-time high of 392 as of July 2023. Furthermore, since the 2022 report, 11 new facilities have commenced operations, and 15 new projects have started construction (GCCSI, 2023). Notably, the United States, the United Kingdom, Canada, China, and Norway are leading the way in new CCS development.

Key Issues Related to CCS Development

In delving into the history of CCS development, it becomes evident that the journey has been fraught with challenges and complexities. Many of the challenges faced by CCS are not unique to this technology; they are common across all climate solutions, largely due to the absence of supportive policies. However, because CCS is specifically a CO_2 control technology rather than an energy technology, the lack of policy frameworks makes its deployment even more difficult. To understand the trajectory of CCS implementation, it is essential to examine key issues, including the economic challenges of CCS and governments' policy response, the need for a comprehensive legal and regulatory framework, social-related issues, and the influence of international climate politics. Through an in-depth analysis of these topics, it is possible to uncover the intricate dynamics surrounding CCS deployment and its implications for achieving global climate goals.

Economic Challenges of CCS and Policy Response

Over the past decades, CCS for decarbonization has grappled with persistent market failures, including unpredictable high upfront capital costs, negative environmental externalities, asymmetric information, and imperfect competition. CCS projects entail substantial upfront construction costs, encompassing investments in capture equipment, transportation infrastructure, and geological storage sites (Budinis et al., 2018). For instance, the Kemper County Integrated Gasification Combined Cycle CCS project in Mississippi, initially budgeted at $2.4 billion with an intended operation by 2014, encountered serious delays and cost overruns, eventually reaching $7.5 billion by 2017 (Swartz, 2017). However, the primary cause of this failure is attributed to the components required before the capture stage (Hawkins, 2017; Wagman, 2017). In addition, the capture system requires additional energy, increasing energy consumption by 13–44%, depending on the application (IPCC, 2022).

While CCS generates positive externalities, the current market fails to reflect these environmental costs (Hardisty et al., 2011; Krahé et al., 2013). Early movers in CCS project development may face challenges in reaping economic benefits from knowledge creation and innovation, as others can easily leverage insights from these early projects, discouraging initial investments (Krahé et al., 2013). In response to these challenges, governments worldwide have implemented various policy approaches to support CCS projects, including establishing carbon pricing, issuing carbon taxes for emitters, offering direct grants, and providing tax credits.

In 1991, Norway emerged as a global pioneer in implementing a carbon tax, initially targeting the oil and gas sector. This policy shift prompted Equinor (formerly Statoil) to strategically initiate the Sleipner Project (operational in 1996) and the Snohvit CCS Project (operational in 2008) (Fattouh et al., 2024). Both projects involve the capture of CO_2 emissions from offshore gas fields, with Sleipner storing them in a saline aquifer deep underground and Snohvit in a subsea gas reservoir. By 2021, these initiatives had successfully stored 26 million tons of CO_2 permanently beneath the seabed (Halland, 2021), serving as early examples of how climate policies can drive CCS projects forward. Zhang et al. (2022) reviewed 20 carbon storage facilities operating between 1996 and 2020, concluding that a total of 197 Mt of CO_2 was stored during this period.

The Kyoto Protocol, established in 1997, marked the first international effort to curb greenhouse gas emissions under the UNFCCC. While not explicitly mentioning CCS, it laid the groundwork for emissions trading between developed and developing nations through the Clean Development Mechanism (CDM). In 2011, CCS was officially recognized as a CDM project by the Conference of the Parties serving as the Meeting of the Parties to the Kyoto Protocol (Dixon et al., 2013), allowing CCS projects to generate Certified Emission Reduction units. These units could be utilized to offset emissions in line with the Kyoto Protocol's mitigation targets for developed nations. However, the CDM market for CCS projects remained limited (JOGMEC, 2023). These domestic and global developments represent some of the earliest forms of policy support for CCS projects, albeit indirect and not specifically dedicated to CCS initiatives.

Comprehensive policy frameworks specifically tailored to promote CCS development emerged later in the 2010s. As CCS regains momentum as a crucial tool for climate change mitigation, governments worldwide are devising diverse policy approaches to bolster its development. These policies are shaped by national priorities, existing infrastructure, and the unique challenges and opportunities encountered by each country.

The United States

CCS enjoys bipartisan support in the United States, with government funding for CCS projects dating back to the 1990s (Jones & Lawson, 2022). Central to the US strategy is the utilization of tax credits to bolster CCS project deployment, notably through the Section

45Q Tax Credit for Carbon Sequestration (Congressional Budget Office, 2023). This credit provides tax incentives for projects that capture CO_2 that is permanently stored underground, including projects where CO_2 is utilized for enhanced oil recovery and permanently stored during the process. Initially introduced under the Energy Improvement and Extension Act of 2008, the 45Q credit underwent subsequent amendments under the Bipartisan Budget Act in 2018 and the Inflation Reduction Act in 2022 (see Table 32.1). This economic incentive directly encourages companies to invest in CCS, CDR, and carbon conversion projects, while also benefiting pipeline and storage infrastructure providers, technology developers, and equipment and service providers.

Moreover, the Department of Energy plays a crucial role by providing funding for research, development, and demonstration projects through appropriations authorized by the US Congress. Annual appropriations for CCS research and related programs totaled $5.3 billion (in nominal dollars) over the 2011–2023 period. In November 2021, the US Congress passed the Infrastructure Investment and Jobs Act, earmarking a historic $8.2 billion funding for CCS programs over FY2022–2026. Specifically, it allocates $2.5 billion for CCS

TABLE 32.1 The Evolution of the 45Q Tax Credit in the United States

	2008	2018	2022	
	The Energy Improvement and Extension Act of 2008	The FUTURE Act of 2018	The Inflation Reduction Act of 2022	
Carbon Capture Tax Credit ($/Metric ton)				
	Point Source	Point Source	Point Source	Direct Air Capture
Storage	$20	$50	$85	$180
Storage in enhanced oil recovery/utilization	$10	$35	$60	$130
Qualified Minimum Size of Capture Facility (Metric tons/year)				
Power generation (For the 2018 version, facilities emitting ≤500,000 tons of CO_2 per year must capture ≥25,000 tons.)	500,000	500,000	18,750	
Industrial	500,000	100,000	12,500	
Direct air capture		100,000	1,000	
Size Cap of Tax Credits				
	75 million tons total	No	No	
Direct Pay Option (Eligible entities can receive a cash payment directly from the IRS for a portion of the 45Q tax credit they would otherwise have claimed against their tax liability.)				
	No	No	Yes	

Source: Organized by the Author.

demonstration projects, $1 billion for large-scale CCS pilot projects, and $3.5 billion for regional direct air capture (DAC) hubs over the next 5 years (Congressional Budget Office, 2023). This combination of tax credits and government funding fosters innovation and aims to enhance the commercial viability of CCS projects.

The United Kingdom

The UK Government has maintained a long-standing commitment to supporting CCS, with initiatives dating back to the early 2000s. There were two rounds of funding competitions for large-scale demonstration projects in 2007 and 2012. However, both programs were discontinued in 2011 and 2017 due to concerns regarding cost risks (Herzog, 2016; Hough et al., 2017). Nevertheless, the United Kingdom's commitment to achieving net-zero emissions by 2050, legislated in 2019, has revitalized interest in CCS. In November 2020, the government unveiled its "Ten Point Plan," setting an ambitious target of capturing and storing 10 million tons of CO_2 annually by 2030 (UK Government, 2020). This ambition was further amplified in December 2023 with a revised target of capturing 20–30 million tons of CO_2 per year by 2030. This revised target is backed by a significant £20 billion investment over 20 years and aims to create 50,000 jobs in the CCS sector (UK Government, 2023).

To facilitate these objectives, the UK Government has established a comprehensive regulatory and legal framework and is providing financial support across the CCS value chain. For emitters, this includes direct capital grants through the CCS infrastructure fund. The £1bn CCS Infrastructure Fund provides upfront financial assistance to developers for project initiation and construction. The Carbon Contract for Difference scheme offers long-term contracts guaranteeing a minimum price for captured CO_2, ensuring operational cost stability for CCS facilities (Fattouh et al., 2024). These initiatives have led to the development of several CCS industry clusters, including HyNet in North West England, the East Coast Cluster in Teesside & the Humber, Acorn in Scotland, and Viking in the Humber (UK Government, 2023).

The European Union

The European Union has consistently championed the global and regional deployment of CCS as part of its commitment to transitioning to a low-carbon economy. CCS plays a pivotal role in the European Commission's climate and energy packages, with the European Union being among the earliest jurisdictions to establish a comprehensive legal framework for CO_2 storage, known as the CCS Directive (2009/31/EC). This directive covers all aspects of CO_2 storage in geological formations within the European Union, including permitting, monitoring, and postclosure liability.

In alignment with its ambitious climate goals, the European Union aims to develop at least 50 million tons per year of CO_2 storage capacity by 2030, as outlined in the 2023 Net-Zero Industry Act proposed by the European Commission (European Commission, 2023). Since 2013, the EU Emissions Trading Scheme has served as a primary driver for CCS investment and has facilitated funding for industrial carbon management projects through the Innovation Fund (European Commission, 2024). In 2022, the European Union allocated €1.8 billion from the Innovation Fund to support 17 large-scale innovative clean tech projects, with seven of these projects involving CCS and spanning various countries including Bulgaria, Iceland, Poland, France, Sweden, and Germany.

The Horizon EU initiative further supports research, development, and innovation for CCS projects, fostering technological advancements in the field. Additionally, the

Connecting Europe Facility for Energy (CEF-E) provides financial support for cross-border energy infrastructure projects, including CCS hubs, with significant investments earmarked for CO_2 transport and storage initiatives. Notably, several CO_2 transport and storage projects across France, the Netherlands, and Norway have received support under this initiative, covering pipeline and terminal development.

Canada

Similar to the United States, the United Kingdom, and the European Union, Canada has long been a proponent of CCS and has implemented a series of policy measures to address climate change. In 2020, the Government of Canada unveiled a comprehensive CCS Policy titled "A Healthy Environment and a Healthy Economy," with the goal of guiding large emitters toward achieving net-zero emissions by 2050. This policy includes a CA$3 billion allocation over 5 years for decarbonization projects and integrates CCS into Canada's broader Hydrogen Strategy. CCS projects play a pivotal role in Canada's Clean Fuel Standard, as they enable the generation of compliance credits. This aspect further incentivizes the adoption of CCS technologies within the country. Building upon these efforts, the 2022 Federal Budget introduced an investment tax credit aimed at bolstering CCS initiatives, representing a cornerstone of Canada's CCS policy. This credit covers 50% of the capital cost of CO_2 capture equipment from 2022 to 2030, with a higher rate (60%) for equipment capturing CO_2 directly from the atmosphere. Additionally, the credit extends to cover 37.5% of the equipment costs associated with transporting, utilizing, and storing captured CO_2 (Kpekou et al., 2024).

In summary, governments are utilizing a variety of policy instruments to address the market failures hindering CCS development (Table 32.2). However, the successful deployment of CCS is also dependent on addressing numerous social issues, such as risk management, public acceptance and trust, and environmental justice. Effectively tackling these social considerations is crucial for increasing the likelihood of successful CCS deployment and ensuring its long-term viability.

Legal and Regulatory Framework for CCS Projects

A critical piece of successful CCS development is a clear and robust legal and regulatory framework, which is vital to address the social issues and to provide stewardship and oversight for CCS activities (Russial, 2011). It ensures environmental protection by setting standards that minimize risks associated with storing CO_2 underground, safeguarding ecosystems and human health. Secondly, regulations establish safety protocols for CCS facilities, ensuring operational safety and the well-being of workers and communities. Additionally, regulatory clarity aids legal compliance, providing guidance on permits and monitoring and reporting obligations. Moreover, a stable regulatory environment instills confidence in investors, driving investment in CCS infrastructure and technology. Public acceptance is fostered through transparent and inclusive regulatory processes, building trust and engaging meaningfully with a wide range of stakeholders and local communities, some of whom oppose CCS or siting of projects nearby. In essence, a robust regulatory framework is indispensable for realizing the potential of CCS in mitigating climate change.

Significant efforts have been made by national and regional legislators to develop CCS regulations since 2005 (Dixon et al., 2015). Generally, two approaches have emerged in this endeavor. The first involves the creation of dedicated CCS legislation, exemplified by the European Commission's CCS Directive on the Geological Storage of CO_2. The second approach builds upon existing regulatory frameworks, augmenting them with new regulations

TABLE 32.2 Existing CCS Policy Mechanisms by Country

Country	Enabling Legislation and Rules	Cost Reduction Measures	Regulation of Industrial Activities	National Target	Revenue Support
	Legal and regulatory frameworks	Grants preferential interest rates, loan guarantees, tax credits	Carbon tax, emission trading scheme, power plant rules	Deployment targets	Carbon contracts-for-difference
United States	X	X	X		
Brazil		X			
Australia	X	X	X		
Canada	X	X	X	X	
China		X	X		
Qatar		X			
Norway	X	X	X		
Saudi Arabia		X			
United Arab Emirates		X			
Japan		X		X	
United Kingdom	X	X	X	X	X
Netherlands	X	X	X	X	
Indonesia	X	X	X		
France	X	X	X	X	
Denmark	X	X	X	X	

Source: Adapted from IEA, 2023a.

tailored to specific components of CCS, as seen in Australia and the United States. However, irrespective of the chosen path, a comprehensive regulatory framework must encompass all stages and key components of CCS projects (Table 32.3) (WRI, 2008). For example, site characterization and selection represent critical initial steps in ensuring the safe and permanent storage of captured CO_2. Selection criteria, including storage capacity, geological favorability, and caprock integrity, inform this process. Environmental impact assessments play a crucial role in evaluating the potential environmental impacts of CCS projects throughout their lifecycle, facilitating risk identification, mitigation strategies, and transparency through public input. Additionally, robust measurement, reporting, and verification (MRV) systems are essential. A transparent MRV framework monitors CO_2 volumes captured, transported, and injected, facilitating CCS integration into carbon markets. Effective MRV ensures the legitimacy of carbon credits and provides vital data for verifying their compliance within carbon markets. The issue of long-term liability is unique to CCS

TABLE 32.3 A Conceptual Legal and Regulatory Framework for a CO_2 Storage Project

	Pre-Injection	During Injection	Post-Injection
Key Legal and Regulatory Issues	Permitting process Site characterization and selection Ownership of pore space Interaction with other surface and subsurface resources Risk assessment Environment impact assessment Measurement, reporting, and verification (MRV)	Storage site inspections Defining operational liabilities and financial security MRV	Site closure process and certification Long-term responsibility post-closure (e.g., maintenance and monitoring of storage site and addressing long-term liabilities) Finance assurances of long-term site stewardship

Source: Adapted from Yang, 2024.

projects, given their involvement in storing CO_2 underground for extended periods. A well-designed regulatory framework must clarify responsibility in the event of CO_2 leaks from storage sites. Various approaches are employed globally to address this issue, ranging from government assumption of liability to financial instruments such as insurance or liability transfer to the government.

Social Issues

Despite its potential, the history of CCS reveals a significant hurdle: social acceptance. Public anxieties and concerns surrounding CCS projects have, at times, stalled progress and cast a shadow on its widespread adoption. This section delves into the key social issues that have impacted the development of CCS, including risk perception, public acceptance, employment opportunities, and environmental justice.

Risk Perception

CCS development is associated with two primary risks. First, there is a concern that its implementation could inadvertently prolong dependence on fossil fuels, resulting in a "lock-in effect" that delays the transition to cleaner energy sources (Vergragt et al., 2011). While CCS can serve as a transitional technology to combat climate change, it is imperative to emphasize to the public that its primary objective is to facilitate the shift toward low-carbon alternatives and to demonstrate good faith efforts that the fossil fuel industry is drastically reducing its carbon emissions (Bandilla, 2020). Given the essential role of hard-to-abate industries[3] in society and the existing carbon emissions in the atmosphere, CCS may prove indispensable in achieving negative CO_2 emissions necessary to meet ambitious climate targets (Haszeldine et al., 2018). Second, operational risks stem from CO_2 transportation and injection underground, potentially leading to heightened seismic activity, groundwater contamination, and the necessity for secure long-term containment (Leiss & Krewski, 2019). Although extensive operational experience and modeling exist, meticulous site selection and rigorous safety management are critical for minimizing these risks (Friedmann, 2007). A 2020 CO_2 pipeline rupture in Satartia, Mississippi, led to respiratory issues among residents (Zegart, 2021). Despite these concerns and such incident, large-scale commercial operations and research-related trial CCS endeavors in the United States and globally have demonstrated effective and safe CO_2 storage, offering valuable insights for future deployments (National Energy Technology Laboratory, 2024).

Public Acceptance and Engagement

Public acceptance remains a significant hurdle for the deployment of CCS. Concerns among the public often center on transparency issues, uncertainties regarding the technology, and potential environmental risks (Nielsen et al., 2022). Due to limited public education and outreach, many individuals lack familiarity with CCS and its purpose, often favoring more well-known renewables like solar and wind (Reiner et al., 2006). Public surveys indicate resistance to higher electricity bills aimed at funding CCS, raising concerns about its economic viability (Whitmarsh et al., 2019). Real-world examples, such as the cancellation of the Vattenfall project in Germany, underscore the impact of public opposition. However, research suggests that framing CCS as a waste-reuse technology can enhance favorable public perceptions (Jones et al., 2017). To overcome these challenges, policymakers and project developers must prioritize public engagement through educational campaigns, transparent communication about risks and benefits, and positioning CCS as complementary to renewables, all while striving to minimize burdens on local communities.

Environmental Justice

Environmental justice groups in the United States continue to voice significant concerns regarding CCS projects, despite the Biden administration's promotion of this technology for climate change mitigation (Chemnick, 2023). These groups articulate several key objections, chief among them the fear that siting CCS projects in disadvantaged communities already burdened by pollution creates a scenario of "burden doubling," exacerbating existing inequalities (see chapter by Farrell and Brostrom in this volume). Additionally, concerns persist regarding potential safety risks associated with CCS, such as leaks or contamination, which could disproportionately impact nearby communities. Finally, some environmental justice groups express a lack of trust in the government and industry to ensure the responsible and equitable implementation of CCS projects, feeling that their voices are not being heard and that potential benefits are not guaranteed to reach their communities (EPRI, 2023).

To address these concerns and increase the potential for CCS projects to gain wider acceptance, it is essential to prioritize transparent communication and meaningful community engagement. Equitable siting practices must be upheld to ensure that projects are not located in communities already overburdened by pollution, disproportionately impacting low-income residents and people of color. Meaningful community engagement, following principles of transparency and inclusivity, is necessary to ensure that local residents are informed about potential risks and benefits, and that their voices shape the project's development (US Department of Energy, 2022). Moreover, CCS projects should provide tangible advantages to host communities, such as job creation and infrastructure improvements. A 2023 analysis by Rodium Group highlights the significant job creation potential associated with scaling up DAC in the United States. Building a DAC plant creates an average of 1,215 jobs annually over a 5-year construction period. Once operational, a DAC plant is estimated to sustain approximately 340 jobs throughout its lifespan (Jones et al., 2023). However, a comprehensive risk and benefit assessment, conducted collaboratively with the community, is necessary to identify potential environmental and health hazards, with mitigation plans and emergency response protocols clearly communicated.

In February 2022, the Council on Environmental Quality of the White House issued a document titled "Carbon Capture, Utilization, and Sequestration Guidance," aimed at federal agencies involved in regulating CCS projects in the United States. This document recommends that agencies employ a transparent process and meaningful public engagement. Agencies should prioritize the development and application of environmental justice best practices for CCS efforts. Actions to be taken include the following (Council on Environmental Quality, 2022):

- Evaluating the impacts of proposed CCS actions on potential host communities early in the planning process
- Providing information about the impacts, costs, and benefits of CCS in advance of Tribal consultation and stakeholder engagement
- Consulting Tribal Nations on potential CCS projects in a manner that strengthens Nation-to-Nation relationships
- Avoiding the imposition of additional burdens on overburdened and underserved communities, including by evaluating direct, indirect, and cumulative effects and identifying and implementing appropriate mitigation and avoidance measures
- Ensuring transparent decisions and accountability to Tribes and communities with respect to any applicable mitigation measures designed to reduce environmental impacts

International Climate Governance and CCS

The history of CCS development underscores the intricate relationship between international climate politics and technological innovation. Scholars in innovation systems emphasize the critical need for robust international cooperation on CCS demonstrations to foster efficient and cost-effective global learning on CCS (De Coninck et al., 2009; Reiner, 2016). Such an approach would not only deepen our understanding of the global capacity and applicability of CCS but also bolster global trust, awareness, and public confidence in the technology. These scholarly debates have also been reflected in international climate governance on CCS.

The formal recognition of CCS as a clean development mechanism (CDM) project activity in December 2011, at the 17th Conference of the Parties serving as the Meeting of the Parties to the Kyoto Protocol (COP17/CMP7) held in Durban, South Africa, marked a significant milestone in its international acceptance as a climate mitigation technology. Integral to shaping this consensus are various international multilateral and bilateral initiatives. Among these, the Carbon Sequestration Leadership Forum (CSLF), established in 2003, has the longest history. Comprising 26 member countries and the European Commission, the CSLF is dedicated to advancing cost-effective technologies for CCS. It serves as a platform for member countries to coordinate research efforts, share best practices, and advocate for CCS as a viable climate change mitigation strategy.

In 2015, the launch of Mission Innovation at COP21 further underscored the international commitment to accelerating the development and demonstration of clean energy technologies, including CCS, through collaborative efforts. This initiative drew participation from 20 countries and the European Commission in the carbon capture challenge, highlighting the global recognition of CCS's importance. In addition, CCS is incorporated within the Clean Energy Ministerial, a high-level global forum to promote policies and programs that advance clean energy technology. All CSLF, Mission Innovation, and Clean Energy Ministerial CCS established formal work mechanisms and dedicated staff, convening Minister-level meetings annually. Most recently, the Carbon Management Challenge (CMC), was launched by US President Biden in March 2023 and convened at COP28 in Dubal UAE. Emphasizing the importance of CCS technologies in achieving the 1.5°C warming target, the CMC advocates for massive scale-up, with gigaton-level CO_2 reduction targeted by 2030. The CMC encourages collaboration among participating countries to set national targets, develop joint initiatives, and collaborate on project demonstrations and policy frameworks.

CCS has also emerged as a critical component of bilateral climate cooperation, evidenced by its inclusion in initiatives such as the US–China Collaboration and the Australia–Japan Collaboration. The history of CCS demonstrates that its success hinges on strong international cooperation. However, a backslide on international climate goals or governance could significantly hinder this technology's progress.

Conclusion

This chapter has provided an in-depth exploration of CCS as a pivotal tool in climate change mitigation and climate action. It has traced the historical trajectory of CCS development, from periods of initial enthusiasm to the current resurgence of interest. Economic challenges were acknowledged, with proposed solutions outlined through government policy responses. Furthermore, the chapter delved into the intricate social issues surrounding CCS, including public perception and environmental justice concerns, while also discussing potential strategies to address these challenges and foster broader social acceptance of the technology. Additionally, the role of international climate governance in promoting research, development, and deployment of CCS was examined.

In conclusion, the widespread deployment of CCS necessitates a multifaceted approach. Continuous technological advancements must be complemented by effective policy frameworks that address economic and other challenges. Proactive measures are required to address social concerns and build public trust in CCS technology. Moreover, international cooperation is vital for achieving the necessary global scale for CCS to significantly contribute to climate change mitigation efforts.

Notes

1. In some regions, the term *carbon capture, utilization, and storage* (CCUS) is preferred. In this article, CCUS will also be referred to as CCS for simplicity.
2. CDR refers to technologies, practices, and approaches that remove and durably store CO_2 from the atmosphere, including DACCS, BECCS, enhanced rock weathering, soil carbon sequestration, and so on (IPCC AR6 WGIII: CDR Factsheet).
3. A hard-to-abate sector is defined as an industry or economic area that faces significant challenges in reducing greenhouse gas emissions. These challenges often stem from complex value chains, the nature of the production process, and the long lifespan of assets (Edelenbosch et al., 2024).

References

Audus, H. (1997). Decarbonization of fossil fuels: Hydrogen as an energy carrier. *Fuel and Energy Abstracts*, *4*(38), 227.

Bandilla, K. W. (2020). 31—Carbon capture and storage. In T. M. Letcher (Ed.), *Future energy* (3rd ed., pp. 669–692). Elsevier. https://doi.org/10.1016/B9780081028865.000311

Budinis, S., Krevor, S., Dowell, N. M., Brandon, N., & Hawkes, A. (2018). An assessment of CCS costs, barriers and potential. *Energy Strategy Reviews*, *22*(22), 61–81.

Chemnick, J. (2023, August 22). "Down your throat": Biden pushes CCS on polluted places. *E&E News by POLITICO*. https://www.eenews.net/articles/down-your-throat-biden-pushes-ccs-on-polluted-places/

Climate Action Tracker. (2023, December 14). CAT net zero target evaluations. https://climateactiontracker.org/global/cat-net-zero-target-evaluations/

Congressional Budget Office. (2023). Carbon capture and storage in the United States. https://www.cbo.gov/publication/59832#_idTextAnchor053

Council on Environmental Quality. (2022). Carbon capture, utilization, and sequestration guidance. https://www.federalregister.gov/documents/2022/02/16/2022-03205/carbon-capture-utilization-and-sequestration-guidance#footnote-9-p8811

de Coninck, H., Stephens, J. C., & Metz, B. (2009). Global learning on carbon capture and storage: A call for strong international cooperation on CCS demonstration. *China Energy Efficiency*, *37*(6), 2161–2165.

Dixon, T., Leamon, G., Zakkour, P., & Warren, L. (2013). CCS projects as Kyoto Protocol CDM activities. *GHGT11 Proceedings of the 11th International Conference on Greenhouse Gas Control Technologies, 1822 November 2012, Kyoto, Japan*, *37*, 7596–7604.

Dixon, T., McCoy, S. T., & Havercroft, I. (2015). Legal and regulatory developments on CCS. *International Journal of Greenhouse Gas Control*, *40*, 431–448. https://doi.org/10.1016/j.ijggc.2015.05.024

Edelenbosch, O. Y., Hof, A. F., van de Chen, H., Daioglou, V., Dekker, M. M., Doelman, J. C., Harmsen, M., Mikropoulos, S., Sluisveld, van Stehfest, E., Tagomori, Zeist, I. S., & Vuuren, V. (2024). Reducing sectoral hard-to-abate emissions to limit reliance on carbon dioxide removal. *Nature Climate Change*, *14*(7), 715–722. https://doi.org/10.1038/s41558024020 25y

EPRI. (2023). EPRI home. https://www.epri.com/research/products/000000003002026035

European Commission. (2023, March 16). *Net-Zero Industry Act*. https://commission.europa.eu/strategy-and-policy/priorities-2019-2024/european-green-deal/green-deal-industrial-plan/net-zero-industry-act_en

European Commission. (2024). *What is the Innovation Fund?* Climate.ec.europa.eu; https://climate.ec.europa.eu/eu-action/eu-funding-climate-action/innovation-fund/what-innovation-fund_en

Fattouh, B., Muslemani, H., & Jewad, R. (2024). *Capture carbon, capture value: An overview of CCS business models*. The Oxford Institute of Energy Studies. https://www.oxfordenergy.org/wpcms/wp-content/uploads/2024/02/CM08-Capture-Carbon-Capture-Value_Final.pdf

Friedmann, J. (2007). Site characterization and selection guidelines for geological carbon sequestration. Lawrence Livermore National Lab. https://www.osti.gov/biblio/915602

Furre, A.-K., Eiken, O., Alnes, H., Vevatne, J. N., & Kiær, A. F. (2017). 20 years of monitoring CO_2-injection at Sleipner. *Energy Procedia*, *114*, 3916–3926. https://doi.org/10.1016/j.egypro.2017.03.1523

G8 Hokkaido Toyako Summit. (2020, June 28). *G8 Hokkaido Toyako Summit Leaders Declaration*. https://www.mofa.go.jp/policy/economy/summit/2008/doc/doc080714__en.html

GCCSI. (2023). Global status of CCS 2023. https://www.globalccsinstitute.com/resources/publications-reports-research/global-status-of-ccs-2023-executive-summary/

Halland, E. (2021, November 6). *CO_2 capture and storage: Success stories and challenges for upscaling deployment*. https://www.c2g2.net/wp-content/uploads/Halland.pdf

Hardisty, P. E., Sivapalan, M., & Brooks, P. (2011). The environmental and economic sustainability of carbon capture and storage. *International Journal of Environmental Research and Public Health*, *8*(5), 1460–1477. https://doi.org/10.3390/ijerph8051460

Haszeldine, R. S., Flude, S., Johnson, G., & Scott, V. (2018). Negative emissions technologies and carbon capture and storage to achieve the Paris Agreement commitments. *Philosophical Transactions of the Royal Society A: Mathematical, Physical and Engineering Sciences*, *376*(2119), 20160447.

Hawkins, D. (2017, July 28). *Kemper County IGCC: Death knell for carbon capture? NOT*. https://www.nrdc.org/bio/david-hawkins/kemper-county-igcc-death-knell-carbon-capture-not#:~:text=The%20key%20fact%20to%20emphasize

Herzog, H. (2016). *Lessons learned from CCS demonstration and large pilot projects: An MIT energy initiative working paper.* https://sequestration.mit.edu/bibliography/CCS%20Demos.pdf
Hough, D., Ares, E., & Delebarre, J. (2017, January 23). *UK decarbonisation and carbon capture and storage.* The UK House of Commons Library. https://commonslibrary.parliament.uk/research-briefings/cdp-2017-0023/
IEA. (2016). *20 Years of carbon capture and storage accelerating future deployment.* https://iea.blob.core.windows.net/assets/24c3d26b-aa44-4b54-b9c0-5201d4d86a04/20YearsofCarbonCaptureandStorage_WEB.pdf
IEA. (2020). *CCS in clean energy transitions.* https://www.iea.org/reports/CCS-in-clean-energy-transitions
IEA. (2023a). *CCS policies and business models building a commercial market.* https://iea.blob.core.windows.net/assets/d0cb5c89-3bd4-4efd-8ef5-57dc327a02d6/CCSPoliciesandBusinessModels.pdf
IEA. (2023b, September). *Net zero roadmap: A global pathway to keep the 1.5 °C goal in reach.* https://www.iea.org/reports/net-zero-roadmap-a-global-pathway-to-keep-the-15-0c-goal-in-reach
IPCC. (2005). *IPCC special report on carbon dioxide capture and storage* (B. Metz, O. Davidson, H. de Coninck, M. Loos, & L. Meyer, Eds.). Cambridge University Press. https://www.ipcc.ch/site/assets/uploads/2018/03/srccs_wholereport-1.pdf
IPCC. (2018). Global warming of 1.5°C. https://www.ipcc.ch/sr15/
IPCC. (2022). Climate Change 2022: Mitigation of Climate Change. https://www.ipcc.ch/report/ar6/wg3/downloads/report/IPCC_AR6_WGIII_FullReport.pdf
IPCC. (2023). Synthesis report of the IPCC Sixth Assessment Report (AR6). https://www.ipcc.ch/report/sixth-assessment-report-cycle/
IPCC AR6 WGIII. (2023). CDR factsheet. https://www.ipcc.ch/report/ar6/wg3/downloads/outreach/IPCC_AR6_WGIII_Factsheet_CDR.pdf
JOGMEC. (2023, March 2). *Handbook for CCS carbon credits.* https://www.jogmec.go.jp/english/news/release/news_10_00026.html
Jones, A. C., & Lawson, A. J. (2022). *Carbon capture and sequestration (CCS) in the United States.* Congressional Research Service. https://sgp.fas.org/crs/misc/R44902.pdf
Jones, C. R., Olfe-Kräutlein, B., & Kaklamanou, D. (2017). Lay perceptions of carbon dioxide utilisation technologies in the United Kingdom and Germany: An exploratory qualitative interview study. *Energy Research & Social Science*, 34, 283–293. https://doi.org/10.1016/j.erss.2017.09.011
Jones, W., Bower, G., O'Rear, E. G., King, B., & Pastorek, N. (2023). Direct air capture workforce development: Opportunities by occupation. *Rhodium Group.* https://rhg.com/research/direct-air-capture-workforce-development/
Kearns, D., Liu, H., & Consoli, C. (2021). *Technology readiness and costs of CCS.* Global CCS Institute. https://www.globalccsinstitute.com/wp-content/uploads/2021/03/Technology-Readiness-and-Costs-for-CCS-2021-1.pdf
Kpekou, R., Nahornick, N., & Scholz, T. (2024, February 1). *Investment tax credit for carbon capture, utilization and storage.* Office of the Parliamentary Budget Officer. https://www.pbo-dpb.ca/en/publications/LEG-2324-022-S--investment-tax-credit-carbon-capture-utilization-storage--credit-impot-investissement-captage-utilisation-stockage-carbone#:~:text=From%202022%20through%202030%2C%20the
Krahé, M., Heidug, W., Ward, J., & Smale, R. (2013). From demonstration to deployment: An economic analysis of support policies for carbon capture and storage. *Energy Policy, 60*(9), 753–763.
Leiss, W., & Krewski, D. (2019). Environmental scan and issue awareness: Risk management challenges for CCS. *International Journal of Risk Assessment and Management, 22*(3/4), 234. https://doi.org/10.1504/ijram.2019.103340
Loria, P., & Bright, M. B. H. (2021). Lessons captured from 50 years of CCS projects. *The Electricity Journal, 34*(7), 106998. https://doi.org/10.1016/j.tej.2021.106998
Marchetti, C. (1977). On geoengineering and the CO_2 problem. *Climatic Change, 1*(1), 59–68.
Marshall, C. (2015, February 4). Clean coal power plant killed, again. *Scientific American.* https://www.scientificamerican.com/article/clean-coal-power-plant-killed-again/
National Energy Technology Laboratory. (2024). Permanence and safety of CCS. https://netl.doe.gov/carbon-management/carbon-storage/faqs/permanence-safety

Natural Resources Canada. (2014, August 27). Evaluation of the Clean Energy Fund (CEF). https://natural-resources.canada.ca/transparency/reporting-and-accountability/plans-and-performance-reports/strategic-evaluation-division/year/evaluation-reports-2014/evaluation-the-clean-energy-fund-cef/16534

Nielsen, J. A. E., Stavrianakis, K., & Morrison, Z. (2022). Community acceptance and social impacts of carbon capture, utilization and storage projects: A systematic meta-narrative literature review. *PLOS ONE*, *17*(8), e0272409.

Reiner, D., Curry, T., De Figueiredo, M., Herzog, H., Ansolabehere, S., Itaoka, K., Akai, M., Johnsson, F., & Odenberger, M. (2006). *An international comparison of public attitudes towards carbon capture and storage technologies*. https://sequestration.mit.edu/pdf/GHGT8_Reiner.pdf

Reiner, D. M. (2016). Learning through a portfolio of carbon capture and storage demonstration projects. *Nature Energy*, *1*(1), 15011. https://doi.org/10.1038/nenergy.2015.11

Russial, T. (2011). *Carbon capture and storage—Legal and regulatory framework*. International Centre for Sustainable Carbon. https://www.sustainable-carbon.org/report/carbon-capture-and-storage-legal-and-regulatory-framework-ccc-179/

Steinberg, M., & Albanese, A. S. (1980). Environmental control technology for atmospheric carbon dioxide. In W. Bach, J. Pankrath, & J. Williams (Eds.), *Interactions of energy and climate: Proceedings of an international workshop held in Münster, Germany, March 3–6, 1980* (pp. 521–555). Springer Netherlands.

Swartz, K. E. (2017, June 22). Southern Co.'s clean coal plant hits a dead end. *E&E News by Politico*. https://www.eenews.net/articles/southern-co-s-clean-coal-plant-hits-a-dead-end/

Tollefson, J. (2008). Carbon burial buried. *Nature*, *451*(7179), 613–613. https://doi.org/10.1038/451612b

UK Government. (2020). *The ten point plan for a Green Industrial Revolution*. https://assets.publishing.service.gov.uk/media/5fb5513de90e0720978b1a6f/10_POINT_PLAN_BOOKLET.pdf

UK Government. (2023, December 20). *New vision to create competitive carbon capture market follows unprecedented £20 billion investment*. https://www.gov.uk/government/news/new-vision-to-create-competitive-carbon-capture-market-follows-unprecedented-20-billion-investment

US Department of Energy. (2022). *Creating a community and stakeholder engagement plan*. https://www.energy.gov/sites/default/files/2022-08/Creating%20a%20Community%20and%20Stakeholder%20Engagement%20Plan_8.2.22.pdf

US Government. (2021). *The long-term strategy of the United States: Pathways to net-zero greenhouse gas emissions by 2050*. https://www.whitehouse.gov/wp-content/uploads/2021/10/US-Long-Term-Strategy.pdf

Vergragt, P. J., Markusson, N., & Karlsson, H. (2011). Carbon capture and storage, bio-energy with carbon capture and storage, and the escape from the fossil-fuel lock-in. *Global Environmental Change*, *21*(2), 282–292.

Wagman, D. (2017, June 13). *The three factors that doomed Kemper County IGCC—IEEE Spectrum*. https://spectrum.ieee.org/the-three-factors-that-doomed-kemper-county-igcc

Whitmarsh, L., Xenias, D., & Jones, C. R. (2019). Framing effects on public support for carbon capture and storage. *Palgrave Communications*, *5*(17). https://doi.org/10.1057/s41599-019-0217-x

WRI. (2008). Guidelines for carbon dioxide capture, transport, and storage. https://www.wri.org/research/guidelines-carbon-dioxide-capture-transport-and-storage

Yang, X. (2024). *A gap analysis of China's regulatory framework for CO_2 geological storage*. Global CCS Institute. https://www.globalccsinstitute.com/resources/publications-reports-research/a-gap-analysis-of-chinas-regulatory-framework-for-co2-geological-storage/

Zegart, D. (2021, August 26). Gassing Satartia: How a CO2 pipeline explosion affected this Mississippi town. *HuffPost*. https://www.huffpost.com/entry/gassing-satartia-mississippi-co2-pipeline_n_60ddea9fe4boddef8boddc8f

Zhang, Y., Jackson, C., & Krevor, S. (2022). An estimate of the amount of geological CO_2 storage over the period of 1996–2020. *Environmental Science & Technology Letters*, *9*(8), 693–698. https://doi.org/10.1021/acs.estlett.2c00296

CHAPTER 33

The Challenges and Opportunities of Landscape Restoration in Latin America

FRANCISCO S. ÁLVAREZ

Forest Cover Dynamics in Latin America: Current Trends and Challenges

FORESTS are an essential ecological component covering almost a third of the Earth's surface (FAO, 2022). However, deforestation and environmental degradation driven by human activities over the last century threaten forests and biodiversity globally (Barlow et al., 2016; Jaureguiberry et al., 2022; Young et al., 2010). Between 1990 and 2020, approximately 420 million hectares of forests were lost by deforestation due to shifting agriculture and livestock farming expansion, as well as urban and industrial development (Curtis et al., 2018; FAO, 2022). Annual deforestation rates in recent years have averaged 10 million hectares (FAO, 2022). These data are alarming and reflect an accelerated pace of the global environmental crisis that affects biodiversity, ecosystem functionality, human development, and global climate (Lenton et al., 2023). Under this context, governments and stakeholders risk failing to meet several international conservation and restoration goals established as climatic actions for the following years, such as the New York Declaration on Forests (NYDF) goal of zero deforestation (see https://forestdeclaration.org/), global conservation goals of the Kunming-Montreal Convention on Biological Diversity (CBD) (30 × 30 Initiative), and the commitments of the United Nations Decade for Ecosystem Restoration (UNDER) by 2030 as part of the Paris Agreement on climate change (see https://www.decadeonrestoration.org/) (Sewell et al., 2020). Therefore, understanding the phenomenon of deforestation and the restoration strategies developed under these commitments will be key to understanding the challenges countries face in the next years.

Latin America is considered a biodiversity hotspot with sizeable tropical forest areas that house high species richness (Mittermeier et al., 2011; Pillay et al., 2022) and work as natural carbon sinks that are key in global efforts to mitigate climate change (Cook-Patton et al., 2020; Harris et al., 2021; Pan et al., 2024). However, over 117 million hectares of forest have been deforested in the last two decades, including more than 45 million hectares of primary forests (see Figure 33.1 and Table 33.1). However, despite the rapid loss of primary forests and generalized deforestation in the region, simultaneously, new forest areas are also being recovered on abandoned agricultural lands (Aide et al., 2013; Pan et al., 2024; Poorter, Craven, et al., 2021; Poorter, Rozendaal, et al., 2021; Rozendaal et al., 2019). Nonetheless, just ~17 million hectares were recovered during the same period (see Hansen et al., 2013; Potapov et al., 2022). This phenomenon shows an apparent compensation deficit between the loss of forest cover and the gain of new forest cover in the region.

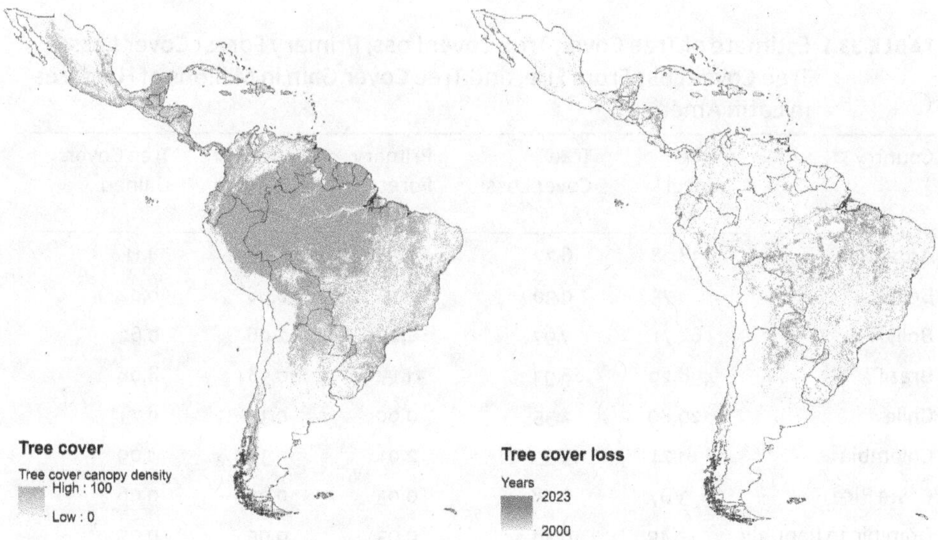

FIGURE 33.1. Tree cover (2010) and tree cover loss in Latin America (2001–2023). The tree cover and tree cover loss dataset comes from Landsat satellite imagery at a resolution of 30 × 30 meters with a >30% tree cover density (see Hansen et al., 2013). The author created the map from cited data sources.

Important efforts have been made to identify the causes of deforestation in Latin America. To date, deforestation from shifting agriculture (small and medium scale) and commercial agricultural expansion (large scale) are the primary drivers of deforestation in the region (Curtis et al., 2018). Countries with high forest cover percentages, such as Brazil, Colombia, Peru, Paraguay, Bolivia, Argentina, and Mexico, report high deforestation rates in the 21st century (see Table 33.1). During 2016 and 2017 alone, tree cover loss in tropical forests increased in the region (Figure 33.2). More than 16 million hectares of this forest cover loss was related to fires, mainly within the Amazon, where many of them were caused by land use changes for agriculture and livestock farming (Pendrill et al., 2022). By 2020, during the COVID-19 pandemic, there was a slight increase in deforestation in the region, despite lockdowns by quarantine and drastic decreases in economic activity worldwide (see Brancalion et al., 2020; Desvars-Larrive et al., 2020). The demand for food during the health crisis possibly drove deforestation during this period. Moreover, the absence of authorities in these forested territories facilitated illegal logging in many countries (see Brancalion et al., 2020). Nevertheless, the most significant impact of the pandemic on forests may occur in the coming years when countries increase resource extraction to try to rebuild their economies, despite that the loss of tropical forests could result in new emerging diseases (FAO, 2022; Guégan et al., 2020).

Understanding the factors driving deforestation in Latin America is crucial to generating better strategies to reduce deforestation, restore land, and mitigate the effects of climate change. It is necessary to understand the social and economic factors that influence decisions about forests at different levels, from decisions made by landowners about cutting trees on farms to macroeconomic factors or public policies favoring land use changes in a country. In Latin America, where many rural communities exist near extensive forested areas (FAO, 2022), commercializing land is usually the only way to generate family income to survive in times of economic crisis. In many cases, this leads to a change in land use, usually from forests to crops such as palm oil, soybeans, pasture, wood fiber, and productive forestry

TABLE 33.1 Estimate of Tree Cover, Tree Cover Loss, Primary Forest Cover Loss, Tree Cover Loss From Fire, and Tree Cover Gain in Millions of Hectares in Latin America

Country	Tree Cover[1]	Tree Cover Loss[2]	Primary Forest Cover[2]	Tree Cover Loss From Fire[2]	Tree Cover Gained[3]
Argentina	38.38	6.77	0.48	0.60	1.11
Belize	1.75	0.30	0.15	0.06	0.02
Bolivia	62.71	7.97	4.22	2.06	0.62
Brazil	498.20	68.93	31.15	10.58	8.06
Chile	20.80	2.35	0.00	0.33	0.73
Colombia	81.74	5.39	2.01	0.37	1.09
Costa Rica	3.87	0.28	0.03	0.02	0.06
Dominican Republic	2.45	0.38	0.03	0.05	0.05
Ecuador	19.23	1.00	0.25	0.07	0.22
El Salvador	0.92	0.08	0.00	0.00	0.01
Guatemala	7.03	1.75	0.57	0.25	0.29
Honduras	7.55	1.40	0.51	0.17	0.14
Mexico	50.35	4.89	0.79	0.75	1.42
Nicaragua	7.62	1.84	0.65	0.12	0.14
Paraguay	20.45	6.96	1.17	0.49	2.12
Peru	78.78	4.08	2.65	0.30	0.43
Panama	5.55	0.51	0.09	0.03	0.08
Uruguay	2.12	0.43	0.00	0.01	0.65
Venezuela	57.27	2.42	0.61	0.34	0.49
Total	**966.78**	**117.72**	**45.38**	**16.59**	**17.74**

[1] The tree cover 2010 dataset comes from Landsat satellite imagery at a resolution of 30 × 30 meters with a >30% tree cover density of woody vegetation with a height of >5 meters (see Hansen et al., 2013).

[2] The tree cover loss dataset comes from 2001–2023 Landsat satellite imagery at a resolution of 30 × 30 meters with a >30% tree cover density of woody vegetation with a height of >5 meters (see Hansen et al., 2013).

[3] The tree cover gained dataset comes from 2000–2020 Landsat satellite imagery at a resolution of 30 × 30 meters of woody vegetation with a height of >5 meters (see Potapov et al., 2022).

plantations (Castellanos-Navarrete et al., 2021; Davis et al., 2020; Goldman et al., 2020; Pendrill et al., 2022). These types of crops or plantations, funded by significant foreign and local investments, have been considered drivers of the deforestation of large areas of tropical forests in Latin America, whose environmental impacts also affect the socio-ecological integrity of rural communities that depend on tropical forests (Davis et al., 2020).

Mining (artisanal and industrial) is also a significant driver of deforestation in many countries in the region (Giljum et al., 2022; Maus et al., 2022). For example, Amazonian

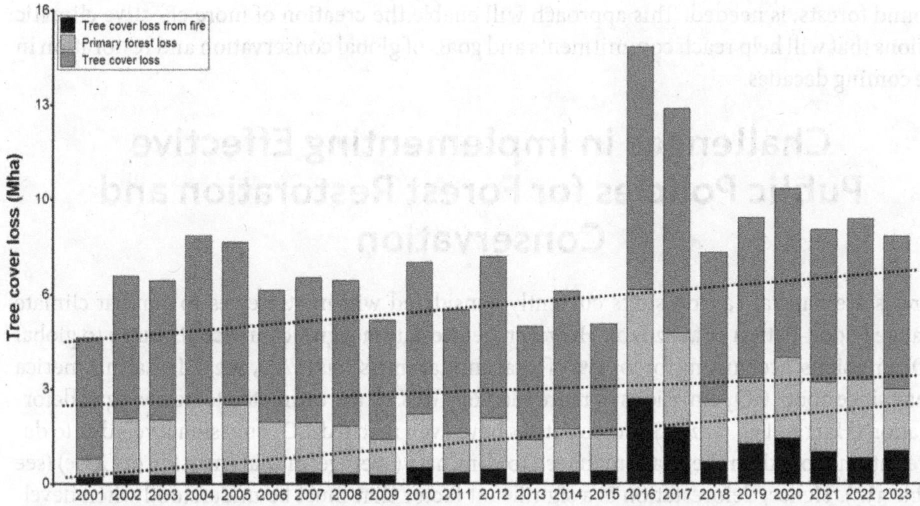

FIGURE 33.2. Estimation of annual deforestation of tree cover, including tree cover loss from fires and primary forest cover loss in Latin America from 2001 to 2023. The tree cover loss dataset comes from Landsat satellite imagery at a resolution of 30 × 30 meters with a >30% tree cover density of woody vegetation with a height of >5 meters (see Hansen et al., 2013).

countries like Brazil, Colombia, and Peru experience high levels of direct deforestation due to mining activities, while many other Latin American countries where mining is developing face significant indirect deforestation due to new urban and agricultural areas emerging near mining sites (Giljum et al., 2022). This activity not only leads to the loss of forests but also causes ecosystem degradation and loss of biodiversity, contributes to CO_2 greenhouse emissions, negatively impacts river and water quality, as well as causes health issues, and generates social conflicts in communities near mining sites (Dethier et al., 2023; Giljum et al., 2022; Maus et al., 2022; Soares et al., 2024).

Another factor that has impacted forest land use change in Latin America is economic globalization (Lambin & Meyfroidt, 2011). In recent decades, international human migration to the Global North has generated remittance growth, mainly in Central American countries, accounting for 20–26% of the country's gross domestic product (GDP) (see Table 33.2). In some cases, this migratory phenomenon has led to two opposing scenarios. On the one hand, the income from remittances replaces the need for agriculture and/or cattle farming as the main economic activities in rural communities, resulting in the abandonment of agricultural lands where new forest cover is now emerging (e.g., Hecht & Saatchi, 2007). On the other hand, it is invested in the extension and/or mechanization of agricultural processes or purchasing new lands to create alternative sources of income through touristic developments or urbanization, including purchasing forest land as an insurance policy for a possible return to their origin country, which in turn results in land use change and more deforestation (Angelsen et al., 2020; Davis & Lopez-Carr, 2014; Mack et al., 2023; Taylor et al., 2016).

Deforestation in Latin America reveals a complex network of social and economic interactions related to land use and land tenure. Therefore, challenges arise to propose unique, feasible solutions to reduce deforestation that are profitable enough to compete with the principal productive activities, such as agriculture and livestock. To effectively reduce deforestation and forest degradation in the region, a comprehensive treatment of the problem, considering social, cultural, and economic factors, including frameworks and public policy around forests, is needed. This approach will enable the creation of more effective climatic actions that will help reach commitments and goals of global conservation and restoration in the coming decades.

Challenges in Implementing Effective Public Policies for Forest Restoration and Conservation

Forests are natural carbon sinks currently considered within strategies to combat climate change (Cook-Patton et al., 2020). However, deforestation significantly contributes to global CO_2 emissions, accounting for 10–15% of total annual emissions (FAO, 2022). In Latin America alone since 2000, CO_2 emissions of more than 69 $GtCO_2e$ are estimated to be owing to deforestation (Harris et al., 2021). Some countries have even reached CO_2 emission levels due to deforestation more than they have managed to eliminate (average annual emission of CO_2e) (see Table 33.3). Hence, deforestation is a significant factor that must be understood when developing actions in the context of climate change. This problem has been addressed in multilateral spaces of the United Nations, where important frameworks for implementing policies to avoid deforestation and promote global restoration have been discussed as a part of strategies aimed to keep global warming below 1.5°C (Sewell et al., 2020). This has triggered considerable interest in international forest-related cooperation and mobilized substantial funding for restoration and forest conservation in many developing countries of Latin America. Despite this, many of these investments and strategies of restoration and conservation have not been implemented correctly or have not had the expected results. In many cases, these difficulties are related to local factors that are not usually considered in the design and implementation of global strategies. Moreover, deforestation is not usually assessed as a multidimensional problem related to local conditions, such as weak policies, low technical and scientific capacity, socioeconomic factors, and cultural dynamics of each country.

Creating and implementing national public policies to protect forests and implementing restoration actions at a global level is a priority. However, these efforts are not equally developed in all countries. Weak or unenforced policies can undermine conservation and restoration strategies. The forest's discontinuity between bordering countries provides a good example: How does a neighboring country protect the forest better than another (Wuepper et al., 2024)? Many Latin American countries face difficulties in producing and applying environmental policies for regulating and protecting forests and promoting forest restoration. Evidence of this is the fluctuating phenomenon of deforestation in the region over the last decades, where many public policies have promoted deforestation to enhance economic growth. For example, some South American countries encouraged colonization and agrarian development in Amazonian lands as a part of developing economic policies, leading to millions of hectares of deforested forest (Carrero et al., 2020; Viteri-Salazar & Toledo, 2020). This shows how public policies may be drivers of deforestation in the region. Even in some cases, only government administration shifts can exacerbate deforestation (e.g., de Area Leão Pereira et al., 2020; Rodrigues-Filho et al., 2015), evidencing institutional weakness within forest management.

On the contrary, there are positive cases of public policies that can reduce deforestation and make land restoration strategies effective. For example, policies promoting economic incentives, such as tax deductions, reforestation loans, and payments for environmental services (PES) can facilitate a progressive transition from agricultural or livestock activities to conserving forest lands (e.g., Alix-Garcia et al., 2018). Through strict public policies, Costa Rica prevented land use change and promoted payment incentives for environmental services to protect forests and promote forest plantations in degraded areas. These public policies allowed Costa Rica to increase its forest coverage in recent years (Arroyo-Mora et al., 2005; Calvo-Alvarado et al., 2009), and now these natural areas contribute to the economy via ecotourism (Courvisanos & Jain, 2006). These contrasts in implementing public policies between countries in the region show challenges that must be addressed. It is necessary to improve the design and implementation of public policies that do not contradict conservation and restoration efforts, guaranteeing their applicability over time and generating legal security in favor of restoration and conservation.

Regulations in international trade or market tendencies may influence the development of new regulations or public policies within the region. Recently, the European Union proposed the Regulation on Deforestation-Free Products (EUDR) to counteract deforestation in many tropical countries, including Latin America (Muradian et al., 2025). This initiative is part of a strategy to mitigate the adverse effects of climate change and decrease the demand for products that contribute to forest loss. Although this restriction can generate undesirable effects, such as disincentives to agroforestry (e.g., low profitability in cocoa and coffee production) or promote trade with other nonrestrictive countries, it is an important step toward creating new sustainable markets with low CO_2 emissions from deforestation. A paradigm shift is essential for creating effective public policies in the region, and for this, it is crucial to strengthen local capacities in most countries. Bilateral and multilateral cooperation will be necessary to develop public policies (fair and rights-based) that ensure forest conservation and restoration in Latin American countries.

Most countries in Latin America have taken important steps in forest conservation and restoration frameworks. However, gaps and barriers make significant progress on these efforts difficult. One of the region's main challenges is the absence of public policies to counter deforestation and promote forest conservation efficiently. On the contrary, public policies are created that contradict the commitments made in restoration and conservation efforts (see Figure 33.3 and Table 33.2). An example is the creation of public policies that promote deforestation through agricultural incentives or extractive activities (Carrero et al., 2020; Pereira et al., 2020; Viteri-Salazar & Toledo, 2020) and simultaneously committing to international climate agreements promoting forest conservation. For instance, Brazil has seen an aggressive agricultural expansion within the Amazonian forest in recent years (Feng et al., 2022; Hoang & Kanemoto, 2021); at the same time, it has promoted efforts to reduce deforestation through international cooperation and carbon offsets. These divergent policies can create conflicts and must be addressed with great attention. Global decisions should influence national policies to encourage investment in conservation and restoration actions or strategies and improve international traceability processes with low carbon emissions from deforestation. Consequently, public policies that facilitate the processes and financing for forest restoration and conservation must be promoted instead of those that may be detrimental to forest conservation. These policies must also avoid conflicts between agricultural policies and seek synergies and balances between sustainable use and forest conservation. Finally, these policies must be articulated among stakeholders, including local communities, to increase citizen participation and promote local governance that facilitates restoration mechanisms, as well as forest conservation and management.

Forest Landscape Restoration as a Climate Action

Over the past decade, many governments have recognized the vital role of forests and the negative impact of the alarming levels of global deforestation. Restoration as a climate action has been proposed in recent decades and has become more popular. Many commitments and promises have emerged from different international forums. Recently, the United Nations Decade on Ecosystem Restoration (2021–2030) was established to avoid or decrease desertification, drought, increasing temperatures, land and forest degradation, loss of biodiversity, and wildfires. This global commitment to restoration is expected to reach 350 million hectares in the following years and stop global deforestation by 2030 (following the New York Declaration on Forests [NYDF]).

However, other global efforts, such as the Bonn Challenge, seek to restore 150 million hectares of the world's deforested and degraded land, adding to 350 million hectares by 2030. Many countries in Latin America and the Caribbean added to this effort through the 20x20 Initiative (see https://initiative20x20.org/), where each country pledged to restore and protect forests. So far, 18 countries in Latin America and the Caribbean have committed to restore and protect 52 million hectares through this Initiative by 2030 (Figure 33.3). Although restoration is a global need, some countries require greater attention, especially countries with greater vulnerabilities to biodiversity and ecosystem services loss or hold key ecosystems where restoration and conservation actions are a priority for global efforts (Lamb et al., 2005). Many authors have asked where restoration opportunities exist and how it should be carried out (e.g., Brancalion, Niamir, et al., 2019; Potapov et al., 2011; Williams et al., 2024). Many platforms and tools now allow for identifying areas with restoration opportunities where restoration would be more efficient using different approaches. For instance, some authors have estimated a restoration potential of at least 2 billion hectares worldwide, considering just climate conditions and land use without human influence (see Figure 33.3) (Potapov et al., 2011), while other authors have estimated close to 215 million hectares with restoration potential by natural regeneration for the tropic regions (Williams

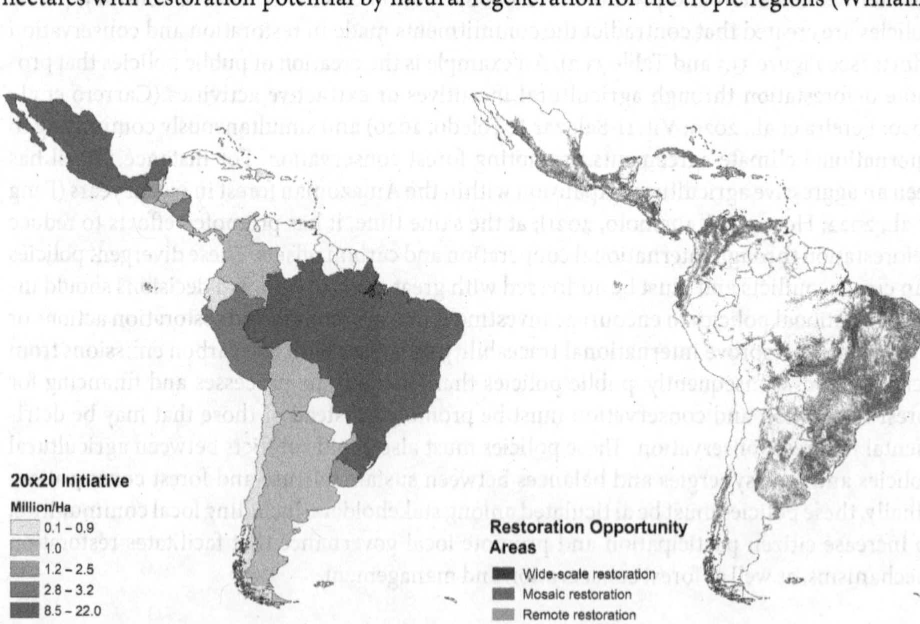

FIGURE 33.3. Number of hectares committed by member countries of the 20x20 Initiative and Restoration Opportunity Areas in Latin America. Restoration Opportunity Areas data come from a spatial resolution of 1 km (see Potapov et al., 2011). Wide-scale restoration: potential areas with less human pressure and potential to support closed forests; mosaic restoration: potential areas with moderate human pressure; remote restoration: potential areas with very low human pressure. The author created the map from cited data sources.

et al., 2024). These examples show favorable scenarios for restoring degraded areas in the region. However, all of these tools must be evaluated cautiously and considered as an approximation of potential restoration sites or as a planning input, where the socioeconomic, political, cultural, ecological, and climatic conditions that operate in each territory or site to be restored must also be considered.

Restoration is understood as a process that intends to recover an ecosystem's health, integrity, and functionality, bringing it closer to its original structure and function (Gann et al., 2019). Considering this, an ecosystem is said to be recovered and restored when it contains sufficient resources to continue its development on its own, maintaining its structure and functionality. In this context, there are many methodologies and strategies to restore lands. Depending on an ecosystem's level of degradation or deforestation, there are different stages of recuperation related to the type of management and actions carried out to contribute to landscape restoration (Di Sacco et al., 2021). These actions can range from assisted processes (active restoration), such as reforestation (commercial plantations, agroforestry, native species), to natural processes, such as natural regeneration (passive restoration). The difference between these methodologies lies in the expected goals at the end of the process and the resources available for the interventions (see Figure 33.4), which can be measured according to the increase in forest cover area, forest attributes (e.g., basal area, stem density, biomass), biodiversity, ecosystem services, and socio-ecological benefits over time (Chazdon, 2008; Gann et al., 2019; Giles et al., 2024; Maniraho et al., 2023).

Reforestation is often confused with restoration; in any case, reforestation contributes to the larger objectives or goals of restoration. Restoration is more than a simplified action of planting trees (reforestation). Therefore, it is essential to distinguish both actions, know their implications, and identify how, when, and where reforestation is necessary for restoration strategies (active or passive restoration). For example, reforesting using unsuitable species (e.g., exotic species) in incompatible sites can generate undesirable results during the restoration process, such as affecting water infiltration, displacement of native biodiversity, rise of invasive species, decrease in pollination, reduction of carbon stock capacity (above and below ground), and increase of forest fuel loads (wildfires/fires), and it could even lead to deforestation by land use conflict (by reforesting areas destined to food production) (Di Sacco et al., 2021). Therefore, restoration efforts should consider multiple factors before the

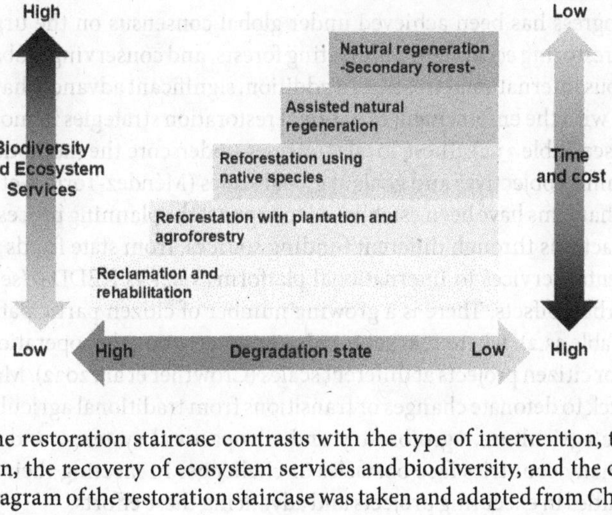

FIGURE 33.4. The restoration staircase contrasts with the type of intervention, the state of ecosystem degradation, the recovery of ecosystem services and biodiversity, and the cost and time of investment. The diagram of the restoration staircase was taken and adapted from Chazdon (2008).

intervention, from the site history, reference ecosystem, land use, anthropogenic factors, or disturbance drivers (e.g., fires, deforestation frequency, agriculture-livestock intensity), and environmental factors such as climate. The degree of disturbance and history of sites determine the complexity of restoration actions that should be carried out (Chazdon, 2008; Poorter et al., 2024).

When financial resources are scarce or the degree of degradation is low, an efficient way to restore is to reduce disturbance drivers and allow natural regeneration, also known as secondary succession (new forests) (Chazdon et al., 2021; Williams et al., 2024). This secondary succession is usually driven by plant regrowth capacity, seed source availability, and soil nutrient availability (Poorter et al., 2024). Currently, secondary succession is known to result in greater structural and functional diversity of the ecosystem at lower costs (Bodin et al., 2022; Brancalion, Meli, et al., 2019; Chazdon, 2008; Chazdon et al., 2021; Chazdon & Guariguata, 2016; Crouzeilles et al., 2017; Lamb et al., 2005; Prach et al., 2007). However, in some cases, management activities or assisted restoration practices, such as fertilization, pruning, enrichment planting, or protection against herbivores, become necessary to improve environmental conditions and achieve the expected results (Castro et al., 2021; Chazdon et al., 2021). Passive restoration, as a product of natural regeneration or secondary succession, can attain 80% of its original biodiversity in about ~20 years, offering multiple ecosystem services very similar to primary forests (Chazdon et al., 2009; Rozendaal et al., 2019). These new forests resulting from natural regeneration (secondary forests) now represent most of the world's tropical forests (FAO, 2022). Therefore, it is crucial to identify the socioeconomic, political, ecological, and cultural factors that facilitate the growth of these new forests in the region and to promote strategies to reduce the threat of their deforestation or reconversion into productive activities (return of agricultural and livestock). Understanding these factors would enable more effective actions to accelerate restoration processes, aiding in accomplishing global biodiversity conservation commitments and restoration pledges aimed at mitigating the effects of climate change.

Obstacles for Developing Countries to Restore Degraded Lands and Protect Their Forests

Significant progress has been achieved under global consensus on the urgency of stopping deforestation, restoring ecosystems, protecting forests, and conserving global biodiversity, reflected in various international treaties. In addition, significant advances have been made on a national scale, with the enforcement of national restoration strategies in most of the countries in the region (see Table 33.2). These local strategies underscore the methodological processes to achieve common objectives and goals at global scales (Méndez-Toribio et al., 2017). Various financing mechanisms have been established to support the planning processes and execution of restoration actions through different funding sources, from state funds such as payments for environmental services to international platforms such as REDD+ (see https://redd.unfccc.int/) or carbon offsets. There is a growing number of citizen participation in restoration projects (see Table 33.2), where market trends and international cooperation favor financing mechanisms for citizen projects at different scales (Crowther et al., 2022). Many of these funds or strategies seek to detonate changes or transitions from traditional agriculture and livestock to regenerative agriculture, agroforestry and silvopastoral systems, and forest protection. Despite this, many countries face obstacles or challenges in meeting their restoration goals, causing difficulties in executing projects and advancing these efforts.

TABLE 33.2 Latin American Restoration Pledges Made Under the New York Declaration on Forests, the Bonn Challenge, the 20 × 20 Initiative, National Restoration Strategies, Citizen Restoration Initiatives, and Economic Indicators by Country

Country	New York Declaration on Forests	Bonn Challenge[1]	20 × 20 (Million/Ha)[1]	Restoration National Strategy, Plans, or Program[2]	RESTOR Projects[3]	GDP (Billions of US$)[4]	Personal Remittance (% of GDP)[4]
Argentina		X	1.00	X	142	589.04	0.2
Belize			0.13	X	66	2.62	4.6
Bolivia			1.00		3,866	39.26	3.2
Brazil	X	X	22.00	X	7,496	1950	0.2
Chile	X	X	0.50	X	1,420	279.68	0.0
Colombia	X	X	1.00	X	1,373	356.79	2.8
Costa Rica	X	X	1.00	X	2,814	73.11	0.8
Dominican Republic			0.89		21	100.37	8.7
Ecuador	X	X	0.50	X	562	112.16	4.6
El Salvador		X	1.00	X	7	28.27	24.1
Guatemala		X	1.20	X	54	80.91	19.6
Honduras		X	1.00	X	260	26.9	26.1
Mexico	X	X	8.50	X	2,626	1,330	3.7
Nicaragua		X	2.80		14	15.4	26.1
Paraguay			ND[5]		22	44.02	1.7
Peru	X	X	3.20	X	334	221.78	1.7
Panama		X	1.00	X	55	72.8	0.6
Uruguay			2.50	X	9	61.99	0.2
Venezuela					11	—	—

[1] Data come from the 20 × 20 Initiative (https://initiative20x20.org/) and the Bonn Challenge Initiative (https://www.bonnchallenge.org/).

[2] Documents were consulted and are available on each country's official website.

[3] Data come from RESTOR platforms, showing citizen restoration initiatives (Crowther et al., 2022).

[4] Data come from the World Bank; GDP data represent the GDP constant 2015 US$. (https://data.worldbank.org/country).

[5] ND: Paraguay is a member, but their commitment is not defined.

Many ambitious restoration initiatives, such as the Bonn Challenge, the UNDER commitments, or national restoration plans goals, involve recovering large degraded land areas. Larger countries like Brazil or Mexico have more areas to restore and are able to commit many hectares in global restoration pledges (see Table 33.2 and Figure 33.3). They also have stronger economies and may have more available funds to meet the challenge. However, some small countries have committed to restoring extensive areas concerning the size of their territory. An example is El Salvador, which pledged about half of its territory as a part of the Boon Challenge Initiative (Table 33.2). These bold challenges by small countries may respond to the need to recover forest coverage as a mitigation and adaptation strategy to climate change and to the need to raise more funds to achieve it (Fagan et al., 2020), even if they cannot fulfill their commitments.

Each country's restoration commitments may face challenges when implementing them, even more so if many approaches to these efforts remain vertical, where decision makers (e.g., the state, carbon offsetting companies, or nongovernmental organizations [NGOs]) plan restoration actions without a participative consultation process involving local stakeholders. Also, it usually does not consider the degraders (e.g., private companies) and restorers (e.g., local communities, private farms, landowners) as crucial actors linked and interrelated, leading to different levels of legitimacy within the restoration process. In addition, the long-term sustainability of these initiatives is generally unclear; often, it only considers the initial restoration process (reforestation) cost but does not include the monitoring and following costs. This lack of resources and proper planning during the restoration process may cause many problems and generate social conflicts, particularly when the (socioeconomic) benefits are unclear for the stakeholders, mainly in the local communities. All of this makes it challenging to advance restoration goals in Latin America, especially in those countries that lack tangible strategies and financial possibilities to sustain these efforts in the long term.

Analyzing the national restoration strategies (including national plans or programs) reveals ambiguous restoration goals. Some restoration strategies use diffuse concepts to reclaim their "restored areas." Many countries do not have mechanisms to guarantee that a restoration intervention has been sustained over time. Only the initial action is often quantified (reforestation: number of planted trees or reforested area), and no transparent mechanisms for monitoring or verifying the actions exist, or data may be overestimated or duplicated (see Imbach & Nello, 2024; see Murcia et al., 2017). Many restoration actions do not prioritize conservation and increased biodiversity as tangible indicators of the reclamation of restored areas. Moreover, it does not consider the estimated time that the efforts and monitoring should last to guarantee that the expected results are achieved. Much of the investment made in many restoration national strategies in the region focuses on generating new productive forest cover (agroforestry and silvopastoral systems, forest plantations) on degraded lands to recover ecosystem services. Yet scarce initiatives promote the restoration of forests despite the need to recover lost ecosystems to conserve global biodiversity (Nicholson et al., 2021; Strassburg et al., 2020).

Establishing new productive forest cover or implementing assisted restoration actions on highly degraded lands can have high execution costs in their initial stages and eventually cause sustainability problems over time. When resources are scarce, as is the case of many projects in the region (Méndez-Toribio et al., 2017), natural regeneration or secondary succession can be a more efficient form of restoration under these circumstances, successfully helping to achieve restoration goals over a relatively short time (Poorter, Craven, et al., 2021; Poorter, Rozendaal, et al., 2021; Rozendaal et al., 2019; Williams et al., 2024). However, many countries have not used this approach and usually focus on assisted restoration (planting trees). Multiple studies consider that the best strategies for restoring land would be those that promote the protection of existing forests, restoration through natural regeneration by secondary succession, greater community participation, and promoting and strengthening the rural livelihoods and governance around forests (Cooke et al., 2019; Di Sacco et al., 2021; Williams et al., 2024; Young, 2000).

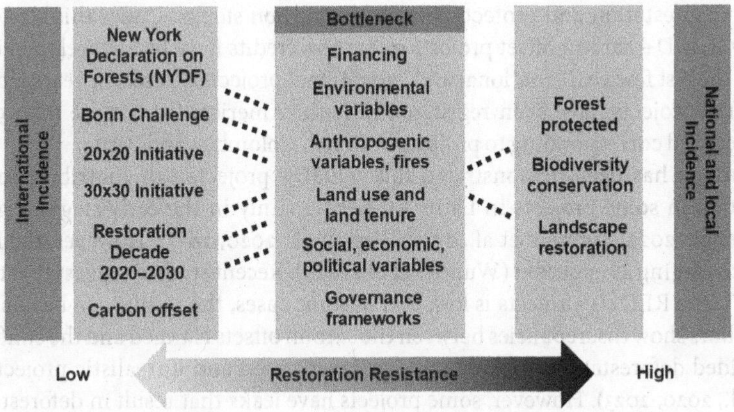

FIGURE 33.5. Social, economic, and environmental factors generate restoration resistance in the commitments in Latin America.

However, many national plans do not consider the factors or disturbances that generate resistance (bottleneck) to restoration (Figure 33.5). The constant pressure of land use change, such as livestock grazing within areas in the recovery process or forest fires, can be negative drivers of restoration, affect the recovery process, or change the dynamics of new forests (e.g., Álvarez et al., 2021). Reducing negative pressures that slow down restoration efforts is necessary. Regulations parallel to national strategies must be enforced to guarantee the success of restoration efforts. Factors related to national public policies that promote deforestation through "perverse incentives," such as agricultural subsidies, low-quality environmental compensation (e.g., carbon offsets), or noncompliance with the law due to weak institutions, can highlight negative pressures that undermine national restoration efforts. An example of this is the fluctuating and constant rate of deforestation in many countries in the region in the last decades. In addition, access to financing for restoration, together with other conditions related to the site, such as socioeconomic context, land use, land tenancy, local idiosyncrasy, poor agricultural practices, forest fires, prolonged droughts, or floods, can affect the success of restoration projects on a local scale (see Figure 33.3). All factors that generate resistance to restoration must be evaluated to develop efficient and realistic strategies in the region. Furthermore, the benefits of restoration must be prevented from dissipating among a few (government, private companies, NGOs), and the empowerment of local actors who will make restoration efforts sustainable over time within the territories must be guaranteed.

Reflections on Carbon Offsets and Restoration Efforts

In recent decades, numerous strategies to mitigate and minimize the effects of global climate change have been discussed in international settings. In 1997, the United Nations Framework Convention on Climate Change (UNFCCC) approved the Kyoto Protocol, an agreement that establishes objectives for reducing greenhouse gas emissions, mainly from developed countries. Consequently, world governments have established new strategies for compensating greenhouse gas emissions through voluntary or regulated carbon offset markets. As part of the Conference of the Parties (COP) agreements, Reducing Emissions from Deforestation and Forest Degradation (REDD+) in Developing Countries was created to reduce deforestation and protect global forest carbon stocks. Under this strategy, many voluntary REDD+ carbon offset projects or carbon credits have been developed in tropical forests in the last few years (national and subnational projects). In recent years alone, nearly 202 REDD+ projects have been registered in Latin America, from more than 93% of the credits recorded corresponding to projects in Brazil, Colombia, and Peru.

Currently, it has been demonstrated that REDD+ projects can contribute to reducing deforestation in some projects in Latin America, mainly in the early stages (e.g., Guizar-Coutiño et al., 2022; Simonet et al., 2019; West et al., 2020, 2023). However, the results are not as encouraging as expected (Wunder et al., 2024). Recent studies suggest that the performance of some REDD+ projects is low, and in some cases, the results can be controversial. Many authors show discrepancies between the carbon offsets claimed and the confirmed impact (avoided deforestation), possibly due to exaggerated and unrealistic projections (e.g., West et al., 2020, 2023). However, some projects have leaks that result in deforestation next to the project areas (West et al., 2020), or the deforestation returns after the payments are over (Carrilho et al., 2022; Roopsind et al., 2019). Also, many of these REDD+ projects have established initiatives in forest areas with higher carbon values located in remote regions far from the real threats of deforestation (Delacote et al., 2022). In contrast, critical areas where greater conservation effort should be focused continue to be deforested. Accountability becomes another issue as many of these projects are not fully transparent when reporting and verifying carbon offsets (Delacote et al., 2024), including a potential overestimation of claimed offsets (West et al., 2020).

Many see carbon credits as a promising tool to mitigate climate change by reducing greenhouse gas emissions from deforestation. The idea of polluting and offsetting in tropical forests in developing countries seems reasonable to some but not easy to carry out to others. For some scholars, evaluating these offsetting initiatives as successful or unsuccessful is still too early. However, evaluating whether we are on the right path is necessary. The Voluntary Carbon Market (VCM) has generated more than 213 million credits in REDD+ projects and more than 253 million credits in Forestry and Land Use in Latin America and the Caribbean (see Haya et al., 2024), mainly in Amazonian countries. An imbalance is evident in the allocation of resources under the context of carbon offsets in the region. Countries with less forest cover, such as some countries in Central America and the Caribbean, may be excluded from these incentives or financing. This inequity creates difficulties for those countries that urgently need to stop deforestation and gain forest cover as a mitigation and adaptation strategy to climate change (Strassburg et al., 2020). This situation raises the question of whether this strategy works equally for everyone.

Many studies have evaluated the implementation costs of REDD+ projects. The results suggest that implementing these projects may be higher than the income received (Luttrell et al., 2018; Nantongo & Vatn, 2019). Results may also be limited by market volatility and resource availability at critical stages of the project (Wunder et al., 2024). In addition, there is criticism regarding project intermediaries who tend to obtain more economic benefits than the local communities or the project itself. There are even questions about the main beneficiaries of carbon credits (Trouwloon et al., 2023). These inequalities are perceived and generate discomfort in the perception of REDD+ and carbon offsets. This concern must be addressed with greater caution, and the environmental impact projected and compliance with the social and environmental safeguards must be verified.

Social conflict or negative impacts during the implementation of these projects should also be avoided, as well as potential greenwashing risks. Institutional procedures should be set into place to guarantee that the benefits of these projects do not end as perverse incentives, creating adverse effects or externalities that have not been considered during and after the

TABLE 33.3 Latin American REDD+ Projects, National Strategy or Action Plan REDD+, Offset Credits, Estimate of Average Annual Values of Emission of CO_2e for Deforestation, and Elimination of CO_2e for Forest Cover

Country	National Strategy or Action Plan REDD+[1]	REDD+ Projects[2]	REDD+ Credits[2]	Forestry and Land Use Projects[2]	Forestry and Land Use Credits[2]	Emission $MtCO_2e/yr$[3]	Elimination $MtCO_2e/yr$[3]	Net Carbon Sink $MtCO_2e/yr$[3]
Argentina	Yes	3	—	15	—	72.00	−164.00	−91.60
Belize	Yes	6	4,281,642	—	4,281,642	6.29	−4.94	1.35
Bolivia	—	4	55,400	6	189,091	160.00	−177.00	−17.50
Brazil	Yes	109	75,372,399	151	79,251,068	1,640.00	−1,760.00	−119.00
Chile	Yes	1	533,654	6	1,927,068	43.80	−164.00	−120.00
Colombia	Yes	39	34,378,176	64	38,513,519	131.00	−241.00	−110.00
Costa Rica	Yes	—	—	6	48,801	6.45	−17.60	−11.10
Dominican Republic	—	—	—	—	—	7.37	−12.00	−4.64
Ecuador	Yes	—	—	3	—	27.70	−62.10	−34.50
El Salvador	—	—	—	—	—	1.57	−4.95	−3.38
Guatemala	Yes	4	7,839,327	8	8,813,366	36.10	−30.00	6.11
Honduras	Yes	1	—	2	18,655	30.00	−33.60	−3.62
Mexico	Yes	2	—	302	6,875,390	87.30	−216.00	−129.00
Nicaragua	—	—	—	7	360,907	44.00	−32.30	11.70
Paraguay	Yes	7	958,773	16	1,768,076	79.90	−101.00	−21.60
Peru	—	24	89,613,220	36	90,081,536	113.00	−170.00	−57.20
Panama	Yes	2	—	10	3,178,398	12.30	−23.40	−11.10
Uruguay	—	—	—	16	17,873,082	15.70	−23.70	−8.08
Venezuela	—	—	—	—	—	48.80	−163.00	−115.00

[1] Data come from REDD + Web Platform (https://redd.unfccc.int/) of government reports.

[2] Data come from the Voluntary Registry Offsets Database (Haya et al., 2024).

[3] Data come from modeled data to annual average (2001–2023) gross emissions, gross removals, and net flux from forests (see Harris et al., 2021).

termination of the projects (Carrilho et al., 2022). Multiple platforms monitor these global initiatives (e.g., Atmadja et al., 2023; Haya et al., 2024). However, efforts must be unified to increase accountability and verification of the implemented actions and to improve adjustment mechanisms that allow adapting strategies when necessary. For now, methodological discrepancies and difficulties in accessing information make it challenging to understand the progress achieved.

Despite multiple strategies promoting REDD+ projects in the region, these initiatives have not been successfully established in all priority areas where forest degradation and deforestation must be addressed. There has been a significant decline in REDD+ projects since they were adopted as a global climate change mitigation strategy at UNFCCC COP13 (Atmadja et al., 2022). Implementation challenges and uncertainties may contribute to the slowdown of REDD+ initiatives in the region. Despite this, it is important to acknowledge the progress made in these efforts and to avoid dismissing or abandoning them. The consensus reached in recent years regarding the need to generate incentives to halt deforestation and forest degradation represents a historic milestone in global climate change mitigation. However, it is essential to address the concerns and problems surrounding the implementation of each initiative, and we must take on the challenge of enhancing our understanding of deforestation and restoration issues while offering solutions to each problem based on science and social justice.

Conclusion

Deforestation remains one of the leading environmental challenges in Latin America. Understanding the drivers of deforestation at different levels will be crucial for effective climate change mitigation and adaptation strategies and for achieving the goals of the Paris Agreement. Currently, many contradictions exist between each country's restoration and conservation commitments and the implementation of public policies that exacerbate deforestation. The role of public policies and the weak institutional capacity of governments to enforce laws or regulations are drivers of environmental degradation and deforestation in many countries across Latin America. Governments must set up appropriate economic and legal incentive structures to tackle deforestation and promote the sustainable use of natural resources. Greater efforts must be made within governments to create coherence between development policies and global efforts to counteract the impacts of climate change, as well as strengthening the institutions in charge of leading these processes and promoting governance and participation in priority territories. Despite this, Latin America remains a region of interest under financing to reverse deforestation and promote restoration as part of the climate actions identified within international discussions. Nonetheless, many of these efforts fail to have the desired impact or are often insufficient. Efforts to understand the progress made in forest protection and restoration in the region in the last decades show many challenges, especially in implementing actions at a national or subnational scale.

Almost all countries in the region have national restoration strategies (including plans or programs). Yet many plans lack proper indicators that guarantee the correct measure of gain in forest cover over time. Regardless of this, there is a significant increase in forest cover in the region. Much of this new forest cover is not the result of massive forest restoration efforts, but rather, in many cases, it is the consequence of smallholders abandoning agricultural land. This natural regeneration in deforested and degraded lands demonstrates the region's restoration potential (passive restoration-secondary forest) (see Williams et al., 2024). This suggests that natural regeneration could be one of the most effective large-scale restoration strategies if properly well-understood and planned. However, in many restoration efforts, natural regeneration has been underutilized, with current strategies often relying on tree planting, which can be highly costly and frequently unsustainable over time.

REDD+ and carbon offset projects have generated many expectations as a global strategy to stop the loss of tropical forests and fight greenhouse gas emissions. However, the results have not been as expected (e.g., West et al., 2020, 2023; Wunder et al., 2024). Evidence of this is the low or modest results obtained from REDD+ projects in recent years. There is evidence that well-planned projects with robust scientific data can have positive impacts (Wunder et al., 2024). Still, many difficulties related to local social and political aspects limit successful execution, including factors related to market volatility that do not guarantee that financial resources will arrive on time and be sufficient or sustainable. Adopting these projects at a national scale could likely guarantee greater political and regulatory commitments from countries, although subnational projects could be handled more efficiently. Currently, there is a consensus on financial mechanisms that allow for avoiding carbon emissions from deforestation and forest degradation through REDD+ and carbon offset projects, which must be improved. Doubts and criticisms must be addressed using robust methods and greater transparency, as well as avoiding leaks and guaranteeing environmental and social benefits (safeguards) based on the rights of local communities.

Latin America's current forest protection and restoration mechanisms must be reviewed and improved. For this, spaces must be created to promote the discussion of genuine and legitimate solutions and alternatives for forest conservation and climate change mitigation. This would address the underlying deforestation problems under an effective holistic and comprehensive approach to facing the climate crisis. Also, it is necessary to understand that forests are more than a "carbon pool" that can trade; forests represent the life of many rural communities and many species' habitats and are part of the people's history. Consequently, it must explore and innovate in other sources of financing and prevent the conservation of forests and biodiversity from depending solely on the funding of large degraders or polluters. New, more participatory financing schemes must be explored for restoration and forest conservation. Also, new offsetting forms and incentives to reduce carbon emissions must be analyzed to create effective and realistic climate actions for the next years.

References

Aide, T. M., Clark, M. L., Grau, H. R., López-Carr, D., Levy, M. A., Redo, D., Bonilla-Moheno, M., Riner, G., Andrade-Núñez, M. J., & Muñiz, M. (2013). Deforestation and reforestation of Latin America and the Caribbean (2001–2010). *Biotropica*, *45*(2), 262–271. https://doi.org/10.1111/j.1744-7429.2012.00908.x

Alix-Garcia, J. M., Sims, K. R., Orozco-Olvera, V. H., Costica, L. E., Fernández Medina, J. D., & Romo Monroy, S. (2018). Payments for environmental services supported social capital while increasing land management. *Proceedings of the National Academy of Sciences*, *115*(27), 7016–7021. https://doi.org/10.1073/pnas.1720873115

Álvarez, F. S., Finegan, B., Delgado, D., Ramos, Z., Utrera, L. P., & Granda, V. (2021). Dispersal limitation, soil, and fire affect functional properties of tropical secondary forests on abandoned cattle ranching landscapes. *Perspectives in Plant Ecology, Evolution and Systematics*, 52, 125632. https://doi.org/10.1016/j.ppees.2021.125632

Angelsen, A., Aguilar-Støen, M., Ainembabazi, J. H., Castellanos, E., & Taylor, M. (2020). Migration, remittances, and forest cover change in rural Guatemala and Chiapas, Mexico. *Land*, *9*(3), 88. https://doi.org/10.3390/land9030088

Arroyo-Mora, J. P., Sánchez-Azofeifa, G. A., Rivard, B., Calvo, J. C., & Janzen, D. H. (2005). Dynamics in landscape structure and composition for the Chorotega region, Costa Rica from 1960 to 2000. *Agriculture, Ecosystems & Environment*, *106*(1), 27–39. https://doi.org/10.1016/j.agee.2004.07.002

Atmadja, S., Komalasari, M., Alusiola, R., Barboza, I., Sartika, L., Theresia, V., & Simonet, G. (2023). The International Database on REDD+ projects and programs Linking Economics, Carbon and Communities (ID-RECCO)—Project tables, V.5.0. https://www.reddprojectsdatabase.org

Atmadja, S. S., Duchelle, A. E., De Sy, V., Selviana, V., Komalasari, M., Sills, E. O., & Angelsen, A. (2022). How do REDD+ projects contribute to the goals of the Paris Agreement? *Environmental Research Letters*, *17*(4), 044038. https://doi.org/https://10.1088/1748-9326/ac5669

Barlow, J., Lennox, G. D., Ferreira, J., Berenguer, E., Lees, A. C., Nally, R. M., Thomson, J. R., Ferraz, S. F. d. B., Louzada, J., Oliveira, V. H. F., Parry, L., de Castro Solar, R. R., Vieira, I. C. G., Aragão, L. E. O. C., Begotti, R. A., Braga, R. F., Moreira Cardoso, T., de Oliveira Jr, R. C., Souza Jr, C. M., ... Gardner, T. A. (2016). Anthropogenic disturbance in tropical forests can double biodiversity loss from deforestation. *Nature*, *535*(7610), 144–147. https://doi.org/10.1038/nature18326

Bodin, B., Garavaglia, V., Pingault, N., Ding, H., Wilson, S., Meybeck, A., Gitz, V., d'Andrea, S., & Besacier, C. (2022). A standard framework for assessing the costs and benefits of restoration: Introducing the economics of ecosystem restoration. *Restoration Ecology*, *30*(3), e13515. https://doi.org/10.1111/rec.13515

Brancalion, P. H. S., Broadbent, E. N., De-Miguel, S., Cardil, A., Rosa, M. R., Almeida, C. T., Almeida, D. R. A., Chakravarty, S., Zhou, M., Gamarra, J. G. P., Liang, J., Crouzeilles, R., Hérault, B., Aragão, L. E. O. C., Silva, C. A., & Almeyda-Zambrano, A. M. (2020). Emerging threats linking tropical deforestation and the COVID-19 pandemic. *Perspectives in Ecology and Conservation*, *18*(4), 243–246. https://doi.org/10.1016/j.pecon.2020.09.006

Brancalion, P. H. S., Meli, P., Tymus, J. R. C., Lenti, F. E. B., Benini, R. M., Silva, A. P. M., Isernhagen, I., & Holl, K. D. (2019). What makes ecosystem restoration expensive? A systematic cost assessment of projects in Brazil. *Biological Conservation*, *240*, 108274. https://doi.org/10.1016/j.biocon.2019.108274

Brancalion, P. H. S., Niamir, A., Broadbent, E., Crouzeilles, R., Barros, F. S. M., Almeyda Zambrano, A. M., Baccini, A., Aronson, J., Goetz, S., Reid, J. L., Strassburg, B. B. N., Wilson, S., & Chazdon, R. L. (2019). Global restoration opportunities in tropical rainforest landscapes. *Science Advances*, *5*(7), eaav3223. https://doi.org/10.1126/sciadv.aav3223

Calvo-Alvarado, J., McLennan, B., Sánchez-Azofeifa, A., & Garvin, T. (2009). Deforestation and forest restoration in Guanacaste, Costa Rica: Putting conservation policies in context. *Forest Ecology and Management*, *258*(6), 931–940. https://doi.org/10.1016/j.foreco.2008.10.035

Carrero, G. C., Fearnside, P. M., do Valle, D. R., & de Souza Alves, C. (2020). Deforestation trajectories on a development frontier in the Brazilian Amazon: 35 years of settlement colonization, policy and economic shifts, and land accumulation. *Environmental Management*, *66*(6), 966–984. https://doi.org/10.1007/s00267-020-01354-w

Carrilho, C. D., Demarchi, G., Duchelle, A. E., Wunder, S., & Morsello, C. (2022). Permanence of avoided deforestation in a Transamazon REDD+ project (Pará, Brazil). *Ecological Economics*, *201*, 107568. https://doi.org/10.1016/j.ecolecon.2022.107568

Castellanos-Navarrete, A., de Castro, F., & Pacheco, P. (2021). The impact of oil palm on rural livelihoods and tropical forest landscapes in Latin America. *Journal of Rural Studies*, *81*, 294–304. https://doi.org/10.1016/j.jrurstud.2020.10.047

Castro, J., Morales-Rueda, F., Navarro, F. B., Löf, M., Vacchiano, G., & Alcaraz-Segura, D. (2021). Precision restoration: A necessary approach to foster forest recovery in the 21st century. *Restoration Ecology*, *29*(7), e13421. https://doi.org/10.1111/rec.13421

Cook-Patton, S. C., Leavitt, S. M., Gibbs, D., Harris, N. L., Lister, K., Anderson-Teixeira, K. J., Briggs, R. D., Chazdon, R. L., Crowther, T. W., Ellis, P. W., Griscom, H. P., Herrmann, V., Holl, K. D., Houghton, R. A., Larrosa, C., Lomax, G., Lucas, R., Madsen, P., Malhi, Y., ... Griscom, B. W. (2020). Mapping carbon accumulation potential from global natural forest regrowth. *Nature*, *585*(7826), 545–550. https://doi.org/10.1038/s41586-020-2686-x

Cooke, S. J., Bennett, J. R., & Jones, H. P. (2019). We have a long way to go if we want to realize the promise of the "Decade on Ecosystem Restoration." *Conservation Science and Practice*, *1*(12), e129. https://doi.org/10.1111/csp2.129

Courvisanos, J., & Jain, A. (2006). A framework for sustainable ecotourism: Application to Costa Rica. *Tourism and Hospitality Planning & Development*, 3(2), 131–142. https://doi.org/10.1080/14790530600938378

Crouzeilles, R., Ferreira, M. S., Chazdon, R. L., Lindenmayer, D. B., Sansevero, J. B. B., Monteiro, L., Iribarrem, A., Latawiec, A. E., & Strassburg, B. B. N. (2017). Ecological restoration success is higher for natural regeneration than for active restoration in tropical forests. *Science Advances*, 3(11), e1701345. https://doi.org/10.1126/sciadv.1701345

Crowther, T. W., Thomas, S. M., van den Hoogen, J., Robmann, N., Chavarría, A., Cottam, A., Cole, R., Elliott, T., Clark, E., Max, S., Danylo, O., & Rowe, C. (2022). Restor: Transparency and connectivity for the global environmental movement. *One Earth*, 5(5), 476–481. https://doi.org/10.1016/j.oneear.2022.04.003

Curtis, P. G., Slay, C. M., Harris, N. L., Tyukavina, A., & Hansen, M. C. (2018). Classifying drivers of global forest loss. *Science*, 361(6407), 1108–1111. https://doi.org/10.1126/science.aau3445

Chazdon, R. L. (2008). Beyond deforestation: restoring forests and ecosystem services on degraded lands. *Science*, 320(5882), 1458–1460. https://doi.org/10.1126/science.1155365

Chazdon, R. L., Falk, D. A., Banin, L. F., Wagner, M., J Wilson, S., Grabowski, R. C., & Suding, K. N. (2021). The intervention continuum in restoration ecology: Rethinking the active–passive dichotomy. *Restoration Ecology*, 32(8), e13535. https://doi.org/10.1111/rec.13535

Chazdon, R. L., & Guariguata, M. R. (2016). Natural regeneration as a tool for large-scale forest restoration in the tropics: Prospects and challenges. *Biotropica*, 48(6), 716–730. https://doi.org/10.1111/btp.12381

Chazdon, R. L., Peres, C. A., Dent, D., Sheil, D., Lugo, A. E., Lamb, D., Stork, N. E., & Miller, S. E. (2009). The potential for species conservation in tropical secondary forests. *Conservation Biology*, 23(6), 1406–1417. https://doi.org/10.1111/j.1523-1739.2009.01338.x

Davis, J., & Lopez-Carr, D. (2014). Migration, remittances and smallholder decision-making: Implications for land use and livelihood change in Central America. *Land Use Policy*, 36, 319–329. https://doi.org/10.1016/j.landusepol.2013.09.001

Davis, K. F., Koo, H. I., Dell'Angelo, J., D'Odorico, P., Estes, L., Kehoe, L. J., Kharratzadeh, M., Kuemmerle, T., Machava, D., Rodrigues Pais, A. d. J., Ribeiro, N., Rulli, M. C., & Tatlhego, M. (2020). Tropical forest loss enhanced by large-scale land acquisitions. *Nature Geoscience*, 13(7), 482–488. https://doi.org/10.1038/s41561-020-0592-3

Delacote, P., L'horty, T., Kontoleon, A., West, T. A., Creti, A., Filewod, B., LeVelly, G., Guizar-Coutiño, A., Groom, B., & Elias, M. (2024). Strong transparency required for carbon credit mechanisms. *Nature Sustainability*, 7, 706–713. https://doi.org/10.1038/s41893-024-01310-0

Delacote, P., Le Velly, G., & Simonet, G. (2022). Revisiting the location bias and additionality of REDD+ projects: the role of project proponents status and certification. *Resource and Energy Economics*, 67, 101277. https://doi.org/10.1016/j.reseneeco.2021.101277

Desvars-Larrive, A., Dervic, E., Haug, N., Niederkrotenthaler, T., Chen, J., Di Natale, A., Lasser, J., Gliga, D. S., Roux, A., Sorger, J., Chakraborty, A., Ten, A., Dervic, A., Pacheco, A., Jurczak, A., Cserjan, D., Lederhilger, D., Bulska, D., Berishaj, D., . . . Thurner, S. (2020). A structured open dataset of government interventions in response to COVID-19. *Scientific Data*, 7(1), 285. https://doi.org/10.1038/s41597-020-00609-9

Dethier, E. N., Silman, M., Leiva, J. D., Alqahtani, S., Fernandez, L. E., Pauca, P., Çamalan, S., Tomhave, P., Magilligan, F. J., & Renshaw, C. E. (2023). A global rise in alluvial mining increases sediment load in tropical rivers. *Nature*, 620(7975), 787–793. https://doi.org/10.1038/s41586-023-06309-9

Di Sacco, A., Hardwick, K. A., Blakesley, D., Brancalion, P. H. S., Breman, E., Cecilio Rebola, L., Chomba, S., Dixon, K., Elliott, S., Ruyonga, G., Shaw, K., Smith, P., Smith, R. J., & Antonelli, A. (2021). Ten golden rules for reforestation to optimize carbon sequestration, biodiversity recovery and livelihood benefits. *Global Change Biology*, 27(7), 1328–1348. https://doi.org/10.1111/gcb.15498

Fagan, M. E., Reid, J. L., Holland, M. B., Drew, J. G., & Zahawi, R. A. (2020). How feasible are global forest restoration commitments? *Conservation Letters*, 13(3), e12700. https://doi.org/10.1111/conl.12700

FAO. (2022). *The state of the world's forests 2022: Forest pathways for green recovery and building inclusive, resilient and sustainable economies*. FAO. https://doi.org/10.4060/cb9360en

Feng, Y., Zeng, Z., Searchinger, T. D., Ziegler, A. D., Wu, J., Wang, D., He, X., Elsen, P. R., Ciais, P., Xu, R., Guo, Z., Peng, L., Tao, Y., Spracklen, D. V., Holden, J., Liu, X., Zheng, Y., Xu, P., Chen, J., ... Zheng, C. (2022). Doubling of annual forest carbon loss over the tropics during the early twenty-first century. *Nature Sustainability*, *5*(5), 444–451. https://doi.org/10.1038/s41893-022-00854-3

Gann, G. D., McDonald, T., Walder, B., Aronson, J., Nelson, C. R., Jonson, J., Hallett, J. G., Eisenberg, C., Guariguata, M. R., Liu, J., Hua, F., Echeverría, C., Gonzales, E., Shaw, N., Decleer, K., & Dixon, K. W. (2019). *International principles and standards for the practice of ecological restoration. Restoration Ecology*, *27*(S1), S1–S46.

Giles, A. L., Schietti, J., Rosenfield, M. F., Mesquita, R. C., Vieira, D. L. M., Vieira, I. C., Poorter, L., Brancalion, P. H., Peña-Claros, M., Siqueira, J., Oliveira Junior, L., do Espírito-Santo, M. M., de Medeiros Sarmento, P. S., Ferreira, J. N., Berenguer, E., Barlow, J., Elias, F., Godinho Cassol, H. L., Silva, R. C., ... Jakovac, C. C. (2024). Simple ecological indicators benchmark regeneration success of Amazonian forests. *Communications Earth & Environment*, *5*(1), 780. https://doi.org/10.1038/s43247-024-01949-9

Giljum, S., Maus, V., Kuschnig, N., Luckeneder, S., Tost, M., Sonter, L. J., & Bebbington, A. J. (2022). A pantropical assessment of deforestation caused by industrial mining. *Proceedings of the National Academy of Sciences*, *119*(38), e2118273119. https://doi.org/10.1073/pnas.2118273119

Goldman, E., Weisse, M., Harris, N., & Schneider, M. (2020). *Estimating the role of seven commodities in agriculture-linked deforestation: Oil palm, soy, cattle, wood fiber, cocoa, coffee, and rubber*. World Resources Institute. https://doi.org/10.46830/writn.na.00001

Guégan, J.-F., Ayouba, A., Cappelle, J., & De Thoisy, B. (2020). Forests and emerging infectious diseases: unleashing the beast within. *Environmental Research Letters*, *15*(8), 083007. https://doi.org/10.1088/1748-9326/ab8dd7

Guizar-Coutiño, A., Jones, J. P., Balmford, A., Carmenta, R., & Coomes, D. A. (2022). A global evaluation of the effectiveness of voluntary REDD+ projects at reducing deforestation and degradation in the moist tropics. *Conservation Biology*, *36*(6), e13970. https://doi.org/10.1111/cobi.13970

Hansen, M. C., Potapov, P. V., Moore, R., Hancher, M., Turubanova, S. A., Tyukavina, A., Thau, D., Stehman, S., Goetz, S. J., Loveland, T. R., Kommareddy, A., Egorov, A., Chini, L., Justice, C. O., & Townshend, J. R. G. (2013). High-resolution global maps of 21st-century forest cover change. *Science*, *342*(6160), 850–853. https://doi.org/10.1126/science.1244693

Harris, N. L., Gibbs, D. A., Baccini, A., Birdsey, R. A., De Bruin, S., Farina, M., Fatoyinbo, L., Hansen, M. C., Herold, M., Houghton, R. A., Potapov, P. V., Suarez, D. R., Roman-Cuesta, R. M., Saatchi, S. S., Slay, C. M., Turubanova, S. A., & Tyukavina, A. (2021). Global maps of twenty-first century forest carbon fluxes. *Nature Climate Change*, *11*(3), 234–240. https://doi.org/10.1038/s41558-020-00976-6

Haya, B. K., Abayo, A., So, I. S., & Elias, M. (2024). *Voluntary Registry Offsets Database v11, Berkeley Carbon Trading Project*. University of California, Berkeley. https://gspp.berkeley.edu/faculty-and-impact/centers/cepp/projects/berkeley-carbon-trading-project/offsets-database

Hecht, S. B., & Saatchi, S. S. (2007). Globalization and forest resurgence: Changes in forest cover in El Salvador. *Bioscience*, *57*(8), 663–672. https://doi.org/10.1641/B570806

Hoang, N. T., & Kanemoto, K. (2021). Mapping the deforestation footprint of nations reveals growing threat to tropical forests. *Nature Ecology & Evolution*, *5*(6), 845–853. https://doi.org/10.1038/s41559-021-01417-z

Imbach, A., & Nello, T. (2024). Aplicando el Barómetro de la restauración. IUCN, Regional Office for Mexico, Central America and the Caribbean (ORMACC).

Jaureguiberry, P., Titeux, N., Wiemers, M., Bowler, D. E., Coscieme, L., Golden, A. S., Guerra, C. A., Jacob, U., Takahashi, Y., Settele, J., Díaz, S., Molnár, Z., & Purvis, A. (2022). The direct drivers of recent global anthropogenic biodiversity loss. *Science Advances*, *8*(45), eabm9982. https://doi.org/10.1126/sciadv.abm9982

Lamb, D., Erskine, P. D., & Parrotta, J. A. (2005). Restoration of degraded tropical forest landscapes. *Science*, *310*(5754), 1628–1632. https://doi.org/10.1126/science.1111773

Lambin, E. F., & Meyfroidt, P. (2011). Global land use change, economic globalization, and the looming land scarcity. *Proceedings of the National Academy of Sciences, 108*(9), 3465–3472. https://doi.org/10.1073/pnas.1100480108

Lenton, T. M., Xu, C., Abrams, J. F., Ghadiali, A., Loriani, S., Sakschewski, B., Zimm, C., Ebi, K. L., Dunn, R. R., Svenning, J.-C., & Scheffer, M. (2023). Quantifying the human cost of global warming. *Nature Sustainability, 6*(10), 1237–1247. https://doi.org/10.1038/s41893-023-01132-6

Luttrell, C., Sills, E., Aryani, R., Ekaputri, A. D., & Evinke, M. F. (2018). Beyond opportunity costs: who bears the implementation costs of reducing emissions from deforestation and degradation? *Mitigation and Adaptation Strategies for Global Change, 23*, 291–310. https://doi.org/10.1007/s11027-016-9736-6

Mack, E. A., Sauls, L. A., Jokisch, B. D., Nolte, K., Schmook, B., He, Y., Radel, C., Allington, G. R., Kelley, L. C., Scott, C. K., Leisz, S., Chi, G., Sagynbekova, L., Cuba, N., & Henebry, G. M. (2023). Remittances and land change: A systematic review. *World Development, 168*, 106251. https://doi.org/10.1016/j.worlddev.2023.106251

Maniraho, L., Frietsch, M., Sieber, S., & Löhr, K. (2023). A framework for drivers fostering social-ecological restoration within forest landscape based on people's participation: A systematic literature review. *Discover Sustainability, 4*(1), 26. https://doi.org/10.1007/s43621-023-00141-x

Maus, V., Giljum, S., da Silva, D. M., Gutschlhofer, J., da Rosa, R. P., Luckeneder, S., Gass, S. L. B., Lieber, M., & McCallum, I. (2022). An update on global mining land use. *Scientific Data, 9*(1), 433. https://doi.org/10.1038/s41597-022-01547-4

Méndez-Toribio, M., Martinez-Garza, C., Cecconc, E., & Guariguata, M. R. (2017). Current ecological restoration plans in Latin America: Progress and omissions. *Tropical Journal of Environmental Sciences, 51*(2), 1–30.

Mittermeier, R. A., Turner, W. R., Larsen, F. W., Brooks, T. M., & Gascon, C. (2011). Global biodiversity conservation: The critical role of hotspots. In F. E. Zachos & J. C. Habel (Eds.), *Biodiversity hotspots: Distribution and protection of conservation priority areas* (pp. 3–22). Springer. https://doi.org/10.1007/978-3-642-20992-5_1

Muradian, R., Cahyafitri, R., Ferrando, T., Grottera, C., Jardim-Wanderley, L., Krause, T., Kurniawan, N. I., Loft, L., Nurshafira, T., Prabawati-Suwito, D., Prasongko, D., Sanchez-Garcia, P. A., Schröter, B., & Vela-Almeida, D. (2025). Will the EU deforestation-free products regulation (EUDR) reduce tropical forest loss? Insights from three producer countries. *Ecological Economics, 227*, 108389. https://doi.org/10.1016/j.ecolecon.2024.108389

Murcia, C., Guariguata, M. R., Peralvo, M., & Gálmez, V. (2017). La restauración de bosques andinos tropicales: Avances, desafíos y perspectivas del futuro. *Center for International Forestry, 170*(4), 1–96.

Nantongo, M., & Vatn, A. (2019). Estimating transaction costs of REDD+. *Ecological Economics, 156*, 1–11. https://doi.org/10.1016/j.ecolecon.2018.08.014

Nicholson, E., Watermeyer, K. E., Rowland, J. A., Sato, C. F., Stevenson, S. L., Andrade, A., Brooks, T. M., Burgess, N. D., Cheng, S.-T., Grantham, H. S., Hill, S. L., Keith, D. A., Maron, M., Metzke, D., Murray, N. J., Nelson, C. R., Obura, D., Plumptre, A., Skowno, A. L., … Watson, J. E. M. (2021). Scientific foundations for an ecosystem goal, milestones and indicators for the post-2020 global biodiversity framework. *Nature Ecology & Evolution, 5*(10), 1338–1349. https://doi.org/10.1038/s41559-021-01538-5

Pan, Y., Birdsey, R. A., Phillips, O. L., Houghton, R. A., Fang, J., Kauppi, P. E., Keith, H., Kurz, W. A., Ito, A., Lewis, S. L., Nabuurs, G.-J., Shvidenko, A., Hashimoto, S., Lerink, B., Schepaschenko, D., Castanho, A., & Murdiyarso, D. (2024). The enduring world forest carbon sink. *Nature, 631*(8021), 563–569. https://doi.org/10.1038/s41586-024-07602-x

Pendrill, F., Gardner, T. A., Meyfroidt, P., Persson, U. M., Adams, J., Azevedo, T., Bastos Lima, M. G., Baumann, M., Curtis, P. G., De Sy, V., Garrett, R., Godar, J., Goldman, E. D., Hansen, M. C., Heilmayr, R., Herold, M., Kuemmerle, T., Lathuillière, M. J., Tyukavina, V. R., … West, C. (2022). Disentangling the numbers behind agriculture-driven tropical deforestation. *Science, 377*(6611), eabm9267. https://doi.org/10.1126/science.abm9267

de Area Leão Pereira, E. J., de Santana Ribeiro, L. C., da Silva Freitas, L. F., & de Barros Pereira, H. B. (2020). Brazilian policy and agribusiness damage the Amazon rainforest. *Land Use Policy, 92*, 104491. https://doi.org/10.1016/j.landusepol.2020.104491

Pillay, R., Venter, M., Aragon-Osejo, J., González-del-Pliego, P., Hansen, A. J., Watson, J. E., & Venter, O. (2022). Tropical forests are home to over half of the world's vertebrate species. *Frontiers in Ecology and the Environment, 20*(1), 10–15. https://doi.org/10.1002/fee.2420

Poorter, L., Craven, D., Jakovac, C. C., Van Der Sande, M. T., Amissah, L., Bongers, F., Chazdon, R. L., Farrior, C. E., Kambach, S., Meave, J. A., Muñoz, R., Norden, N., Rüger, N., van Breugel, M., Almeyda Zambrano, A. M., Amani, B., Andrade, J. L., Brancalion, P. H. S., Broadbent, E. N., … Hérault, B. (2021). Multidimensional tropical forest recovery. *Science, 374*(6573), 1370–1376. https://doi.org/10.1126/science.abh3629

Poorter, L., Rozendaal, D. M., Bongers, F., Almeida, d. J. S., Álvarez, F. S., Andrade, J. L., Arreola Villa, L. F., Becknell, J. M., Bhaskar, R., Boukili, V., Brancalion, P. H. S., César, R. G., Chave, J., Chazdon, R. L., Dalla Colletta, G., Craven, D., de Jong, B. H. J., Denslow, J. S., Dent, D. H., … Westoby, M. (2021). Functional recovery of secondary tropical forests. *Proceedings of the National Academy of Sciences, 118*(49), e2003405118. https://doi.org/10.1073/pnas.2003405118

Poorter, L., van Der Sande, M. T., Amissah, L., Bongers, F., Hordijk, I., Kok, J., Laurance, S. G., Martínez-Ramos, M., Matsuo, T., Meave, J. A., Muñoz, R., Peña-Claros, M., van Breugel, M., Herault, B., Jakovac, C. C., Lebrija-Trejos, E., Norden, N., & Lohbeck, M. (2024). A comprehensive framework for vegetation succession. *Ecosphere, 15*(4), e4794. https://doi.org/10.1002/ecs2.4794

Potapov, P., Hansen, M. C., Pickens, A., Hernandez-Serna, A., Tyukavina, A., Turubanova, S., Zalles, V., Li, X., Khan, A., Stolle, F., Harris, N., Song, X.-P., Song, i.-P., Baggett, A., Kommareddy, I., & Kommareddy, A. (2022). The global 2000–2020 land cover and land use change dataset derived from the Landsat archive: First results. *Frontiers in Remote Sensing*, 3, 856903. https://doi.org/10.3389/frsen.2022.856903

Potapov, P., Laestadius, L., & Minnemeyer, S. (2011). Global map of forest landscape restoration opportunities. *World Resource Institute*. www.wri.org/applications/maps/flr-atlas.

Prach, K., Marrs, R., Pyšsek, P., & van Diggelen, R. (2007). Manipulation of succession. In L. R. Walker, J. Walker, & R. J. Hobbs (Eds.), *Linking restoration and ecological succession* (pp. 121–149). Springer.

Rodrigues-Filho, S., Verburg, R., Bursztyn, M., Lindoso, D., Debortoli, N., & Vilhena, A. M. (2015). Election-driven weakening of deforestation control in the Brazilian Amazon. *Land Use Policy, 43*, 111–118. https://doi.org/10.1016/j.landusepol.2014.11.002

Roopsind, A., Sohngen, B., & Brandt, J. (2019). Evidence that a national REDD+ program reduces tree cover loss and carbon emissions in a high forest cover, low deforestation country. *Proceedings of the National Academy of Sciences, 116*(49), 24492–24499. https://doi.org/10.1073/pnas.1904027116

Rozendaal, D. M., Bongers, F., Aide, T. M., Alvarez-Dávila, E., Ascarrunz, N., Balvanera, P., Becknell, J. M., Bentos, T. V., Brancalion, P. H., Cabral, G. A., Calvo-Rodriguez, S., Chave, J., César, R. G., Chazdon, R. L., Condit, R., Dallinga, J. S., de Almeida-Cortez, J. S., de Jong, B., de Oliveira, A., … Poorter, L. (2019). Biodiversity recovery of Neotropical secondary forests. *Science Advances, 5*(3), eaau3114. https://doi.org/10.1126/sciadv.aau3114

Sewell, A., Van Der Esch, S., & Löwenhardt, H. (2020). *Goals and commitments for the restoration decade: A global overview of countries' restoration commitments under the Rio conventions and other pledges*. PBL Netherlands Environmental Assessment Agency.

Simonet, G., Subervie, J., Ezzine-de-Blas, D., Cromberg, M., & Duchelle, A. E. (2019). Effectiveness of a REDD+ project in reducing deforestation in the Brazilian Amazon. *American Journal of Agricultural Economics, 101*(1), 211–229. https://doi.org/10.1093/ajae/aay028

Soares, M. B., Rodrigues, R. R., Péres, L. O., Cerri, C. E. P., & Alleoni, L. R. F. (2024). Impact of climatic seasons on the dynamics of carbon, nitrogen and mercury in soils of Brazilian biomes affected by gold mining. *Science of the Total Environment, 954*, 176279. https://doi.org/10.1016/j.scitotenv.2024.176279

Strassburg, B. B., Iribarrem, A., Beyer, H. L., Cordeiro, C. L., Crouzeilles, R., Jakovac, C. C., Braga Junqueira, A., Lacerda, E., Latawiec, A. E., Balmford, A., Brooks, T. M., Butchart, S. H. M., Chazdon, R. L., Erb, K.-H., Brancalion, P., Buchanan, G., Cooper, D., Díaz, S., Donald, P. F., … Visconti, P. (2020). Global priority areas for ecosystem restoration. *Nature, 586*(7831), 724–729. https://doi.org/10.1038/s41586-020-2784-9

Taylor, M. J., Aguilar-Støen, M., Castellanos, E., Moran-Taylor, M. J., & Gerkin, K. (2016). International migration, land use change and the environment in Ixcán, Guatemala. *Land Use Policy*, *54*, 290–301. https://doi.org/10.1016/j.landusepol.2016.02.024

Trouwloon, D., Streck, C., Chagas, T., & Martinus, G. (2023). Understanding the use of carbon credits by companies: A review of the defining elements of corporate climate claims. *Global challenges*, *7*(4), 2200158. https://doi.org/10.1002/gch2.202200158

Viteri-Salazar, O., & Toledo, L. (2020). The expansion of the agricultural frontier in the northern Amazon region of Ecuador, 2000–2011: Process, causes, and impact. *Land Use Policy*, *99*, 104986. https://doi.org/10.1016/j.landusepol.2020.104986

West, T. A., Börner, J., Sills, E. O., & Kontoleon, A. (2020). Overstated carbon emission reductions from voluntary REDD+ projects in the Brazilian Amazon. *Proceedings of the National Academy of Sciences*, *117*(39), 24188–24194. https://doi.org/10.1073/pnas.2004334117

West, T. A., Wunder, S., Sills, E. O., Börner, J., Rifai, S. W., Neidermeier, A. N., Frey, G. P., & Kontoleon, A. (2023). Action needed to make carbon offsets from forest conservation work for climate change mitigation. *Science*, *381*(6660), 873–877. https://doi.org/10.1126/science.ade3535

Williams, B. A., Beyer, H. L., Fagan, M. E., Chazdon, R. L., Schmoeller, M., Sprenkle-Hyppolite, S., Griscom, B. W., Watson, J. E., Tedesco, A. M., Gonzalez-Roglich, M., Daldegan, G. A., Bodin, B., Celentano, D., Wilson, S. J., Rhodes, J. R., Alexandre, N. S., Kim, D.-H., Bastos, D., & Crouzeilles, R. (2024). Global potential for natural regeneration in deforested tropical regions. *Nature*, *636*, 131–137. https://doi.org/10.1038/s41586-024-08106-4

Wuepper, D., Crowther, T., Lauber, T., Routh, D., Le Clec'h, S., Garrett, R. D., & Börner, J. (2024). Public policies and global forest conservation: Empirical evidence from national borders. *Global Environmental Change*, *84*, 102770. https://doi.org/10.1016/j.gloenvcha.2023.102770

Wunder, S., Schulz, D., Montoya-Zumaeta, J. G., Börner, J., Ponzoni Frey, G., & Betancur-Corredor, B. (2024). Modest forest and welfare gains from initiatives for reduced emissions from deforestation and forest degradation. *Communications Earth & Environment*, *5*(1), 394. https://doi.org/10.1038/s43247-024-01541-1

Young, J. C., Marzano, M., White, R. M., McCracken, D. I., Redpath, S. M., Carss, D. N., Quine, C. P., & Watt, A. D. (2010). The emergence of biodiversity conflicts from biodiversity impacts: characteristics and management strategies. *Biodiversity and Conservation*, *19*, 3973–3990. https://doi.org/10.1007/s10531-010-9941-7

Young, T. P. (2000). Restoration ecology and conservation biology. *Biological Conservation*, *92*(1), 73–83. https://doi.org/10.1016/S0006-3207(99)00057-9

CHAPTER 34

Labor and Just Transitions

JASMINE KERRISSEY AND VICTORIA COON

THE climate crisis requires bold action. The necessary transition to green energy, however, has the potential to produce and perpetuate social inequities. Historically, major transitions have impacted workers unevenly, leaving workers and communities behind in some industries or geographies (Newell & Simms, 2021). It is possible that new green development could take a low road approach, using expendable labor with low job security and hazardous conditions. Access to the best climate jobs could be uneven, with barriers to entry for women, workers of color, and workers located in the Global South. Around the world, various stakeholders—unions, governments, nongovernmental organizations (NGOs), climate justice organizations, and corporations—are grappling with whether and how to make the energy transition just.

Broadly, *just transition* is the demand that the processes and outcomes of the green energy transition be both worker- and equity-centered. Labor and community organizations are central to the movement for just transition (Cha et al., 2022; Kleinheisterkamp-González, 2023). Decades of organizing led to the inclusion of just transition language in major international climate negotiations and documents, including the preamble of the Paris Climate Agreement (Stevis & Felli, 2015). Although the broad concept of just transition has become mainstream on the global stage, the meaning of "just transition" is contested. From a labor lens, meanings of just transition range from a narrow definition of protecting displaced fossil fuel workers to a transformative vision of equity across multiple axes (Flanagan & Goods, 2022; Vachon, 2023). Due to structural and cultural orientations, labor organizations vary in the approach they take to just transition, including some rejecting the calls for just transition or aligning with carbon-based jobs (Vachon, 2023). Moreover, as the concept diffused globally and transnational corporations adopted the language of just transition, the worker- and equity-centered meaning of just transition has become diluted (Goods, 2022; Moussu, 2020; Rosemberg, 2020).

The burgeoning just transition academic literature is interdisciplinary and rapidly growing. Scholarship from environmental studies, sociology, geography, international relations, political science, management, and other disciplines points to key trends and debates. Several questions animate the literature: How did just transition become a global concept? What does just transition mean—in other words, just transition for *whom*? What institutional, political, and economic factors shape approaches and outcomes of just transition plans? What explains whether and how labor unions organize for just transition? This chapter highlights the major research debates and findings about just transition and labor and concludes with areas for future research.

Making Just Transition a Global Issue

Just transition language has become codified in major international climate negotiations and documents, such as the Paris Climate Agreement, the United Nations Environment Programme, the International Labour Organization, and the World Health Organization (Olsen & Kempter, 2013; Rosemberg, 2020; Stevis & Felli, 2015). Scholars point to the leading organizing role that labor organizations and activists have taken to elevate the concept of just transition to the global stage (Cha, 2024; Rosemberg, 2020; Stevis & Felli, 2015).

The concept of just transition was first advanced by leaders of the Oil, Chemical, and Atomic Workers Union (OCAW) in the United States in the 1970s. Questioning the treatment of OCAW's atomic plant workers, many of whom suffered from chemical exposure on the job, OCAW President Tony Mazzocchi rejected the "jobs versus the environment" framing, which is the idea that there is a tradeoff between climate action and jobs. Rather, Mazzocchi proposed a "Superfund for Workers" to support displaced workers (Mazzocchi, 1993). Following the origins of the just transition framework in the 1970s, a small number of national unions, global unions, and labor-climate groups began to lay the groundwork for the just transition movement. These organizations formed important networks of labor-climate activists (Rosemberg, 2020; Stevis & Felli, 2020). However, early on, many large unions remained hesitant or even opposed to joining international climate efforts. For example, in 2004, the United States' AFL-CIO, which was the largest international affiliate of the International Confederation of Free Trade Unions (ICFTU), applied pressure to the ICFTU to refrain from advocating for climate issues on the international stage, including endorsing the Kyoto Protocol (Rosemberg 2020).

The emerging calls for "green jobs" helped push labor organizations to articulate the type of transition they hoped to secure. Three main critiques of green growth emerged from the labor movement. Rosemberg (2020) notes that green jobs would not improve working conditions, including gender disparities, that green policies would harm workers in the fossil fuel sector, and that green jobs narratives would uphold existing power structures dominated by the Global North. As green growth discussions percolated, labor-climate activists began to find increased space on the global stage for their organizing efforts (Stevis & Felli, 2015). The 2006 merger of the two major international union federations, the ICFTU and the World Confederation of Labour (WCL), created space for a revitalized discussion of the role of international labor federations and brought new voices to the table, which led to the inclusion of environmental protections in the newly merged International Trade Union Confederation's (ITUC) constitution (Rosemberg, 2020). Similarly, the International Labour Organization (ILO) established a Green Jobs Programme in 2009 (Olsen & Kempter, 2013). As the concept gained traction, NGOs and other non-state actors began to pick up the mantle of just transition (Rosemberg, 2020). These organizing efforts were realized in the Paris Climate Agreement, with just transition included in its preamble. In a short two decades, the minority position of labor-climate activists had become mainstream in global environmental negotiations. Despite reaching the global stage, the meanings of just transition remained contested.

Just Transition for Whom?

Just transition calls for the green energy transition to be centered on equitable processes and outcomes. However, the term is far from hegemonic (Flanagan & Goods, 2022). While many stakeholders use the term "just transition," they do not necessarily adopt the same scope and scale in their meanings. A burgeoning literature identifies and describes the various approaches to just transition, which often use different labels and taxonomies depending on their discipline (Flanagan & Goods 2022; Morena et al., 2020; Stevis, 2023; Stevis & Felli, 2015; Wang & Lo, 2022). There are several general approaches to just transition. From a labor perspective, the meanings of just transition range from narrow to expansive in scope and scale. At the narrower end of the spectrum, just transition focuses on the protection of displaced fossil fuel workers. At the other end, just transition is a call for societal transformation on multiple axes. Sociologist Todd Vachon (2023) and others describe these approaches as protective, proactive, and transformative (Cha et al., 2022). Corporations have also begun to use the term, although research suggests that they do not necessarily define just transition to mean equitable processes and outcomes for workers (Goods, 2022; Moussu, 2020).

At the narrow end of the spectrum, *protective* approaches to just transition seek to protect fossil fuel workers who face displacement in the transition to green energy. The transition to green energy is set to upend employment in the energy sector, with the near elimination of fossil fuel jobs. The International Energy Agency (2019) estimates that there are approximately 65 million workers employed in the energy sector, accounting for 2% of global employment. Workers are split evenly among fuel supply, power sources, and end uses. Roughly half of these energy sector workers are in the clean energy sector. The traditional carbon-based energy jobs have been "good" jobs in many ways. On average, energy workers have higher skill levels than other workers, with 45% of the workforce required to have specialized certifications or university credentials (International Energy Agency, 2019). These workers tend to be paid higher wages than the average worker. Established energy sectors, such as nuclear, oil, and gas, tend to have higher wages and are more likely to be unionized. Newer energy sectors, such as solar, are less likely to be unionized (International Energy Agency, 2019).

The employment impacts—or the number of jobs created and lost—during this transition will depend on government policy (United Nations, 2020). Whether displaced workers are cared for is a political process. Scholars, unions, and policy groups have investigated the types of support that existing carbon-intensive energy workers will need during this transition. Applied research tackles how to design just transition programs to support fossil fuel workers, such as the costs of steady versus episodic employment contraction of fossil fuel workers and their related costs (Pollin et al., 2021). Others focus on various elements of just transition practices, from communication of phase-out plans to job guarantees (Pai et al., 2020). Many have focused on skills retraining as a path to transition displaced workers into new sectors (Lim et al., 2023). Retraining outcomes depend in part on geography, since locations of green investments may not align with the locations of displaced fossil fuel workers. Researchers have noted that the success of transition programs has varied. They require planning, time, investment, and government support (Apostolopoulos et al., 2023; Banerjee & Schuitema, 2022; ILO, 2019; Zabin et al., 2020). Recognizing these complexities, some organizations have developed strategic plans to pursue "high-road" approaches to economic development. For example, California's High Road Training Partnership supports high-road regional workforce plans, which centers workers' voices to raise industry standards (UC Merced Community and Labor Center, 2024; Zabin et al., 2020). In addition to training, social safety nets, including access to pensions, health insurance, and unemployment insurance, shape just transition experiences for displaced workers. Although exact plans differ, advocates of the protective approach to just transition insist that these types of protective supports must be funded and accessible to all impacted workers.

In sum, this protective approach focuses on fossil fuel workers and displacement. It does not necessarily attend to the quality of new green energy jobs or to issues of broader equity for those workers and communities beyond displaced fossil fuel workers.

Proactive approaches agree that fossil fuel workers must be protected in the transition to clean energy. Building on the protective approach, proactive approaches insist that green energy jobs must be high-road jobs. The promise of "green and decent jobs" requires planning. The ILO (2019, p. 67) notes these jobs "will not necessarily occur automatically or in the volume required unless just transition planning takes place." This proactive agenda often features state policies that address green futures, such as the Green New Deal in the United States (Vachon, 2023). An interdisciplinary group of scholars have argued that any Green New Deal in the United States depends on a worker- and community-led just transition (Cha et al., 2022). In this view, the promises of a just transition are only able to be realized if workers and communities are leading the way and able to organize the political will to enact a proactive vision (Cha, 2024; Kleinheisterkamp-González, 2023).

A growing body of theoretical and applied research unpacks the pillars for what a high-road clean energy workforce would entail (Cliffton et al., 2021; ILR Climate Jobs Institute, 2024; Mayfield & Jenkins, 2021; UC Merced Community and Labor Center, 2024; Zabin et al., 2020). A proactive approach to just transition calls for the climate jobs sector to be unionized (Cliffton et al., 2021; Vachon, 2023). Union workers have a voice on the job, have equitable personnel practices through grievance procedures, and are more likely to have higher pay and benefits than comparable non-union workers. Unionization efforts will not necessarily be easy, especially in countries with weak labor laws. For instance, Tesla remains committed to thwarting unionization, including actions such as declaring US labor law unconstitutional and refusing to sign collective bargaining agreements with Scandinavian Tesla workers, which ultimately provoked a strike (Chan, 2023). To counter the large non-union sector in the United States, the United Auto Workers (UAW) have dedicated $40 million to new organizing, saying "workers will fight to maintain and raise the standard in the emerging battery industry" (UAW, 2024).

Ensuring equitable access to good-quality climate jobs is another major pillar of the proactive vision for just transition. A large scholarship draws attention to the environmental racism embedded in climate change. The United States is a racialized system, with different risks and rewards to groups along racial lines (Bonilla-Silva, 1997). Imprints of racialized outcomes are systemic in climate change, green economies, and just transition efforts. Communities of color are disproportionately exposed to pollution and risk, and thus bear the burden of climate change consequences (Medwinter, 2021; Mohai et al., 2009). Women, too, face disproportionate impacts of climate change (ILO, 2019; Littig, 2018). Both women and workers of color have been disadvantaged in the labor market, with less access to high-paying, high-skill, or high-autonomy jobs than White men (Browne & Misra, 2003). In the United States, fossil fuel workers are disproportionately White men, and many new climate jobs will be in the traditionally male-dominated construction and manufacturing industries (Pollin & Callavi, 2019). An ILO (2019, p. 48) report on just transition and climate action in Asia and the Pacific cautions that textile and garment supply chains are highly gendered and at risk of deepened gender inequality. Just transition initiatives must consider the precarious position of women within work and household management to ensure that policies do not inadvertently create two-tiered systems.

A proactive approach aims to address these long-standing inequities and exclusions. Pollin and Callavi (2019) argue that "the most effective way to reverse such gender and ethnic disparities is to support affirmative action within the growing clean energy sectors" (2019, p. 95). A report on Massachusetts outlines best practices to ensure both clean energy and workforce standards. The report recommends training and recruitment not only for fossil fuel workers but also for communities at high risk from the effects of climate change, BIPOC and environmental justice populations, and women. Establishing and funding pre-apprenticeship programs that recruit and train women and BIPOC communities is crucial to building an equitable workforce pipeline (ILR Climate Job Institute, 2024).

Transformative approaches to just transition seek to address the root causes of environmental degradation and social injustices. While transformative approaches agree with the concepts of protective and proactive perspectives, this more radical approach advocates for a broad, transformative remaking of the world to address systems of capitalism, colonialism, racism, and patriarchy. Transformation requires a just transition for all, with attention to power differentials (Galgóczi, 2022; Luke, 2023). Process is paramount: Transformative visions require democratic governance and decision making. This approach is most expansive, centering communities rather than focusing on traditional workers. It embraces environmental justice, beyond traditional labor interests only.

Advocates of transformative approaches emphasize that an equitable, green future cannot be achieved by reinforcing current capitalist relations. Transformative just transition envisions a reinvestment in communities and community-based resources, from universal health care to excellent publicly funded schools. It requires that the unequal distribution of energy resources around the world be addressed (Newell & Mulvaney, 2013). It also envisions alternative economic systems. One vein of this approach foregrounds the needs for public and democratic control over energy resources and infrastructure and the slowdown or reversal of privatization. Around the world, some workers' movements have advocated for a socially owned renewable energy sector, including in South Africa and India (Roy et al., 2019; Satgar, 2018). These unions have raised concerns about how private energy firms distribute and use carbon-intensive energies, along with their protections for fossil fuel workers. They insist that a green future must have a mix of various forms of socialized forms of energies, including cooperatives and municipal ownership (Satgar, 2018). Through re-municipalization, vital aspects of social infrastructure, from water to energy systems, can be protected (McDonald, 2024; Roy et al., 2019). Worker-owned cooperatives create another model for building solidarity economies and a just transition (Kleinheisterkamp-González, 2023). An example of this "eco-socialist" program in practice is Cooperation Jackson, which aims to use worker-owned and worker-managed companies to create an economic alternative for Black residents outside of capitalism (Akuno, 2020).

An equitable and transformative approach must also address the power and resource differences between countries and regions. World system theorists and decolonial scholars emphasize that industrialized democracies have taken advantage of countries in the Global South as sources of extractable natural resources, geographic locations to offload waste, and sites of exploitable cheap labor (Brand & Wissen, 2021; Hickel et al., 2022; see chapter by Jorgenson et al. in this volume). These exploitative practices perpetuate social inequalities along the North-South divide by enabling corporate profits and resource-heavy consumer lifestyles in powerful countries, while leaving people of the Global South behind. In this sense, even if some countries were able to achieve a protective and proactive just transition for their citizens, it is still quite possible that the exploitation of the Global South's resources and labor could continue. These types of uneven spatial developments also exist within countries (Tomassetti, 2020). The US South, for example, has a history of racialized exploitation, suppressed wages, lax labor laws, and weak social safety nets. Just transition for displaced fossil fuel workers alone does not address broader concerns of energy and climate justice in the region (Luke, 2023).

Transformative just transition envisions the eradication of deeply entrenched systems of oppression, including racism and patriarchy (Newell, 2021). Equity for racialized minorities, women, immigrants, queer people, and other marginalized groups is necessary for a transformative process. Transitions that do not recognize and address the risk that communities of color face are bound to reproduce racial inequities in future green economies (Newell, 2021). This transformative approach goes beyond affirmative action, instead insisting that those impacted by environmental racism should have community control over energy infrastructure and decision making (Akuno, 2020).

Women, too, are on the frontline of the consequences of climate change. Feminist scholars demonstrate that gender shapes all social structures and is central to any socio-ecological transformation. This literature intervenes in green growth discussions to insist that transformation cannot be gender blind. This scholarship highlights the gender inequities in the care economy, stressing that unpaid labor is largely absent from green economy plans. These discourses are critical of growth narratives, including green growth that relies on technological fixes to the climate crisis (Walk, 2024). They ask who will have access to good climate jobs and whether this new work is able to deconstruct traditional family, work-gender roles, and patriarchy (Cohen, 2017; Littig, 2018). They often propose degrowth and alternative economic systems, including those that will promote local food sovereignty and embrace ancestral knowledge (Walk, 2024; see chapter by Fenelon and Figueroa Helland in this volume).

Although these three approaches—protective, proactive, and transformative—differ in scope and vision, they share several core values (Vachon, 2023). They tend to agree that fossil fuel energy must be eventually phased out and that a low-carbon society is the future. To get there, they all insist that the state must be active in protecting workers and building good climate jobs and communities. Winning state support requires building the political will. This state support entails strategic planning, commitment to funding lines, access to training for disadvantaged communities, and a strong social welfare state. The approaches also agree that while outcomes are important, so are processes. Social dialogue and inclusionary decision making are critical to an equitable transition in democracies. Governments and decision makers must provide a meaningful seat at the table to plan the transition. The United Nations, along with the International Labour Organization (ILO), calls for social dialogue and tripartism as foundational to an equitable transition (United Nations, 2020). In this sense, just transition can also be seen as a governance strategy, though how this style of governance unfolds in autocratic states is less theorized (Wang & Lo, 2021).

Finally, in addition to the protective, proactive, and transformative approaches common to labor and climate activists, new meanings of just transition are proliferating. After just transition language was officially included in the Paris Agreement, the term has been used in a more "sanitized" way, with little connection to equity and voice for workers (Rosemberg, 2020, p. 53). Transnational corporations and their employer associations have increasingly used the language of just transition (Moussu, 2020). Just transition is seen by some business leaders as good for the environment and good for business opportunities (Goods, 2022). Similar to earlier business narratives on "green growth," Moussu (2020) charges that businesses' embrace of just transition language reflects "the never-ending story of corporate sustainability," which does not necessarily align with actual business practices around labor and environmental policies. Some scholars point out that business leaders may be embracing so-called just transition to maintain control in the impending transition (Goods, 2022; Moussu, 2020). Case analyses in Australia finds that business representatives' approach to just transition is shaped by profit motivations, with some businesses reconceptualizing just transition to be less transformative and empowering for workers (Goods, 2022). Despite some business leaders' co-optation of just transition language, Goods (2022) finds that the most visible approach is either dismissal or hardline resistance to just transition.

Just Transition in Context

Institutional, economic, and political landscapes influence just transition planning and outcomes. Because just transition is shaped by context, localities experience just transition processes differently (Wang & Lo, 2021).

Most empirical work on the energy transition focuses on industrialized democracies, with relatively little attention to authoritarian settings (Wang & Lo, 2021, 2022). The political possibilities and constraints in nondemocratic settings shape the contours of just transition. Most strikingly, there is far less space for non-state actors, such as unions and environmental justice groups, and little emphasis on social dialogue. Because the role of civil society groups is minimal or nonexistent in nondemocratic settings, the process of just transition must take a different approach than in democratic settings. In the context of top-down authoritarian settings, local discretion by officials over just transition policy implementation remains important (Lo, 2024). Studying coal closure policy in China, Wang and Lo (2022) analyze the top-down, nonparticipatory approach the Chinese state takes to accelerate climate protections. This approach stands in stark contrast to the slower, more deliberative approaches embraced by many of the Western industrialized democracies. They find that state-owned enterprises used their size and resources to provide a relatively just transition for displaced workers. By contrast, privately owned companies provided far less support for laid-off workers and were generally excluded from government-managed transition plans. More research is needed in nondemocratic settings, especially considering the democratic backsliding that has recently occurred in many countries.

In democratic settings, intersecting factors, such as institutional labor arrangements, political orientations, union strength, and economic structures, help explain transition dynamics. Carper (2022) provides a useful analysis of these dynamics in a comparison of transitions for coal workers in Ruhr Valley in Germany and Appalachia in the United States. The German system of industrial relations is based on tripartism, with voices from labor, industry, and the government. Because German unions were relatively strong, they were able to force democratic decision-making processes in how the coal industry would contract. They were also aided by the "psuedo-nationalization" of the Ruhr Valley coal industry, which was established after a series of strikes and protests in the 1960s. This consolidation meant that just transition could be negotiated with one main entity, rather than multiple firms. The German government made provisions for 3,000 coal miners, including early retirement and retraining. However, it is important to note that as Germany phased out coal production in the Ruhr Valley, it imported coal from the Global South, reinforcing global inequities. The story of Appalachia stands in contrast to the Ruhr Valley. Workers have borne the brunt of the collapse of the coal industry in Appalachia. The power of the union movement, particularly the United Mine Workers of America, dropped substantially over the last 50 years. Unlike in Germany, there is no tripartite seat at the table to represent labor interest in the United States. The state did not intervene to effectively address the harm on workers and their communities. Coal operators were allowed to extract profits as the industry bled, with the workers taking on the risk.

Workers in the Global South face additional challenges and opportunities to just transition. Pucheta and Sańchez (2021) note that the persistence of informal work and high levels of unemployment may amplify just transition tensions, as happened in Chile and Argentina. An ILO (2019) report on just transition efforts in Asia and the Pacific calls for increased state capacity to plan and execute just transition, including attending to workers' needs for climate adaptation, such as energy needs (ILO, 2019). Some studies identify the difficulties of embracing just transition, as energy consumption needs, the perceived job-environment trade-off, and Indigenous and labor conflicts over resources create tensions over just transition pathways (Cock, 2021; Houeland, 2021; Kojola, 2021; Sikwebu & Aroun, 2021). Still, as Acosta and Abarca (2018) describe, from South America to Africa, there are local examples of community-based alternatives to development and growth that serve as useful examples globally. In South Africa, for example, Cock (2021) shows how communities impacted by coal have helped infuse the movement for just transition with new energy and tactics. Further case studies based on the Global South are needed to inform the constraints and possibilities of just transition in a range of political and economic contexts.

Another key point of context for just transition possibilities is the strength and orientation of the broader labor and climate movements. For example, Climate Jobs NY, a climate-labor initiative in New York, has been successful in winning labor standards in climate jobs, including prevailing wage. Skinner (2020) attributes this success to the strength of New York unions in general, and particularly the active role that unions in the building trades, energy, and transport sectors—unions that represent members in the current energy economy—took in spearheading the campaign. However, union presence does not necessarily lead to unions' active organization for just transition, as is the case with Poland and Kenya, where national unions have not been organizing toward just transition (Azzellini, 2023; JTRC, 2018). Moreover, the system in which unions operate shapes how they approach just transition and what they can achieve. Stevis (2023) argues that unions operating under liberal capitalism will see just transition only within the "parameters of liberal capitalism," where market-driven solutions are the norm, unless the system itself is changed.

Organizing for Just Transition

A growing vein of literature analyzes the strategies and tactics that worker organizations and climate-labor activists use to organize for just transition. While individual acts of resistance to harmful transition policies are important, collective action by organized workers plays a fundamental role in just transition (Kleinheisterkamp-González, 2023). As Cha and Pastor (2022) argue, just transition will not be achieved through a set of top-down policy proposals. Rather, advocates must organize toward just transition, understanding it as a long-term political project. In this perspective, building community power is critical to any just transition (Hess, 2019). The distribution of power in societies must shift for there to be a possibility of just transition: This shift requires organizing and building broad based coalitions with similar goals. Such organizing requires a common frame for how to understand or address just transition (Cha & Pastor, 2022; Hess, 2019; Vachon, 2023).

A major organizing aim is to incorporate just transition into state policies and international agreements. States must play a central role in bringing about just transition. The policies, practices, and funding mechanisms for transition depend on a strong, proactive state; they will not emerge from free market forces (Carper, 2022; Cha et al., 2022; Vachon, 2023). Historically, states have played vital roles in managing transitions in short periods of time, though with checkered records of justice for impacted communities (Newell & Simms, 2021; Pulido, 2017). Scholarship examines how unions envision and organize around policies that impact the climate and workers (e.g., Brecher, 2018; Cha et al., 2022; Rosemberg, 2020; Vachon, 2023). Another organizing target has been environmental groups, who may need encouragement around just transition issues and how workers are impacted (Vachon, 2023). In general, the literature stresses that power relations must be changed to win just transition, and labor and community organizing is critical for those changes to occur.

Another axis of organizing for just transition is through strong coalitions. Drawing on Gramsci and Polanyi, Ciplet (2022) theorizes that coalitions are critical to transformative change. Broad coalitions that bring together multiple stakeholders help create momentum for a green transition, including new ideas and networks (Fine, 2011). For instance, the Blue Green Alliance, formed by the Steelworkers Union and the Sierra Club, has shaped US conversations on just transition through identifying broad issues, educating their members and the public, and advocating for suitable policies. In a comparative analysis based in California, Brown and Nelson (2024) found the unions were best positioned for transition when they were able to build broad coalitions between labor and community partners based on workers' agendas, independent from firms. Although coalition work is necessary, tensions and conflicts often surfaced in coalitions (Cha et al., 2022; Ciplet & Harrison, 2019; Fine, 2011). Aligning multiple stakeholders, especially when operating in a difficult political and organizational context, is challenging (Cha et al., 2022). Intentional structures to promote facilitation were helpful for generating discussion and cultivating relationships, which were necessary for building trust between stakeholders (Cha et al., 2022). Coalition partnerships can be complex, and sometimes unions have aligned with fossil fuel companies. Hess (2019), for example, studies the framing used by coalitions when utility unions partnered with the utility firms in an attempt to beat back a green electric initiative. In other cases, unions have chosen not to build coalitions. Fine (2011) identifies the failure of construction unions in New Jersey to enter the emerging new climate jobs space was due to going it alone, without strong community partnerships.

Other research examines the micro level, examining how unions shape attitudes and actions among workers. Survey research finds that union members are more likely to hold pro-environment attitudes than nonmembers (Ringqvist, 2022; Vachon & Brecher, 2016). Almeida and colleagues (2023) argue that civic engagement is key to building the organizational capacity necessary to implement climate resiliency programs. They find that unions are the most important membership that predicts who attends a climate meeting.

Notably, little research analyzes the relationship between just transition and the repertoires of action that are core to the labor movement, including the use of strikes and labor agreements. Labor organizations, especially those with a social movement, bottom-up orientation, often use strikes, protests, boycotts, and other forms of disruption as a collective source of power (Kerrissey & Schofer, 2013). To date, union led strikes over climate issues and just transition have been relatively rare. Labor Notes records that the first US contemporary strike in solidarity with climate issues occurred in 2020 among janitors (Altamirano et al., 2020). The United Auto Workers (UAW) 2023 strike priorities included winning a just transition for EV and battery workers, who were largely working at lower wages and less secure jobs than their union counterparts (UAW, 2023).

Unions also organize for just transition within the workplace and through collective bargaining and sectoral agreements. Some unions have been successful in negotiating just transition climate provisions in agreements (Brecher, 2018). For example, the development of offshore wind in Massachusetts involved a Project Labor Agreement to guide job recruitment and training for previously under-represented communities. These provisions aimed to bring women and workers of color into this new field of work (Markey, 2021). The scholarship analyzing firm or sectoral agreements regarding just transition, especially for new climate jobs, is in its infancy. Understanding whether and how organized labor-climate activists use disruption or collective agreements in ways that intersect with just transition and green futures is a promising area for research.

Explaining Labor Approaches to Just Transition

This range of approaches to just transition grounds a central question for scholars: What factors help explain whether and how specific labor organizations embrace just transition? The labor movement has been far from unified on whether and how growth and just transition should be pursued. Some unions oppose the transition away from carbon-intensive energy, while others propose a slow transition away from carbon-intensive energy (Kalt, 2022; Kojola, 2017; Vachon, 2023). On the other end of the spectrum, some unions embrace the narrative that jobs can be both good and aligned with climate priorities.

Both structural factors and unions' political orientations help explain the approach of specific unions. Unions that work directly with or adjacent to extractive industries have been more likely to be concerned with the environment-versus-jobs narrative (Kalt, 2022; Sikwebu & Aroun, 2021; Vachon, 2023). This narrative suggests that environmental regulations create a zero-sum game, in which climate justice comes at the expense of good jobs. These unions have been more likely to distance themselves from climate causes, seeing them as disrupting their access to work. Some fossil fuel workers are skeptical about the actual outcomes of just transition, believing it to be jargon used by policy elites without much promise for actual action (Cha, 2020). Unions in industries that are further afield from carbon-intensive industries, such as educators and health care workers, have been more likely to reject the climate-versus-jobs narrative. Instead, these unions are more likely to adopt the framework of good jobs *and* a clean environment (Vachon, 2023).

Related to the jobs-versus-environment narrative, unions approach growth differently. The treadmill of production theory depicts labor relations institutions as active promoters of industrial growth. In this theory, Western capitalism runs on a treadmill, where resource extraction for economic development fuels environmental degradation (Hyde & Vachon, 2019; Schnaiberg, 1980; see Gould & Lewis in this volume). This perspective posits that labor relations institutions, namely unions, have sped up this so-called treadmill and thus contributed to ecological destruction. By encouraging production growth, unions bolster the capitalist position that growth is necessary. Because infinite growth is catastrophic for a planet with finite resources, labor institutions are seen as complicit in keeping the treadmill of production running. However, scholarship has shown that union relationships to the treadmill of production are time- and context-dependent. Obach (2004), for example, argues that the role of unions' contribution to the acceleration of the treadmill of production concept should be contested. In the United States, Obach argues, unions tame the treadmill by both using collective action to slow development and by integrating climate goals into their agendas. Studying post–World War II Italy, Tomassetti (2020) shows that practices of just transition and the treadmill of production coexist in the same country. Tomassetti argues that when economic pressures created sharp competition between the jobs-versus-environment dilemma, unions tended to choose jobs—or the treadmill of production. Still, countercurrents to the dominant postwar union logic persisted in unions that had stronger social movement traditions. Factories, in some cases, served as a place of political education and confrontation, which primed workers to make demands for broader societal changes, including just transition. Tomassetti notes that the uneven development in Italy explains some of the variation in union responses to environmental issues. Unions in depressed areas with few job opportunities tended to pragmatically cling to the treadmill of production narrative. In sum, while the treadmill of production might capture some of the conflicts between labor and the environment, research has shown that unions have complicated the treadmill of production theory in important ways.

Unions' position in relationship to the state and their access to organizing in the renewable energy sector also help explain different approaches. Brown and Nelson (2024) find that unions that had a chance to organize solar workers in California were reluctant to rock the boat with employers. Kalt (2022) observes that unions with close ties to powerful political institutions were hesitant to diverge too far from state-sanctioned transition approaches, for fear of giving up their institutional access. This often meant affirming market-driven green transitions.

While unions' structural relationships to extractive industries, the treadmill of production, and to institutional access are important, so are their long-standing political orientations and practices (Tomassetti, 2020; Vachon, 2023). Within extraction-related industries, which are theorized to be structurally aligned with a jobs-first perspective, some workers and unions have rejected the jobs-versus-environment narrative (Loomis, 2016). Union cultures inform how unions respond to the climate crisis and the idea of just transition. Unions that practice business unionism, meaning that they are focused on internal issues of dues-paying members and tend to be staff-driven, are less concerned with broader social justice issues, including climate-labor issues (Tomassetti, 2020). Unions that adhere to a social movement style of unionism are more likely to engage with broader justice issues that impact the working class and to build power through member engagement (Clawson, 2003; Obach, 2004; Vachon, 2023). Because they derive their power though worker organizing and do not rely solely on institutional access, they may be freer to pursue more transformative transition approaches (Kalt, 2022). For example, the National Union of Metalworkers of South Africa (NUMSA) sought to build capacity for the energy transition in South Africa (Satgar, 2015). The union conducted research to analyze energy transition in South Africa and then developed a strategic union position on climate change. Part of this process involved international linkages and study tours, where NUMSA worker-delegates traveled to Latin America to learn how other unions were responding to the climate crisis and just transition.

Union politics around climate-labor issues have shifted over time, due to organizing efforts, the global salience of just transition, and greater climate urgency. While structural analyses of unions' position to extractive industries are useful, it is important to note that unions' approaches to just transition are informed by activists' work to bring issues of just transition and climate action to the forefront of union agendas (Rosemberg, 2020). A project of climate-labor activists has been to shift the narrative away from the jobs-versus-environment framework, and instead to present alternative paths for how good jobs and a clean environment can coexist (Vachon, 2023). Caucuses within unions set agendas and strategic organizing plans to press for their unions to take proactive positions on climate issues, from resolutions to policies on pension divestment from fossil fuel holdings (Brecher, 2018). In cross union networks, activists work together to coordinate and amplify plans (Brecher, 2018). As the language of just transition diffuses globally and becomes codified into international documents, space opens for previously reluctant unions to begin to believe that just transition is a worthy cause that can be won.

Conclusion

Climate change poses existential threats to working people around the world. For decades, labor organizations have led calls for a just transition. The concept of just transition has diffused around the globe, codified into international agreements and documents (Stevis &

Felli, 2015). Accordingly, the academic scholarship on labor and just transition has grown monumentally. In a short two decades, a large interdisciplinary scholarship has emerged to tackle key challenges and opportunities with just transition. Through theory development, case analyses, and systematic reviews, this literature has focused on analyzing the contested meanings of just transition and the local, national, and global politics of organizing for just transition.

The burgeoning just transition literature is ripe for further scholarship and coordination in several areas. First, while the body of case studies is growing, a disproportionate number focus the Global North. Much of the central findings of just transition literature based on industrialized democracies, specifically the role of unions, climate justice organizations, and social dialogue, do not inform the politics of just transition within authoritarian regimes. More studies dedicated to the Global South, to nondemocratic regimes, and to major fossil fuel–producing countries would add important insights into the political economy of just transition (Wang & Lo, 2021). The global trend of retreat away for democracies and toward more authoritarian regimes makes research on just transition and nondemocratic regimes especially important. Second, there is an opportunity to incorporate key mechanisms from the labor studies literature into the growing just transition scholarship. Labor-management agreements and strikes are traditional tools used by labor movements. How might these core tactics for power building be adopted for just transition? Third, the bulk of the labor just transition literature focuses on actions taken by unions and union confederations. The field of just transition would benefit from more scholarship that examines how other types of workers' movements and climate activists organize for just transition, including alignments and tensions between worker organizations and community groups (Räthzel et al., 2021). Finally, a substantial scholarship examines the meanings of just transition and the campaigns to include the concept of just transition in national and global agreements and documents. Less research examines the actual implementation of just transition. As just transition moves from language embedded in resolutions and agreements to actual practice, research on implementation practices will be crucial.

References

Acosta, A., & Abarca, M. M. (2018). Buen Vivir: An alternative perspective from the peoples of the global south to the crisis of capitalist modernity. In V. Satgar (Ed.), *The climate crisis: South African and global democratic eco-socialist alternatives* (pp. 131–147). Wits University Press.

Akuno, K. (2020). Tales from the front lines: Building a people-led just transition in Jackson, Mississippi. In *Just transitions: Social justice in the shift towards a low-carbon world* (pp. 95–114). Pluto Press.

Almeida, P., González, L. R., Flores, E. O., Curry, V., & Padilla, A. (2023). The building blocks of community participation in local climate meetings. *NPJ Climate Action*, *2*(1), 37.

Altamirano, I., Nammacher, G., & Dalal-Whelan, P. (2020). Lessons from the first union climate strike in the U.S. *Labor Notes*. https://labornotes.org/2020/04/lessons-first-union-climate-strike-us

Apostolopoulos, N., Kakouris, A., Liargovas, P., Borisov, P., Radev, T., Apostolopoulos, S., & Anastasopoulou, E. E. (2023). Just transition policies, power plant workers and green entrepreneurs in Greece, Cyprus and Bulgaria: Can education and retraining meet the challenge? *Sustainability*, *15*(23), 16307.

Azzellini, D. (2023). *Sustainable work and just transition*. Rosa Luxemburg Stiftung Geneva Office. https://links.org.au/sites/default/files/Dario_Azzellini_Report_JT_RLS_final-1.pdf

Banerjee, A., & Schuitema, G. (2022). How just are just transition plans? Perceptions of decarbonisation and low-carbon energy transitions among peat workers in Ireland. *Energy Research & Social Science*, *88*, 102616.

Bonilla-Silva, E. (1997). Rethinking racism: Toward a structural interpretation. *American Sociological Review*, *62*(3), 465–480.

Brand, U., & Wissen, M. (2021). *The imperial mode of living: Everyday life and the ecological crisis of capitalism*. Verso Books.

Brecher, J. (2018, January). Climate solidarity: Workers versus warming. *New Labor Forum*, *27*(1), 96–104. https://journals.sagepub.com/doi/full/10.1177/1095796017745039

Brower Brown, K., & Nelson, S. H. (2024). Working sunset to sunrise: Union strategies in three California climate transitions. *Environmental Politics*, *33*(4), 657–677.

Browne, I., & Misra, J. (2003). The intersection of gender and race in the labor market. *Annual Review of Sociology*, *29*(1), 487–513.

Carper, M. (2022). From the Ruhr Valley to Ramp Hollow: Lessons for America from Germany's just energy transition. *Tulane Environmental Law Journal*, *35*, 91.

Cha, J. M. (2020). A just transition for whom? Politics, contestation, and social identity in the disruption of coal in the Powder River Basin. *Energy Research & Social Science*, *69*, 1–9.

Cha, J. (2024). *A just transition for all: Workers and communities for a carbon-free future*. MIT Press.

Cha, J. M., & Pastor, M. (2022). Just transition: Framing, organizing, and power-building for decarbonization. *Energy Research & Social Science*, *90*, 102588.

Cha, J. M., Stevis, D., Vachon, T. E., Price, V., & Brescia-Weiler, M. (2022). A Green New Deal for all: The centrality of a worker and community-led just transition in the US. *Political Geography*, *95*, 102594.

Chan, K. (2023). Musk and Tesla are battling unions across Scandinavia. *AP News*. https://apnews.com/article/tesla-strike-union-sweden-labor-dispute-5124a0074f6b4c5b7f503b5353afe8b3

Ciplet, D. (2022). Transition coalitions: Toward a theory of transformative just transitions. *Environmental Sociology*, *8*(3), 315–330.

Ciplet, D., & Harrison, J. L. (2019). Transition tensions: Mapping conflicts in movements for a just and sustainable transition. *Environmental Politics*, *29*(3), 435–456.

Clawson, D. (2003). *The next upsurge: Labor and the new social movements*. Cornell University Press.

Cliffton, R., Wall, M., Rickets, S. Lee, K., Eckdish K., & Walter J. (2021). The clean energy revolution will be unionized. *Center for American Progress*. https://www.americanprogress.org/article/clean-economy-revolution-will-unionized/

Cock, J. (2021). "Beware of the crocodile's smile": Labour-environmentalism in the struggle to achieve a just transition in South Africa. In *The Palgrave handbook of environmental labour studies* (pp. 177–197). Springer.

Cohen, M. G. (Ed.). (2017). *Climate change and gender in rich countries: Work, public policy and action*. Taylor & Francis.

Fine, J. (2011). When the rubber hits the high road: Labor and community complexities in the greening of the Garden State. *Labor Studies Journal*, *36*(1), 122–161.

Flanagan, F., & Goods, C. (2022). Climate change and industrial relations: Reflections on an emerging field. *Journal of Industrial Relations*, *64*(4), 479–498.

Galgóczi, B. (2022). From a "just transition for us" to a "just transition for all." *Transfer: European Review of Labour and Research*, *28*(3), 349–366.

Goods, C. (2022). How business challenges climate transformation: An exploration of just transition and industry associations in Australia. *Review of International Political Economy*, *29*(6), 2112–2134.

Hess, D. J. (2019). Coalitions, framing, and the politics of energy transitions: Local democracy and community choice in California. *Energy Research & Social Science*, *50*, 38–50.

Hickel, J., Dorninger, C., Wieland, H., & Suwandi, I. (2022). Imperialist appropriation in the world economy: Drain from the global South through unequal exchange, 1990–2015. *Global Environmental Change*, *73*, 1–13.

Houeland, C. (2021). Labour resistance against fossil fuel subsidies reform: Neoliberal discourses and African realities. In N. Rathzel, D. Stevis, & D. Uzzell (Eds.), *The Palgrave handbook of environmental labour studies* (pp. 493–516). Palgrave Macmillan.

Hyde, A., & Vachon, T. (2019). Running with or against the treadmill? Labor unions, institutional contexts, and greenhouse gas emissions in a comparative perspective. *Environmental Sociology, 5*(3), 269–282.

ILR Climate Jobs Institute. (2024). Building the clean energy commonwealth: A climate jobs roadmap for Massachusetts. https://www.ilr.cornell.edu/sites/default/files-d8/2024-05/Building%20the%20Clean%20Energy%20Commonwealth.pdf

International Energy Agency. (2019). World Energy Employment Report. https://www.iea.org/reports/world-energy-employment

International Labour Organisation. (2019). Green jobs and a just transition for climate action in Asia and the Pacific. https://www.ilo.org/publications/green-jobs-and-just-transition-climate-action-asia-and-pacific

Just Transition Research Collaborative. (2018). Mapping just transition(s) to a low-carbon world. https://www.uncclearn.org/wp-content/uploads/library/report-jtrc-2018.pdf

Kalt, T. (2022). Agents of transition or defenders of the status quo? Trade union strategies in green transitions. *Journal of Industrial Relations, 64*(4), 499–521.

Kerrissey, J., & Schofer, E. (2013). Union membership and political participation in the United States. *Social Forces, 91*(3), 895–928.

Kojola, E. (2017). (Re) constructing the pipeline: Workers, environmentalists and ideology in media coverage of the Keystone XL Pipeline. *Critical Sociology, 43*(6), 893–917.

Kojola, E. (2021). Whose labour, whose land? Indigenous and labour conflicts and alliances over resource extraction. In N. Rathzel, D. Stevis, & D. Uzzell (Eds.), *The Palgrave handbook of environmental labour studies* (pp. 365–387). Palgrave Macmillan. https://link.springer.com/chapter/10.1007/978-3-030-71909-8_16

Kleinheisterkamp-González, N. (2023). The case for an environmental labor geography: The role of organized labor in the climate crisis. *Progress in Human Geography, 47*(2), 317–332.

Lim, J., Aklin, M., & Frank, M. R. (2023). Location is a major barrier for transferring US fossil fuel employment to green jobs. *Nature Communications, 14*(1), 5711.

Littig, B. (2018). Good work? Sustainable work and sustainable development: A critical gender perspective from the Global North. *Globalizations, 15*(4), 565–579.

Lo, K. (2024). The politics of just transition: Authoritarian environmentalism and implementation flexibility in forest conservation. *Political Geography, 109*, 103066.

Loomis, E. (2016). *Empire of timber: Labor unions and the Pacific Northwest forests*. Cambridge University Press.

Luke, N. (2023). Just transition for all? Labor organizing in the energy sector beyond the loss of "jobs property". *Annals of the American Association of Geographers, 113*(1), 94–109.

Markey, E. (2021) Building trades unions and vineyard wind sign historic project labor agreement [Press release]. https://www.markey.senate.gov/news/press-releases/building-trades-unions-and-vineyard-wind-sign-historic-project-labor-agreement

Mayfield, E., & Jenkins, J. (2021). Influence of high road labor policies and practices on renewable energy costs, decarbonization pathways, and labor outcomes. *Environmental Research Letters, 16*(12), 1–12.

Mazzochi, T. (1993). An answer to the jobs-environment conflict? *Green Left, 114*. https://www.greenleft.org.au/content/answer-jobs-environment-conflict

McDonald, D. A. (2024). Landscapes of remunicipalization: A critical literature review. *Urban Affairs Review, 60*(6), 1898–1930.

Medwinter, S. (2021). Reproducing poverty and inequality in disaster: Race, class, social capital, NGOs, and urban space in New York City after Superstorm Sandy. *Environmental Sociology, 7*(1), 1–11.

Mohai, P., Pellow, D., & Roberts, J. T. (2009). Environmental justice. *Annual Review of Environment and Resources, 34*, 405–430.

Morena, E., Krause, D., & Stevis, D. (2020). *Just transitions: Social justice in a low-carbon world*. Pluto Press.

Moussu, N. (2020). Business in just transition: The never-ending story of corporate sustainability. In E. Morena, D. Krause, & D. Stevis (Eds.), *Just transitions: Social justice in the shift towards a low-carbon world* (pp. 56–75). Pluto Press.

Newell, P. (2021). Race and the politics of energy transitions. *Energy Research & Social Science*, *71*, 1–5.

Newell, P., & Mulvaney, D. (2013). The political economy of the "just transition." *The Geographical Journal*, *179*(2), 132–140.

Newell, P., & Simms, A. (2021). How did we do that? Histories and political economies of rapid and just transitions. *New Political Economy*, *26*(6), 907–922.

Obach, B. K. (2004). *Labor and the environmental movement: The quest for common ground.* MIT Press.

Olsen, L., & Kempter, D. (2013). The International Labour Organization and the environment: The way to a socially just transition for workers. In N. Rathzel, & D. L. Uzzell (Eds.), *Trade unions in the green economy* (pp. 41–57). Routledge.

Pai, S., Harrison, K., & Zerriffi, H. (2020). *A systematic review of the key elements of a just transition for fossil fuel workers.* Smart Prosperity Institute.

Pollin, R., & Callaci, B. (2019). The economics of just transition: A framework for supporting fossil fuel-dependent workers and communities in the United States. *Labor Studies Journal*, *44*(2), 93–138.

Pollin, R., Wicks-Lim, J., Chakraborty, S., Kline, C., & Semieniuk, G. (2021). A program for economic recovery and clean energy transition in California. Political Economy Research Institute Research Report, University of Massachusetts Amherst.

Pucheta, M., Álvarez Alonso, C., & Silva Sánchez, P. (2021). Just transition and workers' rights in the Global South: Recent Argentine and Chilean nationally determined contributions. *Sustainability*, *13*(17), 9616.

Pulido, L. (2017). Geographies of race and ethnicity II: Environmental racism, racial capitalism and state-sanctioned violence. *Progress in Human Geography*, *41*, 524–533.

Räthzel, N., Stevis, D., & Uzzell, D. (Eds.). (2021). *The Palgrave handbook of environmental labour studies.* Palgrave Macmillan.

Ringqvist, J. (2022). Union membership and the willingness to prioritize environmental protection above growth and jobs: A multi-level analysis covering 22 European countries. *British Journal of Industrial Relations*, *60*(3), 662–682.

Rosemberg, A. (2020). No jobs on a dead planet: The international trade union movement and just transition. In *Just transitions: Social justice in the shift towards a low-carbon world* (pp. 32–55). Pluto Press.

Roy, A., Kuruvilla, B., & Bhardwaj, A. (2019). Energy and climate change: A just transition for Indian labour. In N. K. Dubash (Ed.), *India in a warming world* (pp. 284–300). Oxford University Press.

Satgar, V. (2015). A trade union approach to climate justice: The campaign strategy of the National Union of Metalworkers of South Africa. *Global Labour Journal*, *6*(3), 267–282.

Satgar, V. (2018). The Anthropocene and imperial ecocide: Prospects for just transitions. In The climate crisis: South African and global democratic eco-socialist alternatives (pp. 47–67). Wits University Press.

Schnaiberg, A. (1980). *The environment: From surplus to scarcity.* Oxford University Press.

Sikwebu, D., & Aroun, W. (2021). Energy transitions in the Global South: The precarious location of unions. In N. Rathzel, D. Stevis, & D. Uzzell (Eds.), The Palgrave handbook of environmental labour studies (pp. 59–81). Palgrave Macmillan.

Skinner, L. 2020. Building a pro-worker, pro-union climate movement. In J. Kerrissey, E. Weinbaum, C. Hammonds, T. Juravich, & D. Clawson (Eds.), *Labor in the time of Trump* (pp. 131–148). Cornell University Press.

Stevis, D. (2023). *Just transitions: Promise and contestation.* Cambridge University Press.

Stevis, D., & Felli, R. (2015). Global labour unions and just transition to a green economy. *International Environmental Agreements: Politics, Law and Economics*, *15*, 29–43.

Stevis, D., & Felli, R. (2020). Planetary just transition? How inclusive and how just? *Earth System Governance*, *6*, 1–11.

Tomassetti, P. (2020). From treadmill of production to just transition and beyond. *European Journal of Industrial Relations*, *26*(4), 439–457.

UC Merced Community and Labor Center. (2024). Kern strategic workforce development report. https://clc.ucmerced.edu/sites/clc.ucmerced.edu/files/page/documents/kern_hrtp_report.pdf

United Auto Workers. (2023). Background on Big 3 bargaining. https://uaw.org/backgrounder-big-3-bargaining/

United Auto Workers. (2024). UAW announces $40 million commitment to organizing auto and battery workers over next two years. https://uaw.org/uaw-announces-40-million-commitment-to-organizing-auto-and-battery-workers-over-next-two-years/

United Nations. (2020). Just transition of the workforce and the creation of decent work and quality jobs. Technical paper. Framework Convention on Climate Change.

Vachon, T. (2023). *Clean air and good jobs: U.S. labor and the struggle for climate justice*. Temple University Press.

Vachon, T., & Brecher, J. (2016). Are union members more or less likely to be environmentalists? Some evidence from two national surveys? *Labor Studies Journal 41* (2), 185–203.

Walk, P. (2024). From parity to degrowth: Unpacking narratives of a gender-just transition. *Energy Research & Social Science*, *112*, 103513.

Wang, X., & Lo, K. (2021). Just transition: A conceptual review. *Energy Research & Social Science*, *82*, 102291.

Wang, X., & Lo, K. (2022). Civil society, environmental litigation, and Confucian energy justice: A case study of an environmental NGO in China. *Energy Research & Social Science*, 93, 102831.

Zabin, C., Auer, R., Mijin Cha, J., Collier, R., France, R., MacGillvary, J., Myers, H., Strecker, J., and Viscelli, S. (2020). Putting California on the high road: A jobs and climate action plan for 2030. https://escholarship.org/uc/item/5dx7c5o6

CHAPTER 35

Youth Climate Action and Education

Children's Rights and Epistemic Justice

CARRIE KARSGAARD

Introduction

WHILE global youth climate action takes many names, from Fridays For Future to Youth for Climate, the name of School Strike for Climate—originally *Skolstrejk för klimatet* in Swedish—best encapsulates the activation of schools as political spaces for young people to assert their rights, whether knowingly or unknowingly, as they call for immediate action on climate and for climate justice (see the chapter by Uba in this volume for an overview of the Fridays For Future movement). Politicized through their emptying, schools have also been abandoned by Indigenous youth land protectors drawing attention to the colonial injustices of fossil fuel developments, as when youth from across British Columbia, Canada, occupied the provincial legislature in opposition to a pipeline development across Wet'suwet'en territory. As pedagogical spaces, schools can offer key climate science and political learning through coursework and clubs that support action on climate, toward the climate justice called for by young activists. As students organize for climate action and communication, schools are activated as not only pedagogical spaces but also political ones. Teachers, educational leaders, and the broader public participate as allies and advocates—or as disciplining forces, when they penalize young people for skipping school for strikes or misrecognize youth perspectives in an expression of epistemic injustice.

Most strikingly (pun intended), schools have become sites of refusal of broader political structures and assertion of youth rights, including Indigenous rights, as youth globally have walked out of schools to make a statement about the erosion of their futures due to climate inaction by governments (Thunberg, 2018). While education scholarship has attended to the educational power of youth climate action, the goal of youth activism was not educational change but climate policy change (Biswas & Mattheis, 2022, p. 148). Young people have a limited voice in formal climate (education) policy spaces (Narksompong & Limjirakan, 2015; Thew, 2018), so they transformed schools into political sites through civil disobedience that called for climate action, along with "climate justice and equity" (Fridays For Future, 2022; see also Biswas & Mattheis, 2022). By emptying schools, youth asserted their rights to participation and protest and claimed various rights to life, health, and an adequate standard of living in the face of climate impacts—all of which are protected under the United Nations Convention on the Rights of the Child (UNCRC), and which are arguably within the mandate of formal education to uphold and protect. Notably, these assertions of rights are paired with ongoing legal actions by youth, including by a transnational collective of youth climate activists declaring that "high carbon-emitting economic policies are violating the rights guaranteed to children by the UNCRC, for example the right to life, health and prioritizing the best interest principle" (Earth Justice, 2019, qtd. in Abebe & Biswas, 2021, p. 122).

While many would agree that the school is a site for the amelioration of children's rights, the reality is much more complex. Discourses depicting youth as emotional or idealistic, and decisions of educational leadership to restrict youth participation in school strikes and discipline so-called truants, reveal the "generally autocratic and hierarchical cultures of schooling that have been critiqued by scholars of critical pedagogy (Friere, 2018; hooks, 1994)" (Barrance & Muddiman, 2023, p. 1088). Considering the ways that mainstream education systems both curtail youth voice and contain the terms of sense-making around the climate crisis, the dominant educational response to the school strikes reveals persistent epistemic injustice (Fricker, 2007, 2017), working against young people's rights assertions in the face of climate crisis by both miscomprehending youth expression (testimonial injustice) and shaping the terms of discussion (hermeneutical injustice). At the same time, young people's climate activism perturbs educational restrictions and injustices and presents an alternative politics, offering possibilities and provocations for reorienting education around children's rights and epistemic justice. Such a reorientation, as this chapter will explore, involves not only shifts to the treatment of climate change in schools and the actions by educational leaders but also shifts in the foundations of modern schooling toward expanded hermeneutical space.

Notably, the argument about epistemic justice contained in this chapter is tempered by the fact that most research on the youth climate strikes is situated in the Global North and focuses on the actions of privileged youth in democratic contexts (Porta & Portos, 2023), thus making invisible the various activisms by youth in ranging contexts globally and their accompanying knowledges and rights claims (Karsgaard & Shultz, 2022). While research largely focuses on youth strikes in the Global North, children across the Global South are mobilizing for climate, utilizing a multiplicity of climate action tactics including Indigenous land protection (Wahren, 2021) and antimining efforts (Rahman & Hasan, 2022), tree planting, marches, campaigns like #SaveCongoRainforest, and legal cases in addition to school strikes (Nsah, 2023). However, some contexts create barriers for youth climate strikes, as in Taiwan where youth engage in school-based activism instead of school walkouts due to the "challenging environment" created by the "intersectionality of age and culture" that results in "cultural oppression ... manifested as tangible school regulations and laws, or as intangible social norms" (H.-C. Chang, 2022, p. 18). The Global North is not immune to oppressions and exclusions, as profound power imbalances infuse solidarity in contexts such as New Zealand (Birdsall, 2023) and Canada (Grewal, 2024), and activist groups more broadly "have been critiqued for their own exclusions of certain identities and [for] perpetuating certain inequalities" (Tisdall & Cuevas-Parra, 2022, p. 797). More research is needed to address youth climate activism in marginalized contexts and among marginalized youth if we are to do justice to various youth perspectives, knowledges, and experiences. In the face of this gap, this chapter does not speak to all contexts but rather brings research on the youth climate strikes into dialogue with dominant education systems to consider how education sites might better contribute to epistemic justice and support children's rights in education in the face of climate crisis.

Climate Change, Children's Rights, and Epistemic Justice in Education

The three demands of the Fridays For Future—or School Strikes for Climate—are simple:

1. Keep the global temperature rise below 1.5°C compared to preindustrial levels.
2. Ensure climate justice and equity.
3. Listen to the best united science currently available. (Fridays For Future, 2022)

None of the three demands speaks to education systems, yet the actions of these youth have spurred a flurry of educational research probing the educational value of youth climate activism and how it might speak back to the gaps in education as it stands. At the global policy level, United Nations (UN) member states have responded to youth climate activism by adopting at the 28th Conference of the Parties to the UN Framework Convention on Climate Change a *Declaration on the Common Agenda for Education and Climate Change* (UNESCO, 2023), which addresses adaptation, mitigation, and investment in education. Meanwhile, education scholars are exploring how learning within and from youth climate action might warrant shifts in formal education.

Youth climate action presents a reverberating critique of neoliberalism in education and subsequently calls for "a radical re-evaluation of material taught and a pragmatic focus on pressing social topics across disciplines," including the "topics addressed by school strikes for climate e.g. adultism of societies, climate developments, global inequality and war" (Biswas & Mattheis, 2022, p. 152). Young people are learning through the strikes beyond what they learn in schools, developing a "critical science agency" that connects climate science to issues of justice (White et al., 2022), "critical affective climate justice literacies" that seek to build alternative futures through deep commitment and action (Mayes & Center, 2023), "transformative literacies" that blend the personal and political through youth solidarity and imagination (Bowman & Germaine, 2022), and democratic citizenship education that shifts not only individual worldviews but ideological perspectives (Crouzé et al., 2024). Practiced outside of schools, such learning demonstrates the failures of dominant forms of formal education to engage the affective elements of learning, undo hierarchies that place children on a lower social rung, depoliticize climate education through an overwhelming focus on climate science, and detach education from real-life experiences (Malafaia, 2022, p. 423). Altogether, increasing numbers of education researchers recognize these critiques and draw inspiration from the pedagogies, literacies, and nonformal educational efforts of strikers toward shifts not only in climate change education but education writ large.

While education research is attending to mainstream youth climate action such as the strikes, less scholarship focuses on fossil fuel divestment movements (Belliveau, 2018; Maina et al., 2020) and Indigenous youth climate activism (MacKay et al., 2020; Ritchie, 2021), including through Indigenous land-based movements (Claeys & Delgado Pugley, 2017; Gilio-Whitaker, 2019; Karsgaard, 2023; Whyte, 2014) and human rights movements that link the climate with peace and anticolonialism (Hughes et al., 2023). Such movements address the intersections of climate change with petro-power, extractive capitalism, and colonialism, articulating Indigenous knowledges and advocating for Indigenous and other marginalized peoples to be heard at policy tables. Interestingly, more mainstream youth climate activism has evolved in keeping with these movements "with a more explicit emphasis on post-colonial, anti-capitalist analysis" and is "increasingly … calling for rapid structural transformation in our social, economic and political systems" (Martin, 2023, n.p.). A recent report on the evolution of youth climate activism by the Green Economy Coalition and Climate Vanguard (Herfort et al., 2023), for example, opens with an introduction by Laura Veronica Muñoz, an ecofeminist climate activist from the Colombian Andes:

> We were born into a world already hurtling towards its own destruction, as the individualist choices and capitalist ideologies of those older and supposedly wiser than us play out towards their obvious conclusion. Systemic, structural and collective change is the only way we will avoid the worst outcomes of the coming decades. Young people are aware of this truth. (p. 4)

Despite these shifts, education is far from supporting student learning about economic and political systems necessary to structural change, as will be touched on later. Young people are therefore left with climate change education rooted in environmental education despite its inadequacy to address the complexities of climate crisis (Herfort et al., 2023, pp. 30–31).

As youth climate action continues to evolve, therefore, there is cause for education scholars to attend not only to the demands of mainstream youth activists but also to the critiques of Indigenous, anticolonial, and anticapitalist movements. The ultimate aim of youth climate action is substantive and just action on climate through radical change to economic and political systems, without which young people will continue to unevenly experience the erosion of their rights and futures. States and nongovernmental organizations (NGOs) increasingly point out the adverse effects of climate change on young people's rights (Field et al., 2023; Hall & Tudball, 2023) as encoded in the United Nations Convention on the Rights of the Child (UNCRC) (1989), such as the right to life (article 6), health (article 24), rest and play (article 31), and to both security and an adequate standard of living, including food, water, sanitation, and housing (articles 26 and 27). Youth climate action speaks out directly against the decision making by adults "about energy, extraction, global warming, and guns [that] readily sacrifice the interests of the young for their own in the form of short-term returns and quick profits" (Delgado, 2019, p. 7), as adults do not seriously consider the "best interests of the child," as the UNCRC demands.

Motivated by the efforts of children to draw attention to these issues and recognizing children as agents of change (p. 1), the UN Committee on the Rights of the Child issued General comment No. 26 on children's rights and the environment, with a special focus on climate change (CRC/C/GC/26, 2023). This Committee worked with a children's advisory group and gathered, through surveys, focus groups, and consultations, 16,331 contributions from children from 121 countries. In addition to the rights listed earlier, the General Comment further connects climate change to children's rights to nondiscrimination (article 2) and freedom from violence (article 19) with regard to the structured uneven climate impacts on marginalized children, including Indigenous children and other marginalized children whose rights require unique recognition (article 30). The Comment also speaks to the right of all children to be heard (article 12) in climate decision-making processes—including through "creative means of expression" (p. 5) and digital communication—and to receive access to the best information possible to inform their decision making (articles 13 and 17). Recognizing that children "across the world are taking action, individually and collectively, to protect the environment, including by highlighting the consequences of climate change" (p. 5), the Committee also underscores children's rights to freedom of expression, association, and peaceful assembly (articles 13 and 15), as well as more serious consideration of the "best interests of the child" (article 3).

While climate education scholars have already pointed out how, "beyond the Paris Agreement, there is arguably a moral imperative for schools to improve climate change education based on the rights of children" (Field et al., 2023, p. 159), the fact that education is mentioned 26 times throughout the General Comment points to not only a moral but also a legal imperative for education to provide comprehensive climate change education and support children's rights in the face of climate crisis. Recognizing the right to education (articles 28 and 29) as a transversal right—indeed "one of the cornerstones of a child rights-based approach to the environment" (p. 9)—the Committee highlights that education plays a dual role in both protecting children's rights and providing the education children need to lay hold to their rights. In this way, education is "both a substantive right in itself and also an enabling right," and education must recognize and uphold the "fact that children simultaneously have rights in education and a right to capacity building to enable them to progressively realise their rights" (Jerome & Starkey, 2022, p. 441). When children's rights are not legitimately supported in education, as when dominant groups or institutions define the terms of competence or worthiness to claim rights, it becomes difficult to seriously consider the rights claims of children in the face of more dominant social agendas (see Federle, 2017). Anticipating such potentialities, the Committee asserts that state actors, from decision-makers to police to teachers, are directly implicated in upholding children's rights, and mandates for them "training on children's civil and political rights, including measures to ensure that children can enjoy them safely" (CRC/C/GC/26, 2023, p. 6). Schools therefore have a duty to provide comprehensive climate change education, as well as to uphold and indeed facilitate young people's right to protest through public education and teacher training (see also Barrance, 2020), including by ensuring young people can meaningfully voice their perspectives (Neenan et al., 2021).

Epistemic Injustice and Children's Rights in a Changing Climate

Despite the strength of their rights as upheld by the UNCRC, young people—and particularly marginalized young people—face barriers in asserting their rights in the face of climate change, including through epistemic injustices that play out in sites of education. Epistemic injustice, as conceptualized by Fricker (2007, 2017) references the downgrading or disadvantaging of particular epistemic subjects, with secondary impacts on how the person is "perceived and treated non-epistemically too" (Fricker, 2017, p. 2). For Fricker, there are two types of epistemic injustice: testimonial and hermeneutical. Testimonial injustice is a form of discrimination where the speaker is misjudged or perceived as epistemically inferior, as when young people's knowledges are not attended to because children are assumed to be not fully developed, uninformed, or driven by emotion—a problem that compounds with intersecting oppressions facing young people such as race, disability, class, geopolitical location, or other factors. Hermeneutical injustice, however, involves structural hermeneutical marginalization, resulting in a gap in collective interpretative resources leading to disadvantages when interpreting experiences. Hermeneutic injustice, for instance, puts young people "at an unfair disadvantage in comprehending and/or getting others to comprehend an experience of that kind (a somewhat indirect discrimination)" (Fricker, 2017, p. 2). In both cases—testimonial and hermeneutical—Fricker is clear in pointing out that such injustices do not emerge when a knower is fully understood but deliberately manipulated or undermined; rather, epistemic justice pertains to cases when a knower is simply not recognized as such, even by "conscientious well-meaning others" (p. 8) such as, for instance, educators and school leaders.

Bringing Fricker into dialogue with the UNCRC and General Comment 26 within the context of education may shed light on the inattention to youth voice in climate action—and among well-meaning education leaders and decision-makers who shut down youth activism in efforts to get kids back to class. Through the School Strikes for Climate, young people exhibit complex climate literacies and clearly articulate their demands to leaders; in this case—as with other cases of youth activism—the "issue is not whether children and young people are capable of participating, but whether adults and adult systems *are capable of listening* and giving due weight to children and young people's views" (Tisdall & Cuevas-Parra, 2022, p. 795, italics added). If we read "understanding" for "listening" in Tisdall and Cuevas-Parra's sentence, we can see potential for epistemic injustice—perhaps at both testimonial and hermeneutical levels.

Where dominant education is grounded in colonial and patriarchal knowledge systems, potential for structural epistemic injustice (Fricker, 2017, p. 10) emerges as the "adults, who represent the political and educational world, appear to be the 'we' who silence children by speaking for and about them; and dominate children by exerting power over them (Bradford, 2011, p. 273)" (Rahaman, 2021, p. 54). Grounded in coloniality (Tlostanova & Mignolo, 2012), mainstream education systems in both the Global North and South function in relation to "local and global hierarchies of knowledge, of what is considered worthy of knowing and learning" (Saeed, 2020, p. 5)—hierarchies shaping legitimate knowledge according to European, Western, and White-centric systems of thought. If youth knowledge remains unrecognized in the rights claims of young people within sites of learning, the potential for misrecognition of youth climate knowledge is even greater, especially when that knowledge emerges outside of dominant systems. Writing from the Congo, for example, postcolonial eco-critic Nsah (2023) articulates how foreclosures on knowledge are in keeping with colonial trajectories. Drawing on the writings of what he calls "literary prophets" from the Congo Basin, Nsah reveals how these prophets anticipated structural forms of epistemic injustice against youth climate activists as they depicted "unfounded resistance to the child's contribution to the conservation discourse" (Nsah, 2023, p. 10). Compounded by epistemically limited and hegemonic forms of schooling, the availability of epistemic authority and agency for young people asserting rights in the face of a changing climate appear to be narrow indeed. Can it be otherwise?

Youth Climate Action: Asserting Rights, Facing Epistemic Injustices

Bold youth rights claims through climate action indicate that many young people see the climate crisis differently from formal schooling, as the alternative epistemologies and alternative politics expressed through their activism perturb the dominant frames and restrictions that are part of schooling and have led to the current climate crisis (Komatsu et al., 2020; Lobo et al., 2021; Stein et al., 2020). Despite these alternative expressions, strike shutdowns by educational administrators that disregard youth expressions and curtail participation indicate persistent epistemic injustices in education that limit children's rights even as the climate crisis intensifies the need to support them.

Asserting Rights: The Alternative Epistemologies and Alternative Politics of Youth

Youth expressions in climate action indicate young people's collective ability to see the climate crisis differently, as they introduce alternative epistemologies into formal education through the pedagogies and politics of their movements. Of course, for those who are marginalized or excluded from climate activism, such knowledge may also be marginalized or excluded; here, education has the added responsibility of seeking out those knowledges that are disregarded, hidden, or ignored. As Fricker (2017) discusses, methodologically starting from a marginalized standpoint without essentializing or valorizing that standpoint provides insight into the functioning of power toward possibilities for correcting power imbalances. Often less insulated from climate impacts, young people on the climate crisis's advancing edge, particularly those in the Global South and marginalized communities in the North, experience embodied knowledge of climate impacts (Mendez, 2020). Young people experiencing climate injustice are

> often better in touch with their physical environment compared to their counterparts in industrialized countries and to adults.... They play along riverbeds, in empty lots and in isolated areas, and have a wide knowledge of the physical environment. This kind of knowledge is invaluable and hard to find in the adult world. (Nsah, 2023, p. 12)

Even very young children are attuned to climate impacts and are "carving out new ways of being and doing in the face of climate catastrophe" (Hopkins, 2023, p. 86, qtd. in Mochizuki & Bryan, 2024, p. 3). Oftentimes, such knowledge is transformed through schooling into "hidden knowledge," not recognized within neoliberal competence models and "dismissed as subjective and not scientific, abstract or rational" (Ivinson & Renold, 2022, p. 426). In contrast with the denigration of White students' survivalist knowledge of nature in schools, the knowledges of many youth of color are paradoxically devalued in climate education because they do not have access to natural spaces as children and thus do not "purport a strong relationship to nature and nor did they espouse romantic ideas about 'saving' it" (Neas, 2023, p. 1840). As a result, their experiential knowledges of harm caused by the climate crisis and fossil fuel industry are ignored in education sites, in light of more White, middle-class environmentalism, which "immobiliz[es] youth from substantive climate action" (p. 1840). In both cases, the insights of young people into the climate crisis remain absent in formal learning spaces.

By contrast, youth climate action reflects experimentation with alternative knowledge hierarchies, where their confrontation with the realities of climate change enables students to scrutinize dominant culture and education systems and investigate colonial and industrial causes of the climate crisis. In contrast with educational efforts to foster hope among young people in the face of climate crisis, Bright and Eames (2022) recognize in the strikes a "pedagogy of discomfort" (referencing Boler, 1999), whereby the "strike leaders discovered that by engaging with the discomfort of the climate crisis, acutely evaluating the intersection of their future hopes, their place in history and society's values, enabled them to 'see differently' and consider the change they wish to see" (Bright & Eames, 2022, p. 21). This different sight is articulated through "multiple creative, visual methods" in a "visual youthful politics" (Catanzaro & Collin, 2023, p. 26), where the "political and the creative are mutually constitutive dimensions" (Bowman & Germaine, 2022, p. 81). By tapping into their alternative epistemologies as marginalized knowing subjects on the forefront of climate impacts, and expressing these knowledges in alternative ways, youth activists reflect the "intellectual resistance and eco ethics of the worldwide children who have always been ignored and deemed to be immature about environment, development, politics, climate change and other significant issues" (Rahaman, 2021, p. 38). By freely articulating their climate knowledges, these youth express their "rights to question all the wrongs and injustices perpetrated against their world" (p. 38), including the "social and educational ideologies of the adult world that bring up and socialize children" (p. 38). As knowing subjects, young people interrupt dominant epistemologies that drive both climate change and mainstream, formal schooling, asserting their rights as contributors to how social, political, and educational systems could be shaped otherwise.

Young people's alternative knowledges translate into alternative politics as they refuse the epistemic hegemony of educational institutions and social systems and speak directly to climate decision-makers. Here, young people flip the script of education, which typically functions according to a logic of pedagogical preparation for societal participation. Instead of being prepared for democratic citizenship within the walls of the school, young people are transformed into political actors by leaving behind educational settings (Kvamme, 2019, p. 23), implicitly indicating the school as a hegemonic institution, incapable of meeting the realities of the climate crisis. Walking out of formal education does not imply abandonment of older generations, however; through their calls to action, youth invoke adults, "acknowledging the limits of their activism in an age-stratified society" (Theodorou et al., 2023, p. 70) and calling for a whole-scale shift away from a presentist focus on profits toward a "future-oriented ethic of generational and multispecies responsibility premised on the recognition of multispecies, multispatial, and multigenerational relationality" (p. 70). In asserting their rights—and the rights of all—to healthy futures, and expressing these rights through peaceful assembly and free expression, the youth climate activists exhibit the pedagogical potential of grassroots movements, which "can work as collateral human rights learning situations" (Sainz & Hanna, 2023, p. 15) outside of schools. Returning to Fricker's point that attending to marginalized standpoints can assist in correcting power imbalances, attending to youth political expression calls not for the abandonment of formal education but a purposeful response through education that supports young people's rights claims and political participation.

Epistemic Injustice: Misrecognizing Youth Voice, Perspectives, and Participation

Despite the epistemic and political assertions of young people through climate action, significant testimonial and hermeneutical injustices reveal the epistemic hegemony of dominant schooling that prevents the full expression and support of children's rights. Fricker emphasizes the intention of the hearer in testimonial injustice, articulating that intentional or strategic manipulation is not the same as the kind of unintentional misrecognition that constitutes epistemic injustice. Certainly, intentional misconstruing of youth climate action is evident in media representations linked with coal interests, which have been shown to strategically depict young people as absentee pupils with the purpose of driving young people back to school (Bergmann & Ossewaarde, 2020). At the same time, a range of broader social narratives and images of children seem to dominate even the most well-intentioned responses to the school strikes, as parents and administrators sought the return of young people to formal spaces of learning. Social and educational narratives (Barrance, 2020), commonsense discourses (Mayes & Hartup, 2022), and cultural images (Mochizuki & Bryan, 2024) of children reflect an "adultism bias" (Malafaia, 2022) that results in the inability of young people to be properly heard—a testimonial injustice—resulting in the emphasis of the role of schooling for proper learning and development toward children's full and meaningful participation in society. In this vein, school districts from Australia to Germany reflect the adultism bias whereby young people are represented as "not yet fully educated people who are obligated by law to go back to school" (Bergmann & Ossewaarde, 2020, p. 287).

Adultist social narratives about children take a variety of forms, from developmental to anticipatory and protectionist narratives, all of which negate young people's positions as epistemic agents—and subsequently impede their political participation and rights claims. Developmental and psychologized narratives "provide a biological basis for affirming young people's deficits in exercising rational decision-making and emotional restraint" (Mochizuki & Bryan, 2024, p. 7), treating their activism as merely a "phase," accounted for by hormones, peer influences, or teenage rebellion rather than critical thinking and purposeful political action (Mayes & Hartup, 2022, pp. 1011–1012). In these cases, the climate anxiety articulated by youth is understood not as a logical response to the structural injustices and "defuturing effects" (Fry, 2020) interlinked with the climate crisis (Vamvalis, 2023) but is pathologized or treated as alarmist or hyperbolic. Such developmental narratives place educators in the adjudicatory position of responding to youth activism as a behavioral issue to be managed and trained toward future appropriate political participation by developed adults, once these young people have completed school.

Developmental narratives thus map onto narratives that advocate protection of children and youth from climate realities, maintaining their dependence on adults until which time they are fully educated and prepared to participate in climate decision making. Such prevailing narratives are in line with long-held conceptions of the child as innocent, vulnerable, fragile, and in need of protection, though such conceptions are arguably being "amplified in the present moment, given the complexity, scale, and severity of the climate crisis" (Mochizuki & Bryan, 2024, p. 3) and as youth are becoming increasingly vulnerable to climate impacts. In this context, the school is secured as a key space for the protection and insulation of the child—or at least some children, as protection is not similarly extended to "boys, non-binary children, or children of color in the same way as to white girls" (Buhre, 2023, p. 269, qtd. in Mochizuki & Bryan, 2024, p. 4). Problematically, such narratives "[juxtapose] the adultist figure of the innocent, vulnerable, dependent child with that of the rational adult imbued with knowledge, power, and agency [resulting in] the disempowering effect of denying children's capacities as (already) knowledgeable actors with the right to political expression" (Mochizuki & Bryan, 2024, p. 3). Educators and other adults maintain a position of knowledge and responsibility, while children's knowledge of climate impacts is ignored (Hörter, 2024), their activism becomes unrecognizable, and their agency denied (Barrance, 2020; Jones et al., 2023). If children are recognized as knowers and change agents, it is not on their terms but on the terms of adults in authority over them, who have the power to determine between acceptable and unacceptable forms of political action, the latter being disciplined in school or at the most extreme, arrested (Mochizuki & Bryan, 2024 referencing Tannock, 2021). As a result, young people's knowledge is not deepened through education that builds on youth perspectives and experiences toward meaningful climate learning and the further rights claims of youth in the face of climate injustices but instead closes doors to learning that will support them to come to terms with and act on the climate crisis.

Before moving on to the hermeneutical injustice done to youth, it is worth noting that young people are not merely victims to these misrecognitions, as "school strikers play with and perform stereotypes about children and young people" (Mayes & Center, 2023, p. 527) even as they articulate their rights. Youth climate activists have been shown to undermine and mock, as well as strategically invoke, dominant narratives through their signage, artistic expressions, and calls to action. Most notably, youth "subvert traditional adultist power relations and leverage youth as an ideological position from which to expose adults' inertia" (Malafaia, 2022, p. 430). Indeed, moving against protectionism toward intergenerational understandings and responses to the climate crisis is a powerful mobilizing element of youth climate activism, which holds potential for rethinking education outside of age segregation and industrial schooling, as will be discussed further later.

Despite the efforts of youth to make their voices heard, young people suffer not only testimonial but also hermeneutic injustice due to the social powerlessness of youth in the face of adult authority over the terms of education and climate decision making. Due to the gaps in formal education, young people must struggle to make sense of their experiences without educational support. Despite young people's concern with climate change, only a few locations such as Hong Kong (Jackson & Pang, 2017) and Zambia (Muchanga & Nakazwe, 2015) seem to be advancing climate change education. By contrast, a recent scan of secondary science and social science curriculum in 85 countries shows that 69% contain no references at all to climate change (UNESCO, 2024). Various studies report that Singapore (C.-H. Chang & Pascua, 2017), Chile (Salinas et al., 2022), Norway (Tomren, 2022), Canada (Galway & Field, 2023), Australia, Israel, Finland, Indonesia, and England (Dawson et al., 2022) provide uneven curricular support for climate learning. In the most conservative schools, the climate crisis is left out of the educational conversation altogether (Everth & Bright, 2023, p. 28; Long & Henderson, 2023), leaving young people disillusioned with schools as they seek meaning-making spaces elsewhere. Where climate education is present, petro-pedagogies actively obstruct learning by promoting fossil fuel interests (Dunlop et al., 2021; Eaton & Day, 2020; Tannock, 2020), which erode the possibilities for education to support meaningful hermeneutical development with and for youth in the face of climate crisis.

Traditional schooling practices also contribute to hermeneutical epistemic injustice. Neas (2023) notes how Friere's (2014) "banking concept" of education infuses traditional climate education, which utilizes "an acquisition of knowledge model that focuses on delivering content at the cost of making meaning out of that content. Without this, youth have little opportunity to grapple with the complex emotional and moral dilemmas of climate change" (Neas, 2023, p. 1843). Dominant forms of education rarely provide space for people to confront the "volatility, uncertainty, complexity, and ambiguity" of climate change, including how to "hold space for these complexities and their own multi-layered responses to them" (Suša et al., 2023, p. 991). Considering the deep concerns of youth climate activists with these difficult realities, education has arguably "hermeneutically marginalized" (Fricker, 2007, p. 152) young people, disempowering them from full social participation—and thus the claiming of their rights.

Of course, one might point to the proliferating attention to youth climate activism as an indication that indeed young people are gaining power in shaping collective climate hermeneutics. While young people's perspectives are certainly beginning to change climate conversations, dominant hermeneutics continue to contain the critical force of youth expression. Paradoxically, the valorizing response to youth climate activism, which celebrates young people's heroic efforts for change, treats youth efforts as symbolic rather than substantive, applauding activists' efforts without attending to their knowledges and demands. Here, youth are celebrated and tokenistically displayed for the public without being engaged as equal co-creators of the future and co-participants in decision making (Malafaia, 2022). As a result, "framing of social movements as 'youth-led' can enable elites to dismiss them as a naive and immature approach to politics and social change" (Mochizuki & Bryan, 2024, p. 7), delegitimizing knowledge that could otherwise have spoken into a gap in the collective hermeneutical resources for sense-making through climate crisis. Alternatively, young people's perspectives are at times acknowledged yet positioned within a lower rung on the hierarchies of climate and educational decision making (Bowman, 2020, p. 11). Youth are celebrated for their civic engagement or see their efforts instrumentalized within institutional "greening" programs in schools (Dupuis-Déri, 2021, p. 4) but are themselves denied the capacity to change the terms of the conversation.

With climate education thus maintaining the hermeneutical space, the school strikes open up important questions about hermeneutical inclusion and exclusion in education—as well as the "norms structuring our educational systems situated within larger democratic configurations" (Biswas & Mattheis, 2022, p. 151) with the power to ensure children's rights—including the rights of those most marginalized—in the face of climate crisis. Recognizing that young people are already breaking into the hermeneutical spaces surrounding education and climate change, education arguably holds a responsibility to respond, to "make it easier for children to 'speak' to the adults holding power over their lives ... [and] for them to speak to each other and form solidarity, even across national lines" (Delgado, 2019, pp.

25–26). How might education attend to and support youth knowledge-building in ways that enable greater capacity to respond to the climate crisis in ways that ensure children's rights are upheld?

Reorienting Education for Epistemic Justice and Rights in Education

Young people are already asserting themselves as epistemic agents through the global youth climate strikes, and education can respond by recognizing this agency and supporting epistemic justice, altering education's positioning within the global colonial knowledge order. Mainstream education is arguably complicit in the ongoing climate crisis due to its foundations in colonial modernity, and despite the activism and critiques of young people globally, it persists in reinforcing the status quo, rather than shifting its foundations to address the current climate crisis (Komatsu et al., 2020; Orr, 2004). Transforming mainstream education in relation to climate change therefore involves its epistemological reorientation (Stein et al., 2017; Suša et al., 2023; Tlostanova & Mignolo, 2012) away from the "systemic conditions of structured unsustainability that eliminate possible futures" (Escobar, 2018, p. 16) for young people. To support children's rights in education and open hermeneutical space for sense-making surrounding the climate crisis, educational leaders and decision-makers can ground decisions in the UNCRC toward the collective knowledge-building needed for climate justice.

Bringing together epistemic justice with children's rights, education might work toward supporting children's rights *in* education—rather than simply children's rights *to* education (Abebe & Biswas, 2021). Learning from the ways youth activists and broader communities push back against both hegemonic forms of knowledge and dominant forms of schooling, Abebe and Biswas (2021) bring to light the epistemic agencies of young people that "have been erased and continue to be erased by the universal promotion of mass schooling under the pretext of fulfilling a 'right to education' which, in fact, concerns only the 'right to schooling'" (p. 121). Whereas a focus on the right *to* education results in efforts to ensure access to education and subsequently to keep young people in class, a reorientation around rights *in* education "evokes questions that not only challenge the global knowledge order but also inscriptions of new practices through the mediums of resistance/revolution/activism so as to re-politicize knowledge and re-introduce it in the debates around epistemic justice" (p. 121). Oriented around community formation rather than individual achievement, supporting rights in education is foundationally relational work, invigorating intergenerational knowledge sharing in place of age-segregated schooling and recognizing learning in extra-institutional spaces from homes to virtual spaces to playgrounds and trains. As such, rights in education "realigns the purpose of education with the overall purpose of sustaining life on the planet" (p. 121) and opens space for recognizing and engaging with youth epistemic agency and learning within climate crisis and through the strikes.

Considering the prevailing testimonial injustices facing young people through structural discrimination that deems young people as undeveloped and as-yet incompetent—age discrimination that compounds with other injustices facing young people according to such factors as race, ability, class, gender, sexual orientation, and more—education has the challenge of reconceptualizing rights in education by recognizing the structuring of youth powerlessness (Federle, 2017) or subalterneity (Bowman, 2020) and opening space for legitimate contributions by youth. Such efforts not only involve supporting young people's critical reflection and legitimate spaces for participation (Jerome & Starkey, 2022, p. 446) both inside and beyond schools as part of their fundamental rights but also critical reflexivity on the part of educationists who inhabit unjust structures and ways of knowing. So education professionals and researchers will need to self-reflexively dialogue with youth to ensure they are asking the right questions, using young people's language, and interpreting youth responses on young people's terms (Neenan et al., 2021, p. 2). The centrality of educators as decision-makers may need to be displaced (Tisdall & Cuevas-Parra, 2022, p. 794) in favor of legitimate youth contributions and their relational co-creation of knowledge in relation to the broader human and nonhuman community (Karsgaard et al., 2022)—potentially through an ongoing "'dialectic' between social movement learning and schooling" (McGregor et al., 2018, p. 11). A limit of the chapter you are reading is that it draws on secondary literature—albeit some literature either co-written with youth and much that that amplifies youth perspectives—rather than attending directly to youth knowledge (though such an effort is currently underway). By contrast, co-research and co-creation of educational policy with youth who are aware of the climate crisis and bring informed and experienced expert knowledge to policy settings (Bowman & Germaine, 2022; Casas et al., 2021; Cutter-Mackenzie & Rousell, 2019; Gladwin et al., 2022; Karsgaard & Davidson, 2021) would work to undermine epistemic injustice not only by recognizing the validity of youth knowledge but also by generating youth power through their contributions to decision making, especially collaborations with youth sections of environmental justice organizations.

Educational efforts against testimonial injustice in education can be buoyed by efforts against hermeneutical injustice through expansion of not only the interpretative frames brought to bear on the climate crisis but also through a foundational break with dominant education's foundations in colonial modernity that limit hermeneutical possibilities from the outset. It is nothing new to claim that "formal schooling is a central part of the colonial-capitalist system that strikers are demanding be changed" (Verlie & Flynn, 2022, p. 6). Despite the global youth climate strikes' efforts toward climate justice and the radical challenges they pose to mainstream policies on education for sustainable development (Kowasch et al., 2021, p. 19), mainstream education diminishes collective hermeneutical resources by continuing to overprioritize theories of change that are fundamentally at odds with the systemic transformation necessary to address climate (Trott, n.d.), including technoscientific solutionism, neoliberal growth paradigms, and individual actions. Such dominant outcomes push out learning about alternatives such as degrowth, Indigenous knowledge, and political possibilities for dismantling climate-harming systems (Kowasch et al., 2021, p. 19; Verlie & Flynn, 2022, p. 6). Education's narrow framing of the climate crisis arguably necessitates bringing together climate justice with epistemic justice, as "debates over how the concept [of climate justice] is framed, and by whom, is a justice issue" (McGregor et al., 2018, p. 29) with repercussions for those marginalized youth and communities who are subject to the most intense climate effects and whose knowledges are erased in the face of capitalist growth paradigms. Here, education holds the responsibility to open hermeneutic space for exploration of the functioning of power in relation to knowledge creation and mobilization, including how youth might attend to and articulate "instances of economic, cultural, and epistemic insurgency that aim to re/create and maintain practices of cultural, economic, and ecological difference" (Escobar, 2018, p. 75) from climate-harming capitalist modernity.

While teaching and learning about climate change may thus expand collective hermeneutic capacity for sense-making about the current climate crisis, the structure of modern schooling is counterproductive to such efforts. Many scholars point to the contradictions inherent to education when it articulates concern about climate change through curricular outcomes and school initiatives while "institutionalizing children and youths and not including their interests in current political priorities" (Kvamme, 2019, p. 23). Dominant forms of education tend toward a narrow focus on "individual conformity and success" (McGimpsey et al., 2023, p. 3) to the point of repressing student climate activism as counterproductive to educational aims of developing skilled and competent social contributors. It was against such educational contexts that the "movement for intergenerational climate justice began in a highly privileged Northern context, with White child activists striking a pivotal institution of modern childhood and one of the foundational pillars for sustaining an intergenerationally unjust capitalist economy—*the contemporary school*" (Biswas, 2023, p. 1016, italics added). The dominant model of schooling, grounded in "coloniality, modernity and industrialisation depends on age-based segregation and the breaking down of intergenerational community formation" (Biswas, 2023, p. 1016), in direct contradiction to the relational approach to children's rights *in* education described earlier. Arguably, a movement away from this dominant model is necessary to crack open greater hermeneutical space whereby young people can co-construct knowledge with communities and surrounding lands toward the sustainment of life on a changing planet. Of course, not all education sites and systems are firmly set within the modern structures of schooling—particularly where Indigenous, African, and other alternative land-based cosmologies persist in influencing learning—and such examples may provide alternative footing.

Considering the role of policymakers and educational leaders in mandating young people back to school during the strikes, it is worth considering how such actors can play a role in supporting epistemic justice and children's rights in education as the climate crisis continues to intensify. Certainly, the intensity of political polarization and complexity of entanglements among schools, governments, media, and fossil fuel interests places educational leadership in a difficult position when responding to youth climate strikes (Variyan & Gobby, 2022, p. 8). Furthermore, in some contexts, formal education policy does not provide sufficient grounds to either support or restrict the strikes (Wilmers, 2023), leaving leaders to make decisions in a policy vacuum. Problematically, the controversies and gaps faced by educational leaders result in public discussion reorienting around "*the 'emergency' of student activism* rather than the emergency of the *climate crisis itself*" (Variyan & Gobby, 2022, p. 11, italics original). While the position of school leaders is indeed a challenging one, leaders at some schools have embraced the strikes as learning opportunities and recognized both "teachers' and students' actions as potential lines of exploration that can bring back new knowledge" (Everth & Bright, 2023, p. 29) toward the development of "collective intelligence" (p. 30)—or what we might understand here as an expanded hermeneutical space. Ultimately, this expanded collective intelligence supports "experimentation, student and teacher agency and community engagement to promote the transformation of society towards a sustainable future" (p. 30) with the rights of children in mind. Further, by supporting epistemic justice through engagement with youth activism, educational leaders are effectively supporting children's rights to participation and freedom of expression, as delineated by the UNCRC. In the vacuum created by national policies, therefore, the UNCRC can provide foundations for support of children's rights and epistemic justice while contributing the meaningful climate learning young people need to face their current and future contexts.

Conclusion

By leaving their desks behind and taking to the streets, young people activated schools as political spaces whereby they could assert their rights, articulate their demands for action on climate and climate justice, and assert alternative epistemologies to the dominant ways of understanding the climate crisis, including in their schools. As marginalized contributors to the global climate conversation, young people brought their experiential knowledges of the climate crisis and their advocacy on behalf of other more marginalized youth into public dialogue, advancing climate justice on policy—and education policy—agendas globally.

Building on the long-standing concept of environmental justice, the concept of climate justice recognizes not only scholarly sense-making surrounding the climate crisis but also the "distinctive educative and epistemological contributions of social movements" (McGregor et al., 2018, p. 2; see also Schlosberg & Collins, 2014), positioning itself "critically both 'in and against'" (p. 2) educational policy contexts. While climate justice is gaining steam in education (Trott et al., 2023), there is further to go in attending to the knowledges, experiences, and voices of youth—especially those directly experiencing climate injustice.

Schools have a responsibility to support children's rights and particularly in the face of intensifying climate crisis, as per the UNCRC, but dominant forms of education are constituted by both testimonial and hermeneutical epistemic injustice that downgrade (particular) young people as epistemic subjects and limit youth climate sense-making. Enmeshed with broader adultist social narratives and grounded in the epistemic hierarchies of colonial modernity, education arguably needs a fundamental reorientation away from colonial modernity and industrial schooling toward greater support for children's rights in education. If education is to meet current demands for climate justice, it could do well to attend to the young person informing General Comment No. 26 on the UNCRC (2023, p. 1): "Adults [should] stop making decisions for the future they won't experience. [We] are the key means [of] solving climate change, as it is [our] lives at stake."

References

Abebe, T., & Biswas, T. (2021). Rights in education: Outlines for a decolonial, childist reimagination of the future—Commentary to Ansell and colleagues. *Fennia: International Journal of Geography, 199*(1), Article 1. https://doi.org/10.11143/fennia.107490

Barrance, R. (2020). The right to protest: A children's rights perspective on the school climate strikes. *ORCA–Online Research @ Cardiff.* https://orca.cardiff.ac.uk/id/eprint/129441/63/The%20right%20to%20protest%20A%20children's%20rights%20perspective%20on%20the%20school%20climate%20strikes.pdf

Barrance, R., & Muddiman, E. (2023). Beyond bad behaviour? Towards a broader understanding of school student activism. *British Journal of Sociology of Education, 44*(6), 1087–1107. https://doi.org/10.1080/01425692.2023.2238906

Belliveau, E. (2018). *Climate justice in the fossil fuel divestment movement: Critical reflections on youth environmental organizing in Canada* [Thesis]. University of Victoria. https://dspace.library.uvic.ca/handle/1828/10052

Bergmann, Z., & Ossewaarde, R. (2020). Youth climate activists meet environmental governance: Ageist depictions of the FFF movement and Greta Thunberg in German newspaper coverage. *Journal of Multicultural Discourses, 15*(3), 267–290. https://doi.org/10.1080/17447143.2020.1745211

Birdsall, S. (2023). School Strike 4 Climate in Aotearoa New Zealand: Youth, relationships and climate justice. *Australian Journal of Environmental Education, 39*(2), 139–151. https://doi.org/10.1017/aee.2022.51

Biswas, T. (2023). Becoming good ancestors: A decolonial, childist approach to global intergenerational sustainability. *Children & Society, 37*(4), 1005–1020. https://doi.org/10.1111/chso.12722

Biswas, T., & Mattheis, N. (2022). Strikingly educational: A childist perspective on children's civil disobedience for climate justice. *Educational Philosophy & Theory, 54*(2), 145–157. https://doi.org/10.1080/00131857.2021.1880390

Boler, M. (1999). *Feeling power: Emotions and education*. Routledge.

Bowman, B. (2020). "They don't quite understand the importance of what we're doing today": The young people's climate strikes as subaltern activism. *Sustainable Earth*, *3*(1), 16. https://doi.org/10.1186/s42055-020-00038-x

Bowman, B., & Germaine, C. (2022). Sustaining the old world, or imagining a new one? The transformative literacies of the climate strikes. *Australian Journal of Environmental Education*, *38*(1), 70–84. https://doi.org/10.1017/aee.2022.3

Bradford, C. (2011). The case of children's literature: Colonial or anti colonial? *Global Studies of Childhood*, *1*(4), 271–279.

Bright, M. L., & Eames, C. (2022). From apathy through anxiety to action: Emotions as motivators for youth climate strike leaders. *Australian Journal of Environmental Education*, *38*(1), 13–25. https://doi.org/10.1017/aee.2021.22

Casas, E. V., Pormon, M. M., Manus, J. J., & Lejano, R. P. (2021). Relationality and resilience: Environmental education in a time of pandemic and climate crisis. *Journal of Environmental Education*, *52*(5), 314–324. https://doi.org/10.1080/00958964.2021.1981205

Catanzaro, M., & Collin, P. (2023). Kids communicating climate change: Learning from the visual language of the SchoolStrike4Climate protests. *Educational Review*, *75*(1), 9–32. https://doi.org/10.1080/00131911.2021.1925875

Chang, C.-H., & Pascua, L. (2017). The curriculum of climate change education: A case for Singapore. *Journal of Environmental Education*, *48*(3), 172–181. https://doi.org/10.1080/00958964.2017.1289883

Chang, H.-C. (2022). Climate strike or not? Intersectionality of age and culture encountered by young climate activists in Taiwan. *Childhood*, *29*(1), 7–23. https://doi.org/10.1177/09075682221074869

Claeys, P., & Delgado Pugley, D. (2017). Peasant and indigenous transnational social movements engaging with climate justice. *Canadian Journal of Development Studies / Revue Canadienne d'études Du Développement*, *38*(3), 325–340. https://doi.org/10.1080/02255189.2016.1235018

Convention on the Rights of the Child. (1989). Human Rights Officer of the High Commissioner. United Nations. https://www.ohchr.org/en/instruments-mechanisms/instruments/convention-rights-child

CRC/C/GC/26. (2023). *General comment No. 26 (2023) on children's rights and the environment with a special focus on climate change*. Human Rights Officer of the High Commissioner. United Nations. https://www.ohchr.org/en/documents/general-comments-and-recommendations/crccgc26-general-comment-no-26-2023-childrens-rights

Crouzé, R., Godard, L., & Meurs, P. (2024). Learning democracy through activism: The global climate strike movement and Belgian youth's democratic experience in times of environmental emergency. *Social Movement Studies*, *23*(1), 56–71. https://doi.org/10.1080/14742837.2023.2184792

Cutter-Mackenzie, A., & Rousell, D. (2019). Education for what? Shaping the field of climate change education with children and young people as co-researchers. *Children's Geographies*, *17*(1), 90–104. https://doi.org/10.1080/14733285.2018.1467556

Dawson, V., Eilam, E., Tolppanen, S., Assaraf, O. B. Z., Gokpinar, T., Goldman, D., Putri, G. A. P. E., Subiantoro, A. W., White, P., & Widdop Quinton, H. (2022). A cross-country comparison of climate change in middle school science and geography curricula. *International Journal of Science Education*, *44*(9), 1379–1398. https://doi.org/10.1080/09500693.2022.2078011

Delgado, R. (2019). Children's rights to a livable future. *Alabama Law Scholarly Commons*, *71*, 261–289.

Dunlop, L., Atkinson, L., Stubbs, J. E., & Diepen, M. T. (2021). The role of schools and teachers in nurturing and responding to climate crisis activism. *Children's Geographies*, *19*(3), 291–299. https://doi.org/10.1080/14733285.2020.1828827

Dupuis-Déri, F. (2021). Youth strike for climate: Resistance of school administrations, conflicts among students, and legitimacy of autonomous civil disobedience—The case of Québec. *Frontiers in Political Science*, *3*, 1–11. https://www.frontiersin.org/article/10.3389/fpos.2021.634538

Eaton, E. M., & Day, N. A. (2020). Petro-pedagogy: Fossil fuel interests and the obstruction of climate justice in public education. *Environmental Education Research*, *26*(4), 457–473. https://doi.org/10.1080/13504622.2019.1650164

Escobar, A. (2018). *Designs for the pluriverse: Radical interdependence, autonomy, and the making of worlds*. Duke University Press.

Everth, T., & Bright, R. (2023). Climate change and the assemblages of school leaderships. *Australian Journal of Environmental Education*, *39*(1), 17–36. https://doi.org/10.1017/aee.2022.8

Federle, K. H. (2017). Do rights still flow downhill? *International Journal of Children's Rights*, *25*(2), 273–284. https://doi.org/10.1163/15718182-02502004

Field, E., Spiropoulos, G., Nguyen, A. T., & Grewal, R. K. (2023). Climate change education within Canada's regional curricula: A systematic review of gaps and opportunities. *Canadian Journal of Educational Administration and Policy / Revue Canadienne En Administration et Politique de l'éducation*, *202*, 155–184. https://doi.org/10.7202/1099989ar

Fricker, M. (2007). *Epistemic injustice: Power and the ethics of knowing*. Clarendon Press.

Fricker, M. (2017). Evolving concepts of epistemic injustice. In I. J. Kidd, J. Medina, & G. Pohlhaus Jr. (Eds.), *Routledge handbook of epistemic injustice* (pp. 53–60). Routledge. https://www.routledge.com/9781138828254

Fridays For Future. (2022). Fridays For Future—Our demands. Act now! https://fridaysforfuture.org/what-we-do/our-demands/

Friere, P. (2014). *Pedagogy of the oppressed*. Bloomsbury.

Friere, P. (2018). *Pedagogy of the oppressed: 50th Anniversary Edition*. Bloomsbury Academic.

Fry, T. (2020). *Defuturing: A new design philosophy*. Bloomsbury.

Galway, L. P., & Field, E. (2023). Climate emotions and anxiety among young people in Canada: A national survey and call to action. *Journal of Climate Change and Health*, *9*, 100204. https://doi.org/10.1016/j.joclim.2023.100204

Gilio-Whitaker, D. (2019). *As long as grass grows: The Indigenous fight for environmental justice, from colonization to Standing Rock*. Beacon Press.

Gladwin, D., Karsgaard, C., & Shultz, L. (2022). Collaborative learning on energy justice: International youth perspectives on energy literacy and climate justice. *Journal of Environmental Education*, *53*(5), 1–10. https://doi.org/10.1080/00958964.2022.2113019

Grewal, R. K. (2024). *By no means there yet: Centring the voices of Black, Indigenous, and youth of colour climate activists in Ontario* [Thesis]. Lakehead University.https://knowledgecommons.lakeheadu.ca/handle/2453/5288

Hall, G., & Tudball, L. (2023). Enabling grass roots activism and human rights-based education for sustainability: Case studies of Australian youth organisations. *Human Rights Education Review*, *6*(1), Article 1. https://doi.org/10.7577/hrer.5016

Herfort, N., Johnson, J., Mcintyre, C., & Born, S. (2023). *Mapping the global youth climate movement: Towards a green economic mandate*. Green Economy Coalition & Climate Vanguard.

hooks, b. (1994). *Teaching to transgress: Education as the practice of freedom*. Routledge.

Hopkins, L. (2023). "The ice is melting and I don't want Santa to drown!": Reflections on childhood, climate action, and futurity. *Journal of Childhood Studies*, *48*(1), 85–98.

Hörter, K. (2024). "It makes me sad when so much of the forest burns down!": Childhood and youth in the context of the climate crisis. *The Psychoanalytic Study of the Child*, *77*(1), 389–401. https://doi.org/10.1080/00797308.2023.2287383

Hughes, S. S., Velednitsky, S., & Green, A. A. (2023). Greenwashing in Palestine/Israel: Settler colonialism and environmental injustice in the age of climate catastrophe. *Environment and Planning E: Nature and Space*, *6*(1), 495–513. https://doi.org/10.1177/25148486211069898

Ivinson, G. M., & Renold, E. J. (2022). Emplaced activism: What-if environmental education attuned to young people's entanglements with post-industrial landscapes? *Australian Journal of Environmental Education*, *38*(3–4), 415–430. https://doi.org/10.1017/aee.2022.41

Jackson, L., & Pang, M.-F. (2017). Secondary school students' views of climate change in Hong Kong. *International Research in Geographical and Environmental Education*, *26*(3), 180–192. https://doi.org/10.1080/10382046.2017.1330036

Jerome, L., & Starkey, H. (2022). Developing children's agency within a children's rights education framework: 10 propositions. *Education 3-13*, *50*(4), 439–451. https://doi.org/10.1080/03004279.2022.2052233

Jones, C. A., Davison, A., & Lucas, C. (2023). Innocent heroes or self-absorbed alarmists? A thematic review of the variety and effects of storylines about young people in climate change discourses. *WIREs Climate Change*, *14*(6), e853. https://doi.org/10.1002/wcc.853

Karsgaard, C. (2023). *Instagram as public pedagogy: Online activism and the Trans Mountain pipeline*. Palgrave Studies in Education and the Environment.

Karsgaard, C., & Davidson, D. (2021). Must we wait for youth to speak out before we listen? International youth perspectives and climate change education. *Educational Review*, *75*(1), 74–92. https://www.tandfonline.com/doi/abs/10.1080/00131911.2021.1905611

Karsgaard, C., Pillay, T., & Shultz, L. (2022). Facing colonial Canada through pedagogies of equity for First Nations: An advocacy education project. *Journal of Curriculum and Pedagogy*, *21*(4), 1–25. https://doi.org/10.1080/15505170.2022.2134948

Karsgaard, C., & Shultz, L. (2022, November 22). Youth movements and climate change education for justice. In *Oxford Research Encyclopedia of Education* (pp. 1–23). https://doi.org/10.1093/acrefore/9780190264093.013.1808

Komatsu, H., Rappleye, J., & Silova, I. (2020). Will education post-2015 move us toward environmental sustainability? In *Grading goal four* (pp. 297–321). Brill. https://brill.com/view/book/edcoll/9789004430365/BP000022.xml

Kowasch, M., Cruz, J. P., Reis, P., Gericke, N., & Kicker, K. (2021). Climate youth activism initiatives: Motivations and aims, and the potential to integrate climate activism into ESD and transformative learning. *Sustainability*, *13*(21), Article 21. https://doi.org/10.3390/su132111581

Kvamme, O. A. (2019). School strikes, environmental ethical values, and democracy. *Studier i Pædagogisk Filosofi*, *8*(1), 6–27. https://doi.org/10.7146/spf.v8i1.117967

Lobo, M., Bedford, L., Bellingham, R. A., Davies, K., Halafoff, A., Mayes, E., Sutton, B., Walsh, A. M., Stein, S., & Lucas, C. (2021). Earth unbound: Climate change, activism and justice. *Educational Philosophy and Theory*, *53*(14), 1491–1508. https://doi.org/10.1080/00131857.2020.1866541

Long, D., & Henderson, J. (2023). Climate change as superordinate curriculum? *Research in Education*. https://doi.org/10.1177/00345237231160080

MacKay, M., Parlee, B., & Karsgaard, C. (2020). Youth engagement in climate change action: Case study on Indigenous youth at COP24. *Sustainability*, *12*(16), Article 16. https://doi.org/10.3390/su12166299

Maina, N. M., Murray, J., & McKenzie, M. (2020). Climate change and the fossil fuel divestment movement in Canadian higher education: The mobilities of actions, actors, and tactics. *Journal of Cleaner Production*, *253*, 119874. https://doi.org/10.1016/j.jclepro.2019.119874

Malafaia, C. (2022). "Missing school isn't the end of the world (actually, it might prevent it)": Climate activists resisting adult power, repurposing privileges and reframing education. *Ethnography & Education*, *17*(4), 421–440. https://doi.org/10.1080/17457823.2022.2123248

Martin, B. (2023). *How the youth climate movement has evolved from school strikes to ...*. Green Economy Coalition. https://www.greeneconomycoalition.org/news-and-resources/from-school-strikes-to-systemic-change-how-the-youth-climate-movement-has-evolved

Mayes, E., & Center, E. (2023). Learning with student climate strikers' humour: Towards critical affective climate justice literacies. *Environmental Education Research*, *29*(4), 520–538. https://doi.org/10.1080/13504622.2022.2067322

Mayes, E., & Hartup, M. E. (2022). News coverage of the School Strike for Climate movement in Australia: The politics of representing young strikers' emotions. *Journal of Youth Studies*, *25*(7), 994–1016. https://doi.org/10.1080/13676261.2021.1929887

McGimpsey, I., Rousell, D., & Howard, F. (2023). A double bind: Youth activism, climate change, and education. *Educational Review*, *75*(1), 1–8. https://doi.org/10.1080/00131911.2022.2119021

McGregor, C., Scandrett, E., Christie, B., & Crowther, J. (2018). Climate justice education. In T. Jafry, M. Mikulewicz, & K. Helwig (Eds.), *Routledge handbook of climate justice* (1st ed., pp. 494–508). Routledge. https://doi.org/10.4324/9781315537689-36

Mendez, M. (2020). *Climate change from the streets: How conflict and collaboration strengthen the environmental justice movement*. Yale University Press. https://login.ezproxy.library.ualberta.

ca/login?url=https://search.ebscohost.com/login.aspx?direct=true&db=e000xna&AN=2329368&site=ehost-live&scope=site

Mochizuki, Y., & Bryan, A. (2024). Victims or vanguards? The discursive construction of the Anthropocene generation and climate change education's hopeful, resilient, post-political subject. In J. Wyn, H. Cahill, & H. Cuervo (Eds.), *Handbook of children and youth studies* (pp. 1–16). Springer Nature. https://doi.org/10.1007/978-981-4451-96-3_130-1

Muchanga, M., & Nakazwe, K. M. (2015). Climate change education in the school curricula of Zambia. In G. Nhamo & S. Shava (Eds.), *Climate change education in SADC school curriculum* (pp. 241–260). Africa Institute of South Africa.

Narksompong, J., & Limjirakan, S. (2015). Youth participation in climate change for sustainable engagement. *Review of European, Comparative & International Environmental Law, 24*(2), 171–181. https://doi.org/10.1111/reel.12121

Neas, S. (2023). Narratives and impacts of formal climate education experienced by young climate activists. *Environmental Education Research, 29*(12), 1832–1848. https://doi.org/10.1080/13504622.2023.2193684

Neenan, E. E., Roche, J., & Bell, L. (2021). Time to listen: Children's voice in geoscience education research. *Frontiers in Environmental Science, 9*, 1–6. https://www.frontiersin.org/articles/10.3389/fenvs.2021.669430

Nsah, K. T. (2023). Our children will fight for the climate: How Congo-Basin writers prophesied global youth climate activism. *Electronic Green Journal, 48*, 1–25. https://doi.org/10.5070/G314855358

Orr, D. (2004). Can educational institutions learn? The creation of the Adam Joseph Lewis Center at Oberlin College. In P. F. Barlett & G. W. Chase (Eds.), *Sustainability on campus: Stories and strategies for change* (pp. 159–175). MIT Press.

Porta, D. della, & Portos, M. (2023). Rich kids of Europe? Social basis and strategic choices in the climate activism of Fridays for Future. *Italian Political Science Review, 51*(1), 24–49. https://doi.org/10.1017/ipo.2021.54

Rahaman, H. (2021). Doctor Stockmann and Greta Thunberg: Some implications of intellectual resistance, eco-activism and unschooling. *Journal of Unschooling & Alternative Learning, 15*(29), 36–60.

Rahman, A., & Hasan, M. (2022). From local to global: Networked activism against multinational extractivism. *Review of Communication, 22*(3), 231–255. https://doi.org/10.1080/15358593.2022.2107876

Ritchie, J. (2021). Movement from the margins to global recognition: Climate change activism by young people and in particular indigenous youth. *International Studies in Sociology of Education, 30*(1–2), 53–72. https://doi.org/10.1080/09620214.2020.1854830

Saeed, T. (2020). Reimagining education: Student movements and the possibility of a critical pedagogy and feminist praxis. UNESCO Digital Library. *UNESCO*, 1–13. https://unesdoc.unesco.org/ark:/48223/pf0000374157

Sainz, G. M., & Hanna, A. (2023). Youth digital activism, social media and human rights education: The Fridays for Future movement. *Human Rights Education Review, 6*(1), 116–136. https://doi.org/10.7577/hrer.4958

Salinas, I., Guerrero, G., Satlov, M., & Hidalgo, P. (2022). Climate change in Chile's school science curriculum. *Sustainability, 14*(22), Article 22. https://doi.org/10.3390/su142215212

Schlosberg, D., & Collins, L. B. (2014). From environmental to climate justice: Climate change and the discourse of environmental justice. *WIREs Climate Change, 5*(3), 359–374. https://doi.org/10.1002/wcc.275

Stein, S., Andreotti, V., Suša, R., Ahenakew, C., & Čajková, T. (2020). From "education for sustainable development" to "education for the end of the world as we know it." *Educational Philosophy and Theory, 54*(3), 1–14. https://doi.org/10.1080/00131857.2020.1835646

Stein, S., Hunt, D., Suša, R., & de Oliveira Andreotti, V. (2017). The educational challenge of unraveling the fantasies of ontological security. *Diaspora, Indigenous, and Minority Education, 11*(2), 69–79. https://doi.org/10.1080/15595692.2017.1291501

Suša, R., da Barra, M. T., & Stein, S. (2023). Beyond colonial futurities in climate education. *Teaching in Higher Education*, *28*(5), 987–1004. https://www.academia.edu/104471068/Beyond_colonial_futurities_in_climate_education

Tannock, S. (2020). The oil industry in our schools: From Petro Pete to science capital in the age of climate crisis. *Environmental Education Research*, *26*(4), 474–490. https://doi.org/10.1080/13504622.2020.1724891

Tannock, S. (2021). *Educating for radical social transformation in the climate crisis*. Palgrave Macmillan.

Theodorou, E., Spyrou, S., & Christou, G. (2023). The future is now from before: Youth climate activism and intergenerational justice. *Journal of Childhood Studies*, *48*(1), 59–72. https://doi.org/10.18357/jcs202320466

Thew, H. (2018). Youth participation and agency in the United Nations Framework Convention on Climate Change. *International Environmental Agreements: Politics, Law and Economics*, *18*(3), 369–389. https://doi.org/10.1007/s10784-018-9392-2

Thunberg, G. (2018, November 26). I'm striking from school to protest inaction on climate change—You should too. *The Guardian*. https://www.theguardian.com/commentisfree/2018/nov/26/im-striking-from-school-for-climate-change-too-save-the-world-australians-students-should-too

Tisdall, E. K. M., & Cuevas-Parra, P. (2022). Beyond the familiar challenges for children and young people's participation rights: The potential of activism. *International Journal of Human Rights*, *26*(5), 792–810. https://doi.org/10.1080/13642987.2021.1968377

Tlostanova, M. V., & Mignolo, W. D. (2012). *Learning to unlearn*. Ohio State University Press.

Tomren, T. S. (2022). Climate strikes and curricula: Insights from Norway. *Journal of Teacher Education for Sustainability*, *24*(1), 105–115.

Trott, C. D. (n.d.). Envisioning action-oriented and justice-driven climate change education: Insights from youth climate justice activists. *Children & Society*. https://doi.org/10.1111/chso.12846

Trott, C. D., Lam, S., Roncker, J., Gray, E.-S., Courtney, R. H., & Even, T. L. (2023). Justice in climate change education: A systematic review. *Environmental Education Research*, *29*(11), 1–38. https://doi.org/10.1080/13504622.2023.2181265

UNESCO. (2023). *The declaration on the common agenda for education and climate change at COP28*. https://www.unesco.org/en/articles/declaration-common-agenda-education-and-climate-change-cop28

UNESCO. (2024). *Climate change and sustainability in science and social science secondary school curricula*. https://www.unesco.org/en/articles/climate-change-and-sustainability-science-and-social-science-secondary-school-curricula

Vamvalis, M. (2023). "We're fighting for our lives": Centering affective, collective and systemic approaches to climate justice education as a youth mental health imperative. *Research in Education*, *117*(1), 88–112. https://doi.org/10.1177/00345237231160090

Variyan, G., & Gobby, B. (2022). "The least we could do?": Troubling school leaders' responses to the school strikes for climate in Australia. *Journal of Educational Administration and History*, *56*(4), 1–17. https://doi.org/10.1080/00220620.2022.2153110

Verlie, B., & Flynn, A. (2022). School strike for climate: A reckoning for education. *Australian Journal of Environmental Education*, *38*(1), 1–12. https://doi.org/10.1017/aee.2022.5

Wahren, J. (2021). *Youth activism and climate change in Latin America: Indigenous and peasant youth in defence of their human rights and territories*. Global Campus of Human Rights. https://repository.gchumanrights.org/handle/20.500.11825/2334

White, P. J., Ferguson, J. P., Smith, N. O., & Carre, H. O. (2022). School strikers enacting politics for climate justice: Daring to think differently about education. *Australian Journal of Environmental Education*, *38*(1), 26–39. https://doi.org/10.1017/aee.2021.24

Whyte, K. (2014). Indigenous women, climate change impacts, and collective action. *Hypatia*, *29*(3), 599–616.

Wilmers, A. (2023). From the global to the school level: Connections and contradictions between Fridays for Future and the school context. *Journal of Sustainability Education*, *28*. https://www.susted.com/wordpress/wp-content/uploads/2023/03/Wilmers-JSE-General-March-2023-PDF.pdf

CHAPTER 36

Enacting Transboundary Hydrocommons as a Pathway to Climate Action

AINHOA MONTOYA

Introduction

AVAILABLE predictions estimate that water and hydrological cycles may be significantly impacted by climate change. The 2022 Intergovernmental Panel on Climate Change (IPCC) report dedicated a full chapter to water, indicating that increases in temperature will very likely impact water precipitation and runoff, the quality of water as well as the frequency and severity of extreme weather events resulting in floods, hurricanes, and tsunamis as well as droughts (IPCC, 2022; see also Palmer et al., 2008, UN-Water Expert Group on Water and Climate Change, 2019). Climate factors may therefore affect the temporal and spatial availability of water. These factors may also impact freshwater biodiversity and the biodiverse territories through which water flows (Albert et al., 2021). Meanwhile, 3.6 billion people are estimated to currently inhabit areas affected by water scarcity at least 1 month per year—a figure which could potentially increase up to 5.7 billion by 2050, according to modeling studies (UN-Water Expert Group on Water and Climate Change, 2019, p. 11). The overall fresh water available on Earth has not varied for billions of years, but population growth, increasing human water use, water quality deterioration, and climate-induced water variability will exacerbate water scarcity in some regions and inequities in water distribution overall (McCaffrey, 2019, pp. 5–9, 25). Scholars have thus highlighted the existing threats of tipping points in freshwater ecosystems and their vital hydrological cycles (Hernández Martínez de la Riva et al., 2023, p. 2).

There is global pressure to devise and implement measures that allow addressing current and future climate-driven water variability. Yet such endeavors become particularly challenging when the affected regions are crossed by transboundary water bodies (Cooley & Gleick, 2011; De Stefano et al., 2012, p. 195; Draper & Kundell, 2007). Transboundary river basins or watersheds are shared by two or more nations, thereby transcending international borders. They constitute a hydrological ecosystem composed of surface and underground waters which cut across geopolitical boundaries and flow toward the same finishing point, though the latter trait has been subject to some contention (McCaffrey, 2019, pp. 45–46). In some cases, such as the Amazon River basin, they are also constituted by so-called atmospheric or flying rivers (i.e., corridors or streams of water vapor or moisture that travels from tropical regions toward the poles). The waters of transboundary watersheds such as the Amazon, the Colorado, or the Mekong Rivers flow through geopolitical boundaries, national jurisdictions, and the sovereign territories of different peoples and nations, making it challenging to foster concerted and mutually beneficial climate action by all those concerned and affected.

Some scholars have highlighted the potential for future conflict among co-riparians with regard to transboundary watersheds in the light of climate change–related predictions (De Stefano et al., 2012; Michel et al., 2021). Upstream and downstream co-riparians have often different development agendas and strategic interests, as well as distinct relationships to water bodies, and are unequally affected by environmental impacts on them. Indeed, their interdependency as co-riparians raises the issue of lack of sovereignty over shared water bodies by those located downstream and suffering the impacts (pollution, water extraction, flood control, and so on) of activity in the upper streams of the watershed (see Mason & Khawlie, 2016; Montoya, 2021a, 2024a). It is, however, the case that consequences of activity downstream may be felt upstream as beings like fish may move against the flow, and aquifers and the direction of their flows do not necessarily overlap with surface waters (McCaffrey, 2019, pp. 40–41). Relationships among co-riparians may thus be conflictive and potentially intensify with the water irregularity that is likely to be exacerbated by climate change, even as there is evidence that cooperation tends to prevail over conflict in transboundary watersheds (Schmeier, 2010). Notwithstanding such predictions, there is a lack of climate-proofing initiatives in transboundary watersheds. For instance, most international water treaties do not include provisions for how to deal with climate-induced water variability (Cooley & Gleick, 2011; De Stefano et al., 2012), nor does international water law that regulates the competing interests of co-riparians.

Through political-legal initiatives that have emerged in Central America to govern the Lempa transboundary watershed, this chapter explores the prospect of local, collaborative cross-border environmental action as a pathway to mitigate and manage foreseeable climate change impacts and guarantee the reproduction of the watershed. Two initiatives have stood out: (1) a citizen-led proposal of a transboundary treaty for the management of shared waters (see Montoya, 2021a, 2021b, 2024a); and (2) an institutional initiative that takes the form of the *mancomunidad* (literally translated as "association of communities" or "commonwealth"), an administrative formula that helps coordinate the action of municipalities located within the transboundary watershed but on opposite sides of geopolitical borders. In what follows, I explore the outcomes of enacting transboundary watersheds as "transboundary commons."[1] In this regard, I draw on these two initiatives I have engaged through my own research which contribute to devising forms of cross-border collaborative climate action from which we may learn and replicate methodologies for the governance of water bodies and potentially of other kinds of transboundary commons.

The Challenges of Transboundary Watersheds

The notion of the "river basin" or "watershed" denotes the area drained by a river and its tributaries. Its consideration as the optimal geographic or territorial unit for hydrological planning or management endeavors has early roots in regions such as the Middle East and China, yet it dates back to the 18th century in the West and has since evolved as well as oscillated in importance over time (Molle, 2006, 2009; Teclaff, 1996). Some scholars have traced it back to 18th-century French-led efforts of geographic legibility (Carse, 2014, p. 49). It was not, however, until the early 20th century that integral basin planning was introduced, and it gained momentum after the first quarter of the century (Molle, 2006; Molle & Wester, 2009, p. 1; Teclaff, 1967, p. 1). The notion of the river basin thus emerged as a modernist one premised on the possibility of isolating, taming, and intervening in nature through science, technology, and the development of hydraulic infrastructure (Molle & Wester, 2009, p. 4). In the 1990s, it became established as a fundamental unit for ecosystem planning and management (Omernik & Bailey, 1997; Schlager & Blomquist, 2008, p. 4).

The watershed is conceived as a unit defined by nature and biophysical processes as self-contained, with its climate, soil, topography, and biota constituting an ecological whole (see Teclaff, 1967). Yet its boundaries are relatively arbitrary and unstable; many watershed phenomena and impacts exceed the territory delineated as constituting the watershed or transform it over time, and the watershed may be home to diverse biotopes along its different stretches (Molle & Wester, 2009, pp. 2, 15). Indeed, it is as much an ecological unit as it is a cultural and political one. It is a particular sociocultural construct that serves certain purposes, such as the governance and protection of water. The watershed can indeed work at once as "a legal reality, a geohydrological fiction, and a sociopolitical imaginary" (Carse, 2018). However, there is no questioning that, as an interdependent and connected ecological whole, interventions in any single area of the watershed may have very material and tangible repercussions in others.

Currently, there are over 300 transboundary watersheds, covering about 45% of the planet's territory and being home to roughly 40% of the global population (Mirumachi, 2017, p. 1; Schulze, 2012, p. 49). This number would be even higher if we were to include transboundary aquifers, such as the Guarani.[2] The number of transboundary watersheds increased in the second half of the 20th century as a result of the independence of various African nations and the breakup of the Soviet Union and Eastern Europe, as well as the increasing sophistication of mapping technologies (Abu-Zeid, 2001, p. 26; Cooley & Gleick, 2011, p. 712). Moreover, transboundary watersheds, including both surface and underground waters, amount to 60% of the planet's freshwater flows (United Nations & UNESCO, 2021, p. 1). Such a figure offers a sense not only of the sizeable portion of the globe's freshwater flows that transboundary watersheds represent but also of their importance for sustaining life on the planet.

Considering the relevance of fresh water that is shared across borders and jurisdictions and with the goal of precluding the escalation of conflict over basin control which was prevalent throughout history, conventions, multinational treaties, and other agreements have been devised to govern the navigational and non-navigational uses of shared water bodies. Additional related incentives for transboundary governance have been to enact water diplomacy through stable institutionalized cooperation mechanisms (Schmeier, 2013, 2018) or to cover knowledge gaps on transboundary ecosystems (Milman et al., 2020). The first known legal developments sought to establish multinational agreements on navigational uses, one of river basins' main uses throughout history in addition to irrigation (Cooley & Gleick, 2011, p. 713; Teclaff, 1967, pp. 58–63). Transboundary river basins serve as key transport arteries that have enabled international commerce since antiquity, even though levies and limitations were imposed from the Middle Ages through the 16th century, and it was not until the late 18th century that freedom of navigation extended to basins as integrated wholes (Teclaff, 1967, pp. 56–58, 60). The regulation of navigational uses of river basins then emerged in the beginning of the 19th century, confirming the principle of freedom of navigation and initially prioritizing navigation over other uses (Salman, 2007, pp. 625–627).

With the expansion of industrialization, however, non-navigational uses of rivers gained relevance and from 1921 European nations adopted agreements that reflected this development (Salman, 2007, p. 626). The regulation of non-navigational uses of river basins or watersheds is, however, far more recent and responds to their prioritization over navigational ones in the second half of the 20th century. Non-navigational uses are regulated through a combination of rules and resolutions (customary law, which is not legally binding) and conventions and treaties (legally binding for signatory nations). All of them are founded on the principle of limited territorial sovereignty by which every co-riparian has the right to use the waters that flow through their portion of the watershed's territory and the duty not to cause significant harm to other co-riparians (Salman, 2007, p. 628). Customary law reflected in the 1966 Helsinki Rules on the Uses of the Waters of International Rivers (henceforth the Helsinki Rules) established the principle of reasonable and equitable water use by co-riparian nation-states which has served as the groundwork for further codification (Stoa, 2014, pp. 1326–1327). Although the Helsinki Rules include "rivers" in their name, the text extends their scope of application to the territory of the "international drainage basin." Even as they are not binding, this set of rules constituted an authoritative reference until the UN Convention was ratified.

Customary international water law continued developing through the decades building upon the Helsinki Rules, and these also influenced the only attempt at codifying such law thus far (Salman, 2007, pp. 630–631). The 1997 United Nations Convention on the Law of the Non-Navigational Uses of International Watercourses (henceforth the UN Convention), the only binding instrument to date, was negotiated over 23 years and was not ratified and did not enter into effect until its full ratification in 2014 (McCaffrey, 2019, pp. 409–442; Salman, 2007, p. 625). Although the UN Convention utilizes the term "international watercourse," it defines it in such a systemic way that encompasses the territory of the international drainage basin as do the Helsinki Rules, including as these did interconnected surface water and groundwater. The UN Convention codified the principle of reasonable and equitable water use, and just as the Helsinki Rules, it prioritized it over the duty not to cause significant harm to other co-riparians (Salman, 2007, p. 634). Another important source of regulation is bilateral and multilateral treaties signed by co-riparians that agree on watershed governance, with more than 300 such treaties signed thus far (Cooley & Gleick, 2011, p. 712). Their focus varies depending on the concerns of co-riparians, and not all of them include all co-riparians.

These soft and hard law instruments are still insufficient in many respects. For instance, they lack provisions specific to connected groundwaters even if they acknowledge these to be part of transboundary watersheds, and are not inclusive of confined groundwater or aquifers surrounded by impermeable materials (McCaffrey, 2019, pp. 43–44). This is a crucial gap, given that groundwater represents 97% of the planet's fresh water, and surface water is rapidly decreasing and deteriorating in quality. They also lack any mention of atmospheric rivers that are integral to watersheds and often foster crucial inter-basin connectivity and regional or even planetary stability. Moreover, these instruments do not include specific climate-related provisions. Climate change does not feature in their texts. The Helsinki Rules prioritize current over future uses. While the UN Convention refers to the need to consider future generations in the non-navigational uses of transboundary waters, neither the UN Convention nor customary law makes any mention of climate change. This lacuna is reflected in international water law's neglect of provisions and mechanisms to deal with climate change–related impacts, including increasing water variability such as flooding and droughts that result from extreme weather events (Cooley & Gleick, 2011). The same happens with bilateral and multilateral treaties. Crucially, neither international water law nor bilateral and multilateral agreements include elements of watershed protection, and the latter focus predominantly on concerns pertaining to hydropower production and water allocation (Cooley & Gleick, 2011, p. 713).

Some scholars have highlighted the potential, if adequately harnessed, of institutional arrangements such as river basin organizations or climate policy instruments to climate-proof watershed governance (Blumstein & Petersen-Pelman, 2021; Schmeier & Schulze, 2010). Public participation, other scholars have argued, is critical to make state-negotiated agreements effective on the ground (Bruch et al., 2005). Next, I take the argument further to suggest that to date some of the most interesting and effective initiatives of cross-border watershed governance that have a potential to reverberate in climate action are locally designed and led by citizens, civil society organizations and nongovernmental organizations (NGOs), or local institutions. As I have suggested elsewhere building upon a critique of environmental management, dwelling and being embedded within a territory in a quotidian way allow the interrelationality that characterizes a territory like a watershed to flourish (see Montoya, 2024a). Local public-led initiatives thus have a potential to transform watershed governance based on the management of water-as-a-resource into an approach underpinned by care and protection which involves altogether a different kind of sensibility toward water and the territory of the watershed—a sensibility that only those who are embedded in the watershed can cultivate. The next section expounds how locally elicited imaginaries of transboundary watersheds as commons can foster the commoning and caring practice that is necessary to cooperatively address climate change in the territory of a watershed.

Transboundary Watersheds as Commons

The full waterscape of watersheds—which includes surface water and groundwater—is not easily apprehended and thus represents a challenge to understanding impacts well beyond the immediate territory and to building commoning action that exceeds it and considers the integral biophysical processes of the watershed. For instance, watery interconnections through underground waters within and among river basins are difficult to map out and include (Teclaff, 1967, pp. 9–10). Little is known yet today about groundwater despite being the most extracted water source and even natural resource (Eckstein, 2005; Eckstein & Eckstein, 2005). Floods may also temporarily connect water basins and transfer flows between them. Meanwhile, atmospheric or flying rivers are not even mentioned in international water law or governance, even though they are critical for planetary stability and climate regulation and may connect different basins, as is the case of the atmospheric flows traveling from the Amazon to La Plata in South America (Beveridge et al., 2024). Yet connectivity and interdependence are not only characteristic of watersheds or of the relationship among different river systems but also of the relationships among different kinds of bodies, human and more-than-human, inhabiting such watersheds.

Living beings are aqueous bodies. We are all made of water in significant portions that vary depending on species, gender, and age. Water enters and leaves our bodies, circulating among different kinds of entities, human, nonhuman, and nonliving (e.g. hydraulic infrastructure), connecting us all through a form of "watery kinship" (Neimanis, 2024, p. xxiv). We are *of* and *with* the rivers, aquifers, streams, and other bodies of water that exist or flow within the territories we inhabit (Diaz, 2024). This image of a world made of watery bodies is particularly effective and powerful in disrupting the modernist binaries of human/nonhuman or culture/nature and bringing out their actual entanglements; even as we are reminded by decolonial feminist scholarship of the constant need to interrogate the homogenizing "we" yielded by such an image of a watery world and a "more-than-human hydrocommons" (Neimanis, 2024). In this "hydrocommons," water is in constant circulation, and it is in that circulation that it produces difference as much as sameness (Neimanis, 2024, pp. xxv–xxvi). No water body is equal to another; nor does it have necessarily the same political standing as any other (Chen et al., 2013b, p. 7). Yet this singularity does not preclude being, thinking, and acting in common (Amin & Howell, 2016).

The notion of the commons and its lexicon variants have been employed with various senses in mind (Amin & Howell, 2016; Lloredo Alix, 2020). Among these variants, recent approaches have suggested we do not think of commons as things that have intrinsic qualities that make them apt as resources managed in common (common pool resources), but as practices or processes of commoning (being in common), that is, as forms of being in the world that are not limited to owning in common (Bollier & Helfrich, 2020). Amin and Howell (2016) underline how the precarity of the planet which we inhabit has brought to the fore notions of the commons that transcend different kinds of boundaries. Yet they go on to suggest that addressing the planet or certain regions' precarity requires rethinking how to reclaim the commons in ways that do not erase singularity by drawing on a notion of community that homogenizes and excludes. Holding in common rather than owning, imagining ourselves in common, and thinking about our common responsibilities, without presuming a shared or homogenous identity, is the way to go. "Whatever 'common ground' there is, it cannot be fixed or settled, but is always-emerging, though care, collaboration, and altruism remain the fixture" (Amin & Howell, 2016, p. 12).

In the last few decades, various efforts have emerged to rethink governance in an era of human-induced climate change or the so-called Anthropocene, especially in ways that are reflective of the need to think beyond boundaries. In the latter quarter of the 20th century, a "global commons" approach emerged with a focus on transboundary spheres and regions such as river basins, understanding these as large-scale common pool resources that transcend geopolitical and jurisdictional borders and whose effective governance requires international cooperation (Ostrom et al., 1999; UNESCO, 1980). Some scholars have more recently highlighted the limitations of this concept for the Anthropocene era in which we currently live. The concept of "planetary commons," they argue, is much more attuned to the task ahead of rethinking governance and law in ways that consider how some of those transnational and multi-jurisdictional regions or spheres perform critical regulatory functions for the stability of the planet and therefore their continuity and reproduction into the future is a planetary and systemic concern rather than a regional one (Rockström et al., 2024). This approach, however, is yet to develop what planetary or earth system governance would look like in this light, matching with earth system science's knowledge about the potential of multiple tipping points ahead. Importantly, a planetary commons approach is not free of the limitations and challenges that national sovereignty poses to any form of transboundary governance and requires an important degree of reimagination and reconfiguration.

Another related legal concept is that of "common concerns of humankind" (Biermann, 1996).[3] This concept was employed in the 1990s to juridically address global concerns such as the atmosphere and the need to protect it. Against the notion of "shared resource" that has most commonly underlaid international water law to regulate the exploitation of water by co-riparians, a common concern of humankind acknowledged a particular matter or sphere as critical to sustaining life on earth and therefore "essential for the survival of humankind" (Biermann, 1996, p. 431). This notion would be well suited to a matter like water. A critical aspect to further develop this notion is the need to restrict or redefine state sovereignty (Biermann, 1996, p. 432). Nonetheless, there are yet other aspects that need to be considered, including the need to decommodify water and consider life beyond humanity. Water can no longer be regarded as a resource, given the vital matter it is. Yet water is not just vital for the survival of humankind but for all life on Earth as well as for the hydrological cycles that contribute to the regulation of the planet's climate. The juridical concept of "common concern" might be a way forward to firm such an approach in law insofar as it is extended and conceived as a concern of all forms of life as well as vital for the stability of the planet into the future.

Given the status of water as shared matter that has overwhelmingly made cooperation prevail over conflict and whose qualities allow it to connect different kinds of beings and issues in a political community of sorts (a hydrocommons), it is worth exploring the potential of collaborative initiatives that seek to protect transboundary watersheds to inspire, inform, and stimulate the cross-border collaborations that are critical for climate action and transformation. Water promotes inclusivity in a more-than-human community and raises important questions around the political possibilities of such a community, hence the relevance of joining efforts of "thinking with water" by which water is considered as both metaphor and matter (Chen et al., 2013a). In contrast to approaches to watershed management which retain the underlying consideration of water as a resource or commodity, the initiatives I describe in the following section focus on the protection of the integrity of the watershed, its well-being and its future reproduction. There is thus a potential in these initiatives that enact transboundary commons not only for human collaborations and alliances across geopolitical borders but also for collaborations and alliances among human and other-than-human water bodies (Montoya, 2024a). However, moving away from a perspective of water-as-a-resource toward one of water-as-vital-matter and water-as-relational-matter is critical for these initiatives to be able to foster such transboundary commons.

Transboundary Collaborations as Pathways to Climate Action: The Case of Central America

Water has long been an issue of concern in some areas of Central America. While availability of water in the region is high, it is geographically and temporally unevenly distributed and records substantial differences between the Caribbean and Pacific coasts (CEPAL, 2011, p. 94). Surface water and aquifers are very abundant in the Caribbean coast, whereas the Pacific coast lacks water yet concentrates much of the population of the region and has thus the highest demand for water. Central America relies heavily on rainfalls for the replenishment of its water bodies. While the Caribbean coast registers elevated records of rainfall throughout the year, the Pacific coast experiences an extended dry season that can last about half of the year. Moreover, the seasonality and volume of rainfalls vary annually, with the Pacific coast facing some years of extended periods of droughts. This is especially so in the so-called Dry Corridor, a region beginning in the south of Mexico and cutting across Central America down to Panama, characterized by long dry seasons and severe droughts during the wet season (Díaz Bolaños, 2019, p. 300). El Salvador is located on the Pacific coast and is transected by the Dry Corridor, which explains why many areas of the country risk being water stressed. Yet availability of quality water is also complicated by the fact that 54% of it is used by the country's agro-industries, and that the country has the shortest and less water-abundant rivers in the region which flow toward the Pacific (CEPAL, 2011, pp. 95–96).

El Salvador is water dependent. It is located downstream in the transboundary Lempa River basin shared with Guatemala and Honduras and on which it relies heavily for domestic and industrial water consumption (Montoya, 2021a, pp. 300–301; Montoya, 2024a, p. 21). Being located on its lower streams entails that El Salvador is affected by the management of the upper streams of the basin by Guatemala and Honduras. The Lempa River basin contains the Trifinio region within its territory. The Trifinio covers the trinational forest area of the Montecristo mountain massif that is the cradle of some of the most important water basins of the region. It was declared a biosphere reserve in 1987 and a UNESCO World Heritage site in 2011 (Celata et al., 2013, p. 171). While serving as a unique instance of regional integration in Latin America set up in the 1980s to promote the integral and transnational governance of the Trifinio transboundary bioregion, it also has deficits in areas such as limited municipal devolution or limited regional integration (Celata et al., 2013, p. 183). Nor has the integral governance of the Lempa been the focus of the Trifinio cooperation, even though the bioregion sits within the territory of the basin (Artiga, 2003). In the new millennium, however, Salvadorans began to move from concerns about water to those pertaining to the governance of integral river basins. Both citizen-led and institutional initiatives have been developed beyond geopolitical borders, evidencing a focus on the reproduction of water bodies and not simply on human water consumption. In what follows, I describe two such initiatives.[4]

Envisioning Transboundary Cooperation: A Citizen Draft of a Regional Treaty

Since the mid-2000s, Salvadoran citizens participating in NGOs and civil society organizations engaged in environmental issues commenced to raise their concerns about transboundary waters. At the time, the environmental campaigning of Salvadoran NGOs was focused on precluding large-scale industrial extraction of metals in the country due to the imminent exploitation of a gold mine by a Canadian corporation (see Bebbington et al., 2019; Montoya, 2021a; Nadelman, 2015; Spalding, 2018). Yet members of the Centro de Investigación sobre Inversión y Comercio (Centre for Research into Investment and Trade, CEICOM)—a research-focused NGO with offices in a region of El Salvador bordering Guatemala and Honduras which encompasses the Salvadoran territory of the trinational biodiverse Trifinio region—began to call attention to extractive activities, such as mining and agroindustry, that were being developed or had received the permissions to begin their operations in the Guatemalan or Honduran territory of the river basin. Yet, as some argued, Guatemalan and Honduran state officials could not be approached about the impact of their extractive activities on shared territory and waters if El Salvador did not take the steps to enact laws that prioritized the protection of water over extractive ventures in its own territory (see Montoya, 2021a, p. 298).

Nonetheless, organizations like CEICOM considered both steps could and should be taken in tandem. In 2014, they hired an international legal expert to help them translate their ideas for how shared waters should be governed into a multilateral treaty draft—one that would offer a proposal for how to regulate cooperation among Guatemala, Honduras, and El Salvador on shared river basins. Building on similar soft legal instruments that have been devised in other regions among co-riparian states to govern transboundary river basins, they decided to create their own to then lobby the governments of the three countries with it. The treaty draft was the outcome of workshops and discussions with local populations who inhabit the watershed and experience an increasing everyday lack of quality water but also see their surrounding water bodies become ever drier. Likewise, they involved civil society organizations from the three countries, taking advantage of the transnational networks in which Central American NGOs have participated in their environmental campaigning and advocacy through the leading work in building regional linkages of international organizations such as Oxfam América, Friends of the Earth, or Cáritas (see Montoya, 2024b). They together formed the Central American Network for the Defense of Transboundary Waters (Red Centroamericana por la Defensa de las Aguas Transfronterizas, RedCAT), a platform to promote the protection of shared river basins in common.

As I have explained elsewhere, the treaty draft was completed in 2015 and functioned as a "legal prototype," that is, an open-ended model for how the binational and trinational river basins that El Salvador shares with Guatemala and Honduras could be governed cooperatively (Montoya, 2024a).[5] They then shared it with state officials of all three countries to begin the discussion of what such a legal instrument should be like and its political possibilities.

They approached members of the Central American Parliament and the Foreign Office of all three countries. El Salvador was the country in which the treaty had a more active political life. While the left-wing Farabundo Marti National Liberation Front (FMLN) was in office (2009–2019), the legal prototype of the treaty was considered and discussed by a range of ministries, with amended versions circulating through their offices, and moving back and forth between civil society organizations and the ministries; yet never materializing as law. Since the incumbent president Nayib Bukele took office in 2019, the social life of the treaty was abruptly discontinued. Besides, after the pandemic, CEICOM lost its funding from external sources and was dissolved as an NGO. Their members have since continued to promote the ethos of the legal prototype, though with much less resource and support, and other citizen initiatives with related goals that had been active since before the pandemic have continued to promote transboundary commoning initiatives (see Montoya, 2021b).

The recourse to lawmaking by CEICOM and allied NGOs and civil society organizations was not incidental. Salvadoran citizens involved in environmental campaigning have sought to contribute to lawmaking in various ways in the last two decades (see Montoya, 2021a, 2023, 2024a, 2024b) and so have their counterparts in Honduras by, for instance, drafting a mining law (Montoya et al., 2020). Lawmaking has allowed them, among other things, to introduce their concerns in official political agendas and to inflict law with their own perspectives of environmental governance, including the perspectives of local populations that inhabit the territories in question (Montoya, 2024a, 2024b). In the draft of the legal treaty, they have thus foregrounded "the welfare of transboundary waters and transboundary populations" as interdependent and inextricable. Central to their treaty draft is also the incorporation of citizen decision making and custodianship of the watershed, aspects that are absent from international water law and multilateral agreements governing transboundary river basins. There is a paucity of instances of citizen participation in the governance of transboundary watersheds, and indeed international water law does not incorporate mention of it and is limited to interstate negotiations, agreements, and rights and duties (Mostert, 2003; Seqat, 2022). Of course, manifold processes in Latin America and beyond have sought to increase and prioritize citizen participation in decision making on environmental issues before and especially after its institutionalization at the 1992 Conference on Environment and Development in Rio and through its resulting declaration, and the latter offered an important foundation for the proposals incorporated in the treaty.

While seemingly focused on transboundary rivers basins, protecting water and life, and curbing industrial activities with profound environmental liabilities for the shared bioregion, CEICOM's and RedCAT's development of the treaty and lobby activities with it are the manifestation of an environmental agenda concerned with the impacts of climate change in Central America. Their concern with climate change does not emerge explicitly in the treaty draft. It does, however, in related documents such as the submission that the RedCAT prepared together with Franciscans International for the 2019 Universal Periodic Review of El Salvador[6]—a United Nations mechanism that allows examining the human rights performance of states. They acknowledged in it that impacts on the quality and quantity of transboundary waters might be affected by the broader problem of climate change, in addition to political decision making and regional environmental governance. Members of CEICOM have also been active in other environmental campaigns and platforms such as the Mesa Nacional Frente a la Minería Metálica (National Roundtable Against Mining) and the Mesa de Cambio Climático de El Salvador (El Salvador's Climate Change Roundtable). Indeed, prior research has pointed out that nongovernmental climate action in El Salvador is articulated through campaigning on specific issues, such as the human right to water, food security and sovereignty, or risk management, rather than outright as climate action (Cuéllar et al., 2021, p. 27). CEICOM's and RedCAT's initiatives on the regional governance of shared watersheds have thus been attempts at making proposals for how water, a matter largely affected by climate variability, may be addressed regionally irrespective of the boundaries that are politically and historically erected yet transcended by environmental phenomena.

A Transboundary Commonwealth: Developing a *Mancomunidad*

An initiative of a different nature but with a related goal is that of the Mancomunidad Trinacional Fronteriza Río Lempa (Trinational Border Community Río Lempa). A *mancomunidad* is an association of municipalities for the purposes of managing or providing a service in a cooperative manner. Historical work shows that this supramunicipal administrative figure has its roots in 19th-century Spanish public law that sought to overcome the limitations of municipalities due to their limited scale in order to more efficiently provide municipal services and works or to pursue similar goals of common concern (Quintana López, 1990). *Mancomunidades* have continued to exist into the present and have been actively promoted through processes of devolution that have delegated important political functions to municipalities. They have proliferated in Spain and other European countries as well as in Latin America, where European donors have contributed to the development of manifold associations of this kind throughout the Latin American region (Celata et al., 2013; see also Enríquez Villacorta, 2008; Gómez Navas, 2003).

The Mancomunidad Trinacional Fronteriza Río Lempa was founded in 2007 and began to operate in 2009. It was created as an association of municipalities from El Salvador, Guatemala, and Honduras that are located within the territory of the upper Lempa watershed. It emerged out of the will of mayors from various municipalities of these three countries to join efforts to foster sustainable development within the integral territory of the watershed as well as to promote Central America's political and institutional integration—efforts that they considered to be inadequately carried out by the Trifinio Plan that governs part of the upper Lempa territory. As explained by Héctor, the manager of the Mancomunidad, the Trifinio Plan is an interstate body with centralized decision making regarding issues that directly concern the municipalities.[7] It leaves little space for the participation of local institutions. Nor was the activity of Plan Trifinio at the time focused enough on protecting the integrity of the watershed territory even as some projects have more recently strived to guarantee access to water by local populations. This is paradoxical, given that the Plan Trifinio's motto is "water without borders."

Héctor recounted that the Mancomunidad emerged out of a shared concern with the Lempa watershed. "In 2003, 2004, the mayors of that territory began to think about the possibility of having an entity that would represent them as local governments, that would establish their priorities, their agendas, where they would make decisions and where they would be the architects of development. It is in that context that it [the *mancomunidad*] arises […]. The main issue that motivated the interest of the mayors in creating this *mancomunidad* is exactly what is going to happen with the water of the future." He went on to further expound this point: "Under this principle of intergenerational responsibility, they say, let's create an organization, firstly, that unites us mayors; secondly, that allows us to work together to promote the development of this territory; and thirdly, that puts on our permanent agenda the importance of caring for a shared resource such as water and, in particular, the Lempa River."[8] Water, and more specifically the water body of the Lempa, has been the driving force to act in common and promote sustainable development within the integral territory of the watershed.

With this goal in mind, the *mancomunidad* has recently contributed to form and lead a wide network of more than 30 entities, including civil society organizations, NGOs, grassroots movements, and local institutions. Founded in 2020, it has been named Red Trinacional por el Rescate del Río Lempa (Trinational Network for the Rescue of the Lempa River) and is oriented toward raising awareness of the systemic value of the Lempa ecosystem and the need to protect it from the pollution of industrial activities and agribusinesses.[9] In this vein, on June 28, 2024, representatives of the network presented and handed in to El Salvador's vice-president a draft law for the protection, conservation, and restoration of River Lempa. Members of the network had been working for months on drafting this legislative initiative, which was introduced to, and discussed with, member organizations on March 19, 2024, in San Salvador. As stated in the news item published by the *mancomunidad* on the formal handing of the law draft to state officials in June that year, the network's motivation to take this step was "to ensure a sustainable and healthy future for the river and future generations."[10] To the date of writing, the government has not adopted the law draft.

The *mancomunidad* considers the Lempa River basin as a shared territory, a commons of sorts, that transcends geopolitical, cultural, ideological, and generational boundaries. They deem water a resource but also a vital substance for life (human and nonhuman) to continue its reproduction into the future. Once again, as is the case with the citizen treaty draft and the efforts thereof, this institutional initiative is not articulated outright as being focused on climate change in its strategy and policies. Yet by acting to promote an agenda of sustainable development and having the reproduction of the watershed into the future at the heart of its activity, the *mancomunidad* is involved in climate action. In El Salvador, where environmental action has spread out through agendas promoting sustainable development, water protection and conservation, or food security and sovereignty, forms of climate action may not be articulated as such, yet be providing pathways for how to address climate change through different domains.

Conclusion

Water is critical to the stability of the planet's climate system, and climate change is in turn impacting significantly on the availability and quality of water, as well as on fresh water and river basin biodiversity. Interestingly, these changing conditions have not been reflected in the soft and hard international water law or in bilateral and multilateral agreements that regulate the uses of water yet represent important challenges ahead in the conflictive negotiations over transboundary bioregions such as river basins. More importantly, underlying international law and multilateral agreements is a utilitarian conception of water as a resource and a commodity, whereby water is considered matter that may be exploited, transacted, and even speculated with. This political-ethical perspective on water does not facilitate or nurture the kind of relationship with water necessary to promote the reproduction and flourishing of the watershed. Water, as vital matter that is shared by all forms of life and that creates connectivity in its circulation among beings, can, however, cultivate the sensibility that is required. Water conceived as a "common concern" not just of humanity but of life on Earth is a concept that is implicit in much environmental cross-border action at the local level.

While international water law and agreements are not responding thus far to climate-induced transformations, citizens and local institutions in various parts of the world are pursuing climate action or engaging in actions that can address the impacts of climate change. The two initiatives described in this chapter offer a sense of the affordances of transboundary initiatives that bring local actors together to care for the hydrological cycles that connect us all. Their focus on protecting transboundary watersheds, all the while through enacting them as commons, is a way of doing so. Both initiatives consider there is a need to inflect law and policy with notions of the watershed as a shared and integral territory in order to protect it cooperatively. The transboundary watershed, hence conceived as a transboundary commons that requires a joint practice of care or a practice of commoning by beings across boundaries to be reproduced into the future, is thus an imaginary that can foster the transboundary action necessary to address climate change.

Neither of the initiatives is articulated outright as climate action, yet both aim at anticipating and mitigating aspects of shared water bodies, hydrological cycles, and water variability that are largely the outcome of climate change or at least exacerbated by climate change. Both are initiatives that place the emphasis on the integral nature of river basins and hydrological cycles, on the interdependence of local populations and rivers, and on the future and the principle of intergenerational solidarity. In this vein, climate action at a local level (citizen-led or institutional) may take the form of subject-specific agendas that more or less directly seek to address different dimensions of climate change and should thus be included in analyses and discussions of climate action and future pathways and possibilities for it.

Acknowledgments

This book chapter has been written with the support of a Research Consolidation Grant, Award CNS2023-143561 co-funded by MCIN/AEI/10.13039/501100011033 and the European Union NextGenerationEU/PRTR. It has received further support from a Ramón y Cajal Senior Research Fellowship, Grant RYC2021-034772-I co-funded by MCIN/AEI/10.13039/501100011033 and the European Union NextGenerationEU/PRTR. More importantly, this publication has relied on long-lived exchanges and collaborations with various civil society organizations, NGOs, and institutions in El Salvador. I am particularly grateful to members of CEICOM, CRIPDES, ADES Santa Marta, and to the administrator of the Mancomunidad Trinacional Fronteriza Río Lempa. Finally, I would like to extend my gratitude to Paul Almeida for the invitation to contribute to this volume and be part of the important conversation it seeks to promote on climate action.

Notes

1. For a discussion of the notion of "transboundary commons," see Montoya (2024a).
2. Although not the focus of this contribution, transboundary glaciers also exist in Latin America's Southern Cone or the Eastern Himalayas and have similarly raised questions regarding boundary delimitation and governance.
3. I thank my colleague Karen Morrow for bringing that legal concept to my attention.
4. This section relies on data from *The Legal Cultures of the Subsoil Database* (Montoya et al., 2020), an open-access digital and bilingual (English/Spanish) research resource which maps out relevant legal and legal-like actions employed by a range of actors to assert fundamental rights in the context of socio-environmental conflicts over industrial mining in Central America and Mexico.
5. To access the draft of the treaty, visit https://legalculturessubsoil.ilcs.sas.ac.uk/legal-artefacts/2014-proposed-treaty-for-the-integrated-sustainable-management-of-crossborder-waters-focused-on-shared-watersheds-between-the-countries-of-guatemala-honduras-and-el-salvador-spanish/

6. Their submission is available at https://legalculturessubsoil.ilcs.sas.ac.uk/wp-content/uploads/2024/02/EPU-El-Salvador-Informe-Conjunto-FI-RedCAT-1_0.pdf
7. Interview with Héctor Aguirre, managing director of the Mancomunidad Trinacional Fronteriza Río Lempa, San Salvador, El Salvador, March 9, 2024.
8. Interview with Héctor Aguirre, managing director of the Mancomunidad Trinacional Fronteriza Río Lempa, San Salvador, El Salvador, March 9, 2024.
9. For details of the composition of the network, visit http://riolempa.org/miembros-de-la-red/
10. For further details on this action, see the news item: https://www.trinacionalriolempa.org/mtfrl/noticias/825-vicepresidente-de-el-salvador-dr-felix-ulloa-recibe-anteproyecto-de-ley-presentado-por-la-red-trinacional-por-el-rescate-del-rio-lempa-entidades-de-la-sociedad-civil-de-los-tres-paises-para-la-proteccion-conservacion-y-restauracion-del-rio-lempa

References

Abu-Zeid, K. (2001). International water law from Helsinki Rules to the United Nations Convention on the Law of the Non-Navigational Uses of International Watercourses. *Water Resources Impact*, 3(4), 26–31.

Albert, J. S., Destouni, G., Duke-Sylvester, S. M., Magurran, A. E., Oberdorff, T., Reis, R. E., Winemiller, K. O., & Ripple, W. J. (2021). Scientists' warning to humanity on the freshwater biodiversity crisis. *Ambio*, *50*, 85–94. https://doi.org/10.1007/s13280-020-01318-8

Amin, A., & Howell, P. (2016). Thinking the commons. In A. Amin & P. Howell (Eds.), *Releasing the commons: Rethinking the futures of the commons* (pp. 1–17). Routledge.

Artiga, R. (2003). *The case of the Trifinio Plan in the Upper Lempa: Opportunities and challenges for the shared management of Central American transnational basins.* UNESCO. http://unesdoc.unesco.org/images/0013/001333/133304e.pdf

Bebbington, A., Fash, B., & Rogan, J. (2019). Socio-environmental conflict, political settlements, and mining governance: A cross-border comparison, El Salvador and Honduras. *Latin American Perspectives*, *46*, 84–106. https://doi.org/10.1177/0094582X18813567

Beveridge, C. F., Espinoza, J., Athayde, S., Correa, S. B., Couto, T. B. A., Heilpern, S. A., Jenkins, C. N., Piland, N. C., Utsunomiya, R., Wongchuig, S., & Anderson, E. P. (2024). The Andes–Amazon–Atlantic pathway: A foundational hydroclimate system for social–ecological system sustainability. *Proceedings of the National Academy of Sciences (PNAS)*, *121*(22), 1–10. https://doi.org/10.1073/pnas.2306229121

Biermann, F. (1996). "Common concern of humankind": The emergence of a new concept of international environmental law. *Archiv des Völkerrechts*, *34*(4), 426–481.

Blumstein, S., & Petersen-Pelman, J. D. (2021). When the water runs dry: Supporting adaptive governance in transboundary river basins. *Water International*, *46*(3), 306–324. https://doi.org/10.1080/02508060.2021.1877984

Bollier, D., & Helfrich, S. (2020). *Libres, dignos, vivos. El poder subversivo de los comunes.* Icaria Editorial.

Bruch, C., Jansky, L., Nakayama, M., Salewicz, K. A., & Cassar, A. (2005). From theory to practice: An overview of approaches to involving the public in international watershed management. In C. Bruch, L. Jansky, M. Nakayama, & K. A. Salewicz (Eds.), *Public participation in the governance of international freshwater resources* (pp. 3–18). United Nations University Press.

Carse, A. (2014). *Beyond the big ditch: Politics, ecology, and infrastructure at the Panama Canal.* MIT Press.

Carse, A. (2018, June 27). Watershed, theorizing the contemporary. *Fieldsights*. https://culanth.org/fieldsights/watershed

Celata, F., Coletti, R., & Sanna, V. S. (2013). La cooperación transfronteriza en la región del Trifinio y la difusión de modelos europeos de gobernanza de las fronteras en América Latina, *Si Somos Americanos. Revista de Estudios Transfronterizos*, *XIII*(2), 165–189. http://dx.doi.org/10.4067/S0719-09482013000200008

Chen, C., MacLeod, J., & Neimanis, A. (Eds.). (2013a). *Thinking with water.* McGill-Queen's University Press.

Chen, C., MacLeod, J., & Neimanis, A. (2013b). Introduction: Toward a hydrological turn? In C. Chen, J. MacLeod, & A. Neimanis (Eds.), *Thinking with water* (pp. 3–22). McGill-Queen's University Press.

Comisión Económica para América Latina y el Caribe (CEPAL). (2011). *La economía del cambio climático en Centroamérica.* CEPAL.

Cooley, H., & Gleick, P. H. (2011). Climate-proofing transboundary water agreements. *Hydrological Sciences Journal, 56*(4), 711–718. https://doi.org/10.1080/02626667.2011.576651

Cuéllar, N., Díaz, Ó., & Martí, M. (2021). *Mapeo de actores nacionales de cambio climático en El Salvador.* PRISMA. https://prismaregional.org/wp-content/uploads/2021/10/Mapeo-de-actores-nacionales-del-CC_El-Salvador.pdf

De Stefano, L., Duncan, J., Dinar, S., Stahl, K., Strepek, K. M., & Wolf, A. T. (2012). Climate change and the institutional resilience of international river basins. *Journal of Peace Research, 49*(1), 193–209. https://doi.org/10.1177/0022343311427416

Diaz, N. (2024). Fusings. In D. Adeyemo, N. Diaz, N. Yala Kisukidi, & R. Walcott (Eds.), *Borders, human itineraries, and all our relation* (pp. 45–67). Duke University Press.

Díaz Bolaños, R. E. (2019). El Corredor Seco Centroamericano en Perspectiva Histórica. *Anuario de Estudios Centroamericanos, 45*, 297–322. http://dx.doi.org/10.15517/aeca.v45i0.40697

Draper, S. E., & Kundell, J. E. (2007). Impact of climate change on transboundary water sharing. *Journal of Water Resources Planning and Management, 133*(5). https://doi.org/10.1061/(ASCE)0733-9496(2007)133:5(405)

Eckstein, G. E. (2005). Protecting a hidden treasure: The U.N. International Law Commission and the international law of transboundary ground water resources. *Sustainable Development Law and Policy, 5*(1), 5–12.

Eckstein, Y., & Eckstein, G. E. (2005). Transboundary aquifers: Conceptual models for development of international law. *Groundwater, 43*(5), 679–690. https://doi.org/10.1111/j.1745-6584.2005.00098.x

Enríquez Villacorta, A. (2008). Microrregiones y mancomunidades municipales: nuevas formas de gestión del territorio. *PRISMA, 22*, 149–166.

Gómez Navas, Ó. (2003). Las mancomunidades de municipios en la experiencia venezolana. *Ciencias de Gobierno, 7*(13), 97–122.

Hernández Martínez de la Riva, A., Harper, M., Rytwinski, T., Sahdra, A., Taylor, J. J., Bard, B., Bennett, J. R., Burton, D., Creed, I. F., Haniford, L. S. E., Hanna, D. E., Harmsen, E. J., Robichaud, C. D., Smol, J. P., Thapar, M., & Cooke, S. J. (2023). Tipping points in freshwater ecosystems: An evidence map. *Frontiers in Freshwater Science, 1*, 1–16. https://doi.org/10.3389/ffwsc.2023.1264427

IPCC. (2022). *Climate change 2022: Impacts, adaptation and vulnerability.* Contribution of Working Group II to the Sixth Assessment Report of the Intergovernmental Panel on Climate Change. Cambridge University Press. https://doi.org/10.1017/9781009325844

Lloredo Alix, L. M. (2020). Bienes comunes. *EUNOMÍA. Revista en Cultura de la Legalidad, 19*, 214–236. https://doi.org/10.20318/eunomia.2020.5709

Mason, M., & Khawlie, M. (2016). Fluid sovereignty: State–nature relations in the Hasbani Basin, southern Lebanon. *Annals of the American Association of Geographers, 106*(6), 1344–1359. https://doi.org/10.1080/24694452.2016.1213155

McCaffrey, S. C. (2019). *The law of international watercourses* (3rd ed.). Oxford University Press.

Michel, D., Eriksson, M., & Klimes, M. (2021). Climate change and (in)security in transboundary river basins. In A. Swain, J. Öjendal, & A. Jägerskog (Eds.), *Handbook of security and the environment* (pp. 62–75). Edward Elgar. https://doi.org/10.4337/9781789900668.00012

Milman, A., Gerlak, A. K., Albrecht, T., Colosimo, M., Conca, K., Kittikhoun, A., Kovács, P., Moy, R., Schmeier, S., Wentling, K., Werick, W., Zavadsky, I., & Ziegler, J. (2020). Addressing knowledge gaps for transboundary environmental governance. *Global Environmental Change, 64*. https://doi.org/10.1016/j.gloenvcha.2020.102162

Mirumachi, N. (2017). *Transboundary water politics in the developing world.* Routledge.

Molle, F. (2006). *Planning and managing water resources at the river basin level: emergence and evolution of a concept*. IWMI. https://horizon.documentation.ird.fr/exl-doc/pleins_textes/divers15-08/010047208.pdf

Molle, F. (2009). River-basin planning and management: The social life of a concept. *Geoforum*, *40*(3), 484–494. https://doi.org/10.1016/j.geoforum.2009.03.004

Molle, F., & Wester, P. (2009). River basin trajectories: An inquiry into changing waterscapes. In F. Molle & P. Wester (Eds.), *River basin trajectories: Societies, environments and development* (pp. 1–19). CABI.

Montoya, A. (2021a). On care for our common home: Ecological materiality and sovereignty over the Lempa transboundary watershed. *Journal of Latin American Studies*, *53*(2), 297–322. https://doi.org/10.1017/S0022216X21000249

Montoya, A. (2021b). Lo transfronterizo como futuro común. Hacia la co-gobernanza de las cuencas hidrográficas compartidas en Centroamérica. In Y. Bravo-Espinosa (Ed.), *El derecho y los derechos ante la defensa de los territorios, las formas de vida colectiva y los bienes comunes* (pp. 28–37). CLACSO.

Montoya, A. (2023). Post-extractive juridification: Undoing the legal foundations of mining in El Salvador. *Geoforum*, *138*. https://doi.org/10.1016/j.geoforum.2022.103667

Montoya, A. (2024a). Troubling environmental governance: Citizen legal experiments in transboundary watersheds. *Oxford Development Studies*, 52(4), 396–412. https://doi.org/10.1080/13600818.2024.2395008

Montoya, A. (2024b). Citizens as lawmakers: Legal innovation and the competing moralities of environmental juridification. *International Journal of Human Rights*, 1–25. https://doi.org/10.1080/13642987.2024.2410835

Montoya, A., Sieder, R., Bravo Espinosa, Y., Knox, R., & Pauchulo, M. C. (2020). *The legal cultures of the subsoil database*. https://doi.org/10.14296/SLWU8713

Mostert, E. (2003). Conflict and cooperation in international freshwater management: A global review. *International Journal of River Basin Management*, *1*(3), 267–278. https://doi.org/10.1080/15715124.2003.9635212

Nadelman, R. (2015). "Let us care for everyone's home": The Catholic Church's role in keeping gold mining out of El Salvador. *Center for Latin American and Latino Studies (CLALS) Working Paper* , Series 9, December 2015. https://papers.ssrn.com/sol3/papers.cfm?abstract_id=2706819

Neimanis, A. (2024). Foreword: Hydrofeminisms and the desire for a watery "we." In T. Shefer, V. Bozalek, & N. Romano (Eds.), *Hydrofeminist thinking with oceans: Political and scholarly possibilities* (pp. xxiii–xxviii). Routledge.

Omernik, J. M., & Bailey, R. G. (1997). Distinguishing between watersheds and eco-regions. *Journal of the American Water Resources Association*, *33*(5), 935–949. https://doi.org/10.1111/j.1752-1688.1997.tb04115.x

Ostrom, E., Burger, J., Field, C. B., Norgaard, R. B., & Policansky, D. (1999). Revisiting the commons: Local lessons, global challenges. *Science*, *284*(5412), 278–282. https://doi.org/10.1126/science.284.5412.278

Palmer, M. A., Reidy Liermann, C. A., Nilsson, C., Flörke, M., Alcamo, J., Lake, P. S. & Bond, N. (2008). Climate change and the world's river basins: Anticipating management options. *Frontiers in Ecology and the Environment*, *6*(2), 81–89. https://doi.org/10.1890/060148

Quintana López, T. (1990). Las mancomunidades en nuestro derecho local: Evolución histórica. *Revista de estudios de la administración local y autonómica*, *248*, 713–732. https://doi.org/10.24965/reala.vi248.8657

Rockström, J., Kotzé, L., Milutinović, S., Biermann, F., Brovkin, V., Donges, J., Ebbesson, J., French, D., Gupta, J., Kim, R., Lenton, T., Lenzi, D., Nakicenovic, N., Neumann, B., Schuppert, F., Winkelmann, R., Bosselmann, K., Folke, C., Lucht, W., … Steffen, W. (2024). The planetary commons: A new paradigm for safeguarding Earth-regulating systems in the Anthropocene. *PNAS*, *121*(5). https://doi.org/10.1073/pnas.2301531121

Salman, S. M.A. (2007). The Helsinki Rules, the UN Watercourses Convention and the Berlin Rules: Perspectives on international water law. *Water Resources Development*, *23*(4), 625–640. https://doi.org/10.1080/07900620701488562

Schlager, E., & Blomquist, W. (2008). *Embracing watershed politics*. University Press of Colorado.

Schmeier, S. (2010). Governing international watercourse—Perspectives from different disciplines: A comprehensive literature review. *Hertie School of Governance – Working Papers* , 53. https://papers.ssrn.com/sol3/papers.cfm?abstract_id=1658899

Schmeier, S. (2013). *Governing international watercourses: River basin organizations and the sustainable governance of internationally shared rivers and lakes*. Routledge.

Schmeier, S. (2018). Anchoring water diplomacy—The legal nature of international river basin organizations. *Journal of Hydrology*, *567*, 114–120. https://doi.org/10.1016/j.jhydrol.2018.09.054

Schmeier, S., & Schulze, S. (2010). Governing environmental change in international river basins—The role of river basin organizations. Paper presented at the Berlin Conference on the Human Dimensions of Global Environmental Change, October 8–9, 2010. Berlin, Germany. http://dx.doi.org/10.2139/ssrn.1690244

Schulze, S. (2012). Public participation in the governance of transboundary water resources—Mechanisms provided by river basin organizations. *L'Europe en Formation*, *365*(3), 49–68. https://doi.org/10.3917/eufor.365.0049

Seqat, N. (2022). The principle of public participation in transboundary water resources management: A case study of transboundary basin organizations. *Revue Internationale de Droit Économique*, *XXXVI*(2), 55–75. https://doi.org/10.3917/ride.362.0055

Spalding, R. (2018). From the streets to the chamber: Social movements and the mining ban in El Salvador. *ERLACS*, *106*, 47–74. https://doi.org/10.32992/erlacs.10377

Stoa, R. B. (2014). The united nations watercourses convention on the dawn of entry into force. *Vanderbilt Journal of Transnational Law*, *47*, 1321–1370. https://scholarship.law.vanderbilt.edu/vjtl/vol47/iss5/3/

Teclaff, L. A. (1967). *The river basin in history and law*. Martinus Nijhoff.

Teclaff, L. A. (1996). Evolution of the river basin concept in national and international water law. *Natural Resources Journal*, *36*(2), 359–391. https://digitalrepository.unm.edu/nrj/vol36/iss2/7

UN-Water Expert Group on Water and Climate Change. (2019). *Climate change and water: UN-Water policy brief*. UN-Water. https://www.unwater.org/sites/default/files/app/uploads/2019/10/UN_Water_PolicyBrief_ClimateChange_Water.pdf

UNESCO. (1980). The global commons. *The Unesco Courier*, May, 33–35 and 38. https://unesdoc.unesco.org/ark:/48223/pf0000074757

United Nations & UNESCO. (2021). *Progress on transboundary water cooperation: Global status of SDG 6.5.2 and acceleration needs*. https://www.unwater.org/sites/default/files/app/uploads/2021/09/SDG6_Indicator_Report_652_Progress-on-Tranboundary-Water-Cooperation_2021_EN_UNECE.pdf

PART V

The Obstacles and Limits to Climate Action and the Road Ahead

CHAPTER 37

Limits of Climate Action in the Gulf Petrostates

AISHA AL-SARIHI

Introduction: Gulf Climate Conundrum

COMPRISING approximately 30% of confirmed global petroleum reserves and 20% of total natural gas reserves and accounting for around 30% of the world's crude oil exports (BP, 2021), oil and gas wealth has significantly shaped the Gulf economies, political landscapes, international relations, and geopolitics.[1] Once dependent on pearl diving, agriculture, and fishery, oil and gas wealth has swiftly transformed Gulf nations into some of the world's wealthiest countries (Zumbraegel, 2022).

Meanwhile oil and gas resources have also enabled the Gulf countries to weather the severe impacts of climate change. These include adapting to high temperatures: The region is heating up at a rate twice faster than the rest of the world, and, in a worst-case scenario,[2] it is estimated to experience a heating of 5°C by the end of the century (ESCWA, 2017). Heat records have been already experienced in Kuwait (63°C in 2019) and Saudi Arabia (55°C in 2019). In July 2023, 50°C was recorded in the UAE (NCMUAE, 2023) and Saudi Arabia (Saudi News, 2023). Furthermore, water scarcity, desertification, elongated periods of drought, sea-level rise, and biodiversity loss as well as extreme weather events such as dust storms and cyclones are serious challenges faced by the Gulf countries, exacerbated by climate change. According to data from the World Resources Institute's Aqueduct Water Risk Atlas, based in Washington, DC, the six Gulf Arab states are among the 17 most water-stressed countries in the world (WorldAtlas, 2019). The intensity and frequency of tropical storms are evidently increasing. Oman, with a coast on the Arabian Sea, has been hit by at least four cyclones over the last 12 years alone. Cyclone Shaheen was the latest cyclone to hit the country in October 2021, causing at least a dozen deaths and an estimated USD 500 million in damages (Al-manji, 2022). While the region experiences a general reduction in annual rainfall, the Gulf is also prone to intense rainfall, and flooding incidents are becoming more frequent. In Saudi Arabia, for instance, flash floods hit Jeddah in 2009, 2011, 2022, and 2024, resulting in death tolls and major infrastructural and economic damage (Mahmoud, 2017; Davies, 2022; Kalin, 2022; The New Arab, 2024). Government concerns about dust storms are increasing. In a 2021 interview, Crown Prince Mohammed bin Salman said: "When we were very young, the number of sandstorms was 70 to 80 per cent less than what we are suffering from today. This is very dangerous to our children and our lives ... and will cost our healthcare sector a considerable amount of money in the future" (AlArabiya, 2021).

Oil wealth has enabled Gulf countries to adapt to these severe climate challenges including through the importation of technology and know-how needed to adapt to high temperatures and water scarcity, bringing lessons to other petrostates that are gradually faced with similar challenges because of changing climate. Air conditioning makes up 70%

of Gulf peak electricity consumption and air conditioner stock has increased eightfold in the region in recent years between 1990 and 2021, and it is anticipated to quadrable in 2050 (IEA, 2021a, 2021b, 2021c). Water desalination technologies meet more than 50% of water needs in Bahrain, Kuwait, and Qatar and are rapidly expanding in Oman, Saudi Arabia, and the UAE (Al-Zubari et al., 2017). The Gulf accounts for 45% of the world's desalination capacity (Al-Zubari & Alajjawi, 2022). Additionally, oil wealth has enabled Gulf countries to secure nearly 50–90% of their food from imports as agriculture accounts for only 2% or less of Gulf GDP. Qatar, Bahrain, and the United Arab Emirates import 80–90% of their food, Oman approximately 50%, and Kuwait and Saudi Arabia up to 70% (Ghazaly et al., 2020).

Carbon dioxide emissions from existing fossil fuel production (and combustion) will likely result in global warming in excess of the 1.5°C UNFCCC target, reinforcing the need for an even faster phasing-out of all fossil fuels. While the Gulf nations collectively are not major contributors to global greenhouse gas emissions—the six Gulf countries together contribute around 2.8% of global greenhouse gas emissions, as per 2021 data (CAIT, 2021)—global efforts to reduce dependence on fossil fuels puts Gulf economies at risk of looming declines in oil demands and revenues.

Against this backdrop, climate change presents a conundrum for petrostates: Without oil revenues, Gulf states cannot sustain economic growth, and without oil revenues, petrostates will not be able to invest in climate adaptation strategies or social welfare. Hence, the following sections of this chapter focus on the strategies adopted by Gulf countries to adapt to evolving energy landscapes to transition away from fossil fuels; examine structural challenges needed to accelerate Gulf energy transition; and elaborate on the socioeconomic implications that may arise from implementing such structural changes.

A Dual Energy-Transition Approach

As renewable energy investments are not expected to generate the same amount of revenues that are generated from fossil fuel exports (Krane, 2020), and owing to vast oil and gas reserves, which are cheap to exploit, Gulf nations are adopting a dual energy-transition approach: continue fossil fuel production and export, while simultaneously scaling up investments in cleaner energy options, such as solar, wind, nuclear, hydrogen, and carbon capture and storage (CCS) technologies. By adopting such a dual energy-transition approach, Gulf states aspire to be the last persons standing in global fossil fuel production. To enhance the competitiveness of their energy exports in what is widely regarded as the world's last oil boom, the Gulf dual energy-transition approach is characterized by four strategies.

Monetizing Hydrocarbon Reserves

Oil and gas sales remain essential for Gulf economies, but for decades, they have been bracing for this moment, by aiming to diversify their sources of income away from the overwhelming dependence on oil and gas export revenues. These efforts have been spotty, however. In the 2022 Global Economic Diversification Index, for example, the Gulf countries are in the bottom rank out of a sample of 103 countries in terms of economic diversification: UAE (69th), Bahrain (79th), Saudi Arabia (87th), Qatar (89th), and Oman (95th) (EDI, 2024). A closer examination of the Gulf economies reveals that, for all their efforts, oil and gas revenues remain pivotal, both for current spending and for funding their diversification drives. In the coming years, as the world experiences what is widely expected to be the last oil boom, the Gulf countries will thus seek to exploit their vast oil and gas reserves, albeit via differing strategies.

The UAE: Exploit Hydrocarbons as Quickly as Possible

The UAE, unlike Saudi Arabia, seeks to maximize oil and gas export revenues as quickly as possible. As part of a USD 150 billion investment plan (ADNOC, 2023), the UAE revised timeline to boost oil production capacity to 5 million barrels per day by 2027 from the current 4 million barrels per day, compared to the original timeline of 2030. Aspiring to take advantage of oil and gas export revenues as quickly as possible, the UAE has been frequently acquiring higher quotas for its own production under OPEC production curtailment agreements (France 24, 2021). The UAE stands out regionally for having the most diversified economy as well as the highest scale of clean energy uptake. The contribution of non-oil sectors in the country's total economic output accounts for 72.3% of GDP in 2021 (oil rent accounted for 15.7% of GDP in 2021) coupled with a substantial 50.8% contribution from small and medium-sized enterprises (SMEs).

The UAE's divergent approach to fossil fuels was notable during its hosting of COP28 in 2023. Despite the controversial reaction of Dr Sultan Al Jaber, the COP28 president, toward the language of phasing out fossil fuels—"There is no science indicating that a phase-out of fossil fuels is needed to restrict global heating to 1.5°C" (Carrington & Stockton, 2023)—in the second week of negotiations, Dr Al Jaber played a role in correcting course. He hosted a closed-door meeting with other countries' representatives in a special setting of a "majlis": reflecting local customs and traditions, to ensure that countries reached a compromise around fossil fuel language (IISD, 2023). The move was driven by the desire to leave a legacy as a COP host: "We have a language on fossil fuel for the first time," an outcome that while agreeable was not welcomed by other producers such as Saudi Arabia and Kuwait (Lo, 2023). The UAE-led compromise also reflects its oil-free readiness compared to its neighbors. Committing to provide funding for developing countries, including through the UN Loss and Damage fund, was another instance of divergence between the UAE and its neighbors. The UAE COP28 host pledged $100 million to the start-up fund, another move that contrasted with the avoidance by Saudi Arabia of committing to provide future climate finance,[3] prioritizing national interest against any future declines in oil revenues.

The UAE also pioneered clean energy uptake accounting for nearly 60% of regional total renewable energy installed capacity (IRENA, 2023a) and 70% of renewable energy investments in the 2013–2022 period (IRENA, 2023b). Clean energy accounts for 10% of the country's grid-linked electricity generation capacity (IRENA, 2023b). The UAE is the only Gulf country that has one large-scale nuclear power plant, Barakah, supplying up to 60% of the UAE's clean energy, followed by solar PV and solar thermal accounting for 36% and 2% of the country's clean energy generation, respectively. Besides nuclear and solar power, the UAE is also the first country in the Gulf region to invest in waste-to-energy technology. In April 2022, the emirate of Sharjah completed construction of the country's first waste-to-energy plant, producing 30 MW of low-carbon electricity. Despite limited wind power potential, UAE also considers wind as part of its clean energy mix. In October 2023, the UAE launched its first commercial wind project, a 103.5 MW project spanning four locations. The UAE is also the first country in the region to produce clean hydrogen, using solar power. In November 2021, it released its national hydrogen strategy aiming to support the domestic hydrogen industry as well as enhance the country's position in blue and green hydrogen exports, with aspirations to capture 25% of the global low-carbon hydrogen market by 2030.

Saudi Arabia: Prolong the Life of Hydrocarbons as Much as Possible

In comparison, Saudi Arabia seeks to prolong the production and export of hydrocarbons as long as possible. Despite progressive efforts to diversify its economy away from fossil fuel dependence, as instilled in its Vision 2030, Saudi Arabia's economy is still largely dependent on oil export revenues which also remain essential to finance its economic diversification initiatives. For instance, the increase in Saudi non-oil revenues is largely due to fiscal policy reforms, including the introduction of a value added tax in 2018 and its subsequent increase from 5% to 15% in 2020; and the institutional reforms of the Public Investment Fund (PIF). PIF is central to the Saudi diversification strategy as it oversees investments in strategic non-oil sectors, including investment in giga projects[4]—Neom, The Line, Oxagon, Qiddiya entertainment city, Diriyah Gate heritage, The Red Sea Project—as well as international investments such as Lucid (PIF, 2021). And while Saudi oil exports declined from 84% in 2012–2013 to 74% in 2022 of total exports of goods and services, this decrease is mainly due to the expansion of petrochemical exports with petrochemicals' share rising from 9% to 12% over the same period.

This is why hydrocarbons are at the heart of the kingdom's energy transition strategy—the so-called circular carbon economy, which builds on the concept of circular economy but with an emphasis on the circularity of carbon. As Saudi Arabia continues to expand oil and gas production until there are signs that transition is under way and demand is declining, scaling up CCS and carbon dioxide removal is a priority to ensure the circularity of carbon emissions associated with its expanded fossil fuel–based projects. Early this year, Saudi Arabia launched four projects to generate 7,200 MW electricity using natural gas with carbon-capture technology (Arab News, 2024). At COP 27, a number of CCS projects were announced: Aramco's plan to establish one of the largest CCS hubs in the world in Jubail Industrial City to start operations by 2027, aiming to extract and store 9 million metric tons per annum (MTPA) of CO_2 in its first phase, supporting Saudi Arabia's national target to extract, use, and store 44 million MTPA by 2035. Additionally, Saudi Arabian Mining Company (Ma'aden) signed an agreement with Gulf Cryo to build and operate a CO_2 capture plant at its Phosphate Complex in Ras Al Khair Industrial City for 20 years. The plant will capture 300,000 metric tons per annum of CO_2 emissions from Ma'aden's three ammonia plants (Zawya, 2022). Ambitious plans that emphasize producing and exporting blue hydrogen are also seen as a strategic investment to complement the kingdom's aspirations to capitalize on its hydrocarbon resources. The kingdom also has plans to develop two nuclear reactors, aiming for a capacity between 2.2 to 3.3 gigawatts. Meanwhile, the kingdom has ambitious plans to increase its renewable energy share in its electricity generation mix to 50% in 2030, while the other 50% is expected to come from natural gas. The use of gas and renewables in domestic power generation can free more oil for export—currently, an average of 3.6 million barrels per day (b/d) of oil and gas liquids is used for power generation.

Qatar: Banking on Natural Gas as Transitional Fuel

Qatar, the only Gulf country not to commit to net-zero emissions goals, views liquefied natural gas (LNG), its major export, as a bridging fuel in the "transition away" phase. To that end, it aims to expand LNG production (Aljazeerah, 2024). In February 2024, Qatar announced new plans to boost LNG output from expanded "North Field West" from current 77 MTPA to 142 MTPA by 2030. Asian countries led by China, Japan, India, and South Korea have been the main market for Qatari gas and will continue to be favorable destinations for Qatar's LNG due to long-term contracts. While European demand has also grown since Russia's war on Ukraine, European preference for the spot market does not match Doha's affinity for long-term contracts, and other issues concerning EU citizens such as human rights and meeting climate goals may stall any EU-Qatar natural gas agreements.

Kuwait: Reluctant to Transition Away From Fossil Fuels

Owing to more than eight decades of oil reserves, Kuwait reaffirmed its rejection for the inclusion of any call for phasing out fossil fuels consumption and production in the COP28 draft final climate deal, as Oil Minister Saad Al Barrak said in statements to Kuwait's state news agency (Reuters, 2023). Kuwait is keen on increasing its oil and gas production capacities from 3.15 mb/d to 4.0 mb/d by 2035, and from 1.7 to 2.0 bcf/d by 2040, respectively (KPC, 2024). Kuwait remains the laggard among the GCC nations in terms of economic diversification as well as energy mix diversification. Apart from 70 MW pilot projects,[5] Kuwait does not have any operational renewable energy projects. Its only renewable energy project to generate approximately 3.2 gigawatts of electricity using renewable sources by 2030, Shagaya, had been shelved in 2021 due to the negative economic consequences of the COVID-19 crisis.

Oman: A Leap of Faith in Clean Energy

Unlike Saudi Arabia and the UAE, Oman is left with limited oil and gas resources that are expected to last for less than two decades. As of 2022, the oil and gas sectors contribution in Oman's GDP was much lower than the contribution of non-oil sectors, accounted for 39.5% and 64.3%, respectively (NCSI, 2023). The country is taking a leap of faith in clean energy as a potential replacement for oil rent. It is the second largest investor in renewables after the UAE and has implemented bold efforts to harness hydrogen production and export. Hydrogen production, specifically green hydrogen, has topped the government agenda: The government has identified a total of 50,000 square kilometers of land for hydrogen production including Duqm, Dhofar, and Al-Jazir on the southwest coastline of Oman. A National Hydrogen Alliance, a national hydrogen strategy, and a national technical committee (HYDROM) were formulated to accelerate the country's hydrogen race.

Bahrain: On the Fence

With the smallest oil and gas reserves in the Gulf, Bahrain has long established its non-oil sectors and diversified its economy. The non-oil sector accounted for 83.6% of real GDP in 2023, where the financial services sector overtook oil as the largest contributor to real GDP, standing at 18.1% in 2023 (EDB, 2024). A small state island, Bahrain aims to enhance its competitiveness in a decarbonized world, aiming to generate 700 MW of renewable energy capacity by 2030. By 2023, Bahrain renewable energy installed capacity was 59 MW. In November 2023, Bahrain's Sovereign Wealth Fund launched a voluntary carbon offsetting platform, Safa, aiming to be a hub for carbon trading.

Expanding Clean Energy Domestically to Free More Oil and Gas

While the Gulf's current renewable energy installed capacity accounts for only around 3% of total power installed capacity, the scale of and investment in these technologies in the region have increased quickly over the last two decades, from 17 MW in 2011 to 5,131 MW in 2022 (IRENA, 2023a).

Expanding clean energy deployment in the Gulf has economically rational goals. Despite being home to nearly a third of world oil and more than a fifth of global gas reserves, the Gulf countries are also faced with a looming challenge of energy security. The Gulf states' rising population, economic growth, and industrial expansion, together with rapidly rising living standards, have increased the regional energy consumption by nearly six times since the 1980s—faster than in any other part of the world. This has led the likes of Oman, the UAE, and Kuwait to import natural gas in order to meet the growing domestic demand. Kuwait and the Emirates are currently the region's largest importer of natural gas, with around 8.4 and 1.6 billion cubic meter of LNG imports in 2022, respectively (EIA, 2023; Natural Gas World, 2023). Therefore, for the Gulf countries, selling more oil creates energy security challenges. That is why all Gulf countries are pursuing efforts to increase clean energy investments domestically to wean themselves from fossil fuels and free more oil and gas for exports.

Securing Long-Term Oil and Gas Buyers

To protect their market share in a shrinking fossil fuel market, Gulf countries seek bilateral routes to the marketing of oil, pairing with importers in an environment of enhanced competition. Along with long-term supply deals, the other tactic is to integrate ownership of upstream, midstream, and downstream infrastructure, where national oil companies purchase oil and gas infrastructure inside importing countries. Refineries have been the main target. Aramco has created joint ventures that provide influence or even control over the refinery configuration and, by extension, the supply of crude oil to the refinery process. Aramco has bought stakes in refineries in China,[6] South Korea, Japan,[7] the United States,[8] and Malaysia[9] and is in the process of negotiating further purchases in India[10] (with participation of ADNOC). Similarly, the Kuwait Petroleum Co. purchased a 35% ownership and operational stake in a refinery in Vietnam configured around Kuwaiti crude, the UAE's ADNOC has a 30% equity stake in Azerbaijan's Absheronn Gas Field,[11] and Mubadala Petroleum is the second-largest producer of crude oil in Thailand.

To increase the added value of fossil fuels, the conversion of crude oil and natural gas into chemical products—such as precursor resins and polymers that form the basis for finished products ranging from auto parts to cushions, paint, and toothpaste—has been perhaps the Gulf's most promising climate hedge. Gulf companies have made major investments in petrochemical plants, both inside the Gulf and outside. In Saudi Arabia, domestic investment includes the USD 20 billion Sadara joint venture with Dow Chemical, the largest single-phase chemical plant ever built. Internationally, Saudi Aramco and SABIC, the state-held petrochemical giant, have proposed joint venture projects in China, Malaysia, and India. Similarly, the UAE's ADNOC has joint ventures for chemical projects with China[12] and India.[13]

Securing Clean Energy Markets Tied to Fossil Fuels

Along with expanding in clean energy at home, Gulf states are also expanding their clean energy footprint abroad. Apart from investing in renewable energy projects abroad (Hussain, 2023), Gulf countries also tap into clean energy markets tied to fossil fuels. This strategy enables Gulf countries to continue the fossil fuel business albeit in a cleaner form. Examples of climate action via green energy investments include forging bilateral deals on blue hydrogen and blue ammonia, which are produced from fossil fuel sources using low-carbon technologies such as CCS technologies.[14] These deals range from feasibility studies to letters of intent, memorandums of understanding, energy partnerships, and even trial shipments. To date, these deals have mostly involved countries that have an established energy trade relationship. For example, Japan already imports crude oil from Saudi Arabia and has been looking to expand the trade relationship to blue ammonia. In September 2020, Saudi Aramco announced the world's first shipment of blue ammonia—produced from natural gas with CCS in the hydrogen plant in Jubail—to Japan, where it has been used in power stations to produce emissions-free electricity. It is exploring similar opportunities with Indonesia,[15] the

UAE's ADNOC signed similar agreements with Japan's MHI[16] and Korea's POSCO[17] to explore potential opportunities to build supply chain for low-carbon hydrogen and ammonia.

Structural Challenges of Climate Action

The different strategies discussed in the previous section dictate that Gulf states adopt a mix of strategies to adapt to a changing energy landscape. While Gulf countries adopt a dual energy transition with expanding both fossil fuels and clean energy investments, their strategies remain dominated by a focus on fossil fuels. That is because, despite increasing interest in clean energy options, Gulf states are still adopting a cautious approach toward clean energy. Even though clean energy options can spare Gulf countries more oil and gas for exports, they cannot generate the same level of revenues generated by fossil fuel exports. Additionally, expanding clean energy at home requires structural changes that may present significant trade-offs that can come at the expense of political stability. In this section, I discuss three structural reforms necessary to promote Gulf energy transition. These are reductions in fossil fuel subsidies so that the cost of energy services reflects the actual price; the privatization of state-owned companies; and the introduction of tax systems that enhance the competitiveness of clean energy sources such as fossil fuel fees.

Fossil Fuel Subsidy Reforms

Oil and gas export wealth has enabled Gulf governments to generously subsidize many services, including transport fuel, electricity, water, food, education, health, and industrial activities. For decades, Gulf citizens have enjoyed lavish lifestyles with some of the lowest prices of provisions, which consequently encouraged trends of overconsumption behavior, which have led to increasing trends of greenhouse gas emissions—positioning the Gulf states at the highest end of the global scale of carbon footprints on a per capita basis. Meanwhile, subsidized fossil fuel–based services have rendered clean energy technologies uncompetitive.

While the prices of renewable energy technologies have fallen significantly over the past few decades, without governmental intervention, most of the clean energy technologies to date are still not competitive with traditional energy sources. Intermittency, storage, and efficiency of renewable energy technologies, for instance, still do not make clean energy reliable compared to fossil fuel resources. In the same vein, the high cost of CCS technologies and uncertainties behind its profitability create little incentive for investors and financial agents to prioritize investing in clean energy technology projects compared to the well-established hydrocarbon-sector investments.

At the consumer level, Gulf citizens enjoy highly subsidized provisions of electricity and water. So even though most of the regional governments have introduced incentives and regulations that support the installation of clean energy technologies such as solar rooftops or tax breaks for the use of electric vehicles, citizens' access to cheap and dirty energy options creates little incentive for them to consider investing in the alternatives.

To this end, the role of governments is critical in creating incentives that make clean energy competitive with the traditional energy supplies. Price reforms and reducing fossil fuels subsidies are some of the key policies to create incentives for clean energy technology adoption. Meanwhile, governments can consider redirecting subsidies to the clean energy sector. However, the latter may not be a favorable option as it may put further pressure on state budgets, a challenge that Gulf governments have already been struggling to address, especially during times of low oil prices, and an issue that may become more acute as the global community gradually transitions away from fossil fuels, which will ultimately affect Gulf fiscal stability (Al-Sarihi & Mason, 2020; Soummane & Al-Sarihi, 2023).

Subsidies are a crucial element of the bargaining contract between the state and society, and Gulf citizens consider subsidies as their personal right from oil and gas wealth. Therefore, any changes in subsidies have the potential to create a rapture in state-society relations. Following the drop of oil prices in 2014, Gulf governments introduced energy price reforms as means to mitigate such vulnerability and to lower domestic demand for fossil fuels in ways that improve fiscal health. For example, in 2018, the United Arab Emirates (UAE) removed long-standing fossil fuel subsidies (Boersma & Griffiths, 2016), and Saudi Arabia introduced energy price reforms in 2016 and 2018, and increased domestic gasoline prices and linked them with international prices (Aldubyan & Gasim, 2021). These reforms did indeed challenge state–society relations, albeit mildly, either in the form of physical protests against high energy prices, such as the 2017 protests in Oman, or demonstrations in social media.

Privatization of State-Owned Utilities

Gulf utilities are prominently run by state-owned companies such as ACWA Power of Saudi Arabia, the Dubai Electricity & Water Authority, and Oman's Power and Water Procurement company. State-owned utilities have provided secure supplies of, for example, water and electricity provisions. However, these public services rely heavily on state budgets to remain afloat. Because of subsidized provisions, Gulf citizens do not pay the actual price of utility services. Therefore, without governmental subsidies, utility companies may pass the costs to the consumers, making the utility bills unaffordable, especially for middle- or low-income classes. However, periods of economic vulnerability to oil price declines have shown that utilities strain state budgets. It is for this reason that governments in the region have already explored the possibility of privatizing state-owned companies to increase efficiency, reduce pressure on states' budgets, and improve fiscal health.

For instance, following the 2014 fall in oil prices, the Omani government's holding company for electricity sector entities, Nama, sold a 49% stake in the Oman Electricity Transmission Company to China's State Grid for USD 1 billion in December 2019. Nama controls several other electricity providers within Oman, many also slated for partial privatizations, including the expected sale of Muscat Electricity Distribution Company in 2020. Oman's oil minister also announced the intention of an initial public offering of 20% to 25% of shares in the state-owned Oman Oil Company (Young, 2019). Future plans seek to open prominent state-owned enterprises such as the Oman Tourism Development Company, Oman Food Investment Holding Company, Electricity Holding Company, Oman Global Logistics Group, and Oman Oil Company to private-sector investment (Oxford Business Group, 2019).

Similar trends took place in the UAE wherein the Dubai Electricity & Water Authority held an initial public offering in April 2022, and Saudi Arabia has increased private-sector involvement in water desalination and power generation projects. Bahrain recently transformed its governmental water and energy authority into a governmental company, an initial stride toward privatization, and, in 2023, Kuwait conducted a feasibility study on privatizing some of the country's power and water facilities (Alsayed, 2024).

Privatization of state-owned companies may trigger contention between the government and the owners of the utilities, which are mostly owned by merchant families or members of the government itself. In fact, the ownership of state-owned companies has been one of the factors that played a role in delaying the entry of renewable energy companies because of the lobbies of these groups to protect their interest and block competition.

Fossil Fuel Taxes

Along with energy subsidy reforms and privatization of state-owned utilities, fossil fuel taxes or "carbon taxes" can be one way to drive clean energy investments in the Gulf (Al-Sarihi, 2018; Luomi, 2022). Fossil fuel taxes and carbon pricing set direct prices on carbon content[18] in fossil fuels and are implemented as means of bringing down emissions and meanwhile driving investments of cleaner options (World Bank Group, 2024a). Depending on its design, it can raise revenues for the government, carbon "dividends" can be recycled to citizens, or funds can be targeted to support affected industries or fund decarbonization projects.

Carbon pricing sends an economic signal to carbon-intensive industries to either continue polluting and pay carbon price or invest in alternative cleaner options to avoid paying carbon tax. By the time of writing, 17 countries and 22 cities are implementing carbon taxes with a price range of 0.46–167 USD/tCO_2e (World Bank, 2024b).

To date, none of the Gulf countries have set a price on carbon (Luomi et al., 2023). However, by implementing voluntary carbon markets,[19] Gulf countries have taken initial steps that pave the way for carbon pricing. Saudi Arabia's Public Investment Fund and the Saudi Tadawul Group Holding Company created the Regional Voluntary Carbon Market Company in 2022, and in 2024, it announced the Greenhouse Gas Crediting and Offsetting Mechanism (GCOM). While the mechanism does not set a direct price on carbon, it provides credits for carbon-intensive industries that opt for emission reduction and/or removal, across all sectors. The proceeds from credit sales are supposed to fund nature conservation activities. Reported prices for carbon credits in one of the early auctions in Saudi Arabia were $6.27 per ton (Miriri, 2023). Qatar hosted the Global Carbon Council since 2016; the Abu Dhabi Global Market teamed up with AirCarbon Exchange (ACX) and created a carbon trading exchange and clearing house in 2021; Kuwait Finance House created a carbon offset platform in 2021; and Bahrain's sovereign wealth fund, Bahrain Mumtalakat Holding Company (Mumtalakat), established a voluntary carbon offsetting platform called "Safa" in 2023. Oman established a national Green Alliance to promote afforestation and unlock economic opportunities of carbon credits.

On a global level, carbon pricing has experienced a mix of success and failure in both revenue generation and greenhouse gas emissions. For example, Norway's Climate Action Plan 2021–30 proposes to raise the carbon tax from NOK 590 (USD 69) per ton of CO_2-equivalent in 2021 to NOK 2,000 (about USD 233) by 2030. While green taxes have helped Norway to generate a new stream of revenues, the forgone revenues as tax incentives to support other cleaner options such as electric vehicles has led to reduced green tax revenues, presenting an example when taxes are a victim of their own success.

A major concern for Gulf countries—specifically as their economies still depend heavily on oil export revenues—is that setting a price on carbon could discourage foreign direct investments and the expansion of the oil sector. As shown in the third section, all Gulf countries have plans to continue oil and gas production and are employing strategies to secure long-term markets for their hydrocarbon exports. Unlike Norway, Gulf states still have not decoupled the hydrocarbon sector from government spending. In fact, today's Gulf economic diversification projects largely depend on hydrocarbon export revenues, without which these fossil fuel–based funds' alternative development initiatives may be delayed or become obsolete. Therefore, in the short term, implementing carbon pricing counteracts the Gulf states' economic ambitions of expanding the hydrocarbon sector or funding economic diversification projects. And it would not be logical to combine carbon pricing with continuing subsidization of fossil fuels, so subsidy reform is an essential first step before introducing carbon pricing. Significantly raising the cost of fossil fuels in petrostates can create shock for the residents and generate negative political consequences.

Discussion: The Socioeconomic Implications of Climate Action in Petrostates

Climate change presents an existential dilemma for the Gulf states. Without oil revenues, Gulf countries cannot sustain their economies and, without oil revenues, they will not be able to stave off either climate change or a rupture in their social compacts that will likely put the respective regimes at risk.

That is why the Gulf states have adopted a cautious approach to global energy transition, one that eliminates the tradeoffs and meanwhile enhances their competitiveness in the long term. The dual energy-transition approach, as discussed in this chapter, while it considers expanding both fossil fuels and clean energy options, remains largely focused on monetizing fossil fuels: using clean energy resources to prolong the life of hydrocarbons.

However, beyond short-term gains, today's climate strategies by Gulf states cannot guarantee a long-term fiscal and political stability. In other words, without taking a proactive approach to advance climate policies domestically and abroad, the Gulf's future itself is imperiled, for the following reasons. Climate scientists increasingly demonstrate that global coal, oil, and gas production must decline by an average of 95%, 62%, and 42%, respectively, from 2020 to 2050 in order to limit long-term warming to 1.5°C (Achakulwisut et al., 2023). Aligning science with policy will push climate activists, emboldened by the steps taken in the UAE, to make their presence felt at future COP meetings by influencing political decisions on fossil fuel investments. Early this year, climate activists in the United States put pressure on the Biden administration to pause approval for liquified natural gas (LNG) terminal expansion in a bid to cut American methane emissions. To date, nearly 2,500 climate lawsuits have been filed in dozens of countries, putting carbon emitters at risk of growing climate litigation (Wetzer et al., 2024).

The decision to "transition away from fossil fuels" taken at COP28, while not revolutionary, sets the world on a path towards gradually ending their use. The UN climate conference in the Emirates was just the start for the push for more strident language and can be expected at future COP meetings. "Though we didn't turn the page on the fossil-fuel era in Dubai, this outcome is the beginning of the end," said Simon Stiell, the head of the UNFCCC secretariat, after the conclusion of COP28 (UNFCCC, 2023a, 2023b).

The building momentum for a clean energy transition has already sent signals to investors and banks to refrain from investing in fossil fuels by painting such assets as being susceptible to sharp declines in the near future. According to the International Energy Agency, global energy investment in clean energy has consistently exceeded that in fossil fuels since 2016, amounting to USD 1,864 billion in 2023 compared to USD 1,090 billion in 2016 (IEA, 2024).

Mr Stiell's confidence in his "beginning of the end" statement is supported by the fact that global energy outlooks published by different organizations suggest that half to three quarters of this demand will be erased by 2050. Indeed, there are already signs of this. Growing sales of electric vehicles in recent years have removed some 5 million barrels per day of world oil demand, according to IEA, a figure that, despite short-term technical setbacks, is expected to be on the rise due to increasing electric vehicle economies of scale and continued rollout of electric vehicle–supporting policies and mandates to ban internal combustion engines globally, with the vast majority of this coverage coming from China, the European Union, and the United States. As a former Saudi energy minister once admitted, "the Stone Age did not end for lack of stones, and the oil age will end long before the world runs out of oil."

Conclusions

The strategic approaches depict near-term measures adopted by petrostates to adapt to climate change, aiming not only to endure but to reshape their enterprises to gain competitive advantages. However, Gulf strategies toward climate action, which are mostly designed to prolong the life of fossil fuels, will result in short-term and long-term implications. In the short term, it is important to recognize that the fossil fuel industry—comprising oil, gas, and coal—still contributes two thirds of the ongoing greenhouse gas emissions accumulating in the atmosphere. By promoting the use of fuels that continue to be burned without restriction, these strategies prolong harm to the Earth's climate and the impacts of climate change become more severe, threatening the existence of the Gulf countries themselves. In the long term, as nations hasten investments in alternative energy supplies, the oil market will likely grow more competitive, and competing visions between Gulf states—especially the UAE (to sell as quickly as possible) and Saudi Arabia (to sell as long as possible)—are likely to result in two scenarios.

A *competitive scenario*, wherein regional countries compete to diversify their economies away from oil, while maximizing the benefits of oil to achieve these goals, is likely to put Gulf countries in a course of geoeconomic competition and possible tensions within OPEC grouping. Signs of active geoeconomic competition are already present among major Gulf countries such as the UAE, Saudi Arabia, and Qatar, in multiple dimensions. In the race to attract foreign investments, Saudi Arabia recently required multinational companies to move their regional headquarters to Riyadh. The UAE hosted EXPO 2020 in Dubai, as a means of soft power to attract investments, and in response, Saudi Arabia secured the hosting of Expo 2030. After Qatar successfully hosted the 2022 FIFA World Cup, Saudi Arabia will host the 2034 FIFA World Cup and has increased its investment in sports, attracting top football players and golfers. Similar trends are seen in tourism and entertainment: Royal Diriyah Opera House, Jeddah Opera House, and MiddleBeast are recent entertainment developments in Saudi Arabia, which aims to attract 150 million tourists by 2030, almost four times the UAE's target of 40 million. Competition is also heating up in transportation: Qatar airways and soon Riyadh Air are in a constant competition with Emirates airlines, and ports initiatives such as NEOM, Jeddah Islamic Seaport are challenging the UAE's dominance as air and sea transportation hub.

Frequent contentions between Saudi Arabia and the UAE around OPEC+ production cut agreements have been already observed. The UAE has been acquiring a higher quota for its own production under OPEC production cut agreements. UAE's divergent energy strategy, in contrast to Saudi plans to increase oil production and export as quickly as possible, may result into a departure of the UAE from OPEC, a decision already made by former OPEC members, including Qatar in 2019, Ecuador in 2020, and Angola in 2023. Riyadh's recently imposed requirement to relocate regional headquarter to Riyadh may be another factor that pushes the UAE to leave OPEC. As more members leave OPEC, the share of the grouping in the world's oil market declines, eliminating the power of the group to balance market supply and control over oil prices. If this happened, in a climate-constrained world, producers would compete to secure importers and are likely to compete on gaining premium for the shrinking fossil fuel market. Without groupings like OPEC, oil prices may change significantly, increasing price volatility in the market and making it tough for both producers and buyers.

With such anticipated market volatility and the absence of price control by groups like OPEC, Gulf nations will likely compete to secure long-term contracts and pair with importers. This competition will intensify as oil markets face declining demand and oversupply. To ensure they have buyers for their long-term sales, Gulf producers are likely to enforce their traditionally preferred long-term contracts of their oil and gas to guarantee stable and predictable export revenues.

However, too much oil and not enough buyers may prevent Gulf countries from revenues needed to achieve their economic goals. This explains the UAE's strategy to sell oil quickly while prices are still favorable, contrasting with Saudi Arabia's plan to sell more oil over a longer period. In such scenarios, banking on being the low-cost, low-carbon producers, Gulf companies may push for high prices by gaining premium owing to their low-carbon content compared to other producers.

The latter option is a case that encourages OPEC members to pursue a more *cooperative scenario*, which is unlikely. In a shrinking oil market and as more countries set standards on fuel emissions, OPEC can enhance its role by taking into consideration not only producers' equitable shares in a shrinking global oil market but also global climate concerns. In such a case, OPEC's strategy, whose main concern is to manage oil prices, will likely allocate market shares based on both a country's output capacity and the cleanliness of its oil. Therefore, buyers should be cautious, as OPEC could continue to control energy prices. They may push for higher prices not just by cutting supply but also by charging a premium for cleaner, low-carbon oil compared to other producers.

In either scenario, the last person standing while competing in a shrinking fossil fuel market will likely gain more power and possibly charge more for their oil while enforcing long-term contracting, an outcome that exacerbates the exact issue of energy security and affordability for energy importers.

Acknowledgments

I would like to express my gratitude for the constructive feedback provided by Prof. Michael Mason on a first draft of this chapter. I also thank the participants of Workshop 13 at the 14th Gulf Research Meeting for their constructive feedback on the presentation of this paper.

Notes

1. Gulf countries refer to Bahrain, Kuwait, Oman, Qatar, Saudi Arabia, and the United Arab Emirates (UAE).
2. In case climate goals remain unattained.
3. Author's observations during COP28 first-week negotiations.
4. Under the purview of the Public Investment Fund, giga projects are central to Saudi Arabia economic diversification away from oil that spans investments in real estate, infrastructure, and entertainment.
5. Consisting of a 50 mega-watt (MW) concentrated solar power (CSP) plant that was commissioned in December 2018, a 10 MW wind farm that was commissioned in May 2017, and a 10 MW photovoltaic (PV) plant.
6. Aramco owns a share of the Fujian Refining & Petrochemical Co. (http://www.frep.cn/en/view-category?categoryId=ff8008081290od8e0129f00d94e31100). This year, Aramco announced a joint venture with two other Chinese companies to build a new 300,000 b/d refinery and petrochemical complex in China's northeast. Aramco also agreed to acquire 10% of a Chinese oil refiner, Shenzhen-listed Rongsheng Petrochemical Co, in Zhejiang: https://www.ft.com/content/9e091abb-9463-4f7c-a78f-1b95d0e1130b

7. Showa Shell & Sekiyu, K. K. (2016). "Corporate Report 2017," p. 101. December 31, 2016. https://www.idemitsu.com/content/100866140.pdf
8. Gramer, R. (2017). "Saudi Arabia Now Controls the Largest Oil Refinery in North America." *Foreign Policy*, May 3, 2017. https://foreignpolicy.com/2017/05/03/saudi-arabia-now-controls-the-largest-oil-refinery-in-north-america-energy-middle-east-aramco/
9. PRefChem (Pengerang Refining and Petrochemical) is a major joint venture investment program between Aramco Malaysia and PETRONAS, the national oil and gas company of Malaysia. https://malaysia.aramco.com/en/who-we-are/overview/what-we-do
10. In a limbo—story of the mega refinery project India, Saudi Arabia want to resuscitate. https://indianexpress.com/article/explained/explained-economics/india-saudi-arabia-mega-refinery-project-explained-8940807/
11. https://www.adnoc.ae/en/news-and-media/press-releases/2023/adnoc-to-acquire-30-equity-stake-in-absheron-gas-field
12. ADNOC and Wanhua Chemical form Strategic Shipping JV. https://en.whchem.com/cmscontent/485.html
13. ADNOC and Reliance Sign Strategic Partnership for World-Scale Chemical Projects at TA'ZIZ in Ruwais. https://www.adnoc.ae/en/news-and-media/press-releases/2021/adnoc-and-reliance-sign-strategic-partnership-for-world-scale-chemical-projects-at-taziz-in-ruwais
14. As an example, JOGMEC and ADNOC expand technical cooperation to decarbonize upstream value-chains. https://www.jogmec.go.jp/english/news/release/news_10_00040.html
15. https://www.aramco.com/en/news-media/news/2022/aramco-and-pertamina-explore-hydrogen-and-ammonia-value-chain
16. ADNOC and MHI add new link to UAE-Japan low-carbon fuel value chain. https://www.adnoc.ae/en/news-and-media/press-releases/2023/adnoc-and-mitsubishi-heavy-industries-to-explore-opportunities
17. South Korean POSCO Holdings, its subsidiary POSCO International and UAE's Abu Dhabi National Oil Company (ADNOC) have signed a strategic cooperation agreement (SCA) for a joint investigation of clean hydrogen production business. https://www.offshore-energy.biz/posco-and-adnoc-forge-clean-hydrogen-ties/
18. Carbon content refers to the carbon dioxide equivalent that is released through the combustion or oxidation of a fossil fuel, or that is associated with the combustion or oxidation of a fossil fuel, used to generate electricity.
19. Carbon markets are trading systems in which carbon credits are sold and bought. Companies or individuals can use carbon markets to compensate for their greenhouse gas emissions by purchasing carbon credits from entities that remove or reduce greenhouse gas emissions.

References

Achakulwisut, P., Erickson, P., Guivarch, C., Schaeffer, P., Brutschin, E., & Pye, S. (2023). Global fossil fuel reduction pathways under different climate mitigation strategies and ambitions. *Nature Communications*, *14*, 5425. https://doi.org/10.1038/s41467-023-41105-z

ADNOC. (2023). Our strategy. Abu Dhabi National Oil Company. https://adnoc.ae/en/adnoc gas-ipo/the-opportunity

AlArabiya. (2021). *Transcript: Saudi Crown Prince Mohammed bin Salman's full interview on Vision 2030*. Al Arabiya English.

Aldubyan, M., & Gasim, A. (2021). Energy price reform in Saudi Arabia: Modeling the economic and environmental impacts and understanding the demand response. *Energy Policy*, *148*, 111941.

Aljazeerah. (2024). Qatar announces new gas output boost with mega field expansion. https://www.aljazeera.com/news/2024/2/25/qatar-announces-new-gas-output-boost-with-mega-field-expansion

Al-manji, S. (2022). Building resilience to extreme weather events in Oman. Insight. In A. Al-Sarihi (Ed.), *COP 27 and the Middle East*. Middle East Institute. National University of Singapore. https://mei.nus.edu.sg/wp-content/uploads/2022/11/Climate-Change_Compiled-Volume.pdf

Al-Sarihi, A. (2018). How can carbon pricing support economic diversification in the Arab Gulf States? Carbon Pricing Leadership Coalition. https://www.carbonpricingleadership.org/blogs/2018/12/13/how-can-carbon-pricing-support-economic-diversification-in-the-arab-gulf-states

Al-Sarihi, A., & Mason, M. (2020). Challenges and opportunities for climate policy integration in oil-producing countries: the case of the UAE and Oman. *Climate Policy, 20*(10), 1226–1241. https://doi.org/10.1080/14693062.2020.1781036

Alsayed, N. (2024). Privatizing the gulf utility sector. Arab Gulf States Institute in Washington. https://agsiw.org/privatizing-the-gulf-utility-sector/

Al-Zubari, W., Al-Turbak, A., Zahid, W., Al-Ruwis, K., Al-Tkhais, A., Al-Muataz, I., Abdelwahab, A., Murad, A., Al-Harbi, M., and Al-Sulaymani, Z. (2017). An overview of the GCC Unified Water Strategy (2016–2035). *Desalination and Water Treatment, 81*, 1–18. https://doi.org/10.5004/dwt.2017.20864

Al-Zubari, W., & Alajjawi, S. M. (2022). The water sector in the GCC countries: Climate change vulnerability and adaptation. In Al-*Sarihi, Aisha, COP 27, and the Middle East*. Middle East Institute, National University of Singapore. https://mei.nus.edu.sg/wp-content/uploads/2022/11/Climate-Change_Compiled-Volume.pdf

Arab News. (2024). Saudi Arabia launches 4 projects to generate 7,200 MW using carbon-capture tech. https://www.arabnews.com/node/2435581/business-economy

Boersma, T. & Griffiths, S. (2016). Reforming energy subsidies: Initial lessons from the United Arab Emirates. Energy Security and Climate Initiative, Brookings, p. 2. Masdar Institute.

BP. (2021). *BP statistical review of world energy 2021*. BP p.l.c. https://www.bp.com/content/dam/bp/business-sites/en/global/corporate/pdfs/energy-economics/statistical-review/bp-stats-review-2021-full-report.pdf

Carrington, D., & Stockton, B. (2023). COP28 president says there is "no science" behind demands for phase-out of fossil fuels. *The Guardian*. https://www.theguardian.com/environment/2023/dec/03/back-into-caves-cop28-president-dismisses-phase-out-of-fossil-fuels

Climate Watch (CAIT). (2021). *Country greenhouse gas emissions data*. World Resources Institute https://www.climatewatchdata.org/ghg-emissions?end_year=2020&regions=WORLD&start_year=1990

Davies, R. (2022). Saudi Arabia—Severe floods hit Jeddah after 179mm of rain in 6 hours. Floodlist.com. https://www.wsj.com/articles/flooding-in-saudi-arabia-kills-two-stirs-rare-public-anger-11669388176

Economic Diversification Index (EDI). (2024). *Global Governments Summit 2024*. https://economicdiversification.com/wp-content/uploads/2024/02/Global-Economic-Diversification-Index-2024_Full-report.pdf

EDB Bahrain. (2024). Economic Development Board. Bahrain's golden license attracts USD 2.4 billion in investment contributing to economic diversification & growth. https://www.bahrainedb.com/latest-news/bahrains-golden-license-attracts-usd-2-4-billion-in-investment-contributing-to-economic-diversification-growth#:~:text=Bahrain%20also%20diversified%20its%20economic,18.1%25%20in%20Q3%20of%202023

EIA. (2023). *UAE*. The U.S. Energy Information Administration. https://www.eia.gov/international/analysis/country/ARE

France 24. (2021). Disagreement between UAE, Saudi Arabia puts OPEC at impasse. https://www.france24.com/en/business/20210705-disagreement-between-uae-saudi-arabia-puts-opec-at-impasse

Ghazaly, S., Rabbat, R., & Mokhtar, A. (2020). How GCC countries can ensure their food security. *PWC*. https://www.strategyand.pwc.com/m1/en/articles/2020/how-gcc-countries-can-ensure-their-food-security.html#:~:text=GCC%20countries%20import%20about%2085,leaves%20countries%20vulnerable%20to%20shortages.

Hussain, T. (2023). UAE and Saudi Arabia invest in Central Asia renewable energy deals, with Russia's blessing. *South China Morning Post*. https://www.scmp.com/week-asia/economics/article/3207964/uae-and-saudi-arabia-invest-central-asia-renewable-energy-deals-russias-blessing

IEA. (2021a). Energy system of United Arab Emirates. International Energy Agency. https://www.iea.org/countries/united-arab-emirates/energy-mix
IEA. (2021b). Growth in global air conditioner stock, 1990–2050. International Energy Agency. https://www.iea.org/data-and-statistics/charts/growth-in-global-air-conditioner-stock-1990-2050
IEA. (2021c). Net zero emissions by 2050: A roadmap for the global energy sector. International Energy Agency. https://www.iea.org/reports/net-zero-by-2050
IEA. (2024). Global investment in clean energy and fossil fuels, 2015–2024. International Energy Agency. https://www.iea.org/data-and-statistics/charts/global-investment-in-clean-energy-and-fossil-fuels-2015-2024
IISD. (2023). Daily report for 10 December 2023: UN Climate Change Conference—United Arab Emirates Nov/Dec 2023. The International Institute for Sustainable Development. https://enb.iisd.org/united-arab-emirates-climate-change-conference-cop28-daily-report-10dec2023
IRENA. (2023a). Renewable capacity statistics 2023. International Renewable Energy Agency, Abu Dhabi. https://www.irena.org/Publications/2023/Mar/Renewable-capacity-statistics-2023
IRENA. (2023b). Renewable energy markets: GCC 2023. International Renewable Energy Agency, Abu Dhabi. https://www.irena.org/Publications/2023/Dec/Renewable-energy-market-analysis-GCC
Kalin, S. (2022). Flooding in Saudi Arabia kills two, stirs rare public anger. *The Wall Street Journal.*
KPC. (2024). Kuwait Petroleum Corporation. https://www.kpc.com.kw/Strategy2040
Krane, J. (2020). Climate action versus inaction: Balancing the costs for Gulf energy exporters. *British Journal of Middle Eastern Studies*, *47*(1), 117–135. https://doi.org/10.1080/13530194.2020.1714269
Lo, J. (2023). Twitter post. https://twitter.com/joeloyo/status/1734840563369718065?s=20
Luomi, M. (2022). Carbon markets and related opportunities for the Gulf. In Henderson, J. *The evolution of carbon markets and their role in climate mitigation and sustainable development*. Issue 123, pp. 42–45. The Oxford Institute for Energy Studies.
Luomi, M., Bosse, T., & Sergeeva, Z. (2023). Lessons from Gulf Cooperation Council countries' participation in the clean development mechanism. King Abdullah Petroleum Studies and Research Center. DOI: 10.30573/KS--2023-DP04
Mahmoud, M. (2017). *Weathering climate change in the Gulf.* Arab Gulf States Institute in Washington.
Miriri, D. (2023). Saudi companies buy 2.2 million tonnes of carbon credits in Kenya auction. *Reuters.* https://www.reuters.com/sustainability/saudi-firms-bid-2-mln-tonnes-carbon-credits-kenya-auction-2023-06-14/
Natural Gas World. (2023). Kuwait. https://www.naturalgasworld.com/country-focus-kuwait-onwards-and-upwards-for-lng-gas-in-transition-107679
Ncmuae. (2023). Twitter post. https://twitter.com/ncmuae/status/1680559928522469376?s=20 National Centre of Meteorology.
NCSI. (2023). *Statistical year book 2023*. National Centre for Statistical Information, Oman. https://www.ncsi.gov.om/Elibrary/LibraryContentDoc/bar_Statistical%20Year%20Book%202023%20Issue%2051_937cfd51-85c8-49b2-b61b-0e66af8afb4b.pdf
The New Arab. (2024). Dubai-style floods hit Saudi Arabian provinces, heavy rain in holy city of Madinah. https://www.newarab.com/news/floods-heavy-rain-hit-saudi-arabia-holy-city-madinah
Oxford Business Group. (2019). The report: Oman 2019. https://oxfordbusinessgroup.com/reports/oman/2019-report
PIF. (2021). Public Investment Fund Program 2021–2025. Public Investment Fund. https://www.pif.gov.sa/en/strategy-and-impact/the-program/
Reuters. (2023). Kuwait refuses inclusion of fossil fuels phase-out call in COP28 draft deal—oil minister. https://www.reuters.com/sustainability/climate-energy/kuwait-refuses-inclusion-fossil-fuels-phase-out-call-cop28-draft-deal-oil-2023-12-11/#:~:text=CAIRO%2C%20Dec%2011%20(Reuters),agency%20(KUNA)%20on%20Monday.
Saudi News. (2023). Twitter post. https://twitter.com/SaudiNews50/status/1685287314094755841?s=20
Soummane, S., & Al-Sarihi, A. (2023). Impacts of global climate policies on Middle Eastern oil exporters: A review of economic implications and mitigation strategies. King Abdullah Petroleum Studies and Research Center Discussion Papers (ks-2022-dp16).

UNFCCC. (2023a). Accelerating action towards a green, inclusive and resilient economy. Third Update of Second Nationally Determined Contribution for the UAE. The United Nations Framework Convention on Climate Change https://unfccc.int/sites/default/files/NDC/2023-07/Third%20Update%20of%20Second%20NDC%20for%20the%20UAE_v15.pdf

UNFCCC. (2023b). COP 28 agreement signals "beginning of the end" of the fossil fuel era. The United Nations Framework Convention on Climate Change. https://unfccc.int/news/cop28-agreement-signals-beginning-of-the-end-of-the-fossil-fuel-era

United Nations Economic and Social Commission for Western Asia (ESCWA); Arab Center for the Studies of Arid Zones and Dry Lands (ACSAD) of the League of Arab States; Food and Agriculture Organization of the United Nations (FAO); Deutsche Gesellschaft für Internationale Zusammenarbeit GmbH (GIZ); League of Arab States; Swedish Meteorological and Hydrological Institute (SMHI); United Nations Environment Programme (UN Environment); United Nations Educational, Scientific and Cultural Organization (UNESCO); Office in Cairo, United Nations Office for Disaster Risk Reduction (UNDRR); United Nations University Institute for Water, Environment and Health (UNU-INWEH); and World Meteorological Organization (WMO). (2017). Arab Climate Change Assessment Report—Executive summary. Beirut. E/ESCWA/SDPD/2017/RICCAR/Summary.

Wetzer, T., Stuart-Smith, R., & Dibley, A. (2024). Climate risk assessments must engage with the law. *Science*, *383*, 152–154. DOI: 10.1126/science.adj0598

WorldAtlas. (2019). The 10 most water-stressed countries in the world. https://www.worldatlas.com/natural-disasters/the-10-most-water-stressed-countries-in-the-world.html

World Bank. (2024a). What is carbon pricing? https://carbonpricingdashboard.worldbank.org/what-carbon-pricing

World Bank. (2024b). State and trends of carbon pricing dashboard. https://carbonpricingdashboard.worldbank.org/compliance/price

Young, K. (2019). Sell-off in Oman reveals privatization with regional characteristics. *Al-Monitor*. https://www.al-monitor.com/originals/2019/12/sell-off-oman-privatization-mideast.html#ixzz8bgaUaXRC

Zawya. (2022). Saudi climate pledge: Key announcements from Sharm El-Sheikh. https://www.zawya.com/en/projects/industry/saudi-climate-pledge-key-announcements-from-sharm-el-sheikh-ukel9kbo

Zumbraegel, T. (2022). *Political power and environmental sustainability in Gulf monarchies*. Springer Nature Singapore.

CHAPTER 38

Climate Change Denial

DEBORAH BARROS LEAL FARIAS

"HIV is harmless and does not cause AIDS." "COVID-19 is only dangerous to people with preexisting medical conditions." "Tobacco has not been proven to cause cancer." "The Earth is flat." "Vaccines cause autism." "Climate change is a hoax." According to ample scientific evidence, all these statements are false. Yet many continue to embrace such positions.

On the one hand, it is normal and healthy to be skeptical about what is presented as scientific truth. The scientific method is based on this premise. Given "facts" are questioned based on scientific elements such as ontological assumptions, epistemological approaches, data gathering, and interpretation of results. On the other hand, continued belief over (erroneous) statements as those just presented—regardless of all scientific evidence—tends to be something of a different nature. They are examples of what is considered science denialism. Here, the *belief* in a scientific "fact" comes first. If personal belief and scientific evidence are at odds, the latter is dismissed. The only acceptable kind of evidence is the one confirming preexisting beliefs (Diethelm & McKee, 2009; Schmid & Betsch, 2019).

Climate change is "by far the most coordinated and well-moneyed form of science denial" (Björnberg et al., 2017, p. 235). There is a clear consensus among scientists that climate change is both real and human-caused (Cook et al., 2013). In fact, there is a "99.999% probability" that GHG emissions from human activities are the reason for the increase in the global average temperature in the last 60 years (Kokic et al., 2014). From a scientific perspective, there is little doubt over the existence of climate change and the fundamental role played by humans in the process. However, outside of academic circles, doubt is far from uncommon.

This chapter explores climate change denial. This is not a new topic. In fact, Björnberg et al.'s (2017) systematic literature review covered 25 years of academic research on this issue. This chapter focuses mostly on material published on the topic since that review. It starts by asking *what* climate change denialism is. The second section is interested in *who* embraces and/or promotes climate denial, followed by a section on *why* this happens. A fourth section summarizes the common strategies used by climate change denialists, that is, *how* it is built and propagated. Finally, it concludes with the main challenges climate change denialism poses for the future and presents some suggestions for dealing with this issue.

What Is Climate Change Denialism?

Climate change denialism is anchored upon an unwillingness to believe in existing scientific evidence related to climate change (Björnberg et al., 2017, p. 237). The most obvious example is not believing in the existence of climate change altogether. However, climate change denialism is not always a case of full-out rejection of *all* science or scientific evidence (Fischer, 2019). It covers a much wider range of issues and levels of denial, such as the following:

- Accepting that climate change is happening but denying its link to increases in greenhouse gas (GHG) emissions
- Accepting that climate change is happening but denying its link to *human*-caused increases in GHG emissions
- Accepting that climate change is happening (whether human-caused or not) but denying its negative effects
- Accepting that climate change is happening (whether human-caused or not) and accepting its negative effects but denying the effects' severity or need for urgent action
- Accepting that climate change is happening (whether human-caused or not) and accepting its negative effects but denying the usefulness of doing anything about it

These examples show the complexity of climate change denialism. They also illustrate the debates over whether the terms [climate change] "denialism" and "skepticism" represent the same or different phenomena (see Dunlap, 2013; Lubke, 2022; O'Neill & Boykoff, 2010; Weart, 2011). There is no single answer. A skeptic might just be someone who believes in the scientific method but is yet unconvinced about some existing evidence and is open to newer or better evidence. In this case, the two terms should not be used interchangeably. Yet climate change "skepticism" has been known to be purposefully used as a ruse. Claiming to be a skeptic is a much more "palatable" way of pushing a denialist agenda without the negative connotations of being a denialist. In mainstream circles, it is easier to appear to be a moderate and sensible voice—"I'm just asking questions"—while intentionally disseminating and pushing a denialist agenda, than to admit that no evidence will ever be enough. Most of the literature on the topic tends to use skeptic (-ism, -ist) as synonymous with denial (-ism, -ist) because self-proclaimed "skeptics" are known to raise doubts in bad faith. This point will be explored in a later section on strategies used by climate change denialists, that is, *how* climate change denial is promoted.

Who Embraces Climate Denial?

Climate change denialism is a global phenomenon. Yet this does not mean it happens in the same way in every country. There are variances in how "popular" and intense it is, how salient the issue is in the media (traditional or social), over what is mostly commonly denied, and whether the country's leader or the party in power are complicit in fostering denialism—to name a few. There are climate change denialists worldwide, albeit at different levels depending on the region. A global study of public opinion asking questions related to climate change (WEF, 2020) found important variations, with no visible Global North/South pattern (Figure 38.1).

Cultural elements can play a big role in understanding certain differences (Dunlap & McCright, 2010; Hoffman, 2011, 2015). This can be manifested in a tendency for high levels of trust and respect for individuals with recognized academic credentials or more attention toward collective challenges than individual ones (Jylhä et al., 2021; WEF, 2020). For example, when asked, "How much do you trust what scientists say about the environment?" 86% of Indians and 78% of Bangladeshi responded "a great deal" or "a lot," compared to 25% of Japanese and 23% of Russians (WEF, 2020). Even ideology, which is a major element in understanding denialism in the United States, is rather "muted" in others. Hornsey et al. (2018) compared 24 countries and found that in three quarters of cases, there was "no significant relationship between the various indices of ideology and climate scepticism" (p. 616), with the United States standing out as having the highest relationship between these two elements

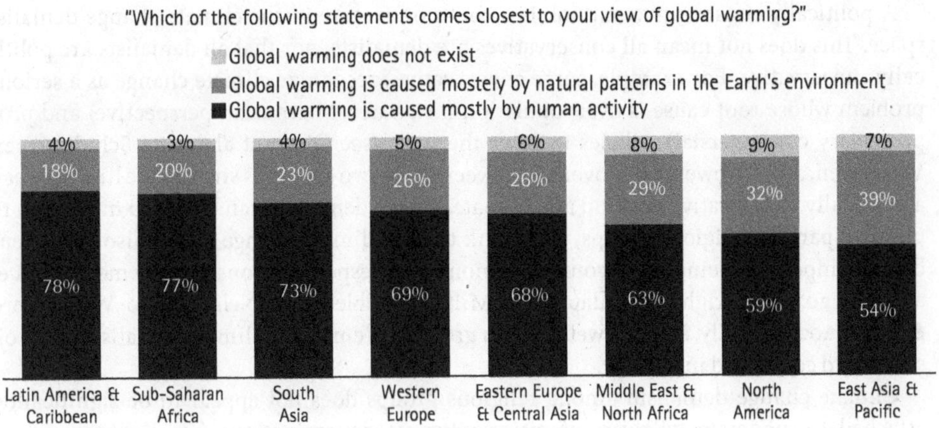

FIGURE 38.1. Questions related to climate change.

(*Source*: Developed by the author based on data from the World Economic Forum [WEF], 2020)

than any other country for all ideological variables analyzed: left–right political leaning, liberal–conservative political ideology, conspiracy beliefs, individualism (as opposed to communitarianism), and hierarchical values (as opposed to egalitarianism). In the United States, research found that climate change denialism was positively and strongly correlated to right-wing political leaning, embrace of conservative values, high levels of belief in conspiracy theories, preference for individualism, and being against reducing inequalities. Still, some countries had very similar beliefs to the US case for measures of ideology, such as Australia, Brazil, and Canada (see Lockwood, 2018). And while conservative ideologies were found to be associated with climate skepticism practically everywhere, the size and the effect in some places (such as Western Europe and the United Kingdom) were relatively modest (pp. 617–618).

It is worth mentioning China's "uniqueness" in this space. Overall, climate change denialism is *not* a very prominent issue (Liu, 2015; Wang & Zhou, 2020; Yang et al., 2021). There is a historically low presence of denialist views in the media (Duan & Miller, 2021; Shen & Wang, 2023) as well as in the country's most popular social media app Weibo (Liu & Wallace, 2023). This overlaps with the government's official position, which accepts climate change as real, human-caused, and an important issue (Jamelske et al., 2015). The biggest points of contention over climate change in China are related to discussions over mitigation and adaptation policies; the main interests are over *what* should be done and *how* (Shen & Wang, 2023). That is, the concern is not over the "realness" of climate change but over the political and economic consequences of dealing with it, both in the domestic and international realms.

There are many explanations for why (some) countries/individuals embrace climate change denialism more commonly or forcefully than others. However, research shows that some actors stand out globally in the climate change denialism "space." Evidently, there are country-to-country variations, with certain groups being more important in some places but not at all in others. Nonetheless, when identifying "who" denies climate change in the world, certain actors are regularly found. Before discussing these cases, it is worth pointing out that few scientists are climate change denialists (i.e., placing beliefs above evidence). Those who are typically do not belong to an established community of researchers working in the climate field and publishing peer-reviewed articles in high-level academic journals, and are not affiliated with an academic institution (Björnberg et al., 2017, pp. 235–236).

A politically conservative outlook is a common feature in the climate change denialist space. This does not mean all conservatives are denialists, nor that all denialists are politically conservative. For example, some conservatives recognize climate change as a serious problem whose root cause is overpopulation (i.e., a neo-Malthusian perspective) and propose (very controversial) policies to solve the issue (see Ojeda et al., 2020; Schultz, 2021; Verhoeven, 2011). However, the overlap between these two groups is strong. The link between a politically conservative position and climate change denialism can be found in relation to political parties, religious groups, and think tanks. Climate change denial also has a gendered component, being more common among men, especially conservative men (Jylhä et al., 2020, 2021; McCright & Dunlap, 2011a; Milfont & Sibley, 2016; Nelson, 2020; Wullenkord & Reese, 2021). Finally, another well-known group that embraces climate denialism is the oil sector and connected industries.

Climate change denialism among religious groups does not appear to be significantly attached to moderate followers of major religious denominations (see Koehrsen, 2021; Sachdeva, 2016). Rather, it seems to appear among conservative congregations and/or fringe groups, which are commonly *science* denialists, who also reject ideas like evolution and stem cell research based on spiritual beliefs. In the United States, in particular, this denialist position is frequently visible among evangelical communities and politicians whose core platform draws from this religious background (Veldman et al., 2021; Zaleha & Szasz, 2015). However, generational differences are emerging in this group, with denialism being less prevalent among younger evangelicals than their elders (Pew, 2022).

How: What Are the Common Strategies of Climate Change Denialists?

This section addresses common tactics used by those who seek to deliberately spread and sow the seeds of climate change denialism, especially among the general public. The next section complements the "how?," focusing on *why* some individuals/groups are more susceptible to these strategies than others.

The strategies used to promote climate change denialism by "merchants of doubt" are akin to those used by science denialism in general and the tobacco industry in particular (Oreskes & Conway, 2010). One common feature they all exhibit is their reliance on misinformation to "support" denialist claims. This can be done through the purposeful creation and/or use of information known to be false, as well as consciously distorting and misinterpreting scientific evidence from a place of bad faith. Misinformation works by lowering people's acceptance of climate change and support for climate policy and action (Cook et al., 2017). In this area, misinformation campaigns have gone beyond targeting scientific findings to also include ad hominem attacks, with climate *scientists* accused of being the "real" creators of climate change misinformation (see Leiserowitz et al., 2013; Maibach et al., 2012).

Some rhetorical techniques and logical fallacies are so commonly found in the context of science and climate misinformation that they have become known by the acronym "FLICC": fake experts, logical fallacies, impossible expectations, cherry-picking, and conspiracy theories (Cook, 2020; Diethelm & McKee, 2009). These techniques and fallacies are used to support denial strategies. Denial strategies—regardless of the topic discussed—can be separated into three main contents: literal, interpretative, and implicatory (Cohen, 2001). As it relates to climate change, literal denial is, as expected, the unwillingness to accept that climate change is real and human-caused. Literal denial is still advocated by "true" believers—those who *really* believe climate change is a hoax—and by those who have to gain

TABLE 38.1 Climate Change Questions and Discourses

Climate Question	Climate Delay Discourse
"Is it our responsibility to take action?"	Redirect responsibility: no, because someone else should take responsibility first
"Are transformative changes necessary?"	Push for *non*transformative solutions: disruptive change is not necessary
"Is it desirable to mitigate climate change, given the costs?"	Emphasize downsides: no, because change will be disruptive
"Is it still possible to mitigate climate change?"	Surrender: it is not possible to mitigate climate change

Source: Adapted from Lamb et al. (2020, p. 2).

with climate change denialism. However, over the years, this strategy has become more difficult to uphold (Fischer, 2019; Shue, 2022). As a result, flat-out climate change literal denial has become relatively less frequent than interpretive and implicatory denial, which have been increasingly amplified. With interpretative denial, there is an acknowledgment that climate change is happening *but* the consequences are being "exaggerated," or the dire urgency to act upon it is being "overblown." In the case of implicatory denial, there is a decoupling of responsibility: climate change is real, human-caused, and bad, *but* "It is not *my* fault/problem" or 'I can't do anything about it' (Cohen, 2001; Wullenkord & Reese, 2021).

Interpretative and implicatory denial are typically linked to an emphasis on a supposed scientific uncertainty surrounding climate change's existence and cause. The goal is to cast doubts over scientific evidence, as opposed to express *certainty* that the science is wrong. The strategy of promoting the "moderate" forms of denial seems to have had at least some success (Cann & Raymond, 2018). They might explain the paradox of many people who acknowledge that climate change is real and human-caused but do not act upon it or, even worse, oppose practical actions to address it (Wullenkord & Reese, 2021).

The specific approach of raising questions over costs and urgency of dealing with climate change is known as *climate delay* (Cann & Raymond, 2018; Lamb et al., 2020; Shue, 2023). The discourse of climate delay can be characterized by pushing a negative response to at least one of the four questions in Table 38.1, where each question has a particular discourse attached to it (Lamb et al., 2020, pp. 1–6).

Conservative Think Tanks

Conservative think tanks (CTTs) are one of the most powerful climate denialist groups (Brulle, 2014, 2023; Brulle et al., 2012; Busch & Judick, 2021; Dunlap, 2013; Dunlap & Jacques, 2013; Oreskes & Conway, 2010). According to a study conducted by Jacques et al., (2008), CTTs were at least partly responsible for 92% of books (published between 1972 and 2005) expressing skepticism toward climate science. CTTs work as an "apparatus for the incubation and diffusion of conservative ideas" (Dunlap & McCright, 2015, p. 304). They are key components of an organized structure set out to create, disseminate, and amplify climate change denialism. The most prominent CTTs in this context are US-based, such as the Cato Institute, Heritage Foundation, Heartland Institute, American Enterprise Institute (AEI), and the Committee for a Constructive Tomorrow (CFACT).

One important reason behind many CTTs' power is having deep pockets. These well-funded organizations belong to the "[climate change] denial machine" (Dunlap, 2013; Dunlap & McCright, 2011) pushed forward by corporate interests, particularly the fossil fuel and petrochemical industries and other high GHG-emitting sectors, such as mining and steel (Busch & Judick, 2021; Oreskes & Conway, 2010). Since the 1990s, many CTTs have acted as a façade for these sectors' interests since they are more likely to come across as independent-looking organizations than their funders, such as ExxonMobil, the American Petroleum Institute, and the US National Coal Association (e.g., Dunlap & Brulle, 2020; Dunlap & McCright, 2010, 2011; Oreskes & Conway, 2010; Union of Concerned Scientists, 2007).

Conservative/right-wing media outlets are important climate denialist actors in many countries (Dunlap & McCrught, 2011; Painter & Ashe, 2012). One common way they participate in this space is by giving denialists time and space, usually with very timid (if any) push-back, featuring individuals with questionable academic credentials on the topic of climate change as if they were "reliable" experts. Assuming politically conservative individuals are more likely to follow conservative media outlets, this reinforces an "echo chamber" arrangement for climate change denialism (Elsasser & Dunlap, 2013). Even when a media venue is not embedded in a conservative agenda, the norm of giving equal space for both sides—in this case, climate change scientists and deniers—is problematic, as it allows deniers to "punch above their weight" by creating the wrongful allure of a valid scientific debate (Painter & Gavin, 2016).

Along with CTTs whose interest topics *include* climate change, there are organizations specifically created to operate in the active and purposeful dissemination of climate change denial. One common strategy used is to give the organization a "neutral" or "academic"-sounding name. This is known as "astroturfing," referencing an artificial turf's brand name, a synthetic surface made to look like real grass. Astroturf organizations look like legitimate grassroots organizations but are actually created and/or sponsored by large corporations to defend their economic interests (Byrne, 2020; Cho et al., 2011)—in this case, by promoting climate change denialism. An example is the now defunct "Global Climate Coalition" (1989–2002), whose members included the US National Association of Manufacturers, the American Petroleum Institute, and several petroleum companies. Its main goal was to *obstruct* climate action in the United States and internationally, particularly in regard to the Intergovernmental Panel on Climate Change (IPCC), relying on four main strategies: (1) undermining climate science, (2) highlighting the costs of climate action, (3) shifting the cultural understanding of climate change to oppose carbon emissions reductions, and (4) conducting aggressive lobbying of political elites to stop meaningful climate action (Brulle, 2023). The Heartland Institute is another interesting case. Here, climate change (denial) is one of its main topics, along with the promotion of "free-market" ideas regarding education, health, and the economy. Climate change science per se is not the main object of dispute. Rather, the framing is how climate change mitigation policies are part of a broader "leftist agenda" that attacks capitalism and wishes to increase government intervention in society (see Klein, 2011, 2015). This strategy is an example of promoting "motivated" denial, discussed in the next section.

Why Is There Denial?

Numerous strategies are used to create and amplify climate change denialism. Yet not all people are affected in the same way. Why are some individuals or groups of individuals more likely to embrace climate change denialism? There is no single answer. In fact, several complementary, correlated, and reinforcing elements are at play when trying to understand why (some) people embrace denial.

As a starting point, climate change denial's main drivers can be seen as belonging to a continuum (Ferkany, 2015, p. 710). On one end, there is "naïve" denial, that is, denial based on pure ignorance of facts or by being "duped" into believing erroneous information. On the other, there is "motivated" denial, when one has access to proper information but a strong commitment to a certain worldview—such as political ideology or religion—does not "allow" them to accept the reality of climate change.

Lack of education in general and of basic science, in particular, tends to correlate with climate change denial (e.g., Hamilton, 2011; Hamilton & Keim, 2009; McCright & Dunlap, 2011a; Zhou, 2015). Those with relatively lower scientific literacy are relatively more likely to adopt denialist positions than those with more education. Nonetheless, science literacy, by itself, is not enough to explain denialism (Lubke, 2022). For example, individuals with low scientific literacy who place high trust in climate experts will probably not embrace denialist positions. On the flip side, high (formal) levels of scientific literacy, such as expected from having a university degree, can be "nulled" by motivated denial (e.g., Fielding et al., 2016; Kahan et al., 2012; Kovaka, 2021; Lockwood, 2018; Ranney & Clark, 2016; Whitmarsh, 2011).

One powerful element underpinning motivated climate change denial seems to be a person's (political) values and worldview (Beiser-McGrath & Huber, 2018; Hornsey et al., 2016)—with this phenomenon being more acute in the United States than in other countries (Hornsey et al., 2018). What someone believes to be right or wrong tends to have a strong direct and indirect impact on how they make sense of climate change (see Campbell & Kay, 2014; Hennes et al., 2016; Kahan et al., 2012). For example, an individual who is highly distrustful of authority and expertise is very likely to extend this view to policymakers and climate scientists' calls for action over climate change (Hobson & Niemeyer, 2013; Huber et al., 2020; Motta, 2018). Similarly, those who are strong believers in individualism, free-market neoliberalism, minimal state interference, and private property rights tend to support a climate delay discourse (e.g., Bohr, 2016; Lo, 2014). Embracing a conservative political ideology stands out as a very strong predictor for embracing climate change denialism (Byrne, 2020; Lockwood, 2018; McCright & Dunlap, 2011a, 2011b). Here, climate scientists and policymakers concerned over climate change tend to be framed as "leftists" (or "communists") who are embedded in pushing a so-called woke agenda that also involves issues such as gender norms, the separation of church and state, and the degree of government regulation (or "interference") in the private sector—to name a few. Within the conservative sphere, climate change denial is even more pronounced among those espousing a far-right conservative ideology, which is, by its own nature, a radical and extreme version of conservatism (Byrne, 2020).

In the context of conservative ideology, especially in its populist and more radical versions, gender also intersects with climate change denialism (Agius et al., 2020; McCright & Dunlap, 2011a). Environmental concerns are framed as "feminine" issues and something self-entitled "real men" are not worried about, or worse, deny them as valid concerns (Anshelm & Hultman, 2014; Jylhä et al., 2020; Krange et al., 2018; Letourneau et al., 2024). The concept of "petro-masculinity" (Daggett, 2018), commonly found among white politically conservative males (especially in the United States), encapsulates a set of beliefs, emotions, and behaviors that combine misogyny and climate change denial. Fossil fuels, under this setting, become a part of the masculine *identity* (Nelson, 2020).

Right-wing *populism* (whether or not "*far*" right) is also prime space for denialism (Huber, 2020; Jylhä & Hellmer, 2020; Lockwood, 2018; Norris & Inglehart, 2018). One of populism's core characteristics is the division of society between "the (good) people" and "the (evil) elite"

(Mudde, 2004, 2007). The most important actors in the (combating) climate change space are those with an international profile, such as the IPCC—which sits under the United Nations umbrella—and the United Nations Framework Convention on Climate Change (UNFCCC). Right-wing populists are known for framing international organizations in general, and the United Nations in particular, as part of a "global elite" that is constantly seeking to expand its reach and power over "the people." This is heightened by populists' general disdain of academic/technocratic expertise and high suspicion of the political establishment, both conceived as part of "the elite" (Hobson & Niemeyer, 2013; Mede & Schäfer, 2020; Motta, 2018). Thus, for populists, climate change and actions to combat climate change are yet another "ploy" from the "evil global elite" to control "the people." Right-wing populists' exclusionary and antiegalitarian preferences also involve a general prejudice toward a range of socially disadvantaged groups, such as the poor, immigrants, and women—which tend to be those affected the hardest by climate change (Jylhä & Hellmer, 2020, pp. 317–318). Additionally, there is also a predisposition to reject initiatives toward climate change mitigation (most likely to benefit marginalized groups) that impose higher taxation upon the wealthy and/or high GHG-emitting industries.

Individuals who embrace conspiratorial ideations are also among those very likely to embrace climate change denialism (Lewandowsky, Gignac, & Oberauer, 2013; Lewandowsky, Oberauer, & Gignac, 2013). Conspiracy believers can be defined as those who seek to explain major political or social events as being derived from some secret plot concocted by a cabal of powerful individuals or organizations (Sunstein & Vermeule, 2009). Conspiracy "logic" is immune to refutation: When evidence disproving the theory is presented, the theory is broadened to also include those who produced the evidence (Lewandowsky et al., 2015). Science and scientific evidence, in general, is probably the most common target of conspiracy theorists. Thus, there is practically no surprise that this applies to climate change as well (see Wood et al., 2012). And while not all populists embrace conspiracy ideations, the combination is not uncommon: It is not difficult to imagine those prone to conspiracy beliefs framing the "[evil] elite" as the master conspirator against the "[good] people" (Castanho Silva et al., 2017).

In addition to political ideology, some researchers have focused on psychological elements that might increase (or decrease) the chances of someone being a climate change denialist (see Gifford, 2011). Individuals with low levels of empathy, altruism, trust, and "openness" (being open to new experiences and ideas) have also been shown to positively correlate with denialism (Jylhä & Hellmer, 2020; Milfont et al., 2013, 2015). Another aspect highlights that when facing an intractable problem (such as climate change), some individuals might prefer denial and self-deception over an "unvarnished" vision of truth (Friedrichs, 2011; Norgaard 2006a, 2006b, 2011). Self-protective strategies are a way for humans to deal with painful emotions and fear, avoid identity conflicts, or maintain positive self-esteem (Wullenkord & Reese, 2021). People try to find ways to distance themselves from uncomfortable information, excluding knowledge of painful facts—such as global crisis—so they can carry on their day-to-day lives (Rayner, 2012). Climate change denial in this context can be manifested as someone who agrees that climate change is real and human-caused but thinks, "There is nothing I can do about climate change, so there is no point being worried about it" or "I have too much going on in my personal life to worry about climate change, too."

Now What?

Climate change is real and human-caused. It is one of humanity's most pressing issues and will not "go away" under a business-as-usual approach. Only with purposeful action, done sooner rather than later, can it be tackled. Coordinating policies at a global scale is essential for success. These are some of the baseline elements of addressing the challenges posed by climate change. Climate change denialism represents the opposite of what should be done. Denialism contributes, both directly and indirectly, to making climate change *worse*. Nonetheless, climate change denial exists globally and is a powerful force in many countries. This phenomenon cannot and should not be easily dismissed, even where denialists are a minority.

Dealing with climate change requires, at the very minimum, a significant reduction in current levels of GHG emissions. To do so means doing many things differently. This includes actions such as replacing fuel sources, changing manufacturing processes, rethinking consumption patterns, and imposing stricter pollution thresholds. However, moving away from the existing status quo creates a direct problem for those who benefit from it: loss in profits, if not a full-out obsolescence (e.g., coal plants). Just like the tobacco industry's reluctance to acknowledge their product's links with cancer because it would make them lose money, sectors yielding high GHG emissions also prefer to stay profitable than have to change their modus operandi or pay (literally and figuratively) for their role in worsening climate change. But while tobacco interests were "only" contained to companies that produced tobacco leaves and cigarettes, there are *several* sectors and governments whose existence and/or wealth are intrinsically tied to high GHG emissions.

In the 1990s, when climate change—then mostly referred to as "global warming"—gained mainstream visibility, denying the existence of climate change and its anthropogenic origin was a common strategy. In the past 20 years, the shift has been more toward promoting doubt, whether over climate science's consensus and/or the urgency and effectiveness of action. The main field of debate has gone from "Is there climate change?" to "Who should pay for it, when, and how much?" (Cann & Raymond, 2018). Still, at their core, all climate change denial strategies seek the same goal: stasis, that is, keeping things as they are for as much and as long as possible.

Denialism leads to inaction over climate change's causes. In doing so, it supports large-scale GHG emitters' interests and financial standing, notwithstanding climate change's high costs elsewhere. Those who benefit from the status quo have actively sought to gather supporters for their cause, especially among policymakers and the general public (electors). In this context, this would be difficult if the main strategy was advocacy for a *pro*-climate change position and *more* GHG emissions (not that it hasn't been tried). There has been a (re)framing of climate change mitigation policies as representing a threat to certain values and worldviews that many people strongly embrace. A common strategy when dealing with an unpopular issue/position is to (re)frame it as a symbol of a highly valued virtue that is under attack. In the case of tobacco, for instance, no one would embrace "Cancer is good" or "Cancer is not a big deal" to defend the industry. But if cigarette smoking is packaged as representing "freedom," government policies to limit smoking could be sold as a symbolic attack on this value, increasing the chance of capturing a wider net of supporters like those already set on contesting government "overreach" upon individual choices. Initiatives that promote climate-friendly actions have also been derailed (or tried to be) using this same "slippery slope" logical fallacy: If one relatively small step is taken in a certain direction, then an unstoppable chain of progressively negative events will take place.

Those interested in *increasing* climate change denialism in society have achieved a good level of success, unfortunately, by adapting their position to more palatable and valid discussions. Conservative think tanks (CTTs) represent the best example of this process. In some cases, CTTs promote and reinforce denialism discourse that matches their preexisting political worldviews—such as "The market should decide over the cost of high GHG emissions" or "Scientists are part of an out-of-touch techno-bureaucratic elite." Here, climate change denialism is one of many topics discussed. In other cases, CTTs are created by those with direct financial interests in maintaining the status quo to specifically promote their denialist position while appearing to be "neutral" and nonideological." Either way, both are interested in achieving the same result: slow/no change to deal with climate change.

The reality of denialism is bleak. This is the case for climate change, but as the COVID-19 pandemic has shown, science denialism, rampant misinformation, and an increasing grasp of conspiracy theories are widespread problems. But there is reason for hope: There are strategies that can help minimize, if not partially reverse, denialism's reach. More education and better communication are some of the most evident paths to counter denial (Lehtonen et al., 2019; Lewandowsky, Gignac, & Oberauer, 2013; Monroe et al., 2019; Ranney & Velautham, 2021). This includes actions such as widespread attention to scientific literacy, the promotion of critical thinking among teenagers, and reaching out to journalists and the general public to promote more effective messaging over what scientific "uncertainty" means. Neutralizing misinformation can be done after people receive it (debunking) or before ("prebunking"). The latter involves "inoculating" people to misinformation and is more effective than debunking, although its protective effects might be conditional (Lewandowsky & Van Der Linder, 2021; Spampatti et al., 2024).

Education and communication are necessary to combat climate change denial but are not sufficient, especially when it is attached to deeply entrenched worldviews (Cook, 2019; McCright & Dunlap, 2011b). In fact, "political conservatives are more likely to *reject* the scientific consensus on climate change if they have more education" (Drummond & Fischhoff, 2017, p. 9587, emphasis added). Directly refuting information among those who view climate change (and related policies) through a strong ideological prism can backfire and *increase* misperceptions among members of the group (Bail et al., 2018; Nyhan & Reifler, 2010). This means addressing denial among highly motivated denialists is not easy. It must be carefully thought of and properly implemented to (1) avoid backlash and (2) have some impact on the targeted group. Research on this topic, which gained even more prominence with the COVID-19 pandemic, has found that "corrections should focus on misinformation that circulates among a substantive portion of the population and carries the potential for harm" (Ecker et al., 2022). Regulation of traditional and social media regarding false and socially harmful misinformation can also help to contain the spread and "normalcy" of blatantly erroneous information, including those linked to conspiracy theories.

Ultimately, countering climate change denialism requires a multidisciplinary approach. As it is anchored upon complex normative grounds, discourses to overcome this denial have to integrate a wide range of actions, attentive to *what* is being denied, *how*, and *why*. Solutions must be integrated at various levels—such as public and private; local, national, and international; cultural, socioeconomic, political, and educational backgrounds; legal, ethical, and moral frameworks—and tailored to specific groups to be most successful.

References

Agius, C., Bergman Rosamond, A., & Kinnvall, C. (2020). Populism, ontological insecurity and gendered nationalism: Masculinity, climate denial and Covid-19. *Politics, Religion & Ideology, 21*(4), 432–450.

Anshelm, J., & Hultman, M. (2014). A green fatwā? Climate change as a threat to the masculinity of industrial modernity. *NORMA: International Journal for Masculinity Studies, 9*(2), 84–96.

Bail, C. A., Argyle, L. P., Brown, T. W., Bumpus, J. P., Chen, H., Hunzaker, M. F., Lee, J., Mann, M., Merhout, F., & Volfovsky, A. (2018). Exposure to opposing views on social media can increase political polarization. *Proceedings of the National Academy of Sciences, 115*(37), 9216–9221.

Beiser-McGrath, L. F., & Huber, R. A. (2018). Assessing the relative importance of psychological and demographic factors for predicting climate and environmental attitudes. *Climatic Change, 149*(3–4), 335–347.

Björnberg, K. E., Karlsson, M., Gilek, M., & Hansson, S. O. (2017). Climate and environmental science denial: A review of the scientific literature published in 1990–2015. *Journal of Cleaner Production, 167*(20), 229–241.

Bohr, J. (2016). The "climatism" cartel: Why climate change deniers oppose market-based mitigation policy. *Environmental Politics, 25*(5), 812–830.

Brulle, R. J. (2014). Institutionalizing delay: Foundation funding and the creation of US climate change counter-movement organizations. *Climatic Change, 122*, 681–694.

Brulle, R. J. (2023). Advocating inaction: A historical analysis of the Global Climate Coalition. *Environmental Politics, 32*(2), 185–206.

Brulle, R. J., Carmichael, J., & Jenkins, J. C. (2012). Shifting public opinion on climate change: An empirical assessment of factors influencing concern over climate change in the US, 2002–2010. *Climatic Change, 114*, 169–188.

Busch, T., & Judick, L. (2021). Climate change—that is not real! A comparative analysis of climate-sceptic think tanks in the USA and Germany. *Climatic Change, 164*(18), 1–23.

Byrne, G. (2020). Climate change denial as far-right politics: How abandonment of scientific method paved the way for Trump. *Journal of Human Rights and the Environment, 11*(1), 30–60.

Campbell, T. H., & Kay, A. C. (2014). Solution aversion: On the relation between ideology and motivated disbelief. *Journal of Personality and Social Psychology, 107*(5), 809.

Cann, H. W., & Raymond, L. (2018). Does climate denialism still matter? The prevalence of alternative frames in opposition to climate policy. *Environmental Politics, 27*(3), 433–454.

Castanho Silva, B., Vegetti, F., & Littvay, L. (2017). The elite is up to something: Exploring the relation between populism and belief in conspiracy theories. *Swiss Political Science Review, 23*(4), 423–443.

Cho, C. H., Martens, M. L., Kim, H., & Rodrigue, M. (2011). Astroturfing global warming: It isn't always greener on the other side of the fence. *Journal of Business Ethics, 104*, 571–587.

Cohen, S. (2001). *States of denial: Knowing about atrocities and suffering.* Polity.

Cook, J., Nuccitelli, D., Green, S. A., Richardson, M., Winkler, B., Painting, R., Way, R., Jacobs, P., & Skuce, A. (2013). Quantifying the consensus on anthropogenic global warming in the scientific literature. *Environmental Research Letters, 8*(2), 024024.

Cook, J., Lewandowsky, S., & Ecker, U. K. H. (2017). Neutralizing misinformation through inoculation: Exposing misleading argumentation techniques reduces their influence. *PloS One, 12*(5), e0175799.

Cook, J. (2019). Understanding and countering misinformation about climate change. In I. Chiluwa & S. Samoilenko (Eds.), *Handbook of research on deception, fake news, and misinformation online* (pp. 281–306). IGI-Global.

Cook, J. (2020). Deconstructing climate science denial. In D. C. Holmes & L. M. Richardson (Eds.), *Research handbook on communicating climate change* (pp. 62–78). Edward Elgar.

Daggett, C. (2018). Petro-masculinity: Fossil fuels and authoritarian desire. *Millennium, 47*(1), 25–44.

Diethelm, P., & McKee, M. (2009). Denialism: What is it and how should scientists respond? *European Journal of Public Health, 19*(1), 2–4.

Drummond, C., & Fischhoff, B. (2017). Individuals with greater science literacy and education have more polarized beliefs on controversial science topics. *Proceedings of the National Academy of Sciences, 114*(36), 9587–9592.

Duan, R., & Miller, S. (2021). Climate change in China: A study of news diversity in party-sponsored and market-oriented newspapers. *Journalism, 22*(10), 2493–2510.

Dunlap, R. E. (2013). Climate change skepticism and denial: An introduction. *American Behavioral Scientist, 57*(6), 691–698.

Dunlap, R. E., & Brulle, R. J. (2020). Sources and amplifiers of climate change denial. In D. C. Holmes & L. M. Richardson (Eds.), *Research handbook on communicating climate change* (pp. 49–61). Edward Elgar.

Dunlap, R. E., & Jacques, P. J. (2013). Climate change denial books and conservative think tanks: Exploring the connection. *American Behavioral Scientist, 57*(6), 699–731.

Dunlap, R. E., & McCright, A. M. (2010). Challenging climate change denial: Sources, actors and strategies. In C. Lever-Tracy (Ed.), *Routledge handbook of climate change and society* (pp. 240–259). Routledge.

Dunlap, R. E., & McCright, A. M. (2011). Organized climate change denial. In S. Brechin & S. Lee *(Eds.), The Oxford handbook of climate change and society* (pp. 144–160). Oxford University Press.

Dunlap, R. E., & McCright, A. M. (2015). Challenging climate change. In R. E. Dunlap & R. J. Brulle (Eds.), *Climate change and society: Sociological perspectives* (pp. 300–332). Oxford University Press.

Ecker, U. K., Lewandowsky, S., Cook, J., Schmid, P., Fazio, L. K., Brashier, N., Kendeou, P., Vraga, E. K., & Amazeen, M. A.(2022). The psychological drivers of misinformation belief and its resistance to correction. *Nature Reviews Psychology, 1*(1), 13–29.

Elsasser, S. W., & Dunlap, R. E. (2013). Leading voices in the denier choir: Conservative columnists' dismissal of global warming and denigration of climate science. *American Behavioral Scientist, 57*(6), 754–776.

Ferkany, M. (2015). Is it arrogant to deny climate change or is it arrogant to say it is arrogant? Understanding arrogance and cultivating humility in climate change discourse and education. *Environmental Values, 24*(6), 705–724.

Fielding, K. S., & Hornsey, M. J. (2016). A social identity analysis of climate change and environmental attitudes and behaviors: Insights and opportunities. *Frontiers in Psychology, 7*, 171214. DOI: 10.3389/fpsyg.2016.00121

Fischer, F. (2019). Knowledge politics and post-truth in climate denial: On the social construction of alternative facts. *Critical Policy Studies, 13*(2), 133–152.

Friedrichs, J. (2011). Peak energy and climate change: The double bind of post-normal science. *Futures, 43*(4), 469–477.

Gifford, R. (2011). The dragons of inaction: Psychological barriers that limit climate change mitigation and adaptation. *American Psychologist, 66*(4), 290–302.

Hamilton, L. C. (2011). Education, politics, and opinions about climate change: Evidence for interaction effects. *Climatic Change, 104*, 231–242.

Hamilton, L. C., & Keim, B. D. (2009). Regional variation in perceptions about climate change. *International Journal of Climatology, 29*, 2348–2352.

Hennes, E. P., Ruisch, B. C., Feygina, I., Monteiro, C. A., & Jost, J. T. (2016). Motivated recall in the service of the economic system: The case of anthropogenic climate change. *Journal of Experimental Psychology: General, 145*(6), 755.

Hobson, K., & Niemeyer, S. (2013). "What sceptics believe": The effects of information and deliberation on climate change scepticism. *Public Understanding of Science, 22*(4), 396–412.

Hoffman, A. J. (2011). The culture and discourse of climate skepticism. *Strategic Organization, 9*(1), 77–84.

Hornsey, M. J., Harris, E. A., Bain, P. G., & Fielding, K. S. (2016). Meta-analyses of the determinants and outcomes of belief in climate change. *Nature Climate Change, 6*(6), 622–626.

Hornsey, M. J., Harris, E. A., & Fielding, K. S. (2018). Relationships among conspiratorial beliefs, conservatism, and climate scepticism across nations. *Nature Climate Change, 8*(7), 614–620.

Huber, R. A. (2020). The role of populist attitudes in explaining climate change skepticism and support for environmental protection. *Environmental Politics, 29*(6), 959–982.

Jacques, P. J., Dunlap, R. E., & Freeman, M. (2008). The organisation of denial: Conservative think tanks and environmental scepticism. *Environmental Politics, 17*(3), 349–385.

Jamelske, E., Boulter, J., Jang, W., Barrett, J., Miller, L., & Han, W. L. (2015). Examining differences in public opinion on climate change between college students in China and the USA. *Journal of Environmental Studies and Sciences, 5*, 87–98.

Jylhä, K. M., & Hellmer, K. (2020). Right-wing populism and climate change denial: The roles of exclusionary and anti-egalitarian preferences, conservative ideology, and antiestablishment attitudes. *Analyses of Social Issues and Public Policy, 20*(1), 315–335.

Jylhä, K. M., Tam, K.-P., & Milfont, T. L. (2021). Acceptance of group-based dominance and climate change denial: A cross-cultural study in Hong Kong, New Zealand, and Sweden. *Asian Journal of Social Psychology, 24*, 198–207.

Jylhä, K. M., Strimling, P., & Rydgren, J. (2020). Climate change denial among radical right-wing supporters. *Sustainability, 12*(10226), 1–15.

Kahan, D. M., Peters, E., Wittlin, M., Slovic, P., Ouellette, L. L., Braman, D., & Mandel, G. (2012). The polarizing impact of science literacy and numeracy on perceived climate change risks. *Nature Climate Change, 2*(10), 732–735.

Klein, N. (2011). Capitalism vs. the climate. *The Nation*, November 28.

Klein, N. (2015). *This changes everything: Capitalism vs. the climate*. Simon and Schuster.

Koehrsen, J. (2021). Muslims and climate change: How Islam, Muslim organizations, and religious leaders influence climate change perceptions and mitigation activities. *Wiley Interdisciplinary Reviews: Climate Change, 12*(3), e702.

Kokic, P., Crimp, S., & Howden, M. (2014). A probabilistic analysis of human influence on recent record global mean temperature changes. *Climate Risk Management, 3*, 1–12.

Kovaka, K. (2021). Climate change denial and beliefs about science. *Synthese, 198*, 2355–2374.

Krange, O., Kaltenborn, B. P., & Hultman, M. (2018). Cool dudes in Norway: Climate change denial among conservative Norwegian men. *Environmental Sociology, 5*(10), 1–11.

Lamb, W. F., Mattioli, G., Levi, S., Roberts, J. T., Capstick, S., Creutzig, F., Minx, J. C., Müller-Hansen, F., Culhane, T., & Steinberger, J. K.(2020). Discourses of climate delay. *Global Sustainability, 3*(e17), 1–5.

Lehtonen, A., Salonen, A. O., & Cantell, H. (2019). Climate change education: A new approach for a world of wicked problems. In J. W. Cook (Eds.), *Sustainability, human well-being, and the future of education* (pp. 339–374). Springer Nature.

Leiserowitz, A. A., Maibach, E. W., Roser-Renouf, C., Smith, N., & Dawson, E. (2013). Climategate, public opinion, and the loss of trust. *American Behavioral Scientist, 57*(6), 818–837.

Letourneau, A., Davidson, D., Karsgaard, C., & Ivanova, D. (2024). Proud fathers and fossil fuels: Gendered identities and climate obstruction. *Environmental Politics, 33*(4), 678–698.

Lewandowsky, S., Cook, J., Oberauer, K., Brophy, S., Lloyd, E. A., & Marriott, M. (2015). Recurrent fury: Conspiratorial discourse in the blogosphere triggered by research on the role of conspiracist ideation in climate denial. *Journal of Social and Political Psychology, 3*(1), 142–178.

Lewandowsky, S., Gignac, G. E., & Oberauer, K. (2013). The role of conspiracist ideation and worldviews in predicting rejection of science. *PloS One, 8*(10), e75637.

Lewandowsky, S., Oberauer, K., & Gignac, G. E. (2013). NASA faked the moon landing—Therefore, science is a hoax: An anatomy of the motivated rejection of science. *Psychological Science, 24*(5), 622–633.

Lewandowsky, S., & Van Der Linden, S. (2021). Countering misinformation and fake news through inoculation and prebunking. *European Review of Social Psychology, 32*(2), 348–384.

Liu, C., & Wallace, J. L. (2023). What's not trending on Weibo: China's missing climate change discourse. *Environmental Research Communications, 5*(1), 011002.

Liu, J. C.-E. (2015). Low carbon plot: Climate change skepticism with Chinese characteristics. *Environmental Sociology, 1*(4), 280–292.

Liu, J. C.-E. (2023). Public opinion on climate change in China—Evidence from two national surveys. *PLOS Climate, 2*(2), e0000065.

Lo, A. Y. (2014). The right to doubt: Climate-change scepticism and asserted rights to private property. *Environmental Politics, 23*(4), 549–569.

Lockwood, M. (2018). Right-wing populism and the climate change agenda: Exploring the linkages. *Environmental Politics, 27*(4), 712–732.

Lubke, C. (2022). Socioeconomic roots of climate change denial and uncertainty among the European population. *European Sociological Review, 38*(1), 153–168.

Maibach, E., Leiserowitz, A., Cobb, S., Shank, M., Cobb, K. M., & Gulledge, J. (2012). The legacy of climategate: Undermining or revitalizing climate science and policy? *Wiley Interdisciplinary Reviews: Climate Change, 3*(3), 289–295.

McCright, A. M., & Dunlap, R. E. (2011a). Cool dudes: The denial of climate change among conservative white males in the United States. *Global Environmental Change, 21*(4), 1163–1172.

McCright, A. M., & Dunlap, R. E. (2011b). The politicization of climate change and polarization in the American public's views of global warming, 2001–2010. *The Sociological Quarterly, 52*(2), 155–194.

Mede, N. G., & Schäfer, M. S. (2020). Science-related populism: Conceptualizing populist demands toward science. *Public Understanding of Science, 29*(5), 473–491.

Milfont, T. L., Milojev, P., Greaves, L. M., & Sibley, C. G. (2015). Socio-structural and psychological foundations of climate change beliefs. *New Zealand Journal of Psychology, 44*(1), 17.

Milfont, T. L., Richter, I., Sibley, C. G., Wilson, M. S., & Fischer, R. (2013). Environmental consequences of the desire to dominate and be superior. *Personality and Social Psychology Bulletin, 39*(9), 1127–1138.

Milfont, T. L., & Sibley, C. G. (2016). Empathic and social dominance orientations help explain gender differences in environmentalism: A one-year Bayesian mediation analysis. *Personality and Individual Differences, 90*, 85–88.

Monroe, M. C., Plate, R. R., Oxarart, A., Bowers, A., & Chaves, W. A. (2019). Identifying effective climate change education strategies: A systematic review of the research. *Environmental Education Research, 25*(6), 791–812.

Motta, M. (2018). The dynamics and political implications of anti-intellectualism in the United States. *American Politics Research, 46*(3), 465–498.

Mudde, C. (2004). The populist zeitgeist. *Government and Opposition, 39*(4), 542–563.

Mudde, C. (2007). *Populist radical right parties in Europe*. Cambridge University Press.

Nelson, J. (2020). Petro-masculinity and climate change denial among white, politically conservative American males. *International Journal of Applied Psychoanalitic Studies, 17*, 282–295.

Norgaard, K. M. (2006a). "People want to protect themselves a little bit": Emotions, denial, and social movement nonparticipation. *Sociological Inquiry, 76*(3), 372–396.

Norgaard, K. M. (2006b). "We don't really want to know": Environmental justice and socially organized denial of global warming in Norway. *Organization & Environment, 19*(3), 347–370.

Norgaard, K. M. (2011). *Living in denial: Climate change, emotions, and everyday life*. MIT Press.

Norris, P., & Inglehart, R. (2018). *Cultural backlash: Trump, Brexit, and the rise of authoritarian-populism*. Cambridge University Press.

Nyhan, B., & Reifler, J. (2010). When corrections fail: The persistence of political misperceptions. *Political Behavior, 32*(2), 303–330.

Ojeda, D., Sasser, J. S., & Lunstrum, E. (2020). Malthus's specter and the anthropocene. *Gender, Place & Culture, 27*(3), 316–332.

O'Neill, S. J., & Boykoff, M. (2010). Climate denier, skeptic, or contrarian? *Proceedings of the National Academy of Sciences, 107*(39), E151.

Oreskes, N., & Conway, E. M. (2010). *Merchants of doubt*. Bloomsbury Press.

Painter, J., & Ashe, T. (2012). Cross-national comparison of the presence of climate skepticism in the print media in six countries, 2007–10. *Environmental Research Letters, 7*(4), 044005.

Painter, J., & Gavin, N. T. (2016). Climate skepticism in British newspapers, 2007–2011. *Environmental Communication, 10*(4), 432–452.

Pew Research Center. (2022). How religion intersects with Americans' views on the environment. https://www.pewresearch.org/religion/2022/11/17/how-religion-intersects-with-americans-views-on-the-environment/

Ranney, M. A., & Clark, D. (2016). Climate change conceptual change: Scientific information can transform attitudes. *Topics in Cognitive Science, 8*(1), 49–75.

Ranney, M. A., & Velautham, L. (2021). Climate change cognition and education: Given no silver bullet for denial, diverse information-hunks increase global warming acceptance. *Current Opinion in Behavioral Sciences, 42*, 139–146.

Rayner, S. (2012). Uncomfortable knowledge: The social construction of ignorance in science and environmental policy discourses. *Economy and Society, 41*(1), 107–125.

Sachdeva, S. (2016). Religious identity, beliefs, and views about climate change. In H. von Storch (Ed.), *Oxford research encyclopedia of climate science*. Oxford University Press.

Schmid, P., & Betsch, C. (2019). Effective strategies for rebutting science denialism in public discussions. *Nature Human Behavior, 3*(9), 931–939.

Schultz, S. (2021). The neo-Malthusian reflex in climate politics: Technocratic, right wing and feminist references. *Australian Feminist Studies, 36*(110), 485–502.

Shen, C., & Wang, Y. (2023). Concerned or Apathetic? Exploring online public opinions on climate change from 2008 to 2019: A Comparative study between China and other G20 countries. *Journal of Environmental Management, 332*, 117376.

Shue, H. (2023). Unseen urgency: Delay as the new denial. *Wiley Interdisciplinary Reviews: Climate Change, 14*(1), 1–6.

Spampatti, T., Hahnel, U. J. J., Trutnevyte, E., & Brosch, T. (2024). Psychological inoculation strategies to fight climate disinformation across 12 countries. *Nature Human Behaviour, 8*(2), 380–398.

Sunstein, C. R., & Vermeule, A. (2009). Conspiracy theories: Causes and cures. *Journal of Political Philosophy, 17*(2), 202–227.

Union of Concerned Scientists. (2007). Smoke, mirrors & hot air: How ExxonMobil uses Big Tobacco's tactics to "manufacture uncertainty" on climate change. https://www.ucsusa.org/resources/smoke-mirrors-hot-air

Veldman, R. G., Wald, D. M., Mills, S. B., & Peterson, D. A. M. (2021). Who are American evangelical Protestants and why do they matter for US climate policy? *Wiley Interdisciplinary Reviews: Climate Change, 12*(2), e693.

Verhoeven, H. (2011). Climate change, conflict and development in Sudan: Global neo-Malthusian narratives and local power struggles. *Development and Change, 42*(3), 679–707.

Wang, B., & Zhou, Q. (2020). Climate change in the Chinese mind: An overview of public perceptions at macro and micro levels. *Wiley Interdisciplinary Reviews: Climate Change, 11*(3), e639.

Weart, S. (2011). Global warming: How skepticism became denial. *Bulletin of the Atomic Scientists, 67*(1), 41–50.

World Economic Forum (WEF). (2020). Toward a more sustainable world: A global study of public opinion. http://www3.weforum.org/docs/WEF_More_Sustainable_World.pdf

Whitmarsh, L. (2011). Scepticism and uncertainty about climate change: Dimensions, determinants and change over time. *Global Environmental Change, 21*(2), 690–700.

Wood, M. J., Douglas, K. M., & Sutton, R. M. (2012). Dead and alive: Beliefs in contradictory conspiracy theories. *Social Psychological and Personality Science, 3*(6), 767–773.

Wullenkord, M. C., & Reese, G. (2021). Avoidance, rationalization, and denial: Defensive self-protection in the face of climate change negatively predicts pro-environmental behavior. *Journal of Environmental Psychology, 77*(101683), 1–15.

Yang, J., Gounaridis, D., Liu, M., Bi, J., & Newell, J. P. (2021). Perceptions of climate change in China: Evidence from surveys of residents in six cities. *Earth's Future, 9*(12), e2021EF002144.

Zaleha, B. D., & Szasz, A. (2015). Why conservative Christians don't believe in climate change. *Bulletin of Atomic Scientists, 71*(5), 19–30.

Zhou, M. (2015). Public environmental skepticism: A cross-national and multilevel analysis. *International Sociology, 30*(1), 61–85.

CHAPTER 39

The Future of Climate Action

PAUL ALMEIDA

Introduction

WITH climate change trends continuing in the wrong direction with accelerated planetary warming, record levels of CO_2 in the earth's atmosphere, and ongoing extreme weather events of wildfires, droughts, hurricanes, flooding, and heat waves, the calls for climate action could not be more urgent. While several forms of climate action gain traction and attempt to place civilization on a climate-friendly and ecologically sustainable trajectory, multiple obstacles remain in place. Much progress has been made in the areas of climate action planning, renewable energy production, and the implementation of nature-based climate solutions. Outside of the institutions and scientific developments that address global warming, over the past two decades, mass mobilization in favor of pro-climate policies has also garnered the attention of the general public and policymakers. After a summary of these achievements, the chapter focuses on the main challenges to overcome in the next decade to deepen the impacts of climate action and avoid the catastrophic scenarios associated with surpassing irreversible earth system tipping points. These obstacles include the lack of civic engagement over the climate crisis, climate financing, an overemphasis on decarbonization, and the growing backlash of climate denialism and authoritarian political mobilization.

Advances in Institutional Climate Action

Three forms of institutional-based climate action appear especially promising in the second quarter of the 21st century. These dominant forms include (1) local-level climate action planning; (2) the expansion of the renewable energy sector and technologies; and (3) natural climate solutions (NCSs).

City Planning

As one of the first extensive initiatives to address climate change in the early 1990s, city-level climate action planning has made significant gains in reducing greenhouse gases (GHGs) as well as developing several other innovative endeavors. City-level programs usually encompass a range of practices, including greenhouse gas inventories, GHG reduction goals and benchmarks, climate-friendly local transportation systems, afforestation, electrification of infrastructure and buildings, and commitments from private industry to reduce GHGs. Originally focused on climate mitigation strategies, climate action plans (CAPs) now also incorporate several concerns with equity, including publishing materials in multiple languages, making green benefits accessible to low-income populations, and enhancing participation of marginalized populations in climate resiliency and adaptation. The combined membership of the major global climate action planning networks at the local level has reached nearly 15,000 cities, including C40, ICLEI, and the Covenant of Mayors. These municipal climate action planning programs are active in over 150 countries (see Leffel, this volume).

City-level CAPs are global in scope and include some of the world's largest cities, including municipalities in authoritarian and semiauthoritarian states. They provide a template on how to mitigate GHGs at the local level and continue to diffuse to more towns and cities across the world. CAPs provide a protective space to develop a range of policies and guidelines relevant to local conditions, even when national governments may be less favorable to addressing the climate crisis. One analysis estimates that between 2008 and 2020, the European Covenant of Mayors Program, across over 1,600 city CAPs, reached their benchmarks of reducing GHG emissions by 20% in 86% of cities (Rivas et al., 2022). The Global Covenant of Mayors Program, representing 13,766 municipalities across the planet, calculates that if city partners continue meeting emission mitigation targets, they will reduce the amount of CO_2 released by 4.2 gigatons (Global Covenant of Mayors, 2024).

Green Energy

The expansion of renewable energy systems and technologies over the past few decades marks another major advance in climate action. This has been especially true for photovoltaics, biofuels, hydropower, and turbine-based energy generation, while the development of geothermal, hydrogen, gravitational, tidal, and other forms of clean energy is increasing. Solar and wind-based energy grew rapidly in the early 2020s (see Hochstetler, this volume). More states continue to subsidize and support such industries while social movements advocate for their implementation (see Vasi, this volume). Nonetheless, most of renewable energy investment is concentrated in the Global North and China, with the lowest income nations receiving less than 1% of global green energy investments (see Hochstetler, this volume). The inequitable distribution of renewable energy financing should place the Global South as a high priority for future growth. It is also critical that in the territorial regions of green energy extraction, production, and distribution that the basic principles of environmental justice are recognized, as discussed in the later section on antiextraction movements (Backhouse & Lehman, 2020; Lehman & Tittor, 2023).

Natural Climate Solutions

NCSs such as urban greening, reforestation, biochar, algae, cover cropping, and other sustainable interventions to trap and absorb carbon continue to accelerate. The term and practices of NCSs became popularized in scientific and policy circles in the 2010s as a means to incorporate natural ecosystems into climate mitigation efforts with minimal impact on food supplies and environmental sustainability (Osaka et al., 2021). By some estimates, if NCSs were deployed on a massive scale, with $400 billion in annual investments, they could account for 30% of all mitigation needed by 2030 to reach net-zero emissions (Ellis et al., 2024). In this handbook, NCS strategies in Africa are discussed by Epule and Poirier and in Latin America by Álvarez. Another mechanism for rapidly expanding NCS programs involves making them integral to national- and local-level climate action planning initiatives. According to Álvarez in this collection, NCS strategies can be especially scaled up when integrated into larger land restoration projects with regenerative ecological benchmarks (including biodiversity) involving local stakeholders.

All three promising forms of climate action outlined earlier would also benefit by prioritizing workforce development opportunities. Workforce development programs with high wages, medical and retirement benefits, career ladders, and reskilling opportunities push climate action on more of a just transitions pathway (Kerrissey and Coon, this volume). These policies enhance just transition outcomes when states recruit large swathes of the low-carbon economy workforce from disadvantaged communities (Zabin, 2020). More research is needed in systematically capturing the preferences of workers in disadvantaged communities (Almeida et al., 2024). Workers traditionally experience "good job" characteristics in the manufacturing sector, but these labor standards also need to be transferred to the service sector in an emerging green economy (Hess, 2012). The kinds of employment conditions are often referred to as "high road" jobs, and local and national governments could provide more incentives to motivate public and private employers working in climate action-related industries to adopt and institutionalize such practices (Brostrom et al., 2024; Pollin et al., 2021).

Advances in Noninstitutional Climate Action

Noninstitutional climate action involves collective mobilization over climate-related issues largely outside of institutions and routine advocacy channels (Almeida, González, and Fonsah, 2024). Two particularly efficacious types of noninstitutional action include the transnational climate movement and local struggles over fossil fuel-related extraction and transportation.

Global Climate Movement

The international climate social movement has made several contributions to addressing the climate crisis. These contributions involve expanding participation and knowledge of planetary warming and its chief causes. Such activities are key to what social movement scholars refer to as diagnostic framing and provide key requisites for social mobilization in terms of blame and attribution of the problem in question (Snow & Benford, 1988). The transnational movement took off in 2005 and continued to diffuse rapidly through 2019. The movement reached nearly every nation with at least one demonstration or pro-climate event and mobilized up to 6 million people during international days of action in the late 2010s (de Moor et al., 2021). Global climate organizing largely targets international institutions (such as the United Nations) and major GHG emitting nation-states (such as the European Union, the United States, Russia, India, and China). In recent years, 350.org, Fridays for Future (FFF), and Extinction Rebellion served as the most prominent global climate movement organizations (see Bates, Boeve, and Henn; Uba; and Carvalho and Gardner, this volume). Additional achievements include placing pressure on the signing of the Paris Agreement in 2015, influencing components of the Green Deal in the European Union (especially the Just Transition Fund, and Popular Participation) (Parks, 2025; Tosun, this volume), and the Loss and Damage fund for Global South nations reached at the 29th Conference of the Parties (COP) in 2024. Others have documented a relationship between global climate movement activity and the growth of local climate action programs in Latin America (Mondragón, 2025). The greatest strength of the global movement occurs when it can consistently mobilize weekly global protests to sustain momentum, as was the case in the second half of 2019 via efforts by FFF.

Antiextraction Movements

Extraction movements form at the point of initial resource removal and mining of fossil fuels and their transportation for processing, consumption, and combustion. Antiextraction struggles provide indispensable resources for the future of climate action in terms of keeping the focus on a just transition trajectory. Communities facing extraction activities are often vulnerable and marginalized, placing them within the larger environmental justice movement (González Marquez, 2025). The conflicts involve oil and gas drilling, fracking, and transportation, especially via pipelines. The Environmental Justice Atlas (EJ Atlas) has documented hundreds of such struggles across the globe (see chapters by Martínez-Alier, Arce, and Ovadia, this volume). Examples of these battles range from the Dakota Access Pipeline (DAPL), the Keystone Pipeline, and Northern Access Pipeline (NAPL) in the United States, to the East African Crude Pipeline (EACOP) running from Uganda to Tanzania, as well as ongoing struggles in the Niger Delta of Nigeria. Excluded communities facing extraction not only demand an immediate cessation or withdrawal of activities but also compensation, environmental restoration, and sustainable economic development investments with high-quality jobs, sovereignty, and autonomy. These kinds of claims help push climate action in general on a just transition trajectory away from climate denialism and carbon reductionism.

As mentioned earlier, territorial conflicts also include green energy extraction, production, and distribution, including mining operations for rare earth minerals and critical minerals such as copper, lithium, and cobalt to drive green manufacturing and production (see Almeida, Fonsah, and González, this volume). Successful antiextraction movements include the case of Arvin in Kern County, California, where a working-class Latino community effectively scaled up its campaign to the state level with the passage of legislation to prevent oil drilling near schools, residential housing, and medical facilities (see Farrell and Brostrom, this volume). In Ecuador, a coalition of Indigenous and environmental groups spearheaded a nation-wide referendum passed in 2023 to prevent oil drilling in the ecologically sensitive Yasuní National Park (Alarcón, 2024). Incorporating antiextraction movements into larger climate action strategies facilitates maintaining climate justice as a primary objective.

Major Impediments to the Future of Climate Action

Despite the successes of institutional and noninstitutional forms of climate action, major obstacles remain. The consequences of these barriers involve accelerated heating of the earth's surface temperature, increased frequency of climate-related shocks and weather events, and the continued growing amount of GHGs in the atmosphere, especially CO_2 (Fisher, 2024). Four major areas of blockage from meeting climate goals in a shorter time horizon include (1) a lack of civic engagement; (2) climate financing and reparations; (3) the dominance of carbon reductionism and decarbonization as strategies in key policymaking circles; and (4) the growing threat of climate delay and denial from authoritarian governing structures and the fossil fuel industry.

Civic Engagement

Cross-national and global surveys show around 60% of the public is concerned about climate change (Leal Farias, this volume). A much smaller percentage of the public is actually engaged or participates in some form of climate action. Most of the scholarly literature tends to concentrate on participation in climate protest events and demonstrations, which is a small percentage of the general population and largely middle class (della Porta & Portos, 2023;

Fisher, 2018). For a rapid expansion in climate action to occur, much more of civil society needs to be involved. In the short term, local and national governments could shape policy mandating civic engagement for climate action projects funded by the state, from subsidies to alternative energy to climate action planning and nature-based climate solutions. Over the long term, a massive implementation of climate science and climate action curriculum in the public school systems across the world would create a new generation of young adults prepared with skills and capabilities to participate in a wide range of climate initiatives.

In terms of priority populations for climate action initiatives, projects should commence in low-income and polluted regions, or what environmental social science scholars refer to as "environmental justice" or "front-line" communities (London et al., 2008), and mentioned earlier as "disadvantaged communities." Working in climate civic engagement with environmental justice communities ensures that other pressing issues will be combined with climate action if participatory principles and local needs are given precedence. The state of California and the national government under President Biden created granular classification schemes to identify front-line communities based on indicators of cumulative pollution and low household income levels. In the United States, where the environmental justice movement initially emerged (Bullard, 2005; Taylor, 2014), policymakers and social science collaborators developed environmental justice classification schemes down to the census tract level. Such geographically based indicators of environmental injustice would be useful to implement in nation-states throughout the world to identify key locations to emphasize climate action led by local front-line communities.

The EJ Atlas is the closest measure we have of documentation of environmental justice regions at the international level with a reported 4,329 conflicts in front-line communities (Martinez-Alier et al., 2016; Martínez Alier, this volume). More pathbreaking and systematic empirical work that includes polluted communities where mobilization has not emerged would be an additional resource for marginalized communities (Auyero & Swistun, 2009). Another mechanism to activate environmental justice communities in climate action civic engagement would be to invest in labor and community-based organizations (CBOs) already working in environmentally compromised regions. Labor unions and CBOs offer training, trust, and capacity building in civic skills to bring more people into public participation involving climate issues (Almeida et al., 2023). In short, a massive expansion of climate action in front-line communities both adds to a mass base of support and sustains focus on just transition type solutions and strategies.

Climate Financing

Another major challenge to advance climate action initiatives involves funding projects in the Global South and environmental justice communities in the Global North. There also needs to be greater recognition and compensation for past climate-induced destruction from more powerful and industrialized nation-states with the highest levels of historic GHG emissions. In the early 2020s, the United Nations COP established a "Loss and Damage" fund for the Global South. Current pledges of around $800 million from 25 nations are viewed by most observers as utterly inadequate, including by the United Nations itself (see Khan and Khan, this volume). Climate financing experts suggest *annual pledges* of hundreds of billions of dollars to make substantial progress in preparing Global South nations for climate resiliency and green technology transfer. Ongoing pressure from Global South environmental justice advocates for increased climate investments from the Global North, and even climate reparations, continue as a top priority, such as in 2010, when an estimated 20,000 people participated in the World People's Conference on Climate Change and the Rights of Mother

Earth in Cochabamba, Bolivia, and made demands for climate reparations (Pellow, 2021). Based on a comprehensive review of the shortcomings of past climate financing, Uri et al. (2024) suggest that the UN Loss and Damage program should be governed with substantial representation from the Global South and build mechanisms and channels for rapid access and delivery of funds. Even in the Global North, renewable energy initiatives and climate action planning programs are more often found in relatively more affluent regions within nation-states.

Carbon Reductionism

The emphasis on decarbonization and climate mitigation alone serves as a major hurdle in expanding climate action (Méndez, 2020). The history of the over focus on carbon reductionism resides in the scientific nature of first detecting the problem of global warming and the greenhouse effect. Highlighting the urgency of the issue gained momentum with atmospheric scientists in the 1980s, and they sought very specific solutions by curbing carbon emissions. Much of the early climate action planning in the 1990s also focused on reducing GHG emissions by achieving benchmarks for cities, states, and national governments. In more recent years, climate advocates demand more equity considerations to be incorporated into climate action planning and renewable energy programs. To achieve these equity goals, placing environmental justice communities at the forefront of green transitions moves away from a strict carbon mitigation effort and acknowledges the intersectional nature of the climate crisis.

In addition, converting climate science into everyday language to raise public awareness remains challenging (Barnes, 2025). Jargon-filled terms such as "decarbonization," "mitigation," "adaptation," and "carbon management" are confined to smaller groups of scholars, policymakers, consultants, and planners. Climate advocates would benefit from using a vocabulary that clearly articulates the urgency of turning back planetary warming with minimal turgidity (Maibach et al., 2023). A listening tour by decarbonization proponents in environmental justice communities would assist in developing such a vocabulary (Cha & Pastor, 2022). At the same time, an engagement with front-line regions would bring in other priorities and pressing needs such as immigration reform, low wages, housing costs, price inflation, child and elderly care, health care access, and the costs of utilities. Incorporating these issues into climate civic engagement would push emphases away from a carbon-only focus.

Climate Denialism and Authoritarian Governance

Perhaps the largest obstacle to advancing climate action centers on the rise of authoritarian populism over the past three decades. From the United States to South America and Europe, a growing number of authoritarian leaders and political parties place antienvironmentalism and anticlimate policies on their electoral campaign platforms. In addition, fossil fuel-based industries and capital independently promote climate misinformation, delay, and denial while also funding anticlimate political parties (Leal Farias, this volume). Some climate countermovement organizations operating at an international level trace their origins back to 1989, such as the Global Climate Council (Brulle, 2023). Industry groups, especially in the oil and gas industry, fund think tanks, front organizations, lobby groups, and commission studies trying to create uncertainty to the established climate sciences (e.g., atmospheric, paleontology, ecology, and geology) on the causes of climate change (Dunlap & Mcright, 2011). Such coalitions sponsored by the fossil fuel industry effectively prevented major climate change legislation from passing in the United States in 2010 (Nanko & Coan, 2024).

Authoritarianism alone may not impede the expansion of green industries such as photovoltaics and electric vehicles in nondemocratic states. However, anticlimate and antienvironmental authoritarian and semiauthoritarian regimes repress pre-existing climate action. Funding and research for climate-related initiatives may become suppressed by the government and activists become targets for state sanctions and harassment, especially in regions characterized by resource extraction. From 2017 to 2020, and especially since 2025, the United States has disinvested in renewable energy and climate action in general and removed climate change data from official websites. Across the world, other leading political figures and political parties engage in climate skepticism or undermine climate action efforts, such as what has occurred in Brazil under President Bolsonaro and partially in Argentina under President Milei. In Europe, this is evident in rightest parties such as Alternative for Germany (AfD), Vox party in Spain, Northern League in Italy, Reform UK, Party for Freedom in the Netherlands, and the Freedom Party of Austria. Rightist parties also at times take an eco-nationalist view on climate issues with a focus on heritage and the local environment within a country's boundaries leading to attempts at implementing xenophobic and neo-Malthusian policies.

Conclusion

While this Handbook has provided detailed examples of cases, groups, and industries engaging in institutional and noninstitutional forms of climate action, tremendous challenges remain as planetary warming intensifies. Since the 1990s, climate scientists, policymakers, and advocates created substantial inroads in terms of how to address the climate crisis. Some of the most promising programs include the thousands of city-level CAPs spread throughout the globe. CAPs are especially effective when they reach their GHG emission benchmarks, while incorporating equity and popular participation in their programs. Green and renewable energy systems also continue to expand globally. These include not only photovoltaics, hydropower, and turbine power but also green hydrogen, geothermal, tidal, gravitational, and other novel forms. A variety of NCSs also appear favorable as an ecologically sustainable means of removing or trapping GHGs. All of these forms of climate action move onto a just transitions trajectory when they provide long-term workforce development opportunities and recruit from environmental justice communities.

The transnational climate movement reached equally impressive outcomes as the institutional forms of climate action. Between 2005 and 2019, the movement coordinated dozens of global days of action, connecting almost every country on the planet and involving millions of participants. The events, and the massive organizing between events, raised awareness and played a substantial role in more binding international climate agreements, a European Union Green Deal, and a United Nations Loss and Damage Fund for the most vulnerable regions of the Global South. At the same time, antiextraction movements on the front lines of visible climate issues of strip mining, oil, and gas extraction (and transportation) added urgency to climate action via an environmental justice lens. Nonetheless, the challenges to expanding climate action and turning back planetary heating appear daunting.

One of the key challenges involves bringing more civil society participation into climate action. Governments and major societal institutions could act as crucial players by encouraging greater involvement in a variety of climate action efforts informing populations of the existential threat and magnitude of continued warming. Climate education in schools, workplaces, and religious centers could be one focus area. Another means of civic engagement expansion over climate includes purposively targeting environmental justice communities for climate action programs. The other competing economic and social needs of front-line communities create an inseparability between climate advocacy and economic equity and redistribution. A massive increase in national and international investment is also required to protect billions of people and ecological systems in the Global South. The UN Loss and Damage fund is a propitious beginning, but the program is woefully underfinanced.

Until recently, most climate action efforts emphasized decarbonization and GHG mitigation. Environmental justice advocates across multiple continents have pushed for a broader perspective, forcing scientists and environmental advocates to place GHG emissions in a proper political and economic context and to consider issues of long-term sustainability as not consistent with an ever-expanding carbon-based treadmill of capitalist production. Again, engaging directly with front-line and working-class communities about decarbonization processes increases the likelihood of moving on more of a just transition pathway, as multiple viewpoints become incorporated into strategies of implementation. Perhaps the largest obstacle of all resides in the growing resonance of various forms of climate denial, delay, skepticism, and misinformation. When powerful nation-states in the world system adopt such views and policies, climate action efforts of all forms take on a defensive posture. Ironically, the growing frequency of "climate shocks" may be one of the routes to move civil society away from the influences of climate denialism (Fisher, 2024). Also, other powerful nations and international bodies still committed to climate action should also immediately push for social and public institutions under their jurisdictions to expand climate action programs and find ways to improve climate communication away from technical jargon to clear and precise language on the urgency to act.

References

Alarcón, P. (2024). What next for supply-side policy in the south: Emerging lessons from Ecuador's Yasuní initiative. *International Environmental Agreements: Politics, Law and Economics*, *24*(1), 75–90.

Almeida, P., Brostrom, I., Pogue, T., & Gonzalez, L. R. (2024). *Climate and jobs: A community survey in South Stockton*. Merced, CA: North Valley THRIVE/UC Merced Community and Labor Center.

Almeida, P., González, L. R., Flores, E. O., Curry, V., & Padilla, A. (2023). The building blocks of community participation in local climate mitins. *npj Climate Action*, *2*(1), 37.

Almeida, P., Gonzalez Marquez, L. R., & Fonsah, E. (2024). The forms of climate action. *Sociology Compass*, *18*(2), e13177.

Auyero, J., & Swistun, D. A. (2009). *Flammable: Environmental suffering in an Argentine shantytown*. Oxford University Press.

Backhouse, M., & Lehmann, R. (2020). New "renewable" frontiers: Contested palm oil plantations and wind energy projects in Brazil and Mexico. *Journal of Land Use Science*, *15*(2–3), 373–388.

Barnes, J. (2025). *Distant concepts, emotional barriers: Framing challenges faced by U.S. organizations engaged in climate change education* [MA thesis, University of California, Merced].

Brostrom, I., Flores, E., Alatriste-Diaz, R., Luna, K., Fonsah, E., Juarez, K., Guerrero, J., Arias Murillo, A., Arana, E., & Almeida, P. (2024). *HRTP: Kern Strategic Workforce development report*. UC Merced Community and Labor Center.

Brulle, R. J. (2023). Advocating inaction: A historical analysis of the Global Climate Coalition. *Environmental Politics*, *32*(2), 185–206.

Bullard, R. D. (Ed.). (2005). *The quest for environmental justice: Human rights and the politics of pollution* (Vol. 19, pp. 32–33). Sierra Club Books.

Cha, J. M., & Pastor, M. (2022). Just transition: Framing, organizing, and power-building for decarbonization. *Energy Research & Social Science*, *90*, 102588.

De Moor, J., De Vydt, M., Uba, K., & Wahlström, M. (2021). New kids on the block: Taking stock of the recent cycle of climate activism. *Social Movement Studies*, *20*(5), 619–625.

della Porta, D., & Portos, M. (2023). Rich kids of Europe? Social basis and strategic choices in the climate activism of Fridays for Future. *Italian Political Science Review/Rivista Italiana di Scienza Politica*, *53*(1), 24–49.

Dunlap, R. E., & McCright, A. M. (2011). Organized climate change denial. In John S. Dryzek, David Schlosberg, & Richard B. Norgaard (Eds.), *The Oxford handbook of climate change and society* (pp. 144–160). Oxford University Press.

Ellis, P. W., Page, A. M., Wood, S., Fargione, J., Masuda, Y. J., Carrasco Denney, V., Moore, C., Kroeger, T., Griscom, B., Sanderman, J., Atleo, T., Cortez, R., Leavitt, S., & Cook-Patton, S. C. (2024). The principles of natural climate solutions. *Nature Communications*, *15*(1), 547.

Fisher, D. R. (2018). Climate of resistance: How the climate movement connects to the resistance. In D. S. Meyer & S. Tarrow (Eds.), *The resistance: The dawn of the anti-Trump opposition movement* (pp. 109–125). Oxford University Press.

Fisher, D. R. (2024). *Saving ourselves: From climate shocks to climate action*. Columbia University Press.

Global Covenant of Mayors. (2024). *The 2024 Global Covenant of Mayors impact report: Banking on change, finance for local climate action*. Brussels.

González Marquez, L. R. (2025). *A clean source for a sustainable future? Hegemonic outcomes of opposition campaigns against renewable energy extraction megaprojects in Central America* [Doctoral dissertation, University of California, Merced].

Hess, D. J. (2012). *Good green jobs in a global economy: Making and keeping new industries in the United States*. MIT Press.

Lehmann, R., & Tittor, A. (2023). Contested renewable energy projects in Latin America: Bridging frameworks of justice to understand "triple inequalities of decarbonisation policies." *Journal of Environmental Policy & Planning*, *25*(2), 182–193.

London, J. K., Sze, J., & Lievanos, R. S. (2008). Problems, promise, progress, and perils: Critical reflections on environmental justice policy implementation in California. *UCLA Journal of Environmental Law & Policy*, *26*, 255–289.

Maibach, E. W., Uppalapati, S. S., Orr, M., & Thaker, J. (2023). Harnessing the power of communication and behavior science to enhance society's response to climate change. *Annual Review of Earth and Planetary Sciences*, *51*(1), 53–77.

Martinez-Alier, J., Temper, L., Del Bene, D., & Scheidel, A. (2016). Is there a global environmental justice movement? *Journal of Peasant Studies*, *43*(3), 731–755.

Méndez, M. (2020). *Climate change from the streets: How conflict and collaboration strengthen the environmental justice movement*. Yale University Press.

Nanko, M. O., & Coan, T. G. (2024). Defeating cap-and-trade: How the fossil fuel industry and climate change counter movement obstruct US climate change legislation. *Global Environmental Change*, *89*, 102919.

Osaka, S., Bellamy, R., & Castree, N. (2021). Framing "nature-based" solutions to climate change. *Wiley Interdisciplinary Reviews: Climate Change*, *12*(5), e729.

Parks, L. (2025). How the climate movement shaped the EU: Protest cycles and democratic spaces in the European green deal. *Journal of European Integration*, *47*(2), 299–318.

Pellow, D. N. (2021). Global environmental and climate justice movements. In Éloi Laurent and Klara Zwickl (Eds.), *The Routledge handbook of the political economy of the environment* (pp. 75–89). Routledge.

Pollin, R., Wicks-Lim, J., Chakraborty, S., Kline, C., & Semieniuk, G. (2021). *A program for economic recovery and clean energy transition in California*. Political Economy Research Institute Research Report, University of Massachusetts Amherst.

Rivas, S., Urraca, R., & Bertoldi, P. (2022). Covenant of Mayors 2020 achievements: A two-speed climate action process. *Sustainability*, *14*(22), 15081.

Rivera Mondragón, E. (2025). Local climate programs in Mexican cities: Analyzing members in the Global Covenant of Mayors for Climate and Energy. Mobilization Conference, San Diego, CA, June 23.

Snow, S., & Benford, R. D. (1988). Ideology, frame resonance, and participant mobilization. *International Social Movement Research*, 1, 197–217.

Taylor, D. (2014). *Toxic communities: Environmental racism, industrial pollution, and residential mobility.* New York University Press.

Uri, I., Robinson, S. A., Roberts, J. T., Ciplet, D., Weikmans, R., & Khan, M. (2024). Equity and Justice in loss and damage finance: A narrative review of catalysts and obstacles. *Current Climate Change Reports*, *10*(3), 33–45.

Zabin, C. (2020). *Putting California on the high road: A jobs and climate action plan for 2030.* University of California, Berkeley Center for Labor Research and Education.

INDEX

For the benefit of digital users, indexed terms that span two pages (e.g., 52–53) may, on occasion, appear on only one of those pages.

Tables, figures, and boxes are indicated by *t*, *f*, and *b* following the page number